Intelligent Ships and Waterways: Design, Operation and Advanced Technology

Intelligent Ships and Waterways: Design, Operation and Advanced Technology

Editors

Chenguang Liu
Wengang Mao
Jialun Liu
Xiumin Chu

Basel • Beijing • Wuhan • Barcelona • Belgrade • Novi Sad • Cluj • Manchester

Editors
Chenguang Liu
Wuhan University of
Technology
Wuhan
China

Wengang Mao
Chalmers University of
Technology
Göteborg
Sweden

Jialun Liu
Wuhan University of
Technology
Wuhan
China

Xiumin Chu
Wuhan University of
Technology
Wuhan
China

Editorial Office
MDPI AG
Grosspeteranlage 5
4052 Basel, Switzerland

This is a reprint of articles from the Special Issue published online in the open access journal *Journal of Marine Science and Engineering* (ISSN 2077-1312) (available at: https://www.mdpi.com/journal/jmse/special_issues/8LU63V096V).

For citation purposes, cite each article independently as indicated on the article page online and as indicated below:

Lastname, A.A.; Lastname, B.B. Article Title. *Journal Name* **Year**, *Volume Number*, Page Range.

ISBN 978-3-7258-2241-6 (Hbk)
ISBN 978-3-7258-2242-3 (PDF)
doi.org/10.3390/books978-3-7258-2242-3

Contents

Journal of
*Marine Science
and Engineering*

Editorial

Intelligent Ships and Waterways: Design, Operation and Advanced Technology

Chenguang Liu [1,2], Wengang Mao [3], Jialun Liu [1,2,4] and Xiumin Chu [1,2,*]

1   State Key Laboratory of Maritime Technology and Safety, Wuhan University of Technology, Wuhan 430063, China; liuchenguang@whut.edu.cn (C.L.); jialunliu@whut.edu.cn (J.L.)
2   Intelligent Transport System Research Center, Wuhan University of Technology, Wuhan 430063, China
3   Department of Mechanics and Maritime Sciences, Chalmers University of Technology, 41296 Göteborg, Sweden; wengang.mao@chalmers.se
4   East Lake Laboratory, Wuhan 420202, China
*   Correspondence: chuxm@whut.edu.cn

Citation: Liu, C.; Mao, W.; Liu, J.; Chu, X. Intelligent Ships and Waterways: Design, Operation and Advanced Technology. *J. Mar. Sci. Eng.* **2024**, *12*, 1614. https://doi.org/10.3390/jmse12091614

Received: 8 August 2024
Accepted: 9 September 2024
Published: 11 September 2024

Intelligent ships have been attracting much attention with the intention of downsizing the number of staff, increasing efficiency, saving energy, etc. With the perspective of a full lifecycle for intelligent ships and waterways, this Special Issue focuses on the advanced technologies for the design and sustainable operation process. Intelligent ships need to have real-time environment perception, human-like decision making, and high-precision motion control. Intelligent waterways should provide real-time and predictive navigation services, e.g., water depth and velocity, ship traffic flow, etc. The advanced methods and technologies, including artificial intelligence, big data analysis, mixed reality, building information modeling, and digital twins, should be introduced to improve the safety, efficiency, and sustainability throughout the lifetime of intelligent ships.

The present Special Issue contains 20 research articles and 2 review articles, with 8 articles related to waterway and port monitoring, 2 articles covering ships' navigation environment perception, 4 articles handling ships' navigation decision making and motion planning, and 8 articles studying ships' motion modeling and control problems.

Navigation safety is the key issue for waterway transportation. Peng et al. [1] investigated the spatiotemporal distribution and evolution characteristics in Asia since 2000 by collecting technological water traffic accident data. The methods of gravity center and standard deviation ellipse analysis were utilized to determine the spatial and data-related characteristics of water traffic accidents. This study provides guidance for improving marine shipping safety, emergency resource management, and relevant policy formulation.

Ship collision risk identification plays a crucial role in the safe navigation and monitoring of ships in inland waterways. Wang et al. [2] proposed a new method for identifying ship navigation risks by combining the ship domain with Automatic Identification System (AIS) data to increase the prediction accuracy of collision risk identification for ship navigation in complex waterways. A ship domain model was constructed based on the ship density map drawn using AIS data. The effectiveness of this method was verified through a simulation of ships' navigation in complex waterways, with correct collision avoidance decisions being able to be made in accordance with the Regulations for Preventing Collisions in Inland Rivers of the People's Republic of China.

In high-traffic harbor waters, marine radars frequently encounter signal interference stemming from various obstructive elements, thereby presenting formidable obstacles in the precise identification of ships. Ma et al. [3] proposed a customized neural network-based ship segmentation algorithm named MrisNet to achieve precise pixel-level ship identification in complex environments. MrisNet employed a lightweight and efficient FasterYOLO network to extract features from radar images at different levels, capturing fine-grained edge information and deep semantic features of ship pixels. MrisNet accurately segments ships with different spot features and under diverse environmental conditions in marine radar images, exhibiting outstanding performance, particularly in extreme scenarios and challenging interference conditions, showcasing robustness and applicability.

There is insufficient research on the mechanisms underlying collision risks specifically related to merchant and fishing vessels in coastal waters. Zhu et al. [4] proposed an assessment method for collision risks between merchant and fishing vessels in coastal waters and validated it through a comparative analysis through visualization. The results indicated that this method effectively evaluated the severity of collision risks, and the identified high-risk areas resulting from the analysis were verified by the number of accidents that occurred in the most recent three years.

To improve the situational awareness ability of ships in busy inland waterways, Lei et al. [5] focused on the situational awareness of intelligent navigation in inland waterways with high vessel traffic densities and increased collision risks. A method based on trajectory characteristics was proposed to determine associations between AIS data and radar objects, facilitating the fusion of heterogeneous data. Through a series of experiments, including overtaking, encounters, and multi-target scenarios, this research substantiated the method, achieving an F1 score greater than 0.95. Consequently, this study furnished robust support for the perception of intelligent vessel navigation in inland waterways and the elevation of maritime safety.

To support ship trajectory prediction at waterway confluences using historical AIS data, Wang et al. [6] proposed a method to improve the recognition accuracy of ships' behavior trajectories, assist in the proactive avoidance of collisions, and clarify ship collision responsibility, ensuring the safety of waterway transportation systems in the event of ship encounters induced by waterway confluences or channel limitations. An improved K-Nearest Neighbor Algorithm considering the sensitivity of data characteristics (SKNN) was put forward to predict the trajectory of ships, which considers the influence weights of various parameters on ship trajectory prediction. The accuracy of the ship trajectory prediction method was above 99%, and the performance metrics of the SKNN surpassed those of both the conventional KNN and NB classifiers, which was helpful for warning ships of collision encounters early to ensure avoidance.

The global demand for oil is steadily escalating, and this increased demand has fueled marine extraction and maritime transportation of oil, resulting in a consequential and uneven surge in maritime oil spills. Xu et al. [7] introduced a methodology for the automated detection of oil spill targets. Experimental data pre-processing incorporated denoise, grayscale modification, and contrast boost. The realistic radar oil spill images were employed as extensive training samples in the YOLOv8 network model. The proposed method for offshore oil spill survey presented here can offer immediate and valid data support for regular patrols and emergency reaction efforts.

Structural damage is a prevalent issue in the long-term operations of harbor terminals. Bi et al. [8] proposed a novel digital twin system construction methodology tailored for the long-term monitoring of port terminals, which elaborated on the organization and processing of foundational geospatial data, sensor monitoring information, and oceanic hydrometeorological data essential for constructing a digital twin of the terminal. Experimental validation demonstrated that this method enabled the rapid construction of digital twin systems for port terminals and supported practical applications in business scenarios. Data analysis and a comparison confirmed the feasibility of the proposed method, providing an effective approach for the long-term monitoring of port terminal operations.

In response to the evolving landscape of maritime operations, new technologies are on the horizon, such as as mixed reality (MR), enhancing navigation safety and efficiency during remote assistance, e.g., in the remote pilotage use case. Ujkani et al. [9] initially tested and assessed novel approaches to pilotage in a congested maritime environment, which integrated augmented reality (AR) for ship captains and virtual reality (VR) and desktop applications for pilots. The efficiency and usability of these technologies were evaluated through in situ tests conducted with experienced pilots on a real ship using the System Usability Scale, the Situational Awareness Rating Technique, and Simulator Sickness Questionnaires during the assessment.

In scenarios such as nearshore and inland waterways, the ship spots in a marine radar are easily confused with reefs and shorelines, leading to difficulties in ship identification. To accurately identify radar targets in such scenarios, Kang et al. [10] proposed a novel algorithm, namely YOSMR, based on a deep convolutional network. The YOSMR uses the MobileNetV3 (Large) network to extract ship imaging data of diverse depths and acquire feature data of various ships. Meanwhile, taking into account the issue of feature suppression for small-scale targets in algorithms composed of deep convolutional networks, the feature fusion module known as PANet was subject to a lightweight reconstruction, leveraging depthwise separable convolutions to enhance the extraction of salient features for small-scale ships while reducing model parameters and the computational complexity to mitigate overfitting problems. As a result, the YOSMR displays a substantial advantage in terms of convolutional computation.

The trajectory planning of multiple autonomous surface vehicles (ASVs) is particularly crucial to provide a safe trajectory. In [11], a swarm trajectory-planning method was proposed for ASVs in an unknown and obstacle-rich environment. Specifically, a kinodynamic path-searching method was used to generate a series of waypoints in the discretized control space at first. Then, after fitting B-spline curves to the obtained waypoints, a nonlinear optimization problem was formulated to optimize the B-spline curves based on gradient-based local planning. Finally, a numerical optimization method was used to solve the optimization problems in real time to obtain collision-free, smooth, and dynamically feasible trajectories relying on a shared network.

To effectively deal with collisions in various encounter situations in open water environments, Zheng et al. [12] established a ship collision avoidance model and introduced multiple constraints into the Velocity Obstacle (VO) method, which was proposed to determine a ship domain by calculating a safe approach distance. Meanwhile, the ship collision avoidance model based on the ship domain was analyzed, and the relative velocity set of the collision cone was obtained by solving the common tangent line within the ellipse. The timing of starting collision avoidance was determined by calculating the ship collision risk.

Electronic Navigational Charts (ENCs) are geospatial databases compiled in strict accordance with the technical specifications of the International Hydrographic Organization (IHO). Facing the urgent demand for high-precision and real-time nautical chart products for polar navigation under the new situation, Jiao et al. [13] systematically analyzed the projection of ENCs for polar navigation. Based on the theory of complex functions, the direct transformations of Mercator projection, polar Gauss–Krüger projection, and polar stereographic projection were derived. A rational set of dynamic projection options oriented towards polar navigation was proposed with reference to existing specifications for the compilation of the ENCs. Taking the CGCS2000 reference ellipsoid as an example, the numerical analysis shows that the length distortion of the Mercator projection is less than 10% in the region up to 74°, but it is more than 80% at very high latitudes.

Sorting out the requirements for intelligent functions is the prerequisite and foundation of the top-level design for the development of intelligent ships. Taking the technical realization of each functional module as the goal, Hao et al. [14] analyzed the status quo and development trend of related intelligent technologies and their feasibility and applicability when applied to each functional module. This clarified the composition of specific functional elements of each functional module, put forward the stage goals of China's inland intelligent ship development and the specific functional requirements of different modules under each stage, and provided a reference for the Chinese government to subsequently formulate the top-level design development planning and implementation path of inland waterway intelligent ships.

Predicting the maneuverability of a dual full-rotary propulsion ship quickly and accurately is the key to manipulate a ship with machines. Yu et al. [15] performed integrated computational fluid dynamics (CFD) and used the mathematical model approach to simulate ship turning and zigzag tests, which were then compared and validated against a full-scale trial carried out under actual sea conditions. The results indicate that the proposed method has a high accuracy in predicting the maneuverability of dual full-rotary propulsion ships, with an average error of less than 10% from the full-scale trial data (and within 5% for the tactical diameters in particular), in spite of the influence of environmental factors, such as wind and waves. It provides reliability in predicting the maneuverability of a full-scale ship during the ship design stage.

Automatic berthing is at the top level of ship autonomy. Zhang et al. [16] introduced the berthing maneuver model, which is able to predict a ship's responses to steerage and external disturbances and provide a foundation for the control algorithm. The similarities and differences between the conventional MMG maneuvering model and automatic berthing maneuvering model were elaborated. Bibliometric analysis on automatic berthing was also carried out to discover common issues and emphasize the significance of maneuver modeling. Furthermore, the berthing maneuver's specifications and modeling procedures were explained in terms of the hydrodynamic forces on the hull, four-quadrant propulsion and steerage performances, external disturbances, and auxiliary devices.

The challenge of accurately modeling and predicting dynamic environments and motion statuses of ships has emerged as a prominent area of research. In response to the diverse time scales required for the prediction of a ship's motion, Zhang et al. [17] explored and analyzed various methods for modeling ship navigation environments, ship motion, and ship traffic flow. Additionally, these motion prediction methods were applied for motion control, collision avoidance planning, and route optimization. Key issues were summarized regarding ship motion prediction, including the online modeling of motion models, real ship validation, and consistency in modeling, optimization, and control. Future technology trends were predicted in mechanism-data fusion modeling, large-scale modeling, multi-objective motion prediction, etc.

The complex navigational conditions, unknown time-varying environmental disturbances, and complex dynamic characteristics of ships pose great difficulties for ship course keeping. Lin et al. [18] proposed a Particle Swarm Optimization (PSO)-based predictive PID-Backstepping (P-PB) controller to realize the efficient and rapid course keeping of ships. The proposed controller took a ship's target course, current course, yawing speed, and predictive motion parameters into consideration. The parameters in the proposed course-keeping controller were optimized by utilizing PSO, which can adaptively adjust the value of parameters in various scenarios, and thus further increase its efficiency.

Cargo Transfer Vessels (CTVs) were designed to transfer cargo from a Floating Production Storage and Offloading (FPSO) unit into conventional tankers. Liu et al. [19] presented a synchronization control strategy based on the virtual leader–follower configuration and an adaptive backstepping control method. The position and heading of the following vessel were proven to be able to globally exponentially converge to the virtual ship via the contraction theorem. Then, the optimization problem of the desired thrust command from the controller was solved through an improved firefly algorithm, which fully considered the physical characteristics of the azimuth thruster and the thrust forbidden zone caused by hydrodynamic interference. The SAF algorithm outperformed the SQP and PSO algorithms in longitudinal and lateral forces, with the R-squared (R2) values of 0.9996 (yaw moment), 0.9878 (sway force), and 0.9596 (surge force) for the actual thrusts and control commands in the wave heading 180°. The experimental results can provide technical support to improve the safe operation of CTVs.

It is challenging to understand and collect correct data about USV dynamics. Zhang et al. [20] proposed a modified Backpropagation Neural Network (BPNN) to address this issue. The experiment was conducted in the Qinghuai River, and the receiver collected the data. The modified BPNN outperformed the conventional BPNN in terms of ship trajectory forecasting and the rate of convergence. The updated BPNN can accurately predict the rotational velocity during a propeller's acceleration and stability stages at various rpms.

To satisfy the needs of autonomous navigation and high-precision control of ship trajectories, Han et al. [21] proposed a fuzzy control improvement method with an Integral Line-Of-Sight (ILOS) guidance principle. A three-degree-of-freedom ship motion model was established with the battery-powered container ship ZYHY LVSHUI 01 built by the COSCO Shipping Group. Then, a ship path-following controller based on the ILOS algorithm was designed. A controller was applied that uses a five-state extended Kalman filter (EKF) to estimate the heading, speed, and heading rate based on the ship's motion model with the assistance of Global Navigation Satellite System (GNSS) position measurements. The research results provided a reference for the path-following control of ships.

The USV is an emerging marine tool due to its advantages of automation and intelligence in recent years. Li et al. [22] proposed a novel prescribed performance fixed-time fault-tolerant control scheme for a USV with model parameter uncertainties, unknown external disturbances, and actuator faults, based on an improved fixed-time disturbance observer. Firstly, the proposed observer not only accurately and quickly estimated and compensated for the lumped nonlinearity, including actuator faults, but also reduced the chattering phenomenon by introducing the hyperbolic tangent function. Then, under the framework of prescribed performance control, a prescribed performance fault-tolerant controller was designed based on a nonsingular fixed-time sliding mode surface, which guarantees the transient and steady-state performance of a USV under actuator faults and meets the prescribed tracking performance requirements.

In conclusion, the articles presented in this Special Issue cover broad research topics related to advancements in the design, operation, and advanced technology of intelligent ships and waterways, guiding readers through the best methods for carrying out analysis.

Author Contributions: Writing—original draft preparation, C.L.; writing—review and editing, J.L., W.M. and X.C. All authors have read and agreed to the published version of the manuscript.

Conflicts of Interest: The authors declare no conflicts of interest.

References

1. Peng, Z.; Jiang, Z.; Chu, X.; Ying, J. Spatiotemporal Distribution and Evolution Characteristics of Water Traffic Accidents in Asia since the 21st Century. *J. Mar. Sci. Eng.* **2023**, *11*, 2112. [CrossRef]
2. Wang, Z.; Wu, Y.; Chu, X.; Liu, C.; Zheng, M. Risk Identification Method for Ship Navigation in the Complex Waterways via Consideration of Ship Domain. *J. Mar. Sci. Eng.* **2023**, *11*, 2265. [CrossRef]
3. Ma, F.; Kang, Z.; Chen, C.; Sun, J.; Deng, J. MrisNet: Robust Ship Instance Segmentation in Challenging Marine Radar Environments. *J. Mar. Sci. Eng.* **2024**, *12*, 72. [CrossRef]
4. Zhu, C.; Lei, J.; Wang, Z.; Zheng, D.; Yu, C.; Chen, M.; He, W. Risk Analysis and Visualization of Merchant and Fishing Vessel Collisions in Coastal Waters: A Case Study of Fujian Coastal Area. *J. Mar. Sci. Eng.* **2024**, *12*, 681. [CrossRef]
5. Lei, J.; Sun, Y.; Wu, Y.; Zheng, F.; He, W.; Liu, X. Association of AIS and Radar Data in Intelligent Navigation in Inland Waterways Based on Trajectory Characteristics. *J. Mar. Sci. Eng.* **2024**, *12*, 890. [CrossRef]
6. Wang, Z.; He, W.; Lan, J.; Zhu, C.; Lei, J.; Liu, X. Ship Trajectory Classification Prediction at Waterway Confluences: An Improved KNN Approach. *J. Mar. Sci. Eng.* **2024**, *12*, 1070. [CrossRef]
7. Xu, J.; Huang, Y.; Dong, H.; Chu, L.; Yang, Y.; Li, Z.; Qian, S.; Cheng, M.; Li, B.; Liu, P.; et al. Marine Radar Oil Spill Detection Method Based on YOLOv8 and SA_PSO. *J. Mar. Sci. Eng.* **2024**, *12*, 1005. [CrossRef]
8. Bi, J.; Wang, P.; Zhang, W.; Bao, K.; Qin, L. Research on the Construction of a Digital Twin System for the Long-Term Service Monitoring of Port Terminals. *J. Mar. Sci. Eng.* **2024**, *12*, 1215. [CrossRef]
9. Ujkani, A.; Hohnrath, P.; Grundmann, R.; Burmeister, H.C. Enhancing Maritime Navigation with Mixed Reality: Assessing Remote Pilotage Concepts and Technologies by In Situ Testing. *J. Mar. Sci. Eng.* **2024**, *12*, 1084. [CrossRef]
10. Kang, Z.; Ma, F.; Chen, C.; Sun, J. YOSMR: A Ship Detection Method for Marine Radar Based on Customized Lightweight Convolutional Networks. *J. Mar. Sci. Eng.* **2024**, *12*, 1316. [CrossRef]
11. Wang, A.; Li, L.; Wang, H.; Han, B.; Peng, Z. Distributed Swarm Trajectory Planning for Autonomous Surface Vehicles in Complex Sea Environments. *J. Mar. Sci. Eng.* **2024**, *12*, 298. [CrossRef]
12. Zheng, M.; Zhang, K.; Han, B.; Lin, B.; Zhou, H.; Ding, S.; Zou, T.; Yang, Y. An Improved VO Method for Collision Avoidance of Ships in Open Sea. *J. Mar. Sci. Eng.* **2024**, *12*, 402. [CrossRef]
13. Jiao, C.; Wan, X.; Li, H.; Bian, S. Dynamic Projection Method of Electronic Navigational Charts for Polar Navigation. *J. Mar. Sci. Eng.* **2024**, *12*, 577. [CrossRef]

14. Hao, G.; Xiao, W.; Huang, L.; Chen, J.; Zhang, K.; Chen, Y. The Analysis of Intelligent Functions Required for Inland Ships. *J. Mar. Sci. Eng.* **2024**, *12*, 836. [CrossRef]

15. Yu, Q.; Yang, Y.; Geng, X.; Jiang, Y.; Li, Y.; Tang, Y. Integrating Computational Fluid Dynamics for Maneuverability Prediction in Dual Full Rotary Propulsion Ships: A 4-DOF Mathematical Model Approach. *J. Mar. Sci. Eng.* **2024**, *12*, 762. [CrossRef]

16. Zhang, S.; Wu, Q.; Liu, J.; He, Y.; Li, S. State-of-the-Art Review and Future Perspectives on Maneuvering Modeling for Automatic Ship Berthing. *J. Mar. Sci. Eng.* **2023**, *11*, 1824. [CrossRef]

17. Zhang, D.; Chu, X.; Liu, C.; He, Z.; Zhang, P.; Wu, W. A Review on Motion Prediction for Intelligent Ship Navigation. *J. Mar. Sci. Eng.* **2024**, *12*, 107. [CrossRef]

18. Lin, B.; Zheng, M.; Han, B.; Chu, X.; Zhang, M.; Zhou, H.; Ding, S.; Wu, H.; Zhang, K. PSO-Based Predictive PID-Backstepping Controller Design for the Course-Keeping of Ships. *J. Mar. Sci. Eng.* **2024**, *12*, 202. [CrossRef]

19. Liu, C.; Zhang, Y.; Gu, M.; Zhang, L.; Teng, Y.; Tian, F. Experimental Study on Adaptive Backstepping Synchronous following Control and Thrust Allocation for a Dynamic Positioning Vessel. *J. Mar. Sci. Eng.* **2024**, *12*, 203. [CrossRef]

20. Zhang, S.; Liu, G.; Cheng, C. Application of Modified BP Neural Network in Identification of Unmanned Surface Vehicle Dynamics. *J. Mar. Sci. Eng.* **2024**, *12*, 297. [CrossRef]

21. Han, B.; Duan, Z.; Peng, Z.; Chen, Y. A Ship Path Tracking Control Method Using a Fuzzy Control Integrated Line-of-Sight Guidance Law. *J. Mar. Sci. Eng.* **2024**, *12*, 586. [CrossRef]

22. Li, Z.; Lei, K. Robust Fixed-Time Fault-Tolerant Control for USV with Prescribed Tracking Performance. *J. Mar. Sci. Eng.* **2024**, *12*, 799. [CrossRef]

Journal of
Marine Science and Engineering

Article

Spatiotemporal Distribution and Evolution Characteristics of Water Traffic Accidents in Asia since the 21st Century

Zhenxian Peng [1,2,3], Zhonglian Jiang [1,2,*], Xiao Chu [1,2,3] and Jianglong Ying [1,2,3]

1. State Key Laboratory of Maritime Technology and Safety, Wuhan University of Technology, Wuhan 430063, China; pengzx@whut.edu.cn (Z.P.); cx330@whut.edu.cn (X.C.); jlying824@whut.edu.cn (J.Y.)
2. National Engineering Research Center for Water Transport Safety, Wuhan University of Technology, Wuhan 430063, China
3. School of Transportation and Logistics Engineering, Wuhan University of Technology, Wuhan 430063, China
* Correspondence: z.jiang@whut.edu.cn

Abstract: As an important mode of transportation for the global trade, waterborne transportation has become a priority option for import and export trade due to its large load capacity and relatively low cost. Meanwhile, shipping safety has been highly valued. By collecting technological water traffic accident data from the EM-DAT database, the spatiotemporal distribution and evolution characteristics were investigated in Asia since 2000. The methods of gravity center and standard deviation ellipse analysis were utilized to determine the spatial and data-related characteristics of water traffic accidents. Temporally, the results indicated that accidents occurred most frequently during the seasons of autumn and winter, leading to a significant number of casualties. Spatially, both South-eastern Asia and Southern Asia emerged as regions with a high frequency of water traffic accidents, particularly along the borders of Singapore, Malaysia, Indonesia, and the Bay of Bengal region. In addition, the Daniel trend test and R/S analysis were conducted to demonstrate the evolution trend of accidents across various regions and seasons. The present study provides guidance for improving marine shipping safety, emergency resource management, and relevant policy formulation.

Keywords: water traffic accidents; water transport safety; spatiotemporal characteristics; Asia; center of gravity analysis; standard deviation ellipse

Citation: Peng, Z.; Jiang, Z.; Chu, X.; Ying, J. Spatiotemporal Distribution and Evolution Characteristics of Water Traffic Accidents in Asia since the 21st Century. *J. Mar. Sci. Eng.* **2023**, *11*, 2112. https://doi.org/10.3390/jmse11112112

Academic Editor: Gerasimos Theotokatos

Received: 24 September 2023
Revised: 31 October 2023
Accepted: 2 November 2023
Published: 5 November 2023

1. Introduction

Waterborne transportation, known for its energy efficiency [1], high cargo capacity, and cost-effectiveness [2], is the primary mode of international trade. Additionally, it is highly concerned as an important transportation mode which can construct a balance between overall economic development [3] and ecological environment protection in developing countries [4,5]. If a ship encounters a safety incident during its journey, the absence of timely and effective rescue efforts would result in substantial economic losses and loss of human lives. To mitigate the frequency of water traffic accidents and ensure maritime safety, the International Maritime Organization (IMO) [6] was established, and is dedicated to promoting a safe, efficient, and sustainable shipping industry.

The potential causes, underlying mechanisms, and risk assessments of water traffic accidents always differ between regions. For example, occurrences of inland shipping accidents in Europe are related to the geography, climate, national economic background, and safety culture characteristics [7]. The risk assessment of waterborne transportation becomes challenging due to the complex mechanisms and multivariate factors. Therefore, great effort has been devoted, both by academia and industry, to improve water transport safety. Cao et al. [8] provided a theoretical basis for the implementation direction of maritime safety development. Huang et al. [9] demonstrated that artificial intelligence-based risk assessment methods could provide systematic and comprehensive results. Ma et al. [10] synthesized DEMATEL (decision-making trial and evaluation laboratory), ISM (interpretative structural modeling method), FBN (fuzzy Bayesian network), and other methods to reveal influencing factors and their weights for maritime dangerous goods transportation accidents, and put forward a new comprehensive risk analysis method.

The analysis of water traffic accidents would also support the decision-making of the authorities, e.g., maritime administrations and waterway management departments. Hanafiah et al. [11] conducted vessel navigation safety analysis to improve the maritime situational awareness in the Malacca Straits which could help authorities to respond more effectively to accidents. Ma et al. [10] identified more than 20 influencing factors according to the reports of maritime dangerous goods transport accidents in China, and established a BN (Bayesian network) model, which provided a quantitative evaluation of risk factors. Fan et al. [12] analyzed the occurrence of shipping accidents from the perspective of seafarers, and put forward an evaluation framework of seafarers' psychological quality based on the neurophysiology society. The framework would provide effective means for the selection of seafarers and their driving behavior evaluations. According to statistical analysis, a series of human-related factors, e.g., operation and psychological quality of crew, contributes to 70% of water traffic accidents [13,14]. In addition, the impact of vessel traffic service operators on the overall water traffic safety cannot be ignored [15].

In this article, the water traffic accidents data were extracted from the EM-DAT database from the beginning of the 21st century for Asia. Through the methods of gravity center analysis and standard deviation ellipse, the spatiotemporal distribution and evolution characteristics of water traffic accidents were analyzed. The incidence periods and locations were explored as well. The Daniel trend test and R/S analysis were adopted to estimate the development trend of water traffic accidents, which could provide a basis for water traffic accidents prevention in Asian countries and ensure maritime traffic safety.

The remaining part of the article is organized as follows. A brief introduction of the EM-DAT database is provided in Section 2. In Section 3, the specific methods are elaborated upon in detail. A comprehensive analysis and discussion on both spatiotemporal characteristics and evolution trend are presented in Sections 4 and 5. The concluding remarks and prospects are provided in Section 6.

2. Data Sources

As a well-known disaster collection database, EM-DAT has recorded natural and other technological disasters worldwide since 1900, including the starting and ending time of the events, the causes (triggering factors), economic losses, secondary disasters, deaths, missing persons, and affected people caused by various events (injuries and homelessness due to disasters, etc.). The database has been widely utilized to reveal the evolution characteristics of disasters such as floods [16], landslides, and extreme weather conditions [17] on different scales in the world, which would be beneficial for risk assessment [18], emergency resourcing management, and disaster prevention [19–21] etc.

Notably, the EM-DAT database upholds stringent criteria governing the incorporation of disaster data; only events meeting specific conditions are deemed eligible for inclusion. Due to the special nature of the EM-DAT database, all water traffic accidents are categorized as one technological group and some specific accidents (e.g., fishing vessel accidents) are not included. This shortcoming might be overcome by introducing some other data sources (e.g., *IMO*, *Lloyd's List Intelligence* [22,23], etc.). In addition, a minimum number of 10 deaths or 100 affected individuals applies for event qualification [24]. Alternatively, if the national government publicly declares a state of emergency during the incident or international assistance is solicited, the event would be recorded. These rigorous standards reinforce the integrity of the EM-DAT database. A total number of 397 accident records were adopted for analysis in the present study.

3. Methodology

3.1. The Methods of Gravity Center and Standard Deviation Ellipse

The center of gravity was put forward in the mechanical field in physics, and is considered as the point where forces are balanced in all directions in the regional space. The center of gravity analysis is widely applied to demonstrate spatial variation characteristics [25], thus their evolving features could be revealed for different time periods [26,27]. ".shp" files with site information (e.g., latitude and longitude) were imported to create an overall visualization of accident data. Scatterplots were thereby generated and the classification tool of ArcGIS was employed to visualize the severity and distribution characteristics. The gravity centers of various regions were calculated, on the basis of which standard deviation ellipses were obtained for different stages. Subsequently, all these elements were integrated to facilitate further analysis. The calculation of the gravity center is given as follows:

$$\overline{x_I} = \sum\nolimits_{j=1}^{n} x_j w_j \Big/ \sum\nolimits_{j=1}^{n} w_j \tag{1}$$

$$\overline{y_I} = \sum\nolimits_{j=1}^{n} y_j w_j \Big/ \sum\nolimits_{j=1}^{n} w_j \tag{2}$$

where (x_j, y_j) is the barycentric coordinate of the sub-region, w_j is element value corresponding to the sub-region, and $(\overline{x_I}, \overline{y_I})$ is the barycentric coordinate of the main region.

The standard deviation ellipse was initially interpreted by Lefever [28] as a summary measure of two-dimensional points. Firstly, the standard deviation of a group of points scattered in two dimensions was obtained as an x axis and y axis, and then these two axes were rotated. Subsequently, the new standard deviation of the x axis and y axis was calculated, after which the rotation angle with the maximum standard deviation could be obtained. A new coordinate axis could thus be created to derive the short and long axes [29]. The short axis indicates the distribution range of elements, the long axis denotes the distribution direction of all elements, and the difference between the long and short axes represents the distribution directivity of elements. The center of the ellipse represents the central position of all elements. The calculation formulas are given below:

$$x = \sqrt{\left(\sum\nolimits_{i=1}^{n} (x_i - X)^2\right)/n} \tag{3}$$

$$y = \sqrt{\left(\sum\nolimits_{i=1}^{n} (y_i - Y)^2\right)/n} \tag{4}$$

where (x_i, y_i) is coordinates of element point which ranks in i place and (X, Y) denotes the regional barycentric coordinate. The rotation angle of ellipse is calculated as follows.

$$\tan\theta = \frac{\left(\sum_{i=1}^{n} x_i'^2 - \sum_{i=1}^{n} y_i'^2\right) + \sqrt{\left(\sum_{i=1}^{n} x_i'^2 - \sum_{i=1}^{n} y_i'^2\right)^2 + 4\left(\sum_{i=1}^{n} x_i' y_i'\right)^2}}{2\sum_{i=1}^{n} x_i' y_i'} \tag{5}$$

where $x_i' = x_i - X$, $y_i' = y_i - Y$.

The calculation of the long and short axes is accomplished using the subsequent formulas:

$$\varphi_x = \sqrt{\left(\sum\nolimits_{i=1}^{n} (x_i' \cos\theta - y_i' \sin\theta)^2\right)/n} \tag{6}$$

$$\varphi_y = \sqrt{\left(\sum\nolimits_{i=1}^{n} (x_i' \sin\theta - y_i' \cos\theta)^2\right)/n} \tag{7}$$

3.2. Daniel Trend Test and R/S Analysis

The Daniel trend test is a statistical analysis method based on the Spearman rank correlation coefficient and can be accomplished by MATLAB. Given a dataset with a length smaller than four, it does not need to consider the data accuracy in the test process, but only sorts the time series [30]. The rank correlation coefficient r is calculated as follows:

$$r = 1 - (6\sum\nolimits_{i=1}^{N} d_i^2)/(N^3 - N) \tag{8}$$

where $d_i = X_i - Y_i$, N is the length of the dataset. Provided that $r > 0$, a rising trend in the data could be demonstrated, and vice versa [31].

R/S analysis is known as a traditional method for time series analysis [32]. By dividing the data of 23 years into T continuous intervals $Dn(n = 1, 2, \ldots, t)$, each interval element is denoted as $y_{k,n}(k = 1, 2, \ldots, n)$, and n is the integer part of $23/T$. The R/S statistic is accomplished by MATLAB and the principal equation is given as follows:

$$Q_n = R_n/S_n \tag{9}$$

In addition, Q_n can be obtained by following the principle of statistics.

$$Q_n = C1 \times n^H \tag{10}$$

The calculated H value (Hurst coefficient) is the index of the correlation and represents the trend of time sequences [30]. The specific values correspond to different evolution features as shown below.

$$\begin{cases} 0 < H < 0.5, \text{ changing trend is contrary to the past} \\ H = 0.5, \text{ unpredictable} \\ 0.5 < H < 1, \text{ continue the changing trend of the past} \end{cases} \tag{11}$$

4. Occurrence Analysis of Water Traffic Accidents

4.1. Time Series Analysis

4.1.1. Seasonal Distribution of Water Traffic Accidents

The accident occurrence features could be explored for different Asian regions, as shown in Figures 1 and 2. It is noted that the accident records greatly exceeded 30 for September, October, and December. In addition, the occurrence frequency of water traffic accidents in South-eastern Asia and Southern Asia are relatively higher than other regions, which might be attributed to the developed shipping industry in these regions. Within South-eastern Asia, the months of January, July, September, and December manifested heightened accident frequencies. Southern Asia, on the other hand, witnessed heightened occurrences in May, July, August, and September. In Western Asia, water traffic accidents were commonly observed in January, September, November, and December, while in Eastern Asia, accidents more frequently occurred in June, October, and December.

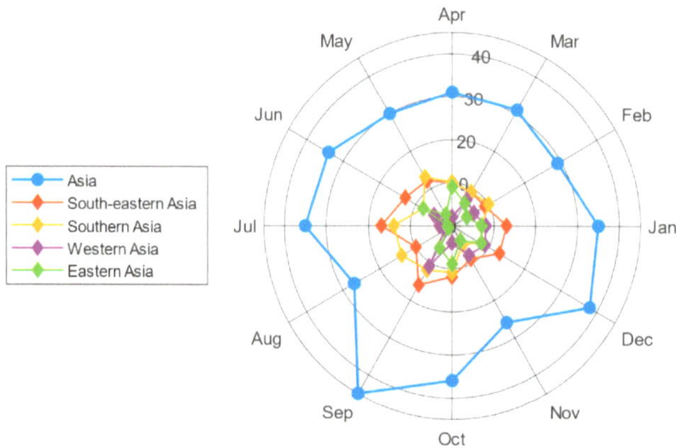

Figure 1. Monthly distribution of water traffic accidents in Asia from year 2000 to year 2022.

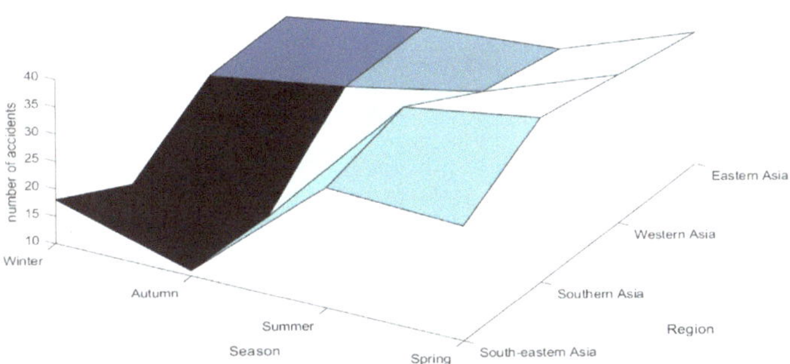

Figure 2. Seasonal distribution of water traffic accidents in Asia from year 2000 to year 2022.

The seasonal distribution characteristics of water traffic accidents are demonstrated in Figure 2. Autumn stands out as the season with the majority of water traffic accidents, tallying 107 occurrences. Following closely, winter documented 102 incidents. Within this context, South-eastern Asia experienced frequent accidents in both summer and autumn, reaching 40 and 37, respectively. Meanwhile, Southern Asia recorded 36 incidents in summer, and Western Asia documented 23 occurrences each in autumn and winter. In Eastern Asia, notable accident frequencies were observed, encompassing 18 incidents in spring and 19 incidents in both autumn and winter.

4.1.2. Yearly Distribution of Water Traffic Accidents, Deaths, and Affected Persons

In general, the temporal variation of water traffic accidents in Asia shows similar patterns with the number of deaths and affected people, i.e., an upward trend at the beginning of the 21st century, a fluctuating decline till 2020, and a minor upward trend in 2022 (as shown in Figure 3a–c). The peak values of water traffic accidents and deaths appeared in 2005 and 2003, respectively, while a maximum of affected people was observed in 2021 with an unprecedented 50,000. After inquiry, it was found that the marine environment was seriously polluted due to the fire and explosion of cargo ships in Sri Lanka [33], which may last for a decade and led to the explosive growth of affected groups. Therefore, this special case was excluded from the present study and the peak point was observed in 2009.

From a regional perspective, water traffic accidents exhibited a heightened frequency within South-eastern Asia and Southern Asia during the early years of the 21st century, as shown in Figure 3a. The accident occurrence in Asia experienced substantial fluctuations since 2007 without distinct discernible patterns for different regions. In terms of fatalities, the Southern Asia predominantly held the highest death tolls from 2000 to 2007 (as shown in Figure 3b).

4.2. Spatial Distribution Analysis

4.2.1. Spring Season

The site information of each water traffic accident was extracted from the EM-DAT database. Records were excluded if exact coordinate information was missing. The distribution diagram and heat map of water traffic accidents in spring are presented, respectively, in Figure 4a,b and Figure 5a.

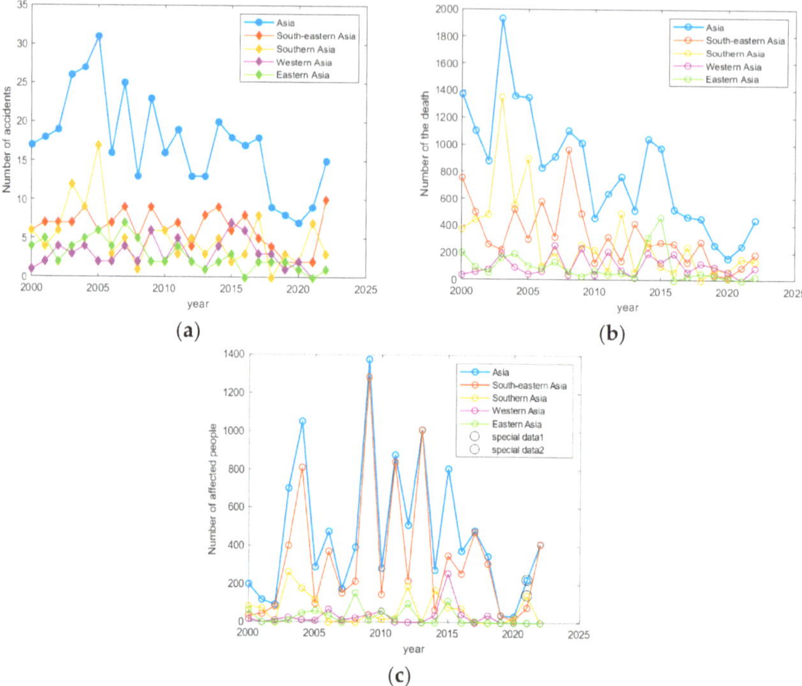

Figure 3. Yearly variation of water traffic accidents (**a**), accident-resulted deaths (**b**), and affected people (**c**) from 2000 to 2022.

Simultaneously, the database revealed a compelling correlation between accidents and the number of deaths attributed to the accidents. This correlation was visually depicted in Figure 4b, where the intensity of the marker color deepens in tandem with the rising death count. Notably, within the nexus of Southern Asia and South-eastern Asia, encompassing countries such as Myanmar, Bangladesh, and the northeastern coastal regions of India, the spring season witnessed 24 recorded accidents. Indonesia reported 16 cases, succeeded by Singapore, Malaysia, and other locales, each registering 13 incidents. The high-incident belt stretches from the central south (Bangladesh) to all directions with heat progressively tapering off, and faint upward trend in the southwest (Yemen), as shown in Figure 5a.

With regards to fatalities (as shown in Figure 4b), the majority remained at relatively lower levels. Bangladesh, the Makassar Strait, and the confluence of the Korea Strait and Tsushima Strait stood out as regions with elevated fatality counts. Most of these locations, where incidents occurred are the coastal areas and bustling port zones. Such areas would be recognized as high-frequency accident locations.

Intriguingly, among the 81 pinpointed accidents, nearly 10 took place in inland waterways; China, situated in Eastern Asia, accounted for seven of these inland incidents. This underscores the imperative of not only concentrating on coastal and open waters but also analyzing the distinct attributes of inland water traffic accidents, especially in a country like China, which actively promotes inland water transportation.

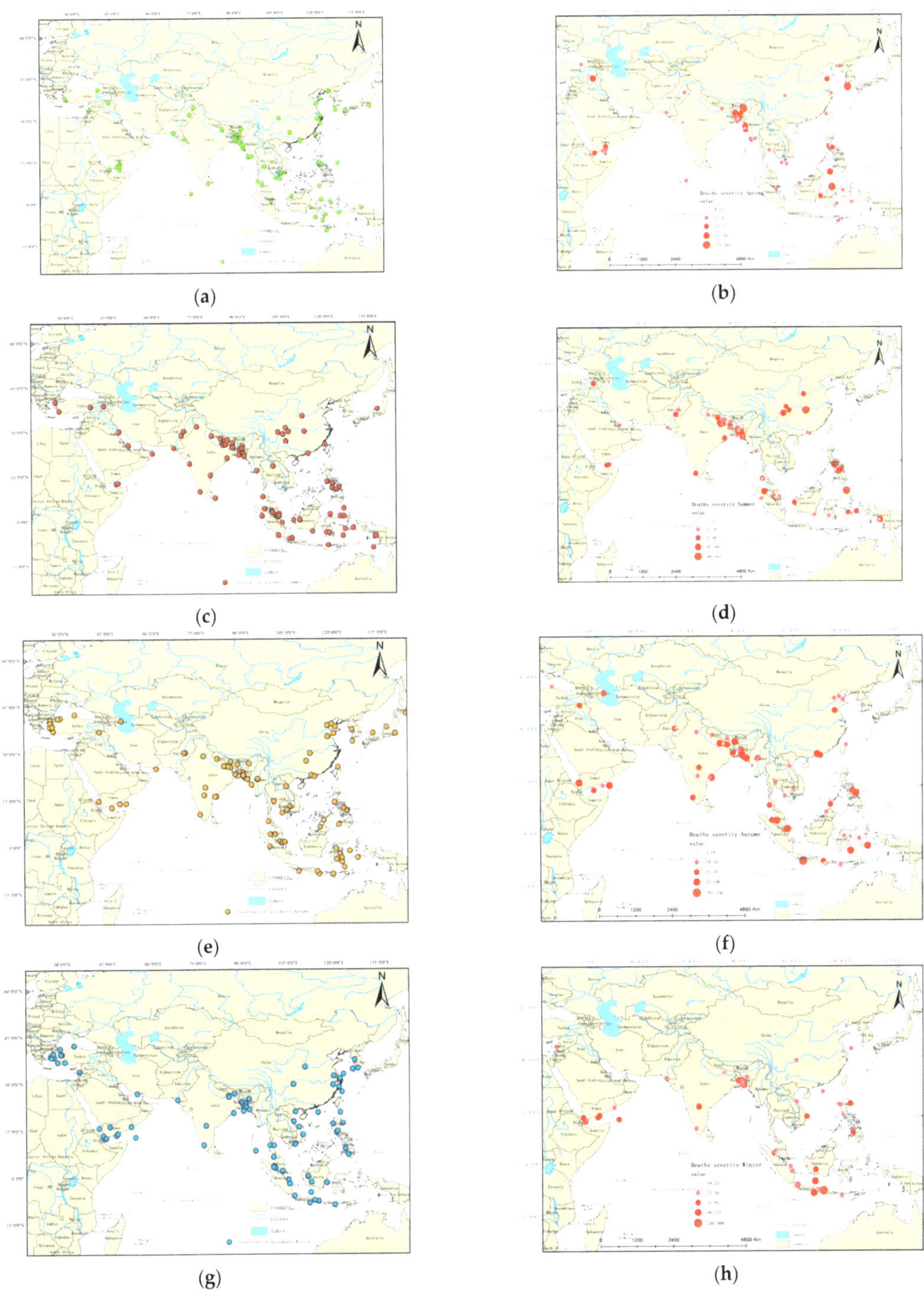

Figure 4. Distribution diagram of water traffic accidents (left panel) and deaths (right panel) in Asia. From top to bottom, Spring (**a,b**); Summer (**c,d**); Autumn (**e,f**) Winter: (**g,h**).

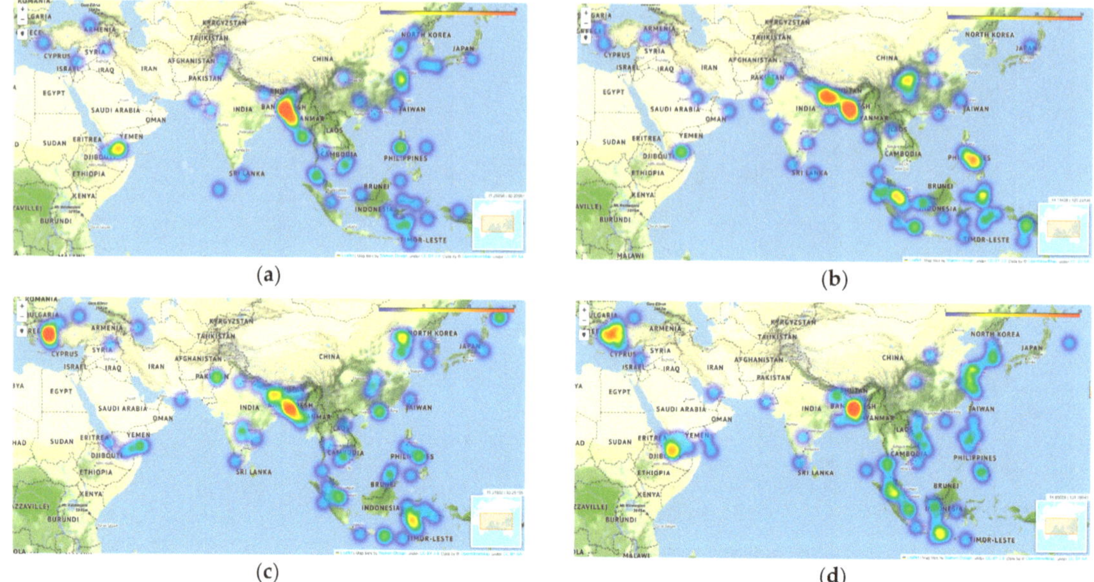

Figure 5. Heat map of water traffic accidents in Asia for different seasons. Spring (**a**); Summer (**b**); Autumn (**c**); Winter (**d**).

4.2.2. Summer Season

In the summer seasons, the same method of latitude and longitude positioning was applied, as illustrated in Figures 4c,d and 5b. The spatial distribution of accidents in summer closely mirrored that of spring. The nexus of Southern Asia and South-eastern Asia remained the focal point of accidents, with Indonesia following closely behind. Singapore, Malaysia, and other regions also figured prominently. However, in comparison to spring, there was an increase of accidents for Indonesia, Singapore, and Malaysia, with increments of 3, 2, and 3, respectively.

The heatmap (Figure 5b) displays a relatively scattered pattern in summer, characterized by a tendency to spread and a larger presence of yellow and red areas. Concerning the fatality count, summer recorded a relatively elevated overall level. Notably, accidents with high fatality counts occurred in Eastern Asia's inland river region in China, as well as in South-eastern Asian countries such as the Philippines, Malaysia, and East Timor. Southern Asia's Bangladesh also witnessed a surge in accidents with substantial fatality rates. The potential causes of these phenomena require further consideration.

Combined with the findings from the spring, it is evident that the water traffic accidents and resulting casualties have increased in China. This underscores the imperative of treating inland waterway accidents with utmost seriousness, even if the navigation conditions for inland water traffic might be excellent.

4.2.3. Autumn Season

As shown in Figure 4e,f, the prevalence of accidents remained consistent between spring and summer. Once again, the Southern Asia and South-eastern Asia junction remains a focal point for accidents, tallying 24 cases. Indonesia followed closely with 13 incidents, while Turkey recorded 12 cases. Singapore, Malaysia, and Philippines all reported 11 cases each. The high mortality points of the death that occurred in southern Indonesia (e.g., Jakarta) are not reflected in the spring and summer seasons.

Notably, the hotspots (as shown in Figure 5c) shifted from the India–Bangladesh–Thailand–Cambodia–Singapore–Indonesia cluster in the spring and summer to a more diversified distribution, concentrating along the Turkey, Indonesia, and the border between

India and Bangladesh. Moreover, areas with high fatality rates encompassed Indonesia, Singapore, Bangladesh, Yemen, and various other locations.

When considering inland water transportation, a notable decrease in the accident number was observed. This phenomenon is quite possibly caused by the onset of the dry season on inland rivers, leading to a reduction in the navigational capacity of rivers and a decrease in vessel traffic flow.

4.2.4. Winter Season

During the winter seasons, as shown in Figure 4g, the highest incidence of accidents was concentrated in South-eastern Asia, particularly in Singapore and Malaysia, which recorded 20 accidents in total. There were 17 records in Bangladesh and the northeastern coastline of India, i.e., at the junction of Southern Asia and South-eastern Asia. In Turkey, 11 records were reported. These regions account for approximately 55% of all winter accidents.

From the hotspot map of accident distribution (Figure 5d), it is evident that, in addition to the aforementioned areas, there was a notable increase in accidents along the eastern coastal regions of China, the coastal zones around Japan and South Korea in Eastern Asia, as well as in Yemen and Saudi Arabia in Western Asia. In addition, a high fatality rate was observed (as shown in Figure 4h) in some countries (e.g., Turkey, Yemen, Bangladesh, Singapore, and Indonesia).

Regarding the continued reduction of inland waterway accidents, this also substantiates the pertinent conclusions drawn in the preceding section (Section 4.2.3).

5. Results and Discussion

5.1. Daniel Trend Test and R/S Analysis

5.1.1. Daniel Trend Test Analysis

To reveal evolution trends of water traffic accidents, the time series were divided into five segments, each spanning 5, 5, 5, 5, and 3 years, respectively. The Daniel coefficients for both water traffic accidents and fatalities were thus calculated and compared with the Spearman rank test values. The magnitude of r signifies the temporal evolution trend, with a positive value indicating an increasing trend and a negative value indicating a decreasing trend [31]. When the absolute value of r surpasses the test value, it signifies a statistically significant trend, and vice versa.

As noted in Table 1, a decline in the number of water traffic accidents and fatalities was witnessed for whole of Asia, with a 95% level of significance. However, there was a noticeable upward trend in accidents during the early period of 21st century and the years following 2020. The former trend can be attributed to the aftermath of the Asian economic crisis in 1997, particularly affecting South-eastern Asia and Eastern Asia, where economic reconstruction was urgently conducted [34]. In this period, the coefficients for South-eastern Asia and Eastern Asia reached 0.850 and 0.750, respectively. The latter trend may be attributed to the global COVID-19 pandemic since 2020. In the post-pandemic era, countries have had to accelerate economic recovery and trade, leading to a substantial increase in shipping vessels and, consequently, a higher likelihood of water traffic accidents.

Generally speaking, the number of accidents in Eastern Asia, South-eastern Asia, Southern Asia, and Western Asia exhibited a declining trend, although the feature in South-eastern Asia is relatively weak. Moreover, the number of fatalities decreased significantly in all regions except Western Asia.

5.1.2. R/S Analysis

The utilization of the Hurst coefficient, similar to the Daniel coefficient, elucidates the evolution patterns of the dataset. However, the Hurst coefficient assumes a dual role; not only does it provide insights into historical trends, but also possesses predictive capabilities regarding future evolution. By computing the Hurst coefficients for variables such as the number of accidents, fatalities across different Asian regions, and the occurrences of water traffic accidents during distinct seasons, the nuanced trajectory of these crucial factors can be further demonstrated.

Table 1. Daniel coefficient of Asia and regions in different periods.

Area	Year	*r* (Accidents/Death)	Trend (Downward-D/Rise-R)	Significance of 95%
Asia	2000–2022	−0.636 (−0.825)	D/D	Y/N
	2000–2004	1.000 (0.100)	R/R	Y/Y
	2005–2009	−0.500 (−0.100)	D/D	N/N
	2010–2014	−0.050 (0.700)	D/R	N/Y
	2015–2019	−0.800 (−1.000)	D/D	Y/Y
	2020–2022	1.000 (1.000)	R/R	Y/Y
Eastern Asia	2000–2022	−0.824 (−0.669)	D/D	Y/Y
	2000–2004	0.750 (−0.100)	R/D	Y/N
	2005–2009	−0.500 (−0.700)	D/D	N/Y
	2010–2014	−0.450 (0.100)	D/D	N/N
	2015–2019	−0.550 (−0.100)	D/D	N/N
	2020–2022	0.250 (−0.500)	R/D	N/Y
South-eastern Asia	2000–2022	−0.225 (−0.649)	D/D	N/Y
	2000–2004	0.850 (−0.400)	R/D	Y/N
	2005–2009	0.350 (0.500)	R/R	N/N
	2010–2014	0.700 (0.500)	R/R	Y/N
	2015–2019	−0.900 (−0.400)	D/D	Y/N
	2020–2022	0.750 (1.000)	R/R	Y/Y
Southern Asia	2000–2022	−0.513 (−0.601)	D/D	Y/Y
	2000–2004	0.250 (0.900)	R/R	N/Y
	2005–2009	−0.300 (−0.300)	D/D	N/N
	2010–2014	−0.500 (0.100)	D/R	N/Y
	2015–2019	−0.050 (−0.300)	D/D	N/N
	2020–2022	0.500 (0.500)	R/R	Y/Y
Western Asia	2000–2022	−0.197 (0.041)	D/R	N/N
	2000–2004	0.350 (0.900)	R/R	N/Y
	2005–2009	0.450 (0.200)	R/R	N/N
	2010–2014	−0.050 (0.000)	D/—	N/—
	2015–2019	−1.050 (−0.600)	D/D	Y/Y
	2020–2022	−0.500 (0.500)	D/R	Y/Y

As shown in Table 2, the calculated Hurst values for water traffic accidents and fatality occurrences within the four regions consistently range between 0.5 and 1.0. This range underscores the enduring correlation embedded within these eight datasets. Given the persistent influence exerted by preceding data and in conjunction with the outcomes deduced from the Daniel trend analysis (2000–2022), it can be reasonably expected that the water traffic accidents and associated fatalities in Asia are poised to exhibit a subtle downward trajectory in the impending years.

Table 2. Hurst coefficient calculations of accidents and deaths in different areas.

	South-Eastern Asia	Southern Asia	Western Asia	Eastern Asia
Accidents	0.65315	0.87011	0.74831	0.73104
Deaths	0.81037	0.80481	0.65312	0.82565

To further explore the ability of this impact to persist, by taking double logarithmic value for R/S statistics and conducting linear fitting, a distinctive set of intersection points of accident frequency were obtained as 1.38, 1.60, 1.79, and 1.60, respectively, in Figure 6. The results show that the dataset of water traffic accidents will sustain its influence on the changing trend for the ensuing 4, 5, 6, and 5 years, respectively. Similarly, for fatality figures, the intersections are noted as 1.71, 1.38, 1.38, and 1.79, indicating a continuous data-driven impact on trend evolution for the subsequent 5.5, 4, 4, and 6 years, correspondingly. These calculated intervals forecast the persisting ripple effect of historical data on the future trajectory of these vital metrics.

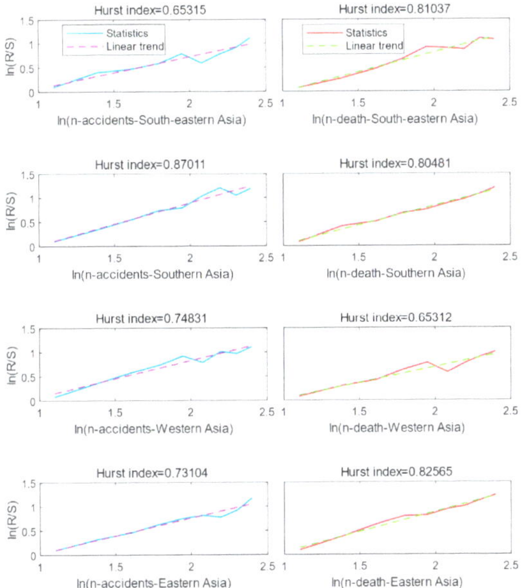

Figure 6. Hurst coefficient of water traffic accidents (**left** panel) and accident-related deaths (**right** panel) in different regions of Asia.

In the context of the Hurst coefficients, it is evident that its range falls between 0.5 and 1.0 (as shown in Table 3), similar to the aforementioned Daniel coefficient. However, a notable distinction emerges where the seasonal index tends to gravitate towards 1.0. For instance, the Hurst index for summer accidents registered a substantial 0.95, underscoring a pronounced level of data correlation.

Table 3. Hurst coefficient calculations of water traffic accidents for different seasons.

Seasons	Spring	Summer	Autumn	Winter
H	0.81288	0.95321	0.68202	0.90795

After being subjected to linear regression analysis (Figure 7), the intersection points for four seasons are derived as 1.65, 1.38, 1.30, and 1.40, respectively. This delineates a noteworthy insight that the data will sustain its influence on seasonal trends over successive timeframes of the next 5, 4, 3.5, and 4 years, respectively.

5.2. Evolution Features of Standard Deviation Ellipse and Gravity Center

The spatial distribution of water traffic accidents was explored in terms of gravity center and the standard deviation ellipse of accident frequencies for Asian and its sub-regions. A comprehensive analysis of the Asia area is presented in Figure 8a, while detailed results are provided for the sub-regions in Figure 8b. The specific coordinates of gravity centers and axes lengths of the standard deviation ellipses are tabulated in Tables 4 and 5.

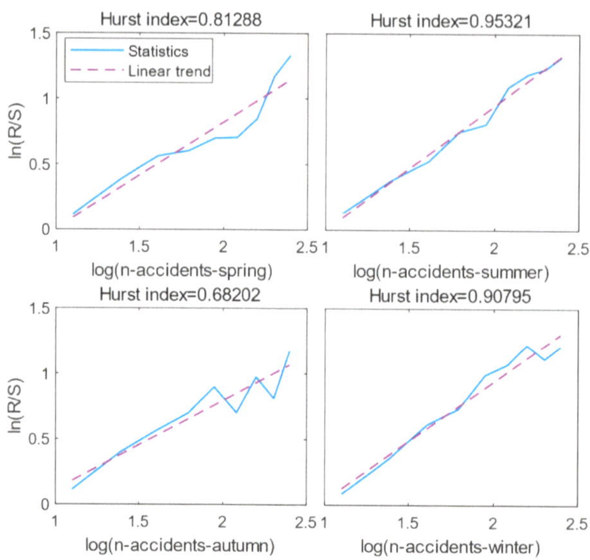

Figure 7. Hurst coefficient calculations of water traffic accidents for different seasons.

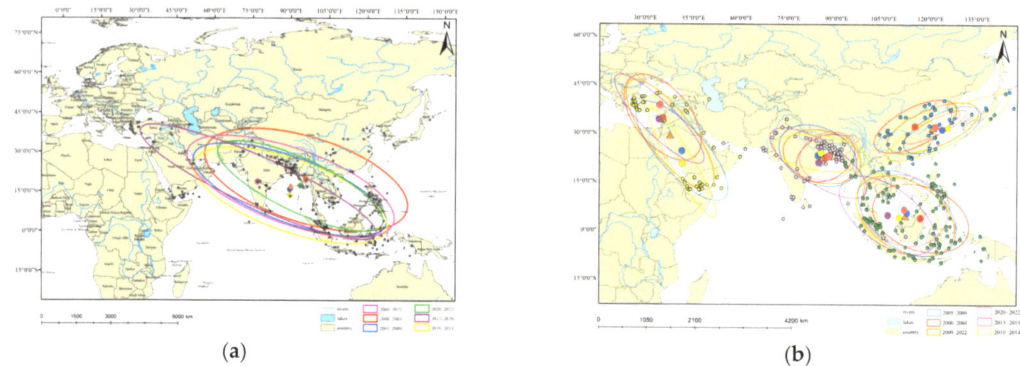

(a)   (b)

Figure 8. Evolution features of water traffic accidents in Asia. (**a**) Gravity centers of the Asia region; (**b**) Gravity centers of four sub-regions in Asia. The triangle represents the gravity center of the whole research period from 2000 to 2022.

Table 4. Coordinates of gravity center in Asia and four sub-regions for different time periods.

Year	Regions				
	Asia	South-Eastern Asia	Eastern Asia	Southern Asia	Western Asia
2000–2022	89.835° E, 16.862° N	112.121° E, 4.523° N	120.785° E, 31.639° N	85.600° E, 22.596° N	37.789° E, 29.207° N
2000–2004	95.808° E, 19.344° N	117.056° E, 3.776° N	115.153° E, 32.086° N	86.736° E, 22.269° N	35.452° E, 34.534° N
2005–2009	89.482° E, 15.562° N	112.690° E, 5.295° N	123.776° E, 31.095° N	84.321° E, 22.955° N	41.490° E, 24.239° N
2010–2014	89.683° E, 13.192° N	110.014° E, 3.908° N	125.611° E, 31.697° N	85.131° E, 23.641° N	41.738° E, 20.441° N
2015–2019	76.816° E, 18.543° N	106.453° E, 4.650° N	121.713° E, 32.022° N	84.493° E, 20.734° N	34.334° E, 34.029° N
2020–2022	93.597° E, 16.996° N	111.765° E, 6.276° N	— —	87.807° E, 22.833° N	34.610° E, 38.495° N

Table 5. Axes lengths of standard deviation ellipse for Asia and four sub-regions in different time periods. (Mm-megameter, km-kilometer; the former data denote the long axis while the latter represent the short axis).

Year	Regions				
	Asia (Mm)	South-Eastern Asia (km)	Eastern Asia (km)	Southern Asia (km)	Western Asia (km)
2000–2022	42.699, 14.323	16.919, 10.617	15.575, 5.887	12.450, 8.050	7.055, 20.689
2000–2004	41.211, 16.251	15.852, 9.768	14.325, 5.938	9.709, 6.747	8.626, 15.743
2005–2009	40.018, 12.360	17.347, 10.866	13.391, 5.268	12.923, 8.706	7.244, 19.703
2010–2014	39.847, 13.640	16.127, 10.261	21.587, 4.909	15.060, 5.452	3.411, 19.659
2015–2019	50.655, 9.363	17.460, 8.703	10.957, 2.446	17.599, 6.590	19.008, 5.090
2020–2022	35.340, 10.134	13.933, 10.746	— —	9.369, 7.348	— —

Over the whole research period of 2000–2022, the center of gravity for water traffic accidents was situated at (89.835° E, 16.862° N). The center of gravity was initially positioned in Myanmar from 2000 to 2004, shifted southwest to the Bay of Bengal waters from 2005 to 2009, continued southward while remaining within the Bay of Bengal from 2010 to 2014, then shifted northwest to regions near India from 2015 to 2019. Subsequently, it migrated southeast and settled on the boundary of Myanmar and the Bay of Bengal. This migration represents the furthest movement among the four shifts in these five periods.

From the perspective of gravity center distribution, the locations of the gravity center differ for five time periods, four of which were located in or near South-eastern Asia. Only the center of gravity for 2015–2019 was positioned in Southern Asia, India. This is attributed to the dense vessel traffic flow through the navigable straits in South-eastern Asia, coupled with the substantial volume of ships, which increases the likelihood of accidents. Additionally, countries such as Bangladesh in Southern Asia and Vietnam in South-eastern Asia have numerous ports (e.g., seventeen in Vietnam, seven in Bangladesh) and rely heavily on exports (export to China, Malaysia, etc.) for their economic development. This leads to a surge in vessel traffic flow through the Straits of Malacca and the South China Sea, potentially resulting in water traffic accidents.

Figure 8b provides a visual representation of the standard deviation ellipses and center of gravity analyses for water traffic accidents across Asia. It was found that accidents frequently occurred (i.e., concentration of gravity centers) in Malaysia and Singapore in South-eastern Asia, the Bay of Bengal in Southern Asia, the western periphery of Turkey in Western Asia, and the southeast region of China in Eastern Asia. The phenomenon of high-frequency water traffic accidents might be attributed the Straits of Malacca, Bay of Bengal region, Mediterranean Economic Belt, and China's Golden Waterway where the shipping industry is well developed. It needs to be emphasized that the Strait of Malacca is a crucial international trade route shared by all countries [35], thus carries a large amount of maritime vessel traffic flow [36]. To ensure navigation safety in this area, maritime supervision efficiency remain focal points of ongoing research [11].

5.2.1. South-Eastern Asia

Between 2000 and 2022, the center of gravity for water traffic accidents in South-eastern Asia resided at (112.121° E, 4.523° N), i.e., the juncture of James Shoal and Malaysia (East). In the initial period of 2000 to 2004, the center of gravity was situated in northern Indonesia near Brunei. It shifted northwest to the James Shoal–Malaysia (East)–Brunei junction during 2005 to 2009. From 2010 to 2014, it migrated southwest to the western waters of Malaysia (East). It moved northwest again to the junction connecting Malaysia (West), Vietnam, and the South China Sea. In the most recent three years, the center of gravity shifted northeastward, nearing the border between James Shoal and the Nansha Islands. Over the entire period, it is evident that the center of gravity in South-eastern Asia consistently hovered around the South China Sea and Malaysia, except the early years (2000–2004) when it was situated in Indonesia as shown in Figure 9a.

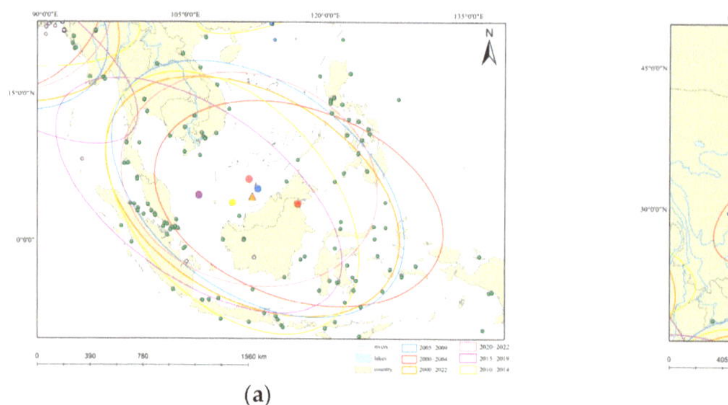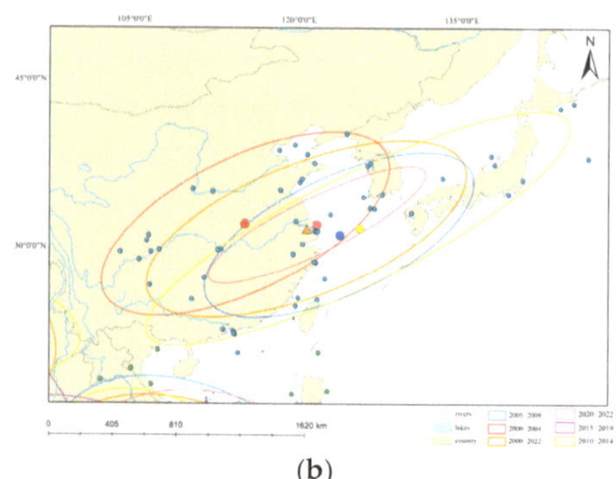

(a)   (b)

Figure 9. Evolution features of water traffic accidents in (**a**) South-eastern Asia and (**b**) Eastern Asia.

Following the standard deviation ellipses analysis within the South-eastern Asia, it was found that the disparities between the long and short axes consistently remained within 10 km during different time periods, which suggests a relatively concentrated data distribution feature. In combination with the center of gravity patterns, it is concluded that the focal area of water traffic accidents in South-eastern Asia is located in the South China Sea because of its important role for the Asian Maritime Silk Road. South-eastern Asia historically constituted the primary trade hinterland of the early Maritime Silk Road [37]. The Maritime Silk Road has built close ties between China and Asian countries, and has promoted all-round cooperation. As an important part of the global shipping routes, thousands of ships pass through this area every day, thereby resulting in a higher risk of water traffic accidents.

5.2.2. Eastern Asia

The water traffic accident data of 2015–2019 were insufficient for center of gravity and standard deviation ellipse analysis, thus the dataset was neglected in the present study.

Over the time period of 2000 to 2022, the center of gravity consistently resided in Suzhou, China, a city situated along the Grand Canal route. The Grand Canal is the earliest and longest artificial waterway in the world, and has played a great role in the economic and cultural development and exchanges between the north and south regions of China. In the time period of 2000 to 2004, and the most recent three years, the centers of gravity were located in the vicinity of the Yangtze River estuary. Conversely, during 2005–2009 and 2010–2014, the centers of gravity were situated in the East China Sea. The evolution features of gravity centers in different periods are correlated with the development strategy of shipping industry in China. China holds a prominent role as the leading nation in the Eastern Asian Maritime Silk Road [38].

Furthermore, results from the standard deviation ellipse analysis consistently exhibited a predominant northeast–southwest directional distribution, with weaker data intensity distributed in the northwest–southeast direction, as shown in Figure 9b.

This phenomenon may be affected by the lengths of coastline in Eastern Asia. The coastline of the Eastern Asia is concentrated around China, Japan, and South Korea, who own approximately 32,000 km (ranked 6th in the world), 29,000 km (ranked 7th in the world,) and 2413 km (ranked 52nd in the world), respectively. The presence of extensive coastlines facilitates the establishment of deep-water ports, allowing for the optimal utilization of coastal advantages for the development of maritime transportation. Additionally, these three nations are major importers and exporters in Asia. Consequently, the number of vessels traversing these coastlines has experienced a notable increase, leading to a higher risk of water transport safety incidents. The analysis also confirmed that a significant portion of these accidents tend to occur at the junctions of these three countries as shown in Figure 4a,e,g.

Throughout the entire research period, the center of gravity was positioned within the inland regions of China. This phenomenon might be attributed to developed inland shipping infrastructures in China, e.g., the Golden waterway of the Yangtze River and the Grand Canal. Simultaneously, the Chinese government has consistently supported the digital development of inland waterway shipping [39]. It is worth noting that inland water traffic accidents might happen occasionally with substantial casualties, such as the sinking accident of the Eastern Star in 2015. This incident underscores the gravity center of water traffic accidents in this specific period.

5.2.3. Southern Asia

The gravity centers in Southern Asia consistently clustered in the eastern region of India, near the Bay of Bengal, as shown in Figure 10a. India and Bangladesh emerged as the primary nations where accidents occurred with high frequency. A notable observation is that the majority of accidents were concentrated in the inland waterways and coastal regions of Bangladesh. In India, the capacity transfer of road and railway transportation to inland waterway transportation has encountered numerous challenges, resulting in suboptimal waterway management and high frequency occurrence of water accidents [40]. Meanwhile, the coastal areas of Bangladesh are prone to natural disasters such as storm surges and hurricanes, which increase accident rates and exert significant impacts on the maritime transportation safety [41].

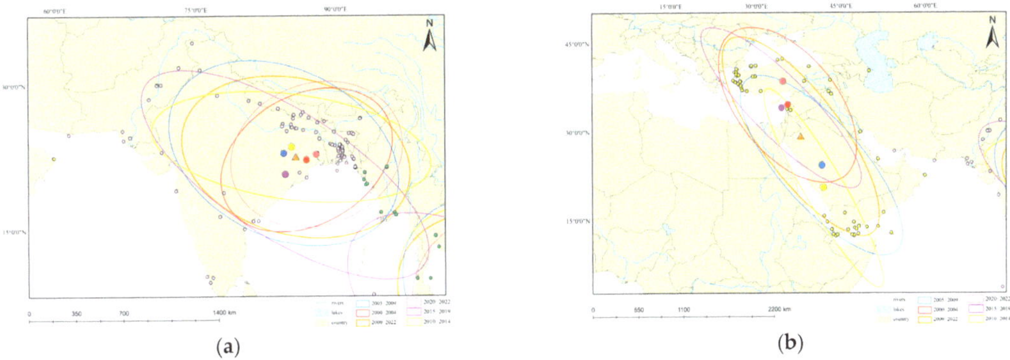

(a) (b)

Figure 10. Evolution features of water traffic accidents in (**a**) Southern Asia and (**b**) Western Asia.

A clear directional distribution pattern of the standard deviation ellipse was observed in South-eastern Asia, Eastern Asia, and Western Asia for different time periods. However, in Southern Asia, there were notable variations in the standard deviation ellipse during different periods. In the time period of 2000 to 2004 and in 2020 to 2022, the data display a northeast–southwest distribution, with minimal disparities between the long and short axes, indicating relatively concentrated data. In the time periods of 2005 to 2009 and 2015 to 2019, a northwest–southeast distribution pattern was observed. The former is characterized by data concentration, while the latter is distributed in a distinct direction and features a significant difference between the long and short axes, extending up to 11 km. The data distribution during 2010–2014 appears primarily horizontal, with a slight rotation toward the northwest–southeast direction.

Interestingly, the gravity center and the ellipse axes of the 2000–2004 period closely resemble those of 2020–2022, as shown in Figure 10a. In these instances, the center of gravity (with a mere 1-degree difference in longitude) and the standard deviation ellipse (featuring only a 0.6-km difference in the short axis) nearly overlap. This suggests that water traffic accidents in this area may follow a periodic pattern. Further data analysis is required to validate this preliminary conclusion.

5.2.4. Western Asia

In the context of gravity center distribution, from 2000 to 2022, the focal point for water traffic accidents in Western Asia consistently gravitated towards the northwest of Saudi Arabia, in proximity to Jordan, as shown in Figure 10b. During 2000 to 2004, the gravity center was situated in Syria along the Mediterranean region. Subsequently, this point shifted southeast to the central part of Saudi Arabia from 2005 to 2009, and continued its journey southwards to the southern region of Saudi Arabia near the Red Sea. Later, the center of gravity shifted significantly northward to an area positioned between the middle of Lebanon and Cyprus in the Mediterranean Sea from 2015 to 2019, marking the most extensive migration distance. Finally, in the most recent three years, the center of gravity has settled in Central Turkey.

An intriguing observation emerges from the gravity center distribution analysis. In all five time periods, the gravity centers consistently aligned along a northwest–southeast axis. This alignment is also clear when examining the standard deviation ellipses. This suggests that water traffic accidents in Western Asia are concentrated along this trajectory, corresponding to the interconnectedness of Turkey, Saudi Arabia, and Yemen at the national level [30]. The rationale behind this phenomenon may be attributed to the Mediterranean Sea facilitating exports for countries like Turkey and Lebanon, while the Red Sea, Gulf of Aden and Arabian Gulf have become vital maritime regions for the export trade of Saudi Arabia, Yemen, and Oman.

5.3. Discussion

The international shipping industry contributes more than 80% of world trade volume. It is imperative to reveal spatiotemporal characteristics of water traffic accidents, thereby more appropriate and effective countermeasures could be proposed to ensure navigational safety. The EM-DAT database has provided a systematic data source for thorough analysis of water traffic accidents worldwide. Both spatial statistical analysis and time series analysis methods were introduced to investigate the features of water traffic accidents in Asia since the 21st century.

The evolution features and intrinsic correlation of water traffic accidents were investigated through time series analysis (i.e., Daniel trend test and R/S analysis). The present study indicates that most of the water traffic accidents occurred in September, October, and December, i.e., in autumn and winter. This might be attributed to the harsh wave conditions and wind speed [42]. An increasing trend has been observed since 2020 (as shown in Figure 3). The calculated Hurst values of water traffic accidents and fatalities fall in a range of 0.5 to 1.0 (as presented in Figures 6 and 7). It is expected that the water traffic accidents and associated fatalities in Asia will show a decreasing trend over the coming years, which agree with the conclusions drawn by Zhou et al. [43].

Following the spatial statistical analysis by gravity centers and standard deviation ellipses, water traffic accidents showed evident spatial variation features during different stages. Water traffic accidents frequently occurred in the South-eastern Asia and Southern Asia (as shown in Figures 4 and 5). This is closely related to the advanced shipping industry of countries within these regions (e.g., Singapore, Malaysia, etc.) and busy shipping routes (e.g., North Pacific shipping line and the Strait of Malacca). The results are consistent with the published literature [43]. With the gradual improvement of inland waterway conditions, China's inland waterway shipping has developed rapidly. The inland river freight volume exceeded 4.4 billion tons in 2022, with an increase rate of 5.1%. The dense vessel traffic flow and trend of larger ships has resulted in a challenging task for the maritime administration of inland shipping. Local extreme weather conditions may cause water traffic accidents, leading to tragic consequences (e.g., The Eastern Star accident in June 2015 [44]). In addition to the primary cause of human error [14], more effort needs be paid to identify key environmental risk factors and evaluate their impacts on water traffic safety [43].

The present study fills the gap in the analysis of characteristics of water transport accidents in Asia, and the current results are of great significance for improving maritime safety services and risk management of shipping companies. It is worth noting that strict criteria have been applied for disaster data inclusion in the EM-DAT database. More sufficient data are thus required to conduct a comprehensive study of risk causes and emergency response strategy in water traffic accidents. Other databases (e.g., *IMO, Lloyd's List Intelligence*, etc.) can provide a good supplement for relevant research. Some interesting results of accident association rules in Artic waters [22], environmental risk factors [43], and regional frequency density [30] have been reported. The incorporation of knowledge graph theories [8] and ergonomics in further studies may provide valuable insights into the underlying mechanisms of water traffic accidents.

6. Conclusions

In the present study, the EM-DAT database was utilized to perform characteristics analysis of water traffic accidents in Asia since the 21st century. Some preliminary conclusions are drawn as follows:

1. Both South-eastern Asia and Southern Asia were identified as high incidence areas of water traffic accidents. Most of the accidents occur in September, October, and December, i.e., in autumn and winter. Overall, the occurrence frequency of water traffic accidents in Asia shows a feature of an upward trend at the beginning of the 21st century, a fluctuating decline till 2020 and a minor increasing trend in 2022.
2. Heat maps and scatter diagrams were presented to demonstrate the distribution patterns of water traffic accidents in different sub-regions. The regional and seasonal evolution trends are anticipated to persist for 4~6 years and 3~5 years, respectively, based on the Daniel trend analysis and Hurst coefficients calculations.
3. The spatial analysis of water traffic accident data demonstrates that the gravity center of Asia is located at the junction between India and Bangladesh. The evolution features of different sub-regions were presented and analyzed. The geographical conditions, industrial planning, and development strategies of Asian countries might have an impact on the distribution and evolution characteristics of water traffic accidents. The potential causes of accidents were also briefly discussed for different sub-regions.

The scope of the present study was confined to a detailed analysis of water traffic accidents in Asia since the 21st century. The results provide a guidance of improving vessel traffic services and disaster prevention. Due to the nature of the EM-DAT database, the potential causes and underlying mechanisms of water traffic accidents were not thoroughly investigated which would be the topic of future studies.

Author Contributions: Conceptualization, Z.P. and Z.J.; methodology, Z.P., X.C. and J.Y.; software, Z.P., X.C. and J.Y.; validation, Z.J.; investigation Z.P. and J.Y.; resources, X.C. and J.Y.; writing—original draft preparation Z.P. and Z.J.; writing—review and editing, Z.J.; visualization, Z.P. and X.C.; supervision, Z.J. All authors have read and agreed to the published version of the manuscript.

Funding: The present study was financially supported by the National Natural Science Foundation of China, Grant Number 52071250 and 51709220.

Institutional Review Board Statement: Not applicable.

Informed Consent Statement: Not applicable.

Data Availability Statement: Access to the data will be considered upon request.

Acknowledgments: We would like to thank the EM-DAT database for the data provided.

Conflicts of Interest: The authors declare no conflict of interest.

References

1. Tolliver, D.; Lu, P.; Benson, D. Comparing rail fuel efficiency with truck and waterway. *Transp. Res. Part Transp. Environ.* **2013**, *24*, 69–75. [CrossRef]
2. Caris, A.; Limbourg, S.; Macharis, C.; Van Lier, T.; Cools, M. Integration of inland waterway transport in the intermodal supply chain: A taxonomy of research challenges. *J. Transp. Geogr.* **2014**, *41*, 126–136. [CrossRef]
3. Dávid, A.; Madudová, E. The Danube river and its importance on the Danube countries in cargo transport. *Transp. Res. Procedia* **2019**, *40*, 1010–1016. [CrossRef]
4. Ministry of Transport of the People's Republic of China. *Notice of the Ministry of Transport on the Issuance of the Outline for the Development of Inland Waterway*; Ministry of Transport of the People's Republic of China: Beijing, China, 2020. Available online: https://www.gov.cn/zhengce/zhengceku/2020-06/04/content_5517185.htm (accessed on 19 October 2023).
5. Ministry of Ports, Shipping and Waterways Government of India. *Maritime India Vision 2030*; Ministry of Ports, Shipping and Waterways Government of India: New Delhi, India, 2021. Available online: https://sagarmala.gov.in/sites/default/files/MIV%202030%20Report.pdf (accessed on 19 October 2023).
6. International Maritime Organization. *30 Years at IMO HQ*; International Maritime Organization: London, UK, 2023. Available online: https://wwwcdn.imo.org/localresources/en/About/HistoryOfIMO/Documents/30%20years%20at%20IMO%20HQ.pdf (accessed on 19 October 2023).
7. Bačkalov, I.; Vidić, M.; Rudaković, S. Lessons learned from accidents on some major European inland waterways. *Ocean Eng.* **2023**, *273*, 113918. [CrossRef]
8. Cao, Y.; Wang, X.; Yang, Z.; Wang, J.; Wang, H.; Liu, Z. Research in marine accidents: A bibliometric analysis, systematic review and future directions. *Ocean Eng.* **2023**, *284*, 115048. [CrossRef]
9. Huang, X.; Wen, Y.; Zhang, F.; Han, H.; Huang, Y.; Sui, Z. A review on risk assessment methods for maritime transport. *Ocean Eng.* **2023**, *279*, 114577. [CrossRef]
10. Ma, L.; Ma, X.; Lan, H.; Liu, Y.; Deng, W. A methodology to assess the interrelationships between contributory factors to maritime transport accidents of dangerous goods in China. *Ocean Eng.* **2022**, *266*, 112769. [CrossRef]
11. Md Hanafiah, R.; Zainon, N.S.; Karim, N.H.; Abdul Rahman, N.S.F.; Behforouzi, M.; Soltani, H.R. A new evaluation approach to control maritime transportation accidents: A study case at the Straits of Malacca. *Case Stud. Transp. Policy* **2022**, *10*, 751–763. [CrossRef]
12. Fan, S.; Blanco-Davis, E.; Fairclough, S.; Zhang, J.; Yan, X.; Wang, J.; Yang, Z. Incorporation of seafarer psychological factors into maritime safety assessment. *Ocean Coast. Manag.* **2023**, *237*, 106515. [CrossRef]
13. Fan, S.; Yang, Z. Towards objective human performance measurement for maritime safety: A new psychophysiological data-driven machine learning method. *Reliab. Eng. Syst. Saf.* **2023**, *233*, 109103. [CrossRef]
14. Galieriková, A. The human factor and maritime safety. *Transp. Res. Procedia* **2019**, *40*, 1319–1326. [CrossRef]
15. Crestelo Moreno, F.; Roca Gonzalez, J.; Suardíaz Muro, J.; García Maza, J.A. Relationship between human factors and a safe performance of vessel traffic service operators: A systematic qualitative-based review in maritime safety. *Saf. Sci.* **2022**, *155*, 105892. [CrossRef]
16. Hu, P.; Zhang, Q.; Shi, P.; Chen, B.; Fang, J. Flood-induced mortality across the globe: Spatiotemporal pattern and influencing factors. *Sci. Total Environ.* **2018**, *643*, 171–182. [CrossRef]
17. Lee, W.V. Historical global analysis of occurrences and human casualty of extreme temperature events (ETEs). *Nat. Hazards* **2014**, *70*, 1453–1505. [CrossRef]
18. De Moraes, O.L.L. Some evidence on the reduction of the disasters impact due to natural hazards in the Americas and the Caribbean after the 1990s. *Int. J. Disaster Risk Reduct.* **2022**, *75*, 102984. [CrossRef]
19. López-Peláez, J. Co-evolution between structural mitigation measures and urbanization in France and Colombia: A comparative analysis of disaster risk management policies based on disaster databases. *Habitat Int.* **2011**, *35*, 573–581. [CrossRef]
20. Mavhura, E. Disaster mortalities and the Sendai Framework Target A: Insights from Zimbabwe. *World Dev.* **2023**, *165*, 106196. [CrossRef]
21. Shi, S.; Yao, F.; Zhang, J.; Yang, S. Evaluation of Temperature Vegetation Dryness Index on Drought Monitoring Over Eurasia. *IEEE Access* **2020**, *8*, 30050–30059. [CrossRef]
22. Fu, S.; Liu, Y.; Xi, Y.; Wan, H. Feature Analysis and Association Rule Mining of Sship Accidents in Arctic Waters. *Adv. Polar Sci.* **2020**, *32*, 102–111. [CrossRef]
23. Lloyd's List Intelligence. *Shipping and Maritime Intelligence*; Lloyd's List Intelligence: London, UK, 2021; Available online: https://www.seasearcher.com/ (accessed on 22 October 2023).
24. Adhikari, P.; Hong, Y.; Douglas, K.R.; Kirschbaum, D.B.; Gourley, J.; Adler, R.; Robert Brakenridge, G. A digitized global flood inventory (1998–2008): Compilation and preliminary results. *Nat. Hazards* **2010**, *55*, 405–422. [CrossRef]
25. Klein, L.R. Measurement of a shift in the world's center of economic gravity. *J. Policy Model.* **2009**, *31*, 489–492. [CrossRef]
26. Grether, J.-M.; Mathys, N.A. Is the World's Economic Center of Gravity Already in Asia? *SSRN Electron. J.* **2008**, *42*, 47–50. [CrossRef]
27. Liang, L.; Chen, M.; Luo, X.; Xian, Y. Changes pattern in the population and economic gravity centers since the Reform and Opening up in China: The widening gaps between the South and North. *J. Clean. Prod.* **2021**, *310*, 127379. [CrossRef]

28. Lefever, D.W. Measuring Geographic Concentration by Means of the Standard Deviational Ellipse. *Am. J. Sociol.* **1926**, *32*, 88–94. [CrossRef]
29. Rogerson, P.A. Historical change in the large-scale population distribution of the United States. *Appl. Geogr.* **2021**, *136*, 102563. [CrossRef]
30. Wei, C.; Guo, B.; Zhang, H.; Han, B.; Li, X.; Zhao, H.; Lu, Y.; Meng, C.; Huang, X.; Zang, W.; et al. Spatial–temporal evolution pattern and prediction analysis of flood disasters in China in recent 500 years. *Earth Sci. Inform.* **2022**, *15*, 265–279. [CrossRef]
31. Liu, J.; Chen, T.; Chi, D. Rainfall trend analysis of the northwest Liaoning Province based on Daniel and Mann-Kendall test. *J Shenyang Agric. Univ. Soc. Ed.* **2014**, *45*, 599–603. [CrossRef]
32. Liu, Y.; Yang, Z.; Huang, Y.; Liu, C. Spatiotemporal evolution and driving factors of China's flash flood disasters since 1949. *Sci. China Earth Sci.* **2018**, *61*, 1804–1817. [CrossRef]
33. International Maritime Organization. *X-Press Pearl—Incident Information Centre*; 9875343; International Maritime Organization: London, UK, 2021. Available online: https://www.x-presspearl-informationcentre.com/ (accessed on 22 October 2023).
34. Hopkins, S. Economic stability and health status: Evidence from East Asia before and after the 1990s economic crisis. *Health Policy* **2006**, *75*, 347–357. [CrossRef] [PubMed]
35. Zaman, M.B.; Kobayashi, E.; Wakabayashi, N.; Maimun, A. Risk of Navigation for Marine Traffic in the Malacca Strait Using AIS. *Procedia Earth Planet. Sci.* **2015**, *14*, 33–40. [CrossRef]
36. Rusli, M. Navigational Hazards in International Maritime Chokepoints: A Study of the Straits of Malacca and Singapore. *J. Int. Stud.* **2020**, *8*, 47–75. [CrossRef]
37. Bellina, B.; Favereau, A.; Dussubieux, L. Southeast Asian Early Maritime Silk Road trading polities' hinterland and the sea-nomads of the Isthmus of Kra. *J. Anthropol. Archaeol.* **2019**, *54*, 102–120. [CrossRef]
38. Schottenhammer, A. The "China Seas" in world history: A general outline of the role of Chinese and East Asian maritime space from its origins to c. 1800. *J. Mar. Isl. Cult.* **2012**, *1*, 63–86. [CrossRef]
39. Chang, Y.-C.; Liu, S.; Zhang, X. The construction of Global Maritime Capital—Current development in China. *Mar. Policy* **2023**, *151*, 105576. [CrossRef]
40. Trivedi, A.; Jakhar, S.K.; Sinha, D. Analyzing barriers to inland waterways as a sustainable transportation mode in India: A dematel-ISM based approach. *J. Clean. Prod.* **2021**, *295*, 126301. [CrossRef]
41. Rose, L.; Bhaskaran, P.K. Tidal Prediction for Complex Waterways in the Bangladesh Region. *Aquat. Procedia* **2015**, *4*, 532–539. [CrossRef]
42. Antão, P.; Soares, C.G. Analysis of the influence of human errors on the occurrence of coastal ship accidents in different wave conditions using Bayesian Belief Networks. *Accid. Anal. Prev.* **2019**, *133*, 105262. [CrossRef]
43. Zhou, X.; Ruan, X.; Wang, H.; Zhou, G. Exploring spatial patterns and environmental risk factors for global maritime accidents: A 20-year analysis. *Ocean Eng.* **2023**, *286*, 115628. [CrossRef]
44. Wang, Y.; Zio, E.; Wei, X.; Zhang, D.; Wu, B. A resilience perspective on water transport systems: The case of Eastern Star. *Int. J. Disaster Risk Reduct.* **2019**, *33*, 343–354. [CrossRef]

Journal of
*Marine Science
and Engineering*

Article

Risk Identification Method for Ship Navigation in the Complex Waterways via Consideration of Ship Domain

Zhiyuan Wang [1,2,3,4,5], Yong Wu [1,3,4,*], Xiumin Chu [1,2,3,4], Chenguang Liu [1,3,4] and Mao Zheng [1,3,4]

[1]   Intelligent Transportation Systems Research Center, Wuhan University of Technology, Wuhan 430063, China; wangzhiyuan@mju.edu.cn (Z.W.); chuxm@whut.edu.cn (X.C.); liuchenguang@whut.edu.cn (C.L.); zhma1987@126.com (M.Z.)
[2]   Fujian Engineering Research Center of Safety Control for Ship Intelligent Navigation, College of Physics & Electronic Information Engineering, Minjiang University, Fuzhou 35021, China
[3]   National Engineering Research Center for Water Transport Safety, Wuhan 430063, China
[4]   School of Transportation and Logistics Engineering, Wuhan University of Technology, Wuhan 430063, China
[5]   Institute of Oceanography, Minjiang University, Fuzhou 35021, China
*   Correspondence: whut_wuyong@whut.edu.cn

Abstract: Collision risk identification is an important basis for intelligent ship navigation decision-making, which evaluates results that play a crucial role in the safe navigation of ships. However, the curvature, narrowness, and restricted water conditions of complex waterways bring uncertainty and ambiguity to the judgment of the danger of intelligent ship navigation situation, making it difficult to calculate such risk accurately and efficiently with a unified standard. This study proposes a new method for identifying ship navigation risks by combining the ship domain with AIS data to increase the prediction accuracy of collision risk identification for ship navigation in complex waterways. In this method, a ship domain model is constructed based on the ship density map drawn using AIS data. Then, the collision time with the target ship is calculated based on the collision hazard detection line and safety distance boundary, forming a method for dividing the danger level of the ship navigation situation. In addition, the effectiveness of this method was verified through simulation of ships navigation in complex waterways, and correct collision avoidance decisions can be made with the Regulations for Preventing Collisions in Inland Rivers of the People's Republic of China, indicating the advantages of the proposed risk identification method in practical applications.

Keywords: collision risk identification; ais data; ship domain; ship navigation; complex waterways

Citation: Wang, Z.; Wu, Y.; Chu, X.; Liu, C.; Zheng, M. Risk Identification Method for Ship Navigation in the Complex Waterways via Consideration of Ship Domain. *J. Mar. Sci. Eng.* **2023**, *11*, 2265. https://doi.org/10.3390/jmse11122265

Academic Editor: Spyros Hirdaris

Received: 7 October 2023
Revised: 16 November 2023
Accepted: 22 November 2023
Published: 29 November 2023

1. Introduction

With the development of waterway transportation, the increase in the number, tonnage, and speed of ships has raised the possibility of ship collision, especially in complex waterways [1,2]. Therefore, more scientific and efficient collision warning and collision avoidance decisions are needed in today's water navigation scenarios. At present, ships have been widely installed with traffic sensing devices such as an automatic identification system (AIS) and a shipborne radar, and most complex or important waterways have been equipped with vessel traffic service (VTS), whose purpose is to obtain more navigation information to ensure the safe navigation of ships. However, ship collisions still exist and have not decreased with the advances in navigation technology [3]. This may be due to a combination of factors such as insufficient experience of ship navigators, sudden collision situations, and high thresholds for advanced equipment analysis [4]. The International Regulations for Preventing Collisions at Sea (COLREG) is implemented by the International Maritime Organization (IMO), and some navigation regulations are implemented in some complex waterways, such as the Regulations for Preventing Collisions in Inland Rivers of the People's Republic of China in the Yangtze River waterway. Nevertheless, these rules are qualitative and do not have quantitative collision prediction thresholds to assist in judgment in order to reduce the subjective judgment of navigators.

Several methods have been proposed to identify the risk of ship collision. In the mid-20th century, concepts such as Distance to Closest Point of Approach (DCPA), Time to Closest Point of Approach (TCPA), and Ship Domain were used to make decisions on ship collision avoidance [5,6]. Zheng and Wu [7–9] proposed the concepts of space collision risk and time collision risk, which considered the main factors that reflect risk levels of collision between ships, such as DCPA, TCPA, safety distance, and the latest avoidance point, to establish their respective collision avoidance risk models. Sotiralis et al. [10] proposed a quantitative risk analysis method based on Bayesian networks by considering human factors more adequately, which integrates elements from the Technique for Retrospective and Predictive Analysis of Cognitive Errors, focusing on analyzing human-induced collision accidents. Abebe et al. [3] proposed a ship collision risk index estimation model based on the Dempster–Shafer theory and its accuracy and fast calculation were verified by comparing it with different machine learning methods. Pietrzykowski et al. [11] proposed an integrated, comprehensive system of an Autonomous Surface Vessel dedicated to ships with various degrees of autonomy, and tests were conducted under the actual operating conditions of ships. Perera and Soares [12] proposed a collision risk detection and quantification methodology that can be implemented in a modem-integrated bridge system. Ozturk and Cicek [13] believe that the risk assessment of wind, current, and wave height on a ship's dynamic cannot be ignored.

More advantageously, ship domain models can promote the rapid identification of ship navigation collision avoidance, and their boundaries determine the accuracy of ships' collision avoidance. As early as 1963, a symmetric elliptical ship domain model centered around the own ship using the ship's radar information was first proposed and widely applied [14]. Goodwin [5] divides the ship domain into three different-sized sectors based on ship lights. Van der Tak and Spaans [15] established a ship domain model based on previous research, in which the center of the elliptical ship domain has a forward offset from the position of the ship, and the bow of the ship was deflected, resulting in an area of approximately equal to the three-sector areas of the Goodwin model. Coldwell [16] established the ship domain model in overtaking and encountering situations by observing the traffic of over 200 ships in the 19 nautical mile waterways of the Humber River in the UK. Zhu et al. [17] proposed a multi-vessel collision risk assessment model based on the Coldwell ship domain model. Wen et al. [18] obtained the shape and size of specific types of ship domains in typical inland waterways by observing AIS-based grid density maps and analyzing grid density data, in which the ship domain in real life is quite different from the theoretical prediction. The shape of the ship domain in typical inland waters takes the form of an asymmetrically shaped ellipse, with its major axis coinciding with the ship's central line.

With the widespread application of modern AIS technology, research on the ship domain based on AIS has become increasingly meaningful. Qi et al. [19] established a ship domain model and obtained boundary curves of the ship domain under different avoidance degrees by utilizing AIS data from Qiongzhou Strait. Hansen et al. [20] estimated the minimum ship domain where a navigator feels comfortable by observing the AIS data of ships sailing in southern Danish waters for four years. Szlapczynski and Szlapczynska [21] proposed a ship collision risk model based on the concept of ship domain and considered the related domain-based collision risk parameters, such as degree of domain violation, the relative speed of the two vessels, combination of the vessels' courses, arena violations, and encounter complexity. Feng et al. [22] proposed a quantitatively evaluated method of the collision risks combining information entropy, which integrated the K-means clustering based on AIS data. Liu et al. [23] proposed a systematic method based on the dynamic ship domain model to detect possible collision scenarios and identify the distributions of collision risk hot spots in a given area. Du et al. [24] proposed a data ship domain concept and an analytical framework based on AIS, in which the ship domain explicitly incorporates the dynamic nature of the encounter process and the navigator's evasive maneuvers. Liu et al. [25] proposed a quantitative method for the analysis of ship collision risk, in which a kinematics feature-based vessel conflict ranking operator was introduced, and both the static and dynamic information of AIS data were considered to evaluate ship collision risk. Silveira et al. [26] introduced a method to define AIS data-based empirical polygonal ship domains, which can fit the empirical domain better. He et al. [27] proposed a dynamic collision avoidance path planning algorithm for complex multi-ship encounters based on the A-star algorithm and the quaternion ship domain from AIS data. Zhang et al. [28] proposed an interpretable knowledge-based decision support method to guide ship collision avoidance decisions with a knowledge base based on the ordinary practice of seamen using AIS data.

In summary, on the one hand, these studies mainly focus on open water areas, and there is a lack of empirical research on ship navigation risks based on complex inland waterways. On the other hand, research methods based on statistical analysis use actual traffic flow data. Early research was mostly based on radar data, and in recent years, it has shifted towards AIS data. However, for research on risk identification with ship domains in complex waterways, qualitative research is often used, and data containing ship navigation information is rarely used for quantitative research. Thus, while taking the risk identification computations in complex waterways, a ship domain model considering the characteristics of navigation separation based on AIS data needs to be developed.

In this paper, a risk identification method is established based on the collision risk detection line and the safety distance boundary of the ship domain to provide real-time and fast guidance for collision avoidance control during navigation in complex waterways. Section 2 provides an overall description of the proposed risk identification method, Section 3 introduces the process of establishing a ship domain model, and Section 4 introduces collision risk classification algorithms based on the ship domain model. Section 5 validates this method using real ship data. In this way, we can evaluate the collision risk of ship navigation based on AIS data and assist ship navigation by providing a collision avoidance decision-making basis for navigators.

2. Procedure of the Proposed Risk Identification Method

The risk identification method for ship navigation in complex waterways consists of two procedures: establishment of ship domain and collision risk classification, as shown in Figure 1.

Due to the complex waterway characteristics and the Regulations for Preventing Collisions, the AIS position data distribution pattern of the target ships around the own ship in the same direction is not the same as that of the target ships in the opposite direction. Therefore, based on historical AIS data, a corresponding ship domain model for the own ship with different encounter situations is established in the complex waterway, which is described in detail in Section 3. When the position data (from AIS, radar, or a fusion of the two) of the own ship and the target ship are collected during real-time navigation, a mathematical equation using Formula (3) in Section 3.5 and Formula (11) in Section 4.1 can be applied for calculating the safety distance boundary in the ship domain. By determining whether the target ship is within the ship domain, whether the collision risk detection line intersects with the safe distance boundary of the own ship, and whether the time (T_{ca}) to the intersection point between the collision risk detection line and the safe distance boundary of the ship domain is sufficient ($T_{ca} < T_{Alarm}$), the collision risk level is described and divided, forming a fast collision risk identification method for ship navigation in complex waterways.

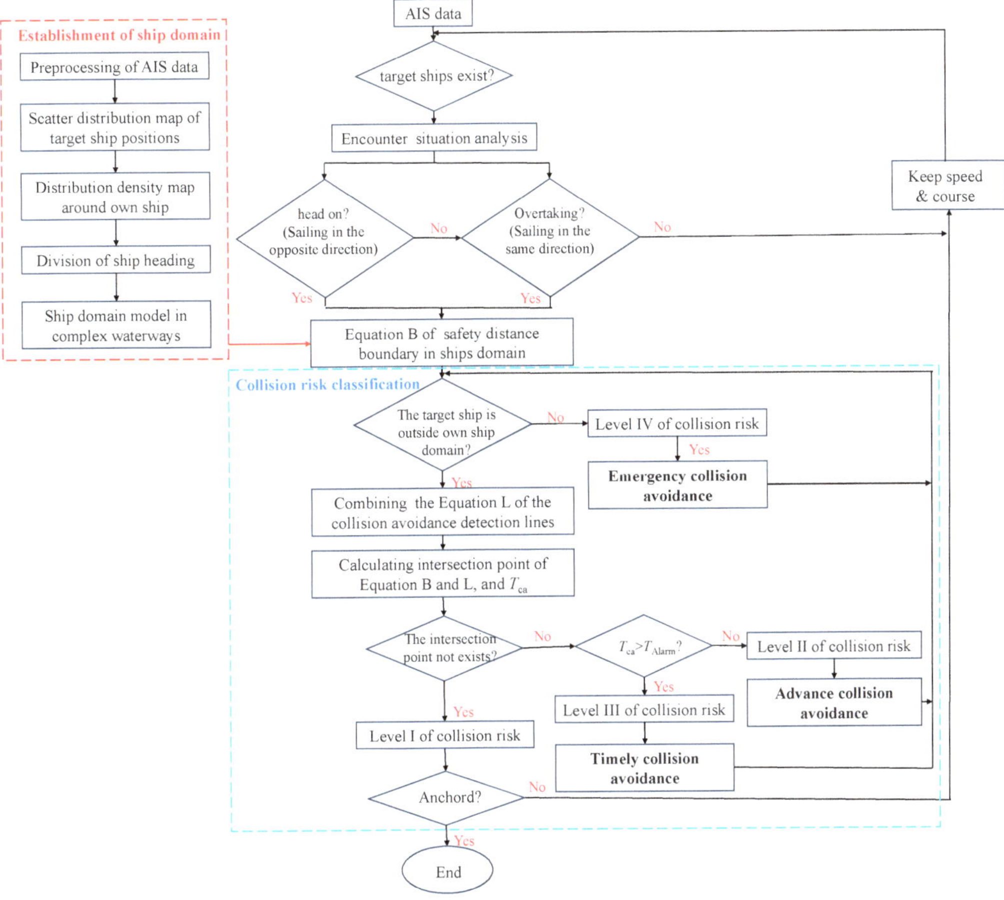

Figure 1. Flow chart of the proposed method.

3. The Ship Domain Model of Navigation in Complex Waterways

The selected data is sourced from AIS onboard equipment on the ship "Channel 1", which includes over 2.7 million AIS and GPS data records of the ships from March 2023 to July 2023 in the Wuhan reach of the Yangtze River.

3.1. Preprocessing of AIS Data

There are some abnormal data in the initial AIS data, so it is necessary to exclude these abnormal data from the selected AIS data, such as ship position anomalies (not within the navigation waterway, with speed but unchanged position), speed anomalies, and heading anomalies, etc. Preprocessing of AIS data mainly includes data cleaning and data repairing. Data repairing mainly involves linear interpolation of the data, with an interpolation interval of 2 s. In addition, the research focuses on the state of the ship during navigation, and it is necessary to determine the current sailing state of the ship based on its speed. Therefore, AIS data with a speed less than 0.5 m/s is considered a parked state, which is excluded. Meanwhile, considering the size characteristics of the ship domain in complex waterways, ships within a 1 km range centered around the own ship are selected as the research object for the encounter situation. The processed AIS data will be used for subsequent calculations of the ship distribution density map.

3.2. Scatter Distribution Map of the Target Ships

Based on historical AIS data, a coordinate system relative to the nearby water area of a ship is established, with the ship's position as the coordinate origin and its heading as the longitudinal axis direction. The relative orientation is calculated from the longitude, latitude, and heading of the ship and the longitude and latitude of the target ship.

Firstly, the relative distance and relative orientation of the two ships at a certain time can be calculated by the longitude and latitude of the ships.

Secondly, to ensure that the own ship's position is the coordinate origin and its heading is the longitudinal axis direction, it is necessary to perform coordinate conversion on the heading of other ships while the own ship is sailing to obtain the relative orientation, which ensures the accuracy of the selected statistical samples.

Finally, based on the calculated relative distance and relative orientation, a scatter distribution map of target ships can be visualized and displayed, as shown in Figure 2.

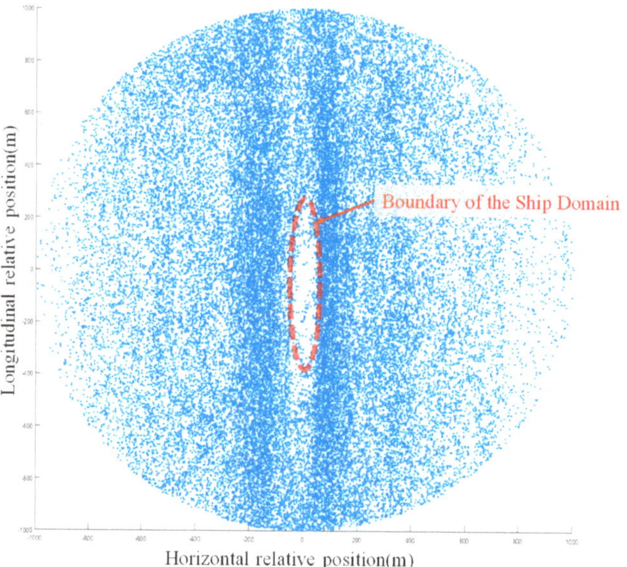

Figure 2. Scatter distribution map of target ships.

Figure 2 shows that there is a sparse distribution area of ship positions near the origin, which is the ship domain of the own ship.

3.3. Distribution Density Map around the Ship

To improve the accuracy of describing the distribution of target ships around the own ship, the Scatter distribution map of target ships was converted into a distribution density map for analysis, in which the density value of coordinate points is the number of ship points within a unit distance divided by the area. When the unit distance is selected as 5 m, the conversion of the distribution density map is shown in Figure 3.

By calculating the density of each coordinate point around the own ship and setting color legends based on the density, a grid density map of the own ship is obtained. The shape and size of the own ship domain can be obtained by analyzing density map data, as detailed in Section 3.5.

Figure 3b shows an area in the center of the image with a density much lower than other surrounding spaces, which is classified as the ship domain of the own ship for safe navigation. The ship domain is all in an asymmetric elliptical shape, with the major axis of the ellipse parallel to the bow direction, and the center of the ellipse deviates from the center of the ship domain (the coordinate origin). The center of the ellipse in the ship domain is located to the left rear of the center of the ship.

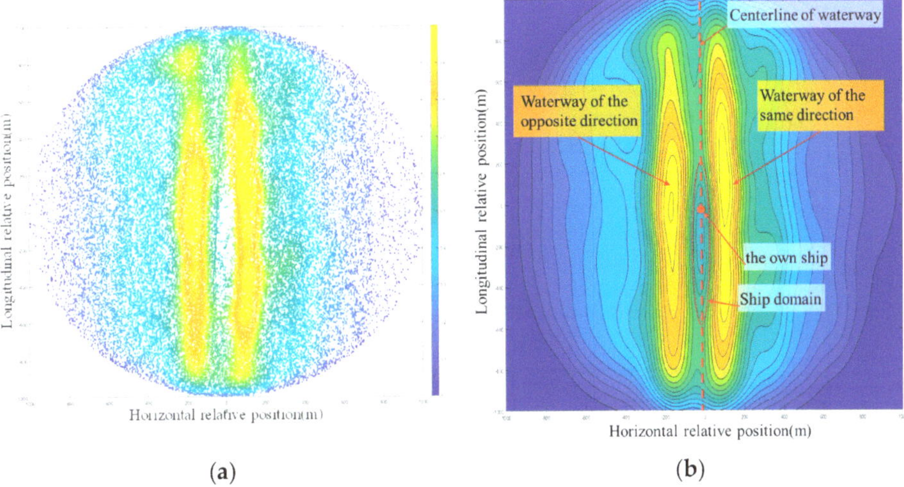

Figure 3. Distribution density map around the own ship. (**a**) Scatter distribution map; (**b**) Distribution density map.

In addition, there are banded high-density areas of ships on both sides of the own ship domain. The "Channel 1" researched is a command ship for channel work, with a total length of 49.60 m, a molded width of 9.3 m, and a molded depth of 3.1 m, which is much faster than other cargo ships sailing in the Yangtze River. The power system of the ship is two C32 twelve-cylinder V-type four-stroke engines produced by Caterpillar in the United States, with a cruising speed of up to 35 km per hour. Therefore, the own ship often navigates close to the centerline of the waterway and is often in a situation of overtaking other ships. When the right chord of the own ship overtook another ship, it should maintain a distance of about 50 m, which is consistent with the banded high-density area of ships within a range of 0 to 200 m on the right side of the origin, and these ships are overtaking ships sailing in the same direction on the right side of the own ship. On the other side, there is also another banded high-density area of ships within the range of 50 to 250 m on the left side of the own ship, which is the encounter of ships sailing in the opposite direction on the left side of the centerline of the waterway. The distance between ships encountered on the port side and the own ship is generally greater than 100 m.

Therefore, the shape and scale of the ship domain in typical complex waterways are significantly different from that in the wide waters of the sea, as well as different from that based on the traditional theoretical analysis. The shape of the ship domain is significantly influenced by its navigation behavior.

3.4. Distribution Density Map with Different Encounter Situations

To ensure the safety of Yangtze River vessel navigation, provide navigation efficiency, and promote shipping development, Yangtze River vessel navigation has implemented segmented waterway navigation and fixed route navigation.

The segmented waterway navigation includes two-way navigation and one-way navigation, and a crossing zone is set up.

During the implementation of two-way navigation, ships follow the prescribed route and try to navigate on the side of the waterway as much as possible in order to maintain sufficient safety distance in case of an encounter. When overtaking, the slow ship should navigate on one side of the channel, while the fast ship should overtake the slow ship on the centerline side of the waterway as parallel as possible and maintain sufficient lateral distance to prevent the occurrence of ship suction. The overtaking ship should first pass through the stern of the overtaking ship and then carry out overtaking. After overtaking another ship, the overtaking ship should not immediately turn to cross the front of the other ship to avoid creating an urgent situation.

During the implementation of one-way navigation, it is stipulated that ships should pass through in one direction and are prohibited from giving way to each other. Ships should navigate laterally within the designated lateral navigation area, strengthen the lookout, and steer with caution.

Due to ships navigating in different lanes along the Yangtze River, the surrounding ships are classified into two categories: in the same and opposite directions for analysis. Figure 4 shows the relative heading distribution of surrounding ships based on the own ship's heading, in which 77.5% of the ships have relative heading between −180 degrees to −160 degrees, −20 degrees to 20 degrees, and 160 degrees to 180 degrees. Therefore, the relative heading of the ship is selected with data from −20 degrees to 20 degrees as the same direction data and data from −180 degrees to −160 degrees and 160 degrees to 180 degrees as the opposite direction data for distribution density statistical analysis of ship.

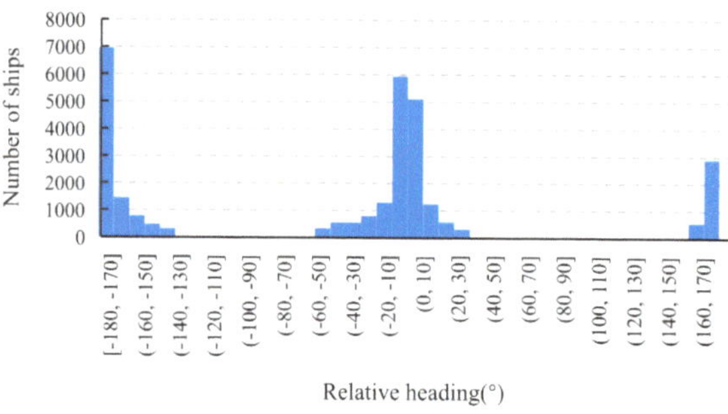

Figure 4. Distribution of relative heading of ships.

Firstly, the scatter distribution map of target ships around the own ship in the same direction can be obtained by selecting data from −20 degrees to 20 degrees for the relative heading of ships as co-heading data, as shown in Figure 5.

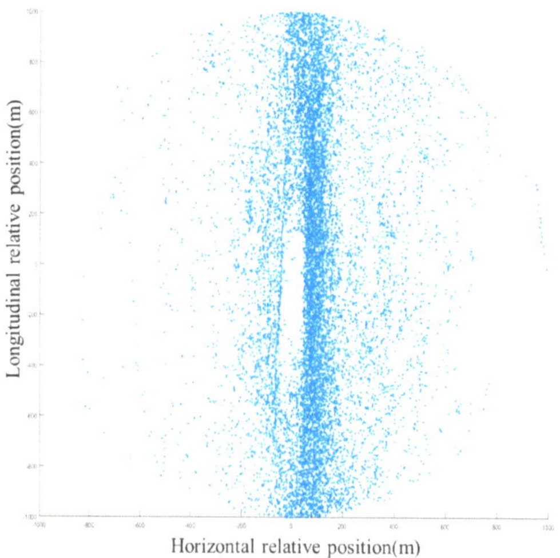

Figure 5. Scatter distribution map of target ships around the own ship in the same direction.

Considering that the navigation diversion of the Yangtze River's inland rivers will change left and right with the flow conditions of the channel, the scatter distribution map will be symmetrically supplemented along the x = 0 line to calculate the distribution density map, as shown in Figure 6.

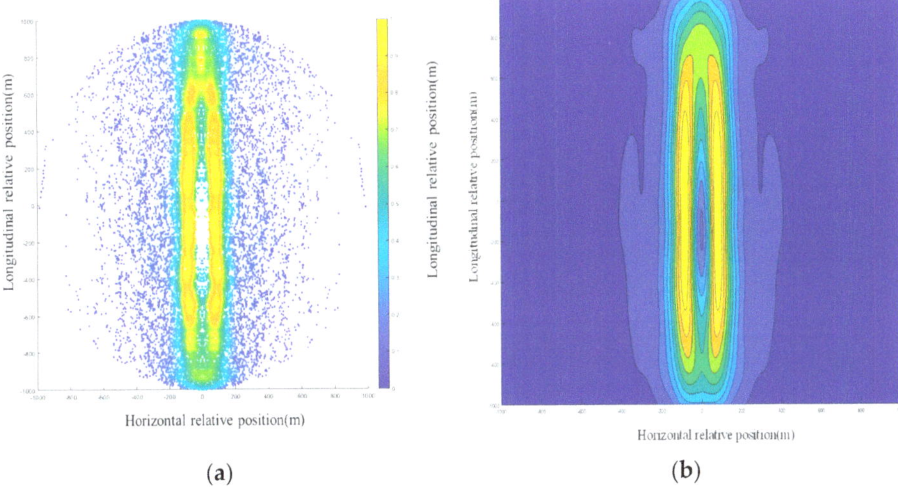

(**a**)　　　　　　　　　　　　　　　(**b**)

Figure 6. Distribution density map around the own ship in the same direction. (**a**) scatter distribution map; (**b**) distribution density map.

Then, the scatter distribution map of target ships around the own ship in the opposite direction can be obtained by selecting data from −180 degrees to −160 degrees and 60 degrees to 180 degrees for the relative heading of ships as opposing navigation data, as shown in Figure 7.

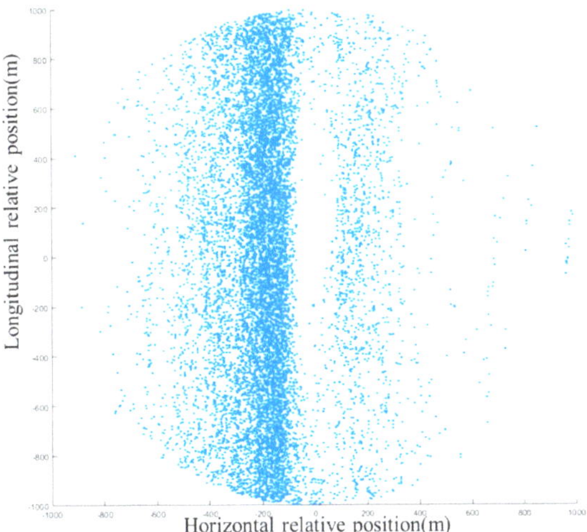

Figure 7. Scatter distribution map of target ships around the own ship in the opposite direction.

Similarly, the scatter distribution map will be symmetrically supplemented along the x = 0 line to calculate the distribution density map, as shown in Figure 8.

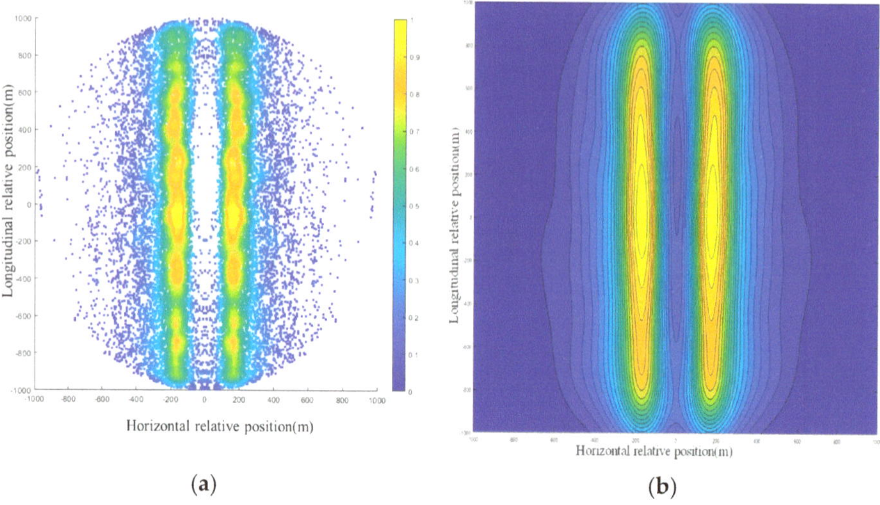

(**a**)   (**b**)

Figure 8. Distribution density map around the own ship in the opposite direction. (**a**) Scatter distribution map; (**b**) Distribution density map.

It can be seen that when sailing in the same direction, the navigation behavior of the own ship is mostly following or overtaking, and the space in the ship domain presents a situation of short front and long back (in a word, the center of the ship domain moves backward), and the space on both sides is relatively narrow due to the influence of lane width;

When sailing in the opposite direction, the navigation behavior of the own ship is mostly encountering behavior, and the space in the ship domain presents a situation of the long front and short back (in a word, the center of the ship domain moves forward), and the space on both sides is relatively wide due to the influence of lane width.

3.5. The Mathematical Models of Ship Domain in Complex Waterways

Considering the characteristics of ship navigation in complex waterways, the elliptical ship domain is chosen as the basic graphic, as shown in Figure 9.

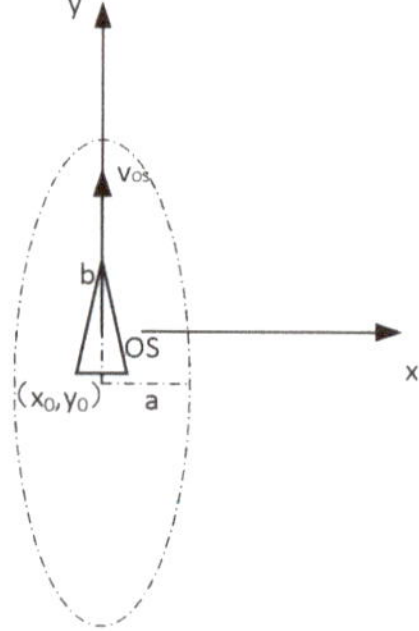

Figure 9. Schematic diagram of an elliptical ship domain model with eccentricity.

The coordinate system is established with the own ship as the origin, the right transverse direction as the *x*-axis direction, and the heading direction of the own ship as the *y*-axis direction.

In this coordinate system, the boundary equation of the own ship domain is given by

$$f(x,y) = \frac{(x + x_0)^2}{a^2} + \frac{(y + y_0)^2}{b^2} = 1 \tag{1}$$

where a is the radius length of the elliptical ship-type field in the positive and negative directions of the x-axis, m; b is the radius length of the elliptical ship-type field in the positive and negative directions of the y-axis; x_0 is the eccentric coordinate in the x-axis direction, m; y_0 is the eccentric coordinate in the y-axis direction, m.

To determine the size of the ship field, the following process was carried out:

① In the distribution density map, select the density data of ships in the same and opposite directions separately and generate the cross-section of $x_0 = 0$ (the center of ship domain) section. Observe the changes in the ship domain under different cutting thresholds to determine the appropriate threshold value. Therefore, a cutting threshold of 40% was selected as the size of the ship domain at the same direction, and the eccentricity points in the same direction ship domain were determined to be (0, −140), with a major axis radius $b = 360$ m; a cutting threshold of 15% was selected as the size of the ship domain in the opposing direction, and the eccentricity points in the same direction ship domain were determined to be (0, 160), with a major axis radius $b = 350$ m, as shown in Figure 10.

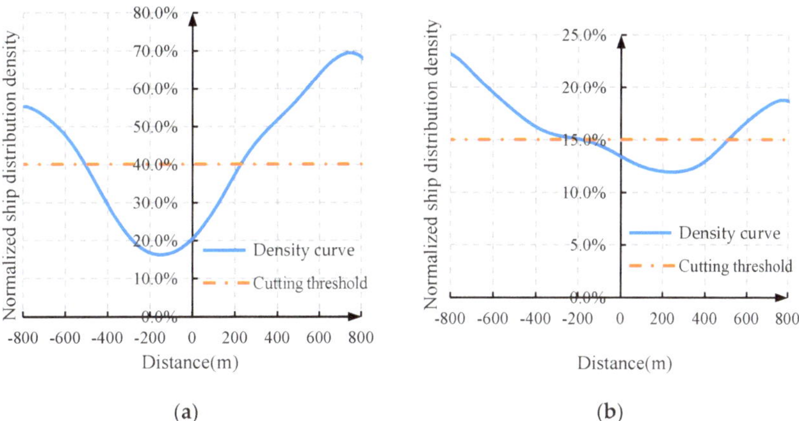

Figure 10. The density curve of the ship under the $x_0 = 0$ section for ship navigation. (**a**) the same direction; (**b**) the opposite direction.

② Select the density data in Figure 6b, and generate the cross-section of $y = -140$ (the center of the same direction ship domain) section; Select the density data in Figure 8b, and generate the cross-section of $y = 160$ (the center of the opposing direction ship domain) section, as shown in Figure 11. A cutting threshold of 40% was selected as the size of the ship domain, and the width a of the same direction sailing ship domain is 30 m, and the width a of the opposite direction sailing ship domain is 70 m.

The mathematical formulas for normalized density in Figures 10 and 11 are given by

$$f(x,y) = \frac{\rho(x,y)}{\max\limits_{-1000<x<1000,-1000<y<1000}\rho(x,y)} \tag{2}$$

where $\rho(x, y)$ is the ship density at the left point of (x, y); select a unit distance of 5 m and calculate the density value of each coordinate point by dividing the number of ship points within the unit distance by the area.

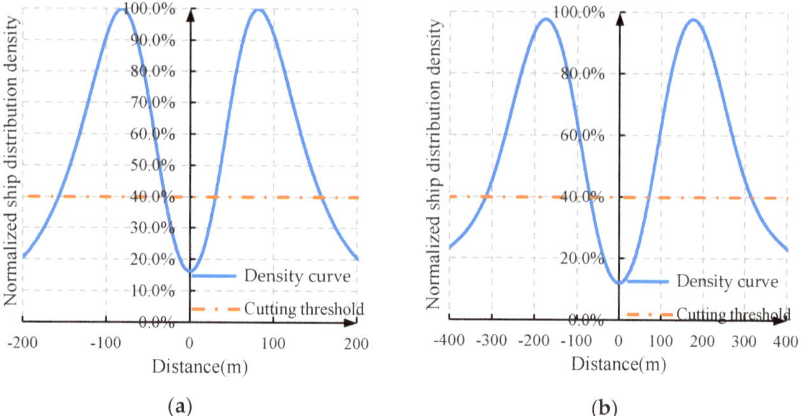

Figure 11. The density curve of the ship under the *y* section for ship navigation. (**a**) the same direction; (**b**) the opposite direction.

③ When judging whether there is a risk of collision solely based on the ship domain, it may result in the two ships being unable to pass at a safe distance in the complex waterway, so expand the values of *a* and *b* in the ship domain by *k* times to increase an external safety distance boundary and improve the safety of ships during encounters. The equation of the safe distance boundary of the ship domain is given by

$$f(x,y) = \frac{(x + x_0)^2}{\left(k_a \cdot a\right)^2} + \frac{(y + y_0)^2}{\left(k_b \cdot b\right)^2} = 1 \tag{3}$$

where k_a and k_b are taken between [1, 2]. By visual comparative analysis, when sailing in the same direction, $k_a = 1.5$ and $k_b = 1$; when sailing in the opposite direction, $k_a = 1.15$ and $k_b = 1.28$, as shown in Figure 12.

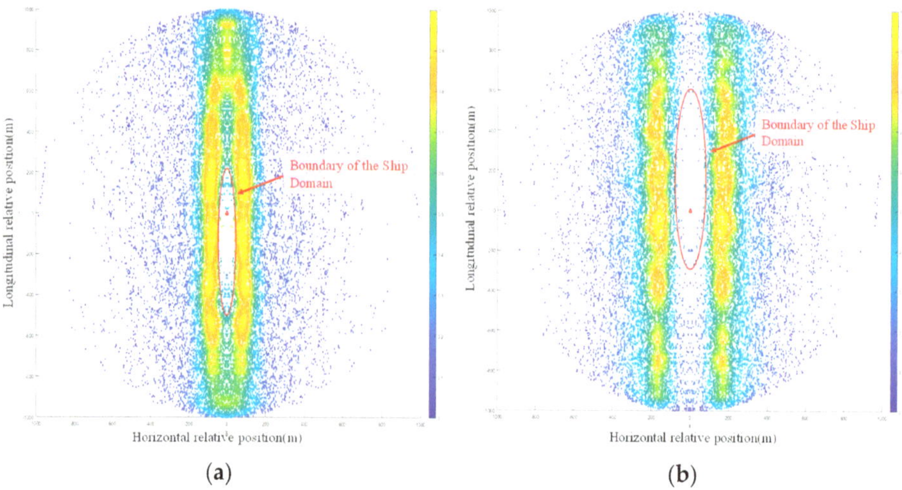

Figure 12. Schematic diagram of the safe distance boundary of the ship domain. (**a**) the same direction; (**b**) the opposite direction.

The equation of the safe distance boundary of the ship domain is defined as Formula (3). According to research conducted with experienced navigators and Wen's research [18], the safe distance between the own ship and the target ship ahead in the same direction is about 200 m in the Yangtze River. In Figure 10a, the Normalized ship distribution density corresponding to a distance of 200 m from the target ship ahead is about 40%. Therefore, we chose a cutting threshold of 40% in Figures 10 and 11. However, in Figure 10b, choosing a cutting threshold of 40% is clearly inappropriate, so a moderate cutting threshold of 15% was chosen to determine the eccentric position of the ellipse. In order to ensure the availability of the final ship domain, adjustments are made in the Formula (3) through k_a and k_b.

4. The Collision Risk Classification Method of Ship Navigation

4.1. Collision Avoidance Collision Model Based on Ship Domain

The collision risk detection line is the straight line where the target ship's direction of its relative velocity vector is located. If the collision risk detection line intersects with the safe distance boundary of the ship domain, there is a risk of collision; otherwise, there is no risk of collision, as shown in Figure 13. Therefore, the equation of the collision risk detection line in the coordinate system with the own ship's position as the origin and its heading direction as the y-axis direction.

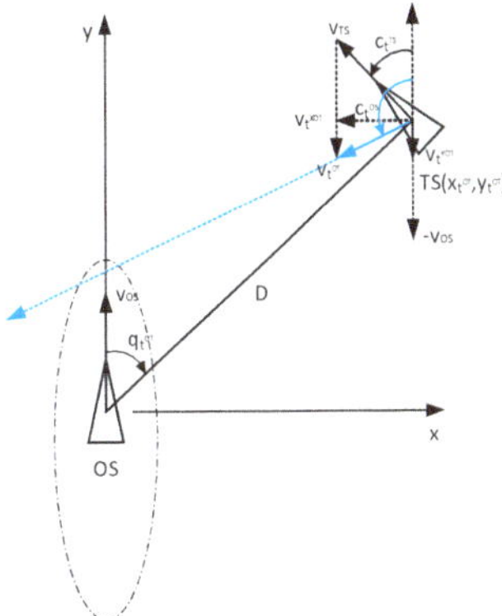

Figure 13. Schematic diagram of collision judgment.

Assuming the coordinate point of the own ship at time t is OS $(0, 0)$, the speed is v^{os}, and the heading is $0°$; the relative distance of the target ship to the own ship at time t is D, the relative orientation is q_t^{OT}, the coordinate point is TS (x_t^{OT}, y_t^{OT}), the speed v^{TS}, and the relative heading is c_t^{TS}. For the convenience of calculation, the range of the relative orientation and the relative heading of the target ship have been changed to between $[-180°, 180°]$.

$$q_t^{OT} = \begin{cases} q_t^{OT}, & 0° \le q_t^{OT} \le 180° \\ q_t^{OT} - 360°, & 180° \le q_t^{OT} \le 360° \end{cases} \tag{4}$$

$$c_t^{TS} = \begin{cases} c_t^{TS}, & 0° \le c_t^{TS} \le 180° \\ c_t^{TS} - 360°, & 180° \le c_t^{TS} \le 360° \end{cases} \tag{5}$$

The coordinate point of the target ship is $TS\ (x_t^{OT}, y_t^{OT})$, which is given by

$$\begin{cases} x_t^{OT} = D \cdot sinq_t^{OT} \\ y_t^{OT} = D \cdot cosq_t^{OT} \end{cases} \tag{6}$$

The velocity components of the target ship's velocity on the x and y axes are as follows:

$$\begin{cases} v_t^{xTS} = v_t^{TS} \cdot sinc_t^{TS} \\ v_t^{yTS} = v_t^{TS} \cdot cosc_t^{TS} \end{cases} \tag{7}$$

The velocity components of the target ship's relative velocity on the x and y axes are as follows:

$$\begin{cases} v_t^{xOT} = v_t^{xTS} \\ v_t^{yOT} = v_t^{yts} - v_t^{OS} \end{cases} \tag{8}$$

When $v_t^{yOT} \neq 0$, the slope k_t of the collision risk detection line is:

$$k_t = \frac{v_t^{xOT}}{v_t^{yOT}} \tag{9}$$

When $v_t^{yOT} = 0$, the slope k_t of the collision risk detection line is:

$$k_t = 0 \tag{10}$$

According to Formulas (9) and (10), the equation of the consolidation risk detection line is given by

$$y = f(x) = k_t x + y_t^{OT} - k_t x_t^{OT} \tag{11}$$

The equation of the consolidation risk detection line is defined as Formula (3).

By using simultaneous Equations (2) and (11), the intersection point between the collision risk detection line and the safe distance boundary of the ship domain can be obtained.

The number of intersection points between Formulas (3) and (11) is determined by the discriminant of a univariate quadratic equation. The discriminant Δ is given by

$$\Delta = \left[2x_0 k_b^2 b^2 + 2k_a^2 a^2 k_t \left(y_t^{OT} - x_t^{OT} k_t + y_0 \right) \right]^2 - 4 \left(k_b^2 b^2 + k_a^2 a^2 k_t^2 \right) \cdot \left[k_b^2 b^2 x_0^2 + k_a^2 a^2 \cdot \left(y_t^{OT} - x_t^{OT} k_t + y_0 \right)^2 - k_a^2 a^2 k_b^2 b^2 \right] \tag{12}$$

When $\Delta < 0$, there is no intersection between Formulas (3) and (11), there is no risk of collision with the target ship; when $\Delta \geq 0$, there are 1 or 2 intersections between Formulas (3) and (11), there is a risk of collision with the target ship. The x-coordinate of the intersection point is given by:

$$\begin{cases} x_1 = \dfrac{-\left[2x_0 k_b^2 b^2 + 2k_a^2 a^2 k_t \left(y_t^{ot} - x_t^{ol} k_t + y_0 \right) \right] + \sqrt{\Delta}}{2 \left(k_b^2 b^2 + k_a^2 a^2 k_t^2 \right)} \\ x_2 = \dfrac{-\left[2x_0 k_b^2 b^2 + 2k_a^2 a^2 k_t \left(y_t^{ot} - x_t^{ol} k_t + y_0 \right) \right] - \sqrt{\Delta}}{2 \left(k_b^2 b^2 + k_a^2 a^2 k_t^2 \right)} \end{cases} \tag{13}$$

And the encounter time to the intersection point T_{cax} is:

$$\begin{cases} T_{cax1} = \dfrac{x_1 - x_0}{v_t^{xOT}}, T_{cax1} \geq 0 \\ T_{cax2} = \dfrac{x_2 - x_0}{v_t^{xOT}}, T_{cax2} \geq 0 \end{cases} \tag{14}$$

Take the minimum between T_{cax1} and T_{cax2} as the value of the encounter time:

$$T_{ca} = min\{T_{cax1}, T_{cax2}\} \tag{15}$$

4.2. Collision Risk Classification

According to the safety distance boundary in the ship domain, refer to Li's risk classification method [29], the collision risk detection line, and the encounter time at the intersection point, the collision risk level of the ship navigation in complex waterways is divided into four levels: safe (Level I of collision risk), unsafe (Level II of collision risk), dangerous (Level III of collision risk), and very dangerous (Level IV of collision risk).

① Level I of collision risk: if the target ship is outside the own ship domain, and the collision risk detection line does not intersect with the safety distance boundary in the ship domain ($\Delta < 0$), it indicates that the two ships can safely encounter each other. In this case, it is considered safe.

② Level II of collision risk: if the target ship is outside the own ship domain, and the collision risk detection line intersects with the safety distance boundary in the ship domain ($\Delta \geq 0$), but the encounter time to the intersection point T_{ca} is greater than the alert time T_{Alarm}, it indicates that the own ship has sufficient time to take avoidance actions to avoid the target ship from entering its domain and reduce the safety. In this case, it is considered unsafe.

③ Level III of collision risk: if the target ship is outside the own ship domain, and the collision risk detection line intersects with the safety distance boundary in the ship domain ($\Delta \geq 0$), but $T_{ca} \leq T_{Alarm}$, it indicates that the own ship must take timely avoidance actions to avoid the target ship from entering its domain, which poses a collision risk. In this case, it is considered dangerous.

④ Level IV of collision risk: if the target ship is inside the own ship domain, it indicates that the target ship has posed a threat to the own ship and poses a significant collision risk, which must implement emergency collision avoidance. In this case, it is considered very dangerous.

5. Simulation Results and Discussion

5.1. Simulation

A complex waterway in the Wuhan reach of the Yangtze River was selected, as ships frequently encounter each other, which significantly increases the demand for collision avoidance warnings. According to the ship "Channel 1", which has been introduced in Section 3.3, and its actual situation of collision avoidance with other ships in this complex waterway, the proposed risk identification method has been verified by mainly taking practical avoidance measures in the change of direction, when "Channel 1" and other ships are in a dangerous or urgent situation of encounter.

From 15:48 to 15:52 on 18 April, the ship "Channel 1" was sailing on the Luoyang reach between the Yangluo Yangtze River Highway Bridge and the Wuhan Qingshan Yangtze River Bridge, which has a width of about 350 m (the up lane width is about 150 m and the down lane width is about 200 m), and a depth of 8 m. Figure 14 shows the electronic chart of the Luoyang reach, in which facing the direction of water flow, the down lane is on the left side of the waterway, while the up lane is on the right side.

Figure 14. Chart with AIS data of the Luoyang reach.

During this navigation, AIS data of the own ship "Channel 1" overtaking the target ship (MMSI: 413786692) was selected for analysis. In the beginning, the target ship was sailing at a speed of 3.4 knots, while the own ship followed at a speed of 13 knots. As the distance gradually narrowed to about 500 m, the own ship turned right and began to overtake the target ship. Then, the own ship was sailing parallel to the target ship, with a parallel distance of approximately 50 m. Finally, the own ship overtook the target ship and then turned left to return to the middle of the up lane of the waterway. The entire overtaking process is shown in Figures 15 and 16.

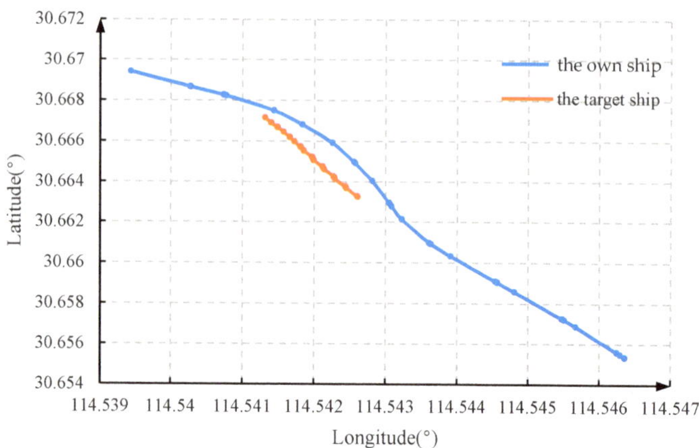

Figure 15. The trajectory map of the own ship and the target ship.

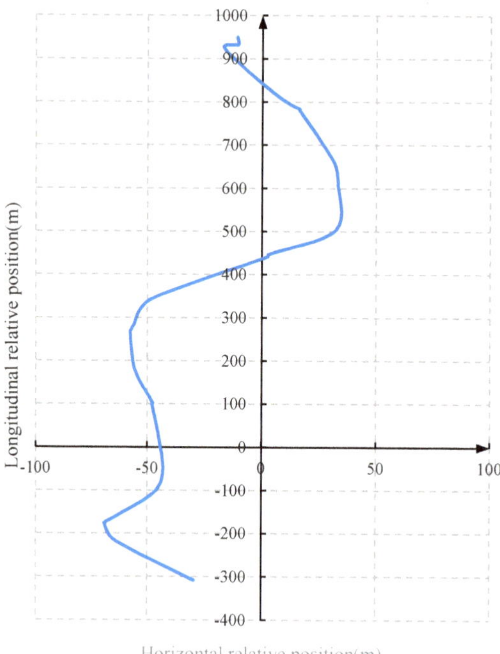

Figure 16. The trajectory map of the target ship relative to the own ship.

5.2. Validation

Figure 17 shows that T_{ca} can effectively describe the collision risk level of the target ship relative to the own ship when the alert time T_{Alarm} is 60 s. In addition, due to the low frequency of sending AIS data from the target ship and the low real-time performance of AIS data, which leads to missed alarms and poor alarm stability when using AIS data alone to calculate the collision risk level, compared to using the fusion data of AIS and radar. Compared with using the fusion data of AIS and radar, the lag time of only using AIS data calculation is about 5–10 s, which may affect the navigator's misjudgment of the navigation risk situation.

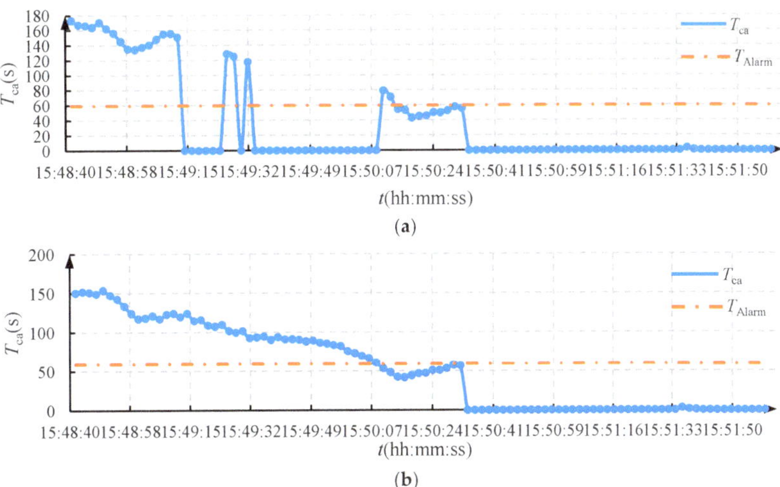

Figure 17. Change rate curve of the encounter time to the intersection point T_{ca}. (**a**) Data from AIS; (**b**) Data from the fusion of AIS and radar.

The change rate of the own ship's heading objectively reflects the behavior of the ship's navigating operation. In other words, when the navigator discovers that there is a collision risk of the target ship, he will take corresponding avoidance actions, resulting in a change in the ship's heading. Therefore, comparing the T_{ca} change curve of the fused data with the change rate curve of the ship's heading can prove whether the collision risk judgment result is consistent with the navigator's risk judgment result, which can also verify the reliability of our risk identification method.

Figure 18 shows that at 15:50:00, our method began to alert collision risks from the target ship, and almost at the same time, the rate of the own ship's heading change gradually increased, indicating that the navigator also judged that there was a risk of collision from the target ship and turned the rudder. Until 15:50:32, our method ends the collision risk alarm for the target ship, and the rate of the own ship's heading change is gradually decreasing, indicating that at this time, the navigator also determines that the collision risk of the target ship is relieved, and turns the rudder in reverse to return to the original heading for navigation. Overall, our method's collision risk alarm judgment is consistent with the navigator's judgment, so our method is effective in identifying the collision risk during navigation and in line with the navigator's cognitive habits.

Therefore, the risk identification method for ship navigation in complex waterways via consideration of ship domain can accurately reflect the collision risk of the target ship relative to the own ship, and the warning data is stable and has good real-time performance, which can effectively assist the navigator in identifying collision risks during navigation, in other words, our proposed method has been validated.

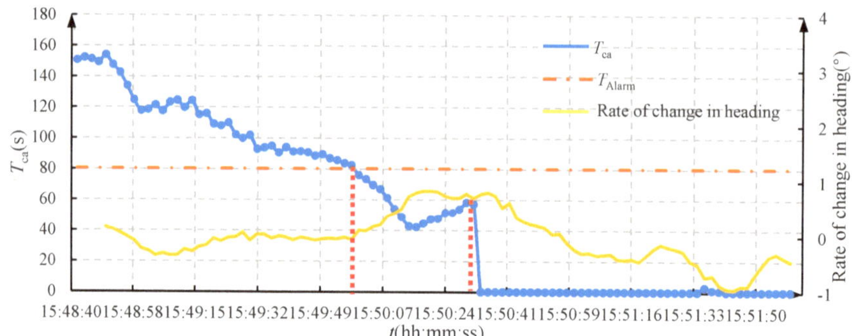

Figure 18. Relationship between T_{ca} and heading change rate (the blue line is T_{ca}, and the yellow line is the actual rate of change in heading).

5.3. Comparison

According to traditional methods, DCPA less than 200 m and TCPA less than 80 s are used as alarm judgment conditions, where DCPA is less than 100 m throughout the entire process, so only the changes in TCPA are considered [30].

Figure 19 shows that our proposed method is more accurate than the traditional DCPA and TCPA judgment methods for early warning. Firstly, the traditional DCPA and TCPA judgment methods lag by 8 s before starting to alarm. Secondly, after the navigation risk alarm, the traditional methods always maintain the alarm state for a long time and only stop the alarm after the ship has passed and cleared, which is inconsistent with the driver's judgment of navigation risk and cannot accurately reflect the navigation situation.

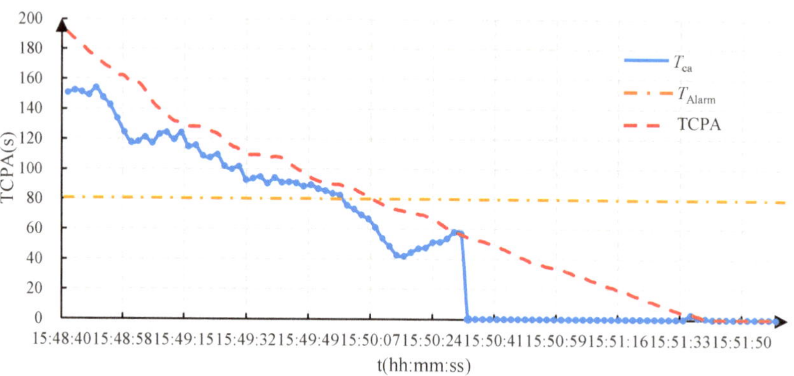

Figure 19. Comparison between our method and TCPA.

5.4. Discussion

In this paper, a new method for identifying ship navigation risks by combining the ship domain with AIS data is proposed, in which the collision time with the target ship is calculated based on the collision hazard detection line and safety distance boundary and formed a method for dividing the danger level of the ship navigation situation.

The research focuses on the Wuhan reach of the Yangtze River, where ships navigate by the Regulations for Preventing Collisions in Inland Rivers of the People's Republic of China. In the complex waterway of the Yangtze River, the simulation results of our method are consistent with the actual navigation behavior and superior to the traditional TCPA method. This also indicates that our method can accurately reflect the consolidation risk of the target ship relative to the own ship, and the warning data is stable and has good real-time performance, which can effectively assist the navigator in identifying consolidation risks during navigation.

However, our method did not consider the hydrodynamic interaction between the ships. External hydrodynamic interference, such as waves and currents, is one of the important influencing factors for ship navigation decisions [31,32]. Our method has limitations by determining a reasonable size of the ship domain and avoiding the impact of hydrodynamic interference through navigators. Due to the lack of actual flow field data, poor ship informatization conditions, and the lack of equipment such as recorders and meteorological instruments, our research cannot consider external disturbances. In addition, hydrodynamic simulation requires a certain amount of computational power support, which may improve the accuracy of early warning but reduce the speed of early warning.

In the future, we will conduct research on rapid simulation and early warning methods that consider hydrodynamic interaction based on our proposed method. Our method will provide a decision-making basis for intelligent ships and assist in the construction of the future New Generation of Waterborne Transportation systems [33].

6. Conclusions

We have considered navigation conditions such as narrow waterways, as well as the characteristics of navigation rules for Yangtze River diversion navigation, and conducted research on methods for estimating the navigation collision risk of ships, and obtained the following conclusions:

(1) In this paper, we combined the ship domain and ship position data from the fusion of AIS and radar to calculate the collision risk level of ship navigation and proposed a new convenient risk identification method for ship navigation in complex waterways.

(2) According to the analysis of the distribution density map around the ship, it can be seen that when sailing in the same direction, the center of the ship domain moves backward, while sailing in the opposite direction, the center of the ship domain moves forward. The space on both sides of the ship is wider in the same direction than that in the opposite direction.

(3) According to the safety distance boundary in the ship domain, the collision risk detection line, and the encounter time at the intersection point, the collision risk level of the ship navigation in complex waterways is divided into four levels: safe (Level I of collision risk), unsafe (Level II of collision risk), dangerous (Level III of collision risk), and very dangerous (Level IV of collision risk). In addition, data analysis was conducted on the real overtaking instance on "Channel 1", verifying the effectiveness, stability, and real-time performance of our risk identification method.

Author Contributions: Conceptualization, Z.W. and Y.W.; methodology, Z.W. and Y.W.; validation, Z.W., Y.W. and C.L.; formal analysis, Z.W. and Y.W.; investigation, Z.W. and Y.W.; resources, X.C. and M.Z.; data curation, Y.W.; writing—original draft preparation, Z.W.; writing—review and editing, C.L. and M.Z.; visualization, Z.W.; supervision, X.C.; project administration, Z.W. and Y.W.; funding acquisition, Z.W., Y.W., X.C., M.Z. and C.L. All authors have read and agreed to the published version of the manuscript.

Funding: This research was funded by the National Natural Science Foundation of China (No. 52172327, No. 52001243, No. 52001240); the Open Project Program of Science and Technology on Hydrodynamics Laboratory (No. 6142203210204); the National Key R&D Program of China (No. 2022YFB4301402); the Fujian Province Natural Science Foundation (No. 2020J01860); the Fujian Marine Economic Development Special Fund Project (No. FJHJF-L-2022-17); the Fujian Science and Technology Major Special Project (No. 2022NZ033023); the Fujian Science and Technology Innovation Key Project (No. 2022G02027, No. 2022G02009); and the Science and Technology Key Project of Fuzhou (No. 2022-ZD-021); and the Scientific Research Foundation for the Ph.D., Minjiang University (No. MJY19032). All authors have read and agreed to the published version of the manuscript.

Institutional Review Board Statement: Not applicable.

Informed Consent Statement: Not applicable.

Data Availability Statement: Data are contained within the article.

Conflicts of Interest: The authors declare no conflict of interest.

References

1. Zhang, M.Y.; Montewka, J.; Manderbacka, T.; Kujala, P.; Hirdaris, S. A big data analytics method for the evaluation of ship-ship collision risk reflecting hydrometeorological conditions. *Reliab. Eng. Syst. Saf.* **2021**, *213*, 107674. [CrossRef]
2. Zhang, M.Y.; Conti, F.; Sourne, H.L.; Vassalos, D.; Kujala, P.; Lindroth, D.; Hirdaris, S. A method for the direct assessment of ship collision damage and flooding risk in real conditions. *Ocean. Eng.* **2021**, *237*, 109605. [CrossRef]
3. Abebe, M.; Noh, Y.; Seo, C.; Kim, D.; Lee, I. Developing a ship collision risk index estimation model based on Dempster-Shafer theory. *Appl. Ocean. Res.* **2021**, *113*, 102735. [CrossRef]
4. Li, S.; Meng, Q.; Qu, X. An overview of maritime waterway quantitative risk assessment models. *Risk Anal. Int. J.* **2012**, *32*, 496–512. [CrossRef] [PubMed]
5. Goodwin, E.M. A statistical study of ship domains. *J. Navig.* **1975**, *28*, 328–344. [CrossRef]
6. Davis, P.V.; Dove, M.J.; Srockel, C.T. A computer simulation of marine traffic using domains and arenas. *J. Navig.* **1980**, *33*, 215–222. [CrossRef]
7. Wu, Z.L.; Zheng, Z.Y. Time collision risk and its model. *J. Dalian Marit. Univ.* **2001**, *27*, 1–5.
8. Wu, Z.L.; Zheng, Z.Y. Space collision risk and its model. *J. Dalian Marit. Univ.* **2001**, *27*, 1–4.
9. Zheng, Z.Y.; Wu, Z.L. New model of collision risk between vessels. *J. Dalian Marit. Univ.* **2002**, *28*, 1–5.
10. Sotiralis, P.; Ventikos, N.P.; Hamann, R.; Golyshev, P.; Teixeira, A.P. Incorporation of human factors into ship collision risk models focusing on human centred design aspects. *Reliab. Eng. Syst. Saf.* **2016**, *156*, 210–227. [CrossRef]
11. Pietrzykowski, Z.; Wołejsza, P.; Nozdrzykowski, Ł.; Borkowski, P.; Banaś, P.; Magaj, J.; Chomski, J.; Mąka, M.; Mielniczuk, S.; Pańka, A.; et al. The autonomous navigation system of a sea-going vessel. *Ocean. Eng.* **2022**, *261*, 112104. [CrossRef]
12. Perera, L.P.; Soares, C.G. Collision risk detection and quantification in ship navigation with integrated bridge systems. *Ocean. Eng.* **2015**, *109*, 344–354. [CrossRef]
13. Ozturk, U.; Cicek, K. Individual collision risk assessment in ship navigation: A systematic literature review. *Ocean. Eng.* **2019**, *180*, 130–143. [CrossRef]
14. Fujii, Y.; Tanaka, K. Traffic capacity. *J. Navig.* **1971**, *24*, 543–552. [CrossRef]
15. Van der Tak, C.; Spaans, J.A. A model for calculating a maritime risk criterion number. *J. Navig.* **1977**, *30*, 287–295. [CrossRef]
16. Coldwell, T.G. Marine traffic behaviour in restricted waters. *J. Navig.* **1983**, *36*, 430–444. [CrossRef]
17. Zhu, K.; Shi, G.Y.; Wang, Q.; Liu, J. Valuation model of ship collision risk based on ship domain. *J. Dalian Marit. Univ.* **2020**, *41*, 1–5.
18. Wen, Y.Q.; Li, T.; Zheng, H.T.; Huang, L.; Zhou, C.H.; Xiao, C.S. Characteristics of ship domain in typical inland waters. *Navig. China* **2018**, *41*, 43–47.
19. Qi, L.; Zheng, Z.Y.; Li, G.P. AIS-data-based ship domain of ships in sight of one another. *J. Navig.* **2011**, *37*, 48–50.
20. Hansen, M.G.; Jensen, T.K.; Lehn-Schiøler, T.; Melchild, K.; Rasmussen, F.M.; Ennemark, F. Empirical ship domain based on AIS data. *J. Navig.* **2013**, *66*, 931–940. [CrossRef]
21. Szlapczynski, R.; Szlapczynska, J. A ship domain-based model of collision risk for near-miss detection and collision alert systems. *Reliab. Eng. Syst. Saf.* **2021**, *214*, 107766. [CrossRef]
22. Feng, H.X.; Grifoll, M.; Yang, Z.Z.; Zheng, P.J. Collision risk assessment for ships' routeing waters: An information entropy approach with Automatic Identification System (AIS) data. *Ocean. Coast. Manag.* **2022**, *224*, 106184. [CrossRef]
23. Liu, K.Z.; Yuan, Z.T.; Xin, X.R.; Zhang, J.F.; Wang, W.Q. Conflict detection method based on dynamic ship domain model for visualization of collision risk Hot-Spots. *Ocean. Eng.* **2021**, *242*, 110143. [CrossRef]
24. Du, L.; Banda, O.A.V.; Huang, Y.; Goerlandt, F.; Zhang, W. An empirical ship domain based on evasive maneuver and perceived collision risk. *Reliab. Eng. Syst. Saf.* **2021**, *213*, 107752. [CrossRef]
25. Liu, Z.; Zhang, B.Y.; Zhang, M.Y.; Wang, H.L.; Fu, X.J. A quantitative method for the analysis of ship collision risk using AIS data. *Ocean. Eng.* **2023**, *272*, 113906. [CrossRef]
26. Silveira, P.; Teixeira, A.P.; Soares, C.G. A method to extract the quaternion ship domain parameters from AIS data. *Ocean. Eng.* **2022**, *257*, 111568. [CrossRef]
27. He, Z.; Liu, C.; Chu, X.; Negenborn, R.R.; Wu, Q. Dynamic anti-collision A-star algorithm for multi-ship encounter situations. *Appl. Ocean. Res.* **2022**, *118*, 102995. [CrossRef]
28. Zhang, J.F.; Liu, J.J.; Hirdaris, S.; Zhang, M.Y.; Tian, W.L. An interpretable knowledge-based decision support method for ship collision avoidance using AIS data. *Reliab. Eng. Syst. Saf.* **2023**, *230*, 108919. [CrossRef]
29. Guo, J.; Li, L.N.; Gao, J.J.; Li, G.D. Early warning of collision and avoidance assistance for river ferry. *Navig. China* **2020**, *43*, 9–14.
30. Hu, Y.J.; Zhang, A.M.; Tian, W.L.; Zhang, J.; Hou, Z. Multi-ship collision avoidance decision-making based on collision risk index. *J. Mar. Sci. Eng.* **2020**, *8*, 640. [CrossRef]

31. Budak, G.; Beji, S. Controlled course-keeping simulations of a ship under external disturbances. *Ocean. Eng.* **2020**, *218*, 108126. [CrossRef]
32. Borkowski, P. Numerical Modeling of Wave Disturbances in the Process of Ship Movement Control. *Algorithms* **2018**, *11*, 130. [CrossRef]
33. Liu, J.L.; Yan, X.P.; Liu, C.G.; Fan, A.L.; Ma, F. Developments and Applications of Green and Intelligent Inland Vessels in China. *J. Mar. Sci. Eng.* **2023**, *11*, 318. [CrossRef]

Journal of *Marine Science and Engineering*

Article

MrisNet: Robust Ship Instance Segmentation in Challenging Marine Radar Environments

Feng Ma [1,2,3], Zhe Kang [1,2,3], Chen Chen [4,*], Jie Sun [5] and Jizhu Deng [6]

[1] State Key Laboratory of Maritime Technology and Safety, Wuhan University of Technology, Wuhan 430063, China; martin7wind@whut.edu.cn (F.M.); kz258852@whut.edu.cn (Z.K.)
[2] National Engineering Research Center for Water Transport Safety, Wuhan University of Technology, Wuhan 430063, China
[3] Intelligent Transportation Systems Research Center, Wuhan University of Technology, Wuhan 430063, China
[4] School of Computer Science and Engineering, Wuhan Institute of Technology, Wuhan 430205, China
[5] Nanjing Smart Water Transportation Technology Co., Ltd., Nanjing 210028, China; sunjie@smartwaterway.com
[6] Nanjing Port Co., Ltd., Nanjing 210011, China
* Correspondence: chenchen0120@wit.edu.cn

Abstract: In high-traffic harbor waters, marine radar frequently encounters signal interference stemming from various obstructive elements, thereby presenting formidable obstacles in the precise identification of ships. To achieve precise pixel-level ship identification in the complex environments, a customized neural network-based ship segmentation algorithm named MrisNet is proposed. MrisNet employs a lightweight and efficient FasterYOLO network to extract features from radar images at different levels, capturing fine-grained edge information and deep semantic features of ship pixels. To address the limitation of deep features in the backbone network lacking detailed shape and structured information, an adaptive attention mechanism is introduced after the FasterYOLO network to enhance crucial ship features. To fully utilize the multi-dimensional feature outputs, MrisNet incorporates a Transformer structure to reconstruct the PANet feature fusion network, allowing for the fusion of contextual information and capturing more essential ship information and semantic correlations. In the prediction stage, MrisNet optimizes the target position loss using the EIoU function, enabling the algorithm to adapt to ship position deviations and size variations, thereby improving segmentation accuracy and convergence speed. Experimental results demonstrate MrisNet achieves high recall and precision rates of 94.8% and 95.2%, respectively, in ship instance segmentation, outperforming various YOLO and other single-stage algorithms. Moreover, MrisNet has a model parameter size of 13.8M and real-time computational cost of 23.5G, demonstrating notable advantages in terms of convolutional efficiency. In conclusion, MrisNet accurately segments ships with different spot features and under diverse environmental conditions in marine radar images. It exhibits outstanding performance, particularly in extreme scenarios and challenging interference conditions, showcasing robustness and applicability.

Keywords: ship segmentation; radar image; lightweight convolution; adaptive attention mechanism; loss function

Citation: Ma, F.; Kang, Z.; Chen, C.; Sun, J.; Deng, J. MrisNet: Robust Ship Instance Segmentation in Challenging Marine Radar Environments. *J. Mar. Sci. Eng.* **2024**, *12*, 72. https://doi.org/10.3390/jmse12010072

Academic Editor: Sergei Chernyi

Received: 24 November 2023
Revised: 22 December 2023
Accepted: 23 December 2023
Published: 27 December 2023

1. Introduction

Real-time monitoring of busy waterways is a critically important task in maritime management, with positive implications for ensuring ship navigation safety, and improving port operational efficiency. Marine radar is widely applied in various domains such as ship collision avoidance, weather forecasting, and marine resource monitoring [1,2]. Similarly, shore-based deployed marine radar plays a significant role as it enables continuous monitoring of ships in expansive water areas, even under adverse weather and low visibility conditions. Compared to detection technologies like Automatic Identification System

(AIS) and Very High-Frequency (VHF), marine radar does not rely on real-time information responses from ships, thereby greatly enhancing the speed of obtaining navigation information.

In general, the images presented by marine radar consist of a two-dimensional dataset composed of a series of spots. These spots represent the intensity and positional information of the radar signal's reflection and echo from objects such as ships, islands, and air masses in space. By analyzing and processing these spots, radar images with different resolutions and clarity can be generated. Additionally, by overlaying and calculating a series of radar images taken at different time intervals, the positions and trajectories of ships can be simulated, resulting in ship spots with varying lengths of trails.

Traditional ship recognition methods for marine radar images mainly involve techniques such as filtering and pattern recognition [3]. These methods demonstrate appropriate suppression capabilities when dealing with relatively simple interference. However, they are relatively inapplicable in tasks that involve low pixel features, slow movement speeds, and strong interference in ship recognition. In recent years, deep learning techniques have rapidly advanced in the field of object recognition, particularly with the widespread application of Convolutional Neural Networks (CNNs). CNN-based methods have shown more competitive performance in complex scenarios, such as equipment defect detection, autonomous driving, and aerospace applications. Compared to traditional methods, these approaches are capable of extracting deep semantic information from images and achieving more accurate object localization.

Both traditional recognition methods and CNN-based object detection algorithms can only provide rough positional information of the targets. As a derivative method in the field of computer vision, instance segmentation offers higher precision in pixel-level target localization and can effectively distinguish multiple instances. For marine radar images, instance segmentation methods also provide richer ship motion information, such as heading and speed. Therefore, utilizing these methods to process marine radar images can provide more accurate and comprehensive ship identification and tracking information for maritime authorities.

Compared to ship instance segmentation in natural images, performing such tasks in radar images faces more challenges. Firstly, there are only small numbers of spots represent actual moving ships. This significantly affects the accurate classification and localization of ship targets, while interferences such as waves, clouds, rain, clutter, etc., further increase the difficulty of interference removal. Secondly, both long-tail and short-tail ships in radar images are considered small or even tiny objects. Particularly, in cases of dense sailing or crossing navigation, distinguishing between ship spots can be challenging. Lastly, due to the extensive use of embedded devices in radar systems, traditional fractal algorithms and micro-Doppler techniques have been commonly employed for radar signal processing for several decades, despite their relatively limited adaptability in many cases. In contrast, deep learning-based recognition methods generally exhibit preferable performance, but they require more computational resources, and only a fraction of mature algorithms can be directly applied to embedded devices.

In response to the characteristics of marine radar images, this paper proposes a customized CNN-based instance segmentation algorithm called MrisNet. Compared to previous research, this method exhibits significant differences in several aspects.

(1)　We enhance the feature network to extract crucial ship features by employing more efficient convolutional modules.

(2)　A convolutional enhancement method that incorporates channel correlations is introduced to further enhance the generalization ability of the feature network.

(3)　An attention mechanism with contextual awareness is utilized to enhance the multi-scale feature fusion structure, enriching the representation of convolutional features at different levels and effectively integrating micro-level and global-level ship features.

(4) The positioning loss estimation of predicted box is optimized to improve the precision of ship localization and enhance the segmentation performance for dense ship scenarios.

(5) To evaluate the effectiveness of various algorithms for ship segmentation in radar images, a high-quality dataset called RadarSeg, consisting of 1280 radar images, is constructed.

The remaining sections of this paper are organized as follows: Section 2 provides a brief overview of relevant research on instance segmentation and detection of ships in different imaging scenarios. In Section 3, a customized CNN-based ship segmentation algorithm is proposed. Section 4 presents a comparative analysis of experimental results using various algorithms in marine radar images. Finally, Section 5 summarizes the main contributions of the proposed method and discusses future directions for development.

2. Related Works

Currently, ship recognition in natural scenes and remote sensing images has been widely explored and developed. Early traditional methods primarily employed techniques such as clustering analysis and filtering for ship recognition [4]. However, these methods have obvious limitations and relatively poor adaptability, making it challenging to achieve satisfactory results in different scenarios. Meanwhile, the rapid development of deep learning and neural network-based technologies has brought significant advancements in tasks like instance segmentation. However, in the field of ship recognition in marine radar images, the application of deep learning-related methods remains relatively limited, presenting vast opportunities for further development.

In this section, we first provided an overview of ship recognition techniques in marine radar images, including neural network-based approaches for object detection and instance segmentation. It is worth noting object detection and instance segmentation share commonalities in feature extraction, object localization, and post-processing, thus the design principles of object detection algorithms can offer valuable insights for the design of instance segmentation algorithms. Additionally, we reviewed and summarized ship recognition techniques and relevant optimization methods in other scenarios.

2.1. Ship Identification Methods under Radar and Other Scenarios

In recent years, significant advancements have been made in object detection and instance segmentation using CNNs. In 2014, the R-CNN algorithm demonstrated a significant advantage on the PASCAL VOC dataset, gradually establishing the dominance of deep learning-based algorithms in the field of object recognition. As is widely known single-stage algorithms is achieving a good balance between computational speed and recognition accuracy, allowing for end-to-end training and widespread adoption in various recognition tasks. However, in certain tasks, single-stage algorithms may exhibit slightly lower accuracy compared to two-stage algorithms such as Faster R-CNN [5] and Mask R-CNN [6]. Therefore, the selection of benchmark algorithms and network architectures requires careful consideration and decision-making based on specific application scenarios and requirements.

In the research on ship recognition in marine radar images, significant progress has been made by scholars who have incorporated deep learning methods into the field of object detection. By designing effective algorithm structure and utilizing techniques such as clutter suppression and feature enhancement, even single-stage algorithms can achieve satisfactory recognition results. Chen et al. [7] proposed a ship recognition algorithm based on a dual-channel convolutional neural network (DCCNN) and a false alarm-controllable classifier (FACC) to suppress clutter and accurately extract ship features in images. Furthermore, some studies have achieved high levels of ship recognition accuracy and effective interference suppression using two-stage recognition algorithms. For instance, Chen et al. [8] made several improvements to Faster R-CNN in multiple aspects, including optimizing the backbone network, sample data balancing, and scale normalization, aiming to enhance the algorithm's accuracy and robustness. Additionally, differential neural architecture search, label reassignment, and various types of feature pyramid structures have been widely applied in algorithm design, yielding promising recognition results [9–11]. In conclusion, deep neural networks have proven effective in extracting ship features. However, current object detection techniques still have limitations in suppressing complex interference, and there is room for improvement in recognizing dense small-sized ships.

Currently, research on ship segmentation under marine radar images is still relatively limited. It is well-known ships in synthetic aperture radar (SAR) images exhibit small scales, limited effective pixels, and significant background interference, which are similar to the challenges faced in marine radar images. In the context of SAR images, there has been extensive research on ship segmentation, and we could draw upon the technical methods from these studies to provide insights for the design of ship segmentation algorithms in marine radar images.

In ship segmentation under complex background conditions of SAR images, a well-structured feature extraction network plays a crucial role in enhancing the model's recognition capability. Zhang et al. [12] proposed methods such as multi-resolution feature extraction networks, and enhanced feature pyramid networks to ensure more adaptable performance in complex scenarios. Moreover, various attention mechanisms can significantly enhance the extraction of ship features and contextual information. Zhao et al. [13] proposed a ship segmentation method based on collaborative attention mechanism, which improved the recognition performance of multi-task branches. Typically, instance segmentation networks employ horizontal bounding boxes to fit the objects, which may include redundant information. Moreover, in dense ship scenarios, it becomes challenging to accurately differentiate between individual targets. Yang et al. [14] proposed a novel ship segmentation network that utilized rotated bounding boxes as the segmentation foundation, enclosing the targets along the direction of the ships. Besides, they designed a dual feature alignment module to capture representative features of the ships. Instance segmentation of small-scale or multi-scale ships has been a focal topic in SAR images. Shao et al. [15] proposed a multi-scale ship segmentation method specifically designed for SAR images. Specifically, they achieved more precise target regression by re-designing the input unit, backbone network, and ROI module of the Mask R-CNN.

The above-mentioned methods provide valuable references for the design of instance segmentation algorithms in marine radar images, particularly in terms of feature network design, attention mechanism application, and multi-scale segmentation.

In other domains, research on ship segmentation primarily focuses on natural images. Researchers have employed various techniques and methods for algorithm design. Some studies have utilized single-stage algorithms, incorporating efficient backbones, feature fusion structures, and prediction networks to achieve high-precision ship segmentation [16]. Others have adopted two-stage algorithm designs, involving the extraction of target candidate regions followed by classification and fine-grained segmentation to enhance accuracy and robustness [17]. These diverse algorithm design strategies cater to the demands and challenges of various domains.

2.2. Optimization Method for Ship Identification Research

The ship recognition in radar images can be enhanced through the utilization of auxiliary techniques. For instance, employing multi-modal fusion techniques and clutter suppression techniques can result in clearer and higher-resolution radar images, thereby providing more accurate ship information. Guo et al. [18] proposed a method that utilized deep learning techniques to identify targets in marine radar images and achieved consistent fusion of electronic chart and marine radar images by treating these targets as reference points. Mao et al. [19] introduced a marine radar imaging framework based on non-uniform imaging theory, which combined techniques such as beam recursive anti-interference, non-uniform sampling models, and dimensionality reduction iterations for improved computational efficiency and higher-quality imaging results. Zhang et al. [20] employed generative adversarial networks to remove noise in radar images and utilized image registration methods to eliminate imaging discrepancies between radar images and chart data, supporting efficient ship recognition through parallel computation of feature data and image fusion.

Currently, there are several techniques available to improve the ship recognition models. Among them, constructing datasets that include rich ship features is an effective method for enhancing recognition performance [21]. In addition, the ship detection can be increased by integrating feature augmentation and key point extraction into the recognition algorithms [22]. Furthermore, through the design of feature enhancement modules and small target attention mechanisms, substantial improvements in the recognition of small-scale ships might be made [23,24]. In certain scenarios, it is necessary to consider the computational limitations of embedded devices, thus requiring the design of efficient lightweight algorithms. Yin et al. [25] addressed this challenge by introducing depthwise separable convolutions and point-wise group convolutions, resulting in a lightweight feature extraction network that balances the accuracy and inference speed of ship recognition.

Based on the above analysis, it can be concluded leveraging deep learning techniques to design feature networks and prediction structures significantly enhances the effectiveness of ship recognition. Furthermore, compared to traditional methods, deep learning approaches exhibit prominent advantages. Given the presence of numerous small-scale targets in marine radar images, this study proposes a ship segmentation algorithm that combines deep convolutional networks, feature attention mechanisms, and multi-scale feature fusion structures. By integrating different hierarchical features from the images, this algorithm effectively suppresses interference and enables accurate localization and precise segmentation of different types of ships.

3. A Proposed Method

The overall framework of the ship segmentation model for marine radar images proposed in this paper is illustrated in Figure 1. The model consists of components such as the feature extraction network, feature fusion network, and prediction structure.

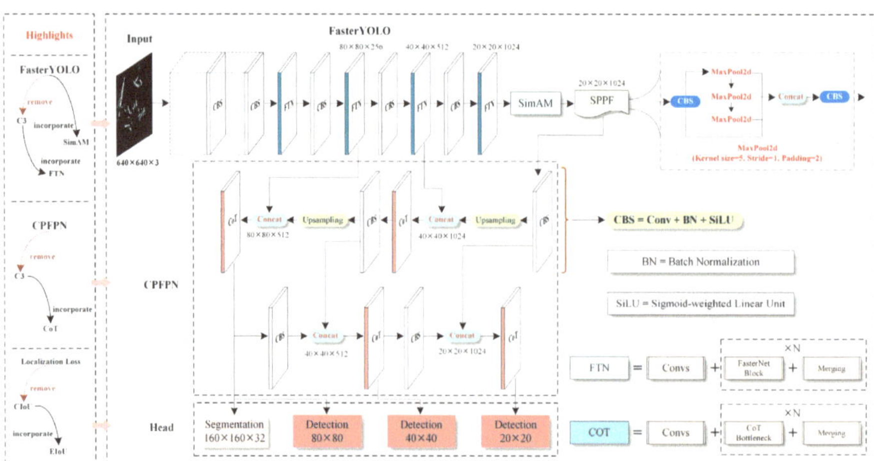

Figure 1. Overall structure of the proposed MrisNet. In comparison to the standard network of YOLOv5(S), MrisNet introduces several improvements and appropriate innovations. Specifically, in the feature extraction network, MrisNet replaces the original C3 (Cross Stage Partial-Darknet53) module with the FTN module and incorporates a SimAM mechanism. In the feature fusion network, the proposed method replaces the original C3 module with the CoT module. In the prediction head structure, MrisNet replaces the localization loss calculation criterion with the EIoU function.

To be specific, the feature network, called FasterYOLO, adopts a novel convolutional structure that combines efficient convolutional units and adaptive attention mechanisms. The FasterYOLO network is composed of four types of structures, namely FTN (Faster-Net) [26], CBS, SimAM (Simple, Parameter-Free Attention Module) [27], and SPPF (Spatial Pyramid Pooling-Fast), which are concatenated in a sequential manner. The FTN structure, an efficient convolutional module, is detailed in Section 3.1.2. CBS, a commonly employed structure in YOLOv5 [28], consists of conventional convolutions, batch normalization layers, and the SiLU function. Additionally, SimAM, a parameter-free attention module, is introduced in Section 3.1.3. Finally, SPPF, a widely adopted feature enhancement method, is utilized. The network reduces redundant computations and improves the extraction of spatial features of the targets, facilitating the model's understanding of the hierarchical pixel and semantic information of ships.

The feature fusion network, CPFPN (Context-sensitive Processing within Feature Pyramid Network), integrates a novel convolutional enhancement method known as the CoT (Contextual Transformer) module, based on the Transformer pattern [29]. CPFPN employs a dual-path feature fusion strategy, combining both bottom-up and top-down approaches, to facilitate the integration of multi-scale features. Notably, the CoT module enhances contextual information of ship targets. It captures global and local dependencies in the feature sequences, thereby avoiding feature loss and degradation.

Finally, in the prediction structure of the proposed algorithm, three object detection branches are designed to target ships at different scales, along with an instance segmentation branch, enabling accurate ship localization and pixel-level segmentation. Especially, the EIoU (Enhanced Intersection over Union) loss [30] is introduced to optimize the loss calculation for predicting bounding box positions. This method improves the convergence speed and accuracy of the predicted boxes, enabling the model to learn feature representations with better generalization ability.

3.1. Feature Extraction Network

As mentioned earlier, marine radar images exhibit unique characteristics such as noise and low resolution, making them significantly different from natural images. Classic feature networks like ResNet-50, ResNet-101, and SENet are typically designed for general natural datasets and may not adapt well to the distinctive image features of radar images, resulting in suboptimal feature extraction performance. Furthermore, these networks often contain a large number of redundant parameters, which can lead to overfitting issues when applied to ship segmentation in marine radar images. Therefore, employing lightweight customized feature networks may be more suitable for such tasks.

Furthermore, due to the generally small-scale of ships in marine radar images, the feature network should possess the ability to perceive small objects. Moreover, to capture different signature features of ships, the feature network should employ suitable convolutional structures and downsampling ratios to obtain feature representations of targets at different scales, thereby enhancing the accuracy of ship segmentation. Based on the aforementioned analysis, the overall framework of the proposed feature extraction network is illustrated in Figure 2. This network is an adaptive neural network that effectively extracts ship features at various levels and achieves more prominent performance.

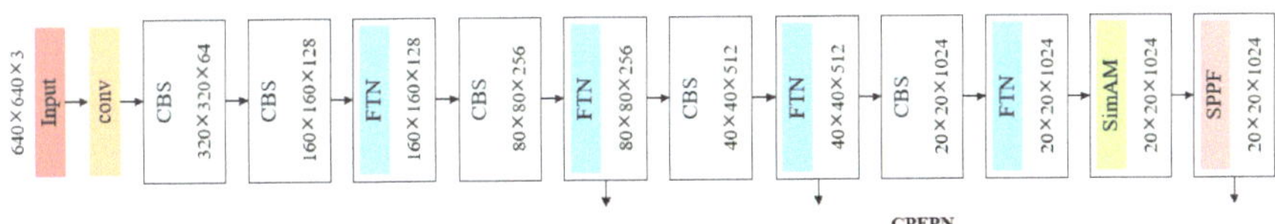

Figure 2. Overall architecture of the feature network. By aggregating convolutional features from three distinct depths, salient information pertaining to different categories of ship spots, with a particular emphasis on small-scale ships, can be acquired.

3.1.1. YOLOv5(S) Network

In this research, we selected the YOLOv5(S) [28] as the baseline feature network. Despite its relatively smaller convolutional scale, YOLOv5(S) is still able to perform effectively. The network is illustrated in Figure 3, YOLOv5(S) utilizes a series of downsampling layers to progressively extract image features, adapting to variations in target scale. It employs different sizes of convolutional kernels and pooling layers to design receptive fields at different levels, capturing multi-scale features of the targets while maintaining a lower number of convolutional parameters and computational complexity. This contributes to improving the accuracy and robustness of ship segmentation, particularly enhancing the recognition of small-scale ships in radar images. It is important to note while YOLOv5(S) has fewer convolutional layers and parameters, as well as lower computational complexity, this lightweight design makes it more suitable for instance segmentation tasks in resource-constrained environments.

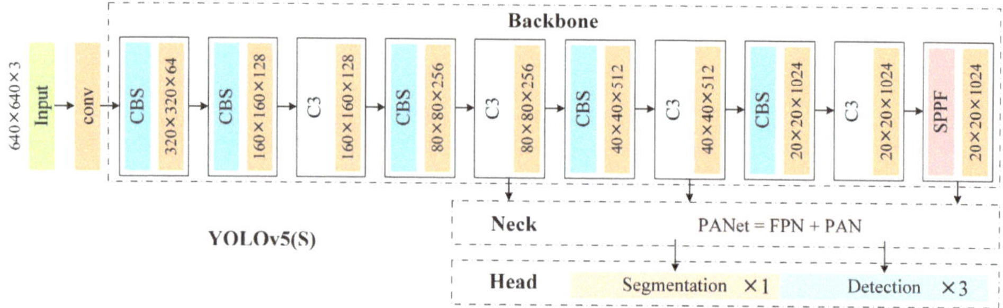

Figure 3. Network structure of instance segmentation of YOLOv5(S). It is apparent the standard YOLOv5(S) exhibits conspicuous disparities from the algorithm proposed in this paper. However, MrisNet still retains the sequential connectivity structure of the feature extraction network and the bidirectional fusion structure of FPN (Feature Pyramid Network) and PAN (Path Aggregation Network) in the feature fusion network.

3.1.2. The Main Architecture of FasterYOLO Network

In marine radar images, image sequences often contain ship regions of various sizes and shapes, with the majority of ships being relatively small in scale. Therefore, it is crucial to capture key ship features with high sensitivity. Additionally, the feature network should possess noise resistance and robustness. To address these requirements, valid convolutional computation units can be employed to enhance network performance. This paper introduces a simple and fast convolutional method, combining standard convolutions and pointwise convolutions, to design a more efficient feature network.

With the advancement of CNNs, classical computation methods represented by standard convolutions often suffer from redundant calculations, resulting in inefficient increases in model parameters and computational costs. Moreover, a significant amount of ineffective convolutional computations can impact the extraction of crucial features, especially in ship recognition under radar images where targets are small in scale and feature information is limited. Excessive convolutional calculations can lead to overfitting issues. Research has revealed convolutional feature maps exhibited high similarity across different channels, and standard convolution unavoidably duplicated the extraction of image features in a per-channel computation. Therefore, simplifying the convolutional computation to reduce feature information not only ensures network performance but also greatly enhances computational speed and efficiency. Specifically, this paper proposes a network structure called FasterYOLO, which incorporates the FasterNet modules [26] after standard convolutions, as illustrated in Figure 4. This improvement aims to enhance the model's ability to extract key ship features and reduce the risk of overfitting.

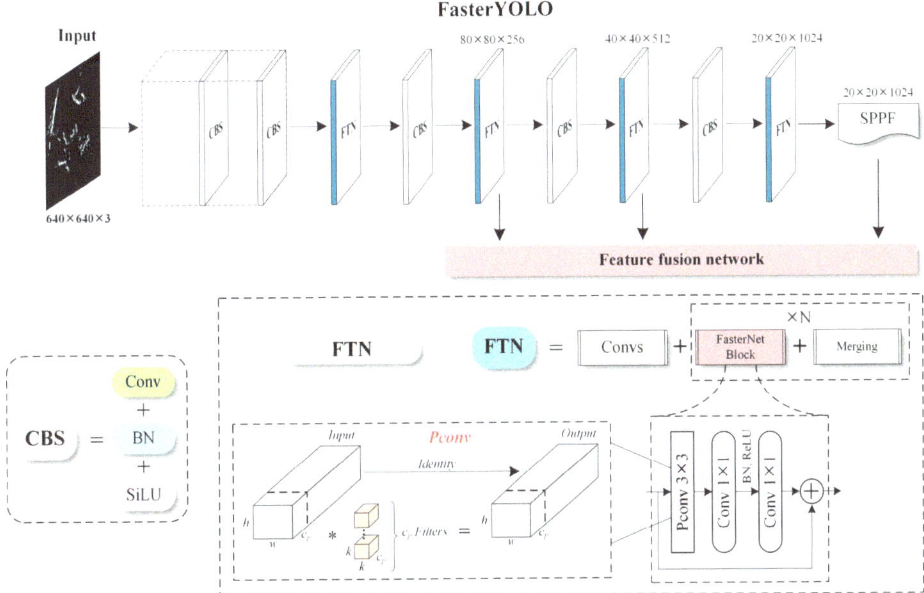

Figure 4. Network structure of FasterYOLO. This research employed four FTN modules to replace the C3 modules in the original YOLOv5, while enhancing the deep-level features through the concatenation of SPPF.

Compared to conventional convolutional networks, integrating the FasterNet modules in the feature network significantly reduces the computational cost of convolutions. The FasterNet employs PConv (Partial Convolution) to optimize per-channel convolutional computations. In practical terms, PConv convolution is a convolutional structure with multiple branches and spatial feature extraction capabilities. It utilizes partial convolution operations, which means it performs conventional convolutions only on a subset of channels in the input feature maps, thereby reducing the computational workload. Additionally, PConv incorporates a parallel branch structure that performs convolutional operations on different spatial positions of the input feature maps, capturing more spatial feature information.

Figure 5 illustrates a brief comparison of the computational processes between PConv convolution, standard convolution, and depthwise convolution [26]. Depthwise convolution, a classical convolutional variant widely used in various neural networks, has shown significant effectiveness in MobileNet series. For an input $I \in R^{c \times h \times w}$, We apply c filters $w \in R^{k \times k}$ to compute the output $O \in R^{c \times h \times w}$. In this case, the computational cost of standard convolution is $h \times w \times k^2 \times c^2$, while depthwise convolution only incurs a cost of $h \times w \times k^2 \times c$. It can be observed depthwise convolution is effective in reducing computational cost. However, it directly leads to a decrease in recognition accuracy and therefore cannot directly replace standard convolution. On the other hand, PConv convolution only applies regular convolution to a subset of input channels, resulting in a computational cost of $h \times w \times k^2 \times c_p^2$. When r is 1/4, the computational cost of PConv is merely 1/16 of that of standard convolution. Consequently, PConv convolution greatly simplifies the computational process of standard convolution while maintaining identical input and output dimensions. Without altering the network hierarchy, standard convolution can be directly replaced with PConv convolution.

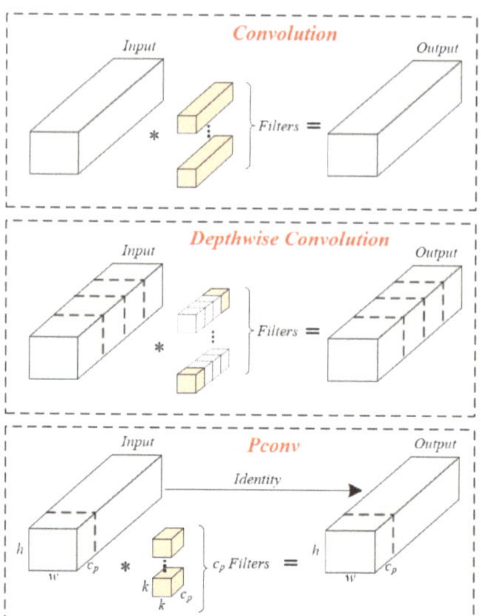

Figure 5. Comparison of several types of convolution methods. It is apparent PConv selectively incorporates conventional computation solely in specific convolutional channels, leading to a substantial reduction in redundant convolutional parameters and computational costs compared to conventional convolutional methods.

3.1.3. Feature Enhancement Mechanism Based on SimAM Attention

For the feature network, it is crucial to fully leverage the contextual information surrounding ships in radar images to enhance the accuracy of ship segmentation. To fulfill this requirement, attention mechanisms can be introduced or convolutional structures with global contextual awareness can be utilized. By capturing the correlations between ships and their surrounding environments, the network can better comprehend the shape, contour, and semantic information of ships, thereby enhancing the effectiveness of ship segmentation.

In this research, the SimAM method [27], which is constructed based on principles derived from visual neuroscience, is appended after the FasterYOLO network. According to the theory of visual neuroscience, neurons carrying more information are more salient compared to their neighboring neurons when processing visual tasks, and thus, they should be assigned higher weights. In ship segmentation, it is equally important to enhance the neurons in the convolutional network that are responsible for extracting crucial ship features. SimAM captures both spatial and channel attention simultaneously and possesses spatial inhibition capabilities that are translation-invariant. Unlike methods such as SENet and CBAM that focus on designing attention mechanisms through pooling and fully connected layers, SimAM evaluates the importance of each feature based on an energy function derived from neuroscientific principles. It offers better interpretability and does not require the introduction of additional learnable parameters. Consequently, SimAM effectively extracts and enhances salient information of ships in marine radar images. Specifically, SimAM evaluates each neuron in the network by defining an energy function based on linear separability, as shown in Equations (1)–(4).

$$e_t^* = \frac{4(\hat{\sigma}^2 + \lambda)}{(t - \hat{\mu})^2 + 2\hat{\sigma}^2 + 2\lambda} \tag{1}$$

$$\hat{\mu} = \frac{1}{M} \sum_{i=1}^{M} x_i \tag{2}$$

$$\hat{\sigma}^2 = \frac{1}{M}\sum_{i=1}^{M}(x_i - \hat{\mu})^2 \tag{3}$$

$$\widetilde{X} = sigmoid(1/e_t^*) \odot X \tag{4}$$

Among them, t is the target neuron, x is the adjacent neuron, M is the number of neurons, and λ is a hyperparameter. A lower energy value of e_t^* indicates higher discriminability between the neuron and its neighboring neurons, implying a higher level of importance for that neuron. In Equation (4), we weight the importance of neurons using $1/e_t^*$. The introduction of SimAM enables the feature network to comprehensively assess feature weights, thereby enhancing the representation of crucial ship information, and reducing reliance on prior information regarding significant variations in the shape of the targets.

3.2. Feature Fusion Network

In marine radar images, the majority of ships exhibit relatively small-scales. Therefore, developing an efficient convolutional architecture to capture fine and effective feature representations of ships becomes crucial. In convolutional networks, shallow convolutional features possess higher resolution, providing more detailed spatial information that aids in the recognition of object edge structures. Moreover, shallow features are more robust to image noise and lighting variations. However, shallow features exhibit limited adaptability to transformations such as translation, scaling, and rotation in images. In contrast, deep convolutional features have lower resolution but contain stronger multi-scale and semantic information, which helps filter out the effects of noise and lighting variations in images. However, deep features have relatively weaker capability in extracting fine-grained details from images. Hence, this paper proposes a more efficient feature fusion network that fully integrates feature information extracted from different scales and depths of receptive fields in the images.

Currently, object detection or instance segmentation algorithms based on deep neural networks commonly employ feature pyramid structures to address the challenge of scale variation. Among them, FPN serves as the most widely used feature structure, delivering more adaptive results in both single-stage and two-stage algorithms. FPN realizes the fusion of features at different scales through a top-down feature propagation path. However, high-level features need to undergo multiple intermediate-scale convolutions and be fused with features at these scales before merging with low-level features. In this process, the semantic information of high-level features may be lost or degraded. In contrast, dual-path fusion structures such as PANet compensate for the shortcomings of FPN in preserving high-level features but also introduce the opposite problem, where the detailed information of low-level features may be degraded during fusion. To address this issue, this paper introduces an attention mechanism that integrates rich feature information into the PANet network. This mechanism adaptively learns the importance of different regions in the feature map and weights the fusion of features at different scales, resulting in more comprehensive feature representations. Based on these theoretical considerations, we refer to the proposed feature fusion network as CPFPN, and its overall structure is illustrated in Figure 6.

Specifically, in this research, we employ an attention mechanism based on CoT [31] to extract crucial features of ships in radar images. This mechanism fully utilizes the contextual information of the input data and enhances the expressive power of key features by learning a dynamic attention matrix. In comparison to traditional self-attention mechanisms, the CoT provides a more comprehensive treatment of contextual information. Traditional self-attention mechanisms only interact information in the spatial domain and independently learn correlation information, thereby overlooking rich contextual information among adjacent features and limiting the self-attention learning capability of feature maps. In terms of specific implementation, the CoT module integrates two types of contextual information about the image. Firstly, the input data is encoded through convolutional operations to capture contextual information and generate static contextual representations. Next, dynamic contextual representations are obtained through concatenation and consecutive convolutions. Finally, the static and dynamic contextual representations are fused to form the final output.

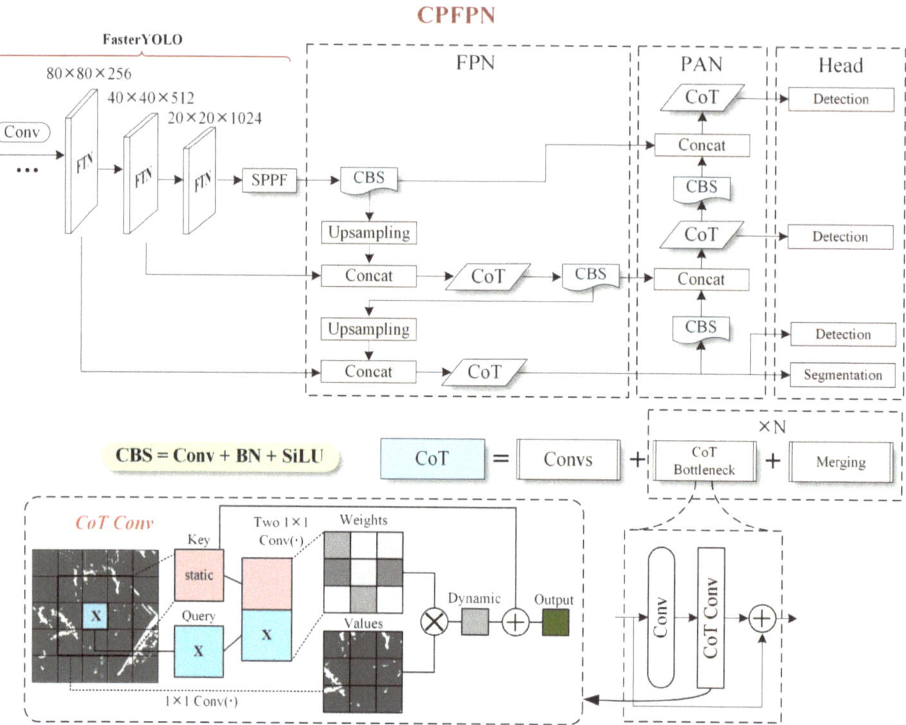

Figure 6. Network structure of CPPFN. In this section, we conducted substitutions of the C3 modules in the original YOLOv5 with four CoT modules. This integration of contextual information derived from image features contributes to the significant feature expression of minute ship spots.

In the context of marine radar images, the CoT module effectively captures global and local contextual features of targets by encoding the input feature maps with contextual information. By leveraging contextual information, the network is able to extract more diverse and meaningful features, thereby obtaining additional information about the background surrounding the ships. Furthermore, CoT adaptively allocates attention weights to different features, enabling the model to flexibly acquire the most relevant and salient ship characteristics from the feature maps. Moreover, CoT exhibits the capability to directly substitute the standard 3×3 convolutional structure, facilitating its seamless integration into other classical convolutional networks.

The convolutional process of the CoT module, as depicted in Figure 7, operates on the input feature map $X \in R^{H \times W \times C}$, with keys ($K = X$), queries ($Q = X$), and values ($V = XW_v$) associated with the convolutional feature map [31]. Unlike traditional self-attention methods, the CoT module employs $k \times k$ groups of convolutional operations to extract contextual information. Through this process, the resulting $K^1 \in R^{H \times W \times C}$ is further utilized as the static contextual representation of the input X.

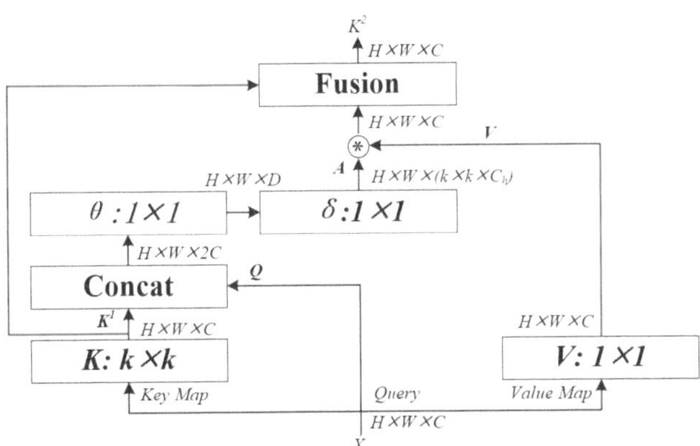

Figure 7. Convolution fusion process of CoT module. Through the concatenation of static and dynamic features, CoT aids the model in capturing appearance information (e.g., texture and color) and motion information relevant to ship spots.

Next, the previously obtained K^1 is concatenated with Q and passed through two consecutive convolutional operations to compute the attention matrix A.

$$A = [K^1, Q]W_\theta W_\delta \tag{5}$$

Based on the given attention matrix A, the enhanced feature K^2 is generated and computed according to the following formula. By leveraging the enhanced feature K^2, it becomes capable of capturing dynamic contextual representations regarding the input X. Ultimately, the fusion of these two contextual representations is achieved through a classical attention mechanism, resulting in the final output.

$$K^2 = V \odot A \tag{6}$$

3.3. Ship Prediction Module

As MrisNet maintains the fundamental structure of YOLOv5, there exist slight variations in the calculation of loss for object detection and instance segmentation. The comprehensive loss for object detection consists of three components, i.e., localization loss, confidence loss, and classification loss. These losses are individually multiplied by corresponding weights and ultimately summed to yield the overall loss. Notably, since this study focuses solely on a single category, the classification loss remains consistently zero. The object detection branch predominantly employs the CIoU (Complete Intersection over Union) [32] metric from the IoU series to compute the localization loss and utilizes binary cross-entropy loss for the calculation of the confidence loss.

In contrast to object detection, the comprehensive loss for instance segmentation comprises four components, i.e., localization loss, confidence loss, classification loss, and mask segmentation loss. Similarly, these losses are weighted and summed to yield the final calculation result for the overall loss. Specifically, the calculation process for localization loss, confidence loss, and classification loss is identical to that in the object detection branch. However, for mask segmentation loss, binary cross-entropy loss is predominantly employed to evaluate the disparity between predicted masks and ground truth masks.

The significance of the localization loss is apparent in both the object detection and instance segmentation branches. In the instance segmentation branch, specifically, the extraction of predicted masks is restricted to the scope of predicted bounding boxes. The predicted bounding boxes, generated through regression using the localization loss, form the basis for calculating the mask segmentation loss. Consequently, this research optimized the calculation process of the localization loss in the instance segmentation branch while preserving the standard calculation of other losses.

Therefore, we introduce a novel evaluation method, referred to as EIoU, as the computation standard for the target's position loss, aiming to further enhance the localization accuracy of ship bounding box predictions. The loss function takes into account multiple factors such as IoU loss, distance loss, and width-height loss, fully considering the relationship between the predicted box and the ground truth box. This leads to a more stable gradient during the algorithm training. Moreover, the EIoU loss helps the model better understand the spatial distribution and relative positions of the targets, which is crucial for achieving high-precision ship segmentation. Specifically, the EIoU loss is defined as follows:

$$Loss_{EIoU} = 1 - (IoU - \frac{\rho^2(b^{gt}, b)}{c^2} - \frac{\rho^2(h^{gt}, h)}{c_h^2} - \frac{\rho^2(w^{gt}, w)}{c_w^2}) \tag{7}$$

$$IoU = \frac{S^{pr} \cap S^{gt}}{S^{pr} \cup S^{gt}} \tag{8}$$

wherein, c represents the Euclidean distance of the minimum bounding rectangle diagonals between the predicted box and ground truth box; $\rho(b^{gt}, b)$ denotes the Euclidean distance between the center points of the two boxes (i.e., ground truth box and predicted box); $\rho(h^{gt}, h)$ represents the difference in length between the two boxes; $\rho(w^{gt}, w)$ represents the difference in width between the two boxes; c_h and c_w are the height and width of the minimum bounding rectangle, respectively; S^{gt} represents the area of the ground truth box; S^{pr} represents the area of the predicted box, and other key indicators are illustrated as shown in Figure 8.

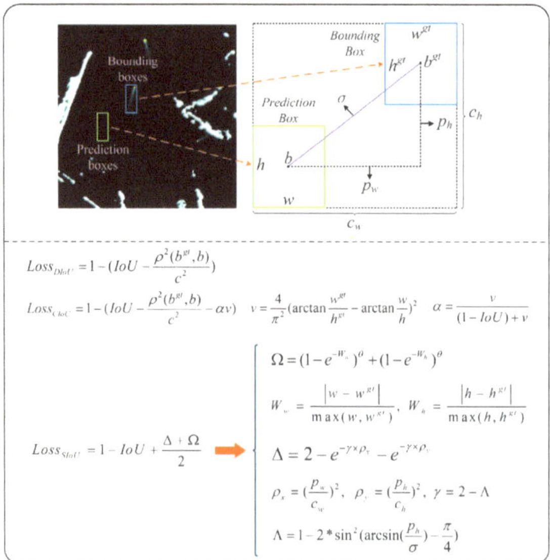

Figure 8. Key indicators of EIoU and calculation comparison of different loss functions. DIoU (Distance Intersection over Union) exclusively takes into account the variation in the distance between the center points of the predicted box and the ground truth box. CIoU extends this concept by incorporating considerations for the difference in aspect ratios between the two boxes. Building upon the considerations for area, shape, and distance differences between the boxes, SIoU (Scylla Intersection over Union) introduces the novel factor of angle difference.

In this section, we compared the EIoU loss with the classical DIoU [32], CIoU, and SIoU [33] and demonstrated their computational processes (refer to Figure 8). Specifically, unlike the IoU metric that solely focuses on the overlapping region, DIoU considers both the distance and the area overlap between the ground truth box and predicted box, thereby enhancing the regression stability of the predicted boxes. CIoU not only takes into account the positional information and localization errors of the predicted box but also introduces shape information to more accurately measure the target accuracy. However, when the aspect ratios of the two boxes are small, there may be issues of gradient explosion or vanishing. Building upon CIoU, SIoU further considers the angle of the regression vector between the two boxes by defining an angle penalty vector, which encourages the predicted box to quickly converge to the nearest horizontal or vertical axis. However, SIoU defines multiple different IoU thresholds, which present challenges in the practical training and evaluation of the model. In comparison, the EIoU function exhibits stronger interpretability in capturing the variations of loss during the regression of the predicted box. This leads to more accurate measurement of position, shape, distance, and width-height losses for small-scale ship predicted boxes in radar images, thereby facilitating the training and convergence of the algorithm.

4. A Case Study

4.1. Dataset

A high-quality dataset can significantly improve the empirical outcomes of CNN-based algorithms. Typically, the performance of the model is heavily reliant on the quality and quantity of the training data. A good dataset should encompass multiple scenarios and situations. This research conducted pre-processing on the real data obtained from the JMA5300 marine radar deployed at Zhoushan Port in Zhejiang, resulting in the creation of a high-quality marine radar image dataset named RadarSeg. The dataset comprises 1280 images. As shown in Figure 9, ships in the images are mainly classified into two categories, i.e., long-wake ships and short-wake ships. To be specific, long-wake ships have distinct features that are easy to extract, while the pixel features of short-wake ships are similar to those of interferences such as reefs, which can interfere with the instance segmentation of ships. Moreover, the RadarSeg dataset covers complex background environments, including different weather conditions, harbor environments, and imaging conditions. Additionally, the dataset also takes into account factors such as the variations in ship heading, and traffic flow. Besides, it also emphasizes an augmentation in the number of images depicting ships of small-scale and miniature sizes. Considering different types of ships exhibit similar spot features in marine radar images, all ships in the dataset are uniformly labeled as "boat" category.

In the RadarSeg dataset, each image has been annotated with ship type labels and tightly fitting bounding boxes along the target edges. These annotations were saved in JSON format files and have been preprocessed to meet the requirements of the YOLO algorithms. We divided all the images into training, validation, and testing sets in an 8:1:1 ratio. Various algorithms were trained on the training and validation sets, and evaluation metrics and algorithm performance were assessed on the testing set. Additionally, cluster analysis revealed the number of pixels occupied by ships accounts for approximately 0.035% of the entire image. Thus, in marine radar images, ships are predominantly small-scale or even miniature targets.

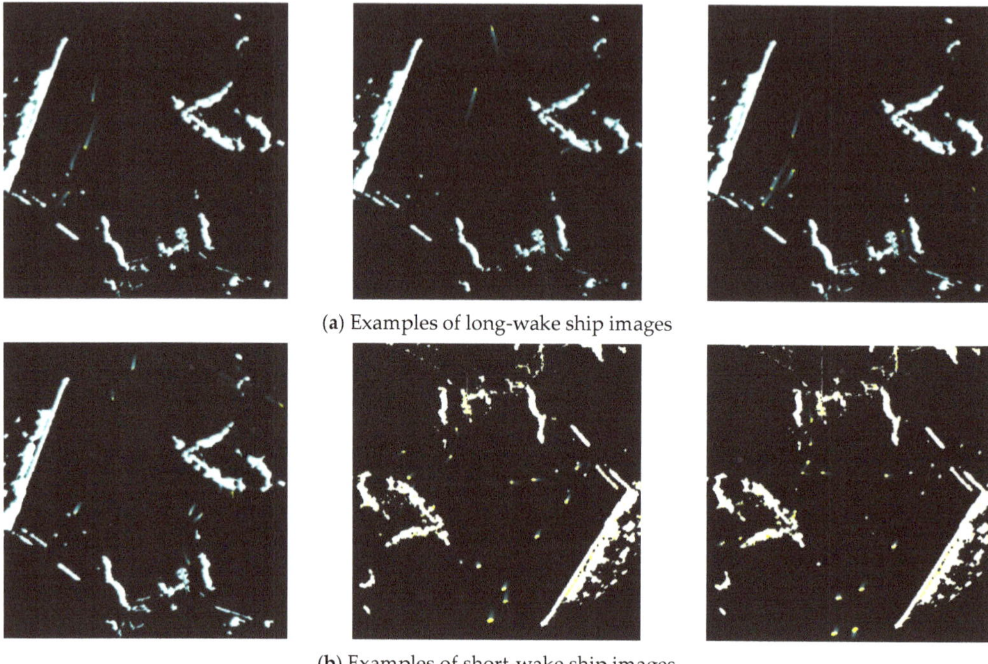

(**a**) Examples of long-wake ship images

(**b**) Examples of short-wake ship images

Figure 9. Marine radar images. Long-wake ships, when compared to short-wake ships, occupy a larger number of pixels in an image, thereby possessing more pronounced and distinctive feature information.

4.2. Training Optimization Methods

This research optimized the training process of the proposed algorithm by improving aspects such as learning rate decay, loss calculation, and data augmentation. Normally, traditional neural network-based algorithms usually employ a series of fixed learning rates for training, which may result in significant learning rate decay at different training stages, leading to unstable changes in model momentum and negatively affecting the algorithm's training effectiveness. To address this issue, the Adam optimizer was introduced to optimize the learning rate, which adaptively adjusted the learning rate without the need for manual tuning, based on the magnitude of parameter gradients. Additionally, label smoothing was introduced to enhance the generalization performance of ship segmentation model. This approach reduces the model's reliance on noise or uncertainty information in the training data. Furthermore, data augmentation methods, including random cropping and horizontal flipping, were employed to increase the diversity of training data and improve the model's robustness and generalization abilities.

4.3. Experimental Environment and Training Results

The experimental setup relied on the Ubuntu 20.04 operating system and utilized the NVIDIA RTX3090 graphics card with an effective memory size of 24GB. The CUDA version used was 11.1.0, PyTorch version was 1.9.1, and the Python environment was 3.8. We compared the performance of different algorithms in the ship segmentation, including the YOLO series and its improved variants, classical two-stage algorithms, hierarchical algorithm based on the Transformer, and MrisNet. All algorithms were evaluated using the same dataset of ship images. During algorithm training, the input image size was set to 640×640 pixels, the momentum was set to 0.9, the batch size was set to 8, and the training was conducted for 800 epochs with an initial learning rate of 10^{-4}.

Under the aforementioned settings, the training of various algorithms and related experiments were carried out. Among them, YOLOv5(S), YOLOv8(S) [34], and MrisNet achieved convergence of the loss values on the training set to 14.91×10^{-3}, 15.09×10^{-3}, and 14.52×10^{-3}, respectively. The experimental results indicated, when evaluated using the same metrics, MrisNet exhibited lower convergence loss compared to the standard YOLO algorithms. This illustrates MrisNet is capable of more accurately identifying ship pixel-level features in radar images. Furthermore, through the analysis of Figure 10, it could be concluded under different threshold conditions, MrisNet has achieved favorable experimental results across various evaluation metrics, further substantiating its ability to minimize misjudgments and omissions in ship segmentation while accurately locating and identifying targets.

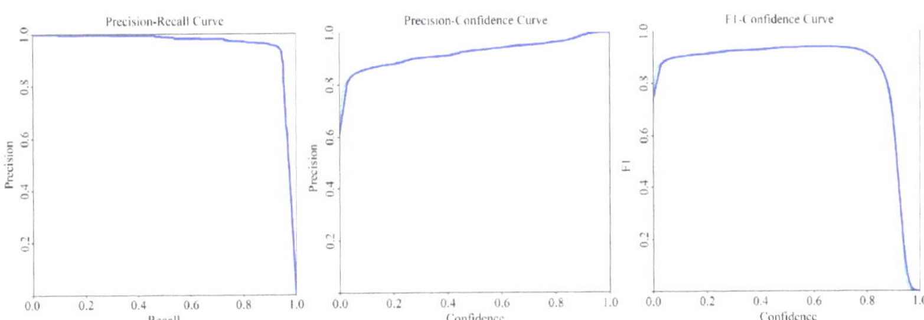

Figure 10. Various verification results of MrisNet. The convergence curves of these three categories manifest the efficacy of the proposed method in extracting positive samples for ship targets and suppressing false positives.

4.4. Comparisons and Discussion

Due to the monolithic nature of the dataset utilized in this research, which consists solely of a single ship category, the calculation of mean Average Precision (mAP) directly corresponds to the Average Precision (AP) value of that specific class. Thus, we selected Recall (R), Precision (P), mAP_{50}, and mAP_{50-95} as metrics to measure the effectiveness of ship segmentation and object detection. Additionally, we evaluated the convolution parameter count and computational cost using parameters (PARAMs) and floating-point operations (FLOPs). Furthermore, five sets of experiments were designed to assess the practical performance of MrisNet. Firstly, comparative experiments were designed to assess the actual performance of various algorithms across different evaluation metrics, thereby validating the effectiveness of the MrisNet. Moreover, ablation experiments were conducted to analyze the individual improvements in MrisNet and verify the specific effects of different methods. Finally, the adaptability of the MrisNet for ship segmentation in marine radar images was examined through the identification of different categories of radar images.

4.4.1. Experimental Analysis of Different Algorithms

On the constructed RadarSeg dataset, a comparison was conducted between several commonly used standard algorithms and the proposed MrisNet. All the algorithms were trained using the same hyperparameters and tested on the same dataset. Moreover, each algorithm underwent cross-testing, and the Recall, Precision, and mAP values were averaged over three experimental trials. As shown in Tables 1 and 2, for the ship segmentation, our proposed method achieved a recall rate of 94.8% and a precision of 95.2% on the testing images. Compared to other algorithms, our method demonstrated satisfactory experimental results. This is attributed to the MrisNet's enhanced capability to accurately identify densely navigated and small-scale ships in radar images. Furthermore, MrisNet demonstrated remarkable performance in the object detection, achieving a recall rate of 98% and precision of 98.6%. This highlights the beneficial ship localization capability of our proposed method. The results also indicate, in the context of ship instance segmentation, MrisNet surpasses other algorithms with its mAP_{50} and $mAP_{50\text{-}95}$ scores of 0.96 and 0.508, respectively. These findings underscore the notable accuracy of MrisNet in recognizing diverse ship instances. Additionally, MrisNet showcased desirable performance in terms of parameter count and real-time computational cost, with values of 13.8M and 23.5 G, respectively. Compared to the standard YOLOv7 [35], MrisNet achieved a reduction of 61.77% and 83.44% in model parameters and computational cost, respectively. These results indicate MrisNet is more suitable for deployment on edge computing devices.

Table 1. Specific experimental results (i.e., Precision, Recall, mAP_{50}, and $mAP_{50\text{-}95}$) of various algorithms.

Algorithms	Box				Mask			
	P	R	mAP_{50}	$mAP_{50\text{-}95}$	P	R	mAP_{50}	$mAP_{50\text{-}95}$
YOLACT	0.901	0.86	0.92	0.492	0.875	0.826	0.871	0.349
SOLOv2	0.897	0.871	0.926	0.508	0.882	0.855	0.884	0.367
Mask R-CNN	0.97	0.975	0.991	0.717	0.938	0.942	0.952	0.495
Deepsnake	0.939	0.94	0.964	0.603	0.917	0.919	0.911	0.424
Swin-Transformer (T)	0.979	0.976	0.991	0.725	0.937	0.94	0.953	0.519
HTC+	0.913	0.883	0.936	0.528	0.903	0.881	0.892	0.373
SA R-CNN	0.946	0.958	0.97	0.594	0.911	0.905	0.919	0.401
YOLOv5(N)	0.957	0.961	0.981	0.618	0.913	0.908	0.92	0.406
YOLOv5(S)	0.962	0.965	0.988	0.655	0.932	0.933	0.94	0.477
YOLOv5(M)	0.966	0.968	0.988	0.669	0.923	0.924	0.933	0.467
YOLOv5(L)	0.965	0.971	0.986	0.668	0.924	0.925	0.937	0.466
YOLOv5(X)	0.967	0.971	0.989	0.671	0.925	0.924	0.939	0.463
YOLOv7	0.968	0.965	0.98	0.675	0.919	0.917	0.924	0.412
YOLOv8(S)	0.961	0.909	0.956	0.608	0.917	0.851	0.914	0.377
YOLOv8(M)	0.967	0.956	0.982	0.662	0.923	0.914	0.935	0.446
YOLOv8(L)	0.958	0.973	0.982	0.663	0.92	0.925	0.927	0.47
YOLOv8(X)	0.969	0.964	0.977	0.659	0.925	0.922	0.94	0.465
Mris_APFN	0.966	0.965	0.981	0.65	0.92	0.923	0.935	0.452
MrisNet	0.986	0.98	0.993	0.737	0.952	0.948	0.96	0.508

Table 2. Specific experimental results (i.e., PARAMs and GFLOPs) of various algorithms.

Algorithms	PARAMs/(M)	GFLOPs
SOLOv2	61.3	232.6
YOLACT	53.72	240.2
HTC+	95.53	1289.5
SA R-CNN	53.79	101.9
Mask R-CNN	62.74	244.8
Swin-Transformer (T)	88	745
Deepsnake	16.37	25.94
YOLOv5(N)	1.8	6.7
YOLOv5(S)	7.1	25.7
YOLOv5(M)	20.65	69.8
YOLOv5(L)	45.27	146.4
YOLOv5(X)	84.2	264
YOLOv7	36.1	141.9
YOLOv8(S)	11.23	42.4
YOLOv8(M)	25.96	110
YOLOv8(L)	43.79	220.1
YOLOv8(X)	68.4	343.7
Mris_APFN	10.3	20.6
MrisNet	13.8	23.5

To evaluate the practical performance of MrisNet, multiple single-stage, two-stage, and contour-based instance segmentation algorithms were employed for comparison. The experimental results demonstrate MrisNet outperforms single-stage algorithms such as YOLACT [36] and SOLOv2 [37] in terms of core evaluation metrics. This indicates MrisNet, through its adaptive design in network architecture, loss function, and feature enhancement, enables more precise extraction of ship features from radar images, resulting in accurate localization and segmentation of ship instances. Due to the introduction of an additional segmentation branch outside the object detection framework, Mask R-CNN achieves relatively robust precision and segmentation quality. Experimental data reveals for ship segmentation, this algorithm surpasses 93% in both accuracy and recall rate. Additionally, it achieves mAP_{50} and $mAP_{50\text{-}95}$ rates of 0.952 and 0.495, respectively, which are very close to the experimental results of MrisNet. However, due to its high parameter count and computational demands, it exceeds the computational capacity of most maritime devices, making it unsuitable for direct deployment on such devices. Additionally, Deepsnake [38], combining deep learning with the Snake algorithm in active contour models, enables real-time instance segmentation in various common scenarios. However, in the marine radar images, Deepsnake achieves a segmentation precision of only 91.7%, significantly lower than the proposed method. The analysis suggests the fine and small ship contours in radar images do not provide sufficient information for accurate learning of vertex offsets, thereby affecting the segmentation accuracy and recall of targets.

The Swin Transformer [39], a novel recognition algorithm based on the evolution of the Transformer architecture, achieves a recall of 94% and precision of 93.7% in ship segmentation. Moreover, the method demonstrates encouraging performance in both the mAP_{50} and $mAP_{50\text{-}95}$ metrics, outperforming the majority of conventional benchmark algorithms. Compared to MrisNet, it experiences a decrease of 0.8% in recall and 1.5% in precision. The analysis suggests due to the limited pixel information of ships in the images and the presence of significant background noise, the fixed-size image patches used in Swin Transformer may struggle to effectively capture and represent targets of different scales. As a result, the algorithm illustrates relatively underwhelming performance in recognizing different types of ships and fails to extract detailed ship features accurately, making it difficult to distinguish ships from interfering objects.

Due to the scarcity of research focused on ship segmentation using CNN or transformer techniques in the context of marine radar images, we selected two instance segmentation algorithms specifically designed for SAR images, namely HTC+ [12] and SA R-CNN [13], to compare against our proposed method, MrisNet, to validate the effectiveness of our approach. As previously discussed, the features of targets in SAR images bear resemblance to the spot-like information of ships in marine radar images. Experimental results indicated the two modified two-stage algorithms, when applied to radar images, exhibited subpar performance. While they demonstrate relatively suitable performance in terms of object detection, their precision, recall, and mAP metrics in ship segmentation are noticeably weaker compared to MrisNet. This discrepancy can be attributed to the fact that ships in SAR images predominantly represent static targets in harbor areas, which starkly contrasts with the characteristics of ships in radar images. Hence, it is likely the algorithms designed for SAR environments may not be well-suited for the scenarios addressed in our research.

Compared to the standard YOLOv5, YOLOv7, and YOLOv8 series, MrisNet exhibits relatively robust segmentation precision and recall rate, surpassing various YOLO algorithms, including the relatively high-performing YOLOv5(S). Moreover, MrisNet exhibits significantly lower parameter count and computational costs compared to many deep-layer algorithms in the YOLO series. Additionally, in terms of recall rate, MrisNet increases 9.7% and 1.5% compared to the lightweight YOLOv8(S) and YOLOv5(S), respectively. Moreover, the method significantly outperforms these two classical lightweight algorithms in terms of both the mAP_{50} and $mAP_{50\text{-}95}$ metrics. This indicates MrisNet experiences fewer instances of ship loss in marine radar images and shows improved efficiency throughout the experiments. Furthermore, the proposed method demonstrates noteworthy capability in suppressing false positives, accurately identifying interferences that resemble ship features, such as coastal objects, reefs, and clouds. This effectively reduces misidentification rates and enhances the precision of ship segmentation in various scenarios.

To investigate the influence of different levels of image features on ship segmentation, we combined a progressive feature pyramid with the MrisNet. This network integrated multi-scale features through a layer-by-layer concatenation approach, preserving deep-level semantic information more effectively compared to standard PANet. In the process of algorithm improvement, the feature network and prediction structure of MrisNet were retained, and the original PANet was replaced with the AsymptoticFPN [40] to construct the Mris_APFN algorithm for comparison. The experimental results demonstrated Mris_APFN achieved a recall rate of 92.3% and a precision of 92%. In addition, the method attains a modest performance with scores of 0.935 and 0.452 in terms of the mAP_{50} and $mAP_{50\text{-}95}$ metrics, respectively. Apparently, it falls short of MrisNet in all evaluation metrics, indicating Mris_APFN suffers from more instances of missing ships and fails to extract features relevant to ship instances effectively. Further analysis suggests deep-level features tend to focus more on abstract semantic information while being relatively weaker in extracting contour details. Therefore, for ship segmentation in radar images, it is necessary to appropriately restrict the influence weight of deep-level features to better express salient features of the targets.

4.4.2. Ablation Experiments

To further validate the practical performance of each improvement method in MrisNet, a comprehensive decomposition analysis was conducted based on the RadarSeg dataset to analyze their impact on ship segmentation. The main experimental process involved step-by-step application of various improvement methods on the standard YOLOv5(S), followed by testing their respective performance metrics. The specific results of the ablation experiments for MrisNet are presented in Table 3.

Table 3. Ablation experiments of MrisNet. The presence of an asterisk denotes the model's adoption of the method corresponding to the leftmost column of the table.

Methods	Model 1	Model 2	Model 3	Model 4	Model 5	Model 6	Model 7
YOLOv5(S)	*	*	*	*	*	*	*
+FasterYOLO		*	*	*	*	*	*
+SimAM			*	*	*	*	*
+CPFPN				*	*	*	*
+SIoU					*		
+DIoU						*	
+EIoU							*
R^{mask}	0.933	0.937	0.94	0.945	0.944	0.94	0.948
P^{mask}	0.932	0.942	0.943	0.946	0.941	0.938	0.952
mAP_{50}	0.94	0.949	0.95	0.957	0.945	0.942	0.96
$mAP_{50\text{-}95}$	0.477	0.489	0.492	0.505	0.49	0.488	0.508

(1) Analysis of the FasterYOLO network. By replacing the feature network with FasterYOLO, experimental results revealed the improved feature network increased the ship segmentation precision by 1.0%, with an improvement in recall as well. Furthermore, the network yields a substantial enhancement of 0.9% and 1.2% in the mAP_{50} and $mAP_{50\text{-}95}$ metrics, respectively. This indicated the algorithm's ability to suppress false targets has been enhanced, reducing the misidentification rate and decreasing the probability of target omissions. Furthermore, through multiple experiments, it has been observed FasterYOLO could accelerate the convergence speed of the algorithm, leading to relatively rapid and stable convergence of the loss values based on the training and validation sets.

(2) Analysis of SimAM. In this paper, a SimAM attention mechanism was appended after the FasterYOLO feature network, enabling the algorithm to weight the convolutional feature maps based on the principle of similarity. This allows the model to focus more on the relevant feature representations associated with the target's spatial location. As a result, the model exhibits enhanced robustness when faced with challenges such as low illumination and occlusion in challenging scenarios. Based on the experimental results, the SimAM module enhances the salient features of ships, resulting in improvements in both precision, recall, and mAP rates of the ship segmentation model.

(3) Analysis of the CPFPN network. As mentioned earlier, applying attention mechanisms based on Transformer structures to feature fusion network enables better modeling of global contextual information in images. According to Table 3, it is evident compared to the standard PANet network, CPFPN improves the recall and precision of ship segmentation by 0.5% and 0.3%, respectively. Concurrently, the network demonstrates an enhancement of 0.7% and 1.3% in the mAP_{50} and $mAP_{50\text{-}95}$ metrics, respectively, leading to a relatively clear improvement in the overall ship recognition capability. However, this improvement comes at the cost of a moderate increase in model parameters. In theory, due to the presence of various interfering factors in radar images, the complexity and diversity of objects such as islands and reefs make it challenging to distinguish ships from the background. The CPFPN network, by adaptively attending to target features and effectively filtering out noise or redundant objects, plays a crucial role in the instance segmentation of small-scale ships.

(4) Analysis of EIoU loss. In neural network-based algorithms, the position loss function aids the model in better accommodating variations in the position and scale of the targets. Experimental results indicate the EIoU loss improves ship segmentation precision by approximately 0.6%, along with improvements observed in other evaluation metrics. To further validate the effectiveness of EIoU, the SIoU and DIoU functions were also compared in this section. Theoretically, these two types of loss functions are also capable of accelerating model convergence. It could be observed the model employing the aforementioned comparative functions achieved somewhat favorable results, yet lower than EIoU in terms of evaluation metrics. Analysis reveals the position loss function, which takes into account spatial and shape factors, exhibits relatively better performance in ship segmentation under radar images.

4.4.3. Comparisons in Radar Images

We presented in Figure 11 the ship segmentation results in marine radar images under different scenarios using MrisNet, with a specific focus on evaluating its performance on ships with various trail features. From Figure 11a, it can be observed MrisNet accurately segments ships with long trails in different water areas. This demonstrates the algorithm's ability to recognize ships with prominent features, as these targets maintain a certain consistency in position and shape across consecutive frames. In radar images, it is noticeable longer ship trails often exhibit trajectory interruptions or significant curvature, posing a significant challenge to the algorithm's adaptability. However, experimental results show MrisNet effectively avoids the degradation of segmentation precision caused by these two scenarios. Figure 11b reveals MrisNet achieves satisfactory segmentation results for short-trail ships under different backgrounds, without missing or misidentifying small-scale or even tiny-scale targets. This indicates the algorithm's relatively positive capability in extracting fine-grained object features and mitigating the influence of terrain, sea conditions, and clutter interference. Figure 11c demonstrates MrisNet accurately segments dense ships in radar images and exhibits good recognition capabilities for complex scenarios such as head-on or crossing trajectories. Analysis suggests the adoption of adaptive attention mechanisms and efficient convolutional computations in MrisNet enables it to capture salient ship features even in scenarios with dense small targets.

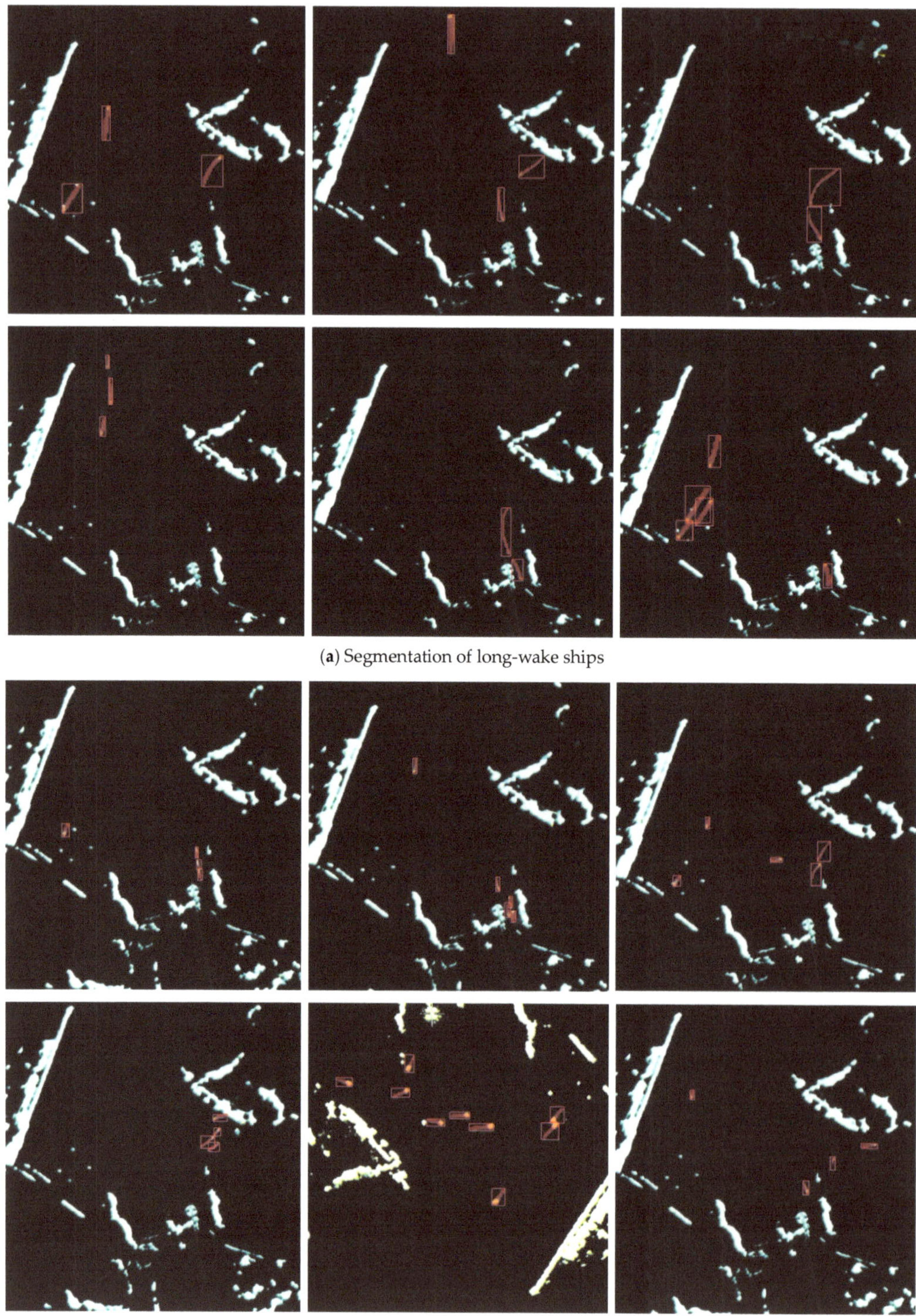

(**a**) Segmentation of long-wake ships

(**b**) Segmentation of short-wake ships

Figure 11. *Cont.*

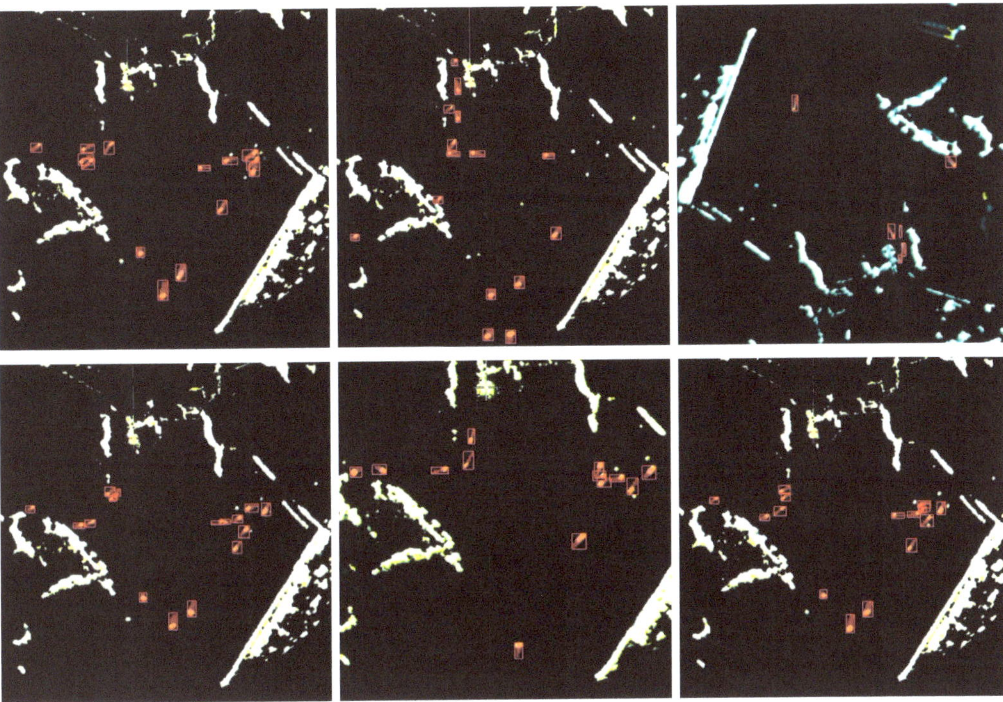

(**c**) Segmentation of dense ships

Figure 11. Segmentation results of MrisNet under marine radar images. The red boxes represent the detected positions of the ship spots, and it can be observed that there is no omission of various types of ships to be recognized. The findings illustrate the proposed method exhibits commendable efficacy across a wide range of radar scenarios.

4.4.4. Comparisons of Small-Scale Ship Segmentation

Instance segmentations of small objects has been a prominent research direction in the field of visual recognition tasks. To evaluate the performance of different algorithms in the segmentation of small-scale ships, this section conducted experiments in two scenarios, i.e., cross navigation and dense navigation. These scenarios involve a large number of ships with short trails, and factors such as islands, reefs, and atmospheric disturbances significantly affect the model's performance. To enhance the credibility of the findings, this experiment compared the performance of MrisNet with the standard YOLOv5(L) and YOLOv8(L). As shown in Figure 12, YOLOv8(L) exhibited a higher misidentification rate but fewer omissions in recognition. YOLOv5(L) demonstrated relatively poor accuracy in recognizing small-scale ships, leading to more omission issues and a tendency to misidentify ships traveling in opposite directions. In comparison to the comparative algorithms, MrisNet achieved more accurate localization of small-scale ships, enabling finer segmentation of ship contours and demonstrating favorable segmentation accuracy for tiny objects. This result highlights lightweight algorithm, through the construction of a rational network structure, can effectively extract ship pixels and contour features from radar images, thereby significantly improving the performance of the model in small-scale instance segmentation tasks.

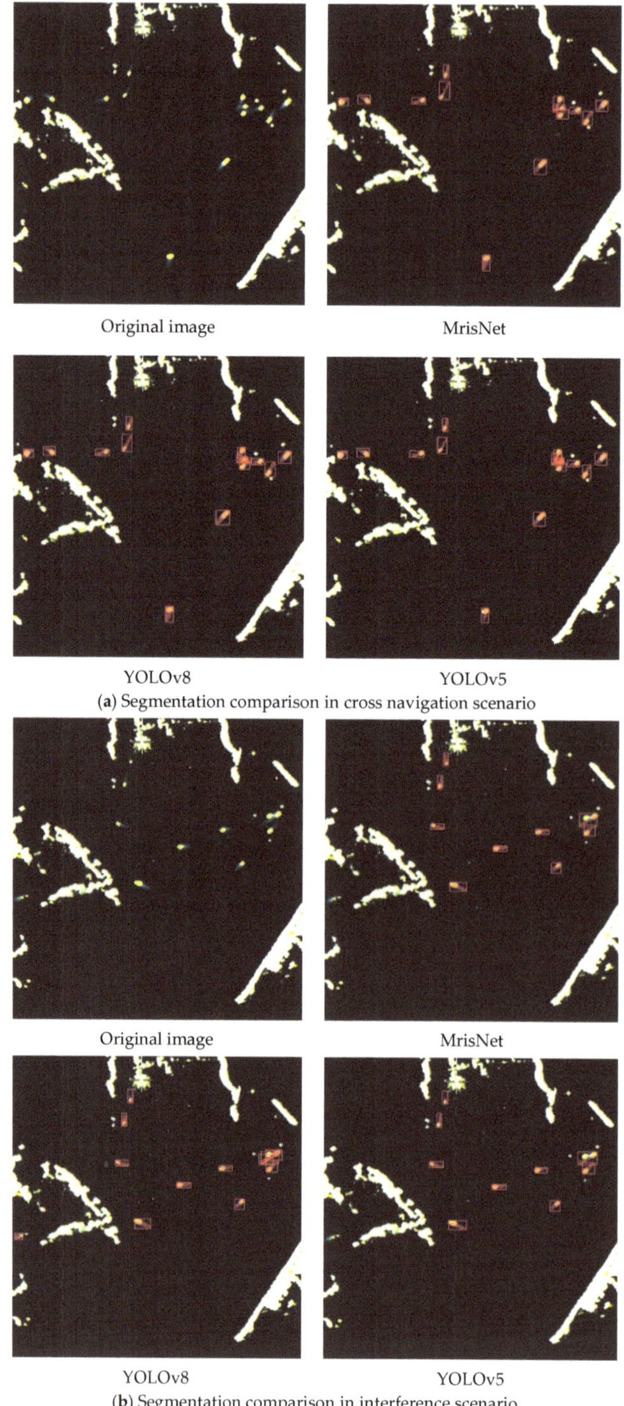

Original image   MrisNet

YOLOv8   YOLOv5

(**a**) Segmentation comparison in cross navigation scenario

Original image   MrisNet

YOLOv8   YOLOv5

(**b**) Segmentation comparison in interference scenario

Figure 12. *Cont.*

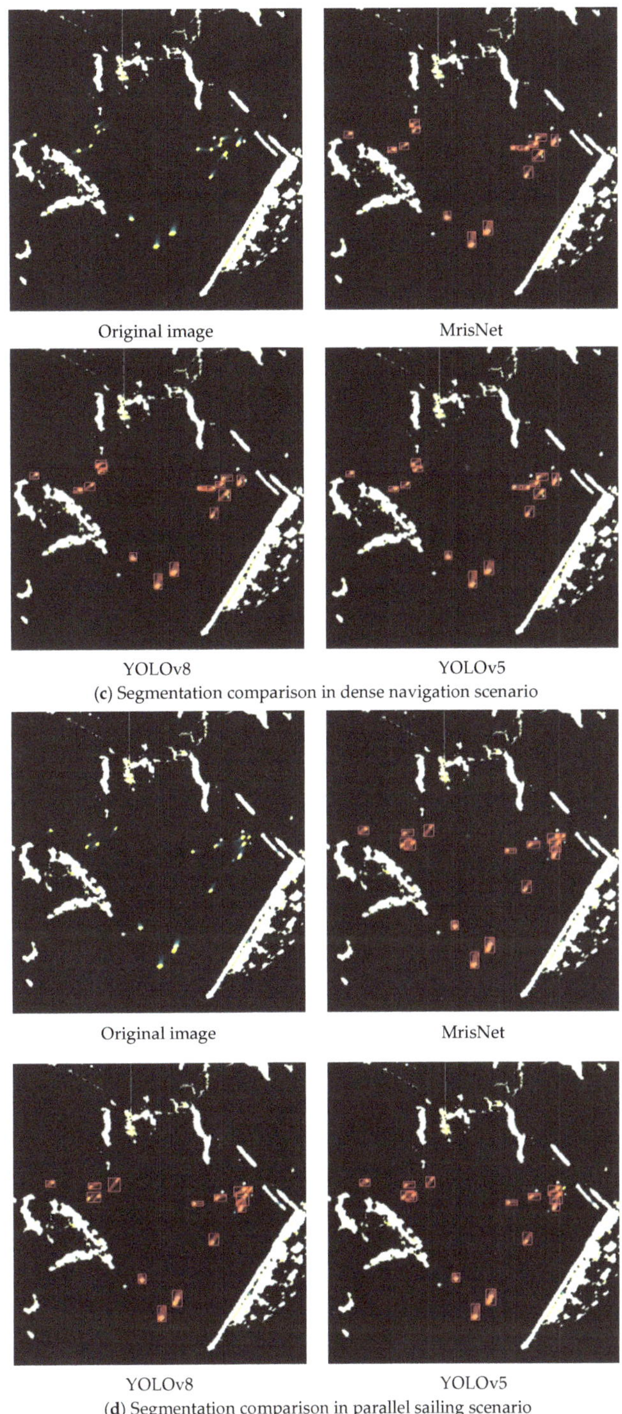

Original image   MrisNet

YOLOv8   YOLOv5

(**c**) Segmentation comparison in dense navigation scenario

Original image   MrisNet

YOLOv8   YOLOv5

(**d**) Segmentation comparison in parallel sailing scenario

Figure 12. Comparisons of various algorithms for small-scale ship segmentation. The positions of the ship spots are indicated by the red boxes, revealing that MrisNet achieves lower false positive and false negative rates compared to the benchmark algorithms.

4.4.5. Ship Identification in Extreme Environments

To evaluate the capability of MrisNet in ship segmentation under extreme environments, a subset of extreme scenario images was constructed by selecting samples from the RadarSeg dataset. This subset consists of 200 images, categorized into two identification scenarios, i.e., dense ship and tiny ship identification. For comparison, YOLACT, YOLOv8(S), and Mask R-CNN were also tested on the same images and threshold settings. The experimental results, as shown in Table 4, indicated MrisNet exhibited encouraging performance. It achieves relatively higher ship segmentation precision and fewer misidentification errors and performs better in terms of recall rate. Therefore, MrisNet demonstrates satisfactory adaptability to ship segmentation in extreme scenarios.

Table 4. Comparisons of experimental data in extreme scenarios.

Algorithms	Detected Ships	True Ships	False Alarms	Recall	Pr
YOLACT	1296	1117	179	0.8007	0.8619
YOLOv8(S)	1288	1170	118	0.8387	0.9084
Mask R-CNN	1389	1299	90	0.9312	0.9352
MRNet	1382	1301	81	0.9326	0.9414

Figure 13 showcases the instance segmentation of MrisNet on ship radar images under various types of interference signals. The experimental images were extracted from the aforementioned subset of extreme scenario images, covering three common target types, i.e., short-trail ships, long-trail ships, and dense ships, which were prevalent in radar images. In this experiment, the original radar images were subjected to three types of noise processing, i.e., salt-and-pepper noise, gaussian noise, and speckle noise. The research found speckle noise and gaussian noise had a more significant impact on radar images, causing noticeable interference with trail features. This led to a decrease in the positional precision of long-trail ships and the potential omission of identification for small-scale ships, as their trail features became confounded with the image background. In contrast, salt-and-pepper noise had a minor impact on the effective ship features. The experimental results demonstrated even under the interference of salt-and-pepper noise, MrisNet was able to achieve precise segmentation of all ships. However, under the influence of speckle noise and Gaussian noise, the confidence of ship segmentation by MrisNet slightly decreased, and the positioning accuracy of ships in dense navigation scenarios also decreased. Nevertheless, in the majority of scenarios, MrisNet maintained an effective ship segmentation performance.

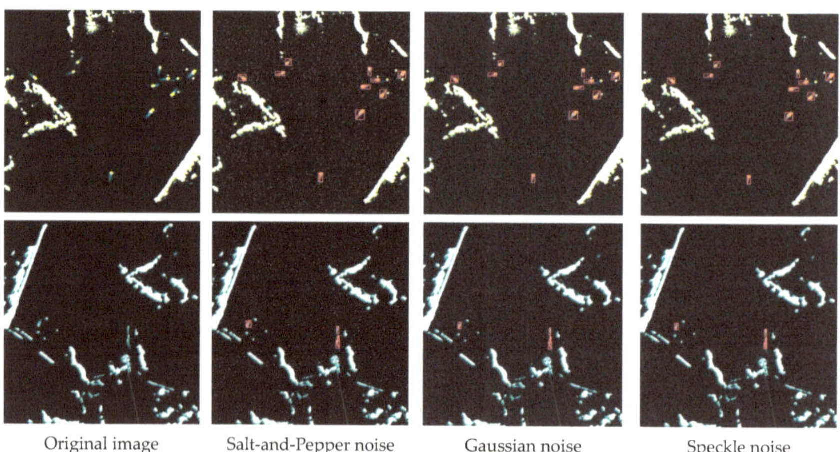

(a) Segmentation of short-wake ships

Figure 13. *Cont.*

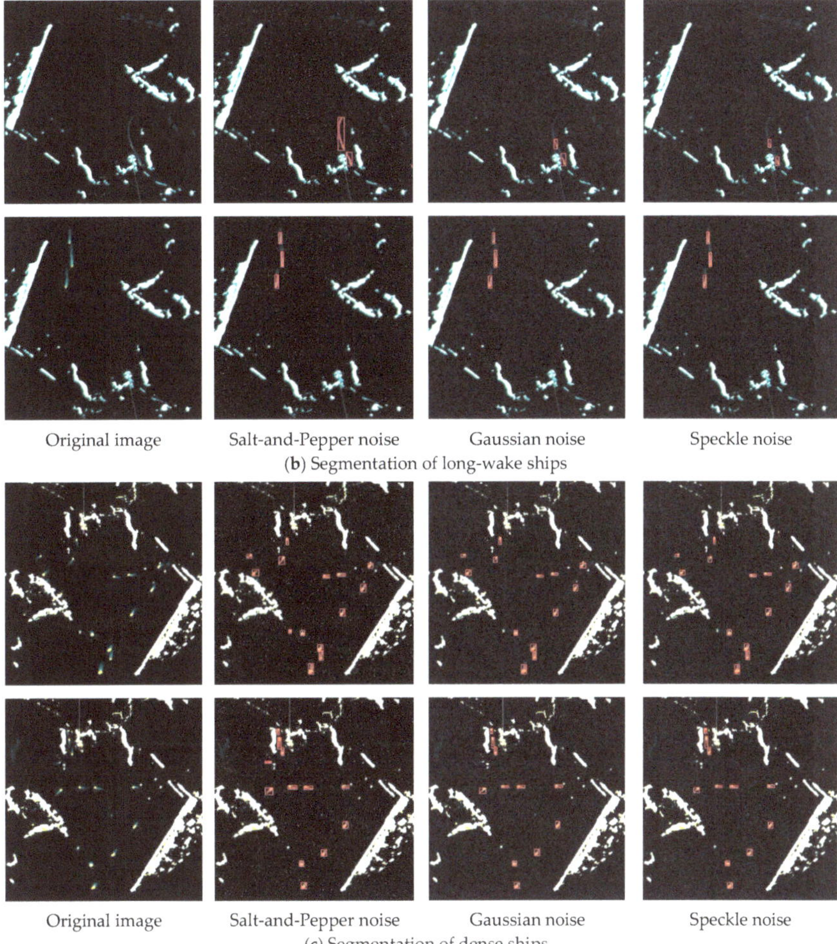

Original image Salt-and-Pepper noise Gaussian noise Speckle noise
(**b**) Segmentation of long-wake ships

Original image Salt-and-Pepper noise Gaussian noise Speckle noise
(**c**) Segmentation of dense ships

Figure 13. Ship segmentation results of MrisNet under various noises. By depicting the detected positions of ship spots using the red boxes, the results underscore the satisfactory ship segmentation performance of MrisNet, which remains robust in the presence of moderate levels of interference.

5. Conclusions and Discussion

This paper presents a ship segmentation algorithm called MrisNet for marine radar images. MrisNet proposes a novel feature extraction network called FasterYOLO, which incorporates efficient convolutional units to accurately extract key features of ships in radar images. Furthermore, a simple, parameter-free attention module is introduced to infer the three-dimensional attention weights of feature maps, optimizing the deep feature output of the network and enhancing the ship segmentation at a higher semantic level. Additionally, the algorithm integrates a self-attention mechanism based on the Transformer structure into the feature fusion network to model and extract long sequence feature of images more accurately, aiding in distinguishing ship instances from the background environment. In the prediction structure, the algorithm improves the calculation of position loss for the predicted boxes by utilizing the EIoU function, thereby extracting more precise positional information of ships.

Experimental results have demonstrated MrisNet outperforms common algorithms in ship segmentation on marine radar images. MrisNet achieves accurate segmentation of long-tail and short-tail ships in various scenarios, with recall and precision reaching 94.8% and 95.2%, respectively. Particularly, it exhibits acceptable performance in scenarios involving tiny ships, dense navigation, and complex interference. Furthermore, MrisNet exhibits advantages in terms of model parameter size and real-time computational cost. It has a parameter size of 13.8 M and a calculation consumption of 23.5 G, significantly reduced compared to deep YOLO series, making it more suitable for deployment in maritime monitoring devices. Considering the relatively limited image samples, future research will focus on expanding the RadarSeg dataset to cover a wider range of ship navigation scenarios. Additionally, the next steps in research will involve introducing more interference factors to enhance the model's robustness.

Author Contributions: Conceptualization, F.M., Z.K. and C.C.; methodology, F.M. and Z.K.; software, Z.K.; validation, Z.K.; formal analysis, F.M. and C.C.; investigation, J.S. and J.D.; resources, J.D.; data curation, J.S.; writing—original draft preparation, Z.K.; writing—review and editing, F.M. and Z.K.; visualization, Z.K.; supervision, F.M. and C.C. All authors have read and agreed to the published version of the manuscript.

Funding: This research was funded by the National Key R and D Program of China, grant number 2021YFB1600400, the National Natural Science Foundation of China, grant number 52171352.

Institutional Review Board Statement: Not applicable.

Informed Consent Statement: Not applicable.

Data Availability Statement: Access to the data will be considered upon request.

Acknowledgments: We are deeply grateful to our colleagues for their exceptional support in developing the dataset and assisting with the experiments.

Conflicts of Interest: Jie Sun is an employee of Nanjing Smart Water Transportation Technology Co., Ltd. and Jizhu Deng is an employee of Nanjing Port Co., Ltd. Other authors declare no conflict of interest.

References

1. Wei, Y.; Liu, Y.; Lei, Y.; Lian, R.; Lu, Z.; Sun, L. A new method of rainfall detection from the collected X-band marine radar images. *Remote Sens.* **2022**, *14*, 3600. [CrossRef]
2. Li, B.; Xu, J.; Pan, X.; Chen, R.; Ma, L.; Yin, J.; Liao, Z.; Chu, L.; Zhao, Z.; Lian, J. Preliminary investigation on marine radar oil spill monitoring method using YOLO model. *J. Mar. Sci. Eng.* **2023**, *11*, 670. [CrossRef]
3. Wen, B.; Wei, Y.; Lu, Z. Sea clutter suppression and target detection algorithm of marine radar image sequence based on spatio-temporal domain joint filtering. *Entropy* **2022**, *24*, 250. [CrossRef] [PubMed]
4. He, W.; Ma, F.; Liu, X. A recognition approach of radar blips based on improved fuzzy c means. *Eurasia J. Math. Sci. Technol. Educ.* **2017**, *13*, 6005–6017. [CrossRef]
5. Ren, S.; He, K.; Girshick, R.; Sun, J. Faster r-cnn: Towards real-time object detection with region proposal networks. *IEEE Trans. Pattern Anal. Mach. Intell.* **2017**, *39*, 1137–1149. [CrossRef] [PubMed]
6. He, K.; Gkioxari, G.; Dollár, P.; Girshick, R. Mask r-cnn. In Proceedings of the IEEE International Conference on Computer Vision (ICCV), Venice, Italy, 22–29 October 2017; pp. 2961–2969.
7. Chen, X.; Su, N.; Huang, Y.; Guan, J. False-alarm-controllable radar detection for marine target based on multi features fusion via CNNs. *IEEE Sens. J.* **2021**, *21*, 9099–9111. [CrossRef]
8. Chen, X.; Mu, X.; Guan, J.; Liu, N.; Zhou, W. Marine target detection based on Marine-Faster R-CNN for navigation radar plane position indicator images. *Front. Inf. Technol. Electron. Eng.* **2022**, *23*, 630–643. [CrossRef]
9. Wang, Y.; Shi, H.; Chen, L. Ship detection algorithm for SAR images based on lightweight convolutional network. *J. Indian Soc. Remote Sens.* **2022**, *50*, 867–876. [CrossRef]
10. Li, S.; Fu, X.; Dong, J. Improved ship detection algorithm based on YOLOX for SAR outline enhancement image. *Remote Sens.* **2022**, *14*, 4070. [CrossRef]
11. Zhao, C.; Fu, X.; Dong, J.; Qin, R.; Chang, J.; Lang, P. SAR ship detection based on end-to-end morphological feature pyramid network. *IEEE J. Sel. Top. Appl. Earth Obs. Remote Sens.* **2022**, *15*, 4599–4611. [CrossRef]
12. Zhang, T.; Zhang, X. HTC+ for SAR ship instance segmentation. *Remote Sens.* **2022**, *14*, 2395. [CrossRef]
13. Zhao, D.; Zhu, C.; Qi, J.; Qi, X.; Su, Z.; Shi, Z. Synergistic attention for ship instance segmentation in SAR images. *Remote Sens.* **2021**, *13*, 4384. [CrossRef]
14. Yang, X.; Zhang, Q.; Dong, Q.; Han, Z.; Luo, X.; Wei, D. Ship instance segmentation based on rotated bounding boxes for SAR images. *Remote Sens.* **2023**, *15*, 1324. [CrossRef]

15. Shao, Z.; Zhang, X.; Wei, S.; Shi, J.; Ke, X.; Xu, X.; Zhan, X.; Zhang, T.; Zeng, T. Scale in scale for SAR ship instance segmentation. *Remote Sens.* **2023**, *15*, 629. [CrossRef]
16. Sun, Y.; Su, L.; Yuan, S.; Meng, H. DANet: Dual-branch activation network for small object instance segmentation of ship images. *IEEE Trans. Circuits Syst. Video Technol.* **2023**, *33*, 6708–6720. [CrossRef]
17. Sun, Y.; Su, L.; Luo, Y.; Meng, H.; Li, W.; Zhang, Z.; Wang, P.; Zhang, W. Global Mask R-CNN for marine ship instance segmentation. *Neurocomputing* **2022**, *480*, 257–270. [CrossRef]
18. Guo, M.; Guo, C.; Zhang, C.; Zhang, D.; Gao, Z. Fusion of ship perceptual information for electronic navigational chart and radar images based on deep learning. *J. Navig.* **2019**, *73*, 192–211. [CrossRef]
19. Mao, D.; Zhang, Y.; Zhang, Y.; Pei, J.; Huang, Y.; Yang, J. An efficient anti-interference imaging technology for marine radar. *IEEE Trans. Geosci. Remote Sens.* **2022**, *60*, 5101413. [CrossRef]
20. Zhang, C.; Fang, M.; Yang, C.; Yu, R.; Li, T. Perceptual fusion of electronic chart and marine radar image. *J. Mar. Sci. Eng.* **2021**, *9*, 1245. [CrossRef]
21. Dong, G.; Liu, H. A new model-data co-driven method for radar ship detection. *IEEE Trans. Instrum. Meas.* **2022**, *71*, 2508609. [CrossRef]
22. Zhang, F.; Wang, X.; Zhou, S.; Wang, Y.; Hou, Y. Arbitrary-oriented ship detection through center-head point extraction. *IEEE Trans. Geosci. Remote Sens.* **2022**, *60*, 5612414. [CrossRef]
23. Zhang, T.; Wang, W.; Quan, S.; Yang, H.; Xiong, H.; Zhang, Z.; Yu, W. Region-based polarimetric covariance difference matrix for PolSAR ship detection. *IEEE Trans. Geosci. Remote Sens.* **2022**, *60*, 5222016. [CrossRef]
24. Qi, X.; Lang, P.; Fu, X.; Qin, R.; Dong, J.; Liu, C. A regional attention-based detector for SAR ship detection. *Remote Sens. Lett.* **2022**, *13*, 55–64. [CrossRef]
25. Yin, Y.; Cheng, X.; Shi, F.; Zhao, M.; Li, G.; Chen, S. An enhanced lightweight convolutional neural network for ship detection in maritime surveillance system. *IEEE J. Sel. Top. Appl. Earth Obs. Remote Sens.* **2022**, *15*, 5811–5825. [CrossRef]
26. Chen, J.; Kao, S.-H.; He, H.; Zhuo, W.; Wen, S.; Lee, C.-H.; Chan, S.-H.G. Run, don't walk: Chasing higher flops for faster neural networks. In Proceedings of the IEEE/CVF Conference on Computer Vision and Pattern Recognition (CVPR), Vancouver, BC, Canada, 18–22 June 2023; pp. 12021–12031.
27. Yang, L.; Zhang, R.-Y.; Li, L.; Xie, X. Simam: A simple, parameter-free attention module for convolutional neural networks. In Proceedings of the International Conference on Machine Learning (ICML), Vienna, Austria, 18–24 July 2021; pp. 11863–11874.
28. Jocher, G. YOLOv5 by Ultralytics. Available online: https://github.com/ultralytics/yolov5 (accessed on 21 September 2023).
29. Vaswani, A.; Shazeer, N.; Parmar, N.; Uszkoreit, J.; Jones, L.; Gomez, A.N.; Kaiser, Ł.; Polosukhin, I. Attention is all you need. In Proceedings of the 31st Conference on Neural Information Processing Systems (NIPS), Long Beach, CA, USA; 2017; pp. 6000–6010.
30. Zhang, Y.-F.; Ren, W.; Zhang, Z.; Jia, Z.; Wang, L.; Tan, T. Focal and efficient IOU loss for accurate bounding box regression. *Neurocomputing* **2022**, *506*, 146–157. [CrossRef]
31. Li, Y.; Yao, T.; Pan, Y.; Mei, T. Contextual transformer networks for visual recognition. *IEEE Trans. Pattern Anal. Mach. Intell.* **2022**, *45*, 1489–1500. [CrossRef] [PubMed]
32. Zheng, Z.; Wang, P.; Ren, D.; Liu, W.; Ye, R.; Hu, Q.; Zuo, W. Enhancing geometric factors in model learning and inference for object detection and instance segmentation. *IEEE Trans. Cybern.* **2021**, *52*, 8574–8586. [CrossRef]
33. Gevorgyan, Z. SIoU loss: More powerful learning for bounding box regression. *arXiv* **2022**, arXiv:2205.12740.
34. Jocher, G. YOLO by Ultralytics. Available online: https://github.com/ultralytics/ultralytics (accessed on 21 September 2023).
35. Wang, C.-Y.; Bochkovskiy, A.; Liao, H.-Y.M. YOLOv7: Trainable bag-of-freebies sets new state-of-the-art for real-time object detectors. In Proceedings of the IEEE/CVF Conference on Computer Vision and Pattern Recognition (CVPR), Vancouver, BC, Canada, 18–22 June 2023; pp. 7464–7475.
36. Bolya, D.; Zhou, C.; Xiao, F.; Lee, Y.J. Yolact: Real-time instance segmentation. In Proceedings of the IEEE/CVF International Conference on Computer Vision (ICCV), Seoul, Republic of Korea, 27–28 October 2019; pp. 9157–9166.
37. Wang, X.; Zhang, R.; Kong, T.; Li, L.; Shen, C. Solov2: Dynamic and fast instance segmentation. *arXiv* **2020**, arXiv:2003.10152.
38. Peng, S.; Jiang, W.; Pi, H.; Li, X.; Bao, H.; Zhou, X. Deep snake for real-time instance segmentation. In Proceedings of the IEEE/CVF Conference on Computer Vision and Pattern Recognition (CVPR), Seattle, WA, USA, 13–19 June 2020; pp. 8533–8542.
39. Liu, Z.; Lin, Y.; Cao, Y.; Hu, H.; Wei, Y.; Zhang, Z.; Lin, S.; Guo, B. Swin transformer: Hierarchical vision transformer using shifted windows. In Proceedings of the IEEE/CVF International Conference on Computer Vision (ICCV), Montreal, QC, Canada, 11–17 October 2021; pp. 10012–10022.
40. Yang, G.; Lei, J.; Zhu, Z.; Cheng, S.; Feng, Z.; Liang, R. AFPN: Asymptotic feature pyramid network for object detection. *arXiv* **2023**, arXiv:2306.15988.

Journal of
Marine Science and Engineering

Article

Risk Analysis and Visualization of Merchant and Fishing Vessel Collisions in Coastal Waters: A Case Study of Fujian Coastal Area

Chuanguang Zhu [1,2,†], Jinyu Lei [1,2,3,†], Zhiyuan Wang [1,2,4,†], Decai Zheng [5], Chengqiang Yu [2], Mingzhong Chen [5] and Wei He [1,2,4,*]

[1] School of Transportation, Fujian University of Technology, Fuzhou 350118, China; guangjie98@outlook.com (C.Z.); j.lei@mju.edu.cn (J.L.); wangzhiyuan@mju.edu.cn (Z.W.)
[2] Fujian Engineering Research Center of Safety Control for Ship Intelligent Navigation, College of Physics and Electronic Information Engineering, Minjiang University, Fuzhou 350108, China; yuchengqiang@mju.edu.cn
[3] College of Computer and Data Science, Minjiang University, Fuzhou 350108, China
[4] Fuzhou Institute of Oceanography, Minjiang University, Fuzhou 350108, China
[5] Fuzhou Aids to Navigation Division of Eastern Navigation Services Center, China Maritime Safety Administration, Fuzhou 350004, China; deacizheng@126.com (D.Z.); mzchen@126.com (M.C.)
* Correspondence: hewei1@mju.edu.cn; Tel.: +86-591-83761115
† These authors contributed equally to this work.

Abstract: The invasion of ship domains stands out as a significant factor contributing to the risk of collisions during vessel navigation. However, there is a lack of research on the mechanisms underlying the collision risks specifically related to merchant and fishing vessels in coastal waters. This study proposes an assessment method for collision risks between merchant and fishing vessels in coastal waters and validates it through a comparative analysis through visualization. First of all, the operational status of fishing vessels is identified. Collaboratively working fishing vessels are treated as a unified entity, expanding their ship domain during operation to assess collision risks. Secondly, to quantify the collision risk between ships, a collision risk index (CRI) is proposed and visualized based on the severity of the collision risk. Finally, taking the high-risk area for merchant and fishing vessel collisions in the Minjiang River Estuary as an example, this paper conducts an analysis that involves classifying ship collision scenarios, extracts risk data under different collision scenarios, and visually analyzes areas prone to danger. The results indicate that this method effectively evaluates the severity of collision risk, and the identified high-risk areas resulting from the analysis are verified by the number of accidents that occurred in the most recent three years.

Keywords: merchant and fishing vessel collisions; coastal waters; ship domain AIS; data visualization

Citation: Zhu, C.; Lei, J.; Wang, Z.; Zheng, D.; Yu, C.; Chen, M.; He, W. Risk Analysis and Visualization of Merchant and Fishing Vessel Collisions in Coastal Waters: A Case Study of Fujian Coastal Area. *J. Mar. Sci. Eng.* **2024**, *12*, 681. https://doi.org/10.3390/jmse12040681

Academic Editor: Md Jahir Rizvi

Received: 27 February 2024
Revised: 26 March 2024
Accepted: 13 April 2024
Published: 19 April 2024

1. Introduction

In recent years, the marine fishing industry has undergone rapid expansion and has emerged as a pivotal facet within the maritime fisheries sector. The proliferation of fishing vessels has led to a discernible increase in channel congestion, making the navigational environment more intricate and increasing the collision risk. Despite the integration of GPS, AIS, and the Beidou positioning system in many fishing vessels, merchant vessels often lack familiarity with the operational practices, methods, and working domains of fishing vessels. Consequently, insufficient attention is devoted to fishing zones and areas with a heightened risk of collisions. And effective communication between merchant and fishing vessels during hazardous situations proves challenging, culminating in collision incidents. Moreover, the visualization of high-risk navigational areas prone to collisions between merchant and fishing vessels holds paramount significance for maritime safety. Firstly, it can issue navigational hazard warnings to vessels in areas characterized by high navigational density, thereby mitigating the occurrence of maritime accidents. Secondly, with a comprehensive understanding of the primary navigation routes and operational domains of merchant fishing vessels, targeted supervision within these areas can be implemented by maritime authorities. However, devising a risk quantification model to assess collision risks between merchant and fishing vessels in different areas remains challenging due to the disparate maneuvering characteristics, navigational environments, and risk standards associated with these vessel types.

With the application requirements of AIS equipment by the IMO and the development of artificial intelligence technology, AIS data become the most powerful "big data" of maritime traffic [1]. Collision risk research based on big data has become a mainstream trend. Liu et al. [2] use static and dynamic information from the automatic identification system (AIS) to calculate the closest points of ship–ship collisions (CPCs) based on ship specifications and geographic positioning, estimate dynamic collision boundaries, and introduce a kinematics-feature-based vessel conflict ranking operator (KF-VCRO) to evaluate collision risk by integrating relative position and velocity information from AIS. Silveira et al. [3] propose a method to calculate the collision risk from the assessment of the number of collision candidates by estimating future distances between ships based on their previous positions, courses, and speeds and comparing those distances with a defined collision diameter. However, the method relies on estimating future distances between ships based on their previous positions, courses, and speeds. So this estimation may not always accurately reflect the actual future positions of the ships, leading to potential inaccuracies in the collision risk assessment.

At present, the analysis of ship collision risk under different navigation conditions can be mainly divided into qualitative analysis and quantitative analysis. Regarding the qualitative analysis of vessel collision risk, Sedov et al. [4] introduce a fuzzy linguistic model to determine collision risk levels for vessels in busy traffic areas. This model comprises 22 terms within five fundamental term sets, each endowed with membership function parameters, and a total of 200 fuzzy rules are generated to delineate collision risk levels. Additionally, Yi et al. [5] leverage fuzzy reasoning and discrete event system specification (DEVS) theory to propose a novel model for predicting vessel collision risks while considering general collision avoidance patterns. This innovative model anticipates collision risks by forecasting changes to future vessel movements, and the authors validate the functionality of the model's structure and fuzzy reasoning module through simulation experiments. In the domain of quantitative analysis for vessel collision risk, scholars often employ the concept of closest point of approach (CPA) for situational analysis. Chin et al. [6] from the National University of Singapore have established a collision risk regression model for port waters based on CPA. In this model, distance to closest point of approach (DCPA) and time to closest point of approach (TCPA) are pivotal factors for determining collision risks.

In order to study the characteristics, causes, and risks of collision accidents between merchant and fishing vessels, scholars have adopted various methods and models for analysis and have put forward corresponding conclusions or strategies, such as risk assessment analysis methods [7,8], navigator collision avoidance behavior models, navigator error development process models [9], machine learning methods [10], and probability risk assessment (PRA) models [11]. Mou et al. [12] constructed a linear regression model employing AIS data collected from collision avoidance scenarios in busy waterways. The study ascertained correlations between vessel size, speed, and heading with distance to closest point of approach (DCPA). Additionally, the researchers proposed a dynamic method for risk assessment based on a Safety Assessment Model for Shipping and Offshore on the North Sea (SAMSON). Uğurlu et al. [13] compiled statistics on the causes of accidents involving fishing vessels during the period from 2008 to 2018. Based on the identified causes of fishing vessel accidents, the researchers constructed a Bayesian network to estimate the probabilities of accidents occurring under various circumstances.

In terms of mitigating the risk of collisions between merchant and fishing vessels, many researchers have put forward related approaches from different perspectives. These include ship-to-ship dialogue and protocols [14], human–machine cooperative collision avoidance systems [15], requirements for fishing vessels [16], safe navigation for merchant vessels [17], and collision avoidance methods during fishing seasons [18]. Obeng et al. [19] systematically summarized the patterns of collisions between merchant and fishing vessels and proposed preventive measures from aspects such as personnel, vessels, environment, and management. Chou et al. [20] conducted studies on collision accidents involving fishing vessels in different water regions. They delineated the risks and causes of collisions between fishing and non-fishing vessels with the aim of reducing incidents involving both types of vessels. In a study focused on preventing fishing vessel collisions, Seo et al. [21] developed a safety navigation system that can simultaneously receive the positions of smartphones and AIS information from vessels. It issues warnings to both merchant and fishing vessels when another vessel approaches within a 500-m range. However, these studies that specifically focused on collision prevention for fishing vessels have limitations because fishing vessels exhibit significant heading deviations, making it challenging to achieve precise risk predictions solely based on distance, DCPA, and TCPA.

Although there is a plethora of research on ship collisions, studies concerning the spatiotemporal characteristics of such collisions extend beyond the events themselves, and the literature lacks comprehensive exploration of the temporal and spatial dynamics from the initiation of ship encounters to the culmination of collision incidents. Moreover, research focusing on collisions between merchant and fishing vessels in coastal waters is relatively scant. Some of these studies often overlook the individual differences between merchant and fishing vessels at the time of collision, failing to consider the impact of fishing vessels in fishing states on collisions. Therefore, this paper conducts a collision risk analysis of merchant and fishing vessels using vessel traffic data during the fishing season in the coastal areas of Fujian. Initially, the study adopts concepts from the ship domain and incorporates considerations for the operational status of fishing vessels to extract potential collision risk events from historical AIS data. Subsequently, a comprehensive visualization analysis of the spatial distribution of encounters between merchant and fishing vessels is undertaken using a weighted kernel density estimation method. Finally, the paper identifies high-risk collision areas between merchant and fishing vessels based on the density of spatial distributions. This study utilizes the concepts of ship domains and collision risk to calculate collision risks in real-time throughout the entire process of ship encounters, which serves as the basis for visualizing collision risks. The innovations of this research are as follows:

- When assessing collision risks, we innovatively adjust the ship domain for fishing vessels during their operations.
- We conduct a visual analysis of the temporal and spatial distribution of encounter risk data.
- We validate the accuracy of our risk assessment methodology using ship collision incidents from the past three years.

The subsequent organization of the paper is as follows: Section 2 introduces the preprocessing of historical AIS trajectories and defines encounter risks based on ship domains and the operational statuses of fishing vessels. Section 3 systematically extracts encounter risk data between merchant and fishing vessels from historical records and conducts spatial and temporal analyses using a weighted kernel density method. Section 4 examines accident data from the past three years in the study area and conducts validations of the collision risk assessment algorithm proposed in this paper. Section 5 provides a summary and highlights the shortcomings of this paper.

2. Methodology

This paper utilizes historical AIS data from the coastal areas of Fujian to establish a ship collision risk model for the research and analysis of collision risks between merchant and fishing vessels. Firstly, the historical AIS data are processed, and the encounter risk data are extracted on the basis of ship domains and the operational status identification of fishing vessels. Subsequently, a weighted kernel density estimation visualization method is employed to display encounter risk data, facilitating the identification of navigational areas prone to collision risks. Lastly, the approach is validated using the locations of ship collisions that occurred within the research area over the past three years. The workflow of this paper, as illustrated in Figure 1, primarily comprises four modules: (1) AIS data pre-processing, (2) identification of fishing vessel operational states, (3) extraction of encounter risk data, and (4) visualization of encounter risk data.

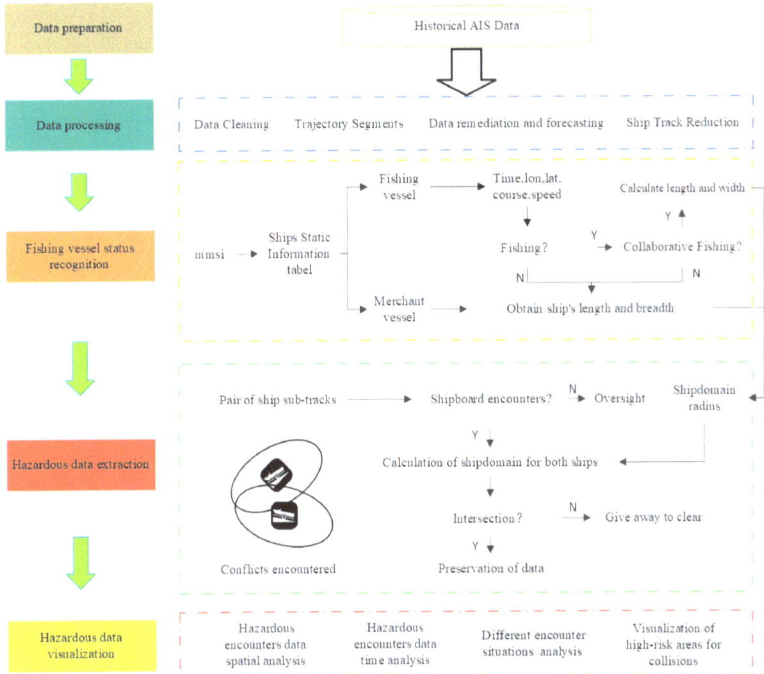

Figure 1. Visualization and assessment method for collision risks between merchant and fishing vessels in coastal waters.

2.1. Data Preprocessing

Due to issues such as location offsets and data loss in the raw AIS data, the analysis of collision risk areas between merchant and fishing vessels based on historical AIS data necessitates preprocessing of the raw data. Processing historical AIS data for merchant and fishing vessels includes data cleaning, trajectory separation, and the repair and prediction of AIS data. Finally, vessel static information is queried based on MMSI to distinguish between merchant and fishing vessels, providing an analytical foundation for assessing collision risks between these two types of vessels.

2.1.1. Data Cleaning

The data cleaning process primarily involves removing abnormal and duplicate data from AIS data. Abnormal data include the following situations:

1. Geographical coordinates located on land;
2. Values exceeding the normal range (refer to Table 1 for data range);
3. MMSI not conforming to specifications.

The occurrence of duplicate data is primarily attributed to the repetitive transmission of AIS data by anchored vessels. As the experiment primarily focuses on the potential risks associated with moving vessels, it is necessary to eliminate AIS data for vessels at anchor.

Table 1. Data range of the study area.

Parameters	Type	Range
Time	timestamp	1 August 2022–1 December 2022
Maritime Mobile Service Identity (MMSI)	string	200,000,000–799,999,999
Longitude	float	0–180°
Latitude	float	0–90°
Speed Over Ground (SOG)	float	0–20 KN
Course Over Ground (COG)	float	0–360°
True Heading	float	0–360°

2.1.2. Trajectory Separation

When analyzing collision risks between merchant and fishing vessels, it is crucial to identify different behavior patterns of vessels accurately by considering key features such as location, time, heading, and speed. In order to reconstruct the historical navigation behavior of vessels, trajectory separation is necessary. Trajectory separation involves two main tasks: one is to separate the trajectory data of different ships according to MMSI; the second is to separate different trajectories from the same ship. Firstly, according to the data characteristics of AIS data, MMSIs are the unique identifications of different ships and can be used as the basis for separating different ship trajectories. Here, we directly group AIS data according to MMSIs to obtain the trajectory data of different ships. Secondly, different trajectories of the same ship are separated according to the AIS data receiving time interval and the distance between adjacent points. The algorithm process for trajectory segmentation is shown in Figure 2.

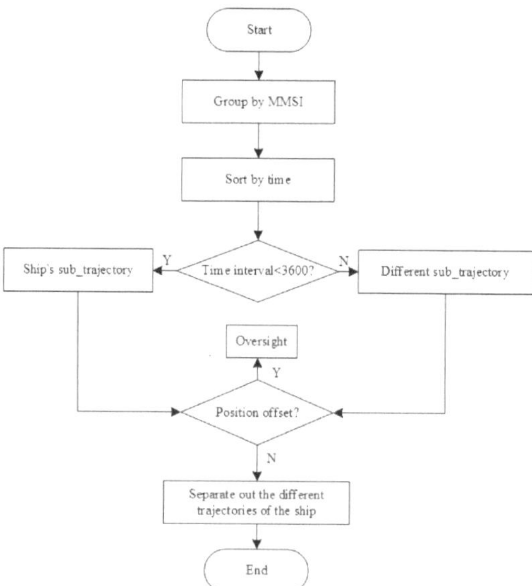

Figure 2. Schematic diagram of trajectory segmentation algorithm.

The experiment selects 1 h as the time interval to determine whether the trajectories before and after the interval are consistent. In addition, the distance between two adjacent points is also used as the basis for judgment. According to the calculation of the ship speed of 0–20 knots, the sailing distance every three minutes is within 0–1 nautical miles. If the distance is excessively large, it is deemed to indicate equipment failure or latitude and longitude drift, leading to the direct exclusion of such points. The latitude and longitude data used in this paper are based on the WGS84 coordinate system. Formula (1) is used to calculate the cosine distance between two points on the sphere, where x_i and y_i are the longitude and latitude of the I-th point on the sphere, with i ranging from 2 to m. In Formula (2), D represents the distance between the two points being calculated, and R denotes the radius of the Earth, which is taken as 6371 km. Formula (3) is used to convert speed values from knots to meters per second.

$$C = \sin(y_{i-1})\sin(y_i) + \cos(y_{i-1})\cos(y_i)\cos(x_{i-1} - x_i) \ i = 2, 3, \ldots, m - 1, m \qquad (1)$$

$$D = R * \arccos(C) * \left(\frac{\pi}{180}\right) \qquad (2)$$

$$v_m = v_k * 0.518 \qquad (3)$$

During processing, the separated AIS trajectory points are grouped into sub-trajectories and connected based on time in ascending order. This process further reconstructs the historical navigation path of the vessels, as shown in Figure 3. The processed data clearly reveal the main channels of vessel navigation and points of trajectory intersection.

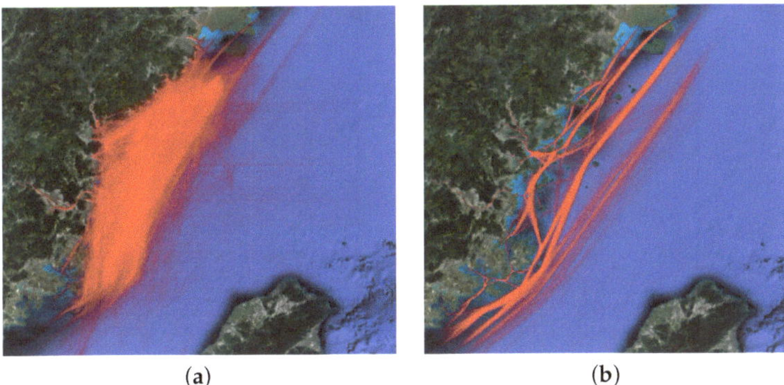

(a)   (b)

Figure 3. Comparison of before and after data pretreatment: ((**a**) before preprocessing; (**b**) after preprocessing).

2.1.3. Data Restoration

In the actual data collection process, AIS data reception can be affected by sensor noise and external environmental factors, leading to issues such as data loss. To ensure the temporal synchronization and high frequency of AIS data for various vessels, interpolation and resampling of AIS data are necessary. The proposals in references [22,23] suggest that, in the absence of prior experience, using piecewise cubic hermite (PCH) interpolation yields minimal errors when reconstructing missing data. Therefore PCH interpolation is applied in this study to restore AIS data.

2.2. Identification of Fishing Vessel Operation Statuses

The statuses of the fishing vessels are crucial in order to define the avoidance distance of passing ships, and the operating status of the fishing vessels should be taken into account when judging the risk of collision. In this experiment, the analysis of AIS trajectory data in the research area during fishing operations reveals distinctive features. Different rules are established to identify the operational states of fishing vessels, and varying safety encounter distance thresholds are set based on the specific state of each fishing vessel.

2.2.1. Characteristics of Fishing Vessel Operational Trajectories

Different fishing methods exhibit distinct trajectory characteristics during fishing operations. Fishing vessel operations can be classified into three main categories, which mainly include trawling operations, purse seine operations, and gillnet operations. Trawling operations are the trawling of fishing gear along the seabed to catch fish. Purse seining entails surrounding a school of fish with a net to capture them; gillnet operations employ a long net suspended vertically in the water to entangle fish. The trajectory characteristics for each of these fishing methods are as follows (Figure 4):

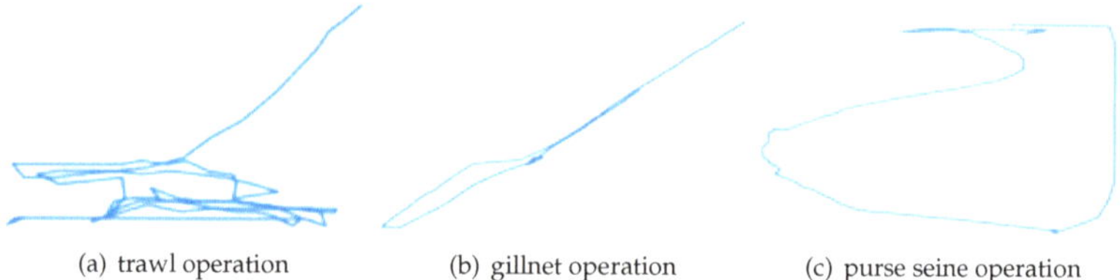

(a) trawl operation   (b) gillnet operation   (c) purse seine operation

Figure 4. Characteristics of fishing vessel operational trajectories under different operating modes.

The typical trajectory characteristics for different fishing operations are as follows:

1. Trawling operation: characterized by frequent turns during operation to tow the net back and forth in a specific area.
2. Gillnet operation: marked by dropping numerous drifting gillnets along a trajectory and returning along the same path to retrieve the nets.
3. Purse seine operation: distinguished by deploying a net to encircle a target area, returning to the starting point, and then retrieving the net.

2.2.2. Fishing Boat Operation Status Judgment

In the study, the operational states of fishing vessels are categorizes into three modes: normal navigation, anchorage, and active fishing. The criteria for each state are defined as follows:

1. Normal navigation: Fishing vessels engaged in normal navigation exhibit relatively stable speed and direction, with no abrupt changes. Vessels with speeds greater than 2 knots and that maintain stable speed and heading within a 10 min interval are identified to be in a normal navigation state.
2. Anchorage: Vessels at anchor may exhibit some speed due to factors like sea currents. Therefore, during processing, vessels with speeds below 2 knots and a substantial number of AIS data positions in close proximity—exhibiting a movement distance less than 0.1 nautical miles within a 10 min interval—are identified as vessels at anchor.
3. Fishing: Identifying whether a fishing vessel is actively fishing relies on distinguishing trajectory characteristics that differ from normal navigation. Fishing vessels typically operate at speeds ranging from 2 to 5 knots. We identify vessels within this speed range and examine their trajectories to determine if they are engaged in fishing. If nearby vessels are also in a fishing state and are within 0.5 nautical miles, this is considered to be cooperative fishing. During this operation, the fishing vessel's maneuverability is limited, and passing merchant vessels are advised to maintain a distance of at least 1 nautical mile.

2.3. Encounter Risk Data Mining and Visualization

Establishing an effective model for identifying vessel navigation risks is crucial for extracting encounter risk data. This paper constructs a model for identifying vessel encounter risks based on ship domains. Then, this model is employed to extract data related to vessel encounter risks, and it further quantifies the collision risk by calculating the CRI. Firstly, DCPA and TCPA are computed for pairs of vessels that may encounter each other. According to the experience of relevant maritime regulatory personnel, the minimum encounter distance between large-tonnage vessels during daytime and good weather conditions with high visibility should satisfy a DCPA of at least 1 nautical mile. In adverse weather conditions or during nighttime, this distance should be at least 2 nm [24]. To obtain a sufficient number of potential collision encounter samples for analysis, we selected the maximum DCPA. After determining the DCPA threshold and considering that fishing vessels typically operate at speeds of 2–5 knots, we consider vessels to be at risk of collision if their TCPA is less than 20 min. The encounter risk is further determined by calculating whether there is an intersection in the vessel domains. Finally, the data related to encounter risks are saved, and visualization tools such as QGIS and Folium are employed for the analysis of encounter risk data.

2.3.1. Ship Domain Model

In the early 1960s, Japanese scholars Fujii et al. [25] proposed the concept of a ship domain when studying the traffic capacity in the waters near Japan, and they tried to establish a calculation method based on ship domains when ships navigated in narrow waters. According to this model, the ship domain is an elliptical area centered around the vessel and with the major axis aligned with the vessel's heading and the minor axis perpendicular to the vessel's heading. The variables a and b represent the major and minor axes, respectively, and determine the elliptical domain. As illustrated in Figure 5, specific values for a and b can be chosen based on the actual circumstances.

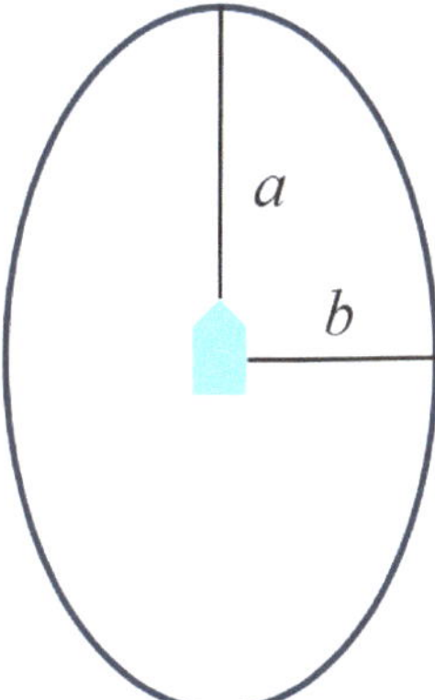

Figure 5. Fujii ship domain model.

This paper employs the Fujii model to determine the values of the axis (a) and the axis (b) for the elliptical ship domains of vessels during normal navigation, as expressed in Equation (4). Here, l represents the vessel's length, and w represents its width.

$$\begin{cases} a = 1.5 \times l \\ b = 1.5 \times w \end{cases} \tag{4}$$

In the case of fishing vessels engaged in operations, occurrences such as obstructive fishing nets render the classical elliptical ship domain inapplicable. The ship domain radius for fishing vessels involved in operations was determined by considering the lengths of the fishing nets commonly used by fishermen. According to information from the Fuzhou Aids to Navigation Division of Eastern Navigation Services Center, the maximum length of trawl nets and gillnets should not exceed 600 m. To ensure an adequate safety distance between passing merchant ships and fishing vessels engaged in operations, we expanded the length of the fishing nets by one-third to establish a safety domain for fishing vessels. Therefore, the ship domain radius for fishing vessels was set to 0.5 nm, as depicted in Figure 6a. Furthermore, for fishing vessels engaged in coordinated operations, they are treated as points on a polygon. The circumscribed circle of the polygon is considered as a whole when calculating the ship domain, ensuring that the encounter distance for each fishing vessel is not less than 0.5 nm, as shown in Figure 6b.

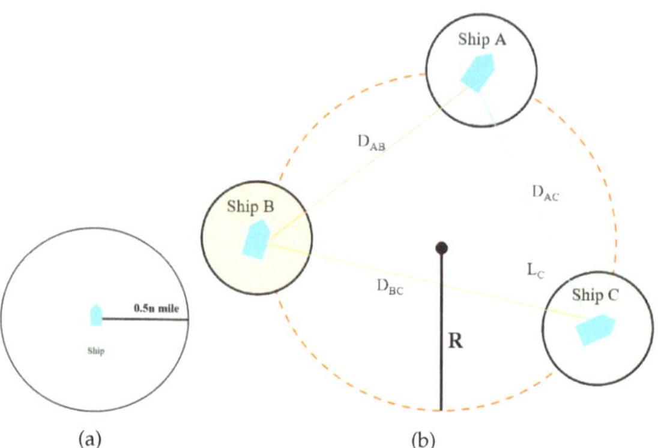

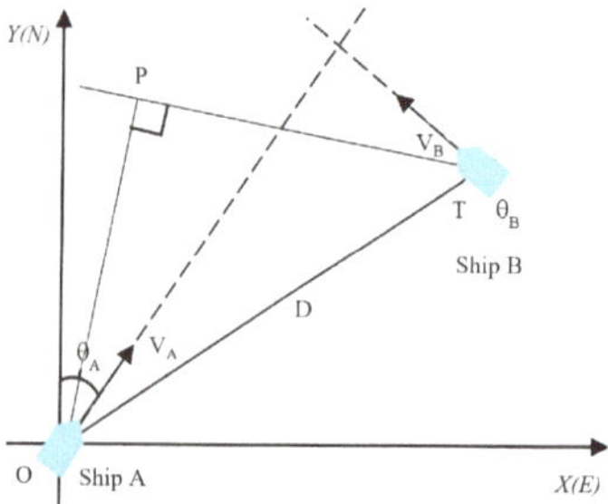

Figure 6. Schematic diagrams of fishing vessel domains in operation; (**a**) single-vessel operation; (**b**) multi-vessel collaborative operation.

2.3.2. Calculation of DCPA and TCPA

DCPA and TCPA are crucial parameters for determining the risk of ship collisions when vessels encounter each other. During calculation, one of the vessels is chosen as the reference vessel in order to establish a coordinate system as illustrated in Figure 7.

Figure 7. Schematic diagram of the encounter of two ships.

Assume that A is the target ship; $lonA, latA, V_A$, and θ_A are the longitude, latitude, speed, and course, respectively, of ship A. B is the encountering ship; $lonB, latB, V_B$, and θ_B are, respectively, the longitude, latitude, speed, and course of ship B. Then the relative speed of the two ships is shown in Formula (5):

$$V_r = \sqrt{V_A^2 + V_B^2 - 2|V_A \cdot V_B| \cos\left(\theta_B \cdot \theta_A\right)} \tag{5}$$

DCPA and TCPA can be calculated by Formulas (6)–(8):

$$\mathrm{DCPA} = D \times \sin\left(\angle OTP\right) \tag{6}$$

$$|TP| = \sqrt{|OT|^2 - |OP|^2} \tag{7}$$

$$\mathrm{TCPA} = \frac{|TP|}{V_r} = D \times \frac{\cos\left(\angle OTP\right)}{V_r} \tag{8}$$

where D represents the relative distance between two trajectory points, which can be obtained from Formula (2). Then, DCPA and TCPA are utilized for assessment of collision risk during vessel navigation and for visualization of risk data.

2.3.3. Collision Risk Index

According to the definition of ship domains, a dangerous encounter is considered when the domain of a navigating vessel is breached. However, this cannot be used to measure the magnitude of the collision risk. In order to further quantify the risk after the invasion of the ship domain during the encounter, the CRI is introduced to quantify the collision risk. This paper evaluates the magnitude of collision risk throughout the entire ship encounter process by considering both the intersection of ship domains and the collision risk index.

Among the calculation methods of CRI, Kearon [26] comprehensively considered the influence of DCPA and TCPA and took into account both factors by using a weighted combination of DCPA and TCPA to determine the CRI, as shown in Equation (9).

$$\rho_i = (a \cdot \text{DCPA}_i)^2 + (b \cdot \text{TCPA}_i)^2 (i = 1, 2, \cdots, n) \tag{9}$$

However, the equation has severe deficiencies and even mistakes. DCPA has dimensions of length; TCPA has dimensions of time. However, the weighted summation of them only involves a value without consideration of the dimensions, which are in accordance with the actual situation. To resolve this issue, we referenced the paper [27] and improved the calculation method of the CRI.

$$\rho = \sqrt{\text{DCPA}s^2 + (\eta V_0 \times \text{TCPA}s)^2} / \sqrt{\text{DCPA}^2 + (\eta V_0 \times \text{TCPA})^2} \tag{10}$$

In Equation (10), V_0 is the speed, $V_0 = \text{DCPA}s/\text{TCPA}s$, DCPAs are the secure approach distances, TCPAs are the secure approach times, and η is the weighting coefficient between DCPA and $V_0 * \text{TCPA}$, which is related to the relative speed of two ships and also varies with the encounter situation. When the ship is coming from the starboard side, then $\eta = 0.5 \times \frac{\text{TCPA}-\text{TCPA}s}{\text{TCPA}s}$, and when the ship is coming from the port side, then $\eta = 0.55 \times \frac{\text{TCPA}-\text{TCPA}s}{\text{TCPA}s}$. Generally speaking, the condition whereby the ship is coming from the starboard sie is more dangerous than the condition whereby the ship is coming from the port side. Therefore, when the ship is coming from the port side, η increases; then, the security vector increases, and the situation is less dangerous.

2.3.4. Risk Data Extraction

Based on the ship domain theory, navigational encounter risk can be defined as a situation where two or more vessels come into contact or overlap when encountering each other. The COLREGs defined three encounter situations: head-on, crossing, and overtaking. However, the COLREGs only quantitatively define overtaking situations and do not provide quantitative definitions for crossing encounters and head-on encounters. To extract risk data from different encounter scenarios, the experiment divides ship encounter situations based on the difference in heading between two ships [28], as illustrated in Figure 8.

The process of extracting encounter risk data involves three primary steps. Firstly, it is essential to determine whether an encounter situation has arisen between the reference ship and the approaching vessel. According to the COLREGs, the masthead light visibility distance is specified as 6 nautical miles for vessels with a length exceeding 50 m. Therefore, we only compute the DCPA and TCPA for vessels within a 6 nautical mile radius of the reference ship. It is considered that an encounter situation exists when the closest vessel's DCPA is less than 2 nautical miles. Secondly, further calculations are performed to determine whether there is an overlap in the ship domains. If ship domains intersect, the encounter data are stored in different encounter scenario datasets based on their relative positions during the encounter. Finally, computation of the CRI for the vessels involved in the encounter is performed. The algorithm flow is illustrated in the diagram below (Figure 9).

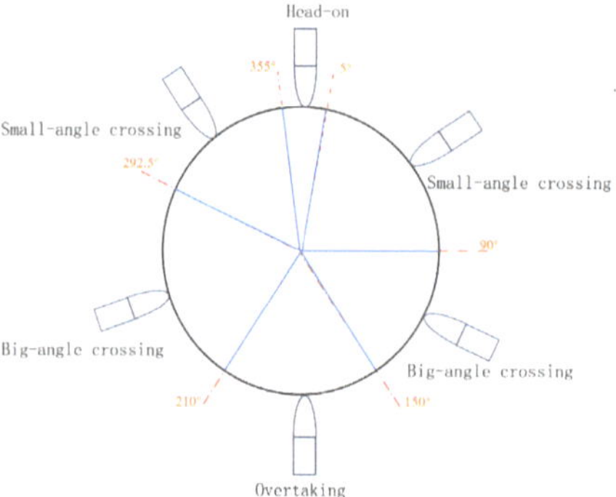

Figure 8. Classification of encounter situations.

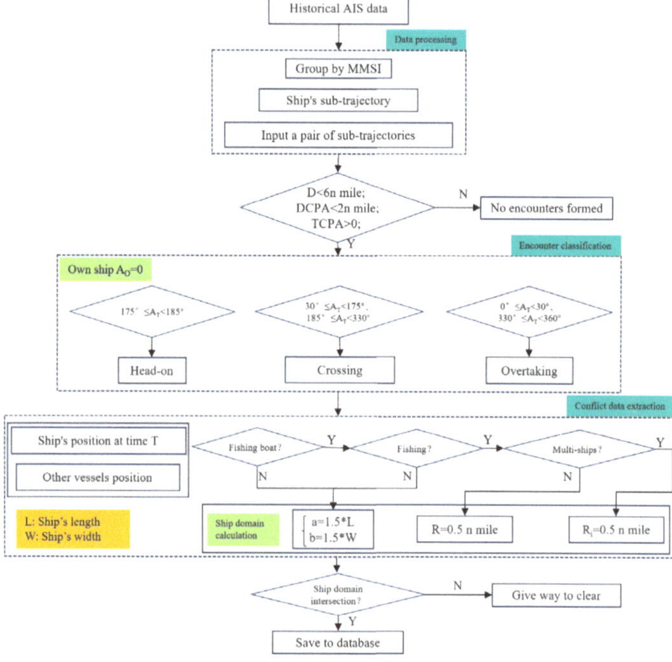

Figure 9. Schematic diagram of dangerous encounter extraction algorithm.

2.3.5. Risk Data Visualization

Risk data visualization enables the concise and intuitive display of spatial distribution patterns and reveals deeper insights. QGIS (Quantum GIS) is user-friendly open-source desktop GIS3.28.5 software with powerful analysis capabilities that make it suitable for conducting visual analyses of vessels' encounter risk data. Figure 10 is based on the historical AIS track data of the coastal area of Lianjiang County, Fujian Province, combined with the navigation environment of the coastal area of Fujian Province and the status quo of customary shipping routes, traffic flow rules, conflict risk of merchant fishing vessels, and ship track clustering analysis to obtain a visualization of merchant and fishing vessel routes. Figure 10 is a visualization of vessels' trajectories using the QGIS software. This visualization is based on historical AIS trajectory data along the coastal waters combined with the navigational environment and the analysis of traffic flow.

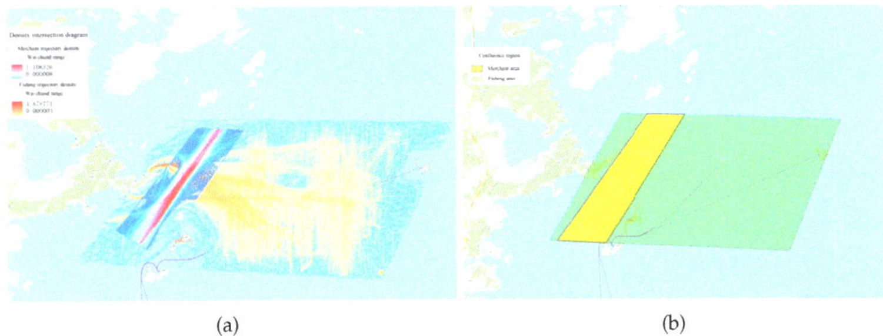

(a)   (b)

Figure 10. Overlapping channels of merchant and fishing vessels; (**a**) Trajectory density maps of merchant and fishing ships; (**b**) Area of convergence of merchant and fishing ships activities.

Kernel density analysis of ship trajectory data can also be performed using QGIS and can be used to display different colors by calculating the density in each raster pixel; this can more intuitively show the navigational density situation and risk severity in different encounter situations. Figure 11 is a kernel density visualization of the original trajectory data in the coastal waters. For typical fishing vessels engaged in operations, a circular area with a radius of 0.5 nautical miles is utilized as the safety domain; enlarging the safety radius of fishing vessels during operations can unearth more potential collision risk vessels. For vessels with potential collision risks, real-time calculation of collision risk will be conducted, and the CRI will be used as the weighting factor for risk visualization.

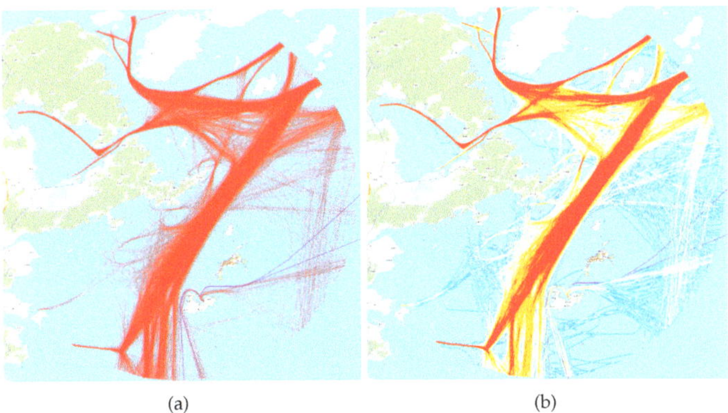

(a)   (b)

Figure 11. Comparison of Original and weighted kernel density trajectory visualizations; (**a**) Original trajectories data; (**b**) Weighted kernel density trajectories data.

3. Experiments

3.1. Analysis of the Characteristics of Dangerous Vessels

3.1.1. Ship Type Distribution in Encounter Scenarios

The experiment statistically analyzed the histogram of the distribution of ship lengths and widths involved in dangerous encounters, as illustrated in Figure 12. The histogram revealed that the lengths of fishing vessels involved in dangerous encounters were predominantly concentrated in the range of 30–50 m, with widths between 5–10 m. On the other hand, merchant ships involved in dangerous encounters showed a concentration of lengths between 50–150 m and widths between 8–20 m, comprising small- to medium-sized merchant ships. Typically, some small- to medium-sized merchant and fishing vessels studied in the research area did not strictly adhere to regulations in the same manner as large cargo ships, often as a result of discrepancies in crew equipment and instances of over-ranking of officers. Some inland individuals engage in fishing operations as temporary laborers without valid employment certificates or systematic safety training. This results in a lack of strong safety awareness, collision avoidance knowledge, and navigational skills, leading to the occurrence of collision risks. Additionally, the experiment conducted a statistical analysis of the distribution of ship types involved in dangerous encounters, as illustrated in Figure 13. Out of the 6442 vessels involved in dangerous encounters, fishing vessels accounted for the largest proportion at approximately 61.8%, followed by merchant vessels at around 28.86%. This highlights that cargo and fishing vessels are the primary types contributing to maritime accidents in this area.

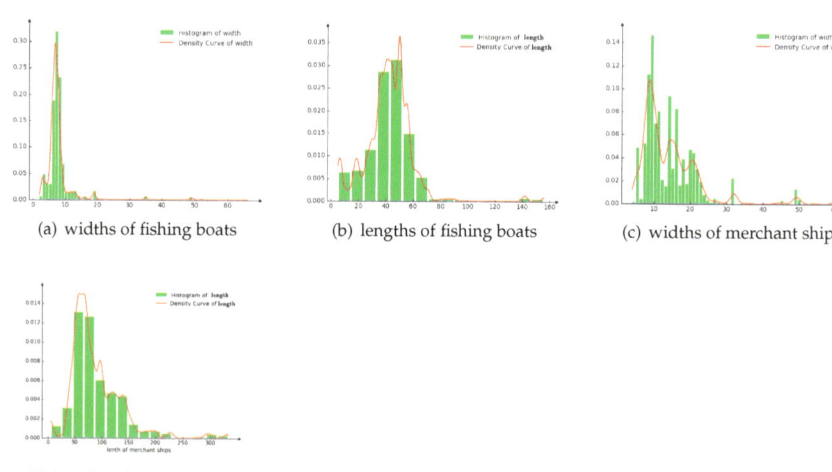

(a) widths of fishing boats

(b) lengths of fishing boats

(c) widths of merchant ships

(d) lengths of merchant ships

Figure 12. Histogram of frequency distribution of ship lengths and ship widths.

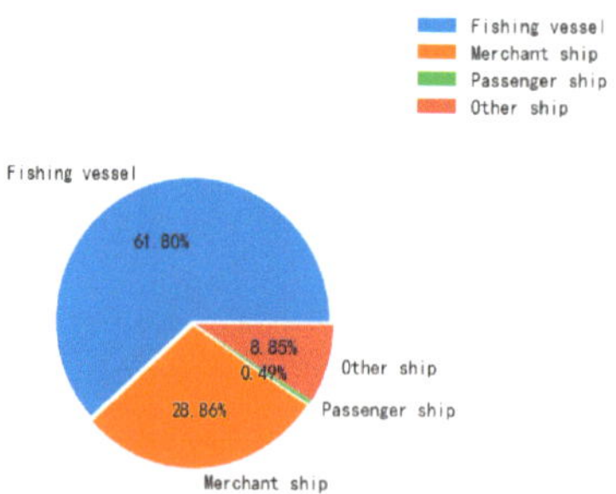

Figure 13. Distribution of vessel types involved in hazardous encounters.

3.1.2. Speed Distribution of Encounters

A statistical analysis of vessel speeds is presented in different encounter situations in Figure 14. Overall, there is a similar trend in the distribution of speeds, with a higher proportion falling within the 7–10 knots range. Specifically, for vessels involved in head-on encounters, their speed distribution follows a Weibull distribution, as depicted in Figure 14a. The majority of speeds are in the 8–10 knot range, accounting for 50.47%, with an average speed of approximately 8.23 knots. In the case of vessels involved in crossing encounters, their speed distribution follows the distribution shown in Figure 14b. The predominant speeds are in the 8–10 knots range, constituting 39.7%, with an average speed of 7.5 knots. For vessels involved in overtaking encounters, their speed distribution is illustrated in Figure 14c. The majority of speeds are between 8–10 knots, making up 46.1%, with an average relative speed of approximately 7.94 knots. Notably, vessels involved in overtaking and crossing encounters exhibit lower average speeds compared to vessels in head-on encounters in hazardous encounter scenarios.

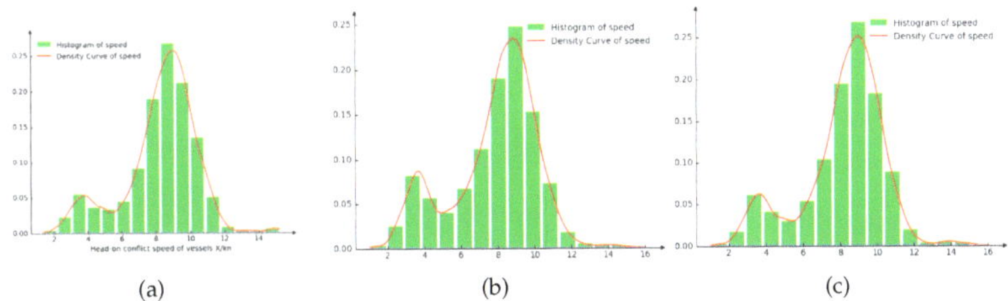

(a)   (b)   (c)

Figure 14. Histogram of collision velocity distribution of different collision situations; (**a**) Head-on conflict speed of ships X/kn; (**b**) Crossing conflict speed of ships X/kn; (**c**) Overtaking conflict speed of ships X/kn.

A heat map illustrating the velocity distribution of hazardous vessels within the study waters is depicted in Figure 15. It can be seen from the figure that the ship speed in the hazardous encounter scenarios from the internal route to the external route along the coast of Fuzhou presents an increasing distribution. The speed of the external route is significantly higher than that of the internal route; this is closely related to the wide sea depth of and fewer obstacles in the external route.

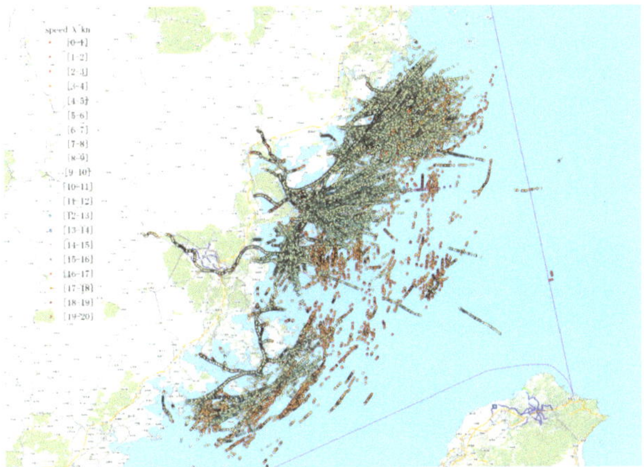

Figure 15. Heat map of speed distribution in dangerous situations.

3.2. Spatial Distribution of Hazardous Encounter Events

Based on the spatial distribution of hazardous ship encounter events depicted in Figure 16, hazardous encounters along the Fujian coast are predominantly concentrated in four regions. These regions are the fishing port areas (such as Tailu Fish Port), the entrances to bays (such as Sansha Bay), the estuary of the Minjiang River, and Guanbei Island. A conflict peak is observed at the entrance waters of Sansha Bay and forms due to the fact that this is the sole entrance and exit of Sansha Bay. During the fishing season, numerous fishing vessels navigate through this area, and the complex navigation conditions, including obstacles such as reef islands in the middle of the channel, contribute to the dense distribution of collision conflicts. The secondary conflict areas are located from Guanbei Island to the main channel of the Minjiang River estuary. This area is a mandatory passage for fishermen from the town of Tailu entering and leaving the fishing port and intersects with the traffic flow to and from the Minjiang River estuary. With a higher navigation density, the likelihood of maritime collision accidents is significantly increased.

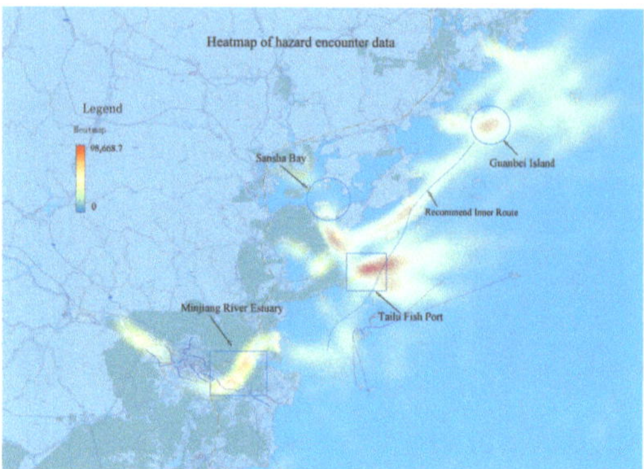

Figure 16. Spatial distribution of risk encounter data.

3.2.1. Spatial Distribution of Hazardous Head-On Encounter Scenarios

The most intensive area of head-on hazardous events is situated near the entrance to Sansha Bay, as illustrated in Figure 17. This area serves as the exclusive entry and exit point for fishermen heading out to sea from Sansha Bay. The presence of navigational constraints such as reefs on the southern side creates a traffic bottleneck, making it a high-traffic area prone to head-on encounter conflicts between commercial and fishing vessels. The conflicts are particularly pronounced within the primary shipping lane, emphasizing the prevalence of head-on encounter issues in this region. The second-highest density of head-on encounter hazardous events is observed in the nearshore waters. This area functions as a major transportation hub for fishermen from the town of Tailu and intersects with the recommended inland waterway connecting Guanbei Island and the Min River estuary. The notable conflict area arises from encounters between southbound and northbound commercial vessels and fishermen heading out for fishing activities. The conflicts are widespread in this area due to the high volume of fishing vessels.

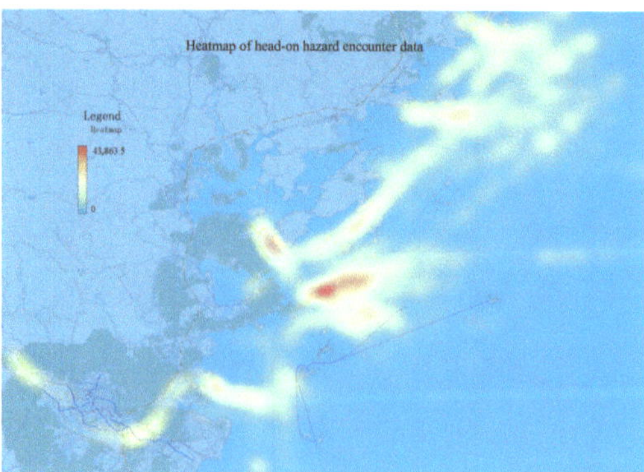

Figure 17. Spatial distribution of hazardous head-on encounter scenarios.

3.2.2. Spatial Distribution of Hazardous Crossing Encounter Scenarios

Hazardous crossing encounter events constitute the predominant type of maritime conflicts in the research area, accounting for a substantial 63.14%. This highlights the significance of crossing encounter hazards as a primary concern for vessels navigating the strait. Figure 18 illustrates the spatial distribution of hazardous crossing encounter events in the research area. The maximum concentration of crossing encounter hazards is located near the waters of the Minjiang River estuary. This area serves as a convergence point for river vessels from the Minjiang River and ocean-going commercial vessels entering the Minjiang River estuary. It is a crucial turning point for vessels navigating from Guanbei

Island to the main channel leading to the Minjiang River, creating a region with a broad distribution and high density of crossing encounter hazards. The second dense area of crossing encounter hazards is situated in the Sansha Bay. Vessels navigating through this area tend to make significant turns near Jigongshan, leading to crossing encounter situations. Limited visibility and the necessity for vessels to make substantial maneuvers increase the potential for dangerous encounters with passing vessels, posing a significant risk of collision.

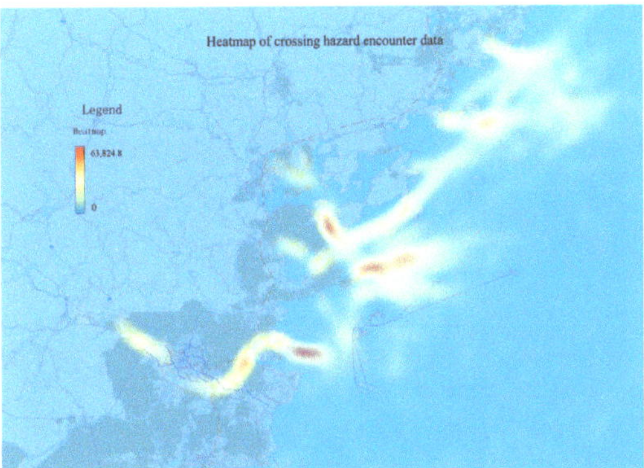

Figure 18. Spatial distribution of hazardous crossing encounter scenarios.

3.2.3. Spatial Distribution of Dangerous Overtaking Scenarios

The spatial distribution of hazardous overtaking encounter events within the studied area is illustrated in Figure 19. The primary hotspot for hazardous overtaking encounter events is situated at the entrance and exit points of Sansha Bay and Luoyuan Bay, especially in the vicinity of Jigongshan when navigating the channel to and from Sansha Bay. The deceleration and acceleration of vessels during course changes in this area are likely the primary contributing factors to the occurrence of hazardous overtaking encounter events. The second dense area of hazardous overtaking collision events is in the region connecting the towns of Tailu and Haidao. This area serves as a crucial intersection for fishermen from Tailu heading north to the open sea utilizing the recommended waterway, and it also serves as a convergence area for fishermen changing course to return to port. The likelihood of vessels accelerating and decelerating in this waterway is significant, making it a key factor in the numerous hazardous overtaking encounter events observed in this area.

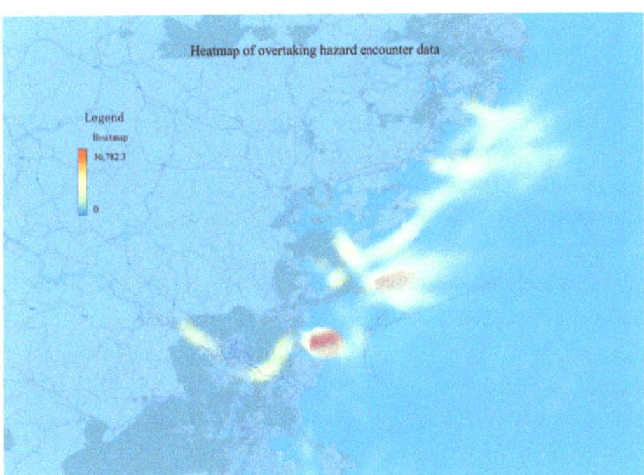

Figure 19. Spatial distribution of dangerous overtaking scenarios.

3.3. *Temporal Distribution of Hazardous Encounter Data*

The study area's fishing season is divided into two distinct periods: spring and autumn. The spring season spans from March to May, while the autumn season extends from August to November. The data utilized cover the period from August to November, corresponding to the autumn flood fishing season. Fishing vessels typically navigate without a fixed course during the operational process, involving both departure and return. These movements are generally determined by the tide and the abundance of fish. Typically, fishing vessels are small, limiting the use of large nets for deep-sea fishing. Consequently, they engage in small-scale fishing in the shallow waters near the sea. Due to the abundance of predators in shallow waters during the day, fish in these areas disperse, hide, or migrate to the deep sea. Consequently, fishermen opt to fish at night. To analyze the temporal distribution of dangerous events involving fishing vessels, statistical analysis was applied to the AIS data of the study waters.

3.3.1. Distribution of Fishing Vessels in Hazardous Situations

The temporal distribution of hazardous encounter events, as shown in Figure 20, reveals a notable concentration during the evening hours from 6:00 PM to 12:00 AM. This period is often considered the peak period for fishing activities. Navigational conditions become more complex during this time due to the onset of darkness coupled with the fact that certain fish species exhibit increased activity during the night. Additionally, nighttime weather conditions at sea are typically more intricate than those during the day. The reduced visibility for crew members, coupled with fatigue from continuous work, significantly impacts their vigilance and responsiveness, thereby elevating the risk of maritime accidents [29].

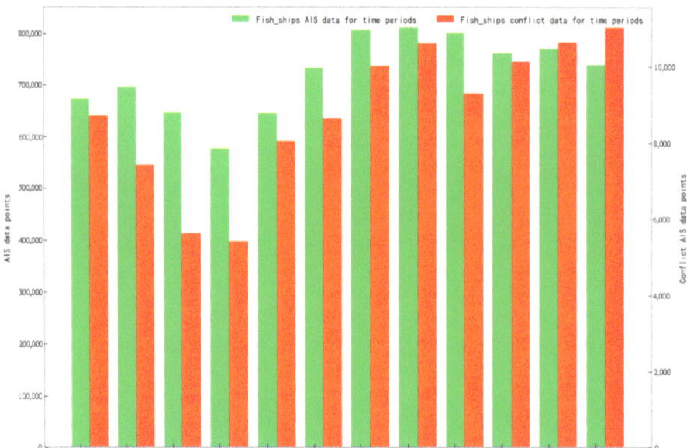

Figure 20. Data distribution of fishing boats during different time periods.

3.3.2. Temporal Distribution of Hazardous Encounter Events under Different Scenarios

The temporal distribution of hazardous encounter events under various encounter scenarios is illustrated in Figure 21. Across all encounter scenarios, the primary time frame for encountering risks is from 8:00 AM to midnight, with the least data recorded between 4:00 AM and 6:00 AM. However, the peak periods for hazardous encounter events slightly differ among the encounter situations. For head-on encounters, the peak time occurs between 2:00 PM and 4:00 PM, while for crossing encounters, it is between 10:00 PM and midnight, and for overtaking encounters, it is between 12:00 PM and 4:00 PM.

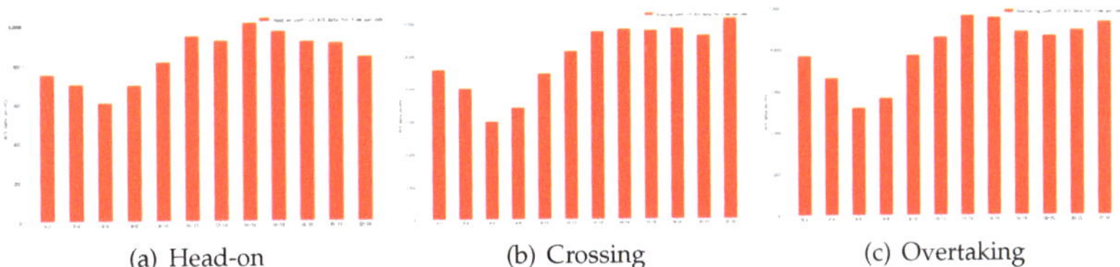

Figure 21. Distributions of risk encounter times under different encounter situations.

3.3.3. Spatial and Temporal Distribution of Hazardous Encounter Events

The experiment incorporated the time dimension into the heat map. Figure 22 illustrates the thermal maps representing hazardous data during six time intervals: 0–4, 4–8, 8–12, 12–16, 16–20, and 20–24. The figure reveals a similar spatial distribution of hazardous encounter events across different time periods, with relatively dense concentrations along the east coast of the Lianjiang River, the Shacheng Port, the Sansha Port, and the Minjiang estuary. It exhibits a higher density during the period from 12 to 24, corresponding to the busy hours for fishermen. Typically, merchant ships and fishing boats navigate during this period, resulting in the intersection of their sailing paths and elevating the risk of ship encounters and potential collisions.

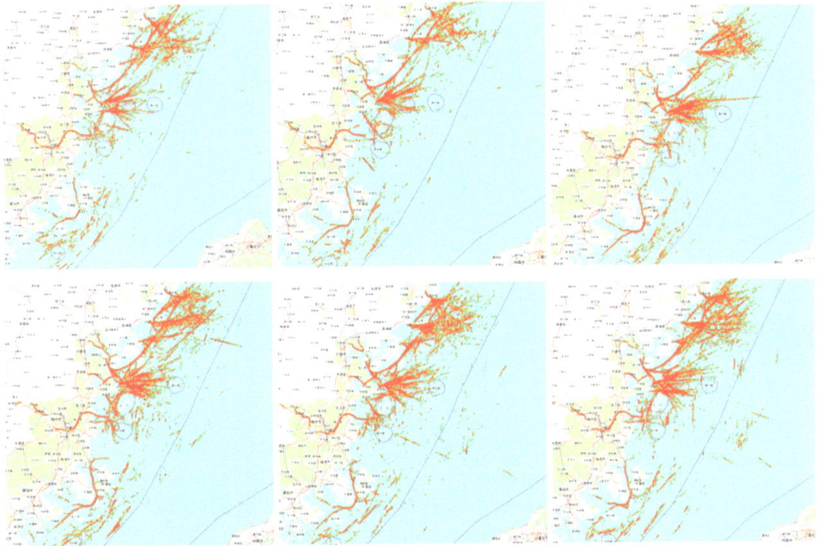

Figure 22. Heat maps of hazardous encounter data during different time periods.

4. Analysis and Discussion

4.1. Case Analysis of Collision Accidents

To validate the accuracy of the high-risk collision zones identified based on historical AIS data, this study gathered maritime accident data from the study area over the past three years. The analysis of accident occurrence regions was coupled with an examination of the causes behind these incidents. Furthermore, a comparison was made between the locations of six maritime collision incidents that occurred in the study area over the last three years and the high-risk collision zones identified above.

4.1.1. Marine Accident Data Statistics

According to the International Maritime Organization (IMO), maritime accidents are classified into the following major categories: (1) Very serious marine casualty: a very serious casualty means a marine casualty involving the total loss of the ship, death, and severe damage to the environment. (2) Serious marine casualty: serious casualties are ship casualties that do not qualify as very serious casualties and that involve a fire, explosion, collision, grounding, contact, heavy weather damage, ice damage, hull cracking, suspected hull defect, etc., resulting in immobilization of main engines, extensive accommodation damage, severe structural damage such as penetration of the hull under water, etc., rendering the ship unfit to proceed or pollution (regardless of quantity) and a breakdown necessitating towing or shore assistance. (3) Less serious marine casualty: less serious casualties are ship casualties that do not qualify as very serious casualties or serious casualties. (4) Marine incident: a marine incident means an event or sequence of events other than a marine casualty that occurred directly in connection with the operations of a ship that endangered or, if not corrected, would endanger the safety of the ship, its occupants, or any other person or the environment. Marine incidents include hazardous incidents and near misses.

We conducted a statistical analysis of maritime accidents that occurred in the study area from 2020 to 2023 following IMO standards. In Figure 23b, serious accidents in coastal waters made up 12.12% of the total incidents, major accidents accounted for 27.27%, while general accidents were the most frequent, representing 60.61%. Among these incidents, self-sinking accidents were the most prevalent, constituting 30.30% of the total, followed by collision accidents at 21.21%. Fire/explosion accidents had the lowest proportion, accounting for a mere 3.03%, as depicted in Figure 23a.

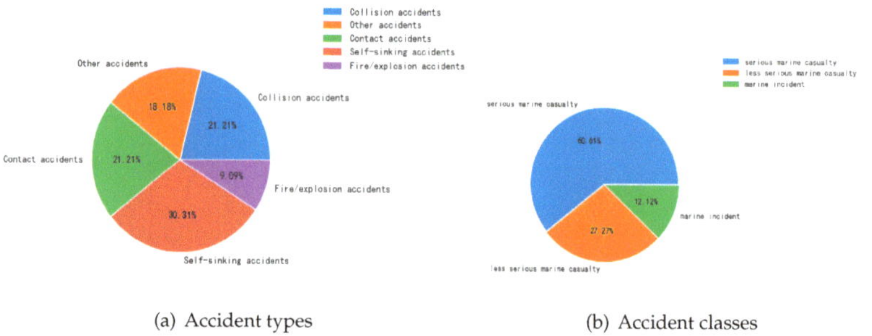

(a) Accident types   (b) Accident classes

Figure 23. Statistical diagram of ship accident data.

4.1.2. Analysis of High-Risk Areas of Collisions between Merchant Ships and Fishing Vessels

Merchant ships and fishing boats serve distinct purposes, resulting in varied sailing areas. To delve deeper into the analysis of high-risk collision areas for these vessels, the experiment presents heat maps depicting the frequency of encounter risks for both types. The identified high-risk areas for merchant and fishing vessels are illustrated in Figure 24.

In the case of fishing vessels, the predominant incident-prone region is around the line connecting the town of Tailu and Yangyu Island. This is attributed to the high volume of fishing vessels navigating through this area and overlapping with the recommended inner shipping route for merchant vessels. The complex interweaving of trajectories, narrow channels, numerous obstacles, and intricate water currents contribute to frequent accidents. Particularly, small- to medium-sized wooden fishing vessels with poor radar reflection and limited visibility are prone to accidents in this region, as passing merchant vessels may struggle to detect their radar echoes effectively.

For merchant vessels, the primary incident-prone area is the exit of the Minjiang River channel. This region is characterized by numerous bends, and many merchant vessels tend to overlook radar blind spots and fail to exercise sufficient caution. There is a tendency to excessively rely on radar functionality and neglect proper lookout procedures. Lack of essential experience and knowledge of utilizing radar echoes to detect smaller fishing vessels contributes to collisions in this area.

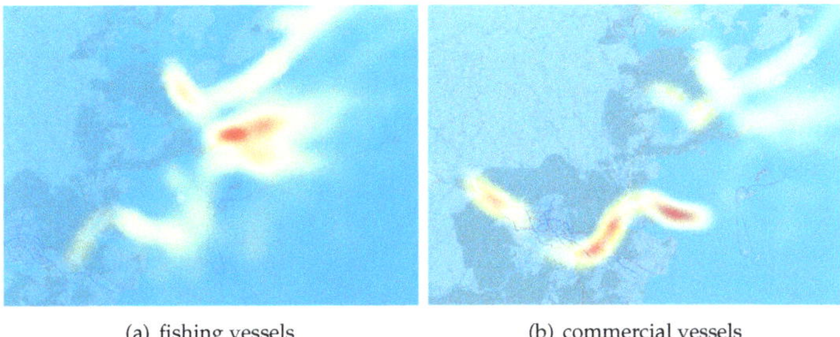

(a) fishing vessels   (b) commercial vessels

Figure 24. Commercial fishing vessel risk encounter data heat map display.

4.1.3. Comparison with Actual Collision Incidents

The analysis conducted in this study focuses exclusively on collision incidents. Therefore, a comparative analysis is performed between the areas where six collision incidents occurred in the past three years and the high-risk collision areas identified in this study. Figure 25 illustrates the comparison between the experimentally derived high-risk collision areas for merchant and fishing vessels and the actual regions where ship collisions occurred. Moreover, five out of the six collision incidents align with the areas identified as high-risk collision zones in this experiment. This correspondence validates the effectiveness of the experiment's methodology at identifying high-risk collision areas for merchant and fishing vessels.

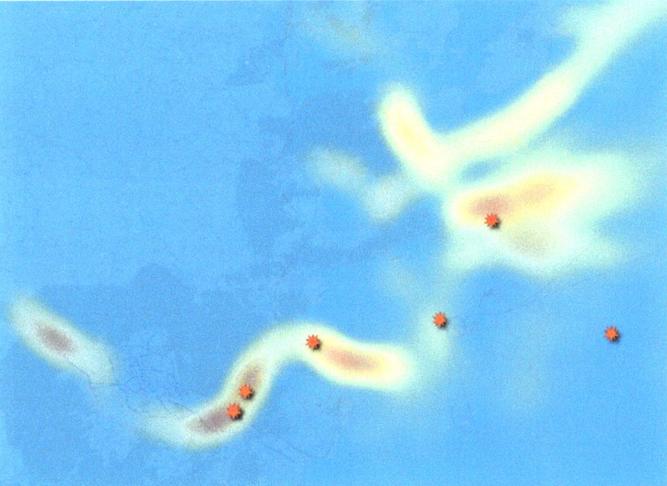

Figure 25. Comparison of merchant fishing vessel collisions and high collision hazard areas. (The red star is the accidents place.)

5. Conclusions

Based on historical AIS data collected during the autumn 2022 fishing season in the study waters, this paper proposes a method for visualizing high-risk collision areas for commercial and fishing vessels. In this method, the safety domain of fishing vessels operating is redefined by considering fishing vessels engaged in cooperative operations as a whole entity. Additionally, the process of vessel intrusion into ship domains is visualized based on the degree of collision risk using different weights.. The aim is to identify areas prone to collisions between merchant and fishing vessels in order to provide a warning for vessels to exercise sufficient caution when navigating through high-risk zones and to reduce the occurrence of maritime collisions. The primary research findings are summarized below:

1. Identification of fishing vessel operational status: The paper proposes a method to identify the operational status of fishing vessels. Recognizing the operational status of fishing vessels is crucial for collision prevention, particularly when encountering fishing vessels engaged in operations. The experiment distinguishes between the navigation trajectories of fishing vessels under different operational characteristics to effectively identify the operational statuses of fishing vessels, with a focus on those engaged in operations for encounter risk analysis.

2. Evaluation of hazardous encounter events for merchant and fishing vessels: The paper introduces a method to assess encounter risk data for merchant and fishing vessels. By calculating the CPA and TCPA, the collision risk is quantified. The experiment sets the vessel domain for fishing vessels engaged in operations to 0.5 nautical miles, ensuring a safe distance of at least 0.5 nautical miles between each vessel in a collaborative operation. The paper categorizes and assesses risk data for different encounter situations.

3. Visualization of high-risk collision areas for merchant and fishing vessels: The paper conducts a visual analysis of high-risk collision areas for merchant and fishing vessels in the research area under different encounter situations. The identified high-risk collision areas include the eastern nearshore waters of Lianjiang, with latitudes ranging from 26°17′56.40″ to 26°25′12″ and longitudes between 119°24′25.20″ and 119°48′7.20″, and the Minjiang River estuary. The paper validates the identified high-risk collision areas using data on the locations of maritime collisions in the research area over the past three years, demonstrating the reasonableness of the experiment's outcomes.

Although the research in this paper revealed some interesting findings, there are still some limitations that need to be further studied in the future. First, this paper does not consider the impact of ship maneuverability and ship tonnage for quantifying ship collision risk, which are needed in further research. Second, the method does not consider the influence of objective conditions such as adverse weather conditions like typhoons, visibility, and traffic density on collision risks. Finally, the assumption of dynamic collision boundaries (polygon and ellipse) may influence the results of the collision risk evaluation, which will be further explored. In future research, we aim to research or refine more advanced models to better suit the characteristics of our study area. Additionally, further collaboration with relevant authorities can be pursued to equip multifunctional navigation aids within high-risk collision areas, providing real-time maritime traffic alerts and information on hazardous zones. By establishing a comprehensive maritime safety system within these high-risk collision areas, vessels' perception of danger can be heightened, thus reducing the occurrence rate of collision incidents and ensuring safe and smooth maritime traffic flow. Additionally, the construction of vessel trajectory prediction models within high-risk collision areas can be explored to estimate the time and location of potential collisions during encounters.

Author Contributions: Conceptualization, W.H., C.Z. and J.L.; methodology, J.L. and Z.W.; validation, M.C. and D.Z.; formal analysis, W.H.; investigation, M.C. and W.H.; resources, D.Z.; data curation, C.Z.; writing—original draft preparation, C.Y.; writing—review and editing, C.Z. and J.L.; visualization, J.L. and Z.W.; supervision, M.C. and W.H.; project administration, C.Y.; funding acquisition, W.H. All authors have read and agreed to the published version of the manuscript.

Funding: This research was funded by name the National Natural Science Foundation of China grant number 52172327, the Fujian Province Natural Science Foundation, grant number 2022J05236, the Science and Technology Planning Project of Fuzhou grant number 2022-P-006, the Fujian Province Key Science and Technology Innovation Project grant number 2022G02009, 2022G02027, and the Science and Technology Key Project of Fuzhou grant number 2022-ZD-021.

Institutional Review Board Statement: Not applicable.

Informed Consent Statement: Not applicable.

Data Availability Statement: Data are contained within the article.

Acknowledgments: The authors wish to thank the Fuzhou Aids to Navigation Division of Eastern Navigation Services Center for providing ship AIS data for this study.

Conflicts of Interest: The authors declare no conflicts of interest.

References

1. Nieh, C.Y.; Lee, M.C.; Huang, J.C.; Kuo, H.C. Risk assessment and traffic behaviour evaluation of inbound ships in Keelung harbour based on AIS data. *J. Mar. Sci. Technol.* **2019**, *27*, 2.
2. Liu, Z.; Zhang, B.; Zhang, M.; Wang, H.; Fu, X. A quantitative method for the analysis of ship collision risk using AIS data. *Ocean Eng.* **2023**, *272*, 113906. [CrossRef]
3. Silveira, P.; Teixeira, A.; Soares, C.G. Use of AIS data to characterise marine traffic patterns and ship collision risk off the coast of Portugal. *J. Navig.* **2013**, *66*, 879–898. [CrossRef]
4. Sedov, V.A.; Sedova, N.A.; Glushkov, S.V. The fuzzy model of ships collision risk rating in a heavy traffic zone. *Vibroeng. Procedia* **2016**, *8*, 453–458.
5. Yi, M. Design of the model for predicting ship collision risk using fuzzy and DEVS. *J. Korea Soc. Simul.* **2016**, *25*, 127–135. [CrossRef]
6. Chin, H.C.; Debnath, A.K. Modeling perceived collision risk in port water navigation. *Saf. Sci.* **2009**, *47*, 1410–1416. [CrossRef]
7. Liu, K.; Yu, Q.; Yuan, Z.; Yang, Z.; Shu, Y. A systematic analysis for maritime accidents causation in Chinese coastal waters using machine learning approaches. *Ocean Coast. Manag.* **2021**, *213*, 105859. [CrossRef]
8. Yu, Y.; Chen, L.; Shu, Y.; Zhu, W. Evaluation model and management strategy for reducing pollution caused by ship collision in coastal waters. *Ocean Coast. Manag.* **2021**, *203*, 105446. [CrossRef]
9. Shinoda, T.; Shimogawa, K.; Tamura, Y. Applied Bayesian Network Risk Assessment for Collision Accidents between Fishing Vessels and Cargo Vessels. *J. Inst. Navig.* **2012**, *127*, 165–174. [CrossRef]
10. Takemoto, T.; Sakamoto, Y.; Yano, Y.; Furusho, M. Collision Avoidance Actions and Backgrounds in Fishing Vessels Collision Accidents. *J. Jpn. Inst. Navig.* **2010**, *122*, 113–120.
11. Xiao, F.; Ma, Y.; Wu, B. Review of Probabilistic Risk Assessment Models for Ship Collisions with Structures. *Appl. Sci.* **2022**, *12*, 3441. [CrossRef]
12. Mou, J.M.; Van der Tak, C.; Ligteringen, H. Study on collision avoidance in busy waterways by using AIS data. *Ocean Eng.* **2010**, *37*, 483–490. [CrossRef]
13. Uğurlu, F.; Yıldız, S.; Boran, M.; Uğurlu, Ö.; Wang, J. Analysis of fishing vessel accidents with Bayesian network and Chi-square methods. *Ocean Eng.* **2020**, *198*, 106956. [CrossRef]
14. Argüelles, R.P.; Maza, J.A.G.; Martín, F.M.; Bartolomé, M. Ship-to-ship dialogues and agreements for collision risk reduction. *J. Navig.* **2021**, *74*, 1039–1056. [CrossRef]
15. Huang, Y.; Chen, L.; Negenborn, R.R.; Van Gelder, P. A ship collision avoidance system for human-machine cooperation during collision avoidance. *Ocean Eng.* **2020**, *217*, 107913. [CrossRef]
16. Jung, C.H. A study on the requirement to the fishing vessel for reducing the collision accidents. *J. Korean Soc. Mar. Environ. Saf.* **2014**, *20*, 18–25. [CrossRef]
17. Yang, J.; Sun, Y.; Song, Q.; Ma, L. Laws and preventive methods of collision accidents between merchant and fishing vessels in coastal area of China. *Ocean Coast. Manag.* **2023**, *231*, 106404. [CrossRef]
18. Wang, X.; Zhou, Z.; Liu, Y.; Song, H. Discussions about the best methods of collision between merchant ships and fishing boats during fishing seasons near chinese coast. In Proceedings of the International Conference on Computer Information Systems and Industrial Applications, Bangkok, Thailand, 28–29 June 2015; pp. 758–761.
19. Obeng, F.; Domeh, V.; Khan, F.; Bose, N.; Sanli, E. Capsizing accident scenario model for small fishing trawler. *Saf. Sci.* **2022**, *145*, 105500. [CrossRef]
20. Chou, C.C. Application of ANP to the selection of shipping registry: The case of Taiwanese maritime industry. *Int. J. Ind. Ergon.* **2018**, *67*, 89–97. [CrossRef]
21. Seo, A.; Hida, T.; Nishiyama, M.; Nagao, K. Comparison and evaluation on user interface of small ship navigation support system. *Proc. Jpn. Inst. Navig.* **2018**, *6*, 1–4.
22. Ma, F.; Chu, X.; Yan, X.; Liu, C. Error Distinguish of Ais Based on Evidence Combination. *J. Theor. Appl. Inf. Technol.* **2012**, *45*, 83–90.
23. Ma, F.; Chu, X.; Liu, C. The error distinguishing of automatic identification system based on improved evidence similarity. In Proceedings of the ICTIS 2013: Improving Multimodal Transportation Systems-Information, Safety, and Integration, Wuhan, China, 29 June–2 July 2013; pp. 715–722.
24. Qin, Z. The Study on Motion Control and Swarm Coordinated Planning for Unmanned Surface Vehicles (USV). Ph.D. Thesis, Harbin Engineering University, Harbin, China, 2018.

25. Fujii, Y.; Tanaka, K. Traffic capacity. *J. Navig.* **1971**, *24*, 543–552. [CrossRef]

26. Kearon, J. Computer programs for collision avoidance and traffic keeping. In Proceedings of the Conference on Mathematical Aspects on Marine Traffic, London, UK, 15–16 September 1977; Academic Press: London, UK, 1979.

27. Xu, X.; Geng, X.; Wen, Y. Modeling of ship collision risk index based on complex plane and its realization. *TransNav Int. J. Mar. Navig. Saf. Sea Transp.* **2016**, *10*. [CrossRef]

28. Gao, M.; Shi, G.Y.; Liu, J. Ship encounter azimuth map division based on automatic identification system data and support vector classification. *Ocean Eng.* **2020**, *213*, 107636. [CrossRef]

29. Hasanspahić, N.; Vujičić, S.; Frančić, V.; Čampara, L. The role of the human factor in marine accidents. *J. Mar. Sci. Eng.* **2021**, *9*, 261. [CrossRef]

Journal of
Marine Science and Engineering

Article

Association of AIS and Radar Data in Intelligent Navigation in Inland Waterways Based on Trajectory Characteristics

Jinyu Lei [1,2,3,†], Yuan Sun [3,†], Yong Wu [4], Fujin Zheng [5], Wei He [1,3,*] and Xinglong Liu [1,3,*]

1   Fuzhou Institute of Oceanography, Minjiang University, Fuzhou 350108, China; j.lei@mju.edu.cn
2   College of Computer and Data Science, Minjiang University, Fuzhou 350108, China
3   Fujian Engineering Research Center of Safety Control for Ship Intelligent Navigation, Minjiang University, Fuzhou 350108, China; sunyuan@mju.edu.cn
4   Navigation College, Jiangsu Maritime Institute, Nanjing 211170, China
5   School of Civil Engineering and Architecture, Wuhan Institute of Technology, Wuhan 430063, China
*   Correspondence: hewei11@mju.edu.cn (W.H.); xinglong_liu@foxmail.com (X.L.)
†   These authors contributed equally to this work.

Abstract: Intelligent navigation is a crucial component of intelligent ships. This study focuses on the situational awareness of intelligent navigation in inland waterways with high vessel traffic densities and increased collision risks, which demand enhanced vessel situational awareness. To address perception data association issues in situational awareness, particularly in scenarios with winding waterways and multiple vessel encounters, a method based on trajectory characteristics is proposed to determine associations between Automatic Identification System (AIS) and radar objects, facilitating the fusion of heterogeneous data. Firstly, trajectory characteristics like speed, direction, turning rate, acceleration, and trajectory similarity were extracted from ship radar and AIS data to construct labeled trajectory datasets. Subsequently, by employing the Support Vector Machine (SVM) model, we accomplished the discernment of associations among the trajectories of vessels collected through AIS and radar, thereby achieving the association of heterogeneous data. Finally, through a series of experiments, including overtaking, encounters, and multi-target scenarios, this research substantiated the method, achieving an F1 score greater than 0.95. Consequently, this study can furnish robust support for the perception of intelligent vessel navigation in inland waterways and the elevation of maritime safety.

Keywords: trajectory association; data fusion; inland waterways; situational awareness

Citation: Lei, J.; Sun, Y.; Wu, Y.; Zheng, F.; He, W.; Liu, X. Association of AIS and Radar Data in Intelligent Navigation in Inland Waterways Based on Trajectory Characteristics. *J. Mar. Sci. Eng.* **2024**, *12*, 890. https://doi.org/10.3390/jmse12060890

Academic Editor: Marco Cococcioni

Received: 6 April 2024
Revised: 15 May 2024
Accepted: 21 May 2024
Published: 27 May 2024

1. Introduction

With the continuous evolution of maritime logistics, the integration of intelligence and technology has gained significant attention. Within this domain, intelligent navigation plays a pivotal role, and vessel navigation situational awareness has emerged as a critical element. In recent years, inland waterway transport has experienced rapid development, resulting in the accumulation of a substantial repository of foundational data resources. These resources encompass various aspects, such as channel surveying, lock scheduling, operational vessel information, AIS data, radar images, and more. However, the intricate nature of inland waterways, characterized by complex shorelines, winding channels, high vessel traffic densities, and frequent vessel encounters, poses substantial constraints on situational awareness, especially on perception data association and fusion. In addition, inland intelligent ship navigation systems primarily focus on intelligent ships as the central element [1]. The key to intelligent ships lies in the association of multi-source data in navigation situational awareness [2]. Conducting research on the association of perception data in inland vessel navigation is essential for enhancing intelligent situational awareness. Therefore, this research endeavor contributes to the advancement of industrial technology in inland intelligent vessels.

In maritime navigation situational awareness, perception data primarily originate from various sensors, such as radar, AIS, remote sensing satellites, and BeiDou [3], which is a global navigation satellite system (GNSS) similar to other GNSS systems like GPS (Global Positioning System) and GLONASS (Global Navigation Satellite System). BeiDou provides precise positioning, navigation, and timing services to users worldwide and especially prevails in passenger ships and fishing ships in China. Radar is one of the primary sensor technologies for improving navigation safety. While small plastic or wooden boats without AIS systems may not have as strong a radar signature as metal-hulled vessels, radar is still capable of detecting plastic or wooden vessels, such as fishing boats, within a certain range, typically up to 3 km. Multi-source data invariably contain a variety of noise and interruptions in trajectory continuity that arise from signal disruptions, introducing uncertainties in navigation situational awareness. Therefore, achieving more precise target association and perception is of the utmost importance when dealing with multi-source trajectory data. Singer introduced the nearest-neighbor method, which employs a distance gating approach to eliminate spurious targets. This algorithm measures the similarity between different trajectories, enabling the determination of trajectory associations [4]. Bar-Shalom et al. proposed a probabilistic data association approach for trajectory association in single-target scenarios [5]. These methods are characterized by their simplicity and low computational loads. However, their performance tends to degrade in areas with complex traffic patterns and high levels of noise [6].

In addressing the intricate and multi-track fusion scenarios present in maritime surveillance data, Ming et al. introduced a weighted trajectory fusion algorithm leveraging local information entropy for the integration of AIS and X-band radar data [7]. And, based on fuzzy theory, Liu proposed a trajectory association method for AIS and surface wave radar (SWR) based on a fuzzy dual-threshold approach. This method utilizes fuzzy membership to quantify the degree of association between trajectories and employs dual-threshold detection to determine associated trajectory pairs [8]. In addition to AIS and SWR data, synthetic aperture radar (SAR) data and satellite images are employed to facilitate trajectory association for the objective of ship traffic monitoring in open seas [9,10]. With the advancement of deep learning, relevant techniques have also been applied to ship trajectory association. Jin et al. integrated track and scene features to estimate the probability of track association by deep learning [11]. Simulation results reveal the method's superior scene adaptability and association accuracy compared to traditional approaches. And Yang et al. developed a multi-target association algorithm for AIS–radar tracks using a graph matching-based deep neural network [12]. The above-mentioned method primarily relies on shore-based equipment and is commonly applied in vessel perception data association research in coastal areas and validated through simulation to assess its effectiveness in real-world scenarios. However, the navigational environment in inland waterways is significantly distinct, exhibiting intricate shorelines, convoluted channels, and diverse inland electronic interference factors. Consequently, the practical applicability of these methods in such environments necessitates further validation.

In inland waterways, closed-circuit television (CCTV) is prevalent in management to enhance traffic situational awareness and monitor abnormal vessel behavior due to its remote and real-time capabilities [13]. Guo et al. incorporated a dynamic time warping algorithm that calculates the similarity of AIS- and CCTV-based vessel trajectories to improve vessel traffic surveillance in inland waterways [14]. Huang et al. established a ship information fusion model based on CCTV images and AIS data, specifically focusing on the tracking of ships [15]. By employing the YOLOv3 algorithm, Gan et al. presented a vision-based data fusion approach for enhancing environmental awareness in ship navigation [16]. In addition to the fusion of CCTV and AIS data, it is also employed for radar data integration to facilitate ship target detection. Liu proposed a method of multi-scale matching vessel recognition (MSM-VR) by fusing CCTV and marine radar to ensure navigation safety [17]. CCTV surveillance systems are susceptible to adverse weather conditions, such as rain, fog, and strong winds. These conditions can result in blurred or obstructed visibility, thereby potentially compromising their detection performance. However, advanced CCTV systems with multi-spectral or thermal imaging technology often outperform human vision, particularly in challenging visibility conditions and nighttime operations. Additionally, their capabilities for accurately measuring distances and sizes are limited, making the recognition of distant or smaller vessels challenging. As a result, the identification of remote or compact vessels may prove to be a challenging task.

Maritime ship track association methods are mainly based on statistical methods and fuzzy mathematics, including the nearest-neighbor (NN) method, fuzzy double thresholds, fuzzy comprehensive functions, etc. Nearest-neighbor data association is a relatively simple method and mainly suitable for situations in which there is little noise and scenarios with a small number of targets [18]. Due to factors such as random noise and the inconsistent detection ranges of different sensors, there is ambiguity in the similarity between their tracks, and fuzzy mathematics has been applied to judge track associations [19,20]. However, existing vessel navigation perception techniques, primarily designed for coastal and open-water areas, require validation and refinement for inland waterway applications. Furthermore, inland waterways primarily emphasize the fusion of video images and AIS data, which have the capability to detect objects not discernible by radar or lacking AIS data, while research into AIS and radar trajectory association methods for shipborne perception systems is lacking. Moreover, AIS and radar systems are usually mandatory equipment for vessels according to maritime regulations, so it is essential to enhance the accuracy and reliability of vessel position and motion information to compensate for the limitations of each system and provide more accurate vessel positions, aiding in real-time adjustment of course and speed to maintain safe distances at sea. Therefore, this study focuses on vessel association of shipborne data in inland waterway vessels by leveraging trajectory features. By employing machine learning and harnessing trajectory information, this research endeavors to enhance the precision and efficacy of vessel situational awareness, thereby contributing to the safety and intelligence of inland waterway navigation. There are two specific contributions of this paper:

- We propose a novel classifier approach that incorporates trajectory characteristics to solve data association issues in inland waterways.
- We propose a dataset construction method to build positive and negative sample datasets for data association using labeled shipborne perception data.

The rest of the paper is organized as follows: the methodology is introduced in Section 2, the computation of trajectory features is detailed in Section 3, the dataset construction is described in Section 4, and the experiments conducted for the method analysis and validation are described in Section 5.

2. Methods

This paper employs a binary classification method based on trajectory features to achieve ship target association from AIS and radar data in intelligent navigation perception systems, as illustrated in Figure 1. The approach encompasses trajectory feature calculation, positive and negative sample dataset construction, and support vector machine (SVM) model development. Initially, trajectory feature calculation is performed on the training dataset. Subsequently, a training set is constructed with radar and AIS trajectory features extracted from the targets, encompassing both positive and negative samples. Then, trajectory feature calculation is performed on the prediction dataset to establish the features of the trajectories that require association. Finally, an SVM model is built using the constructed dataset for training and predicting newly received AIS and radar data. In cases where AIS data are missing, interpolated AIS data along with radar data are utilized to make predictions.

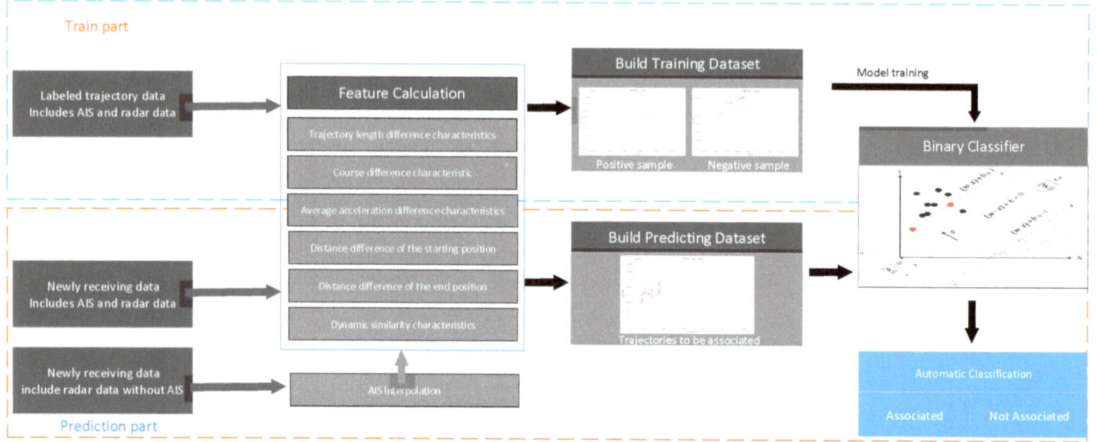

Figure 1. Flowchart of association process.

2.1. SVM Model

Ship radar and AIS are two prevalent ship monitoring technologies that capture ship position and movement data via radio waves and signals. However, the integration of data from these two distinct sources poses a noteworthy challenge stemming from their unique characteristics and inherent incompleteness. The Support Vector Machine (SVM), a supervised learning algorithm, was originally proposed by Vladimir Vapnik [21]. Since its inception, it has undergone continuous development and refinement, emerging as a prominent algorithm in machine learning with wide-ranging applications in pattern recognition, classification, and regression tasks. This section will delve into the fundamental principles of the SVM and explore its utilization in the association and classification of ship radar and AIS track data.

The core idea of the SVM is to find the optimal hyperplane that effectively separates different classes of samples in a feature space. In the case of linear separability, a hyperplane exists that perfectly separates two classes of samples and maximizes the margin, as shown in Figure 2.

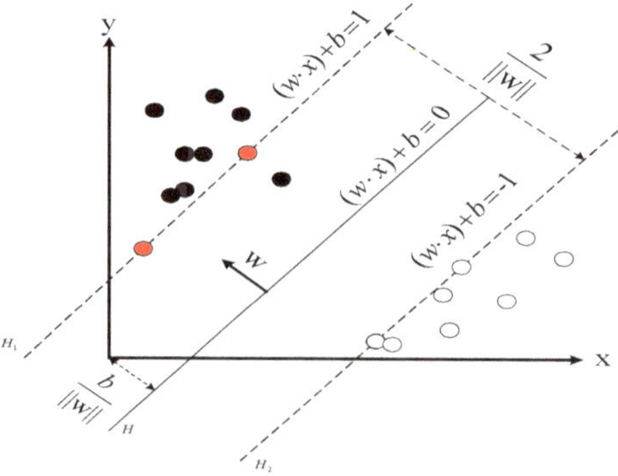

Figure 2. Separating hyperplane.

The training set is set to the following form:

$$D = (x_1, y_1), (x_2, y_2), ..., (x_n, y_n) \tag{1}$$

$x_i \in \mathbb{R}^m$ represents the input feature vector, and $y_i \in \{-1, +1\}$ represents the corresponding class label.

The hyperplane can be represented by a linear equation, $w * x + b = 0$, where w is the weight vector, x is the feature vector, and b is the bias. For any sample point (x_i, y_i), the relationship between its class label and the hyperplane can be expressed as $y_i[(w \cdot x) + b] - 1 \geq 0$.

When the training data are linearly inseparable, a non-linear SVM can be learned by using the kernel function to transform the data and combining it with the margin maximization method. The main components include margin maximization, kernel functions, and the solution of the SVM.

2.2. Margin Maximization

The objective of the SVM is to ascertain a hyperplane that maximizes the margin, that is, the distance between samples of different classes. The distance between two different samples and the hyperplane is defined as the sum of the distances from each sample to the hyperplane. The optimization problem to maximize the distance can be formulated as a convex optimization problem.

$$\max_{w,b} \frac{2}{\| w \|}, \ subject \ to \ y_i[(w \cdot x) + b] - 1 \geq 0, \ i = 1, 2, \ldots, l \tag{2}$$

The norm of the weight vector, denoted as $\| w \|$, represents the magnitude of the weight vector, while b represents the bias term. The label of the i-th sample point is denoted as y_i, and the corresponding feature vector is denoted as x_i. The objective of the optimization problem is to maximize the margin, which refers to the distance between the hyperplane and the two class sample points. Maximizing the margin helps improve the generalization ability of the classifier and its accuracy for new samples.

Meanwhile, the constraint $y_i[(w \cdot x) + b] - 1 \geq 0$ ensures that each sample point is classified on the correct side. These constraints require all sample points to meet the correct classification requirements, thereby ensuring that the margin is not affected by misclassified samples.

2.3. Kernel Function

In practical applications, the data may be linearly non-separable, making it impossible to directly use a linear hyperplane for classification. To address this issue, the concept of kernel functions is introduced to map the data to a higher-dimensional feature space, making them linearly separable in the new feature space.

A kernel function is a function used to calculate the inner product between two sample points in the feature space. Common kernel functions include linear kernels, polynomial kernels, and radial basis function (RBF) kernels. By introducing kernel functions into the optimization problem of the SVM, non-linear decision boundaries can be obtained. For linearly non-separable cases, we can modify the optimization problem to the following form:

$$\min_{w,b} \frac{1}{2} \| w \|^2, \ subject \ to \ y_i[(w \cdot x) + b] - 1 + \xi_i \geq 0, \ i = 1, 2, \ldots, l \tag{3}$$

In the above formula, the constraint condition is softened by introducing slack variables to allow some samples to be misclassified. With $y_i[(w \cdot x) + b] - 1 + \xi_i \geq 0$, the slack variables, $\xi = (\xi_1, \xi_2, \ldots, \xi_l)^1$, represent the degree to which the training set is incorrectly classified. A larger slack variable indicates a higher degree of misclassification.

2.4. Solving the Support Vector Machine

The optimization problem of the SVM is a convex optimization problem, which can be transformed into a dual problem through Lagrange duality.

By constructing the Lagrange function:

$$L(w, b, \alpha) = \frac{1}{2} \| w \|^2 - \sum_{i=1}^{l} \alpha_i[y_i(w \cdot x_i + b) - 1] \tag{4}$$

where α_i is the Lagrange multiplier vector, the dual problem can be formulated as follows:

$$Maximize\ F(\alpha) = \sum_{i,j=1}^{l} \alpha_i - \frac{1}{2}\alpha_i\alpha_j v_i v_j (x_i \cdot y_j),\ subject\ to\ \sum_{i=1}^{l} \alpha_i v_i = 0, \alpha \geq 0 \qquad (5)$$

Solving the dual problem yields the optimal weight vector, w, and the bias term, b. In addition, according to the KKT (Karush–Kuhn–Tucker) conditions, only the Lagrange multipliers of the support vectors, α_i, are non-zero, and they are located on the margin boundary, which determines the position of the optimal hyperplane.

In the practical association of AIS and radar track data, optimization algorithms such as SMO (Sequential Minimal Optimization) and QP (Quadratic Programming) are used to solve the dual problem and obtain the optimal solution for the support vector machine. In the context of ship radar and AIS track data association, the SVM demonstrates effective handling of high-dimensional and complex data, thereby enhancing the accuracy and reliability of vessel perception data association.

3. Trajectory Characteristic Calculation

Track characteristics denote the meaningful information extracted from various data sources, such as ship radar and AIS, which aid in the association of track data and the understanding of ship movement patterns. When calculating track characteristics, it is essential to preprocess the ship's movement data and extract relevant characteristics. This preprocessing involves tasks like data cleaning, denoising, and handling missing data to guarantee the accuracy and completeness of the input data. Subsequently, by extracting features like speed, direction, turning rate, acceleration, and trajectory similarity from ship radar and AIS data, a feature vector representing the ship's trajectory can be established. The following content explains the method of constructing each characteristic.

3.1. Trajectory Length Difference Characteristics

The utilization of the characteristics of length difference between radar tracks and AIS tracks aims to compare any discrepancies in the length of target trajectories captured by the two distinct data sources. By calculating the overall lengths of the radar track and the AIS track, the consistency of the target trajectory information across the different data sources can be evaluated. The length difference characteristic, Len_{diff}, is expressed by the following formula:

$$Len_{diff} = \left| Len(\text{Radar\_Trajectory}) - Len(\text{AIS\_Trajectory}) \right| \qquad (6)$$

where $Len(\text{Radar\_Trajectory})$ represents the length of the radar trajectory sample within a certain time span and $Len(\text{AIS\_Trajectory})$ represents the length of the AIS trajectory sample within a certain time span. If the difference in length between radar and AIS trajectories is small, this indicates that the target observed by the two data sources is more likely to be the same target, and vice versa. The length difference feature serves as a valuable indicator in detecting variations in trajectory length between radar and AIS data, which, in turn, aids in determining whether a target is associated with both data sources.

3.2. Course Difference Characteristic

Course denotes the direction of a ship's movement trajectory relative to the ground, and we exclusively utilized course over ground for our analysis. The characteristics of course differences between radar tracks and AIS tracks are employed to compare any discrepancies in the course of targets observed by the two distinct data sources. By quantifying the difference between the average courses of radar tracks and AIS tracks, we can evaluate the consistency of target course information across diverse data sources. The course difference characteristic, $Course_{diff}$, can be expressed by the following formula:

$$Course_{diff} \left| = Course(\text{Radar\_Trajectory}) - Course(\text{AIS\_Trajectory}) \right. \qquad (7)$$

where Course(Radar_Trajectory) signifies the course of the radar trajectory and Course(AIS_Trajectory) denotes the course of the AIS trajectory. If the difference in course is minimal, this implies that the target detected by the two data sources is highly likely to be the same target. Conversely, significant differences in the course information could indicate inconsistencies.

3.3. Average Acceleration Difference Characteristics

The characteristics comparing the average acceleration difference between radar tracks and AIS tracks are employed to analyze any discrepancies in the acceleration of targets detected by the two distinct data sources. Acceleration pertains to the rate of change in ship target speed with respect to time, and the acceleration range varies among different ships. During the construction of the dataset, we standardize the sample length to align with the temporal span of three AIS data points, ensuring consistency across samples. Consequently, the initiation time for calculating acceleration corresponds to the timestamp of the first AIS data point, while the termination time corresponds to that of the third AIS data point. Then, the average acceleration of AIS and radar data is computed within this designated timeframe. By calculating the difference between the average acceleration of radar track points and AIS track points, we can assess the consistency of target acceleration information across diverse data sources. Acceleration difference characteristics, Avg_Accelration, can be expressed by the following formula:

$$\text{Avg\_Accelration} = \left| \frac{1}{n} \sum_{i=1}^{n} \text{Accelration}(Radar\_\text{Trajectory})_i - \frac{1}{m} \sum_{j=1}^{\infty} \text{Accelration}(AIS\_Trajectory)_j \right| \tag{8}$$

where Accelration$(Radar_\text{Trajectory})_i$ represents the acceleration of radar trajectory point i, Accelration$(AIS_Trajectory)_j$ represents the acceleration of AIS trajectory point j, n is the number of data points in the radar trajectory, and m is the number of data points in the AIS trajectory. If the average acceleration difference is small, this indicates that the target accelerations observed by the two data sources are relatively consistent, indicating that the two observed trajectories are likely to be from the same target.

3.4. The Distance Difference in Starting Positions

The characteristics of the initial-position distance differences between radar tracks and AIS tracks are employed to compare any discrepancies in distance between the starting positions of targets observed by the two distinct data sources. By quantifying the differences in the distance between the initial points of radar tracks and AIS tracks, we can evaluate the consistency of targets' initial-position information across diverse data sources. The distance difference characteristics of the starting position can be expressed by the following formula:

$$\text{Diff\_Start} = \| \mathbf{P}_{\text{start\_Radar}} - \mathbf{P}_{\text{start\_AIS}} \| \tag{9}$$

where $\mathbf{P}_{\text{start\_Radar}}$ represents the starting-position point of the radar trajectory and $\mathbf{P}_{\text{start\_AIS}}$ represents the starting-position point of the AIS trajectory. If the initial-position distance difference is small, this indicates that the target's starting positions observed by the two data sources are relatively consistent, indicating that the two observed trajectories are likely to be from the same target.

3.5. The Distance Difference in End Positions

The feature of the end-position distance difference between radar tracks and AIS tracks serves as a crucial metric to compare the variance in the distance between the end positions of a target tracked by the two distinct data sources. By quantifying the distance difference between the end points of a radar track and an AIS track, we can assess the conformity of the target's end-position information across different data sources.

$$\text{Diff\_End} = \| \mathbf{P}_{\text{end\_Radar}} - \mathbf{P}_{\text{end\_AIS}} \| \tag{10}$$

where $\mathbf{P}_{\text{end\_Radar}}$ represents the end point of the radar trajectory and $\mathbf{P}_{\text{end\_AIS}}$ represents the end point of the AIS trajectory. If the end-position distance difference is small, this indicates that the target's starting positions observed by the two data sources are relatively consistent, indicating that the two observed trajectories are likely to be from the same target.

3.6. Dynamic Similarity Characteristics

Dynamic time warping (DTW) is a method used to compare the similarity between two time series, which is widely applied in trajectory feature calculation. By treating a radar trajectory and an AIS trajectory as time series, the similarity between them can be calculated using the DTW algorithm, which allows us to quantitatively measure the dynamic similarity between the two trajectories.

Given two time series of radar and AIS trajectories, $X = \{x\_1, x\_2, \ldots, x\_m\}$ and $Y = \{y\_1, y\_2, \ldots, y\_n\}$, where x_i and y_j represent the elements at timepoints i and j, respectively, firstly, construct an m x n cumulative distance matrix D, where D[i][j] represents the distance between the first i elements of sequence X and the first j elements of sequence Y. This distance can be calculated based on Euclidean distance metrics.

Secondly, compute the optimal path through dynamic programming to find the best alignment between sequence X and sequence Y.

$$D[i][j] = \text{dist}(x_i, y_i) + \min(D[i-1][j], D[i][j-1], D[i-1][j-1]) \tag{11}$$

where $\text{dist}(x_i, y_j)$ denotes the distance between the sequence elements x_i and y_j. Finally, the DTW similarity between sequence X and sequence Y can be obtained by accumulating the lower-right element, D[m][n], of the distance matrix, D.

The primary advantage of DTW similarity lies in its ability to handle scenarios in which the lengths of time series are inconsistent and the speeds vary. In ship trajectory analysis, the speeds of ships may vary and the sampling frequencies of radar and AIS data may differ, leading to a mismatch between the two trajectories in the time dimension. The DTW algorithm utilizes dynamic programming to determine the optimal time alignment, effectively addressing these challenges and enabling more accurate similarity assessments.

This characteristic holds a crucial role in ship trajectory matching and association problems. By comparing the DTW similarity among diverse target trajectories, it aids in determining whether the radar and AIS data correspond to the identical ship target, thereby facilitating data association and consistency analysis.

4. Trajectory Dataset Construction

The aim of trajectory dataset construction is to extract the trajectory characteristics, as previously described, from labeled radar and AIS trajectory data. By extracting these features from labeled radar and AIS trajectory data and creating positive and negative samples, we can effectively train an SVM classifier, which facilitates the automatic classification and association of unlabeled data. These samples include both positive and negative instances, with positive samples comprising radar and AIS trajectory features labeled as the same target and negative samples consisting of radar and AIS trajectory features labeled as different targets. To maintain sample consistency, we standardized the sample length to correspond to the time span of three AIS data points. In the end, to address the issue of imbalanced distribution between positive and negative samples, we conducted imbalanced preprocessing to create the final dataset for model training.

4.1. Data Preprocessing

The AIS data of ships in a navigation environment are collected by the on-board AIS terminal, and then the location information of surrounding ships is extracted through protocol analysis. Due to the quality problem of AIS data, they usually need to be polished in historical data analysis. However, to mimic real-time scenarios, in which AIS reports sent from other ships are decoded and applied for association directly, raw AIS data are collected in the dataset construction. When vessels are not equipped with an AIS device, they can only be detected by radar. And, in data association, radar data will not match any AIS data. Therefore, track data from radar will be used for collision avoidance and radar data will be employed to build the negative samples.

Radar data preprocessing mainly includes shoreline elimination, connected component detection, and coordinate transformation. Firstly, to acquire radar targets, it is necessary to eliminate shorelines from the original radar images to obtain radar images solely containing the navigation areas of ships. Eliminating shorelines can remove the influence of riverbank objects on radar target detection. Based on the acquired shoreline positions, the intersection of the radar image and the area enclosed by the shoreline can be taken to eliminate the shoreline.

Subsequently, connected component detection is performed on these images to extract the targets of ships. In this paper, the two-pass scanning method [22] was chosen for connected component detection. Through two scans, the connected components in an image can be detected, thereby identifying the radar targets within these connected components, as shown in Figure 3. Then, ship objects are filtered according to the pixel values of each of the connected components.

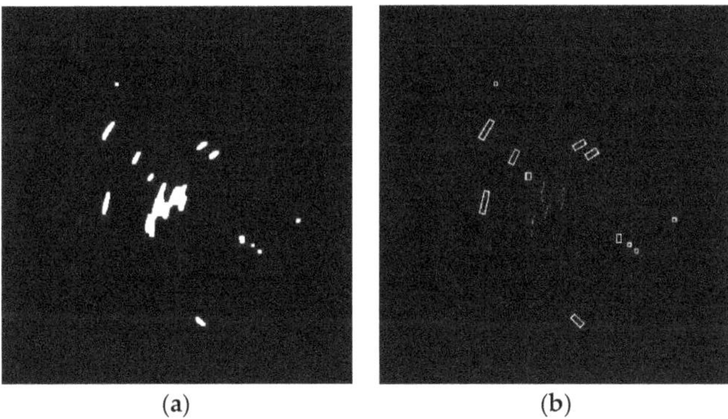

(a)   (b)

Figure 3. Radar target recognition. (**a**) Radar image. (**b**) Connected component analysis.

Finally, in the domain of waterway transportation, the trajectory data generated by radar and AIS exhibit different data formats, which requires the transformation of coordinates. The conversion of radar polar coordinates to AIS coordinates fundamentally involves transforming the geodetic coordinate system into the polar coordinate system. The radar coordinate system operates in a polar fashion, with the radar device as its origin, measuring both distance, ρ, and rotational angle, θ. Meanwhile, AIS target position data (lon_i, lat_i) are originally in the form of longitude and latitude in the geodetic coordinate system. Therefore, to integrate AIS and radar data effectively, conversion from the polar coordinates of radar to geodetic coordinates is necessary, as demonstrated by the equation below:

$$lat\_r = \arcsin(\sin(lat\_o) \times \cos(d/R) + \cos(lat\_o) \times \sin(d/R) \times \cos(\theta)) \tag{12}$$

$$lon\_r = lon\_o + \arctan\left(\cos\left(\tfrac{d}{R}\right) - \sin(lat\_o)\right) \times \sin(lat\_r), \sin(\theta) \\ \times \sin(d/R) \times \cos(lat\_o)) \tag{13}$$

where (lon_r, lat_r) denote the geodetic coordinates of the radar data, (lon_o, lat_o) denote the coordinates of the radar device, θ represents the relative angle, d signifies the radar detection distance, and R denote the radius of the Earth.

4.2. Positive Sample Construction

The process of constructing positive samples aims to establish a feature model for the target vessel, allowing the radar and AIS trajectory features belonging to the same target to be correctly associated. The specific steps are as follows:

Step 1—Data Preparation: Initially, manual labeling of radar and AIS trajectory data is performed. These data encompass vessel motion information along with labeling information, indicating which radar and AIS trajectories correspond to the same vessel. For each pair of radar and AIS trajectories labeled as the same target, they are combined to form a positive sample trajectory pair, facilitating subsequent trajectory feature calculations. Positive sample trajectories are illustrated in Figure 4.

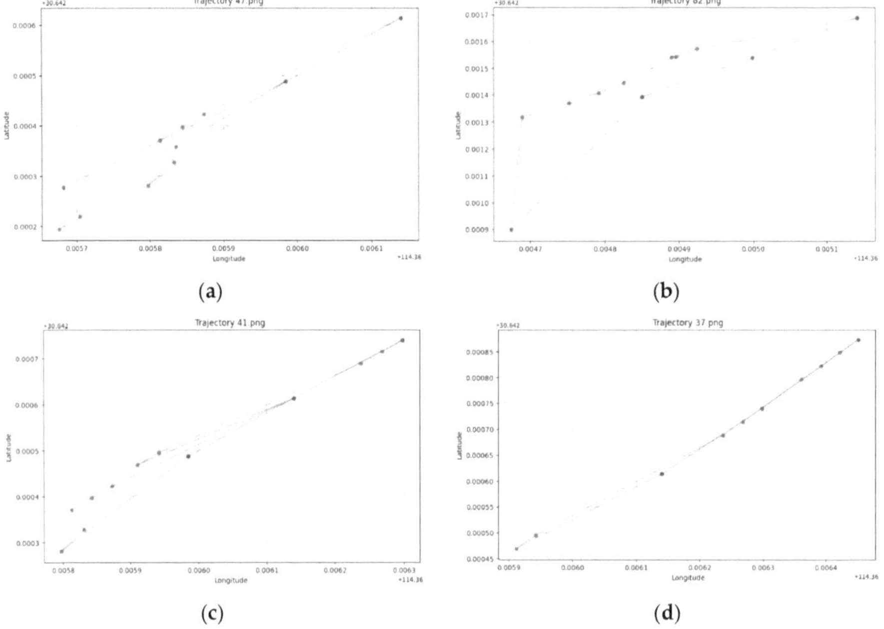

Figure 4. Positive sample trajectory pair (red represents radar trajectory; blue represents AIS trajectory; subfigure (**a**–**d**) represents four positive sample trajectory 47, 82, 41 and 37 respectively).

Step 2—Time Alignment: Due to potential differences in the sampling frequency of radar and AIS data, there may be time discrepancies in the sample's time dimension. To ensure data continuity and consistency, time alignment is carried out when constructing positive samples. Typically, the radar's sample length is set to match the time span of three consecutive AIS data points, which helps mitigate issues related to inconsistent time intervals.

Step 3—Feature Extraction: Trajectory features, such as distance differences, course differences, and average acceleration differences, are extracted from radar and AIS data. These features reflect crucial characteristics of vessel motion, aiding in the establishment of ship identification and association models.

Step 4—Sample Labeling: For each constructed positive sample, a label of "1" is assigned, indicating that they belong to the same target vessel. These labels serve as training data for supervised learning, assisting the model in comprehending the characteristics of the target vessel.

4.3. Negative Sample Construction

The process of negative sample construction aims to establish a feature model capable of distinguishing between different vessels. Negative samples are composed of radar and AIS trajectory features labeled as different targets, assisting the model in understanding the trajectory differences between various vessels from AIS and radar data. The detailed procedure for negative sample construction is as follows:

Step 1—Data Preparation: In contrast to the positive sample construction process, we initially select radar and AIS trajectory data that are not labeled as the same target during the same time interval. These datasets contain vessel motion information and labeling information, indicating which radar and AIS trajectories correspond to different target vessels. For each pair of radar and AIS trajectories labeled as different targets, they are combined to form a negative sample trajectory pair. Negative sample trajectory pairs are illustrated in Figure 5.

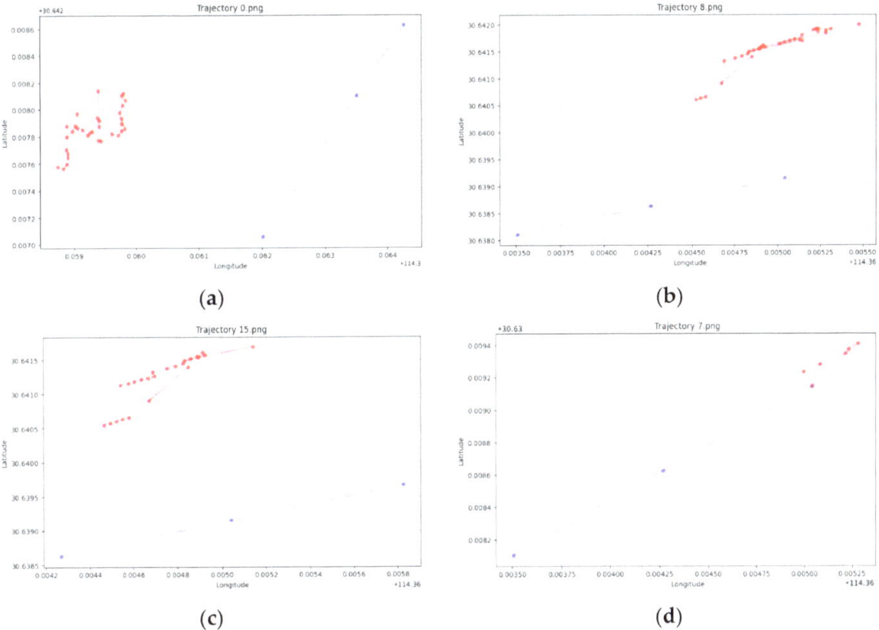

Figure 5. Negative sample trajectory pair (red represents radar trajectory; blue represents AIS trajectory; subfigure (**a**–**d**) represents four negative sample trajectory 0, 8, 15 and 7 respectively).

Step 2—Time Alignment: Similar to the positive sample construction process, time alignment is crucial for radar and AIS trajectories to guarantee the synchronization of heterogeneous data within a consistent temporal interval.

Step 3—Feature Extraction: Similar to the process of constructing positive samples, trajectory features, such as distance differences, course differences, and average acceleration differences, are extracted from radar and AIS trajectory data. These features serve to characterize the differences in AIS and radar data between different vessels.

Step 4—Sample Labeling: For each constructed negative sample, a label of "0" is assigned, indicating that the features of this negative sample trajectory pair belong to different vessels.

4.4. Imbalanced Preprocessing

The process involves combining the constructed positive sample set with the negative sample set to create a comprehensive dataset. In tasks associated with associating ship radar and AIS data, the number of positive samples representing the same target vessel trajectories is relatively limited, while the number of negative samples corresponding to different target vessels is more substantial. Given that the SVM algorithm is significantly affected by sample distribution within the dataset, this imbalance can potentially lead to a reduction in the model's performance during both training and testing phases. This is primarily because the model tends to favor predicting the class with a higher sample count while neglecting the one with fewer samples.

To guarantee the precision and resilience of model training, it is paramount to ensure a balanced distribution of both positive and negative samples across the entire dataset. In this research, we harness the SMOTE (Synthetic Minority Over-Sampling Technique) algorithm to synthesize additional samples, thereby augmenting the minority class representation and mitigating the imbalance in sample category distribution. The generation of these synthetic samples occurs within the feature space and leverages the inherent similarity among samples in the minority class, thereby improving the original data's class distribution imbalance. Consequently, this effectively addresses the problem of sample category imbalance, enhancing the model's performance and generalization capabilities.

5. Experiments and Results

5.1. Data Sources

In our experiments, we employed radar and AIS trajectory data collected from the perception-integrated system installed on the vessel "HANG DAO 1 HAO" within the Yangtze River inland waterway. The shipborne perception system incorporates SIMRAD solid-state radar, which is widely used in maritime field. The detection range of the radar is between 1/32 nm and 36 nm. The AIS device used in the system meets the relevant standards of AIS Class B and can receive data related to ship navigation safety in real time. The dataset encompasses radar and AIS data for various target vessels, along with corresponding labeling information, as shown in Figure 6 and illustrated in Table 1. In the figures, own ship is the "HANG DAO 1 HAO" vessel with the MMSI 413835537, and the straight lines in front of the vessel icons represent their headings. This dataset was collected under sunny weather conditions with high visibility. The AIS data contained data decoded from AIS reports with static and dynamic information. Radar data included radar IDs, labeled MMSIs (Maritime Mobile Service Identities), and the other features were the same as in the AIS dataset, which had 9307 records with labels. From this dataset, we extracted several sample features, including distance differentials, course differentials, average acceleration differentials, starting-point distance differentials, end-point distance differentials, and DTW similarity features. To facilitate our experiments, we divided the constructed dataset into training and testing sets in a 7:3 ratio, allowing for comprehensive testing and evaluation of our proposed methods. This division ensured the independence of the test data from the training data, enabling us to assess the effectiveness and performance of our approaches accurately. The utilization of real-world ship monitoring data from the Yangtze River inland waterway added authenticity and applicability to our experimental framework, contributing to the robustness and relevance of our research outcomes.

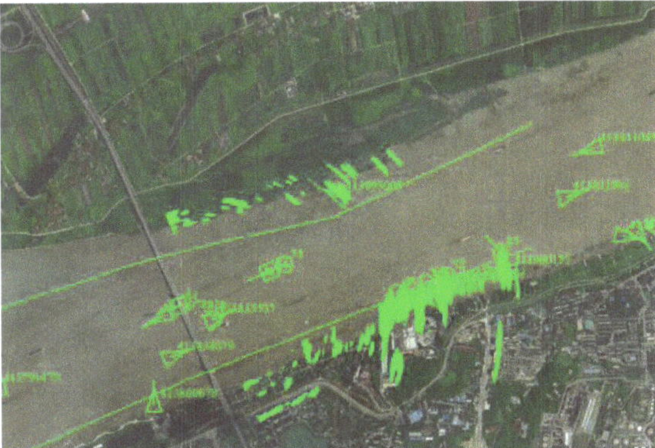

Figure 6. Experimental scenarios.

Table 1. Dataset information.

Features	Type	Values
MMSI	String	0–799,999,999
Longitude	Float	0–180°
Latitude	Float	0–90°
Time	Timestamp	17 April 2022
Course	Float	0–360°
Speed	Float	0–20 kn
Radar ID	String	0–1992

5.2. Evaluation Criteria

Experimental evaluation metrics were utilized to assess the performance of the SVM model in the task of associating ship radar and AIS trajectories, specifically its ability to accurately identify AIS and radar trajectories as belonging to the same vessel. The primary experimental evaluation metrics were precision, recall, and F1 score.

Precision: Precision refers to the proportion of samples that are predicted as "true" by a model and are indeed true positives. It is particularly relevant when dealing with binary classification problems, where the goal is to classify instances into one of two classes, typically referred to as the positive class and the negative class. Here, a high precision value indicates that the model is good at identifying the heterogeneous data belonging to one vessel and does not make many false-positive errors.

Recall: Recall refers to the ratio of positive samples correctly predicted by a model to true-positive samples. Specifically, in the present study, recall is defined as the number of true-positive classifications (correctly identified instances of data from the same vessel) divided by the sum of true positives and false negatives. It indicates the model's capacity to capture and correctly classify data instances that truly belong to the same vessel, which is essential in vessel tracking, navigation, and various maritime applications. A high recall score means that the model is effective at finding and classifying most of the heterogeneous data belonging to the same vessel, reducing the risk of missing important information.

F1 Score: The F1 score is a metric used in classification tasks, including the classification of AIS and radar data belonging to one vessel [23]. It is a valuable measure that combines both precision and recall into a single value to provide a more comprehensive evaluation of a model's performance, especially in scenarios with imbalanced class distributions. The F1 score is calculated as follows:

$$F1 = 2 \times \text{Precision} \times \text{Recall} / (\text{Precision} + \text{Recall}) \tag{14}$$

F1 scores range from 0 to 1, with a high F1 score suggesting, here, that the model achieves a balance between correctly classifying data as belonging to one vessel while minimizing the risk of missing relevant data points. Therefore, we can conduct a comprehensive evaluation of the model's performance on the ship radar and AIS trajectory association task in the test dataset using F1 scores. High precision, recall, and F1 score will substantiate the capability of our proposed method to accurately discern whether AIS and radar trajectories pertain to the same target.

5.3. Results

In inland waterways, where there is a high density of traffic flow, frequent cross encounters, and substantial diversity in vessel trajectories, the challenge of data association becomes particularly intricate and complex. Therefore, we conducted experiments categorized into four groups: vessels moving with the same heading, vessels moving close together with the same heading, vessel encounter scenarios, and multiple vessel encounter scenarios. These experiments allowed us to conduct a comparative analysis of the performance of the ship radar and AIS trajectory data association method based on the SVM in different scenarios.

(1)   Vessels moving with the same heading

The purpose of this experimental group was to explore situations in which vessels move in the same direction, observed by both radar and AIS. Specifically, we selected two typical situations within this group for analysis. We extracted and processed the data to obtain a total number of 174 trajectory samples for further analysis. In this situation, the vessels' movements are in the same direction, albeit with noticeable distances between them, as illustrated by target 92 and target 57 in Figure 7a and target 1532 and target 1548 in Figure 7b. The experiment aimed to confirm the effectiveness and accuracy of our approach in addressing these same-direction forward- and backward-movement scenarios.

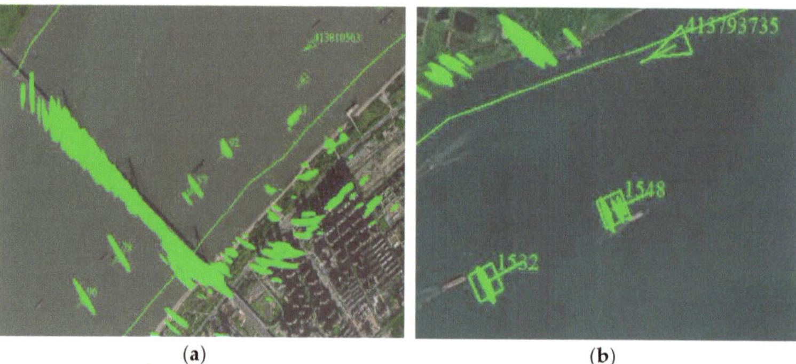
(a)   (b)

Figure 7. Vessels moving with same heading. (**a**) Target 92 and target 57. (**b**) Target 1532 and target 1548.

An F1 score of 0.96 signifies a balanced trade-off between precision and recall delivering accurate classification results. The model effectively discriminates between the radar and AIS trajectories of the target vessels, aligning precisely with the actual labels.

(2)   Vessels moving close together with the same heading

This experimental grouping was designed to replicate scenarios in which vessels closely follow the same course and are ready to overtake in both radar image and AIS data. Specifically, we selected two typical situations within this group for analysis. We extracted and processed the data to obtain a total number of 415 trajectory samples for further analysis, as illustrated by target 7 and target 8 in Figure 8a and target 1880 and target 1881 in Figure 8b. In this case, multiple vessels are navigating near each other while maintaining a consistent direction of movement. This suggests a scenario in which the vessels may be in the process of overtaking one another, with one vessel gradually moving past another while maintaining a similar course. In these circumstances, the radar and AIS trajectories of the target vessels displayed distinct temporal and spatial similarities and were characterized by minimal differences in distance, heading, and speed.

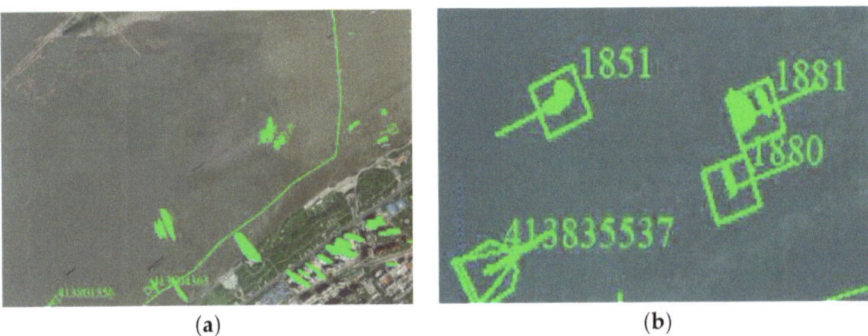

(a)   (b)

Figure 8. Vessels moving closely with same heading. (**a**) Target 7 and target 8. (**b**) Target 1880 and target 1881.

Through calculations, we obtained an F1 score of 0.95. These results signify the performance of the data association method in the scenarios in which vessels 7 and 8 were moving close together in the same direction.

(3)   Vessel encounter scenarios

In this group, we considered scenarios in which target vessels encounter each other in both radar image and AIS data. An encounter refers to a situation in which vessels approach each other in close proximity or along intersecting paths. Specifically, we selected two typical situations within this group for analysis. We extracted and processed the data to obtain a total number of 77 trajectory samples for further analysis. In such cases, the trajectories of the target vessels may exhibit significant differences in terms of distance and heading, as illustrated by the trajectories of vessels 1960 and 1981 in Figure 9a and target 1751 and target 1782 in Figure 9b. This experiment was designed to evaluate the performance of our method in situations involving vessel encounters.

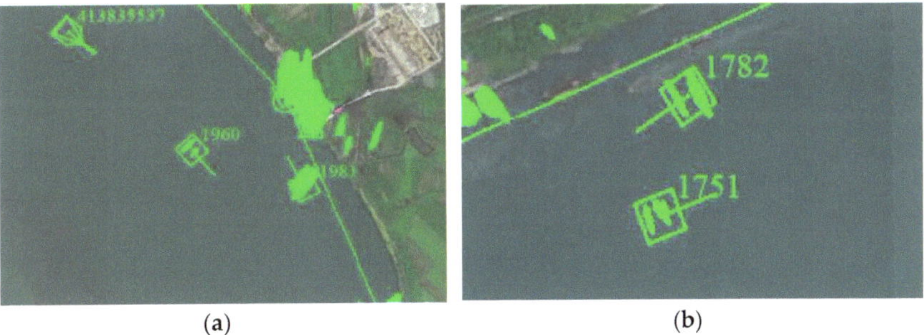

(a)   (b)

Figure 9. Vessel encounter scenarios. (**a**) Target 1960 and target 1981. (**b**) Target 1751 and target 1782.

By computation, an F1 score of 0.98 was obtained, indicating that the proposed method for associating ship radar and AIS trajectory data performs accurately in scenarios in which vessels encounter each other.

(4)   Multiple vessel encounter scenarios

In this set of experiments, we explored scenarios in which multiple vessels simultaneously encounter one another in both radar image and AIS data. Specifically, we selected a typical situation within this group for analysis. We extracted and processed the data to obtain a total number of 230 trajectory samples for further analysis. Multiple-target association requires simultaneous associations across multiple sets of radar and AIS trajectories, as depicted by the examples involving vessels 158, 144, and 136 in Figure 10a and vessels 1805,

1807, 1810, and 1812 in Figure 10b. The experiments aimed to investigate the applicability and efficiency of our proposed association method in multi-target scenarios.

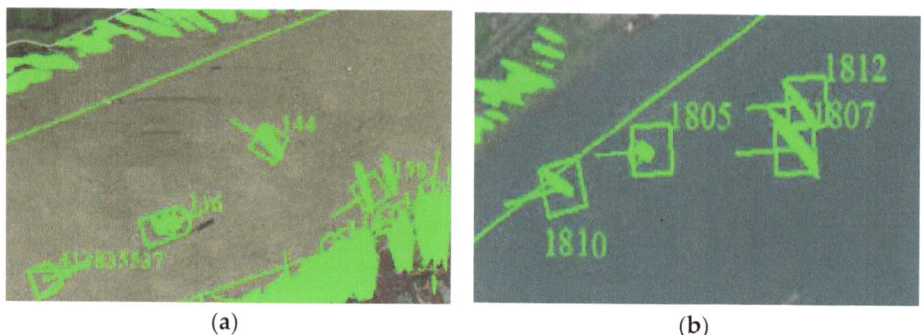

(a)　　　　　　　　(b)

Figure 10. Multiple vessel encounter scenario. (**a**) Target 144, 158, and 136. (**b**) Target 1805, 1807, 1810, and 1812.

Through calculations, we obtained an F1 score of 0.97. These results confirmed the performance of the classifier-based association method in the scenario involving vessels encountering one another, which indicates the model's ability to distinguish between different vessels in multiple vessel encounter scenarios.

6. Discussion

We conducted a comparison with the nearest-neighbor method to provide a more comprehensive evaluation of our classifier approach. The comparison results are presented in Tables 2–5, which detail the evaluation metrics for the different scenarios. For scenarios involving vessels moving with the same heading (Tables 2–4), both our classifier approach and the NN method demonstrated high precision, recall, and F1 score values. In the multiple vessel encounter scenarios (Table 5), our classifier approach consistently achieved higher precision, recall, and F1 score values than the NN method, highlighting its robustness and effectiveness in identifying vessel encounters.

Table 2. Evaluation of vessels moving with same heading scenario.

	Precision	Recall	F1 Score
Classifier Approach	0.97	0.96	0.96
Nearest Neighbor	0.98	0.95	0.96

Table 3. Evaluation of vessels moving close together with same heading scenario.

	Precision	Recall	F1 Score
Classifier Approach	0.96	0.95	0.95
Nearest Neighbor	0.98	0.98	0.98

Table 4. Evaluation of vessel encounter scenario.

	Precision	Recall	F1 Score
Classifier Approach	0.99	0.98	0.98
Nearest Neighbor	1.00	1.00	1.00

Table 5. Evaluation of multiple vessel encounter scenario.

	Precision	Recall	F1 Score
Classifier Approach	0.99	0.95	0.97
Nearest Neighbor	0.87	0.86	0.86

Our analysis focused on a representative set of data collected by on-board perception systems in several classical scenarios, including overtaking and encounters. In these simpler scenarios, both existing NN models and the proposed model exhibited satisfactory trajectory association performance. This is attributable to the relatively straightforward nature of these scenarios, in which ship movement patterns are more uniform and thus easier for models to associate. However, when confronted with more complex scenarios, such as multiple vessel encounters, the proposed method demonstrated its distinct advantage. In multiple vessel encounter scenarios, multiple vessels interact within a limited space, resulting in more intricate and variable trajectory characteristics. In the multiple vessel encounter scenario (Table 5), the nearest-neighbor method exhibited a precision, recall, and F1 score of 0.86. Compared to our classifier approach, the NN method demonstrated a lower performance across all evaluation metrics in this scenario. The NN method, which relies solely on proximity-based matching, may struggle to accurately identify and associate trajectories in such dense and intricate scenarios. And this is attributable to the complexity of the situation, in which distinguishing between multiple overlapping vessel trajectories poses a challenge. In contrast, because it utilizes a comprehensive set of features derived from trajectory characteristics, which enable it to capture nuanced patterns and relationships in the data, our classifier approach ensures accurate identification of positive instances.

Overall, while both methods performed well, the results shown in Table 5 highlight the superior performance of our classifier approach, particularly in scenarios involving multiple vessel encounters. Incorporating multiple trajectory characteristics to solve data association issues makes it more reliable than approaches like the NN method which just take trajectory distance into account. However, its accuracy relies heavily on the training dataset, which may not effectively generalize to complex or open-sea scenarios not adequately represented in the training data mainly obtained from an inland waterway. And as we focused on the association method for AIS and radar data, cases in which radar or AIS are not employed were not considered in this study. Furthermore, the data primarily originated from a single vessel's perception system, and the scenarios selected were relatively limited, potentially affecting the model's generalization capabilities. In future research, we aim to collect a more diverse and extensive dataset, encompassing data from various types of vessels, different weather conditions, and across diverse geographical locations. This will enhance the model's performance and generalization.

Meanwhile, video sensors play a crucial role in enhancing vessel perception, especially in challenging visibility conditions and nighttime operations, where AIS and radar may fall short. After the inclusion of video, we can extend our approach to utilize target detection algorithms to identify vessels present in a video at first. Subsequently, coordinate transformation can be performed to align the coordinates of detected vessels with AIS and radar. Furthermore, leveraging the calibrated relationships between video targets and those identified by radar and AIS, positive and negative sample sets could be constructed for vessel trajectory features in video data. These sample sets will serve as the basis for training a data association classifier, enabling the correlation of vessel perception data from video, AIS, and radar sources. Sensor fusion techniques allow us to leverage the strengths of each sensor type while compensating for their individual limitations. By combining AIS, radar, and video sensor data, we can enhance the accuracy and reliability of vessel motion perception.

7. Conclusions

In this study, a trajectory characteristic-based SVM binary classifier approach is proposed to achieve effective association between ship radar image and AIS data. Based on the data captured from a perception system installed on a vessel named "HANG DAO 1 HAO", we extracted trajectory characteristics of different vessels from radar and AIS data. Then, positive and negative training sets were constructed to feed them into the classifier for association analysis. The research results demonstrate that the trajectory characteristic-based SVM binary classifier excels in ship radar and AIS data association. Through a series of experiments that included two typical situations for each of the following: overtaking, encounters, and multi-target groups, which are common situations in inland waterway traffic contexts, this research substantiated the method, which achieved an F1 score greater than 0.95, with the aim of enhancing the precision and reliability of ship monitoring and navigation information.

In a future study, more diverse and extensive datasets will be collected to enhance the model's performance and generalization. This could involve using data from different types of vessels, varying weather conditions, and different geographical locations. Moreover, the integration of data from other sources, such as CCTV or BEIDOU data, which have the ability to detect objects without radar or AIS data in challenging visibility conditions, will be applied to expand and compensate for the detection capacities of current ship monitoring and navigation. And in future work, data fusion methods, such as covariance intersection, will be implemented after data association to provide more accurate position information about surrounding vessels and enhance maritime situational awareness.

Author Contributions: Conceptualization, J.L., X.L. and Y.W.; formal analysis, Y.S.; methodology, J.L. and W.H.; validation, X.L. and F.Z.; writing—original draft preparation, J.L. and F.Z.; writing—review and editing, Y.S. All authors have read and agreed to the published version of the manuscript.

Funding: This research was funded by the Natural Science Foundation of Fujian Province, grant number 2022J05236; the National Natural Science Foundation of China, grant number 52172327; the MinJiang University Science Project, grant number MJY21037; the Fujian Province Key Science and Technology Innovation Project, grant number 2022G02009; the Fuzhou Science and Technology Planning Project, grant numbers 2022-P-006 and 2022-S-005; and the Fuzhou Institute of Oceanography Science and Technology Project, grant number 2021F07.

Institutional Review Board Statement: Not applicable.

Informed Consent Statement: Not applicable.

Data Availability Statement: The data presented in this study are available on request from the corresponding author.

References

1. Liu, C.; Chu, X.; Wu, W.; Li, S.; He, Z.; Zheng, M.; Zhou, H.; Li, Z. Human–machine cooperation research for navigation of maritime autonomous surface ships: A review and consideration. *Ocean Eng.* **2022**, *246*, 110555. [CrossRef]
2. Wu, Y.; Chu, X.; Deng, L.; Lei, J.; He, W.; Królczyk, G.; Li, Z. A new multi-sensor fusion approach for integrated ship motion perception in inland waterways. *Measurement* **2022**, *200*, 111630. [CrossRef]
3. Hu, J.; Hong, Y.; Jin, Q.; Zhao, G.; Lu, H. An Efficient Dual-Stage Compression Model for Maritime Safety Information Based on BeiDou Short-Message Communication. *J. Mar. Sci. Eng.* **2023**, *11*, 1521. [CrossRef]
4. Singer, R.; Sea, R.; Housewright, K. Derivation and evaluation of improved tracking filter for use in dense multitarget environments. *IEEE Trans. Inf. Theory* **1974**, *20*, 423–432. [CrossRef]
5. Bar-Shalom, Y.; Daum, F.; Huang, J. The probabilistic data association filter. *IEEE Control Syst.* **2009**, *29*, 82–100. [CrossRef]
6. Lu, Q.; Wu, L.; Chen, Z.; Wang, Q.; Xu, Y. A Review of multi-source trajectory data association for marine targets. *J. Geo-Inf. Sci.* **2018**, *20*, 571–581. [CrossRef]
7. Chi, M.W.; Li, Y.; Min, L.; Chen, J.; Lin, Z.; Su, S.; Zhang, Y.; Chen, Q.; Chen, Y.; Duan, X.; et al. Intelligent marine area supervision based on AIS and radar fusion. *Ocean Eng.* **2023**, *285*, 115373.
8. Liu, G.W.; Liu, Y.X.; Ji, Y.G.; Wang, C. Track association for higj-frequency surface wave radar and AIS based on fuzzy double threshold theory. *Syst. Eng. Electron.* **2015**, *38*, 557–562.
9. Vivone, G.; Millefiori, L.M.; Braca, P.; Willett, P. Performance Assessment of Vessel Dynamic Models for Long-Term Prediction Using Heterogeneous Data. *IEEE Trans. Geosci. Remote Sens.* **2017**, *55*, 6533–6546. [CrossRef]
10. Liu, Z.; Xu, J.; Li, J.; Plaza, A.; Zhang, S.; Wang, L. Moving Ship Optimal Association for Maritime Surveillance: Fusing AIS and Sentinel-2 Data. *IEEE Trans. Geosci. Remote Sens.* **2022**, *60*, 5635218. [CrossRef]
11. Jin, B.; Tang, Y.; Zhang, Z.; Lian, Z.; Wang, B. Radar and AIS Track Association Integrated Track and Scene Features Through Deep Learning. *IEEE Sens. J.* **2023**, *23*, 8001–8009. [CrossRef]
12. Yang, Y.; Yang, F.; Sun, L.; Xiang, T.; Lv, P. Multi-target association algorithm of AIS-radar tracks using graph matching-based deep neural network. *Ocean Eng.* **2022**, *266*, 112208. [CrossRef]

13. Chen, X.; Wei, C.; Xin, Z.; Zhao, J.; Xian, J. Ship Detection under Low-Visibility Weather Interference via an Ensemble Generative Adversarial Network. *J. Mar. Sci. Eng.* **2023**, *11*, 2065. [CrossRef]
14. Guo, Y.; Liu, R.W.; Qu, J.; Lu, Y.; Zhu, F.; Lv, Y. Asynchronous Trajectory Matching-Based Multimodal Maritime Data Fusion for Vessel Traffic Surveillance in Inland Waterways. *IEEE Trans. Intell. Transp. Syst.* **2023**, *24*, 12779–12792. [CrossRef]
15. Huang, Z.; Hu, Q.; Mei, Q.; Yang, C.; Wu, Z. Identity recognition on waterways: A novel ship information tracking method based on multimodal data. *J. Navig.* **2021**, *74*, 1336–1352. [CrossRef]
16. Gan, X.; Wei, H.; Xiao, L.; Zhang, B. Research on Vision-based Data Fusion Algorithm for Environment Perception of Ships. *Shipbuild. China* **2021**, *62*, 201–210.
17. Liu, X.; Li, Y.; Wu, Y.; Wang, Z.; He, W.; Li, Z. A Hybrid Method for Inland Ship Recognition Using Marine Radar and Closed-Circuit Television. *J. Mar. Sci. Eng.* **2021**, *9*, 1199. [CrossRef]
18. Aziz, A.M. A new nearest- neighbor association approach based on fuzzy clustering. *Aerosp. Sci. Technol.* **2013**, *26*, 87–97. [CrossRef]
19. Qiu, J.; Cai, Y.; Li, H.; Han, Y. A New Track Association Model Based on Uncertainty Distribution. *IEEE Access* **2023**, *11*, 58890–58900.
20. Jing, P.L.; Liu, F. Track association method using modified fuzzy membership. *J. Appl. Sci.* **2012**, *30*, 181–186.
21. Vapnik, V.; Golowich, S.; Smola, A. Support vector method for function approximation, regression estimation and signal processing. In Proceedings of the Advances in Neural Information Processing Systems, Denver, CO, USA, 2–5 December 1996; Volume 9.
22. Wu, K.; Otoo, E.; Suzuki, K. Optimizing two-pass connected-component labeling algorithms. *Pattern Anal. Appl.* **2009**, *12*, 117–135. [CrossRef]
23. Salem, M.H.; Li, Y.; Liu, Z.; AbdelTawab, A.M. A Transfer Learning and Optimized CNN Based Maritime Vessel Classification System. *Appl. Sci.* **2023**, *13*, 1912. [CrossRef]

Journal of
*Marine Science
and Engineering*

MDPI

Article

Ship Trajectory Classification Prediction at Waterway Confluences: An Improved KNN Approach

Zhiyuan Wang [1,2,3], Wei He [1,2,*], Jiafen Lan [3,*], Chuanguang Zhu [2], Jinyu Lei [1,2] and Xinglong Liu [1,2]

[1] Fuzhou Institute of Oceanography, Minjiang University, Fuzhou 350121, China; wangzhiyuan14@mails.ucas.edu.cn (Z.W.)

[2] Fujian Engineering Research Center of Safety Control for Ship Intelligent Navigation, College of Physics & Electronic Information Engineering, Minjiang University, Fuzhou 350121, China

[3] Intelligent Transportation Systems Research Center, Wuhan University of Technology, Wuhan 430063, China

* Correspondence: heweil1@mju.edu.cn (W.H.); lanjiafen@whut.edu.cn (J.L.)

Abstract: This study presents a method to support ship trajectory prediction at waterway confluences using historical Automatic Identification System (AIS) data. The method is meant to improve the recognition accuracy of ship behavior trajectory, assist in the proactive avoidance of collisions, and clarify ship collision responsibility, to ensure the safety of waterway transportation systems in the event of ship encounters induced by waterway confluence or channel limitation. In this study, the ship trajectory based on AIS data is considered from five aspects: time, location, heading, speed, and trajectory by using the piecewise cubic Hermite interpolation method and then quickly clustered by regional navigation rules. Then, an improved K-Nearest Neighbor Algorithm considering the sensitivity of data characteristics (SKNN) is proposed to predict the trajectory of ships, which considers the influence weights of various parameters on ship trajectory prediction. The method is trained and verified using the AIS data of the Yangtze River and Han River intersection in Wuhan. The results show that the accuracy of SKNN is better than that of conventional KNN and Naive Bayes (NB) in the same test case. The accuracy of the ship trajectory prediction method is above 99% and the performance metrics of the SKNN surpass those of both the conventional KNN and NB classifiers, which is helpful for early warning of collision encounters to ensure avoidance.

Keywords: waterway transportation safety; ship trajectory prediction; AIS; KNN; trajectory classification

Citation: Wang, Z.; He, W.; Lan, J.; Zhu, C.; Lei, J.; Liu, X. Ship Trajectory Classification Prediction at Waterway Confluences: An Improved KNN Approach. *J. Mar. Sci. Eng.* **2024**, *12*, 1070. https://doi.org/10.3390/jmse12071070

Academic Editor: Xinqiang Chen

Received: 15 May 2024
Revised: 21 June 2024
Accepted: 21 June 2024
Published: 26 June 2024

1. Introduction

With the continued growth of the economy, waterway transportation is becoming increasingly congested, and more and more channels are becoming saturated, especially at the confluence of rivers, putting increased demands on waterway navigation management [1]. The popularity of the AIS not only improves the efficiency and safety of navigation but also makes it possible to collect a huge volume of ship motion data for waterway transportation studies. Ship motion data can be extracted to understand the navigation status and they can be applied to many maritime fields, such as ship collision avoidance, maritime monitoring, ship trajectory prediction, and maritime accident investigation. The analysis and application of AIS data are mainly reflected in research on traffic flow [1], ship encounter characteristics analysis [2–5], ship trajectory prediction [6–8], and navigation anomaly identification [9,10]. The following technologies are used as the foundation: data recovery, trajectory clustering, and trajectory prediction modeling.

AIS data cleaning is the basis of ship trajectory data preprocessing, usually based on the ship's position, speed, and heading data to identify anomalies and repair them [11]. On this basis, clustering methods are often used for ship trajectories, such as DBSCAN, K-means, GMM, EM, and other algorithms. Lee et al. (2007) [12] presented the TRACLUS algorithm, which has been widely used, and many researchers have developed the algorithm to improve the clustering effect. Among them, Rong et al. (2020) [13] achieved trajectory clustering through the improved TRACLUS algorithm. Gudmundsson et al. (2015) [14] proposed a distance-based trajectory clustering algorithm using the Frechet distance to define the similarity between trajectories. Further, trajectory prediction has been based on data preprocessing and trajectory clustering, such as the Markov model, the Naive Bayesian model, SVM, neural networks, the particle swarm algorithm, the Kalman filter, and other algorithms for prediction. Perera (2012) [15] uses the K-means algorithm to classify the historical trajectory of the ship and establishes an ANN model to predict the trajectory of the ship based on the ship's information and grouping results. Experiments show that the algorithm has an accuracy of more than 70%. Hu et al. (2021) [16] used an artificial neural network to study AIS data and realized the prediction of ship trajectories. The results show that the accuracy of route prediction was 76.15%. Although the above methods are effective, the accuracy of trajectory prediction may not meet requirements due to the lack of in-depth research on ship trajectory changes. Murray et al. (2021) [17] used the Gaussian mixture model to cluster AIS historical data, and then predicted ship trajectory based on this. The results showed that the accuracy of ship trajectory prediction based on clustering was higher than that of un-clustered ships. The effect of clustering directly affects the accuracy of trajectory prediction. Wang (2019) [18] used the improved TRACLUS to cluster the AIS trajectory and adjust the internal parameters, but because the clustering effect was limited by the choice of internal parameters, under the influence of waterway confluence, the final clustering effect did not fully reflect the actual situation.

Artificial Neural Networks (ANNs) are widely used in ship trajectory prediction research. Qian et al. (2022) used a deep Long Short-term Memory Network Framework (LSTM) and Genetic Algorithm (GA) to predict ship trajectories in inland water [19]. But LSTM models have high complexity, long training time, and require a large amount of historical data. Tian et al. (2023) proposed a ship trajectory prediction model using a Difference Long Short-term Memory (D-LSTM) neural network, which more effectively processes differential sequences, improves prediction accuracy, and stability [20], but also requires high computational resources and high-quality logarithmic data. Gan et al. (2016) combined clustering and an ANN to predict ship trajectory, which reduce data complexity and dimensions through clustering to improve the training efficiency of the ANN [21]. However, this method requires adjusting multiple parameters to tune the model and is sensitive to the initial clustering results. Recurrent Neural Networks (RNNs) are suitable for time-series prediction and have dynamic prediction capabilities, but traditional RNNs suffer from gradient vanishing problems, limited ability to predict long time series, and require a large amount of computational resources [22,23]. Zhao et al. (2023) proposed a ship trajectory prediction method based on GAT and LSTM, which has high robustness [24]. Li et al. (2023) analyzed five classical machine learning methods and eight deep learning methods on ship trajectory prediction [25,26]. Zhou et al. (2024) proposed a ship trajectory prediction method based on Optuna–BILSTM [27]. Li et al. (2024) proposed a ship trajectory prediction method based on ACoAtt–LSTM [28]. Jiang et al. (2023) proposed a ship trajectory prediction method based on an attention mechanism model [29].

Due to the large volume of ship traffic and complex interactions in intersecting water areas, machine learning methods such as KNN and Naive Bayes (NB), which have strong adaptability and can quickly respond to environmental changes, are used to predict ship trajectory intentions in intersecting water areas, achieve real-time trajectory category prediction, and improve navigation safety and efficiency.

The confluence area of main and branch waterways is one of the most complicated channel circumstances [30]. Due to waterway confluence, the traffic flow of ships often forms a complex situation of intersection, such as serious traffic conflicts, irregular navigation order, and heavy pressure on the safety of drivers, which increases the risk of collisions between ships in the confluence area [31,32]. It has been shown that more than 80% of ship collision accidents are caused by human error, including misidentification of the ship's behavior trajectory without correctly perceiving the risk of collision, failing to take appropriate collision avoidance measures, and ultimately causing collision accidents [33].

Therefore, we present a simple method to support ship trajectory prediction using historical AIS data. In our method, a clustering method based on regional navigation rules is used to achieve fast and accurate clustering of trajectories, and an improved K-Nearest Neighbor Algorithm considering the sensitivity (SKNN) of data characteristics is proposed to establish the mapping relationship between ship navigation characteristics and ship trajectory, to then predict the ship trajectory reflecting perceived collision risks while ensuring allocation of avoidance responsibilities. The exploration of the Ship K-Nearest Neighbors (SKNN) algorithm within the realm of maritime trajectory forecasting should support advancements in various pivotal sectors. These include enhancing the efficacy of collision avoidance warning systems, optimizing navigational efficiency, refining energy management strategies for navigation [34], and innovating in the domain of navigational route planning [35].

In this paper, a machine learning approach is proposed to predict ship trajectory at waterway confluences using an improved K-Nearest Neighbor Algorithm considering the sensitivity of data characteristics. Section 2 provides a description of the proposed method for ship trajectory prediction, Section 3 presents a comparison between the proposed method and a conventional KNN, NB, based on the same AIS data of the Yangtze River and the Han River intersection in Wuhan, and Section 4 concludes the article.

2. Proposed Method for Ship Trajectory Prediction

Taking the AIS data of the Wuhan Bridge section in the Yangtze River from October 2020 to November 2020 as an example, a total of 148,325 trajectories, a method for ship trajectory prediction using historical AIS data is established based on the K-Nearest Neighbor Algorithm (KNN). The main scheme of the methods (Figure 1) primarily includes three parts: (1) Ship trajectory data preprocessing; (2) Ship trajectory data clustering; and (3) Ship trajectory prediction.

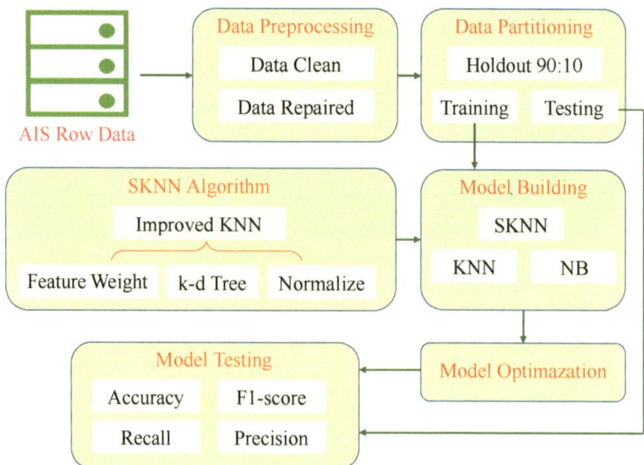

Figure 1. Flowchart for prediction of ship trajectory based on AIS data.

2.1. Ship Trajectory Data Preprocessing

As AIS-based raw data are the research object of ship trajectory, there can be sampling data errors due to the abnormal operation of software or hardware equipment. Hence, these data need to be preprocessed first to clean the wrong AIS data dynamic information and repair the missing information. Figure 1 shows that there are a large number of erroneous data in the AIS-based raw data of the Wuhan Bridge section in the Yangtze River from October 2020 to November 2020, and there are many ferry trajectories to and from Jijiazui Wharf, Qingchuan Wharf, Wuhan Guan Wharf, Zhonghua Road Wharf, etc., which are useless for the study of ship trajectory data on the main routes between the Han River and the Yangtze River. Therefore, the AIS-based raw data need to be cleaned and repaired before studying it, such as with regard to ship latitude and longitude, speed, heading, etc.

2.1.1. AIS Data Cleaning

The parsed dynamic information is sorted according to the ship's MMSI, and if the ship's MMSI does not match, it is classified as the sub-trajectories of different ships; for trajectories with the same MMSI, when the time difference between two consecutive ship trajectory points is greater than 900 s, they are divided into sub-trajectories of the ship and data points with MMSI equal to 0 are deleted. In this paper, we mainly use the time, longitude, latitude, speed, and heading characteristics of the AIS data dynamic information, which are denoted t, x, y, v, and θ respectively, as shown in Table 1.

Table 1. AIS data dynamic information.

Parameter	Data Type	Data Range
Time (t)	timestamp	01/10/2020–30/11/2020
MMSI	text	(200000000–799999999) Nine-digit integer
Longitude (x)	float	0–180°
Latitude (y)	float	0–90°
SOG (v)	float	0–20 kn
Heading (θ)	float	0–360°

According to observation of the AIS data, we find that there are three types of error data: abnormal latitude and longitude, abnormal heading, and abnormal speed. Therefore, the following cleaning rules are formulated for error data types.

(1) Area restriction: Defining the research area and deleting data points outside the area.

(2) Removing the duplicate data: Repeated AIS data are the same AIS data sent by a ship continuously while underway. For this kind of data, set rules are as follows: if the speed of the i-th track point is greater than 2 knots, but the data of this track point i are the same as the data of the next track point, then they are deleted.

(3) Position offsetting: If the coordinates of the trajectory points in the AIS data have changed, it will cause the distance between the two adjacent coordinate points to suddenly become larger, which will lead to the calculated vessel sailing speed being clearly unrealistic. Since the coordinates in the AIS data are from the WGS84 coordinate system in the differential GPS-DGPS, the distance between two track points can be calculated by the following:

$$C = \sin(y_{i-1})\sin(y_i) + \cos(y_{i-1})\cos(y_i)\cos(x_{i-1} - x_i) \tag{1}$$

$$D = R \times \arccos(C) \times \left(\frac{\pi}{180}\right) \tag{2}$$

where R is the radius of the Earth.

The average speed between the two trajectory points is as follows:

$$\bar{v}_i = \frac{D}{t_i - t_{i-1}}, \; i = 2, 3, \ldots, m - 1, m \tag{3}$$

When the average speed is greater than 10 knots, the corresponding track points are deleted as error data.

(4) Processing excessive acceleration: According to the Inland Waterway Navigation Standards, the acceleration distance L_m is 20 times the ship's length in theoretical conditions, while the distance can be reduced to 10–14 times the ship's length when the ship is empty. To prevent deleting correct AIS data, the acceleration distance L_m is set to be 10 times the ship's length in the unloaded case in this paper. According to the equation of uniform acceleration motion, the maximum acceleration of the ship is calculated as follows:

$$L_m = 10 \times L = 0.5 \times t_m \times V_m \tag{4}$$

$$a_{\max} = \frac{V_{\mathrm{m}}}{t_{\mathrm{m}}} = \frac{V_{\mathrm{m}}^2}{20 \times L} \tag{5}$$

where L is the length of the ship and V_{m} is the maximum speed of the ship sailing in the Wuhan section.

In the calculation, the value of V_{m} is taken as the maximum inland vessel speed of 12 knots (6.17 m/s), and the length L is set to 100 m. The maximum acceleration of 0.02 m/s² is obtained using the above formula and is removed when the acceleration between two trajectory points exceeds 0.02 m/s².

(5) Removing ferry data: To eliminate the ferry data, this paper adopts the following rules: the trajectory passes through Region 1 and Region 2 at the same time, Region 1 and Region 3 at the same time, or Region 2 and Region 3 at the same time, as shown in Figure 2.

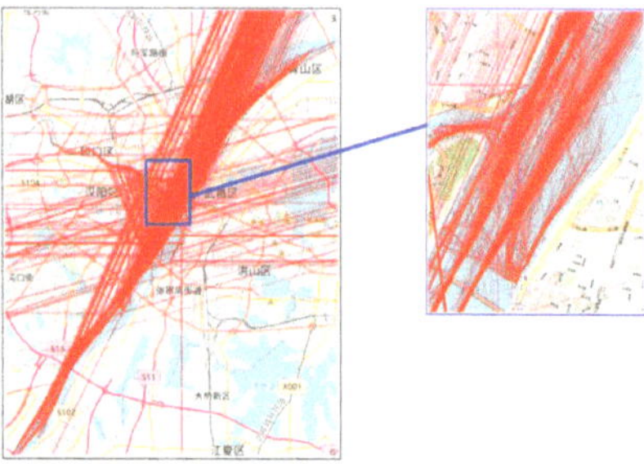

Figure 2. Prediction of ship trajectory process based on AIS-based raw data.

After processing the data as described above, the cleaned trajectories are as shown in Figure 3, and it can be seen that the wrong AIS dynamic data have been removed. However, anomalous data with little position deviation or speed not exceeding the cleaning criteria still exist, and personal judgement may be required if the anomalous data are to be completely cleaned.

Figure 3. Schematic diagram of AIS trajectory after cleaning.

2.1.2. AIS Data Repairing

After data cleaning, there are some problems such as missing track points and inconsistent data time intervals in the AIS-based ship trajectory. To ensure the subsequent clustering and prediction, the segmented cubic Hermite interpolation is selected to repair the missing trajectory points. Its expression is as follows:

$$f(x) = \begin{cases} f_1(x), x \in [x_0, x_1] \\ f_2(x), x \in [x_1, x_2] \end{cases} \tag{6}$$

$$f_i(x) = \frac{[h_i + 2(x - x_{i-1})](x - x_i)^2}{h_i^3} y_{i-1} + \frac{[h_i - 2(x - x_{i-1})](x - x_{i-1})^2}{h_i^3} y_i$$
$$+ \frac{(x - x_{i-1})(x - x_{i-1})^2}{h_i^2} y'_{i-1} + \frac{(x - x_i)(x - x_{i-1})^2}{h_i^2} y'_i \tag{7}$$

where $h_i = x_i - x_{i-1}, i \in [1, 2]$.

2.2. Ship Trajectory Data Clustering

When a ship is navigating in the Yangtze River Channel, its trajectory clusters are obvious due to the constraints of the "Regulations for Navigation in the Middle Reaches of the Yangtze River". Hence, we set the five sailing areas to quickly obtain each route and mark six training labels, as shown in Figure 4.

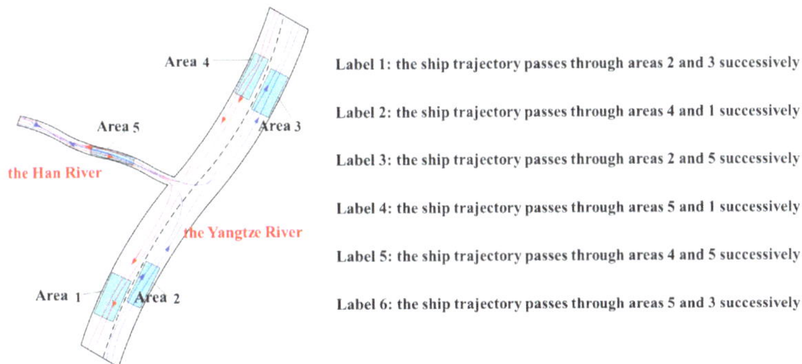

Figure 4. Schematic diagram of sailing area division and route label marking.

2.3. Ship Trajectory Prediction

The proposed method for ship trajectory prediction by SKNN is described in this section to predict which route the ship will sail on according to the current trajectory and determine appropriate collision avoidance measures that can be taken in time to avoid collision accidents. In this method, we improved KNN by considering the sensitivity.

2.3.1. K-Nearest Neighbor Algorithm Considering the Sensitivity (SKNN)

The K-Nearest Neighbor Algorithm is a common classification algorithm whose basic approach is to find the individual that differs the least from the predicted sample and consider the class of the predicted sample to be the same as the class of that individual. The Euclidean distance is often used as a distance metric function in KNN algorithms. The Euclidean distance assigns the same weight to the different characteristic quantities of the sample, as shown in Equation (8).

$$D(x, y) = \sqrt{\sum_{i-1}^{n} (x_i - y_i)^2} \tag{8}$$

where x is the test sample, and y is the training sample.

However, different feature quantities have different effects on the accuracy of the classification results in practical situations. For this reason, the sensitivity method is introduced to improve the KNN, and the improved Euclidean distance equation is as follows:

$$D(x, y) = \sqrt{\sum_{i-1}^{n} w_i (x_i - y_i)^2} \tag{9}$$

where w is the feature weight. Its calculation process is as follows:

(1) The test samples are classified using the conventional KNN algorithm and counting the number of misclassified samples as n.

(2) The samples are removed i (i = 1, 2, 3, 4, . . ., l) feature vectors at a time, and then the conventional KNN method is used to classify the test samples and count the number of misclassified samples as n_i.

(3) Stipulating when n_i = 0 or n = 0, u_i = 1; The larger the n_i, the greater the error of classification, and the greater the contribution of the i-th feature to classification; The smaller n_i is, the smaller the classification error is and the smaller the effect of the i-th feature is.

The weight coefficient of the i eigen weight is defined as:

$$w_i = \frac{u_i}{\sum\limits_{k=1}^{l} u_k}, i = 1, 2, \ldots, l \tag{10}$$

where the fulfillment of conditions is $\sum\limits_{i=1}^{l} u_i = 1$.

2.3.2. Data Training of Ship Trajectory Prediction

In our method, the labels are regarded as the prediction results, and the sub-trajectories of all ships are taken as samples. Then, the k sub-trajectories with the closest distance to the current ship's trajectory data are found, and the labels of these k sub-trajectories are counted, and the highest number of labels is the final prediction result. The conventional KNN algorithm assigns equal weights to the labels of all k sub-trajectories, i.e., the weights of all the neighboring points are equal. However, in instances of sample imbalance, where one class has a significantly larger number of samples compared to others, there is a high probability that the classification model will be skewed towards the majority class. This bias can lead to misclassification errors for test samples that belong to the minority classes. For this reason, this paper uses the method of weights to improve this problem. By using the reciprocal of the distance as the weight of the label, the neighbors with a small distance from the sample have large weights, while the neighbors with a large distance from the sample have relatively small weights; thus, the factor of distance is also taken into account to avoid misclassification due to too large a sample.

To avoid the effect of the difference in the data magnitude of longitude, latitude, speed, and heading on the model results, this paper uses Min-Max scaling to normalize the data, as shown in Equation (11).

$$y_i = \frac{x_i - \min\{x_j\}}{\max\{x_j\} - \min\{x_j\}} \tag{11}$$

where, $1 \leq i \leq$ n, $1 \leq j \leq n$, $\max\{x_j\}$ is the maximum value of the sample data, $\min\{x_j\}$ is the minimum value of the sample data, and the transformed data are all within [0, 1].

In addition, this paper uses the data structure of the k dimension tree algorithm (kd-tree) to save the training data; using the kd-tree for nearest neighbor search reduces the number of calculation times. The main steps are as follows:

(1) Find the leaf node containing x in the k-d tree.

(2) Take this leaf node as the current nearest point and calculate the distance from the current nearest point to the target point, which is D.

(3) Recursive upward backtracking, with the following operation at each node: If the distance D_{cur} from the instance point is saved by that node to the target point, and if $D_{cur} < D$, the current node is taken as the current nearest point. Check whether the region corresponding to another child node of the parent of this child node intersects the hypersphere with the target point x as the center of the sphere and D as the radius; (i) intersect: there may be a point closer to the target point in the region corresponding to

the other child node, move to the other child node and recursively perform the nearest neighbor search; (ii) do not intersect, backtrack upwards.

(4) When it returns to the root node, the search ends, and the current nearest neighbor is the X' nearest neighbor.

3. Results and Discussion

3.1. Data Preprocessing Results

The trajectories of the ship with MMSI 413210950 are analyzed before and after restoration for a total of 1000 s from 21/10/2020 14:41:00 to 21/10/2020 14:57:50, comparing linear interpolation, quadratic interpolation, and segmented triple Hermite interpolation methods for the cleaned trajectories in five aspects: time, longitude, latitude, speed, and track direction. The restoration effect is considered below. The red circles in Figure 3 are the original trajectory points; it can be seen from the figure that the original AIS trajectory data have a lot of data loss and the data sending period is unstable.

Figure 5 shows the restoration results of longitude (lon), latitude (lat), velocity (SOG), and heading (COG). Figure 5a shows that the three interpolation methods have comparable longitude restoration effects when the data are missing at short distances, while the secondary interpolation and segmented triple Hermite interpolation methods have closer restoration effects when the data are missing at long distances. Figure 5b shows that the latitude restoration results of the three interpolation methods are comparable at both long and short distances. Figure 5c,d show that the speed and heading fixes of the three interpolation methods are comparable in the case of missing data at short distances, while the secondary interpolation is poor in the case of missing data at long distances, and the linear interpolation and segmented three-time Hermite interpolation methods are close to each other. In summary, the three interpolation methods are comparable when the data are missing at short distances, the segmented three-times Hermite interpolation method is better when the data are missing at long distances, while the original trajectory characteristics are well maintained. Therefore, this paper selects the segmented three-times Hermite interpolation method to repair the trajectory.

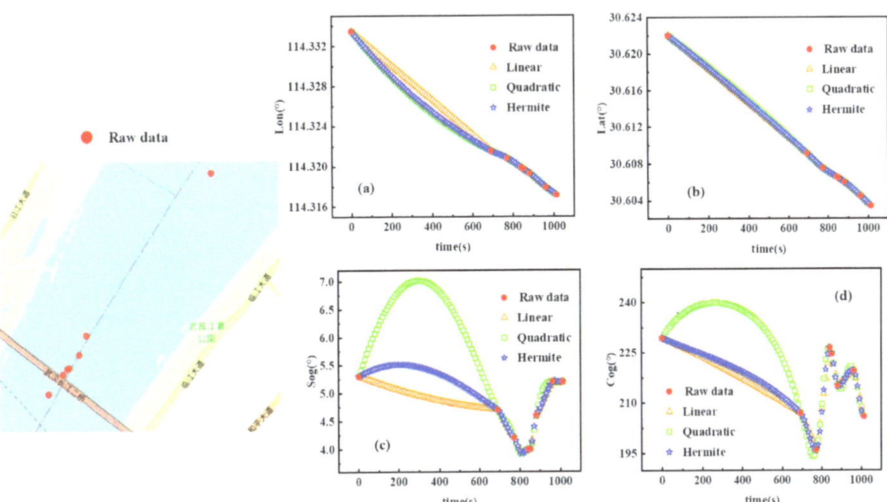

Figure 5. Schematic diagram of sailing area division and route label marking (**a**: longitude; **b**: latitude; **c**: speed; **d**: heading).

3.2. Analysis Effect of Clustering

To verify the effectiveness of the above clustering method, this paper clusters the cleaned AIS data. The clustering results are shown in Figure 6. As can be seen from the figure, the clustering method proposed in this paper can cluster the ship routes well.

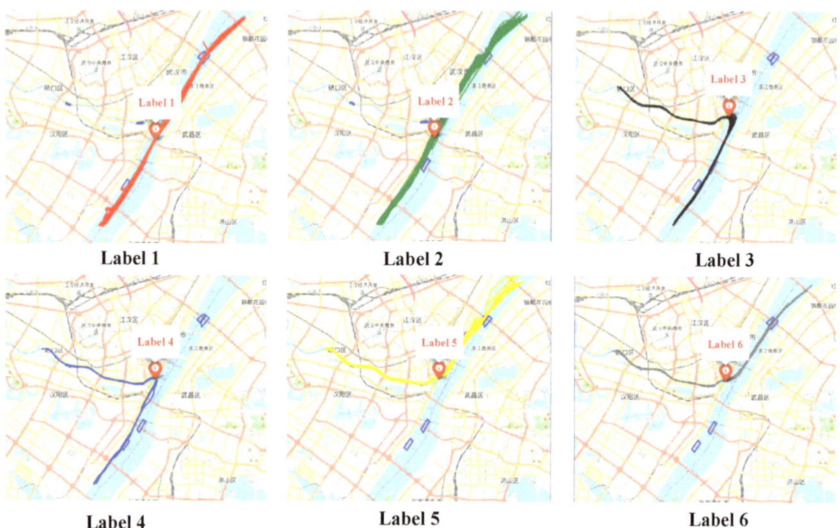

Figure 6. The result of ship trajectories clustering based on AIS data.

To gain a deeper understanding of the navigational patterns of vessels within inland waters, this study employs the speed and heading information from AIS data corresponding to label-1 and label-2 as indicative features of vessel behavior. This analysis aims to assess the impact of both upstream and downstream conditions on these navigational behaviors.

The Figures 7a and 8a show the ship heading distribution of this trajectory class, and Figures 7b and 8b shows the ship speed distribution of this trajectory class. As can be seen from the graph, the heading of upbound vessels is concentrated between 144°~288° and the speed is concentrated between 2 m/s~6 m/s, while the heading of downbound vessels is concentrated between 0°~72° and the speed is concentrated between 6 m/s~10 m/s. The heading and speed of the two labels have obvious regional characteristics. In addition, the vessel speed of tag 2 is significantly greater than that of tag 1. This is because the inner speed of river vessels is affected by the current; downstream vessels (vessels in tag 2) need to travel against the current and their speed is affected by the current. The impact of the ship is weakened; in contrast, the upstream ship (the ship in label-1) sails with the current, and the speed is equal to the sum of the ship's speed and the current speed, which is necessarily relatively faster. It can be seen that the behavioral pattern recognition under the trajectory clustering algorithm based on this paper can effectively explore the navigation pattern of ships, thus helping maritime managers to perceive the traffic form in the waterway.

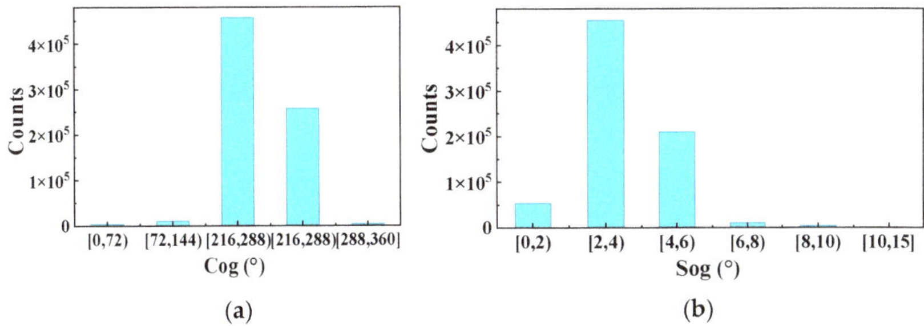

Figure 7. Frequency distribution of heading (**a**) and speed (**b**) of label-1.

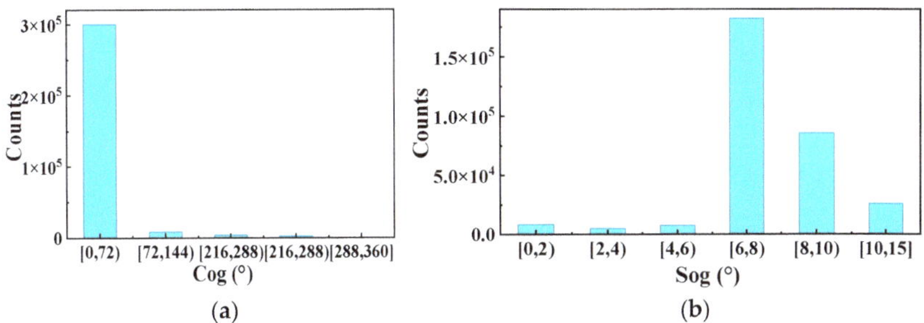

Figure 8. Frequency distribution of heading (**a**) and speed (**b**) of label-2.

3.3. Prediction Accuracy

3.3.1. Subtract the Length

In our method, 95% of the ship sub-trajectories from label 1–6 are taken as training sets and 5% as test sets, respectively. The length of the current trajectory of the ship has an important influence on the route prediction. The ship may predict different results at different stages, and if the ship has sailed only a short distance, the route prediction result may be very inaccurate and difficult to predict by the current trajectory; however, if the ship has sailed a relatively long distance, the prediction result is more accurate. For this reason, sub-trajectories with different trajectory lengths are used as input quantities.

Figure 9 shows that the accuracy of the SKNN, the NB (Naive Bayesian algorithm), and the KNN gradually increase with increase in the sub-trajectory length. Under each sub-trajectory segment, the accuracy of the SKNN is the highest, the KNN is the second, and the NB is the worst, which indicates that the SKNN is effective.

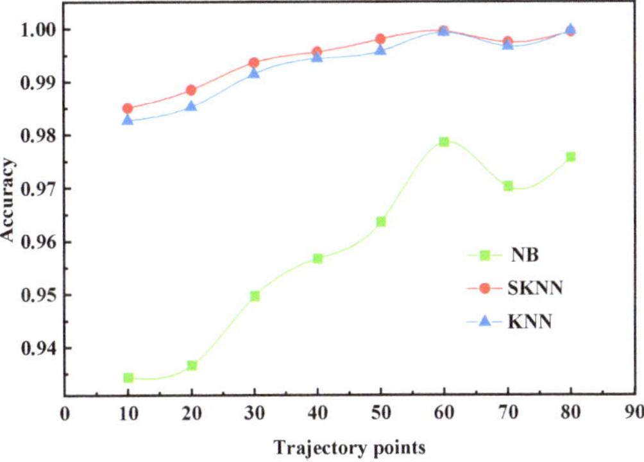

Figure 9. The effect of trajectory with test sample length.

Figure 10 shows that the three methods can achieve high accuracy when the input trajectory length n is 60 trajectory points. For this reason, 60 trajectory points are selected as the input quantity in this paper, and the feature weights of longitude, latitude, speed, and heading are [0.17, 0.23, 0.3, 0.3].

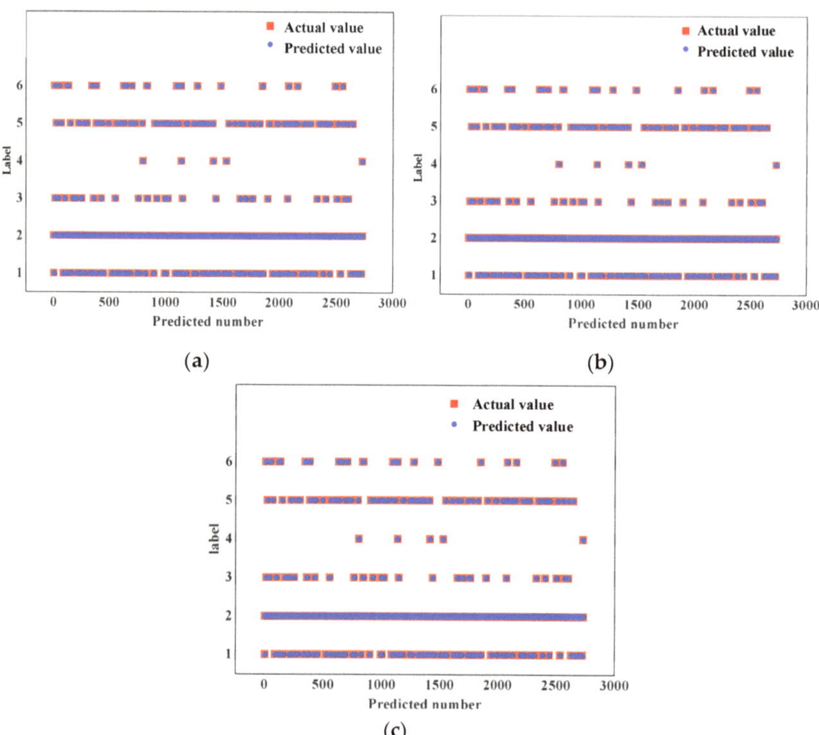

Figure 10. Comparison of accuracy from different algorithms. (**a**) Accuracy of NB = 0.9762; (**b**) accuracy of KNN = 0.9967; (**c**) accuracy of SKNN = 0.9996.

3.3.2. Optimal k-Value for Comparing SKNN and KNN

Selecting an appropriate value of k is crucial for the effectiveness of the KNN. If the value of k is too small, the model has a high squared error problem, leading to overfitting; If the value of k is too large, there is a high bias problem in the model, resulting in underfitting. The k-value can be adjusted based on the distribution of data points in the feature space. If the data points are relatively dense in a certain area, a smaller k-value can be used to avoid overfitting; And in areas with sparse data points, larger k-values are used to increase the model's generalization ability.

Figure 11 shows that when k takes the same value, the accuracy of SKNN is less than or equal to that of KNN, and the accuracy of both SKNN and KNN decreases when k increases. When k equals 6, the accuracy of SKNN and KNN is the same, which provides a reasonable benchmark for evaluating the effectiveness of SKNN relative to KNN. To further compare the accuracy of the two algorithms under different test samples, the value of k was selected as 6 for subsequent calculations. Therefore, choosing a k-value of 6 helps to fairly compare the performance of the two algorithms under different testing conditions.

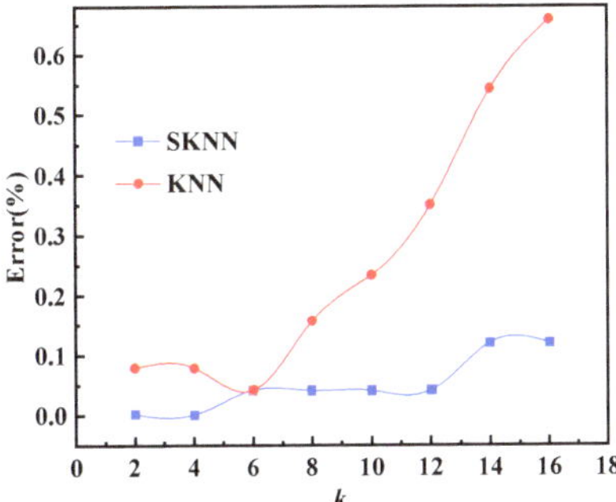

Figure 11. Influence of *k*-value selection on accuracy.

3.3.3. Number of Training Samples

The percentage of test samples is gradually increased from 5% to 30%, with each growth step being 5%. How number of test samples relates to the error rate at the time is shown in the figure below.

Figure 12 shows that the error of both SKNN and KNN gradually increases with increase in the training samples. The error of the SKNN algorithm is smaller than that of the KNN algorithm for different numbers of test samples, and the advantage becomes more obvious as the number of test samples increases. Due to its weighting mechanism, the SKNN can better adapt to this change and maintain a lower error rate. In contrast, the KNN is affected by noise and outliers when the data features change significantly, resulting in an increase in error due to treating all neighbors equally. Therefore, the SKNN is more stable and more accurate than the KNN algorithm with increase in the training sample proportion under the condition that the k-value is determined. The SKNN algorithm can maintain more than 99% accuracy when the percentage of test samples is below 40%.

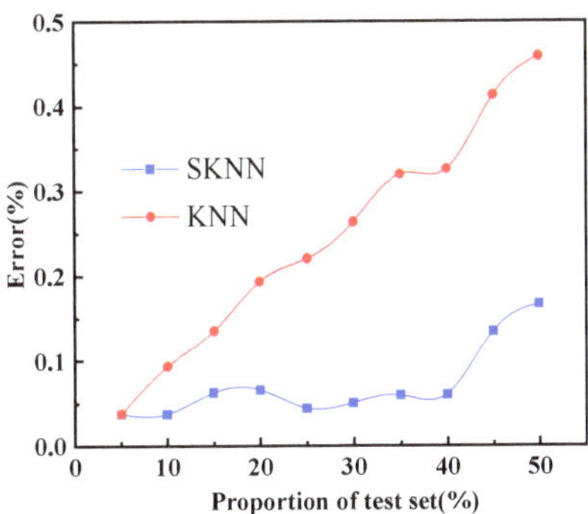

Figure 12. Effect of the number of training samples on algorithm accuracy.

Table 2 shows that SKNN performs the best. The macro avg does not consider the number of samples in a category and assigns the same weight to all categories. The performance metric SKNN > KNN > NB using the macro avg indicates that the SKNN has a very balanced performance across various categories, while the NB performs poorly on small categories.

The weighted avg considers the number of samples in a category, assigning greater weight to categories with larger sample sizes. The performance metric SKNN > NB > KNN using the weighted avg indicates that the SKNN has good adaptability to imbalanced datasets. Using the weighted avg, the precision, recall, and F1-score of all algorithms improved due to the large sample size of certain categories in the dataset, which have better performance and thus improve the overall performance indicators. The weighted avg performance index of the NB significantly improves, indicating that the NB may perform better in categories with a larger sample size.

Table 2. Comparison of performance metrics of SKNN, KNN, and NB.

	Performance Metrics	SKNN	KNN	NB
Macro avg	precision	0.99	0.93	0.62
	recall	0.99	0.87	0.72
	F1-score	0.99	0.9	0.64
Weighted avg	precision	0.99	0.93	0.96
	recall	0.99	0.92	0.94
	F1-score	0.99	0.92	0.95

In short, the SKNN has good adaptability to imbalanced datasets; the KNN performs better when dealing with categories with a larger sample size; the NB algorithm performs better in categories with larger sample sizes.

3.4. Discussion

In general, the length of the navigation trajectory for training has a significant impact on trajectory prediction. Compared to the conventional KNN and NB, the proposed SKNN has better generalization ability on new data, especially when there are smaller lengths of the navigation trajectories in the predicted samples, so, the prediction effect of the ship navigation trajectory is better. Although there is only a slight improvement in accuracy, this reflects the model's more stable performance on unknown data.

Based on the above research results, the proposed SKNN improves the limitations of conventional KNN on training large datasets by enhancing the sensitivity of close-range points, thereby reducing time complexity. However, the SKNN algorithm still faces challenges in real-time prediction when fast response is required.

The proposed SKNN still has sensitivity to data quality and distribution. If the dataset contains noise or outliers, it may affect the accuracy of the prediction results. Although the training effect of SKNN is better than that of KNN, the influence of the selection of k still exists, and improper selection of k may lead to overfitting or underfitting.

Therefore, further research and improvement are needed in the field of ship trajectory prediction by SKNN and support is needed for research in areas such as ship navigation efficiency, energy management, and navigation path planning.

4. Conclusions

A method to support ship trajectory prediction using historical AIS data is established to address the problem of waterway transportation system safety in the event of ship encounters induced by waterway confluence or channel limitation, which includes three parts: ship trajectory data preprocessing, ship trajectory data rule-based clustering, and ship trajectory prediction by SKNN. The research conclusions for each part are as follows:

(1) For the preprocessing of AIS data, the segmented cubic Hermite interpolation method is more suitable for repairing the AIS data of ships, especially when long-distance data are missing.

(2) Aiming at the characteristics of the waterway confluence, a rule-based clustering method is proposed to cluster the AIS data from the converging waters. The results show that the clustering algorithm in our method can effectively determine the ship's navigation law and ship heading distribution.

(3) To predict ship trajectory, an improved K-Nearest Neighbor Algorithm considering the sensitivity is proposed. Based on the clustered ship trajectory data, our KKNS, the conventional KNN, and the Naive Bayes algorithm are used to compare the effect of ship trajectory prediction. The results show that under various input trajectory lengths, the

accuracy of our SKNN is higher than that of the KNN and the Naive Bayes algorithm. In the case of different k-values and proportion of test set samples, the accuracy of our SKNN algorithm is better, being above 99%. Furthermore, the SKNN has good adaptability to imbalanced datasets.

The research conclusions can assist maritime management departments in formulating traffic management policies and providing technical support for the identification of abnormal ship trajectories and the analysis of ship navigation risks.

Author Contributions: Conceptualization, Z.W.; methodology, Z.W. and J.L. (Jiafen Lan); software, J.L. (Jiafen Lan) and C.Z.; validation, Z.W. and C.Z.; formal analysis, Z.W. and C.Z.; resources, Z.W. and X.L.; data curation, W.H.; writing—original draft preparation, J.L. (Jiafen Lan); writing—review and editing, Z.W. and C.Z.; visualization, J.L. (Jiafen Lan); supervision, J.L. (Jinyu Lei); project administration, W.H. and J.L. (Jinyu Lei); funding acquisition, W.H. and Z.W. All authors have read and agreed to the published version of the manuscript.

Funding: This research was funded by the National Natural Science Foundation of China (No. 52172327); the Fujian Marine Economic Development Special Fund Project (No. FJHJF-L-2022-17); the Fujian Science and Technology Major Special Project (No. 2022NZ033023); the Science and Technology Key Project of Fuzhou (No. 2022-ZD-021) and the Fuzhou Marine Research Institute's "Talent Recruitment for Project Leaders" Science and Technology Project (No. 2024F04); and the Scientific Research Foundation for the Ph.D., Minjiang University (No. MJY19032).

Institutional Review Board Statement: Not applicable.

Informed Consent Statement: Not applicable.

Data Availability Statement: Data are contained within the article.

Conflicts of Interest: The authors declare no conflicts of interest.

References

1. Kim, K.; Lee, D.; Essa, I. Gaussian process regression flow for analysis of motion trajectories. In Proceedings of the IEEE International Conference on Computer Vision, Barcelona, Spain, 6–13 November 2011.
2. Qi, L. Ship Encounter Intention Identification and Navigation Aid Application in Intersection Waters. Master's Thesis, Wuhan University of Technology, Wuhan, China, 2020.
3. Ma, J.; Li, W.K.; Zhang, C.W.; Zhang, Y. Ship encounter situation recognition by processing AIS data from traffic intersection waters. *Navig. China* **2021**, *44*, 7.
4. Cho, Y.; Han, J.; Kim, J. Intent inference of ship maneuvering for automatic ship collision avoidance. *IFAC* **2018**, *51*, 384–388. [CrossRef]
5. Cho, Y.; Kim, J.; Kim, J. Intent Inference of Ship Collision Avoidance Behavior Under Maritime Traffic Rules. *IEEE Access* **2021**, *9*, 5598–5608. [CrossRef]
6. Luo, Y.H. Ship Trajectory Prediction Based on AIS Data. Master's Thesis, South China University of Technology, Guangzhou, China, 2017.
7. Gao, D.W.; Zhu, Y.S.; Zhang, J.F.; He, Y.K.; Yan, K.; Yan, B.R. A novel MP-LSTM method for ship trajectory prediction based on AIS data. *Ocean Eng.* **2021**, *228*, 108956.1–108956.16. [CrossRef]
8. Chen, Y.C. Research on Ship Trajectory Prediction Based on Data Mining. Harbin Engineering University, 2020. Master's Thesis, South China University of Technology, Harbin, China, 2020.
9. Wang, W.G.; Chu, X.M.; Jiang, Z.L.; Liu, L. Classification of Ship Trajectory Based on the Weighted Naive Bayes Algorithm. *Navig. China* **2020**, *43*, 20–25.
10. Karata, G.B.; Karagoz, P.; Ayran, O. Trajectory pattern extraction and anomaly detection for maritime vessels. *Internet Things* **2021**, *2021*, 100436. [CrossRef]
11. Zhang, L.Y.; Zhu, Y.A.; Lu, W.; Wen, J.; Cui, J.Y. A detection and restoration approach for vessel trajectory anomalies based on AIS. *J. Northwestern Polytech. Univ.* **2021**, *39*, 7. [CrossRef]
12. Lee, J.G.; Han, J.; Whang, K.Y. Trajectory clustering: A partition-and-group framework. In Proceedings of the 2007 ACM SIGOD International Conference on Management of Data, Beijing, China, 20–25 June 2007.
13. Rong, H.; Teixeira, A.P.; Soares, C.G. Data mining approach to shipping route characterization and anomaly detection based on ais Data. *Ocean Eng.* **2020**, *198*, 106936. [CrossRef]
14. Gudmundsson, J.; Valladares, N. A GPU approach to subtrajectory clustering using the fréchet distance. *IEEE Trans. Parallel Distrib. Syst.* **2015**, *26*, 924–937. [CrossRef]
15. Perera, L.P.; Oliveira, P.; Soares, C.G. Maritime traffic monitoring based on vessel detection, tracking, state estimation, and trajectory prediction. *IEEE Trans. Intell. Transp. Syst.* **2012**, *13*, 1188–1200. [CrossRef]

16. Hu, Y.D.; Gao, C.S.; Li, J.L.; Jing, W.X.; Li, Z. Novel trajectory prediction algorithms for hypersonic gliding vehicles based on maneuver mode on-line identification and intent inference. *Meas. Sci. Technol.* **2021**, *32*, 115012. [CrossRef]
17. Murray, B.; Perera, L.P. Ship behavior prediction via trajectory extraction-based clustering for maritime situation awareness. *J. Ocean Eng. Sci.* **2021**, *7*, 1–13. [CrossRef]
18. Wang, L.X. Research on Ship Adaptive Trajectory Prediction and Application Based on GPR Model. Master's Thesis, Wuhan University of Technology, Wuhan, China, 2019.
19. Qian, L.; Zheng, Y.Z.; Li, L.; Ma, Y.; Zhou, C.H.; Zhang, D.F. A new method of inland water ship trajectory prediction based on long short-term memory network optimized by genetic algorithm. *Appl. Sci.* **2022**, *12*, 4073. [CrossRef]
20. Tian, X.; Suo, Y. Research on Ship Trajectory Prediction Method Based on Difference Long Short-Term Memory. *J. Mar. Sci. Eng.* **2023**, *11*, 1731. [CrossRef]
21. Gan, S.J.; Liang, S.; Li, K.; Deng, J. Ship trajectory prediction for intelligent traffic management using clustering and ANN. In Proceedings of the 2016 UKACC 11th International Conference on Control (CONTROL), Belfast, UK, 31 August–2 September 2016; pp. 1–6.
22. Mehta, N. Ship Trajectory Prediction in Confined Waters. Master's Thesis, Norwegian University of Science and Technology, Trondheim, Norway, 2023.
23. Suo, Y.F.; Chen, W.K.; Claramunt, C.; Yang, S.H. A ship trajectory prediction framework based on a recurrent neural network. *Sensors* **2020**, *20*, 5133. [CrossRef]
24. Zhao, J.S.; Yan, Z.W.; Zhou, Z.Z.; Chen, X.Q.; Wu, B.; Wang, S.Z. A ship trajectory prediction method based on GAT and LSTM. *Ocean. Eng.* **2023**, *289*, 116159. [CrossRef]
25. Li, H.H.; Jiao, H.; Yang, Z.L. Ship trajectory prediction based on machine learning and deep learning: A systematic review and methods analysis. *Eng. Appl. Artif. Intell.* **2023**, *126*, 107062. [CrossRef]
26. Li, H.H.; Jiao, H.; Yang, Z.L. AIS data-driven ship trajectory prediction modelling and analysis based on machine learning and deep learning methods. *Transp. Res. Part E* **2023**, *175*, 103152. [CrossRef]
27. Zhou, Y.; Dong, Z.; Bao, X. A Ship Trajectory Prediction Method Based on an Optuna–BILSTM Model. *Appl. Sci.* **2024**, *14*, 3719. [CrossRef]
28. Li, M.; Li, B.; Qi, Z.; Li, J.; Wu, J. Enhancing Maritime Navigational Safety: Ship Trajectory Prediction Using ACoAtt–LSTM and AIS Data. *ISPRS Int. J. Geo-Inf.* **2024**, *13*, 85. [CrossRef]
29. Jiang, J.; Zuo, Y. Prediction of Ship Trajectory in Nearby Port Waters Based on Attention Mechanism Model. *Sustainability* **2023**, *15*, 7435. [CrossRef]
30. Kasyk, L. Intensity of vessel traffic after crossing a waterway intersection. *Reliab. Eng. Syst. Saf.* **2007**, *3*, 2705–2708.
31. Li, B.; Pang, F.W. An approach of vessel collision risk assessment based on the D-S evidence theory. *Ocean Eng.* **2013**, *74*, 16–21. [CrossRef]
32. Maria, R.; Giuliana, P.; Michele, V. Maritime anomaly detection: A review. *Wiley Interdiscip. Rev. Data Min. Knowl. Discov.* **2018**, *8*, e1266.
33. He, Y.X.; Jin, Y.; Huang, L.W.; Xiong, Y.; Chen, P.F.; Mou, J.M. Quantitative analysis of COLREG rules and seamanship for autonomous collision avoidance at open sea. *Ocean Eng.* **2017**, *140*, 281–291. [CrossRef]
34. Chen, X.; Liu, S.; Zhao, J.; Wu, H.; Xian, J.; Montewka, J. Autonomous port management based AGV path planning and optimization via an ensemble reinforcement learning framework. *Ocean Coast. Manag.* **2024**, *251*, 107087. [CrossRef]
35. Chen, X.; Lv, S.; Shang, W.l.; Wu, H.; Xian, J.; Song, C. Ship energy consumption analysis and carbon emission exploitation via spatial-temporal maritime data. *Appl. Energy* **2024**, *360*, 122886. [CrossRef]

Journal of
*Marine Science
and Engineering*

Article

Marine Radar Oil Spill Detection Method Based on YOLOv8 and SA_PSO

Jin Xu [1,2,3,4,5,†], Yuanyuan Huang [1,2], Haihui Dong [1,2,*], Lilin Chu [1,2,3,4,†], Yuqiang Yang [1,3,4], Zheng Li [1,2,3,4], Sihan Qian [1,2], Min Cheng [1,2], Bo Li [1,2,3,4,5], Peng Liu [6] and Jianning Wu [2]

1   Shenzhen Institute of Guangdong Ocean University, Shenzhen 518116, China; jinxu@gdou.edu.cn (J.X.); huangyuanyuan@stu.gdou.edu.cn (Y.H.); 18861913315@stu.gdou.edu.cn (S.Q.); 13625632485@stu.gdou.edu.cn (M.C.); boli@gdou.edu.cn (B.L.)
2   Naval Architecture and Shipping College, Guangdong Ocean University, Zhanjiang 524091, China; sk.rt2@stu.gdou.edu.cn
3   Guangdong Provincial Key Laboratory of Intelligent Equipment for South China Sea Marine Ranching, Guangdong Ocean University, Zhanjiang 524088, China
4   Technical Research Center for Ship Intelligence and Safety Engineering of Guangdong Province, Zhanjiang 524088, China
5   Key Laboratory of Philosophy and Social Science in Hainan Province of Hainan Free Trade Port International Shipping Development and Property Digitization, Hainan Vocational University of Science and Technology, Haikou 570100, China
6   Navigation College, Dalian Maritime University, Dalian 116026, China; liupeng@dlmu.edu.cn
*   Correspondence: donghh@gdou.edu.cn; Tel.: +86-138-2827-8906
†   These authors contributed equally to this work.

Abstract: In the midst of a rapidly evolving economic landscape, the global demand for oil is steadily escalating. This increased demand has fueled marine extraction and maritime transportation of oil, resulting in a consequential and uneven surge in maritime oil spills. Characterized by their abrupt onset, rapid pollution dissemination, prolonged harm, and challenges in short-term containment, oil spill accidents pose significant economic and environmental threats. Consequently, it is imperative to adopt effective and reliable methods for timely detection of oil spills to minimize the damage inflicted by such incidents. Leveraging the YOLO deep learning network, this paper introduces a methodology for the automated detection of oil spill targets. The experimental data pre-processing incorporated denoise, grayscale modification, and contrast boost. Subsequently, realistic radar oil spill images were employed as extensive training samples in the YOLOv8 network model. The trained detection model demonstrated rapid and precise identification of valid oil spill regions. Ultimately, the oil films within the identified spill regions were extracted utilizing the simulated annealing particle swarm optimization (SA-PSO) algorithm. The proposed method for offshore oil spill survey presented here can offer immediate and valid data support for regular patrols and emergency reaction efforts.

Keywords: oil spill; marine radar; YOLO v8

Citation: Xu, J.; Huang, Y.; Dong, H.; Chu, L.; Yang, Y.; Li, Z.; Qian, S.; Cheng, M.; Li, B.; Liu, P.; et al. Marine Radar Oil Spill Detection Method Based on YOLOv8 and SA_PSO. *J. Mar. Sci. Eng.* **2024**, *12*, 1005. https://doi.org/10.3390/jmse12061005

Academic Editor: Merv Fingas

Received: 16 May 2024
Revised: 10 June 2024
Accepted: 13 June 2024
Published: 16 June 2024

1. Introduction

Oil is one of the most vital energy sources for humans, and as economic development advances, so does the worldwide demand for oil [1]. This has already led to larger-scale offshore oil extraction and transportation activities. Oil spills can result from ship collisions during transport or from explosions on extraction platform [2]. Following an oil spillage incident on the ocean apparent, bumper oil sheen will quickly drift and spread due to the influence of waves and sea breezes, wreaking havoc on the marine ecology and the surrounding natural environment. In addition to gravely harming the marine environment and causing the deaths of numerous marine animals, plants, birds, fish, and mammals, oil spills worldwide also pose a grave risk to human health [3]. The cleanup process for oil spill pollution is time-consuming and labor-intensive, requiring a substantial investment of financial, material, and human resources. Additionally, ocean regions contaminated by oil become unsuitable for tourism and fishing [4]. They also affect maritime traffic, complicating the transportation of shipping operations. A notable oil spill occurred at the Bohai *Penglai 19-3* oilfield on 4 June 2011, which was being jointly developed by ConocoPhillips and CNOOC. This spill impacted 6200 square kilometers of ocean region, with 870 square kilometers experiencing severe pollution [5]. On 27 April 2021, approximately 9419 tons of cargo oil leaked into the Yellow Sea due to a collision between the Panamanian-flagged cargo ship *Sea Justice* and the Liberian-flagged oil tanker *Symphony*. This incident contaminated 786.5 km of coastline and 4360 square kilometers of ocean regions [6].

The nation will suffer the least amount of damage if oil spill monitoring data can be accessed quickly for emergency action plans [7–14]. Marine radar can be mounted on cleanup vessels for oil spill monitoring, which is ideal for real-time assisted pollution control. This oil spill monitoring technology comprises three main aspects: image preprocessing, extraction of meaningful oil spill surveillance regions, and oil film separation [15,16]. Image preprocessing involves denoise, gray scale modification, and contrast boost. Methods for extracting effective oil spill monitoring regions are mainly divided into two kinds: gray-scale distribution matrix and texture feature classification. Oil film segmentation currently involves adaptive thresholding and machine learning classification methods. Oil spill radar image treatment is a tremendous question, apart from the above, and many scholars have conducted research on various aspects [17–20]

The application of deep learning in marine radar oil spill monitoring technology is infrequent. We employed the YOLOv8 model here for marine radar oil spill region detection. The experimental results illustrated that the YOLOv8 model can accurately and swiftly extract the effective oil spill monitoring regions. In this paper, the oil film segmentation is conducted by using the simulated annealing particle swarm optimization (SA_PSO) method. While the particle swarm optimization (PSO) method offers numerous advantages, including simple implementation with few parameters, straightforward convergence with strong robustness, and wide applicability, it does have a drawback: a tendency to settle into local optimal solutions. To address this limitation, the simulated annealing (SA) algorithm's competence to flee locally optimal solutions is incorporated into the PSO algorithm, facilitating faster escape from local optima while searching for the best solution. The experimental results corroborated the practicality of the SA_PSO methodology for oil film segmentation.

2. Materials and Methods

2.1. Materials

The analysis data were acquired by a marine radar transceiver, which was integrated with a computer system featuring a monitor that displayed the processed imagery, as shown in Figure 1. The short wavelength of the X-band radar produces finer echo signals, resulting in high resolution and the ability to detect targets at longer distances. Additionally, X-band radar has strong penetration capabilities, enabling it to penetrate rain and snow. The X-band marine radar was chosen as the platform for experimental data collection here.

With the rotation of the X-band horizontal polarization radar antenna, the system was capable of capturing and storing digital representations of the clutter signals. It could capture between 28 and 45 images in one minute. The radar measurement distance is adjustable through variations in pulse width. To enhance the image resolution of oil film targets, the experimental data detection range was specifically set to 0.75 nautical miles, resulting in images with a resolution of 1024 × 1024 pixels. The type of oil spills in the experimental data were crude oil spilled from oil tankers and storage tanks at oil terminals.

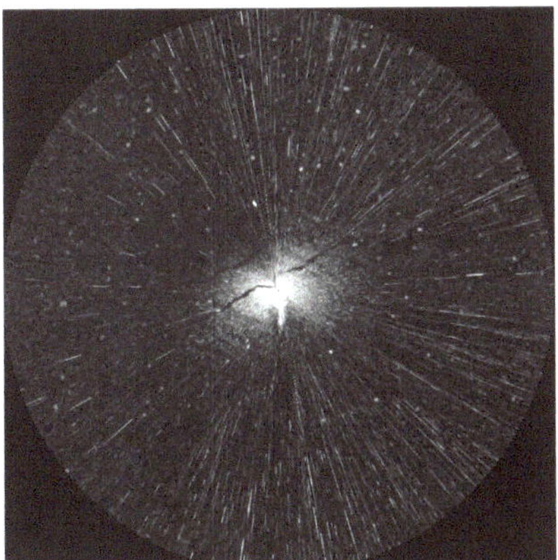

Figure 1. The experimental marine radar remote-sensing image in the polar coordinate system.

2.2. Experimental Process

The experimental process is shown in Figure 2. First, the original marine radar images were preprocessed. Afterwards, the preprocessed images were loaded into the YOLOv8 model for training to generate an oil film prediction model. Furthermore, the new preprocessed images were input into the prediction model to obtain the oil film detection result images. Then, the SA_PSO algorithm was used to preliminarily segment oil film targets. Finally, speckle noises were removed to obtain the final segmentation results.

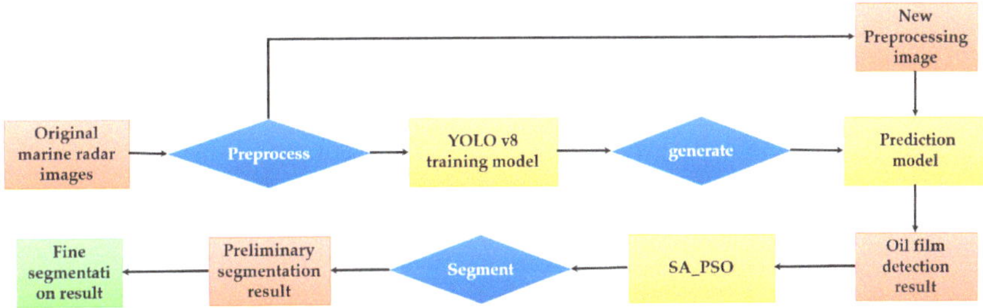

Figure 2. Experimental process.

2.3. Data Preprocessing

The data preprocessing process is shown in Figure 3. The detailed process includes:

a. The original image under the polar coordinate system was converted into Cartesian coordinate system, as shown in Figure 4a.

b. The data in Cartesian coordinate system were convolved with the row vectors of $[-1,-1,4,-1,-1]$. The co-frequency interference noise pixels in the Cartesian coordinate system exhibit bright features compared to adjacent horizontal pixels. Thus, the above row vector that highlights the central pixel was used to enhance co-frequency interference noises.

c. The co-frequency interferences were extracted according to the gray threshold. After this, the mean filter was accessed to smooth the co-frequency interference noises [15].

d. The smoothed co-frequency interference image was binarized again using the gray threshold [16].

e. The speckle noises were extracted by the pixel-quantity threshold.

f. The median filter of the 20×20 window was used to remove speckle noises, as shown in Figure 4b.

g. The noise reduction image was processed for gray correction, as shown in Figure 4c.

h. An overall grey contrast enhancement was applied to the image, as shown in Figure 4d.

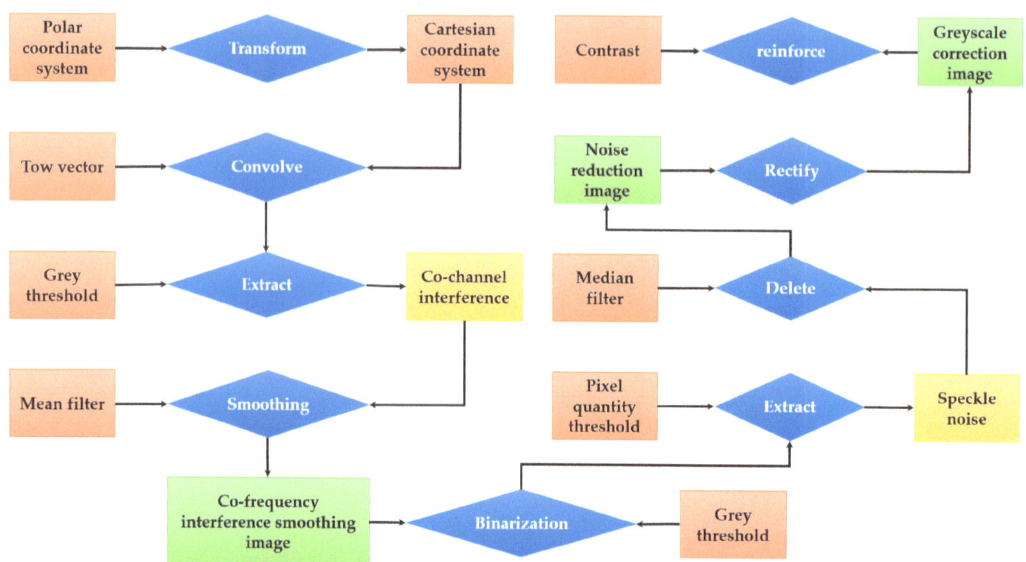

Figure 3. Data preprocessing scheme.

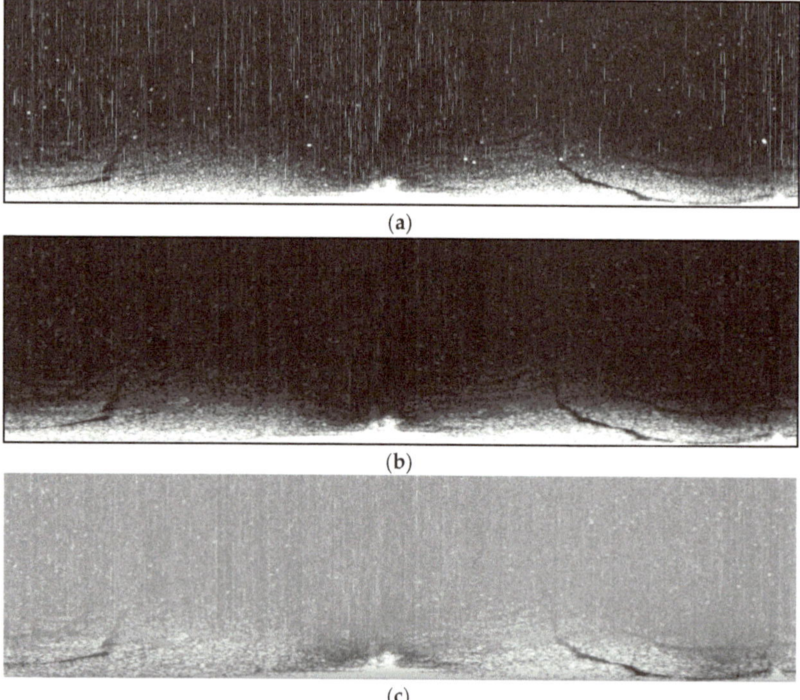

(a)

(b)

(c)

Figure 4. *Cont.*

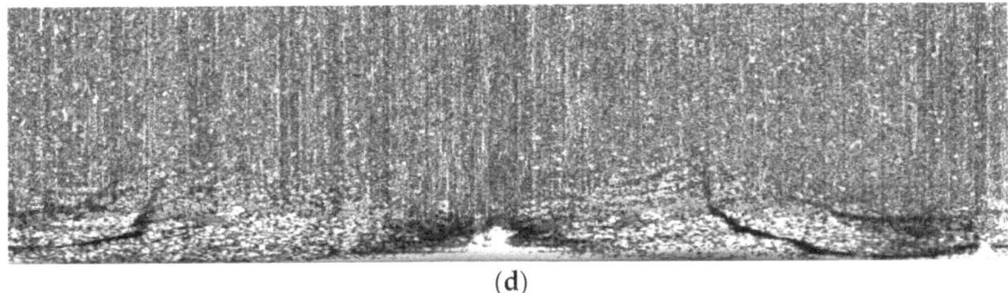

(d)

Figure 4. The preprocess: (**a**) Cartesian coordinate system conversion; (**b**) noise reduction; (**c**) gray correction; and (**d**) local contrast enhancement.

2.4. YOLO

Spaceborne and airborne oil spillage detection utilizing deep learning techniques has emerged as a dominant methodology [21]. Nevertheless, the implementation of deep learning methods for marine radar oil spillage detection remains limited. Deep learning technology has the capability to automatically extract oil film characteristics from remote sensing data, eliminating the need for intricate algorithmic designs. The development of a target detection network mainly focuses on two directions: two-stage algorithms and one-stage algorithms [22]. The main difference between them is that two-stage algorithms need to use a feature extractor to generate a series of pre-selected boxes that may contain the objects to be detected, then apply certain filtering rules to filter the pre-selected frames [23]. By contrast, the one-stage algorithm can extract features directly in the network to predict object classification and location. Because of the constraints related to computer memory usage and communication expenses, the two-step approach is less favored compared to the one-stage approach. The one-step approach is suitable for real-time dynamic monitoring due to its precision and low resource consumption, which can meet the significant demand for oil spill cleanup and management. Hence, the one-step YOLOv8 model was employed here for marine radar oil spillage detection.

The YOLOv8 structure consists of the following four primary components: Input, Backbone Network, Neck, and Output (Figure 5). The input side is responsible for receiving raw images and pre-processing them. In YOLOv8, the input side usually uses Mosaic data enhancement technique to stitch multiple images for increasing the diversity of the training data. In addition, preprocessing operations, such as adaptive image scaling and gray-scale filling, are performed to resize the image into the input size and format required by the network. The backbone network is the core part of YOLOv8 and is responsible for extracting features from the input images. In the backbone network, a series of convolutional layers, pooling layers, and other operations are usually employed to gradually extract feature information from the image. Common backbone network structures in YOLOv8 include Convolutional Layer (Conv), Connect to Fusion (C2F), and Spatial Pyramid Pooling (SPPF), which can effectively capture semantic and spatial information in the image. The Neck end is responsible for further processing and fusion of the features extracted from the backbone network. In YOLOv8, the Neck side is usually designed based on the Path Aggregation Network (PAN) structure, which fuses feature maps at different scales through operations such as up-sampling, down-sampling, and feature splicing, so as to improve the network's ability to detect and recognize targets. The output side is the last part of YOLOv8 and is responsible for generating detection results and outputting them. In the output side, a decoupled head structure is usually used to decouple the classification and regression processes to improve the efficiency and accuracy of the model. In addition, the output side includes steps, such as positive and negative sample matching and loss calculation, in order to evaluate and optimize the detection results. The YOLOv8 network uses the Task Aligned Assigner method to weight the classification scores and the regression scores. The loss calculation contains classification calculation and regression loss calculation. Binary

Cross-Entropy (BCE) is used to calculate classification loss and Complete Intersection over Union (CIoU) loss functions.

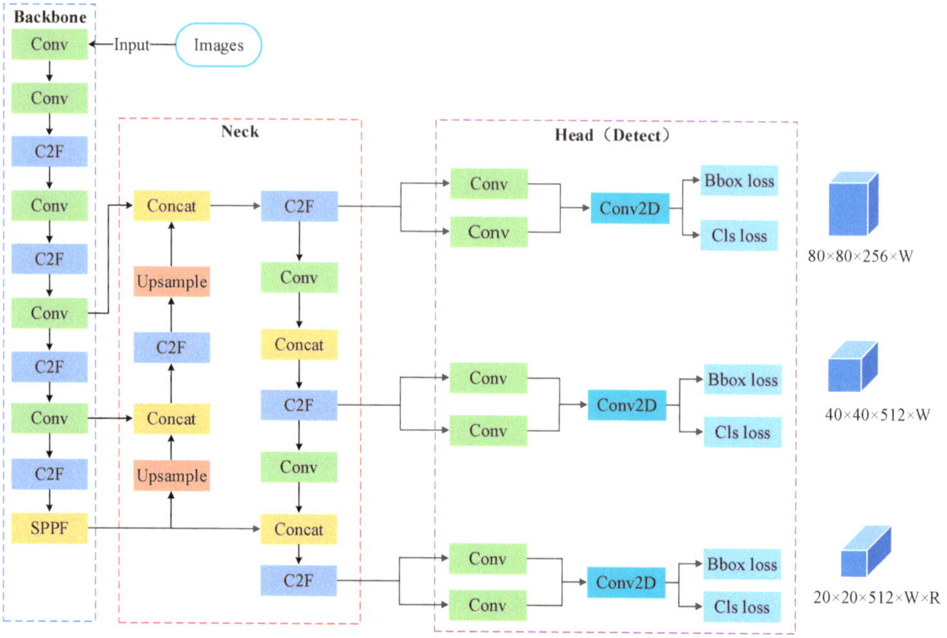

Figure 5. The YOLOv8 model architecture.

2.5. Sample Labeling Tool

LabelImg is an open-source image annotation tool designed specifically for creating bounding boxes and assigning labels to objects in images. Developed primarily in Python and utilizing the Qt framework for its Graphical User Interface (GUI), LabelImg provides a friendly operating platform for efficiently annotating large datasets for object detection and localization tasks. Bounding boxes around objects can be quickly and precisely drawn with LabelImg. The annotations in various formats, including the PASCAL VOC XML format, YOLO, and CreateML, can also be exported correspondingly. Labelimg was used here to mark oil film targets in marine radar images, as shown in Figure 6.

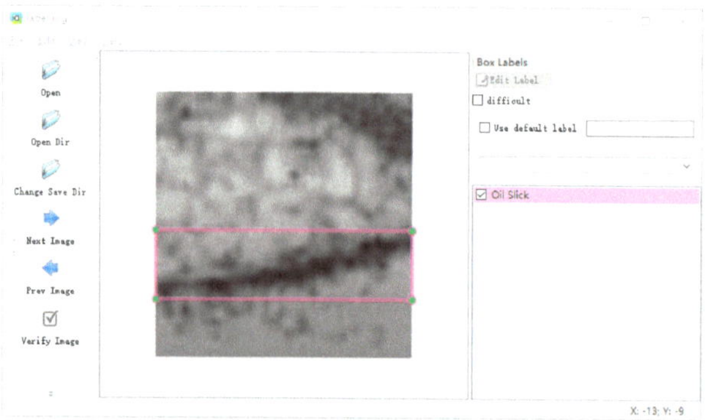

Figure 6. Sample labeling method.

2.6. Model Training Evaluation Indicators

The model training evaluation indicators include the Precision-Confidence Curve and the Recall-Confidence Curve.

a. The Precision-Confidence Curve

YOLO assigns a confidence score to each predicted bounding box, representing the model certainty. The Precision-Confidence Curve in YOLO illustrates the connection between the precision of predicted bounding boxes and their confidence scores. Precision in object detection refers to the percentage of true positive predictions among all positive predictions, indicating how accurately the model predicts the existence of an object inside a bounding box. As the confidence threshold increases, the model becomes more selective, predicting fewer bounding boxes. This may result in lower recall but higher precision. Conversely, decreasing the threshold leads to the model predicting more bounding boxes, potentially increasing precision due to more false positives but also increasing recall;

b. The Recall-Confidence Curve

The Recall-Confidence Curve captures the relationship between the model confidence in its predictions and recall, or ability to detect true objects. The recall gauges the percentage of real objects that the model correctly identifies. The Recall-Confidence Curve displays the recall value against the confidence threshold. As the confidence threshold rises, the model becomes more selective and only predicts bounding boxes with higher confidence scores. However, this may lead to certain real objects being missed due to lower confidence scores, resulting in a decrease in recall.

2.7. Particle Swarm Optimization

In PSO models, a group of particles, firstly randomly initialized without volume or mass, are considered as feasible solutions to an optimization problem [24]. Subsequently, the algorithm iteratively searches for the optimal solution. During each iteration, particles update themselves by tracking two extreme values: the best solution found by the particle itself, known as individual optimal solution $P_{best.i}$ and the second the best solution found by the entire population, known as the global best G_{best}. The particle performs the selection of velocity v and position x by the following:

$$v_i(k+1) = \omega v_i(k) + c_1 r_1 (P_{best.i}(k) - x_i(k)) + c_2 r_2 (G_{best} - x_i(k)) \tag{1}$$

$$x_i(k+1) = x_i(k) + v_i(k+1) \tag{2}$$

where ω is the inertia weight coefficient. c_1 and c_2 are used to control the influence of individual best position and global best position on particle movement. By adjusting the values of c_1 and c_2, a balance between local and global search can be achieved, thus enhancing the algorithm performance and convergence speed $v_i(k)$ and $x_i(k)$ denote the velocity and position of the particle at the kth iteration. r_1 and r_2 are random numbers uniformly distributed in [0, 1]. If any dimension of a particle position exceeds the specified upper bound (Upbound) or lower bound (Low bound), it is directly assigned the value of Upbound or Low bound, respectively. Due to the normalization of the grayscale values of the image, the Upbound was set to '1' and the Low bound was set to '0' here. The ω determines the tendency of particles to maintain their current velocity. A higher ω value makes particles more inclined to retain their previous velocities, which helps the algorithm move quickly across the search space to explore new regions. However, if ω is set too high, it can lead to instability in the search process, and the algorithm may even miss the global optimal solution. When $\omega = 1.1$, the PSO algorithm exhibits faster exploration ability through experimental comparisons. It enables the PSO algorithm to find candidate solutions closer to the global optimal solution more quickly, and a stable oil film threshold can be obtained.

2.8. Simulated Annealing

SA is a probability based global optimization algorithm that simulates the process of material annealing and cooling at high temperatures [25]. It is widely used in the field of signal processing [26]. The principle of the classic simulated annealing algorithm is as indicated below:

a. The beginning temperature T_0 is set for the initial state of the object.

b. Given E_g represents the internal energy of the current best point in the population, T_i represents the current temperature, and the energy of a new state i after iteration is E_i. If E_g is greater than E_i, the new state is accepted as the current state. Then $E_g = E_i$, Otherwise, the state is accepted with a certain probability determined by the acceptance probability formula p_i. The expression for p_i is as follows:

$$p_i = \begin{cases} 1 & E_i < E_g \\ \exp\left(-\frac{E_i - E_g}{T_g}\right) & E_i \geqslant E_g \end{cases} \tag{3}$$

c. If the new state is accepted, the new temperature decreases gradually as:

$$T_{t+1} = \alpha T_t \tag{4}$$

where α is the cooling attenuation factor, $0 < \alpha < 1$, T_{t+1} is the new temperature, T_t is the temperature of last state.

d. Back to step 2 and repeat the iteration process until the termination condition is met, such as several consecutive new solutions not being accepted, reaching the preset number of iterations, or temperature threshold, etc.

e. When the algorithm stops, the current solution E_g is output as the optimal solution.

2.9. Simulated Annealing Particle Swarm Optimization

As an efficient intelligent optimization algorithm, the PSO algorithm requires minimal parameter tuning, which is easily implementable. It avoids complex operations and can swiftly solve intricate optimization problems. However, when dealing with optimization problems featuring multiple local optima, the algorithm is prone to getting stuck in local optima, resulting in slow convergence. To address this issue, when particles in the population become trapped in local optima, efforts should be made to facilitate their escape from these local optima, thereby enhancing the diversity of the entire population. Considering that the SA algorithm can probabilistically accept suboptimal solutions during the optimization process, it can effectively prevent the algorithm from getting trapped in local optima during iterative searches. Therefore, we integrated the core principles of the SA into the PSO algorithm for finding the segmentation threshold between real and suspected oil spills, as shown in Figure 7.

a. Set an initial temperature T_0 and cooling attenuation factor α based on the initial state of particles in the population. The T_0 is a crucial parameter that determines the hotness of the algorithm during the search process. Typically, the initial temperature needs to be set high enough to ensure that the algorithm can escape from local optima during the initial search phase and prevent premature convergence. The value of '100', as a relatively large number, often satisfies this requirement. The value of α determines the rate at which the temperature decreases in each iteration step. A smaller value of α, such as '0.95', implies a relatively slower decrease in temperature, which allows the algorithm to explore the search space for a longer time, thereby increasing the possibility of finding the global optimal solution. So, the T_0 was set to '100', the α was set to '0.95' here.

b. Particle i computes its velocity v_i and position x_i through Formula (1).

c. Calculate the fitness $f(x_i)$ of the current position x_i based on the evaluation objective function.

d. If $f(x_i) < f(P_{best.i})$, then set $P_{best.i} = x_i$. If $f(x_i) > f(P_{best.i})$, utilizing the above-mentioned simulated annealing acceptance probability p_i to apply the Metropolis criterion [27]. If the probability p_i is greater than a random number within the range [0, 1], then the state $P_{best.i}$ is still accepted. If this condition is not met, then the velocity v_i and position x_i will be recomputed.

e. If $P_{best.i}$ is accepted as the new value, then $f(P_{best.i})$ and $f(G_{best})$ are compared according to the evaluation objective function. If $f(P_{best.i}) > f(G_{best})$, then $G_{best} = P_{best.i}$.

f. If the number of iterations reaches the maximum iteration J (The J was set to '100' here), G_{best} is determined as the threshold for the particle swarm optimization algorithm. If the condition is not met, proceed back to step (2) for further iterations.

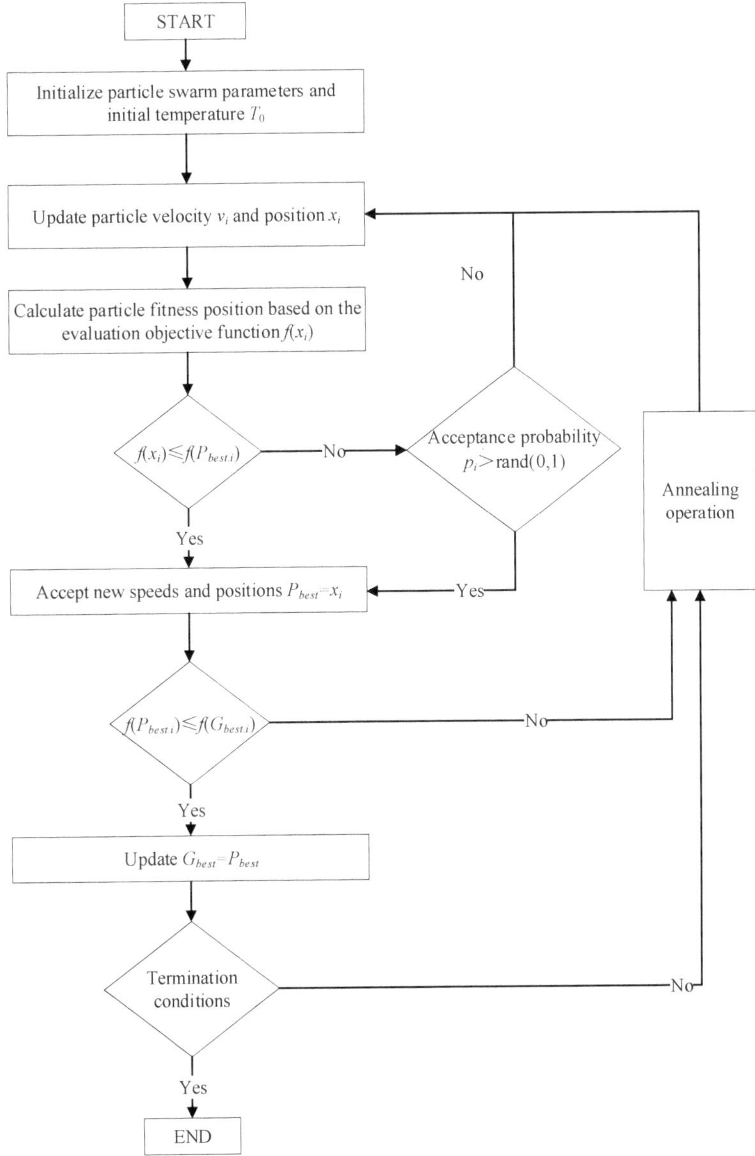

Figure 7. SA PSO algorithmic process.

Ultimately, based on SA_PSO method, dual thresholds were calculated iteratively for separating the real and the suspected oil films here, respectively.

3. Results

3.1. Model Training Curve Analysis

The Precision-Confidence Curve and the Recall-Confidence Curve are shown in Figure 8. As the confidence level increases, so does the accuracy of the model in detecting oil sheen. At a confidence threshold of '0.846', the precision rate stood at '100%', as reflected in Figure 8a. Due to insufficient sample size, the model may not be able to fully learn the distribution of data, resulting in unstable performance on validation or testing sets. This is the reason why the confidence curve oscillates between '0.6' and '0.8'. The collection of oil spill data from maritime radar is an ongoing work in the future. The model's ability to identify oil slicks improves as the single recall value increases. The model had a single recall of '0.92', indicating that the model exhibits high accuracy in oil identification and localization, as shown in Figure 8b.

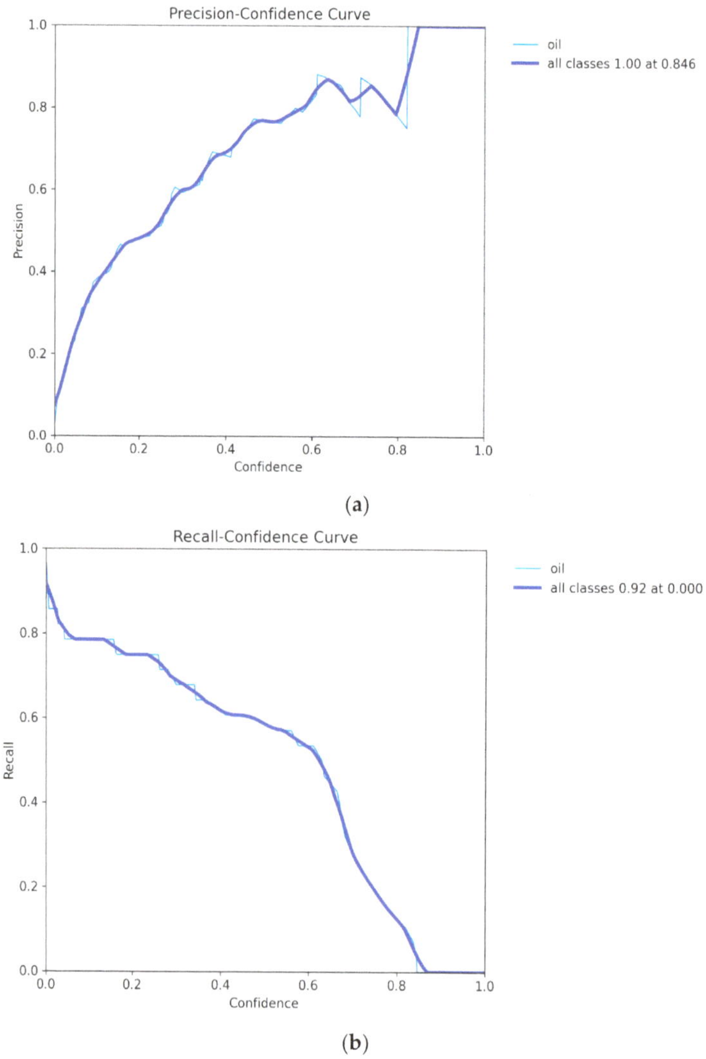

(a)

(b)

Figure 8. YOLOv8 model training curve: (**a**) the Precision-Confidence Curve; and (**b**) the Recall-Confidence Curve.

3.2. Oil Film Prediction

The YOLOv8n model, which was trained here, was employed to predict the oil slick targets within the newly preprocessed marine radar image and the results obtained were shown in Figure 9. The location and size of the effective oil spill monitoring area marked in red in Figure 9a were subsequently extracted in Figure 9b.

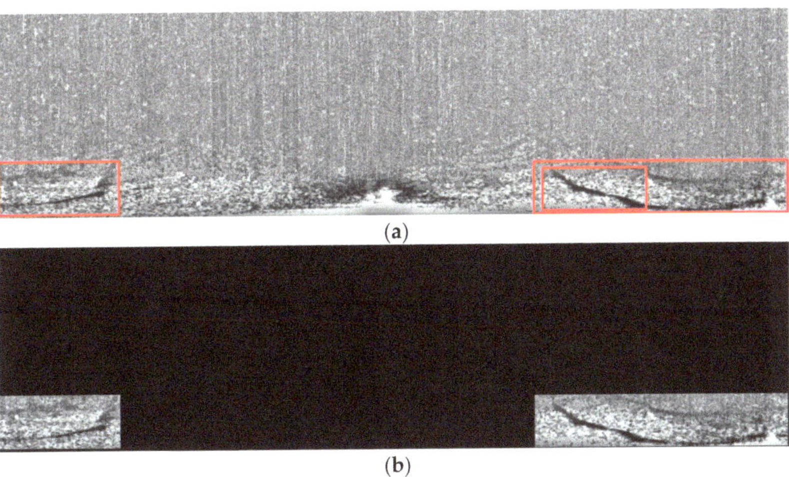

Figure 9. YOLOv8 model prediction results: (**a**) preliminary detection result; and (**b**) the oil film regions were preserved.

3.3. Oil Film Segmentation

The classification result was obtained by using SA_PSO, as shown in Figure 10a. Then, the true oil spills were preserved by removing the sparkles, as shown in Figure 10b. Finally, the oil film identification image was converted from the Cartesian coordinate to the Polar coordinate, as shown in Figure 10c.

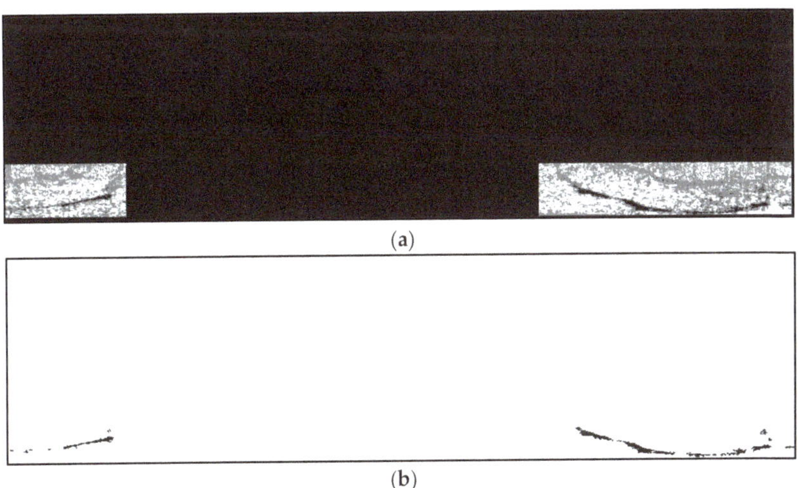

Figure 10. *Cont.*

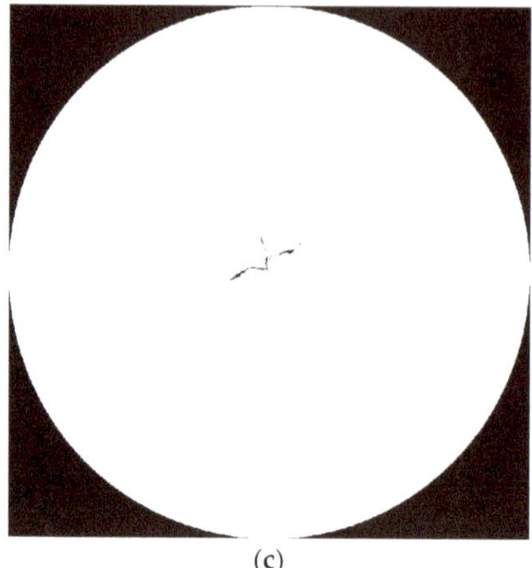

(c)

Figure 10. The oil spill segmentation results: (**a**) the segmentation result of SA_PSO; and (**b**) the real oil pill result; (**c**) The polar coordinates result.

4. Discussion

4.1. Comparison of Prediction Results of Different Training Models

There are five official versions of YOLOv8: 8n, 8s, 8m, 8l, and 8x. Since the 8l and 8x models were too large, only the 8n, 8s, 8m models were trained here for comparison. The three training models were used to predict oil spill targets on preprocessed marine radar images, as shown in Figure 11. The prediction times are displayed in Table 1. The average detection speed of the 8n model for each image was 24.9 ms, with fine prediction results. The effective oil spill monitoring regions were detected, while the ship wake region was not misidentified in Figure 9b. The 8s model exhibited an even detection tempo of 49.8 ms. The detection performance of the oil spill effective monitoring area was excellent, but the ship wake region was misidentified as the oil spill region in Figure 11a. The 8m model exhibited an even detection tempo of 49.8 ms. Although the ship wake region was not misidentified, some effective oil spill detection regions were missed (blue box), as shown in Figure 11b. Therefore, the YOLOv8n model was chosen for subsequent oil film segmentation work in this paper.

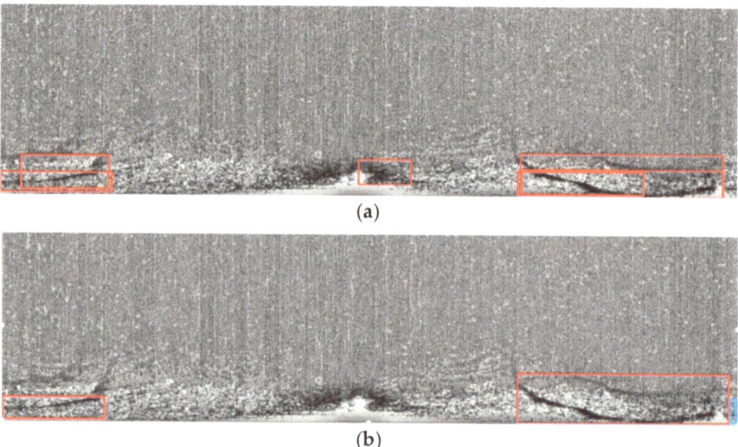

Figure 11. The prediction results of different training models in red color: (**a**) YOLOv8s model; and (**b**) YOLOv8l model.

Table 1. The prediction times of different training models.

Training Model	Forecast Time
YOLOv8n	24.9 ms
YOLOv8s	49.8 ms
YOLOv8m	109.6 ms

4.2. Comparison with the Prediction Result of YOLOv5n Model

The YOLOv5n model, after being trained, was utilized for predicting the oil spill targets in Figure 4d with a computational time of 1.3 ms, as shown in Figure 12. The YOLOv5n model detects the ship wake as oil spill targets which marked in blue box in the middle of Figure 12. The YOLOv8n model used here did not misidentify the ship wake region as oil spill targets in Figure 9a. The trained YOLOv5n model did not detect the small oil spill target in the left corner of Figure 12, whereas the method in this paper was more comprehensive.

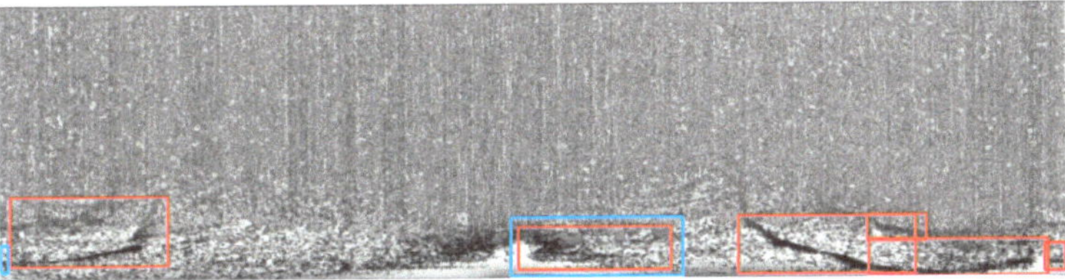

Figure 12. The YOLOv5n model prediction result in red color.

4.3. Comparison with U-Net Semantic Segmentation Network

The U-Net of PyTorch is a deep learning architecture for image semantic segmentation tasks. It combines encoder and decoder for semantic segmentation of high-resolution input images down to the pixel level. The encoder extracts the features through convolution and pooling operations while reducing the spatial resolution. By employing up-sampling and inverse convolution techniques, the decoder diminishes the eigenmaps output from the encoder to match the dimensions of the original input image. This section combines the eigenmaps from the respective encoder layers to produce the segmentation outcomes. The uniqueness of U-Net is that it introduces jump connections, which connect the feature maps of each layer in the encoder to the feature maps of the corresponding layer in the decoder, helping the network to better recover detailed information.

In reference [28], the number of U-Net training epochs was set to '200', the batch size was set to '1', the learning rate was set to '10^{-5}', the validation dataset accounted for 10%, the number of classes was set to '3', the prediction computing time was 2.56 s. The same parameter settings were used here for comparative discussion. The experimental results are shown in Figure 13.

The target detection network is faster than the semantic segmentation network, but it must be combined with other algorithms or models to segment the oil films. The semantic segmentation network directly assigns image pixels to different categories for achieving accurate segmentation of the oil spills at once. However, the semantic segmentation network has very high requirements for the sample labelling, and the suspected oil slicks are the most difficult to mark in the marine radar training images. Thus, many suspected oil slick segmentations were generated in green during our experiment. In terms of the number of real oil film pixels detected in red, due to Figure 13b identifying false positive targets in the ship wake region and sparkles as real oil films, many more real oil film image pixels were identified, as shown Table 2.

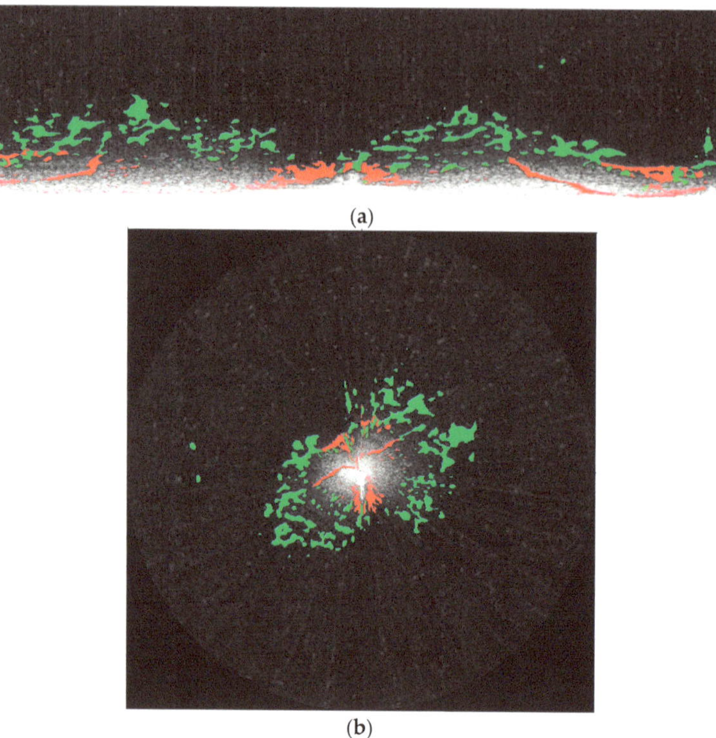

Figure 13. The U-Net segmentation results: (**a**) Cartesian coordinate system; and (**b**) Polar coordinate system.

Table 2. Comparison with the detection pixel numbers of real oil films.

Method	Detection Pixel Number of Real Oil Films
Proposed method	304
U-Net	5384

4.4. Comparison of Performance with Other Machine Learning Threshold Segmentation Methods

Three machine learning methods were used to compare with the SA-PSO method. The first one is the traditional PSO method [29]. The second one is the FCM-based marine radar oil spill segmentation method [16]. The last one is the traditional K-Means method [30]. The segmentations of 4 methods of Figure 9b shown similar results. Because there were little wave echoes in the far range in the experimental data, the upper 2/3 region of the contrast-enhanced image (Figure 4d) was uniformly assigned the same gray value '128', as shown in Figure 14.

Figure 14. Contrast enhanced image with no-wave-region information removed.

Figure 14 was segmented by the above four methods and the results obtained were shown in Figure 15. All the methods identified some ship wake regions as oil slick targets. Among them, there were fewer errors of SA-PSO method. The initial segmentation of SA_PSO method also contained fewer noise results. The corresponding results of PSO and FCM methods were similar with more noises. However, the K-Means method became excessively noisy. The K-Means method segmented suspected oil slicks with weak image features into real ones (in red boxes), while also adding many blocky false targets. The PSO and FCM methods also determined some of the suspected oil films as real ones (in red boxes). The SA_PSO preliminary segmentation were relatively accurate and can show better results by simply excluding speckle noises.

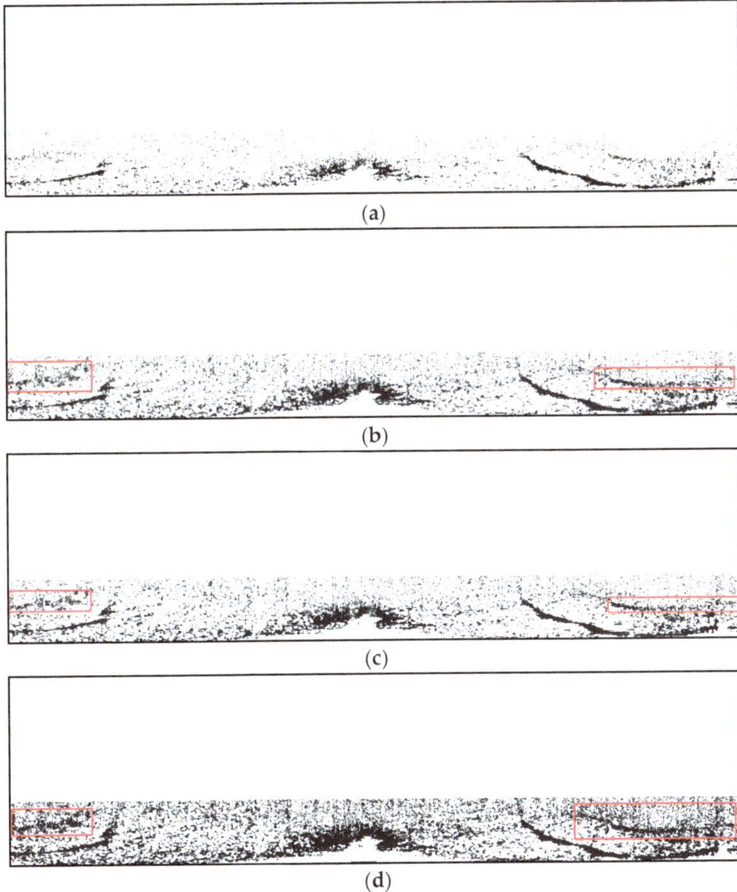

Figure 15. Comparison of four machine learning threshold segmentation methods: (**a**) SA_PSO; (**b**) PSO; (**c**) FCM; and (**d**) K-means.

4.5. Validation of Experimental Results

During the daytime, the accuracy of marine radar oil spill detection can be judged by visual observation or unmanned aerial vehicle (UAV) visible light data, as shown in Figure 16a. During the nighttime, infrared or laser fluorescence data can be used to verify the accuracy of marine radar oil spill monitoring method. Our marine radar experimental data were collected at night, and the corresponding thermal infrared images (Figure 16b) were obtained. In the thermal infrared images, the gray value of the oil film is slightly lower than that of water [31]. The thermal infrared images corresponding to the oil spill locations in the experimental data also show the characteristics of the oil films, which can assist in verifying the feasibility of our method.

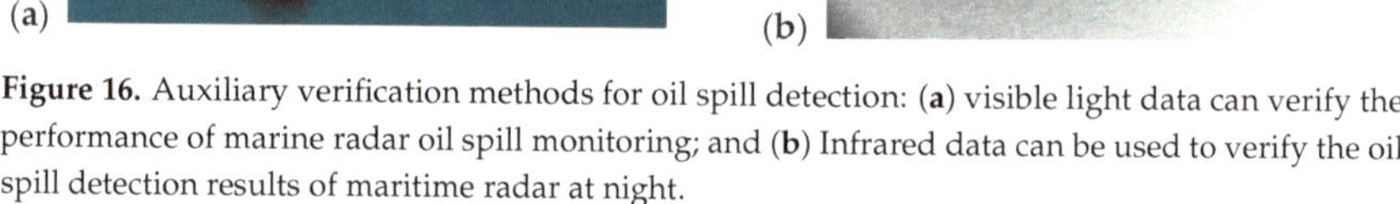

(a) (b)

Figure 16. Auxiliary verification methods for oil spill detection: (**a**) visible light data can verify the performance of marine radar oil spill monitoring; and (**b**) Infrared data can be used to verify the oil spill detection results of maritime radar at night.

4.6. The Impacts of Weather Conditions on Marine Radar Oil Spill Detection

The marine radar oil spill monitoring technology has the capability to detect oil spills in a large range and has a relatively low construction cost. Despite its promising application prospects, this technology is still constrained by weather conditions. When the weather causes excessive ocean waves, ships equipped with marine radar are unable to go out to sea to perform pollution clean-up tasks. When the sea surface is too calm, the radar cannot retrieve certain wave echoes, making it difficult to identify the oil films. Moreover, the ability of the oil films to absorb waves is weak in both cases. In addition, for deep learning networks, training and detection data under the same sea conditions (including sea breeze) can achieve better results. To establish a comprehensive oil spill detection model, continuous collection of marine radar oil spill data under different sea conditions is needed.

5. Conclusions

In this study, the preprocessed images of marine radar oil spills underwent training using the YOLOv8 model. In contrast to prior methodologies, the proposed approach exhibited a notable improvement in detection efficiency. Solely tasked with pinpointing the location and dimensions of oil spills, the YOLOv8 deep learning network was exclusively employed. The SA_PSO algorithm was applied for the final segmentation. As more marine radar oil spill images are gathered across diverse oceanic conditions and deep learning networks are refined, enhanced techniques for oil spill detection will be realized. The future research focus will focus on how to distinguish offshore oil films from false positive targets such as shadows, wind, and plumes, etc.

Author Contributions: Conceptualization, J.X. and Y.H.; methodology, J.X., Y.H. and H.D.; software, L.C. and Y.Y.; validation, L.C. and Z.L.; formal analysis, S.Q., M.C. and B.L.; investigation, J.X., Y.H. and L.C.; resources, J.X., B.L. and P.L.; data curation, J.X. and H.D.; writing—original draft preparation, Y.H. and J.X.; writing—review, B.L. and H.D.; visualization, L.C.; supervision, Y.H. and J.W.; project administration, H.D. and J.X.; funding acquisition, H.D. and J.X. All authors have read and agreed to the published version of the manuscript.

Funding: This research was funded by the Natural Science Foundation of Guangdong Province, grant numbers 2022A1515011603, 2023A1515011212, the Special Projects in Key Fields of Ordinary Universities in Guangdong Province, grant numbers 2021ZDZX1015, 2022ZDZX3005, the Natural Science Foundation of Shenzhen City, grant numbers JCYJ20220530162200001, JCYJ20210324122813036, Program for scientific research start-up funds of Guangdong Ocean University, grant numbers 060302132009, 060302132106, Postgraduate Education Innovation Project of Guangdong Ocean University grant numbers 202421, 2023XSLT_032, National Natural Science Foundation of China, grant number 52271359.

Institutional Review Board Statement: Not applicable.

Informed Consent Statement: Not applicable.

Data Availability Statement: Data are contained within the article.

Conflicts of Interest: The authors declare no conflicts of interest.

References

1. Armaroli, N.; Balzani, V. The Future of Energy Supply: Challenges and Opportunities. *Angew. Chem. Int. Ed. Engl.* **2007**, *46*, 52–66. [CrossRef] [PubMed]
2. Prabowo, A.R.; Bae, D.M. Environmental Risk of Maritime Territory Subjected to Accidental Phenomena: Correlation of Oil Spill and Ship Grounding in the Exxon Valdez's Case. *Results Eng.* **2019**, *4*, 100035. [CrossRef]
3. Salomidi, M.; Katsanevakis, S.; Borja, A.; Braeckman, U.; Damalas, D.; Galparsoro, I.; Mifsud, R.; Mirto, S.; Pascual, M.; Pipitone, C.; et al. Assessment of Goods and Services, Vulnerability, and Conservation Status of European Seabed Biotopes: A Stepping Stone towards Ecosystem-Based Marine Spatial Management. *Mediterr. Mar. Sci.* **2012**, *13*, 49. [CrossRef]
4. Sherman, E.F. Specific Oil Spill Incidents. In *Managing the Risk of Offshore Oil and Gas Accidents*; Edward Elgar Publishing: Cheltenham, UK, 2019; pp. 408–446. ISBN 9781786436740.
5. Yin, J.; Feng, J.; Wang, Y. Social Media and Multinational Corporations' Corporate Social Responsibility in China: The Case of ConocoPhillips Oil Spill Incident. *IEEE Trans. Prof. Commun.* **2015**, *58*, 135–153. [CrossRef]
6. Cao, R.; Rong, Z.; Chen, H.; Liu, Y.; Mu, L.; Lv, X. Modeling the Long-Term Transport and Fate of Oil Spilled from the 2021 A Symphony Tanker Collision in the Yellow Sea, China: Reliability of the Stochastic Simulation. *Ocean Model.* **2023**, *186*, 102285. [CrossRef]
7. Krig, S. Image Pre-Processing. In *Computer Vision Metrics*; Springer International Publishing: Cham, Switzerland, 2016; pp. 35–74. ISBN 9783319337616.
8. Ullah, Z.; Farooq, M.U.; Lee, S.-H.; An, D. A Hybrid Image Enhancement Based Brain MRI Images Classification Technique. *Med. Hypotheses* **2020**, *143*, 109922. [CrossRef] [PubMed]
9. Sonali; Sahu, S.; Singh, A.K.; Ghrera, S.P.; Elhoseny, M. An Approach for De-Noising and Contrast Enhancement of Retinal Fundus Image Using CLAHE. *Opt. Laser Technol.* **2019**, *110*, 87–98. [CrossRef]
10. Al-Ruzouq, R.; Gibril, M.B.A.; Shanableh, A.; Kais, A.; Hamed, O.; Al-Mansoori, S.; Khalil, M.A. Sensors, Features, and Machine Learning for Oil Spill Detection and Monitoring: A Review. *Remote Sens.* **2020**, *12*, 3338. [CrossRef]
11. Conceição, M.R.A.; de Mendonça, L.F.F.; Lentini, C.A.D.; da Cunha Lima, A.T.; Lopes, J.M.; de Vasconcelos, R.N.; Gouveia, M.B.; Porsani, M.J. SAR Oil Spill Detection System through Random Forest Classifiers. *Remote Sens.* **2021**, *13*, 2044. [CrossRef]
12. Li, K.; Yu, H.; Xu, Y.; Luo, X. Detection of Oil Spills Based on Gray Level Co-Occurrence Matrix and Support Vector Machine. *Front. Environ. Sci.* **2022**, *10*, 1049880. [CrossRef]
13. Misra, A.; Balaji, R. Simple Approaches to Oil Spill Detection Using Sentinel Application Platform (SNAP)-Ocean Application Tools and Texture Analysis: A Comparative Study. *J. Ind. Soc. Remote Sens.* **2017**, *45*, 1065–1075. [CrossRef]
14. Chehresa, S.; Amirkhani, A.; Rezairad, G.-A.; Mosavi, M.R. Optimum Features Selection for Oil Spill Detection in SAR Image. *J. Ind. Soc. Remote Sens.* **2016**, *44*, 775–787. [CrossRef]
15. Liu, P.; Li, Y.; Liu, B.; Chen, P.; Xu, J. Semi-Automatic Oil Spill Detection on X-Band Marine Radar Images Using Texture Analysis, Machine Learning, and Adaptive Thresholding. *Remote Sens.* **2019**, *11*, 756. [CrossRef]
16. Li, B.; Xu, J.; Pan, X.; Ma, L.; Zhao, Z.; Chen, R.; Liu, Q.; Wang, H. Marine Oil Spill Detection with X-Band Shipborne Radar Using GLCM, SVM and FCM. *Remote Sens.* **2022**, *14*, 3715. [CrossRef]
17. Ajadi, O.A.; Meyer, F.J.; Tello, M.; Ruello, G. Oil Spill Detection in Synthetic Aperture Radar Images Using Lipschitz-Regularity and Multiscale Techniques. *IEEE J. Sel. Top. Appl. Earth Obs. Remote Sens.* **2018**, *11*, 2389–2405. [CrossRef]
18. Chen, G.; Li, Y.; Sun, G.; Zhang, Y. Application of Deep Networks to Oil Spill Detection Using Polarimetric Synthetic Aperture Radar Images. *Appl. Sci.* **2017**, *7*, 968. [CrossRef]
19. Guo, Y.; Zhang, H.Z. Oil Spill Detection Using Synthetic Aperture Radar Images and Feature Selection in Shape Space. *Int. J. Appl. Earth Obs. Geoinf.* **2014**, *30*, 146–157. [CrossRef]
20. Gil, P.; Alacid, B. Oil Spill Detection in Terma-Side-Looking Airborne Radar Images Using Image Features and Region Segmentation. *Sensors* **2018**, *18*, 151. [CrossRef]
21. Huang, X.; Zhang, B.; Perrie, W.; Lu, Y.; Wang, C. A Novel Deep Learning Method for Marine Oil Spill Detection from Satellite Synthetic Aperture Radar Imagery. *Mar. Pollut. Bull.* **2022**, *179*, 113666. [CrossRef]
22. Wen, H.; Dai, F.; Yuan, Y. A Study of YOLO Algorithm for Target Detection. *Proc. Int. Conf. Artif. Life Robot.* **2021**, *26*, 622–625. [CrossRef]
23. Du, L.; Zhang, R.; Wang, X. Overview of Two-Stage Object Detection Algorithms. *J. Phys. Conf. Ser.* **2020**, *1544*, 012033. [CrossRef]
24. Wang, C.-F.; Liu, K. A Novel Particle Swarm Optimization Algorithm for Global Optimization. *Comput. Intell. Neurosci.* **2016**, *2016*, 9482073. [CrossRef] [PubMed]
25. Delahaye, D.; Chaimatanan, S.; Mongeau, M. Simulated Annealing: From Basics to Applications. In *Handbook of Metaheuristics*; Springer International Publishing: Cham, Switzerland, 2019; pp. 1–35. ISBN 9783319910857.

26. Chen, S.; Luk, B.L. Adaptive Simulated Annealing for Optimization in Signal Processing Applications. *Signal Process.* **1999**, *79*, 117–128. [CrossRef]
27. Liu, S.; Huang, F.; Yan, B.; Zhang, T.; Liu, R.; Liu, W. Optimal Design of Multimissile Formation Based on an Adaptive SA-PSO Algorithm. *Aerospace* **2021**, *9*, 21. [CrossRef]
28. Li, B.; Xu, J.; Chu, L.; Yang, Y.; Huang, X.; Liu, P. Oil Film Semantic Segmentation Method in X-Band Marine Radar Remote Sensing Images. *IEEE Geosci. Remote Sens. Lett.* **2023**, *20*, 1503205. [CrossRef]
29. van Zyl, J.P.; Engelbrecht, A.P. Set-Based Particle Swarm Optimisation: A Review. *Mathematics* **2023**, *11*, 2980. [CrossRef]
30. Hartigan, J.A.; Wong, M.A. Algorithm AS 136: A K-Means Clustering Algorithm. *J. R. Stat. Soc. Ser. C Appl. Stat.* **1979**, *28*, 100. [CrossRef]
31. Fingas, M.; Brown, C. Review of Oil Spill Remote Sensing. *Mar. Pollut. Bull.* **2014**, *83*, 9–23. [CrossRef]

Journal of
Marine Science and Engineering

Article

Research on the Construction of a Digital Twin System for the Long-Term Service Monitoring of Port Terminals

Jinqiang Bi [1,2,3], Peiren Wang [4], Wenjia Zhang [1,2,*], Kexin Bao [1,2,3] and Liu Qin [1,2]

[1] Ministry of Transport Tianjin Research Institute of Water Transport Engineering, Tianjin 300456, China; bijq@tiwte.ac.cn (J.B.); baokx@tiwte.ac.cn (K.B.); qinliu@tiwte.ac.cn (L.Q.)
[2] National Engineering Research Center of Port Hydraulic Construction Technology, Tianjin 300456, China
[3] School of Marine Science and Technology, Tianjin University, Tianjin 300192, China
[4] School of Biomedical Engineering, Tianjin Medical University, Tianjin 300070, China; wangpr@tiwte.ac.cn
* Correspondence: zhangwenjia@tiwte.ac.cn

Abstract: Structural damage is a prevalent issue in long-term operations of harbor terminals. Addressing the lack of transparency in terminal infrastructure components, the limited integration of sensor monitoring data, and the insufficient support for feedback on service performance, we propose a novel digital twin system construction methodology tailored for the long-term monitoring of port terminals. This study elaborates on the organization and processing of foundational geospatial data, sensor monitoring information, and oceanic hydrometeorological data essential for constructing a digital twin of the terminal. By mapping relationships between physical and virtual spaces, we developed comprehensive dynamic and static models of terminal facilities. Employing a "particle model" approach, we visually represented oceanic and meteorological elements. Additionally, we developed a multi-source heterogeneous data fusion model to facilitate the rapid creation of data indexes for harbor elements under high concurrency conditions, effectively addressing performance issues related to scene-rendering visualization and real-time sensor data storage efficiency. Experimental validation demonstrates that this method enables the rapid construction of digital twin systems for port terminals and supports practical application in business scenarios. Data analysis and comparison confirm the feasibility of the proposed method, providing an effective approach for the long-term monitoring of port terminal operations.

Keywords: digital twin; data-driven; port infrastructure; long-term service monitoring; multi-source heterogeneous data

Citation: Bi, J.; Wang, P.; Zhang, W.; Bao, K.; Qin, L. Research on the Construction of a Digital Twin System for the Long-Term Service Monitoring of Port Terminals. *J. Mar. Sci. Eng.* **2024**, *12*, 1215. https://doi.org/10.3390/jmse12071215

Academic Editor: Xinqiang Chen

Received: 20 June 2024
Revised: 18 July 2024
Accepted: 18 July 2024
Published: 19 July 2024

1. Introduction

The harsh marine environment in which harbor terminals operate is characterized by complex factors such as wind, waves, currents, temperature, salinity, and load conditions, leading to pervasive structural damage during long-term operations [1–3]. The terminal components are numerous and highly complex, with current safety and early warning measures primarily reliant on periodic manual inspections. However, existing inspection methods face challenges such as adverse working conditions, concealed internal damage, and low data accuracy, making it difficult to dynamically, continuously, and comprehensively reflect the safety status of the terminals. The deployment of sensors for automated and digital health monitoring has emerged as a new trend in recent years [4–6], offering a more effective approach to addressing these issues.

Digital twins are used in many disciplines to support engineering, monitoring, controlling, and optimizing cyber-physical systems [7]. Digital twin technology, a virtual representation of a physical entity, has revolutionized various industries, including port terminals [8–11]. Digital twin technology is a digital and technological concept, which means that is based on the integration and fusion of data and models. It accurately constructs physical objects in real-time in digital space and simulates, verifies, predicts, and controls the entire lifecycle process of physical entities based on data fusion and analysis prediction [12,13]. Digital twins can capture the combined usage of heterogeneous models and respective evolving data for the entire lifecycle [14]. The development of digital twin technology provides new solutions for the integration of virtual and real, real-time interaction, and iterative operation and optimization, as well as full factor digital transformation and the intelligent upgrading of port terminals. The key to the application of digital twins to ports lies in IoT (Internet of Things) perception and full element digital expression, massive data visualization rendering and data fusion, spatial analysis calculation and simulation deduction, etc. The core lies in the interaction and collaboration between the real scene and twin scene of the port, as well as the transmission and automatic construction of real-time data. Also key is the acquisition of data such as water depth measurements, sensor perception, and maintenance and operation history at the front of the port, ultimately achieving a multi-physical quantity, multi-scale, and multi-probability simulation process of the port.

By creating dynamic models of port operations, digital twin technology has become instrumental in advancing smart water transportation by offering real-time monitoring and predictive maintenance capabilities for port structures. A digital twin (DT) creates a revolutionary opportunity for smart ports' authorities, with the capability of high-fidelity digital representation of real-world things [15,16]. At present, digital twin technology has already been applied to numerous domestic ports [17–20]. Yu P. analyzed the key problems and technologies that need to be solved and used, respectively, in the specific implementation of digital twins of integrated port energy systems [21]. Hofmann proposed a digital twin for truck-dispatching operator assistance, which enables the determination of optimal dispatching policies using simulation-based performance forecasts [22]. Li Y proposed a framework integrating DT with the AdaBoost algorithm to realize the real-time optimization and security of the ACT(Automated Container Terminals) [23]. Martínez-Gutiérrez proposed a new DT design concept based on external service for the transportation sector [24]. Shanghai Port has employed digital twin technology to monitor the health of its infrastructure. By creating a virtual model of the port's structures, including piers and docks, the system can detect structural anomalies and predict maintenance needs, thereby enhancing safety and operational efficiency. Guangzhou Port has integrated digital twin technology with advanced sensors and IoT devices to detect structural integrity issues. The system analyzes data from these devices to provide insights into potential weaknesses and degradation patterns, enabling timely interventions. Shenzhen Port has developed a comprehensive digital twin-based health monitoring system. This system combines real-time data acquisition, simulation models, and predictive analytics to monitor the structural health of port facilities, ensuring long-term stability and reliability. Ningbo–Zhoushan Port has deployed a digital twin to monitor real-time operations and predict maintenance needs. The technology aids in efficient berth allocation, reducing vessel waiting times and improving overall port throughput. The Port of Rotterdam has implemented a digital twin system to continuously monitor the health of its infrastructure. The system uses real-time data to detect structural issues and predict future maintenance requirements, ensuring the safety and efficiency of port operations. Antwerp Port's digital twin framework integrates with various detection technologies, including LIDAR (Light Detection and Ranging) and sonar, to analyze the structural health of underwater and above-water components. This integration allows for the detailed inspection and timely maintenance of critical infrastructure. The Port of Los Angeles has constructed a robust digital twin-based health monitoring system that incorporates machine learning algorithms to analyze structural data. This system provides real-time insights and predictive maintenance recommendations, enhancing the port's operational resilience. Singapore Port has developed a comprehensive digital twin to manage complex logistics and improve service levels. The system supports real-time decision-making, asset management, and operational planning, leading to increased port productivity [25–27].

Structural health monitoring of terminals through sensor data collection allows for real-time safety assessments based on empirical values. However, there is a lack of effective support for feedback on the long-term service performance of terminal structures. This paper proposes a digital twin system methodology tailored for the long-term service monitoring of harbor terminals. Utilizing multi-source heterogeneous data from long-term service monitoring, we establish monitoring data resources encompassing terminal structures, environment, materials, and loads. By extracting characteristic parameters that represent the terminal's performance under marine environmental influences and using deployed sensors to collect status data, we investigate techniques for constructing a comprehensive digital twin of the terminal. The system is designed to handle large volumes of data with low latency and intelligent processing. This digital twin system enables data-driven visual management of structural safety through historical backtracking, monitoring and early warning, and future predictions, thereby illustrating the structural safety status of the terminal at different stages of its service life.

2. Methodology

The framework for the digital twin system construction method proposed in this paper is illustrated in Figure 1.

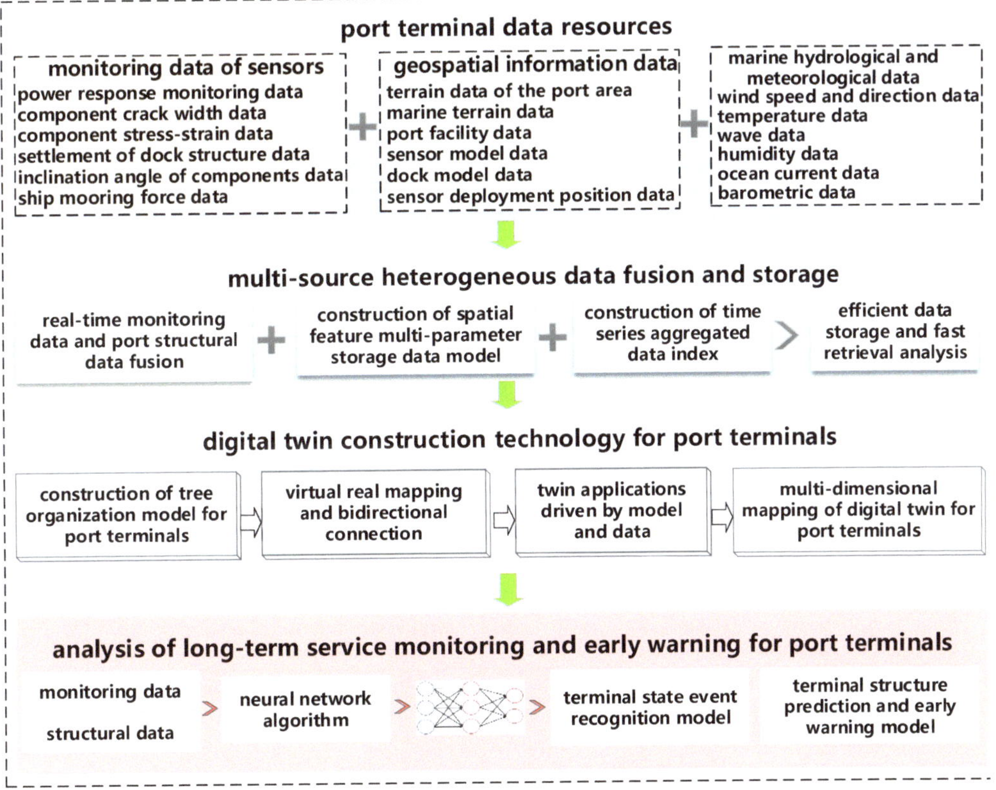

Figure 1. Construction framework of digital twin system.

Based on the physical information–mapping relationship between the physical space scene and the virtual space scene, the logic block diagram is shown in Figure 2.

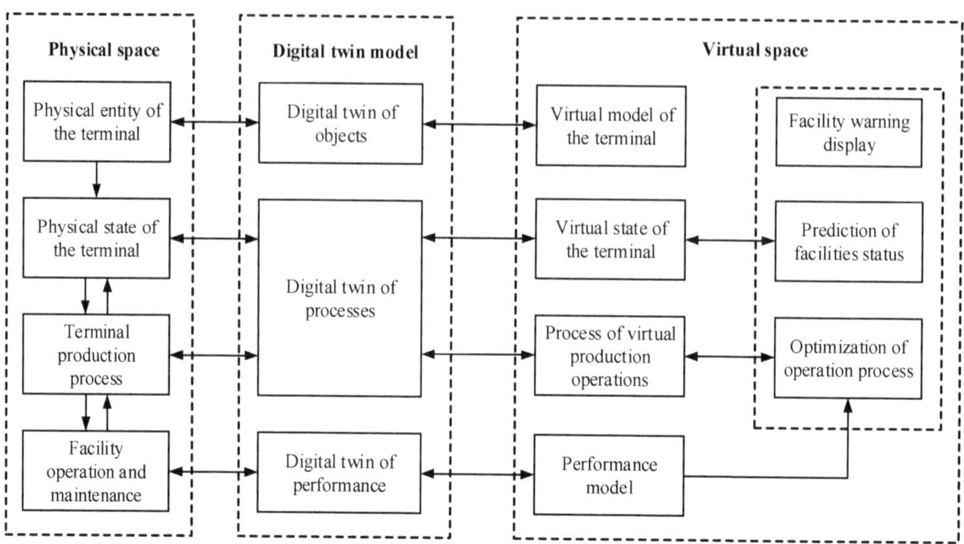

Figure 2. Terminal digital twin logic block diagram.

2.1. Port Terminal Data Resource

The construction of a digital twin for a seaport terminal encompasses foundational geospatial data, sensor monitoring data, and marine hydrometeorological data. These data resources exhibit semantic independence and heterogeneity, varying across temporal and spatial dimensions and storage formats. This includes unstructured data such as vector and raster maps of the port area, and detailed models of the terminal (as shown in Table 1), as well as structured data such as terminal stress–strain data, structural settlement data, and marine hydrometeorological information (as detailed in Table 2). The diverse acquisition methods, data content, and spatiotemporal characteristics necessitate varied approaches to data processing, structure, representation, and storage.

Table 1. Unstructured data.

Serial Number	Data Item	Data Description
1	Topographic data of port area	Including port infrastructure, elevation, and other data.
2	Port infrastructure data	Including data of docks and berths.
3	Marine topographic data	Including data such as the water depth at the wharf front.
4	Wharf model data	Including model data such as wharf components.
5	Sensor model data	Include sensor model data such as stress and strain.
6	Sensor layout position data	Includes sensor point, line, and area data.

Table 2. Structured data.

Serial Number	Data Item	Data Description
1	Dynamic response monitoring data	Including fundamental frequency, amplitude, acceleration, sensor position, and other data.
2	Stress and strain data of components	Including stress value, strain value, sensor position, and other data.
3	Dynamic inclination data of components	Including vibration, angle, sensor position, and other data.
4	Component crack width data	Include data such as crack length, width, and sensor position.
5	Settlement data of wharf structure	Include data such as settlement depth and sensor position.
6	Ship mooring force data	Including data such as stress and sensor position.
7	Marine hydrological element data	Including ocean wind, waves, currents, and other data.
8	Meteorological element data	Including temperature, humidity, air pressure, and other data.

2.2. Multi-Source Heterogeneous Data Fusion and Storage

Terminal monitoring and related data have the characteristics of large data volume, high concurrency, and strong real-time performance. For unstructured data, this study employs an XML-based (Extensible Markup Language) distributed spatial data organization and storage method. This approach supports deep-level nested expressions and tree-like storage structures, systematically managing metadata concerning data sources, types, information content, structure, and access methods. This framework establishes a unified spatial reference for terrain scenarios above, at, and below the waterline at the terminal's edge. On this basis, an integrated spatiotemporal dataset of port infrastructure is constructed, correlated by geographic location and attribute information, and stored using a combination of file systems and relational databases. Regarding structured data, the study first analyzes sensor monitoring data in terms of acquisition frequency, data format, and volume. Then, a time-series-based multi-parameter column storage database model for terminal monitoring is developed, specifying database entity attributes, data types, lengths, and precisions. A distributed storage system database is utilized for storing these data.

Following the organization and processing of these data, the CGCS2000 (China Geodetic Coordinate System 2000) geographic coordinate system is adopted as the reference to create a foundational geographic information basemap of the port. This basemap integrates terminal and sensor models, allowing for the real-time monitoring and collection of terminal safety status data, ultimately forming a unified digital twin data resource for the terminal.

2.3. Digital Twin Construction Technology for Port Terminals

2.3.1. Construction Method for Wharf Digital Twin

The construction of a digital twin for the terminal integrates physical models, simulation models, and data models through mutual coupling. Based on the mapping relationship between physical and virtual spatial scenarios, the process begins by constructing comprehensive models of static elements within the terminal's physical space in the virtual environment. Sensor perception data and structural state data are then fused. Subsequently, dynamic elements of the terminal are simulated, based on kinematic and dynamic characteristic models. Finally, data organization and the optimization of the digital twin model are performed, achieving the construction of a terminal structure digital twin scenario driven by both models and data. The specific steps are as follows:

(1) Construction of static element scenarios: This phase involves four stages: model creation, model establishment, texture mapping, and model baking. Utilizing 3ds Max for three-dimensional modeling, we develop high-precision 3D models of the natural environment within a 3 km radius of the terminal, including marine, terrain, and topographic features, as well as the relevant port infrastructure. Structural parameters, geometric parameters, material parameters, and state parameters are embedded within these 3D models. A multidimensional terminal scene model compatible with sensors is constructed, analyzing the rotation, translation, and orientation of the terminal scene model to determine the relationships between variables such as sensor inclination, angles, and positions. Real-time updates of model data are achieved through data interfaces, facilitating the information interaction and data synchronization between the physical entities and the digital twin.

(2) Construction of dynamic element scenarios: Building on the static model scenarios, dynamic elements of the terminal, such as movable equipment and facilities with action attributes, are modeled according to their motion characteristics. Corresponding animation frames are created, and a graphics engine is employed to execute related operational actions and commands. For complex motion models, elements are categorized and grouped based on the relative motion relationships of terminal components in a tree structure. Each dynamic element of the terminal is then modeled according to its structural characteristics, using kinematic and dynamic properties to construct physical simulation models. These models simulate kinematic and dynamic behaviors, providing parameters for the operational states of the terminal's digital twin (Figure 3).

Figure 3. Scene construction of dynamic element model of wharf.

Organization and rendering optimization of the digital twin model: This study employs a scene tree directory structure for the logical management and optimization of massive model data. The constructed terminal digital twin scene comprises billions of model nodes, each containing vast amounts of data such as triangles and polygons, materials, and textures. The scene graphics are organized using a top-down, hierarchical tree data structure. The root node is at the top, extending downward, with each group node containing geometric information and rendering state information to control its appearance. At the bottom of the scene graph, leaf nodes contain the actual geometric information constituting the scene's objects. Utilizing the rendering tools in 3ds Max, models are rendered in sequence according to the directory structure. This process involves adding materials, defining surface color, transparency, roughness, and texture, and incorporating the physical model's material parameters, structural data, and geometric data, along with optimizing boundary conditions. Additionally, rendering optimization is applied to the edges of the models.

2.3.2. Expression of Marine Hydrometeorological Elements

During the actual operation of a seaport terminal, risks to safety arise from adverse weather conditions, including wind, waves, and currents. To effectively integrate and display marine hydrometeorological data within the digital twin scenario, a dynamic evolution simulation method using "particle models" is proposed. This method involves performing coordinate transformation operations on particles and setting parameters such as color, texture, lighting, and fogging to achieve a three-dimensional dynamic visualization of marine hydrometeorological elements. The specific steps are as follows:

(1) Index Construction: Taking the ocean current field data of the demonstration port area as an example, the raw data are parsed to obtain the basic information of the grid. The values in the u and v directions are stored using two arrays. The data grid spacing is $0.03°$, with each grid cell storing the parsed (u, v) values and the calculated flow direction and velocity information based on these (u, v) values.

(2) Parameter Initialization: The fundamental structure of a particle includes variables x, y, dx, dy, age, birthAge, and path, which are initialized to define position and velocity parameters. The initialization of particle position parameters involves generating random floating-point numbers within specified ranges. The initial lifespan (birthAge) is randomly generated. Once the particle position is determined, the corresponding velocity values from the flow field are obtained via bilinear interpolation. This interpolation method establishes a function based on two variables and linearly extends the function. It determines the value of function $f(x)$ at point P using the known values at points Q_{11}, Q_{12}, Q_{21}, and Q_{22} (Figure 4).

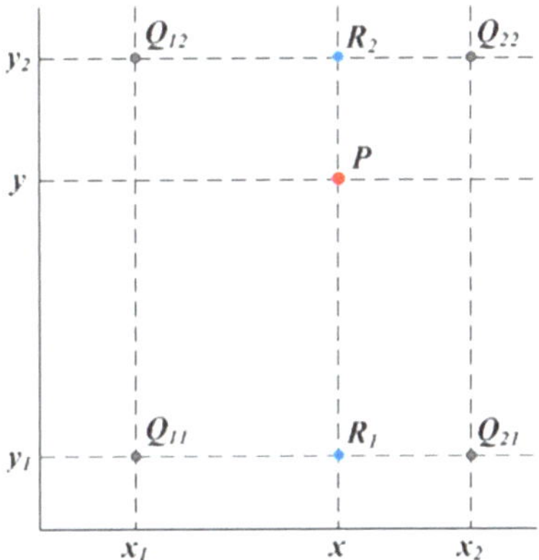

Figure 4. Bilinear interpolation diagram of flow field particles.

Firstly, bilinear interpolation is performed in the x-direction twice according to Formulas (1) and (2), as follows:

$$f(R_1) \approx \frac{x_2 - x}{x_2 - x_1} f(Q_{11}) + \frac{x - x_1}{x_2 - x_1} f(Q_{21}) \tag{1}$$

$$f(R_2) \approx \frac{x_2 - x}{x_2 - x_1} f(Q_{12}) + \frac{x - x_1}{x_2 - x_1} f(Q_{22}) \tag{2}$$

Subsequently, linear interpolation is conducted in the y-direction according to Formula (3), as follows:

$$f(P) \approx \frac{y_2 - y}{y_2 - y_1} f(R_1) + \frac{y - y_1}{y_2 - y_1} f(R_2) \tag{3}$$

(3) Data Computation Update: Based on the velocity values of the vector field, integration is performed to calculate the next position of the particle, continuing iteratively until the termination condition for integration is met. The lifespan of the particle corresponds to the period when integration of the particle's trajectory ceases. Assuming the velocity of the vector field is represented as $\vec{v}(\vec{r}(t))$, where t is the integration variable or time variable and $\vec{r}(t)$ is the position vector of the particle's trajectory, differential functions are utilized for computational solution. According to the integral equation of formula 4, the position of the particle at the next moment can be determined as follows:

$$\vec{r}(t) = \vec{r}(t_0) + \int_{t_0}^{t} \vec{v}(\vec{r}(t)) dt \tag{4}$$

(4) Streamline Generation: After updating the particle position data, particles with different lifespans generate a series of points. These points can be connected using piecewise linear segments to form streamlines. To enhance the dynamic effect of the streamlines, transparency is adjusted based on the order of particle generation. This approach allows all particle points to be connected into line segments, forming cohesive streamlines.

2.3.3. Driven by the Fusion of Digital Twin Scenes and Feature Data

A complete digital twin of the port is established, based on the integration of digital models, sensor data collection, and marine hydrological and meteorological elements that reflect the authenticity of physical entities. Applications such as data perception, intelligent identification, and control prediction need to be driven by the fusion of twin scene and element data. Through multi-source sensor technology, real-time monitoring data of the port structure, environment, materials, loads, and other operational processes are collected.

Through multi-source heterogeneous data analysis, system functional integrity verification, and the iterative optimization of control algorithms, decision-making control, performance evaluation, and health predictions of the digital twin of the port are achieved, as well as real-time feedback and autonomous learning of the physical and historical states of physical entities. A digital twin driving model is established for the data-driven analysis and interactive mapping of digital twin data at the dock, integrating scene construction, element data access, and prediction and deduction. The model mainly includes the following:

(1) The description and precise characterization of dock equipment, equipment components, and system integration, achieving component coupling simulation and the precise characterization of digital twins.

(2) The dynamic data regarding virtual–real interaction at the dock is updated and visualized in real time. Through real-time updates of physical models, state parameters, operational data, and marine environmental data of in-service dock operation facilities, data-driven and interactive iterations, process monitoring, performance evaluation, and health predictions of the digital twin of all dock elements are achieved.

(3) The consistency mapping detection of dock digital twins and application systems is achieved through parallel intelligence, compressive sensing, and deep learning to generate and optimize digital twin interaction mapping, completing accurate data analysis and consistency mapping. We use data-driven models to express the characteristic elements of port terminal structural safety monitoring for different scene objects and application requirements, and map them to the application system through visualization.

2.4. Analysis of Long-Term Service Monitoring and Early Warnings for the Wharf

Based on the constructed digital twin model of the port, driven by real-time monitoring data, deep learning neural network algorithms are used to deeply analyze and mine the multimodal data collected from real-time monitoring of the port, in order to establish an accurate data analysis model, reveal the relationship and temporal characteristic trends between the structural data indicators of the port, improve the analysis ability of the digital twin of the port, and achieve accurate identification and warning of the safety status of the port.

By applying deep learning algorithms to real-time monitoring data such as settlement, displacement, and acceleration in dock structures, utilizing long short term memory neural networks (LSTM) and gated recurrent units (GRU), complex relationships and nonlinear features between data can be discovered. Extracting valuable information from massive monitoring data allows us to provide more accurate intelligent analysis support for digital twin structures at docks. By analyzing the characteristics of structural state monitoring data for specific events such as ship collisions, heavy load operations, extreme weather, etc., a deep neural network model is trained to identify these events and analyze their impact on structural state. We integrate the predictive analysis results into the digital twin system to achieve real-time control, historical review, and prospect prediction of port structural safety, thereby enhancing the safety analysis capability of port infrastructure.

3. Performance Optimization, Comparative Experiments, and Results

3.1. Efficiency Optimization Methods for Scene Data

Optimizing digital twin scenarios for port operations is a critical challenge, requiring the further enhancement of efficiency for the effective deployment of digital twin applications. Efforts have been directed towards optimizing the organization, management, scheduling, and visualization of scene data, incorporating multi-level detail, view frustum culling, asynchronous data loading, and request prediction. Additionally, a semantic-indexing model oriented towards the organization, transmission, and storage of heterogeneous data collected by sensors has been proposed. Furthermore, for the access and interactive visualization of massive ship data, a density clustering approach is employed to manage densely packed AIS (Automatic Identification System) data symbols, thereby reducing data display latency.

(1) Efficiency optimization of port scene data involves several key strategies. Firstly, employing multi-level detail (LOD) optimization gradually simplifies the surface details of models, in order to reduce geometric complexity. Thresholds are set and evaluated based on the ratio of view height to data size, in order to determine whether further decomposition into higher resolution data is necessary, thereby enhancing graphical rendering efficiency. Secondly, view frustum culling optimization controls the range of front and back clipping planes, defined by the field of view angles and projection matrix, to ascertain the size and depth of visible content within the visible view frustum. Thirdly, asynchronous data loading divides data among different threads; the main thread calculates required data through LOD algorithms and clipping. If not cached, data are added to the request queue. Once loaded, the main thread is notified for tile utilization and view refresh (Figure 5). Lastly, predictive request optimization forecasts data needs during continuous scene movement. Data are preloaded into buffers based on the predicted requirements, maintaining the invariant graphic projection frustum. Adjustments are made to frustum size and shape according to viewpoint movement direction and speed, thereby improving the accuracy and specificity of data retrieval predictions.

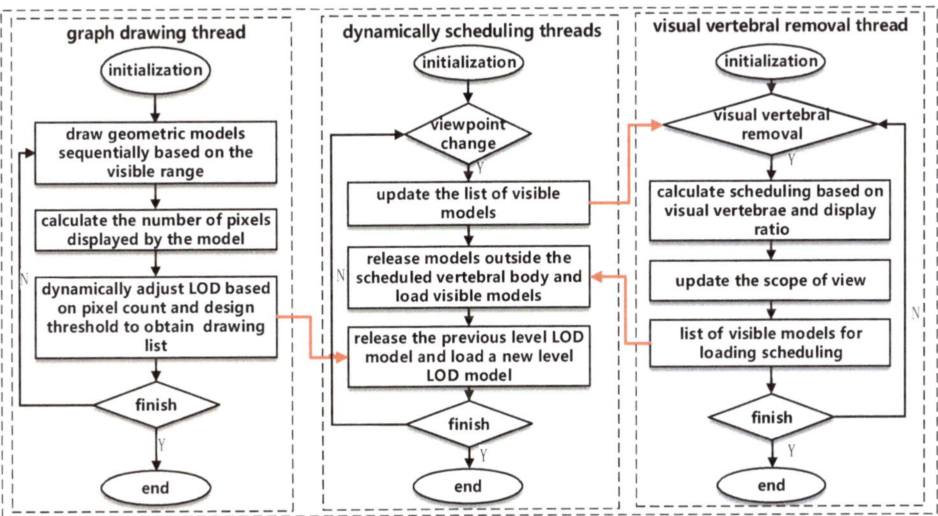

Figure 5. Optimal design of asynchronous data loading.

(2) The efficiency optimization of sensor monitoring data involves several strategic steps. Initially, the storage space for sensor monitoring data is partitioned into N grid regions, abstracting data entities and their attribute relationships into nodes and edges within a model. This enables the indexing of multidimensional features and associative relationships. Subsequently, prior to storage, sensor monitoring data undergoes normalization, where raw data are linearly distributed to achieve a normalized distribution within the [0, 1] interval. This transformation converts raw data into semantic vectors, establishing a semantic indexing structure. Furthermore, during the query–response phase, query requests are forwarded to an integrator module. The integrator module distributes the query to multiple relay modules for the parallel execution of searches. Each relay module is designed with multiple identical instances to ensure load balancing and fault tolerance. Results from queries are merged and returned to the integrator module. Lastly, for sensor monitoring data such as stress–strain and structural settlement data, feature extraction is performed. Multiple datasets are classified and their features correlated to construct a data classification and attribute inverted index. This facilitates the establishment of vector connectivity graphs for different types of data at the same moment in time.

(3) Ship data simulation and efficiency optimization in port twin scenarios involve the real-time integration of ship position data. Each update of ship data necessitates re-rendering to enhance the display's efficiency and reduce latency. This study employs density clustering to optimize dense ship AIS data, ensuring the efficient presentation of ships within the twin port scenario.

Building upon the DENCLUE (Density-based clustering) algorithm, this study optimizes the integration of ship AIS data in twin port scenarios (Figure 6a). For ship position points where the distance between two ships is d and the interaction range is set to ε, if $d \leq \varepsilon$, these points are considered to be within the same clustering range. Using Pi as the center and ε as the radius, the algorithm calculates the number of ship position points u within this neighborhood circle. It then calculates the number of points in the neighborhood of each point within the neighborhood of Pi and takes the maximum value v. If $u > v$, Pi is saved as a density attractor point for that region; if $u \leq v$, then Pi (where (x_2, y_2) are coordinates) with the maximum neighborhood density becomes the attractor point, and all ship position points within Pj's neighborhood, including Pi, are clustered around Pj. To prevent the sequential clustering of point coordinates, clustering is only performed if $d \leq \varepsilon$. This process is repeated for all ship position points to complete one clustering iteration. The pseudo code of the DENCLUE algorithm for ship density clustering is shown in Figure 6b. Based on the previous layer of clustering points, the threshold is incrementally increased, and clustering is repeated until the desired density point count is achieved. As more ship position points are clustered, the density attractor points gradually become the local maxima of global density. When the required density point count is reached, clustering is considered complete. Addressing the issue of low efficiency and display latency in rendering three-dimensional ship models, this method integrates the efficiency optimization techniques proposed earlier for port scene data. Starting from rendering, view frustum culling optimization dynamically loads ships within the view range as the viewpoint moves, simultaneously removing ship models outside the view frustum.

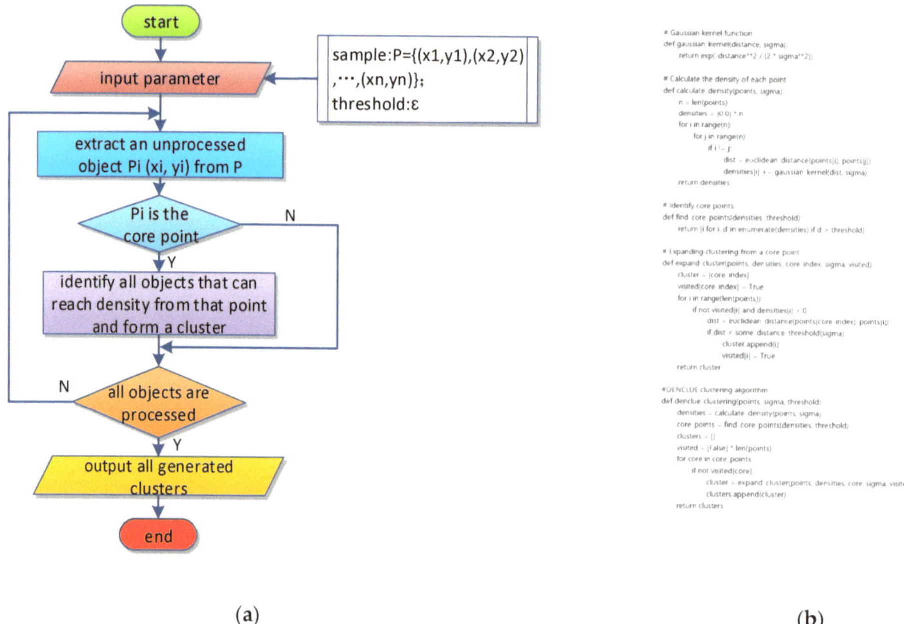

(a)   (b)

Figure 6. Optimization design of ship density clustering and pseudo code. (**a**) Optimization design process for ship density clustering. (**b**) Pseudo code of DENCLUE algorithm.

Through the aforementioned efficiency optimization design for scene data, several performance issues related to the organization, display, and scheduling of three-dimensional scenes have been addressed. Simultaneously, the real-time storage and query efficiency problems of sensor monitoring data have also been resolved. This integrated approach not only enhances the rendering and interaction capabilities of complex three-dimensional scenes but also ensures that sensor data can be stored, accessed, and queried efficiently in real time, thereby improving the overall system performance and user experience in digital twin applications for port environments.

3.2. Twin Scenes and Platform Construction

The demonstration utilizes the ore terminal berth of the demonstration port area. It employs unmanned aerial vehicles for photogrammetric data collection, integrating oblique photography with satellite imagery and spatial elevation data. This process establishes a spatial geographic information repository. Utilizing the methodology proposed in this study, a digital twin model of the port area (Figure 7a) is constructed. Surface and vertex/fragment shaders are employed to create realistic shadows, enhancing the fidelity of the terminal model imported into the Unity model library. These digital twins provide a perspective on the internal components of the terminal and the status of sensors (Figure 7b). The reason for choosing Unity in this article is that it has a good visual rendering effect, supports both Client/Server architecture for easy deployment and an efficient display, and supports publishing Browser/Server architecture for browser access when the display efficiency is not high.

(a)   (b)

Figure 7. Construction of digital twin model. (**a**) Twin scene model of port area. (**b**) Refined dock model.

The development of the dock digital twin system (Figure 8) using the Unity3D base platform expands functionalities with the WebGL (Web Graphics Library) 3D graphics module. This integration of JavaScript and OpenGL for Embedded Systems provides hardware-accelerated rendering for HTML5 (HyperText Markup Language 5) Canvas, enabling the smooth display of dock digital twin scenes and models within web browsers. The system consists of four main components: a data layer, a service layer, a rendering layer, and a presentation layer. The data layer is structured according to dock feature data characteristics, encompassing geographic spatial data, sensor monitoring data, and hydrometeorological data. The service layer centrally manages basic geographic information services, dock feature services, hydrometeorological model services, and interfaces for sensor monitoring and perception data access. The rendering layer focuses on simulating and rendering dock component models, vessels, sensors, and marine hydrometeorological elements. Meanwhile, the presentation layer implements the functionalities of the digital twin system platform, including twin scene display, monitoring data presentation, monitoring alerts, and intelligent decision-making capabilities.

Figure 8. Digital twin system platform.

The paper applies a digital twin system for monitoring the safety status of port terminals. The monitoring data collected by sensors (stress and strain data, structural settlement data, siltation depth data, meteorological environment data, etc.) and spatial data (terminal refinement data, vector and grid map data, etc.) are used as input values. The characteristics of sensor data, collection frequency, and data volume are comprehensively analyzed, and the safety status of port terminals is used as the target output. A deep learning model for evaluating the structural status of ports is constructed to achieve an impact analysis of port structures. Based on monitoring data and numerical analysis results, we extract characteristic elements such as the settlement (Figure 9a) and stress (Figure 9b) of the dock structure, and use multi-dimensional simulation special effects technology to visualize and enable interaction with early warnings, using aspects such as particle effects, light and shadow materials, and grading colors.

(**a**) (**b**)

Figure 9. Application of digital twin platform. (**a**) Simulation of settlement of structure. (**b**) Simulation of stress of structure.

3.3. Efficiency Optimization Experimental Analysis

In the experimental setup, the raw data for constructing the dock digital twin model amounted to 4.56 GB. Rendering and analysis of the scene were conducted using various computational servers, where CPU (Central Processing Unit) usage rates encompassed the operational overhead of internal system processes. Findings indicated that, for a system with 16 GB of memory, CPU usage reached 58.7%, whereas, for a system with 64 GB of memory, CPU usage was 21.6%. Overall, these results suggest that the dock digital twin model construction is adaptable to lower-tier computational resources. However, for practical applications, consideration should be given to provisioning higher-tier computational resources.

In the experiment integrating monitoring data into the digital twin scene, comparisons were made between the initial rendering time and the subsequent rendering time after changes in the view. Before optimization, the initial rendering time was prolonged, increasing exponentially with larger datasets. Conversely, rendering times after changes in the view area were relatively faster and less affected by dataset size increments. After optimization, there was minimal difference between the initial rendering time and the rendering time after view changes. However, overall processing time significantly decreased (as detailed in Table 3). The formula for efficiency improvement rate is to subtract the efficiency before improvement from the efficiency after improvement, and then compare it with the efficiency before improvement.

Table 3. Comparative analysis of scene-rendering efficiency (ms).

Number of Access Data		100	865	1985	7582	13,500	21,000
Before optimization	First rendering time	150.14	681.84	1146.1	6095.36	15,652.38	36,034.96
	View change rendering time	130.22	319.32	461.38	641.96	1351.74	2977.05
After the optimization	First rendering time	148.82	345.96	393.22	667.04	1450.74	3036.52
	View change rendering time	146.3	270.24	273.34	558.56	1243.26	2560.84
Comparative analysis before and after	Efficiency improvement in first rendering	0.88%	49.26%	65.69%	89.06%	90.73%	91.57%
	Efficiency improvement in rendering after view changed	−12.35%	15.37%	40.76%	12.99%	8.03%	13.98%

In the experiment on the efficiency of displaying dynamic ship data, the real-time performance and smoothness of the digital twin scene model were analyzed before and after clustering optimization. This analysis focused on two aspects: monitoring the average frame rate after moving the scene view and comparing the time taken to switch between different ship symbols. The experiment involved 2896 ship data entries of various types, using the frame rate before loading ship models as a baseline. Compared with the situation before loading the ship data, the average frame rate showed a slight decrease post-loading, but the difference was minimal and did not significantly affect the smoothness of the visuals in practical applications. Regarding the time taken to switch between different symbols, the transition time between clustered ship positions and their three-dimensional models was only 368 milliseconds after optimization. As shown in Table 4, this demonstrates that the optimized three-dimensional visualization of ship data performs well in terms of display efficiency, meeting the requirement for smooth visuals.

Table 4. Rendering efficiency of multilevel symbolic modeling.

Display Model	Average Frame Rate/fps	Switching Rate/(ms)
Digital twin scene	18.53	-
Ship point (original)	34.49	0.183
Ship location (clustering)	23.94	0.245
Three-dimensional model of ship	21.65	0.159

4. Conclusions

This study explores the construction of a digital twin system for harbor terminals in an exploratory manner. Firstly, it designs the architecture of the digital twin system focusing on four aspects: data resources for harbor twin scenes, the integration and storage of heterogeneous data from multiple sources, the construction of the harbor digital twin body, and long-term service monitoring and early warning analysis. Secondly, it analyzes the data format of the harbor digital twin body, elaborates on data organization and processing methods, and proposes methods for constructing and simulating visualizations of harbor digital twin bodies and marine hydrometeorological data. Solutions are presented for organizing and scheduling massive data, optimizing twin scene efficiency, and addressing issues related to the visualization performance of scenes and the real-time storage efficiency of sensor monitoring data. Lastly, a digital twin system tailored for the long-term monitoring of ports is developed and experimentally applied at the demonstration berth of an ore terminal in a harbor area. Comparative efficiency analyses of twin scenes using the methods proposed in this study demonstrate their capability to integrate real-time sensor monitoring data, enhance the display of digital twin system data elements, and provide essential data support for practical applications of digital twin scene models.

This study has certain limitations in the interactivity of twin scenarios and deep learning-based analysis and prediction. It requires iterative training and upgrading of the model through long-term accumulated data, ultimately forming a warning analysis model suitable for the long-term service monitoring of dock structures, and achieving efficient collaboration between physical and twin scenarios.

The research outcomes presented in this paper are applicable to port operations and the analysis of vessel arrivals and departures. They provide technical support for the intelligent management of port infrastructure and the optimization of infrastructure health monitoring. The rapidly constructed comprehensive digital twin scenes of docks enhance cognitive efficiency and accuracy, facilitating the integration of next-generation information technologies with the maritime transport industry. The next focus of research includes two aspects. Firstly, in-depth research will be conducted on the interactivity of twin scenes and deep learning-based analysis and prediction. Secondly, research will be conducted on complex scene perception, collaborative control, scheduling organization, information security interaction, etc., to form ubiquitous, interconnected, efficient, and intelligent applications in port digital twin scenes. In the future, the rapid development of the shipping trade will drive automated container terminals towards intelligence, safety, and efficiency. Future research results can be further applied to the formulation of job scheduling tasks and the safety and stability of transportation paths [28,29].

Author Contributions: Methodology, J.B.; Software, P.W., W.Z. and L.Q.; Investigation, K.B. All authors have read and agreed to the published version of the manuscript.

Funding: This work was financially supported by the National Key R&D Program of China (grant number 2022YFB3207400), the Tianjin Transportation Technology Development Plan Project (grant number 2023-47), the Tianjin Natural Science Foundation of China (grant number 22JCQNJC00290), and the Fundamental Research Funds for the Central Public Welfare Research Institutes (grant number TKS20240305).

Institutional Review Board Statement: Not applicable.

Informed Consent Statement: Not applicable.

Data Availability Statement: Data are contained within the article.

Conflicts of Interest: The authors declare no conflict of interest.

References

1. Vytiniotis, A.; Panagiotidou, A.I.; Whittle, A.J. Analysis of seismic damage mitigation for a pile-supported wharf structure. *Soil Dyn. Earthq. Eng.* **2019**, *119*, 21–35. [CrossRef]
2. Zhou, Y.; Zheng, Y.; Liu, Y.; Pan, T.; Zhou, Y. A hybrid methodology for structural damage detection uniting FEM and 1D-CNNs: Demonstration on typical high-pile wharf. *Mech. Syst. Signal Process.* **2022**, *168*, 108738. [CrossRef]
3. Su, L.; Wan, H.P.; Dong, Y.; Frangopol, D.M.; Ling, X.Z. Seismic fragility assessment of large-scale pile-supported wharf structures considering soil-pile interaction. *Eng. Struct.* **2019**, *186*, 270–281. [CrossRef]
4. Sony, S.; Laventure, S.; Sadhu, A. A literature review of next-generation smart sensing technology in structural health monitoring. *Struct. Control Health Monit.* **2019**, *26*, e2321. [CrossRef]
5. Yau KL, A.; Peng, S.; Qadir, J.; Low, Y.C.; Ling, M.H. Towards smart port infrastructures: Enhancing port activities using information and communications technology. *IEEE Access* **2020**, *8*, 83387–83404.
6. Abdulkarem, M.; Samsudin, K.; Rokhani, F.Z.; ARasid, M.F. Wireless sensor network for structural health monitoring: A contemporary review of technologies, challenges, and future direction. *Struct. Health Monit.* **2020**, *19*, 693–735. [CrossRef]
7. Botín-Sanabria, D.M.; Mihaita, A.S.; Peimbert-García, R.E.; Ramírez-Moreno, M.A.; Ramírez-Mendoza, R.A.; Lozoya-Santos JD, J. Digital twin technology challenges and applications: A comprehensive review. *Remote Sens.* **2022**, *14*, 1335. [CrossRef]
8. Kirchhof, J.C.; Michael, J.; Rumpe, B.; Varga, S.; Wortmann, A. Model-driven digital twin construction: Synthesizing the integration of cyber-physical systems with their information systems. In Proceedings of the 23rd ACM/IEEE International Conference on Model Driven Engineering Languages and Systems, Montreal, QC, Canada, 18–23 October 2020; pp. 90–101.
9. Liu, M.; Fang, S.; Dong, H.; Xu, C. Review of digital twin about concepts, technologies, and industrial applications. *J. Manuf. Syst.* **2021**, *58*, 346–361. [CrossRef]
10. Zheng, Y.; Yang, S.; Cheng, H. An application framework of digital twin and its case study. *J. Ambient Intell. Humaniz. Comput.* **2019**, *10*, 1141–1153. [CrossRef]
11. Qi, Q.; Tao, F.; Hu, T.; Anwer, N.; Liu, A.; Wei, Y.; Wang, L.; Nee, A.Y.C. Enabling technologies and tools for digital twin. *J. Manuf. Syst.* **2021**, *58*, 3–21. [CrossRef]
12. Tuegel, E.J.; Ingraffea, A.R.; Eason, T.G.; Spottswood, S.M. Reengineering Aircraft Structural Life Prediction Using a Digital Twin. *Int. J. Aerosp. Eng.* **2011**, *2011*, 154798. [CrossRef]
13. Korth, B.; Schwede, C.; Zajac, M. Simulation-ready digital twin for realtime management of logistics systems. In Proceedings of the 2018 IEEE International Conference on Big Data (Big Data), IEEE, Seattle, WA, USA, 10–13 December 2018.
14. Eramo, R.; Bordeleau, F.; Combemale, B.; van Den Brand, M.; Wimmer, M.; Wortmann, A. Conceptualizing digital twins. *IEEE Softw.* **2021**, *39*, 39–46. [CrossRef]
15. Wang, K.; Hu, Q.; Zhou, M.; Zun, Z.; Qian, X. Multi-aspect applications and development challenges of digital twin-driven management in global smart ports. *Case Stud. Transp. Policy* **2021**, *9*, 1298–1312. [CrossRef]
16. Yao, H.; Wang, D.; Su, M.; Qi, Y. Application of digital twins in port system. *J. Phys. Conf. Ser.* **2021**, *1846*, 012008. [CrossRef]
17. Ding, Y.; Zhang, Z.; Chen, K.; Ding, H.; Voss, S.; Heilig, L.; Chen, Y.; Chen, X. Real-Time Monitoring and Optimal Resource Allocation for Automated Container Terminals: A Digital Twin Application at the Yangshan Port. *J. Adv. Transp.* **2023**, *2023*, 6909801. [CrossRef]
18. Zhou, C.; Zhu, S.; Bell, M.G.; Lee, L.H.; Chew, E.P. Emerging technology and management research in the container terminals: Trends and the COVID-19 pandemic impacts. *Ocean Coast. Manag.* **2022**, *230*, 106318. [CrossRef] [PubMed]
19. Cheng, J.; Lian, F.; Yang, Z. The impacts of port governance reform on port competition in China. *Transp. Res. Part E Logist. Transp. Rev.* **2022**, *160*, 102660. [CrossRef]
20. Wang, L.; Lau, Y.Y.; Su, H.; Zhu, Y.; Kanrak, M. Dynamics of the Asian shipping network in adjacent ports: Comparative case studies of Shanghai-Ningbo and Hong Kong-Shenzhen. *Ocean Coast. Manag.* **2022**, *221*, 106127. [CrossRef]
21. Yu, P.; Zhaoyu, W.; Yifen, G.; Nengling, T.; Jun, W. Application prospect and key technologies of digital twin technology in the integrated port energy system. *Front. Energy Res.* **2023**, *10*, 1044978. [CrossRef]
22. Hofmann, W.; Branding, F. Implementation of an IoT-and cloud-based digital twin for real-time decision support in port operations. *IFAC-PapersOnLine* **2019**, *52*, 2104–2109. [CrossRef]
23. Li, Y.; Chang, D.; Gao, Y.; Zou, Y.; Bao, C. Automated Container Terminal Production Operation and Optimization via an AdaBoost-Based Digital Twin Framework. *J. Adv. Transp.* **2021**, *2021*, 1936764. [CrossRef]
24. Martínez-Gutiérrez, A.; Díez-González, J.; Ferrero-Guillén, R.; Verde, P.; Álvarez, R.; Perez, H. Digital twin for automatic transportation in industry 4.0. *Sensors* **2021**, *21*, 3344. [CrossRef] [PubMed]
25. Keegan, M.J. Driving digital transformation: A perspective from Erwin Rademaker, program manager, port of Rotterdam authority. *Digit. Transform.* **2019**, *5*, 68–72.
26. Neugebauer, J.; Heilig, L.; Voß, S. Digital twins in seaports: Current and future applications. In Proceedings of the International Conference on Computational Logistics, Berlin, Germany, 6–8 September 2023; Springer: Cham, Switzerland, 2023; pp. 202–218.
27. Klar, R.; Fredriksson, A.; Angelakis, V. Digital twins for ports: Derived from smart city and supply chain twinning experience. *IEEE Access* **2023**, *11*, 71777–71799. [CrossRef]

28. Chen, X.; Liu, S.; Zhao, J.; Wu, H.; Xian, J.; Montewka, J. Autonomous port management based AGV path planning and optimization via an ensemble reinforcement learning framework. *Ocean Coast. Manag.* **2024**, *251*, 107087. [CrossRef]
29. Chen, X.; Lv, S.; Shang, W.-L.; Wu, H.; Xian, J.; Song, C. Ship energy consumption analysis and carbon emission exploitation via spatial-temporal maritime data. *Appl. Energy* **2024**, *360*, 122886. [CrossRef]

Journal of
*Marine Science
and Engineering*

Article

Enhancing Maritime Navigation with Mixed Reality: Assessing Remote Pilotage Concepts and Technologies by In Situ Testing

Arbresh Ujkani *, Pascal Hohnrath, Robert Grundmann and Hans-Christoph Burmeister

Fraunhofer Center for Maritime Logistics and Services CML, Blohmstraße 32, 21079 Hamburg, Germany;
pascal.hohnrath@cml.fraunhofer.de (P.H.); robert.grundmann@cml.fraunhofer.de (R.G.);
hans-christoph.burmeister@cml.fraunhofer.de (H.-C.B.)
* Correspondence: arbresh.ujkani@cml.fraunhofer.de

Abstract: In response to the evolving landscape of maritime operations, new technologies are on the horizon as mixed reality (MR), which shall enhance navigation safety and efficiency during remote assistance as, e.g., in the remote pilotage use case. However, up to now, it is uncertain if this technology can provide benefits in terms of usability and situational awareness (SA) compared with screen-based visualizations, which are established in maritime navigation. Thus, this paper initially tests and assesses novel approaches to pilotage in the congested maritime environment, which integrates augmented reality (AR) for ship captains and virtual reality (VR) and desktop applications for pilots. The tested prototype employs AR glasses, notably the Hololens 2, to superimpose the Automatic Identification System (AIS) data directly into the captain's field of view, while pilots on land receive identical information alongside live 360-degree video feeds from cameras installed on the ship. Additional minimum functionalities include waypoint setting, bearing indicators, and voice communication. The efficiency and usability of these technologies are evaluated through in situ tests conducted with experienced pilots on a real ship using the System Usability Scale, the Situational Awareness Rating Technique, as well as Simulator Sickness Questionnaires during the assessment. This includes a first indicative comparison of VR and desktop applications for the given use case.

Keywords: remote pilotage; navigation; augmented reality; virtual reality; in situ test; situational awareness; usability assessment

Citation: Ujkani, A.; Hohnrath, P.; Grundmann, R.; Burmeister, H.-C. Enhancing Maritime Navigation with Mixed Reality: Assessing Remote Pilotage Concepts and Technologies by In Situ Testing. *J. Mar. Sci. Eng.* **2024**, *12*, 1084. https://doi.org/10.3390/jmse12071084

Academic Editors: Chenguang Liu, Wengang Mao, Jialun Liu, and Xiumin Chu

Received: 14 May 2024
Revised: 24 June 2024
Accepted: 25 June 2024
Published: 27 June 2024

1. Introduction

The primary task of maritime pilots is to ensure the safe passage of vessels through challenging or congested waters by providing expert navigation guidance [1]. This remains important despite technology advancements in navigation technologies. Pilots possess in-depth knowledge of local waterways, including currents, tides, depth variations, and potential hazards. While pilots provide guidance, the ultimate responsibility for the safety and navigation of the vessel rests with the ship's captain. Pilots serve in an advisory capacity, offering recommendations and assistance based on their expertise. Pilotage is still a dangerous profession with several casualties happening each year. Especially the process of entering and leaving a vessel is associated with risks [2]. This can be one of the benefits of enabling remote pilotage: by removing the need for a pilot to be physically present onboard a vessel, remote pilotage can reduce the risks associated with accidents and injuries during pilot transfer operations. Additional benefits are scheduling flexibility and time savings according to a Finnish study [3].

However, changing from onboard to remote pilotage comes with challenges, specifically with regard to human factor-oriented topics. Communication remains crucial, as the pilot and master must still communicate efficiently to ensure smooth and safe navigation, despite being geographically separated and lacking non-verbal communication capabilities. Additionally, with the pilot not being on the bridge, it is essential to ensure high situational awareness for the pilot ashore. Addressing these challenges must be supported by remote pilotage technologies in the future to ensure smooth and safe operations. Those challenges must be addressed by any future Remote Pilotage Technology, to ensure that its benefits can be realized without countering effects on navigational safety. Hereby, new technology and interaction concepts should be also investigated with regard to their applicability. Thus, different user interface concepts for remote pilotage are tested and assessed within this paper with regard to their principal usability for remote pilotage operations. This answers the question if they are suitable visualization and interaction technologies for Remote Pilotage Systems of the future. A significant innovation of this paper is the on-site testing of these technologies on an actual ship vessel, a pioneering approach in the field. Our research uniquely evaluates the practical application and usability of these systems in real-world maritime conditions. This hands-on testing not only provides more realistic insights, but also bridges the gap between theoretical concepts and practical implementation, offering valuable contributions to the current literature and industry practices. The tested technologies are classical desktop visualizations for shore-based pilots, immersive virtual reality (VR) technology for shore-based pilots, and immersive augmented reality (AR) technology for onboard masters being piloted.

For the immersive technologies, prototype systems introduced in the concept presented in [4] have been used. By transmitting sensor data and a 360° video stream from the ship to shore and displaying it in a VR environment, a high level of SA for the pilot shall be assured in combination with an integrated electronic nautical chart application. On the ship, an AR system is used to superimpose the view of the master with essential data about the traffic situation. Voice communication between shore and ship is supported by a marker and hint system with the aim of achieving an efficient and unambiguous communication. The prototype of the concept was tested in an in situ trial on a ferry during a voyage in the Baltic Sea. For the desktop visualization, a similar prototype has been created having a comparable level of technological readiness. As the focus was on testing the principal usability of the different technologies rather than testing a specific system implementation, all systems had a pre-commercial implementation standard.

The Introduction is followed by a brief recap of the state of the art in remote pilotage and maritime mixed reality in Section 2. The system prototypes are described in Section 3. The testing procedure and the assessment are presented in Sections 4 and 5, respectively. Finally, Section 7 concludes the paper, followed by a discussion in Section 6.

2. State of the Art

2.1. Remote Pilotage

Remote pilotage has been the subject of scientific publications for at least 20 years [5]. To date, however, the most advanced practical efforts to establish remote control have been limited to only parts of the fairway [6]. An example of this is the port of Rotterdam, where remote pilotage (here called shore-based pilotage) is offered between the Maas Centre pilot station and the Hoek van Holland traffic center. In this port, the pilot communicates with the ship via a VTS-monitored VHF channel. The pilot is supported by a VTS monitoring view and a land-based radar image [7]. Apart from a few exceptional cases, the pilot still enters the ship at one point in time. Complete remote pilotage does not take place [8].

Implementations of remote pilotage in the context of research projects are often limited to implementation in simulation environments [6]. One exception is the Sea4Value research project. In 2022, remote piloting was tested in Finnish waters as part of the research project. To support the pilot, sensor data are transmitted from the ship to the shore, including the ship's foresight as a video stream. The above-mentioned research projects and implementations are based on conventional display technology, with communication between the shore station and the ship taking place exclusively verbally via a radio link.

There is a broad consensus in the scientific community that communication and trust are the two most important factors for the successful execution of pilot maneuvers [9,10]. In addition, communication is considered essential for successful decision-making and for building SA [6]. A survey of pilots confirms that communication is a key factor for successful piloting and that advanced technologies that facilitate communication are seen as a prerequisite for remote pilotage [6]. Besides human factors, communication stability, as well as proper officer qualification are seen as key enablers for such services [3].

2.2. Maritime Mixed Reality

The approach of developing a remote piloting system based on VR and AR technology has not yet been implemented. Modern ship navigation on a bridge is already supported by a variety of sophisticated digital and automated tools such as radar, automatic radar plotting aids, and electronic navigation systems. However, the International Regulations for Preventing Collisions at Sea (COLREGS) apply. Rule 5 requires that *every vessel shall at all times keep a proper lookout by sight and hearing and by any other available means appropriate to the circumstances and conditions, giving a full view of the situation and the possibility of a collision* [11]. It is expected that AR technology will allow bridge personnel to fulfill their duty to keep a lookout while having the information from the modern tools at their disposal.

There are already several research projects investigating AR in ship navigation [12,13] and maritime traffic control [14]. In most cases, augmentation is limited to overlaying the navigator's field of view with AIS data. Furthermore, most implementations have a low technical readiness level (TRL). In a systematic literature review on AR in maritime collaboration, Van den Oever et al. [15] concluded that it would be more advantageous to develop prototypes with a higher TRL and criticized the lack of scientific evaluation of the prototypes developed so far.

Until now, virtual reality in the maritime sector is primarily used for trainings. An exception is the project FernSAMS, where a VR setup is used to steer a tugboat from shore with a 360-degree video stream as the primary sensor input [16,17]. This has led to the MR demonstrator for remote pilotage demonstrated in [4], which is used here for the technology assessment. To the knowledge of the authors, there exist no further approaches for VR-based remote pilotage stations at the moment.

3. System Overview

This section outlines the development and deployment of a mixed reality infrastructure aimed at facilitating remote pilotage operations. An in depth analysis of the infrastructure can be found in the paper *Use Case Remote Pilotage—Technology Overview* [4]. Focusing on augmenting SA and interaction between the ship's captain and remote pilots, this infrastructure integrates augmented reality on the ship side with a desktop and a virtual reality application for the shore side. Notably, all system validations, including the shore-side application, were conducted onboard a vessel, reflecting a unique approach to evaluating the interaction dynamics under real navigational conditions. All applications were developed in the Unity game engine (Unity Technologies, San Francisco, CA, USA).

3.1. System Infrastructure

Data acquisition is systematically facilitated through a comprehensive network of onboard sensors, capturing vital navigational inputs such as GNSS and AIS signals. These sensors channel data to a central processing unit. A general overview can be seen in Figure 1.

It is worth mentioning that the setup for the user study was different from the previously mentioned conceptional setup. For logistical and organizational reasons, the shore-side equipment was also placed on the ship during testing. Therefore, the data exchange was performed via a WiFi router instead of a 4G/5G connection. This paper will continue to refer to the respective interfaces as the shore UI/shore application and the ship UI/ship application for clarity and consistency with the initial conceptual framework.

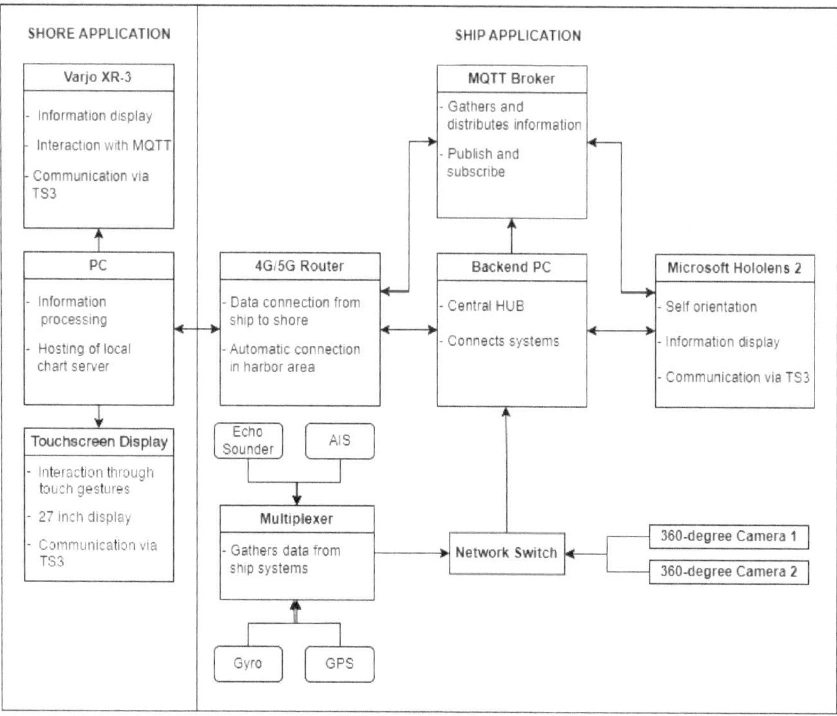

Figure 1. Conceptional system overview: for the user study, a WiFi-connection was used instead of 4G/5G (adaption of [4]).

3.2. Information Exchange between Ship and Shore

To ensure a high SA of the pilot, sensory data are transmitted from ship to shore. These data include AIS and a 360° video stream. In our setup, the GeoVision VR360 camera was used. It produces a H264-encoded RTSP video stream with a resolution of 3840*2160 pixels. One of the installed cameras faces forward from its position directly above the ship's bridge. The second camera is mounted on the port side at the forwardmost corner, also above the ship's bridge.

Our system offers verbal communication between ship and shore via a VoIP connection. Additionally, we aim to compensate non-verbal communication with a system of markers and clues that can be placed by the pilot and give the master additional information. These are as follows:

- Markers for guiding attention towards specific coordinates in the environment.
- Bearing line for guiding attention towards a specific direction.
- Highlighting for guiding attention towards specific vessels in the environment.
- A message system with predefined messages for ensuring safe communication about a chosen target.

3.3. Shore-Side Implementation

The shore-side system was implemented in two variations: a desktop solution and a VR solution. The former runs on a 27-inch touch screen for interaction purposes, while the later uses the Varjo XR-3 headset with hand tracking capabilities (see the *Shore Application* column in Figure 1). A VR desktop application requires significant computational power to deliver a smooth and immersive experience. For our application, we used an Alienware laptop equipped with an NVIDIA RTX 3070 graphics card and an Intel CPU from the 13th generation. Communication with the ship's system was established via local WiFi, as both systems were placed on the ship. As the focus was on usability, this modification was acceptable during testing. The available information in both variants is the same, and the user interfaces are designed to be as similar as possible. Both variants present the live video stream from the 360° cameras on the ship and an electronic sea chart[1]. It is possible to change the displayed chart section and to alter the scale of the map. Essential information about the ship, namely heading, course, and speed, is always visible. Symbols for the vessels in the environment are shown on the chart according to the received AIS data. These symbols can be selected to open a panel with detailed information about a vessel. The info panel also offers the options for tagging a vessel with a highlight and attach it with a message. Markers for the purposes of highlighting specific locations for the master can be placed on the chart.

3.3.1. Desktop Application

The layout of the UI of the desktop application can be seen in Figure 2. The interface is divided between a section for the video stream and a section for the chart. The chart can also be closed so that the video stream is shown as full screen. The interface shows a maximum 140° section of the 360° video stream. On top of the video stream, the compass element and the ship information are placed.

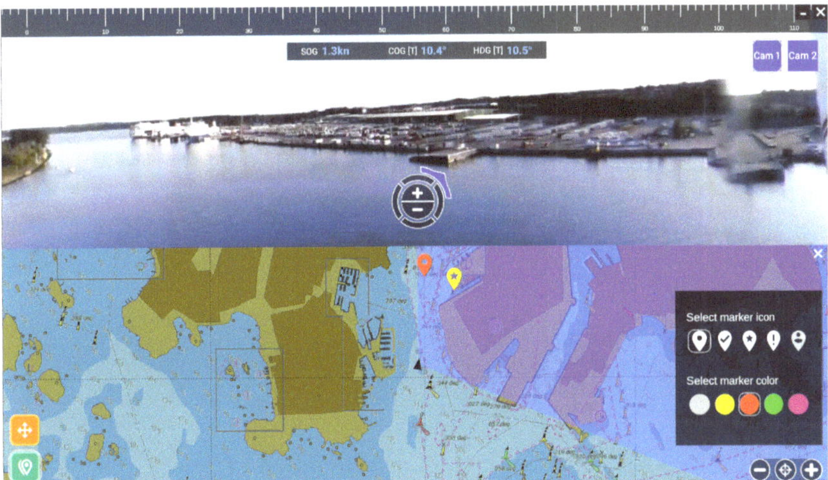

Figure 2. Shore-side desktop UI in split-screen mode.

All interactions with the application are performed via touch gestures on the display surface. The selection of UI elements is performed via a single touch; changing the chart section and the displayed section of the video stream is performed via swipe gestures. Zooming in and out of the chart and the video stream is performed via two-finger gestures.

Figure 2 shows the UI in split-screen mode.

3.3.2. Virtual Reality Application

In the virtual environment, the user is surrounded by a sphere on which the video stream is projected. The chart is placed as a 3D element in front of the user; its position can be changed between predefined options or be disabled completely if needed. The compass element is implemented as a ribbon in the upper part of the users field of view.

Figure 3 shows the user interface of the VR application. Interaction with the environment is performed via the hand-tracking capabilities of the headset. Rays from the shoulder to the hand intersecting with the UI elements in the environment define the cursor position. A finger pinch gesture selects the element the cursor is currently on.

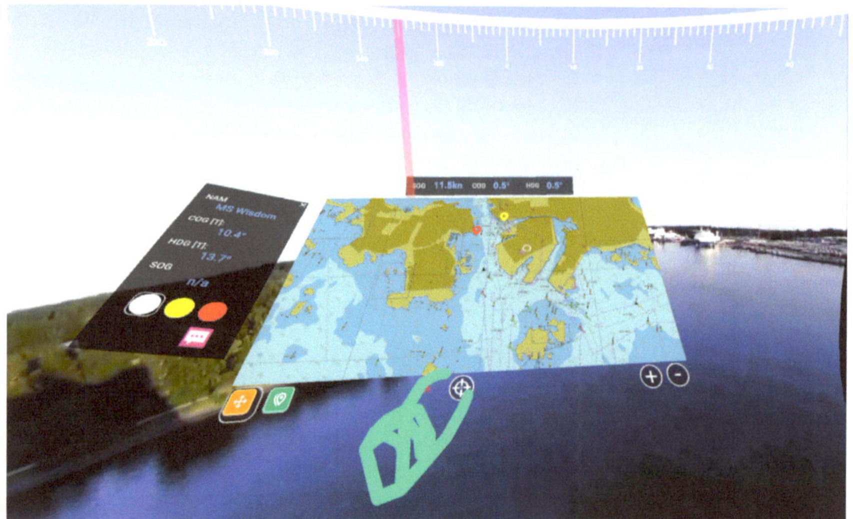

Figure 3. Shore-side virtual reality user interface inside 360° video environment with hand-tracking rig.

3.4. Ship-Side Implementation

The ship-side system harnesses the capabilities of the Microsoft Hololens 2 (Microsoft Corporation, Redmond, WE, USA) to superimpose critical navigational data, including AIS information and route specifics, directly into the captain's field of vision. This integration is designed to augment the physical maritime environment with digital data.

Figure 4 shows an AR overlay with navigational data for a vessel called "Testship". On the left side is the vessel information collapsed behind an AR button. After interacting with the button, the information on the right side is displayed. Essential information such as heading, course, speed, and status is presented alongside the vessel's position. A 3D wireframe model provides visual context. Additionally, collision avoidance details like CPA and TCPA are shown.

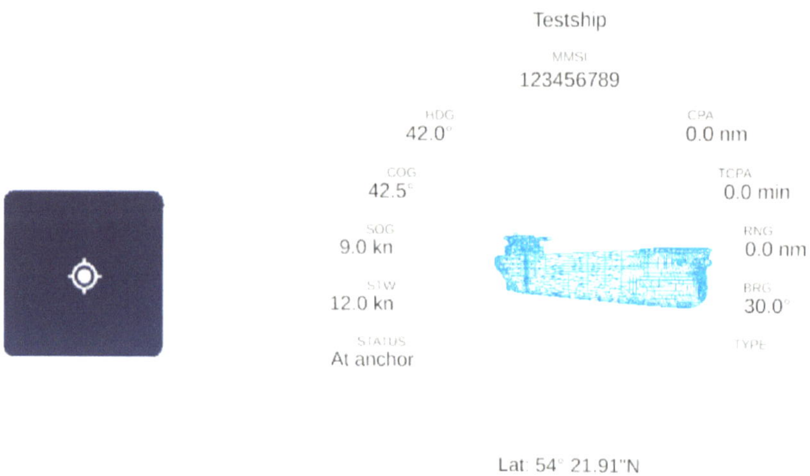

Figure 4. Ship information visualized in AR.

Interaction with the Hololens 2 system is intuitively designed to occur through hand tracking and hand gestures, alongside voice commands. Specific gestures allow for the expansion or collapse of ship information by pinching the thumb and pointer finger. Additionally, looking at the right-hand palm activates the display of the vessel's own navigational information, while glancing at the left palm summons a menu for layer management (Figure 5). This feature permits the toggling of various data layers, such as ships, markers, and waypoints.

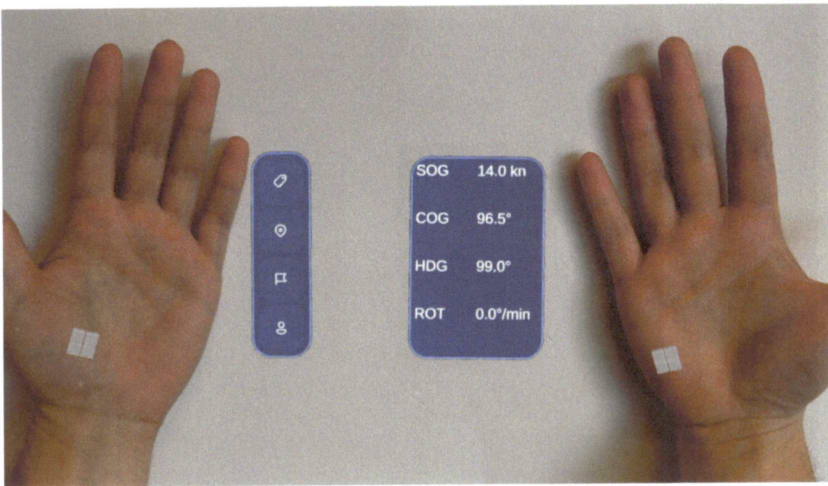

Figure 5. Layer menu and ship information as seen through the Hololens 2.

Figure 6 delineates the overlays in greater detail. Besides AIS data, the overlays include route information and markers, with AIS details encircling identifiable objects. Depending on the type of vessel, a 3D wireframe model is also presented, enhancing the spatial understanding of nearby maritime traffic. Ships that are highlighted via the shore-side application are marked with arrows at the field of view edges if positioned outside the captain's immediate visual range. Moreover, a bearing indicator from the shore-side application can be visualized within the Hololens 2, ensuring synchronized navigation and planning between the ship and remote pilots.

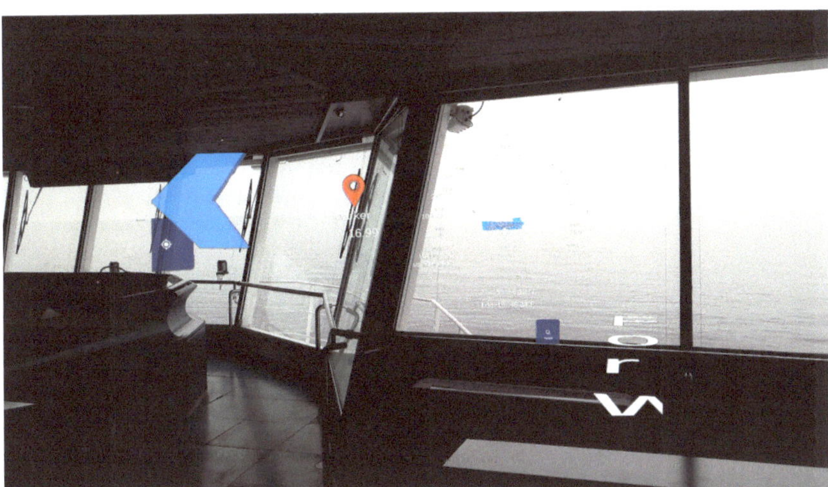

Figure 6. View through the Hololens 2.

The technical setup was already installed and tested in voyages previous to the user study. During these voyages, informal feedback was collected from the ship's crew and integrated into the prototype. To improve the system further, it was deemed necessary to collect systematic feedback from pilots in an in situ test run.

4. Testing and Evaluation

The testing procedure consists of several campaigns in which the usability, intuitive operation, effectiveness, and accuracy of the systems are evaluated. The results aim to determine the potential of each system to establish maritime SA for safe and robust remote pilotage and, where possible, provide insight into which enhancements could improve these systems. The test setup is divided into four phases aimed at assessing different aspects of technology application in a maritime context (see Figure 7).

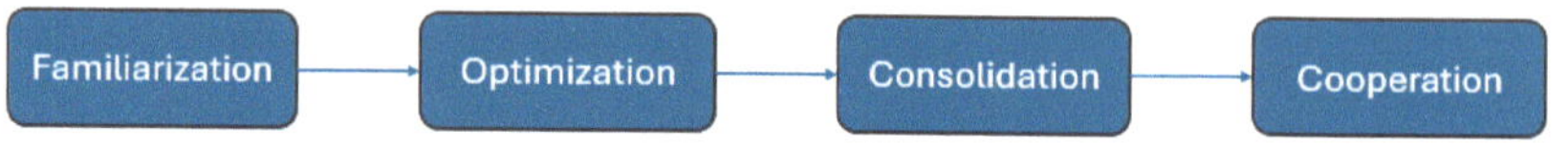

Figure 7. The four phases of the testing procedure.

4.1. Survey Methods Applied

During testing, three survey methods were employed that are utilized in human–computer interaction research, particularly with respect to immersive systems such as AR and VR. These methods aid in understanding and evaluating user experience, system performance, and the physical and psychological impacts of these technologies on users:

- *System Usability Scale (SUS):* The SUS is an effective tool for assessing the usability of a system. It consists of a brief questionnaire with 10 items, providing a quick gauge of how user-friendly a system or product is [18]. The questions are designed to be generic, making them applicable to a wide range of products or systems, including AR and VR. It scores on a scale from 0 to 100, though it is not a percentage. Scores can be categorized into ranges: scores from 85 to 100 indicate excellent usability; scores from 68 to 84 reflect good usability; scores from 51 to 67 are considered marginal; scores below 50 are deemed poor, highlighting significant usability issues. It is critical to acknowledge that the SUS does not provide detailed insights into the specific issues or potential enhancements of a system. Rather, the SUS serves as a general metric for a system's usability and can be employed as a benchmark for comparing different systems or iterations of the same system.
- *Situational Awareness Rating Technique (SART):* SART was developed to assess the level of SA experienced by users during system interaction [19]. It quantifies SA [20]. This is particularly critical in AR and VR environments, as these technologies aim to seamlessly integrate digital information into the user's visual surroundings. A high level of SA implies that users can effectively perceive, comprehend, and respond to information provided by the system. When using this method, subjects rate their own awareness using a questionnaire consisting of ten questions. This questionnaire uses a 7-point scale, with the lowest score being 1 and the highest score being 7. Based on these scores, the questions are divided into three categories (understanding, demand, and supply) to calculate SA. The SART scores range from -14 to 46.
- *Simulator Sickness Questionnaire (SSQ):* The SSQ is used to assess symptoms of simulator sickness (also known as cybersickness) that may occur when using MR systems. Symptoms include nausea, disorientation, and discomfort [21]. The SSQ assists developers and researchers in identifying system aspects that may cause discomfort, to improve them and optimize user experience.

In addition to the traditional surveys, application-specific questionnaires were incorporated and discussed with the participants during the testing campaign, i.e., AIS Data and Efficiency, Live Video Stream, and General Questions.

4.2. Phase 1: Familiarization

The introduction of new technologies requires an adjustment period in which users can become acquainted with the functionalities and possibilities of these technologies. This is crucial for acceptance and effective use. In the first test run, the participants were familiarized with the AR, VR, and desktop systems. This allowed them to become comfortable with the functions and operating methods of the systems. The benefits of this familiarization are that the participants develop a better understanding of the functionality of the systems and can assess their user-friendliness:

- Objective: determination of the adjustment time and comfort of the participants with the various technologies (AR, VR, desktop).
- Methodology: Participants are systematically guided using the AR, VR, and desktop systems. The SUS and SSQ surveys provide quantitative data on usability and potential physical impairments caused by the technology. Here, an initial baseline of usability and user comfort is examined, as well as the effects of the technologies on the users.

4.3. Phase 2: Optimization

In this testing campaign, the usability and intuitive operation of the systems were evaluated. Participants were asked to perform tasks with the various systems and then conduct the SUS and SSQ surveys. Intuitive operation and user-friendliness are crucial for the effectiveness of the systems in critical situations. High usability reduces the risk of operating errors and increases acceptance among users. This phase focuses on how intuitively and effectively the systems can be applied by end-users, which is crucial for the practical implementation of the technologies. Smaller features and changes have been implemented in the applications for an optimized routine in the following tests:

- Objective: evaluation of the usability and intuitive operation of the shore-side and ship-side systems.
- Methodology: performing tasks using the various systems and subsequent evaluation through the SUS and SSQ to quantify usability. SSQ surveys are used to monitor the well-being of the users.

4.4. Phase 3: Consolidation

This phase was designed to evaluate how effectively and accurately the mixed reality technologies could deliver specific, critical information and support users in executing precise maneuvers. By guiding participants through a series of predefined tasks, the tests assessed the capacity of both the VR and AR infrastructure, as well as the desktop and AR setup to provide real-time assistance in navigational decision-making. With developers on the application's opposite end, the evaluation was structured to simulate a realistic pilotage scenario where developers relayed tasks to the pilots via VoIP communication. The interaction of the pilots with the systems has been consolidated in this phase to prepare for the next one:

- Objective: to investigate the effectiveness and accuracy of both VR and AR systems, as well as the desktop and AR setup in handling specific pilotage tasks.
- Methodology: Each pilot participant was required to test every system once to evaluate the systems' efficiency and accuracy. The developers, serving as the counterparts in these tests, assigned a set of tasks that the pilots had to perform using the systems. These tasks included the following:
  - Highlighting a specific ship within the visual field.
  - Setting a status message for a ship to communicate its operational condition.
  - Placing a marker in proximity to their vessel or on top of other ships to designate points of interest or navigational relevance.
  - Adjusting the bearing indicator to aid in the navigation and orientation process.

This phase aimed to provide insights into the user's ability to complete navigation-specific tasks effectively with each system and to determine the operational accuracy of the mixed reality infrastructure.

4.5. Phase 4: Cooperation

Phase 4 is pivotal in assessing the independent operational capability of the mixed reality systems by examining the direct interaction between pilots without developer intervention. This phase focuses on the core of maritime operations: the effective collaboration and communication between ship masters and pilots under authentic conditions:

- Objective: to evaluate the independent usability and efficiency of the shore-side and ship-side applications for cooperative maritime tasks.
- Methodology: Pilots enacted typical maritime scenarios, including vessel entry and departure from a harbor, by collaboratively performing tasks using the mixed reality systems. These tasks were designed to simulate the coordination required during actual pilotage without external assistance, ensuring the systems facilitate effective pilot-to-pilot interaction.

This phase of testing emphasizes real-world applicability and the self-sufficiency of the pilots in utilizing the systems' collaborative tools. The execution and repetition of these scenarios contribute to the iterative enhancement of system performance, mirroring the dynamic and sometimes unpredictable conditions of maritime navigation. Through these exercises, the systems' potential to reinforce safety and improve operational fluidity in maritime navigation is rigorously examined.

4.6. Test and Interpretation Notes

In total, 42 tests with 3 test participants and 3 systems have been conducted. Table 1 gives an overview about the executed test schedule. With regard to the interpretation, it must be noted that the test faced limitations concerning the diversity and number of test persons. While all participants had professional maritime backgrounds either as pilots or navigator, they were all male and in the age range from 45 to 65. This is representative of a pilotage peer group as of today, but of course, the scope for evaluating the system's usability and effectiveness across a broader demographic spectrum was consequently constrained. Additionally, as testing took place under in situ conditions, the external maritime traffic was representative of real-world situations, but of course not controllable and comparable between all phases of testing, which could leave room for different interpretations, as system and traffic assessment can be interlinked by human test participants.

Table 1. Executed test schedule according to phases.

Phase	Date	Time	Test Person and Topic	Test Person and Topic
Phase 1	11.03.	11:00–11:10	Person A (AR)	Person B (Desktop)
Phase 1	11.03.	11:20–11:30	Person B (AR)	Person C (Desktop)
Phase 1	11.03.	11:40–11:50	Person C (AR)	Person A (Desktop)
Phase 1	11.03.	12:10–12:20	Person A (VR)	Person B (AR)
Phase 1	11.03.	12:30–12:40	Person B (VR)	Person C (AR)
Phase 1	11.03.	12:50–13:00	Person C (VR)	Person A (AR)
Phase 2	11.03.	15:00–15:15	Person A (AR)	-
Phase 2	11.03.	15:15–15:30	Person A (VR)	-
Phase 2	11.03.	15:30–15:45	Person A (Desktop)	-
Phase 2	11.03.	16:00–16:15	Person B (AR)	-

Table 1. *Cont.*

Phase	Date	Time	Test Person and Topic	Test Person and Topic
Phase 2	11.03.	16:15–16:30	Person B (VR)	-
Phase 2	11.03.	16:30–16:45	Person B (Desktop)	-
Phase 2	11.03.	17:00–17:15	Person C (AR)	-
Phase 2	11.03.	17:15–17:30	Person C (VR)	-
Phase 2	11.03.	17:30–17:45	Person C (Desktop)	-
Phase 4-1	12.03.	08:00–08:10	Person A (AR)	Person C (Desktop)
Phase 4-1	12.03.	08:20–08:30	Person C (AR)	Person B (Desktop)
Phase 4-1	12.03.	08:40–08:50	Person B (AR)	Person A (Desktop)
Phase 4-1	12.03.	09:10–09:20	Person A (Desktop)	Person C (AR)
Phase 4-1	12.03.	09:30–09:40	Person C (Desktop)	Person B (AR)
Phase 4-1	12.03.	09:50–10:00	Person B (Desktop)	Person A (AR)
Phase 4-2	12.03.	15:00–15:10	Person A (VR)	Person B (AR)
Phase 4-2	12.03.	15:20–15:30	Person B (VR)	Person C (AR)
Phase 4-2	12.03.	15:40–15:50	Person C (VR)	Person A (AR)
Phase 4-2	12.03.	16:10–16:20	Person A (VR)	Person B (AR)
Phase 4-2	12.03.	16:30–16:40	Person B (VR)	Person C (AR)
Phase 4-2	12.03.	16:50–17:00	Person C (VR)	Person A (AR)
Phase 3	13.03.	10:30–10:45	Person B (AR)	-
Phase 3	13.03.	10:45–11:00	Person B (VR)	-
Phase 3	13.03.	11:00–11:15	Person B (Desktop)	-
Phase 3	13.03.	11:15–11:30	Person C (AR)	-
Phase 3	13.03.	12:00–12:15	Person C (VR)	-
Phase 3	13.03.	12:15–12:30	Person C (Desktop)	-
Phase 3	13.03.	12:30–12:45	Person A (AR)	-
Phase 3	13.03.	12:45–13:00	Person A (VR)	-
Phase 3	13.03.	13:00–13:15	Person A (Desktop)	-
Phase 4-3	13.03.	19:00–19:10	Person A (AR)	Person C (Desktop)
Phase 4-3	13.03.	19:20–19:30	Person C (AR)	Person B (Desktop)
Phase 4-3	13.03.	19:40–19:50	Person B (AR)	Person A (Desktop)
Phase 4-3	13.03.	20:10–20:20	Person A (VR)	Person C (AR)
Phase 4-3	13.03.	20:30–20:40	Person C (VR)	Person B (AR)
Phase 4-3	13.03.	20:50–21:00	Person B (VR)	Person A (AR)

5. System Assessment and User Feedback

During the testing phase, participants provided valuable insights into the usability, functionality, and practical limitations of the systems for remote pilotage. Their feedback is instrumental in identifying areas for improvement and potential enhancements to the system.

5.1. User Interface and Usability

Participants reported that both the AR and VR headsets, along with the screen interface, were generally user-friendly and intuitive to operate. However, they noted that accurately selecting options using the AR headset required a period of adjustment. Concerns were raised about the image quality provided by the 360° video stream, specifically mentioning that it was insufficient for operational needs. Additionally, the zoom functionality did not meet expectations, and the system's performance during night-time operations was deemed inadequate. Prolonged use of the VR headset was found to cause eye fatigue, suggesting a need for further ergonomic optimization to ensure user comfort during extended periods of use.

Despite these challenges, the overall system architecture aligns with standard nautical procedures and workflows, making it usable in principle. It was acknowledged, however, that the system prototype has not yet achieved full operational readiness, but served as a technology demonstrator. In terms of usability, the overall SUS score with an average of 68.33 across participants indicates that the system is on the threshold of above-average usability (see Figure 8). Given the small group of test participants, this is, however, only a first indication and not representative, as it can also be seen that the assessment differs between the participants in height, as well in order of preference.

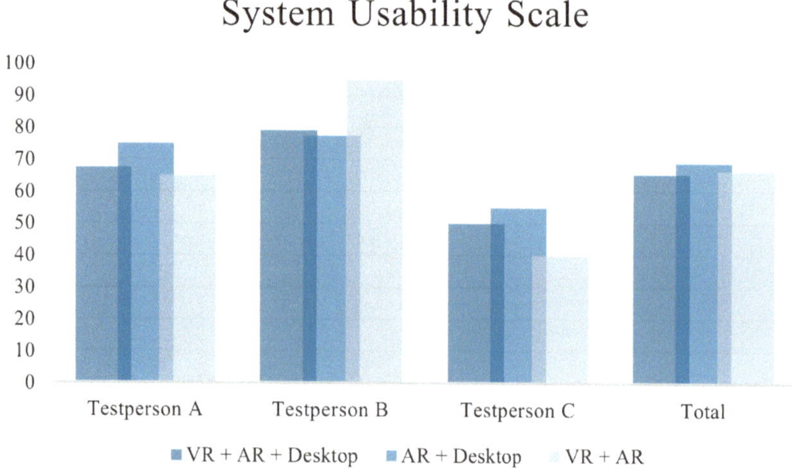

Figure 8. System Usability Scale Test Persons A, B, and C.

5.2. Situational Awareness

The average values of the respondents' ratings for the individual questions are shown in a graph (Figure 9). The computed SA scores of all the participants were above the middle of this range, indicating that they had good SA during the test scenario.

The overall SA score is a calculated value from the three dimensions.

It is described as $SA = Understanding - (Demand - Supply)$ [22]:

- "Demand" quantified the extent of human awareness processes during the simulation. (represented by Question 1, 2, and 3).
- "Supply" represented the available cognitive capacity and uncommitted attention available to the subject during the simulation. (represented by Question 4, 5, 6, and 7).
- "Understanding" indicated the extent to which the individual grasped the situational circumstances during the simulation. (represented by Question 8, 9, and 10).

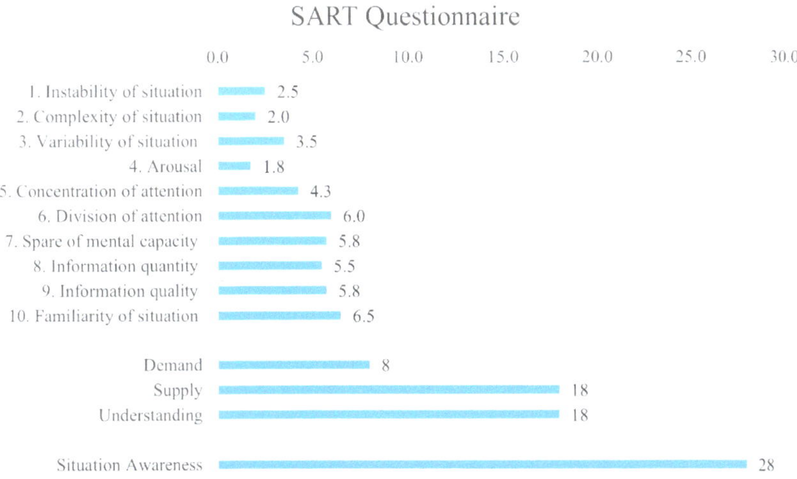

Figure 9. SART questionnaire statistical results.

SA describes the recognition, understanding, and anticipation of environmental factors and events within defined time frames, particularly in dynamic and complex contexts. SA is typically divided into three hierarchical levels [22]:

- Level 1—perception (perception of the surrounding elements).
- Level 2—Comprehension (Comprehension of the current situation).
- Level 3—projection (projection of future states).

The characteristics of effective situational awareness include the following:

- Complete and accurate perception: all relevant information is recorded completely and without error.
- Correct understanding of the situation: the meanings and interactions of the various elements are correctly interpreted.
- Effective projection of future developments: future states and developments are reliably predicted, enabling informed and proactive decisions.

In contrast, the characteristics of inadequate situational awareness are incomplete or erroneous perception, misunderstanding of the situation, and inadequate projection of future states. Endsley's theory [22] provides a central model of SA that emphasizes the importance of the three levels and shows how effective SA is achieved through a combination of environmental perception, processing, and cognitive prediction.

Having effective SA is essential for decision-making and the ability to act in dynamic and complex environments, while inadequate SA can lead to poor decisions and ineffective actions.

5.3. VR and AR Sickness

In the context of VR, subjective symptom reports were gathered from participants during testing sessions. On March 12th, during the late session, Test Persons A and B both experienced mild fatigue prior to engaging with the VR prototype. Following the test, Test Person A reported symptoms of mild eye strain and a sensation of fullness in the head. Test Person B experienced only mild eye strain. Test Person C, who initially felt moderately fatigued, did not report any post-testing issues.

On the morning of March 13th, Test Person A commenced the VR application without any pre-existing simulator sickness symptoms, but subsequently reported "difficulty in focusing". Test Persons B and C did not report any simulator sickness symptoms either before or after their VR tests.

Later that day, Test Person A maintained an absence of simulator sickness symptoms both before and after the VR test. Conversely, Test Persons B and C both noted mild fatigue related to simulator sickness in the pre- and post-testing phases.

These findings indicate variability in the manifestation of simulator-related symptoms among participants within the VR environment, with some experiencing mild discomfort and others reporting no adverse effects. The symptoms observed, such as eye strain and difficulty focusing, align with known indicators of simulator sickness and highlight the need for further research into their etiology and potential mitigation within VR development.

The evaluation of the SSQ is based on the assessment of the three primary dimensions of simulator sickness: nausea, ophthalmic (visual) symptoms, and discomfort. The interpretation of the SSQ should be considered in conjunction with other factors such as the test environment, duration of exposure, and individual experiences. Since the test participants conducted the VR sessions standing for a relatively short duration (10–15 min), the results were, as expected, positive. For future tests, longer standing durations should be considered. Additionally, tests should be planned that involve sitting and the combination with desktop development within a single test session. It must be further noted that the VR system was presented in a more challenging way, as the shore application would normally be on hard ground and not on a moving vessel, the movements of which are neither aligned nor synchronized with the movements within the VR system.

5.4. Comparative Analysis of VR, AR, and Desktop

A comparison between the VR and AR headsets and the desktop interface highlighted the distinct presentation styles of each platform. Notably, the VR headset, while offering an immersive experience, was considered to have limited practical viability for prolonged use due to eye strain. The AR headset, on the other hand, posed fewer issues regarding extended wear, though it was observed to diminish overall sensory perception. Such limitations could potentially lead to disadvantages in nautical practice, raising questions about the system's effectiveness in critical scenarios, such as a fire alarm on the bridge.

In the course of the test campaign, the test subjects had the opportunity to test all three systems. As part of these tests, there was constant feedback on the comfort of the systems, how intuitive they were to use, and comparisons with similar conservative systems. The results are shown in the following graphs. It can be seen that the comfort of the systems and the intuitive operation were always above average. For the desktop system, touch operation was preferred to a conventional setup.

The VR glasses (Figure 10) are characterized by several functions, in particular the availability of the electronic chart applications, which allows navigation information to be integrated into the VR environment. The abilities to enter data and access ECDIS-like data are also important features. The 360-degree overview of the situation and the intuitive user interface improve SA. The full view in the camera supports the visual capture of the surroundings.

The availability and use of chart information within the VR environment were seen as particularly useful. Although the functions are not yet fully developed in the current development phase, the potential for the use of VR is considered to be conceivable and promising for the future. This could even lead to the integration of existing systems and proven applications like PPU software in the VR environment.

The Hololens 2 AR glasses (Figure 11) can make work more difficult in certain situations, especially when several ships are on the same bearing. Problems can arise here if there is an overflow of data and too many visualizations block the view. A key point mentioned was that radar and ECDIS are considered more reliable information systems and the AR glasses only have a supplementary function. The glasses can also interfere with normal vision and make it difficult to find relevant information quickly, which is particularly problematic during maneuvers.

Useful functions of the setup are the display of ship data, interaction with the AIS system, the ability to deactivate information, the three-dimensional representation of ships, the distance to waypoints, and the color coding of traffic vessels for better differentiation.

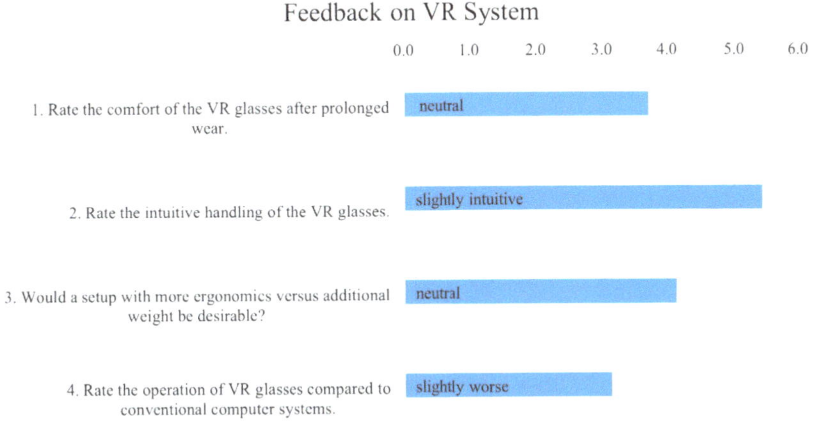

Figure 10. Feedback on the VR system regarding comfort and interactivity.

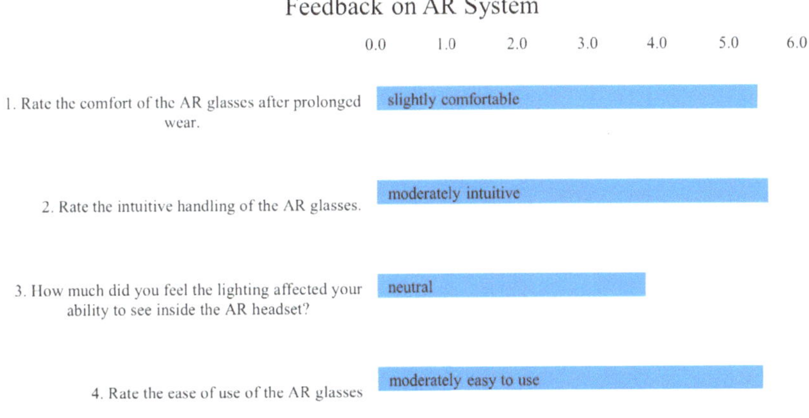

Figure 11. Feedback on the AR system regarding comfort and interactivity.

The desktop solution (Figure 12) offers a number of features that were highlighted for users. These features include the ability to interact between the map and live view, including centering the camera, improved overall operation on the desktop, selecting and tagging other ships, and intuitive handling.

Sending messages appears to be particularly useful, although according to the test subjects, it involves certain risks that still need to be assessed. However, it was noted that the available functions may not be sufficient to make well-founded decisions. This is due to the early stage of development.

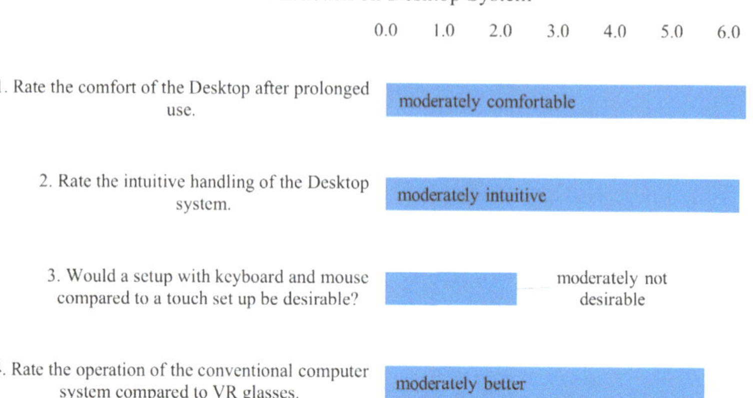

Figure 12. Feedback on the desktop system regarding comfort and interactivity.

5.5. Interaction Design

The system's interactions were designed to be user-friendly, with a specific emphasis on minimizing the need for a physical controller. However, feedback suggested that foregoing a controller entirely should only be considered after comprehensive and effective training to ensure complete operational proficiency and safety.

6. Discussion

The development process during the testing phase was characterized by iterative enhancements and responsive adaptations. An integral finding was the inconsistency in hand tracking within low-light conditions, a common scenario on the bridge at night. The precise positioning of the user's hand relative to the Hololens 2 camera proved crucial; it was noted that users had to slightly tilt their hands to ensure visibility for the camera to register the interaction, which was not entirely intuitive. Moreover, the standard settings of the Hololens 2 presented challenges in dynamic maritime environments. The system, which relies on built-in cameras and a gyroscope for spatial tracking, faced difficulties with the ship's movement, especially when dealing with ocean waves. The gyroscope's detection of tilting conflicted with the camera's perception of a stationary bridge. To address this, enabling the *Moving Platform Mode* was a critical adaptation, minimizing the gyroscope's influence and stabilizing the AR experience amidst the vessel's motion. Consequently, a maritimization of AR and VR equipment is needed, if commercial onboard operation is intended.

While the participants refrained from evaluating whether the prototypes could fully replace onboard pilots at this stage of the project, they concurred with the paper's outlined advantages and disadvantages. Looking ahead, the addition of a night vision feature by including infrared/thermal imaging was proposed as a beneficial enhancement, particularly for operations in foggy conditions. This indicates potential directions for future development in the VR application.

The quality and communication speed over VoIP was considered very good by all participants.

The journey to integrate mixed reality technologies in maritime navigation is ongoing, and this analysis indicates that at least a similar usability level with respect to desktop solutions can be achieved already today on a prototype level. However, the ultimate objective remains clear: to create a system that not only aligns with, but enhances the natural workflows of maritime professionals, thereby promoting safety, efficiency, and precision in the complex domain of seafaring considering the best technology options for the respective tasks.

7. Conclusions and Future Work

Enabling remote pilotage requires solutions addressing the communication and SA challenges inherit to this novel concept. Within this paper, different technology options for visualization and interaction have been initially tested to support technology scouting when setting up these systems in a safe and interactive manner. Given the results with the peer group of three experienced nautical persons, it can be noted that the classical desktop, as well as immersive mixed reality reached a comparable usability on the SUS score. All passed the SUS threshold for *good usability* on average and reached a good SA level in SART. Consequently, this indicates that all three technology options are suitable in principle. However, the limitation of the peer group prevents a comprehensive understanding of how the system would perform or be received by users of varying ages, experience levels, and roles within the maritime industry.

Despite the principle usability of the prototype, several technical constraints have also been identified that need ergonomic enhancements to fully exploit the technologies for the remote pilotage use case properly. First, high-resolution cameras with better specifications than the GeoVision VR360 are necessary to provide clear, detailed images with minimal latency and high frame rates. This is essential for maintaining accurate situational awareness (SA). The zoom function sufficiency is another important aspect. The system's zoom capabilities must maintain image clarity and detail even when magnified. High-quality digital zoom functions are required to avoid significant loss of resolution and minimize noise, ensuring that small details remain visible and actionable. Regarding augmented reality (AR) and virtual reality (VR), reducing eye fatigue for prolonged use is crucial. The hardware must incorporate advanced display technologies to mitigate eye strain. This includes implementing high refresh rates, adaptive brightness, and ergonomic design to facilitate extended use without causing discomfort to the users. Hand tracking consistency is also vital for immersive AR and VR systems. The Varjo XR-3 and Hololens 2 must ensure precise and responsive gesture recognition, allowing users to interact intuitively with virtual elements. This level of accuracy and responsiveness is critical. Lastly, the stability and reliability of AR overlays must be ensured. The Hololens 2 must deliver stable and accurate overlays that provide essential navigational data reliably. This includes superimposing AIS information, route specifics, and collision avoidance details onto the captain's field of view. The system must be robust enough to withstand maritime environmental challenges, such as ship motion and varying light conditions, ensuring continuous and reliable support for safe navigation. By addressing these technical constraints, the Remote Pilotage Technology system can effectively support remote pilotage operations, enhancing both safety and efficiency without compromising navigational integrity.

Future work shall include extended test durations. The combination of desktop and immersive technologies within a single test session also needs to be examined to offer pilots the flexibility to choose the interface that best suits the task at hand.

Author Contributions: Conceptualization, A.U., P.H., R.G. and H.-C.B.; methodology, A.U., P.H. and R.G.; state of the art, P.H.; supervision, H.-C.B.; technical implementation, A.U. and P.H.; testing, A.U. and P.H.; assessment, R.G.; figures, R.G.; writing—original draft preparation, A.U., P.H. and R.G.; writing—review and editing, H.-C.B.; introduction, P.H. and H.-C.B.; conclusion and discussion, R.G. and H.-C.B. All authors have read and agreed to the published version of the manuscript.

Funding: This research was internally funded within the Fraunhofer Innovation Platform for Smart Shipping FIP-S2@Novia collaboration by Fraunhofer CML.

Institutional Review Board Statement: Not applicable.

Informed Consent Statement: Not applicable.

Data Availability Statement: The SUS, SART, and SSQ data sets from the test runs are available upon request from the corresponding author.

Acknowledgments: The authors would like to thank the three test participants for their willingness to join a three-day ferry ride in winter during these tests, as well as the ferry operator for facilitating these tests. Specifically, the authors would also like to thank Niclas Seligson for his support in facilitating this research in practice.

Conflicts of Interest: The authors declare no conflicts of interest.

Abbreviations

The following abbreviations are used in this manuscript:

AIS	Automatic Identification System
AR	augmented reality
COLREGs	Convention on the International Regulations for Preventing Collisions at Sea
CPA	Closest Point of Approach
ECDIS	Electronic Chart Display and Information System
MR	mixed reality
RTSP	Real-Time Streaming Protocol
SA	situational awareness
SART	Situational Awareness Rating Technique
SSQ	Simulator Sickness Questionnaire
SUS	System Usability Scale
TCPA	Time to Closest Point of Approach
TRL	technological readiness level
UI	user interface
VHF	very high frequency
VoIP	Voice over Internet Protocol
VR	virtual reality
VTS	Vessel Traffic Service

Note

[1] Within this prototype, basic sea chart implementations have been used. The authors are aware of the existence of special screen-based pilotage software, so-called Portable Pilot Units (PPUs), that offer pilotage-specific chart applications. For a final commercial system, the PPUs' functionalities or even the software itself could be integrated here so that the pilots can benefit from their established set of known functionalities also remotely, but as this was not the focus of this usability tests, simplified prototypes have been used. It is, however, important to note that Remote Pilotage Systems are not a substitute for PPUs, but that the authors recommend fully integrating PPUs into such systems in the future for improved SA and to smooth the transition between onboard and remote pilotage execution.

References

1. Wild, C.R.J. The Paradigm and the Paradox of Perfect Pilotage. *J. Navig.* **2011**, *64*, 183–191. [CrossRef]
2. Sakar, C.; Sokukcu, M. Dynamic analysis of pilot transfer accidents. *Ocean. Eng.* **2023**, *287*, 115823. [CrossRef]
3. Heikkilä, M.; Himmanen, H.; Soininen, O.; Sonninen, S.; Heikkilä, J. Navigating the Future: Developing Smart Fairways for Enhanced Maritime Safety and Efficiency. *J. Mar. Sci. Eng.* **2024**, *12*, 324. [CrossRef]
4. Grundmann, R.; Ujkani, A.; Weisheit, J.; Seppänen, J.; Salokorpi, M.; Burmeister, H.C. Use Case Remote Pilotage—Technology Overview. *J. Phys. Conf. Ser.* **2023**, *2618*, 012007. [CrossRef]
5. Hadley, M.; Pourzanjani, M. How remote is remote pilotage? *WMU J. Marit. Aff.* **2003**, *2*, 181–197. [CrossRef]
6. Lahtinen, J.; Valdez Banda, O.A.; Kujala, P.; Hirdaris, S. Remote piloting in an intelligent fairway—A paradigm for future pilotage. *Saf. Sci.* **2020**, *130*, 104889. [CrossRef]
7. Rotterdam Port Authority. Port Information Guide. 2023. Available online: https://www.portofrotterdam.com/sites/default/files/2023-01/port-information-guide_0.pdf (accessed on 1 March 2024).
8. Verbeek, E. Shore Based Pilotage, a matter of trust. *Seaways* **2021**, *10*, 6–8.
9. Bruno, K.; Lützhöft, M. Shore-Based Pilotage: Pilot or Autopilot? Piloting as a Control Problem. *J. Navig.* **2009**, *62*, 427–437. [CrossRef]
10. Hadley, M.A. Issues in Remote Pilotage. *J. Navig.* **1999**, *52*, 1–10. [CrossRef]
11. IMO. *COLREG—Collision Regulations 1972*; IMO: London, UK, 2019. Available online: https://www.imo.org/en/About/Conventions/Pages/COLREG.aspx (accessed on 1 July 2021).

12. Rowen, A.; Grabowski, M.; Rancy, J.P. Moving and improving in safety-critical systems: Impacts of head-mounted displays on operator mobility, performance, and situation awareness. *Int. J.-Hum.-Comput. Stud.* **2021**, *150*, 102606. [CrossRef]
13. Okazaki, T.; Kitagawa, R.; Matsubara, K.; Kashima, H. Development of maneuvering support system for ship docking. In Proceedings of the 2017 Joint 17th World Congress of International Fuzzy Systems Association and 9th International Conference on Soft Computing and Intelligent Systems (IFSA-SCIS), Otsu, Japan, 27–30 June 2017; pp. 1–5.
14. Nađ, Đ.; Mišković, N.; Omerdic, E. Multi-Modal Supervision Interface Concept for Marine Systems. In Proceedings of the OCEANS 2019, Marseille, France, 17–20 June 2019; pp. 1–5. [CrossRef]
15. Floris van den Oever, M.F.; Sætrevik, B. A Systematic Literature Review of Augmented Reality for Maritime Collaboration. *Int. J. Human–Computer Interact.* **2023**, *2023*, 1–16. [CrossRef]
16. Burmeister, H.C.; Grundmann, R.; Schulte, B. Situational Awareness in AR/VR during remote maneuvering with MASS: The tug case. In Proceedings of the Global Oceans 2020: Singapore—U.S. Gulf Coast, Biloxi, MS, USA, 5–30 October 2020; pp. 1–6. [CrossRef]
17. Byeon, S.; Grundmann, R.; Burmeister, H.C. Remote-controlled tug operation via VR/AR: Results of an in-situ model test. *TransNav Int. J. Mar. Navig. Saf. Sea Transp.* **2021**, *15*, 4. [CrossRef]
18. Brooke, J. SUS: A quick and dirty usability scale. *Usability Eval. Ind.* **1995**, *189*, 4–7.
19. Bolton, M.; Biltekoff, E.; Humphrey, L. The Level of Measurement of Subjective Situation Awareness and Its Dimensions in the Situation Awareness Rating Technique (SART). *IEEE Trans.-Hum.-Mach. Syst.* **2021**, *52*, 1147–1154. [CrossRef]
20. Taylor, R. Situational Awareness Rating Technique (SART): The Development of a Tool for Aircrew Systems Design. In *Situational Awareness in Aerospace Operations (AGARD-CP-478)*; NATO-AGARD: Neuilly Sur Seine, France; Routledge: London, UK, 1990.
21. Kennedy, R.S.; Lane, N.E.; Berbaum, K.S.; Lilienthal, M.G. Simulator Sickness Questionnaire: An Enhanced Method for Quantifying Simulator Sickness. *Int. J. Aviat. Psychol.* **1993**, *3*, 203–220. [CrossRef]
22. Endsley, M.R. Measurement of situation awareness in dynamic systems. *Hum. Factors* **1995**, *37*, 65–84. [CrossRef]

Journal of
*Marine Science
and Engineering*

Article

YOSMR: A Ship Detection Method for Marine Radar Based on Customized Lightweight Convolutional Networks

Zhe Kang [1,2,3], Feng Ma [1,2,3,*], Chen Chen [4] and Jie Sun [5]

[1] State Key Laboratory of Maritime Technology and Safety, Wuhan University of Technology, Wuhan 430063, China; kz258852@whut.edu.cn
[2] National Engineering Research Center for Water Transport Safety, Wuhan University of Technology, Wuhan 430063, China
[3] Intelligent Transportation Systems Research Center, Wuhan University of Technology, Wuhan 430063, China
[4] School of Computer Science and Engineering, Wuhan Institute of Technology, Wuhan 430205, China; chenchen0120@wit.edu.cn
[5] Nanjing Smart Water Transportation Technology Co., Ltd., Nanjing 210028, China; sunjie@smartwaterway.com
*   Correspondence: martin7wind@whut.edu.cn

Abstract: In scenarios such as nearshore and inland waterways, the ship spots in a marine radar are easily confused with reefs and shorelines, leading to difficulties in ship identification. In such settings, the conventional ARPA method based on fractal detection and filter tracking performs relatively poorly. To accurately identify radar targets in such scenarios, a novel algorithm, namely YOSMR, based on the deep convolutional network, is proposed. The YOSMR uses the MobileNetV3(Large) network to extract ship imaging data of diverse depths and acquire feature data of various ships. Meanwhile, taking into account the issue of feature suppression for small-scale targets in algorithms composed of deep convolutional networks, the feature fusion module known as PANet has been subject to a lightweight reconstruction leveraging depthwise separable convolutions to enhance the extraction of salient features for small-scale ships while reducing model parameters and computational complexity to mitigate overfitting problems. To enhance the scale invariance of convolutional features, the feature extraction backbone is followed by an SPP module, which employs a design of four max-pooling constructs to preserve the prominent ship features within the feature representations. In the prediction head, the Cluster-NMS method and α-DIoU function are used to optimize non-maximum suppression (NMS) and positioning loss of prediction boxes, improving the accuracy and convergence speed of the algorithm. The experiments showed that the recall, accuracy, and precision of YOSMR reached 0.9308, 0.9204, and 0.9215, respectively. The identification efficacy of this algorithm exceeds that of various YOLO algorithms and other lightweight algorithms. In addition, the parameter size and calculational consumption were controlled to only 12.4 M and 8.63 G, respectively, exhibiting an 80.18% and 86.9% decrease compared to the standard YOLO model. As a result, the YOSMR displays a substantial advantage in terms of convolutional computation. Hence, the algorithm achieves an accurate identification of ships with different trail features and various scenes in marine radar images, especially in different interference and extreme scenarios, showing good robustness and applicability.

Keywords: marine radar; ship identification; lightweight convolution; feature fusion network; maritime management

Citation: Kang, Z.; Ma, F.; Chen, C.; Sun, J. YOSMR: A Ship Detection Method for Marine Radar Based on Customized Lightweight Convolutional Networks. *J. Mar. Sci. Eng.* **2024**, *12*, 1316. https://doi.org/10.3390/jmse12081316

Academic Editor: Marco Cococcioni

Received: 3 July 2024
Revised: 24 July 2024
Accepted: 31 July 2024
Published: 3 August 2024

1. Introduction

The continuous surveillance of ships within ports and designated navigational waterways represents a pivotal undertaking, serving to provide regulatory personnel and ship operators with instantaneous insights into the state of maritime passages [1]. As a widely deployed monitoring apparatus, a shore-based marine radar enables the continuous detection of ship targets in wide-ranging water areas under adverse weather conditions (e.g., rain or fog) and poor nighttime visibility. It offers expansive imaging coverage and demonstrates stable imaging performance at relatively close observation distances. Compared to detection methods like the Automatic Identification System (AIS) and Very High Frequency (VHF) radio, marine radar does not require a real-time response from ships, significantly improving the speed of obtaining navigation information. Consequently, marine radar has emerged as a critical means for ship identification in open water domains.

In the context of shore-based surveillance, marine radar systems intended for regulatory purposes commonly incorporate diverse tail display modes of varying lengths. By discerning the distinctive features of ship trails, maritime regulators and navigators can approximate absolute or relative speeds, enabling them to assess the existence of potential risks. Consequently, in this particular scenario, the prompt and effective detection of ship targets assumes paramount importance. The ship identification in marine radar images can be effectively categorized into two distinct classes, i.e., long-wake ship identification and short-wake ship identification. Long-wake ships typically display conspicuous spot and trail features that exhibit clear differentiation from the background, thereby facilitating their localization and feature extraction with relative ease. Conversely, short-wake ships present image characteristics akin to environmental elements like rocky formations and coastlines within the maritime domain. Therefore, the discriminative process for such targets becomes susceptible to numerous background interferences, resulting in substantial difficulties in their distinguishment.

In comparison to object classification conducted on natural images, ship detection in the context of marine radar introduces a relatively higher degree of complexity. Primarily, within congested waterways, only a small portion of the detected spots genuinely correspond to mobile ships, leading to a heightened presence of interference in marine radar images. This, in turn, exerts a significant influence on the precise classification and localization of ships, alongside the formidable challenges associated with clutter removal, encompassing the elimination of sea waves, atmospheric elements such as clouds and rain, as well as extraneous noise. Moreover, the observational angle between the radar system and the ships imparts a substantial impact on the resulting ship imaging, while the irregular shapes of ship trails and spots further impede the efficacy of ship identification efforts. Meanwhile, ship targets within radar images exhibit a relatively diminished occupancy of absolute pixels, usually numbering in the hundreds. This results in a reduced pool of discernible features. Furthermore, when ships draw closer to coastlines, the backscattered spot from ships and the coastal backdrop often prove indistinguishable. Thus, the extraction of effective features and subsequent detection of small-scale ships become pivotal areas of focus and complexity within the research domain. Additionally, given the widespread adoption of embedded devices within radar systems, ship identification methods designed for radar images necessitate careful consideration of the practical computational limitations imposed by the deployed hardware.

Traditional methodologies for ship detection in radar images encompass a range of techniques, including reference object calibration, filtering algorithms, and pattern identification methods. The investigation of reference object calibration approaches primarily capitalizes on image processing methods, such as thresholding and connected component extraction, to extract salient contour features pertaining to ships and the surrounding coastal boundaries. These extracted features are subsequently employed in tandem with target feature-matching techniques to effectuate the calibration of ships within the surveilled water domain [2,3]. Furthermore, filtering algorithms are judiciously applied to distill the authentic trajectories of ship spots from a multitude of traces within the image, thus enabling the acquisition of essential ship attributes encompassing temporal and positional information [4,5]. Based on an exact modeling of ship motion trajectories, the judicious application of filtering algorithms may yield efficacy. However, conventional filtering algorithms, prevalent in radar systems, may not be amenable to ship detection manifesting diminutive target scales, occlusions, or intricate environmental perturbations [6,7].

Conversely, pattern identification methods have engendered commendable results in the classification and localization of ship spots within marine radar images [8]. Notably, several non-probabilistic models have gained wide acclaim, engendering improvements in ship identification efficacy. Nevertheless, the efficacy of the aforementioned techniques may exhibit limitations when confronted with ship identification characterized by a paucity of pixel features, languid motion velocities, and formidable background interferences.

In recent times, the utilization of Convolutional Neural Networks (CNNs) in object detection algorithms has witnessed notable advancements, encompassing both single-stage algorithms [9–12] and two-stage algorithms [13,14]. These CNN-based techniques have demonstrated conspicuous advantages over traditional methodologies, particularly concerning the extraction of deep semantic information from images and the attainment of precise object localization [15]. Of particular significance, researchers have made seminal contributions to ship identification by harnessing the potential of CNNs within the domains of Synthetic Aperture Radar (SAR) and remote sensing images, yielding satisfactory outcomes [16,17]. The prevailing ship instances within the aforementioned contexts predominantly manifest as diminutive object types, invariably accompanied by the conspicuous presence of background interferences. This intrinsic similarity shared with ship identification in marine radar images thus furnishes invaluable inspiration for the formulation of the identification algorithm elucidated within this present research.

Presently, radar image-oriented ship identification methods grounded in CNN architectures have attained a moderate degree of advancement. The judicious design of backbone networks and attention mechanisms aids in suppressing clutter interferences and ameliorating ship confidence levels [18]. Concurrently, certain investigations have achieved an efficient extraction of pivotal ship features within radar images through the implementation of two-stage algorithms [19,20]. Nonetheless, the aforementioned approaches still exhibit inadequacies in effectively mitigating complex interferences and precisely identifying dense small-scale ships. Moreover, it is imperative to elucidate that models based on deep neural networks are conventionally deployed on high-performance computing devices, which entail elevated computational costs due to the excessive convolutional layers and parameters encompassed within. This predicament engenders a diminished sensitivity of higher-level features within the network towards diminutive targets, consequently yielding unsatisfactory outcomes in terms of feature extraction for small-scale objects. Ergo, the adoption of lightweight algorithms, characterized by optimized computational efficiency and a reduced number of convolutional layers [21,22], may be deemed more appropriate for small target identification, thereby augmenting the identification effectiveness pertaining to small-scale ship instances characterized by truncated wake signatures within marine radar images.

In light of the distinctive attributes associated with ship identification in marine radar images, this research designs a novel algorithm, denoted as YOSMR, which leverages a tailored lightweight convolutional network. This approach deviates significantly from prior research in several key aspects.

[1] Adoption of a more efficient lightweight network for extracting crucial ship spot features;
[2] Introduction of a deep feature enhancement method that integrates multi-scale receptive fields to enhance the generalization capability of the feature network;
[3] Incorporation of convolution methods with higher parameter efficiency into a bidirectional feature fusion network, enabling effective learning of spatial and channel features from input data and facilitating the fusion of ship features at both micro and global levels;
[4] Improvement of prediction box formation through advanced non-maximum suppression (NMS) and localization loss estimation, leading to improved ship localization accuracy in dense scenarios;
[5] Design of a more robust ship identification method by utilizing a lightweight convolutional neural architecture to address the computational limitations of embedded devices in radar systems.

It is noteworthy that the ship detection undertaken in this effort is based on the output images from a shore-based radar system. As the imaging foundation, fundamental signal processing techniques within radar systems, including the CFAR (Constant False Alarm Rate) operating modes, will profoundly influence the resulting radar imaging quality. For the shore-based radar used in this research, the echo acquiring and processing technique has undergone targeted adjustments in prior work, with the aim of enhancing the efficiency and stability of the radar imaging in the specific region. It is only upon this foundation that the image-oriented ship detection method possesses true practical value. Additionally, traditional radar's multi-target tracking generally employs the Track-While-Scan (TWS) model, wherein the core technologies are tracking filters and data association. Techniques such as Least Squares Filtering (LSF), Kalman Filtering (KF), Extended Kalman Filtering (EKF), and Unscented Kalman Filtering (UKF) have long played a pivotal role in tracking filters. For data association, the Nearest Neighbor algorithm is suitable for environments with low clutter interference, while for scenarios with medium to high-density clutter and multiple targets, the Joint Probabilistic Data Association (JPDA) algorithm and Multiple Hypothesis Tracking (MHT) algorithm have been proposed and utilized, though these methods also exhibit limitations in high real-time computational demands and large computational loads.

In this work and the upcoming research, we are exploring a novel image-based ship detection-tracking pipeline. Specifically, we will build upon the proposed YOSMR model and seamlessly integrate a customized tracking algorithm tailored for ship targets in radar images, which will enable the development of an innovative working paradigm for shore-based radar detection and tracking. This will serve as a complementary technique to the traditional TWS method, thereby enhancing the real-time performance, accuracy, and robustness of radar perception.

The subsequent sections of this manuscript are structured as follows: Section 2 elucidates a CNN-based customized ship identification algorithm. Section 3 juxtaposes and scrutinizes the experimental outcomes of diverse algorithms employed in marine radar images. Lastly, Section 4 encapsulates the principal contributions of the proposed method and deliberates upon prospective avenues for future advancement.

2. A Proposed Method

The overall framework of the proposed method is depicted in Figure 1, consisting primarily of a feature extraction network and a lightweight feature fusion network. The feature extraction network utilizes the MobileNetV3(Large) architecture [21], which possesses a deeper convolutional layer structure and demonstrates adaptable capability in extracting features at different levels from radar images. The lightweight design of the feature fusion network incorporates three prediction channels, each encompassing the identification process for ships at different scales. Additionally, this network employs depthwise separable convolutions (DSC) [23] as a replacement for standard 3×3 convolutions, significantly reducing the model's parameter and computational complexity. Furthermore, an SPP module [24] is introduced between the feature extraction network and the feature fusion network. The SPP module employs multiple pooling layers to transform feature maps of arbitrary sizes into fixed-size feature vectors, which enhances the capability of extracting ship features and reduces overfitting issues. Lastly, in the prediction head of the algorithm, the non-maximum suppression process is improved using the Cluster-NMS method [25]. This enhancement elevates the accuracy and confidence level of the predicted boxes. Additionally, the α-DIoU loss function is introduced to optimize the calculation of position loss for the predicted boxes, thereby improving the convergence speed and accuracy of the predicted boxes [26].

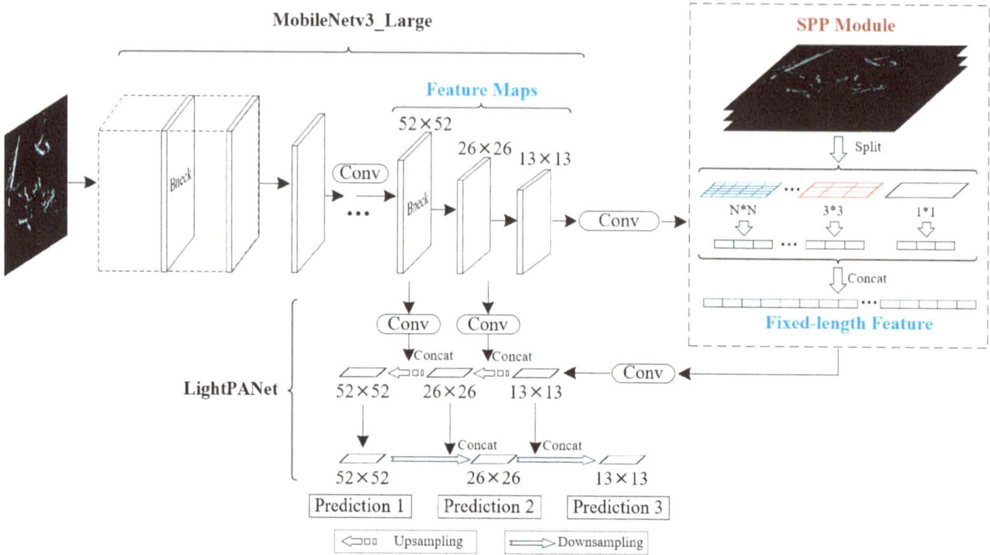

Figure 1. The pipeline of the proposed algorithm. We present a novel detection algorithm grounded in the YOLO architecture, which we term YOSMR. The holistic architecture of YOSMR can be delineated into three core components: Backbone, Neck, and Head. Furthermore, the algorithm also encompasses loss functions and training strategies as pivotal elements. Relative to the standard YOLO framework, YOSMR has undertaken adaptive adjustments across its Backbone, Neck, Head, Loss function, and NMS components to better cater to the unique characteristics of radar-based applications. (a) Within the Backbone, YOSMR has integrated a mature feature extraction network, MobileNetV3(Large), and appended a feature enhancement module known as the Spatial Pyramid Pooling (SPP). (b) We leverage the efficient Depthwise Separable Convolution (DSC), a lightweight convolutional unit, to reconstruct the feature fusion network. This not only ensures the effective extraction of small-scale object features but also significantly reduces the convolution parameters. (c) In the Head structure, we have introduced three prediction channels of diverse scales to encompass the detection of various target types. (d) we have incorporated Cluster NMS and designed the α-DIoU loss to optimize the algorithm's training and accelerate convergence.

2.1. Feature Extraction Network

MobileNetV3(Large) is a remarkable lightweight network renowned for its favorable identification performance on general datasets involving multiple object categories. It comprises 15 so-called Bneck modules with diminishing feature map sizes. In addition, the inverse residual structures within the network exhibit an increment in channel numbers and feature layer quantities. This architectural configuration effectively tackles the issue of feature networks losing object salient information in deep convolutional layers. The Bneck module, depicted in Figure 2, utilizes an inverse residual structure with linear bottlenecks to enhance model dimensionality. Meanwhile, it incorporates residual edge structures for convolutional feature fusion. Additionally, the module employs depthwise separable convolution and lightweight feature attention mechanisms to perform feature extraction, resulting in reduced parameter count and computational requirements. This approach enhances the network's ability to capture significant features of small objects. Research [21] has demonstrated that MobileNetV3(Large) exhibits efficient convolutional computation capabilities and a relatively deep network architecture, enabling robust extraction of essential object features. Moreover, the network has significantly fewer parameters compared to general network architectures such as RCNN series, ResNet-101, SENet, and Darknet53, striking a good balance between accuracy and speed.

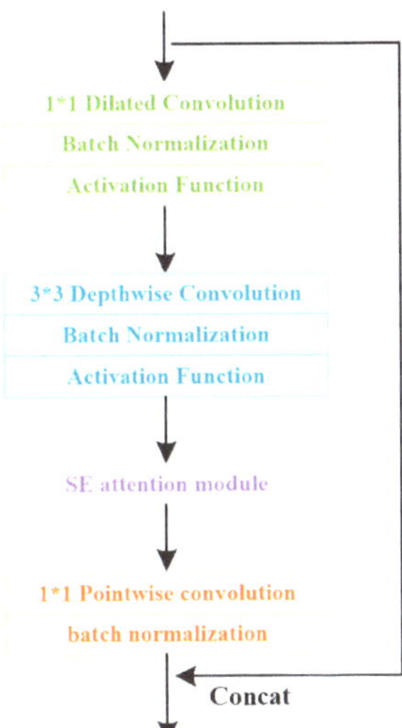

Figure 2. Structure of the Bneck module. In the process of forward convolution, this module employs an attention calculation mechanism and a residual edge structure to enhance the extraction of crucial features relevant to the targets.

In marine radar images, the small scale of ship targets and their limited distinctive features result in the representation of only a limited number of abstract features in deep convolutional networks. This can give rise to the challenge of confounding ship features with background information. Therefore, ship identification in radar images imposes elevated demands on feature extraction networks. MobileNetV3(Large), with its efficient feature extraction architecture for small targets, can capture more comprehensive ship information and enhance accuracy. By incorporating convolutional heatmaps, it is observed that MobileNetV3(Large) primarily leverages the trailing features of ships for object localization. As shown in Figure 3, MobileNetV3(Large) is capable of effectively extracting ship features and accurately distinguishing ships from the background environment, even in scenarios involving minute scales or extreme conditions. This significantly improves ship detection performance and ensures increased accuracy with relatively fewer convolutional parameters.

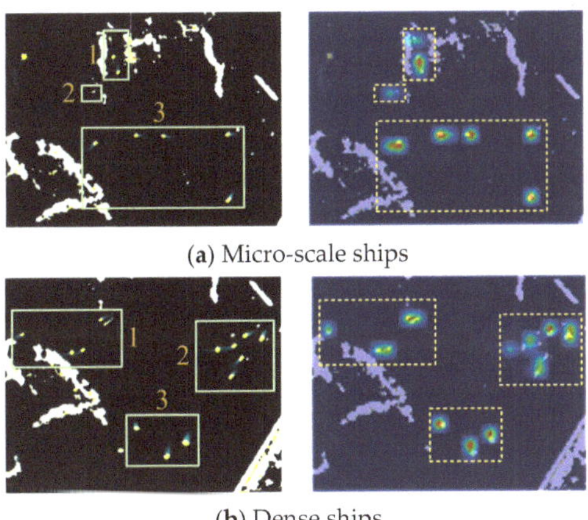

(**a**) Micro-scale ships

(**b**) Dense ships

Figure 3. *Cont.*

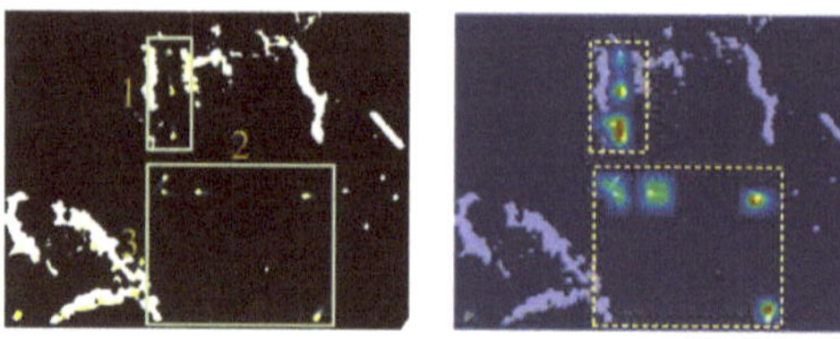

(**c**) Ships in cross-traveling

Figure 3. Convolutional heatmaps. It is apparent that MobileNetV3(Large) leverages the detection of radar spot features to discern the validity of targets. The heatmaps substantiate the remarkable precision of the feature network in localizing ships while effectively mitigating false positives.

2.2. Receptive Field Expansion Module

When performing convolutional computations in the feature network for radar image inputs, variations in the input image size often necessitate operations like stretching and cropping, resulting in the loss of pixel information from the original image. Moreover, small-scale ships tend to have fewer preserved effective features in deep convolutions, leading to lower accuracy in their identification by the model. To address these challenges, concatenating an SPP module after the feature network can preserve more comprehensive image features. This is because SPP utilizes pyramid-like pooling operations, which increase the network's receptive field without altering the resolution of the feature maps. Consequently, it better captures object-related features at different scales, thereby enhancing the model's capability. The SPP structure, as depicted in Figure 4, employs the concatenation of multiple max pooling modules with different sizes to transform the multi-scale feature maps into fixed-size feature vectors [10]. Through numerous experiments, it has been observed that concatenating four max-pooling layers, with sizes of 1×1, 5×5, 9×9, and 13×13, yields optimized detection results for ship identification in radar images.

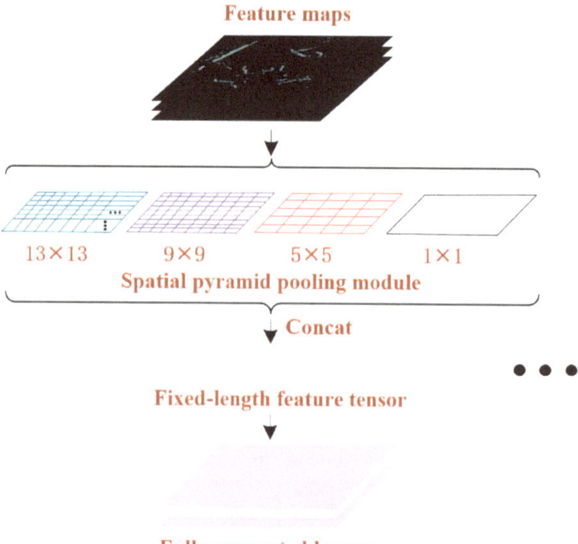

Figure 4. Structure of the SPP. Through the concatenation of multiple scales of maximum pooling layers, this module captures the relatively prominent feature representations from diverse local regions of the feature map. This strategy ensures the positional invariance of the feature data and contributes to mitigating the risk of overfitting.

Drawing upon prior knowledge, it is widely acknowledged that the inclusion of the SPP module plays a critical role in effectively integrating features from diverse scales, thereby enhancing the identification efficacy for small targets and mitigating overfitting concerns. In ship detection using marine radar images, the dataset encompasses diverse scenes and scales of ships, which exert an influence on the complexity and learning capacity of the model, as well as the convergence of the feature network's parameters. The SPP module, in addressing these challenges, enhances the extraction effectiveness of significant features from small-scale ships by expanding the receptive field range of the feature maps. This expansion, in due course, positively contributes to the overall practical capability of the model.

By conducting evaluations within an appropriate training environment and utilizing relevant radar images, the practical performance of the YOSMR algorithm with the incorporation of the SPP module was assessed. In certain marine radar images, a notable presence of ships with short wakes is observed. These ships are commonly considered small-scale targets, posing significant challenges for accurate identification, as depicted in Figure 5. Experimental observations indicate that in situations involving dense and intersecting ships, the SPP module demonstrates robust adaptability to various types of ship targets. It exhibits high precision in ship localization without any instances of mistakes. In contrast, when the SPP module is not utilized, the detection results fail to differentiate between densely packed ships and intersecting ships, leading to multiple erroneous results. Analysis suggests that the SPP module effectively mitigates the common issue of misidentification in small target detection, thus enhancing the actual effectiveness of the algorithm.

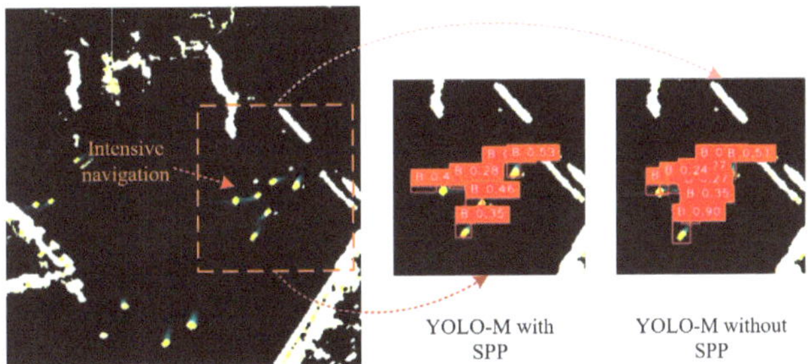

(**a**) Comparison of dense scenarios

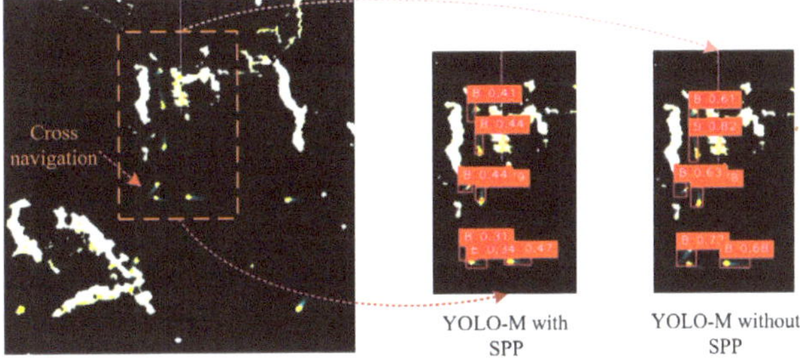

(**b**) Comparison of cross-traveling scenarios

Figure 5. Comparison of identification results with and without SPP. The SPP module, by extracting finer-grained target features, enables effective discrimination of adjacent spots in dense scenes, reducing the probability of misidentification and enhancing the model's robustness.

2.3. Feature Fusion Network

In the field of ship detection and identification in marine radar images, it is typical to observe multiple ship spots with varying sizes and distinct shapes within a single image. Furthermore, the majority of these ships tend to possess relatively diminutive scales, thereby potentially leading to overlapping ship pixels or a striking resemblance to the background. These factors significantly compound the challenges associated with ship identification. To address these aforementioned issues, the extraction of salient features, such as ship contour morphology and the distinguishing characteristics of the bow and stern, can furnish the model with precise discriminative information. Consequently, this augmentation serves to enhance the accuracy and robustness of ship detection.

In general, within deep convolutional networks, shallow convolutions tend to possess higher resolution, capturing more detailed spatial information that aids in improving the precision of object localization. Conversely, deep convolutions have lower resolution but encapsulate stronger multi-scale and semantic information. This research aims to fully integrate feature information extracted from different scales of receptive fields in a single image, thereby devising a more efficient feature fusion network.

With the advancement of convolutional neural networks, feature fusion structures, exemplified by the standard Feature Pyramid Network (FPN) [27], often employ convolutional units that entail redundant computations. Moreover, a plethora of ineffective convolutional parameters can impede the extraction of salient features. Additionally, the single-level top-down feature fusion structure within the FPN architecture fails to concatenate shallow features with deep convolutions, leading to varying fusion effects for different levels of convolutional features. Therefore, the feature blending effect of certain prediction channels is compromised. Research has shown that when PANet [28] is employed as the feature fusion network, its bidirectional fusion structure facilitates a secondary fusion of convolutional features from different levels. This reinforces the outputs of each convolutional level, consequently enhancing the algorithm's estimation and classification capabilities. Simultaneously, the feature fusion network devised in this research incorporates three prediction channels, encompassing the prediction processes for large, medium, and small-scale targets. Consequently, the convolutional computations of the feature fusion network far exceed those of a single prediction channel. This, in turn, leads to a more complex network structure with a larger parameter count and computational overhead, resulting in a surplus of redundant information. Experimental findings indicate that when PANet employs standard 3×3 convolutions, the identification model experiences a significant increase in ineffective parameters. Therefore, simplifying the convolutional computations of feature data serves as a direct approach to reduce the parameter count of the model and alleviate overfitting issues in deep networks.

It has been established through research that depthwise separable convolution profoundly simplifies the computational process of standard convolution while maintaining identical input and output dimensions [23]. Compared to standard convolution, depthwise separable convolution exhibits heightened precision in extracting features pertaining to small targets due to a substantial reduction in redundant parameters. In the context of marine radar images, diminutive-scale ships possess fewer discernible features that distinguish them from background pixel information within the convolutional network. Consequently, such targets gradually fade or even vanish within deep-level feature maps. In actuality, standard convolution tends to suppress the feature expression process of such targets, consequently influencing actual outcomes. Conversely, depthwise separable convolution augments the diversity of convolutional features, thereby ameliorating the algorithm's proficiency in recognizing small-scale ships. Consequently, within the framework of feature fusion networks, this research enhances the PANet network by incorporating five modules of depthwise separable convolution, resulting in the design of the LightPANet network. This endeavor aims to curtail superfluous parameters within the network, enhance computational efficiency, and elevate identification performance for small targets. The LightPANet structure, as devised in this paper, is depicted in Figure 6.

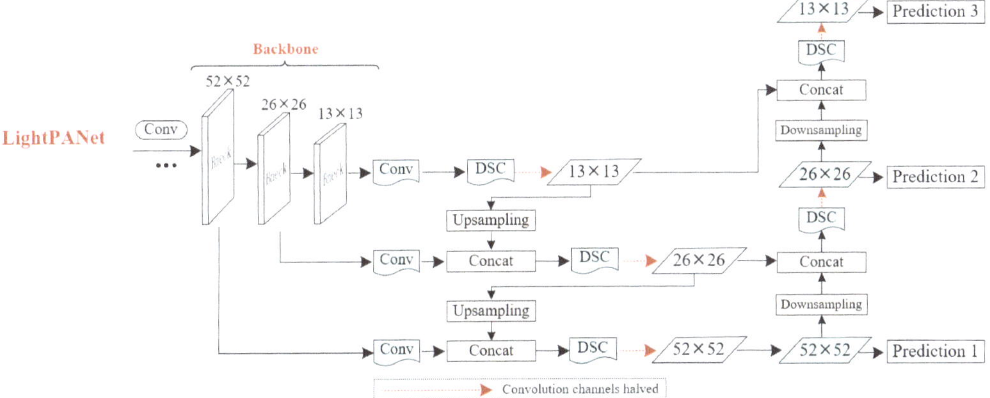

Figure 6. Structure of the LightPANet. In this network, the employment of Depthwise Separable Convolution (DSC) results in a remarkable reduction in parameter count. By decomposing the convolution operation into depthwise convolution and pointwise convolution, DSC achieves a significant decrease in parameters, thereby reducing model complexity and computational demands. Moreover, the independent processing of each input channel during the depthwise convolution allows for the extraction of highly discriminative features. This facilitates the network's ability to capture spatial information within the input data and enhances its generalization capabilities.

To compare the practical capabilities of LightPANet and PANet for ship detection in radar images, the aforementioned networks were employed as feature fusion networks within the YOSMR algorithm. While keeping other structures constant, their practical performance was assessed. Referring to the inference results of radar images in Figure 7, it was observed that the crossing of ship trajectories in regions 1 and 2 posed significant challenges, resulting in substantial interference with the algorithm. The PANet-based model encountered issues of misidentification, leading to a higher rate of false output for ships. Conversely, the LightPANet-based model achieved accurate localization of all ship spots. This experiment validates the detrimental impact of excessive convolution on the detection of small targets in radar images, impeding the effective acquisition of ship pixel features. In contrast, a lightweight feature fusion structure, such as LightPANet, effectively mitigates this issue.

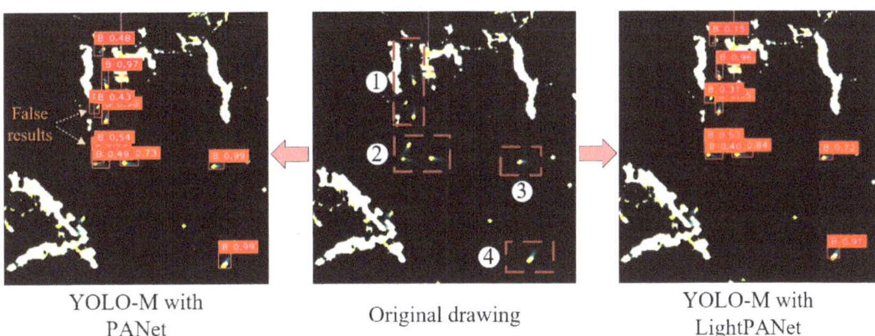

Figure 7. Comparison of identification results between LightPANet and PANet. The utilization of the optimized feature fusion network, empowered by the integration of the DSC module, yields higher precision in localizing ship spots and enhances the accuracy of identifying small-scale targets. Consequently, this leads to a reduced occurrence of false positive predictions.

2.4. Non-Maximum Suppression

Within the YOSMR algorithm, this research employed the LightPANet network to construct a target prediction structure with three channels, enabling the presence of multiple bounding boxes for the same ship target. Typically, the non-maximum suppression (NMS) method is utilized to retain the optimal detection results by filtering out redundant predictions. In conventional algorithms, the IoU metric is commonly employed to filter out redundant predictions, preserving only the bounding box with the highest overlap ratio. However, the IoU metric solely considers the degree of overlap between target bounding boxes, neglecting other crucial target attributes such as shape, size, and orientation. Thus, in scenarios where similar but not entirely overlapping targets exist, the IoU metric may fail to accurately assess the similarity between targets. Moreover, when there is a significant difference in scale between targets, the IoU metric may inadequately measure the similarity between targets, leading to the selection of inappropriate bounding boxes during non-maximum suppression.

This research introduces Cluster-NMS [25] as a solution to the aforementioned issues, serving as a metric for performing non-maximum suppression. In comparison to the IoU metric, the Cluster-NMS method incorporates the DIoU metric [26] and prediction box fusion strategy [29] to achieve weighted adjustments of the bounding box positions and confidence values. This approach enhances the prediction accuracy for small-scale and dense ships. Particularly, the application of a weighted fusion strategy combines multiple prediction boxes, where the weight of each prediction box is determined based on its confidence score. Prediction boxes with higher confidence scores carry greater weights, thereby exerting a more significant influence on the final fusion result. This method better captures the position and shape information of targets, providing more accurate bounding boxes and improving the quality and accuracy of object detection.

As depicted in Figure 8, YOSMR generates multiple bounding boxes for a single ship. Conventional NMS or Soft-NMS methods solely filter the bounding boxes, which may result in lower localization accuracy for the retained boxes. In contrast, the Cluster-NMS method captures more precise ship position and shape information and performs a fusion of multiple prediction boxes. This enables the method to handle ships of different scales, shapes, and scenes, leading to a significant improvement in ship prediction accuracy.

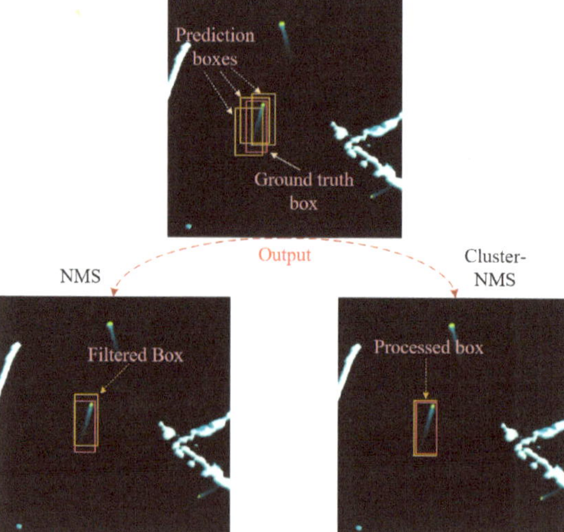

Figure 8. Comparison between Cluster-NMS and other methods. By comparison, Cluster-NMS stands out by utilizing an innovative weighted fusion approach to process candidate prediction boxes, leading to satisfactory precision in target localization.

2.5. Position Loss Function

For the purpose of refining the localization accuracy of ship-bounding boxes, this research introduces the α-DIoU loss function as a computational measure for evaluating the positional error of predicted boxes. On general datasets, this method significantly outperforms standard loss functions such as IoU, DIoU, and CIoU. Moreover, in different scenarios, by adjusting the α coefficient, the detection model exhibits greater flexibility in achieving regression accuracy at various levels, making it easier to find more adaptive threshold settings. Experimental results demonstrate that the α-DIoU loss function exhibits stronger robustness to marine radar images and various types of noise. The overall calculation of the α-DIoU function is illustrated in Equations (1) and (2), with partially key metrics of the function also explained in Figure 9.

$$Loss_{\alpha-DIoU} = 1 - IoU^{2\alpha} + \frac{\rho^{\alpha}(b, b^{gt})}{c^{\alpha}} \tag{1}$$

$$IoU = \frac{S^{pr} \cap S^{gt}}{S^{pr} \cup S^{gt}} \tag{2}$$

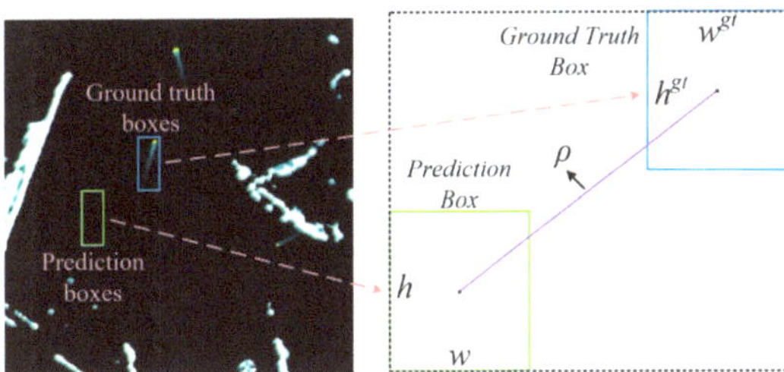

Figure 9. Key indicators of α-DIoU function. Through separate adjustments of a hyperparameter, this method effectively modifies the impact weight of the center point distance metric compared to the standard DIoU metric. This adjustment, made during the loss calculation, facilitates faster convergence of prediction boxes for small-scale targets.

In this context, c represents the Euclidean distance between the minimum bounding rectangles of the prediction box and ground truth bounding box, while $\rho(b, b^{gt})$ denotes the Euclidean distance between their center points. After multiple tests, it has been observed that when the α coefficient is set to 1/2, the α-DIoU function is better suited for ship identification in radar images.

3. A Case Study

3.1. Dataset

The present research focuses on the waters adjacent to Mount Putuo in Zhoushan, China, a large coastal passenger terminal, where the primary ship types encompass passenger ships and maritime auxiliary craft. The JMA5300-MKII marine radar (source from Furuno Electric Co., Ltd., Hyogo, Japan) was selected as the data collection instrument. Following the preprocessing of raw data, a dataset named Radar3000 comprising 3000 images of high quality was constructed for the purpose of training and validating various algorithms. It is noteworthy that passenger terminals prohibit ship operation during inclement weather conditions like heavy rain, dense fog, and strong winds. As such, radar images acquired under these adverse environments would hold limited practical research value. Consequently, the Radar3000 dataset does not consider the aforementioned situations but instead includes other factors such as daytime, dusk, nighttime, and electromagnetic interference, which represent the majority of real-world scenarios. As depicted in Figure 10, the ships in the images can be primarily classified into two categories, i.e., long-wake ships and short-wake ships. The edge features of long-wake ships are more salient. Conversely, it is readily apparent that the pixel characteristics of short-wake ships bear resemblance to interferences such as islands and reefs. This issue is particularly pronounced during ships in dense environments, where mutual interference among short-wake ships tends to manifest.

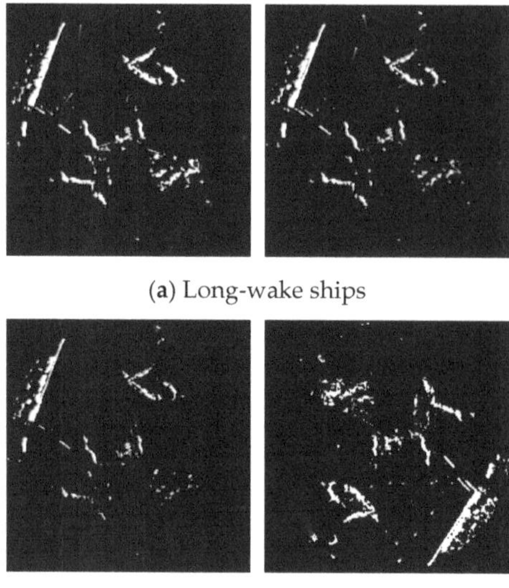

(**a**) Long-wake ships

(**b**) Short-wake ships

Figure 10. Marine radar images. The radar spots present in the image are characterized by their minuscule scale, while small islands and atmospheric clusters, due to their high feature similarity, significantly interfere with the accurate recognition of actual ships.

Furthermore, the Radar3000 dataset encompasses a diverse range of complex background environments, including varying weather conditions, harbor settings, and imaging conditions. It also takes into account factors such as ship heading, distance variations in imaging, and angle transformations. Moreover, to address the issue of mistakes during cross-traveling and busy traffic environments of ships, the Radar3000 dataset incorporates an increased number of ship images specifically tailored to these scenarios. Given the highly similar characteristics exhibited by different ship types in marine radar images, all ships in the Radar3000 dataset have been uniformly labeled and designated as "B", with corresponding XML annotation files generated to adhere to the format requirements of the Pascal VOC dataset.

In the entirety of experiments conducted in this research, all the images in the Radar3000 dataset were partitioned into training, validation, and testing sets in an 8:1:1 ratio. Various algorithms were trained on the training and validation sets, and the actual identification accuracy and effectiveness of the algorithms were evaluated on the testing set. Additionally, through K-means clustering analysis, it was discovered that the average size of the bounding boxes for ship targets was 21 × 25 pixels, accounting for approximately 0.05% of the entire image area. This finding indicates that ship targets in marine radar images predominantly belong to the categories of small-scale and miniature-scale objects.

3.2. Experimental Environment and Training Results

This research conducted algorithm training and testing using a computational platform equipped with an NVIDIA RTX3090 24G (source from NVIDIA Corporation, Santa Clara, CA, USA) graphics card, operating under the Ubuntu 20.04 operating system. The experimental comparisons encompassed conventional algorithms, the YOLO series, and the YOSMR algorithm. The experiments were conducted using the same set of ship images.

Furthermore, transfer learning techniques were employed to optimize the training process of the different algorithms. Specifically, the pre-trained network weights of different YOLO models and MobileNetV3 on the ImageNet dataset were utilized as initial weights for algorithm training. Subsequently, several rounds of algorithm iterations were performed to improve training stability and enhance the learning capabilities of the algorithms in capturing ship features under different scenarios.

During model training, the input image size was set to 416×416 pixels, and the momentum was set to 0.9. In the transfer learning phase, where only the prediction layers were activated, the batch size was set to 10 for 15 iterations, with an initial learning rate of 10^{-4}. When all convolutional layers were enabled, the batch size was set to 4, and the initial learning rate was again set to 10^{-4}. This phase consisted of a total of 300 predetermined iterations. If the algorithm's loss value did not decrease in 15 consecutive iterations, the training process was terminated early. Based on the aforementioned settings, the convergence of the loss during training for different algorithms is illustrated in Figure 11. Specifically, the YOLOv3, YOLOv4, YOLOv5(L)-6.0, and YOSMR algorithms underwent 141, 125, 137, and 133 iterations, respectively. The losses in the training set converged to 9.38, 8.55, 10.31, and 9.26, while the ones in the validation set converged to 10.02, 9.03, 9.82, and 9.4, respectively. It is evident that the training process of YOSMR exhibits smoother loss convergence, with a smaller difference between the two types of losses, thereby contributing to improved training efficacy of the algorithm.

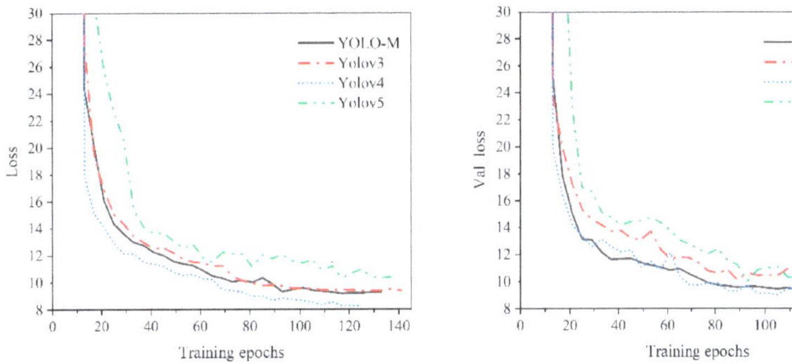

Figure 11. Comparison of training processes of various algorithms. Throughout the entirety of the training process, the proposed algorithm exhibits an expedited and consistent convergence compared to standard methods, resulting in obviously lower overall loss computation values.

3.3. Comparisons and Discussion

This paper quantitatively evaluates the model using five metrics, namely recall, accuracy (Ac), precision (Pr), model parameters (PARAMs), and floating-point operations (FLOPs) [30]. Additionally, a series of experiments were devised to examine the identification performance of the YOSMR algorithm on ship images from marine radar. Firstly, based on the testing images from the Radar3000 dataset, comparative experiments were designed to scrutinize the actual performance of various algorithms across different evaluation metrics, thus validating the effectiveness of the YOSMR algorithm. Moreover, ablation experiments were devised to analyze the different improvement methods within YOSMR, thereby verifying the specific effects of each method. Lastly, by identifying ships from different scenarios, the adaptability of the YOSMR algorithm to different types of tasks was assessed.

A.   Experimental analysis of different algorithms

In the constructed Radar3000 dataset, a comparative analysis was undertaken to assess the performance of various generic algorithms in comparison to the proposed YOSMR algorithm in this paper. Uniformly, the algorithms were trained using the same approach and tested on an identical dataset. Furthermore, cross-validation was employed for all the algorithms, resulting in the recall, accuracy, and precision values being the average of three experimental runs. According to Table 1, the proposed YOSMR achieved a recall rate of 0.9308, accuracy of 0.9204, and precision of 0.9215 for the testing set images. Compared to the top-performing comparative algorithm, YOSMR exhibited a 0.64% increase in recall rate. This improvement can be attributed to YOSMR's adequate ability to accurately identify dense and small-scale ships in radar images. Additionally, in relation to model parameter size and real-time computational consumption, YOSMR exhibited favorable performance, with figures of 12.4 M and 8.63 G, respectively. In comparison to the standard YOLOv3, YOSMR achieved significant reductions of 80.18% and 86.9% in the respective metrics mentioned earlier. Meanwhile, YOSMR has achieved an inference throughput of 122 frames per second (FPS) when deployed on a server equipped with a 3090 24 GB GPU (source from NVIDIA Corporation, Santa Clara, CA, USA) in our research lab. Furthermore, we have integrated the trained model into the existing radar-based surveillance system at Zhoushan Port, where a 1660 6 GB GPU (source from NVIDIA Corporation, Santa Clara, CA, USA) is utilized, delivering a real-time inference speed of 82 FPS and satisfying the requirements of real-time computation.

Table 1. Specific experimental results of various algorithms.

Different Types	Algorithms	Recall	Ac	Pr	PARAMs/(M)	FLOPs/(G)
Conventional methods	CV + GHFilter	0.8910	0.8815	0.8744	N/A	N/A
Standard algorithms	YOLOv3	0.9217	0.9195	0.9233	62.57	65.88
	YOLOv4	0.9209	0.9286	0.9225	63.94	59.87
	YOLOv5(S)	0.8944	0.9089	0.9236	7.2	16.5
	YOLOv5(L)	0.9177	0.912	0.9216	46.5	109.14
	YOLOv7	0.9195	0.9177	0.9281	37.21	104.7
	YOLOv8(S)	0.911	0.9057	0.921	11.2	28.6
	YOLOv8(L)	0.9149	0.9107	0.93	43.28	165.67
Lightweight algorithms	YOLOv3-MobileNetV3(Large)	0.9019	0.9001	0.9127	25.74	20.77
	YOLOv3-MobileNetV3(Small)	0.8565	0.8462	0.8504	7.72	4.73
	YOLOv3-Ghostnet	0.8973	0.8916	0.9003	25.44	20.23
Other algorithms	SRDet	0.8962	0.9025	0.9098	35.1	/
	AFSar	0.8936	0.8813	0.8612	6.52	9.86
YOSMR	YOSMR(PANet)	0.9244	0.9135	0.9095	42.11	31.69
	YOSMR	0.9308	0.9204	0.9215	12.40	8.63

The conventional methods [31,32], incorporating CV and GHFilter, exhibited a recall of 0.8910, accuracy of 0.8815, and precision of 0.8744 in the identification of ships from radar images. These figures were lower by 3.98%, 3.89%, and 4.71%, respectively, compared to the performance achieved by YOSMR. Analysis indicates that when dealing with small-scale ships, difficulties arise due to their diminutive dimensions, absence of prominent characteristics, and limited distinguishability from the surrounding background. These factors pose challenges in terms of target association and data integration, rendering the processes more intricate. Within complex backgrounds, traditional methods are susceptible to false or missed associations, resulting in imprecise detection outcomes. Therefore, experimental results demonstrate the advantages of CNN-based approaches in ship detection under marine radar images.

YOSMR attains a level of identification precision that closely rivals several YOLO algorithms, encompassing YOLOv3, YOLOv4, YOLOv5(L)-6.0, YOLOv7 [33], and YOLOv8(L) [34]. It closely approaches the performance of the best-performing YOLOv8(L) and even surpasses them in terms of recall. Furthermore, YOSMR outperforms YOLOv8(L) by a significant margin in terms of model parameter size and computational consumption. Compared to the lightweight YOLOv5(S) and YOLOv8(S) algorithms, YOSMR demonstrates an enhancement of 3.64% and 1.98% in the recall, respectively, while sustaining relatively superior overall performance. In comparison to optical images, radar images exhibit lower resolution and higher levels of noise. These factors hinder the clear depiction of details and features of small targets in radar images, posing challenges for standard YOLO algorithms in accurately localizing and recognizing such targets. Consequently, the experimental performance of YOLO algorithms in this context often falls short of the desired expectations. Conversely, the proposed YOSMR demonstrates competent capability in suppressing false targets, accurately distinguishing coastal objects, reefs, clouds, and other interferences that bear resemblance to ship features. This effectively reduces mistaken rates and enhances adaptability for radar images.

To extract more accurate ship features from radar images while simultaneously balancing identification capability and model parameter size, this experiment explored several combinations of lightweight feature networks with YOLO architecture. By utilizing the prediction head structure of YOLOv3, the MobileNetV3(Large), MobileNetV3(Small), and Ghostnet networks are successively employed to replace the original Darknet53 network, resulting in three categories of lightweight algorithms for comparison. As shown in Table 1, the YOLOv3-MobileNetV3(Large) algorithm achieved recall, accuracy, and precision rates of 0.9019, 0.9001, and 0.9127, respectively. It outperformed the YOLOv3-Ghostnet algorithm in all evaluation metrics and significantly outperformed the YOLOv3-MobileNetV3(Small) algorithm, particularly in terms of recall. This underscores its effectiveness across diverse radar images. Moreover, YOLOv3-MobileNetV3(Large) exhibits minimal deviation from various standard YOLO series in terms of detection precision in radar scenarios. Given its desirable performance in all aspects, YOLOv3-MobileNetV3(Large) has been selected as the benchmark algorithm in this research.

To validate the practical performance of the LightPANet network, the standard PANet network was applied as the feature fusion structure in YOSMR, resulting in the YOSMR(PANet) algorithm. It is worth noting that the only difference between YOSMR(PANet) and YOSMR lies in the feature fusion structure, while the other components remain consistent. According to the data in Table 1, YOSMR(PANet) exhibits a slight decline in overall performance compared to YOSMR, particularly in terms of recall and precision, which decreased by 0.64% and 1.2%, respectively. This suggests that excessive convolution calculations can lead to a noticeable decrease in the model's accuracy in predicting ship-positive samples and discerning interference objects. The experiments confirm that reducing redundant convolutions is beneficial for improving the feature extraction of deep convolutional networks for small targets. Furthermore, in terms of convolution parameter count and real-time computational consumption, YOSMR achieves a reduction of 70.57% and 72.77%, respectively, compared to YOSMR(PANet). This provides evidence that the 3×3 convolutions in the PANet network contribute significantly to the presence of ineffective parameters in the model. In conclusion, the lightweight LightPANet significantly enhances the algorithm's performance for ship detection in radar images and is better suited for designing lightweight algorithms.

Amidst the current lack of CNN or transformer-based ship detection algorithms specifically developed for marine radar images, through a comparative assessment, we have determined that the target features in SAR images exhibit similarities to the small-scale characteristics of ships in radar scenes. As such, we have selected two detection algorithms designed specifically for SAR images, namely SRDet [35] and AFSar [36], to benchmark against our proposed YOSMR, thereby validating the efficacy of our approach. The experimental results indicate that the comparative algorithms demonstrate unconvincing adaptability when applied to radar images. Although they exhibit better performance in sparse target detection, the precision, recall, and accuracy metrics for the identification of radar spots are markedly weaker in comparison to YOSMR. This disparity can be attributed to the fact that ship targets in SAR images primarily represent static objects in port or near-shore environments, which contrasts sharply with the dynamic characteristics of ships in radar scenes, thus accounting for their suboptimal real-world performance.

B. Ablation Experiments

To further validate the practical performance of the proposed method, a decomposition validation was conducted on the constructed Radar3000 dataset to analyze the influence of each method on ship identification results. The experimental process primarily involved applying various improvement methods step by step on the basis of YOLOv3-MobileNetV3(Large) and testing their respective metrics. The ablation experiments for YOSMR are presented in Table 2.

Table 2. Ablation experiments of YOSMR.

Methods	Model 1	Model 2	Model 3	Model 4	Model 5	Model 6
YOLOv3 + MobileNetV3(Large)	★					
+SPP		★				
+Cluster-NMS			★			
+α-DIoU				★		
+FPN					★	
+LightPANet						★
Recall	0.9019	0.9098	0.9138	0.9224	0.9229	0.9308
Ac	0.9001	0.8936	0.9040	0.9142	0.9126	0.9204
Pr	0.9127	0.9053	0.9182	0.9160	0.9097	0.9215

(1) The incorporation of the SPP module following the MobileNetV3(Large) network yields a noteworthy increase of 0.79% in the recall rate of ship targets in radar images. This enhancement signifies an improved capability of the algorithm to suppress false targets, reduce mistaken rates, and diminish the likelihood of ship omissions. Furthermore, through multiple experiments, it has been observed that the integration of the SPP module accelerates algorithm convergence, resulting in an average reduction of 11 iterations in the YOSMR training process. Additionally, the discrepancy between the training set loss and the validation set loss decreases from 0.7 to 0.14. The empirical findings unequivocally demonstrate that SPP significantly enhances the training quality of the algorithm and improves the identification ability of the model. Notably, in the later stages of algorithm training, the model loss continues to steadily decline with increasing training iterations. Therefore, the SPP module aids in mitigating the overfitting issue in YOSMR and enhances the model's generalizability across diverse ship detection scenarios.

(2) This research focuses on optimizing two modules, namely non-maximum suppression (NMS) for bounding box prediction and the calculation of object localization loss, in the conventional YOLO's prediction structure. Firstly, the Cluster-NMS method is introduced to optimize the candidate results of bounding boxes. By preserving more accurate ship prediction boxes, experimental results demonstrate improvements of 0.4% in recall, 1.04% in accuracy, and 1.29% in precision. Secondly, the α-DIoU loss function is incorporated to calculate the localization loss of bounding boxes, providing a more accurate evaluation of the positioning accuracy of predicted ships. Experimental findings reveal that this approach increases recall by 0.86% and accuracy by 1.02%.

(3) As previously elucidated, the application of depthwise separable convolutions to the feature fusion network presents a notable avenue for reducing convolutional parameters and computational costs, thus facilitating the development of lightweight algorithms. Analysis of the data in Table 2 reveals a significant enhancement in the identification capability of the YOSMR algorithm with the incorporation of the LightPANet network. In comparison to the original FPN network in YOLO, this architecture achieves a 1.18% increase in precision and a 0.79% improvement in recall. Theoretically, assigning higher impact factors to critical feature information is of paramount importance for the detection of small targets, such as ships, in marine radar images. Consequently, LightPANet effectively

improves the representation of crucial target features by reducing redundant computations, ultimately bolstering the model's ability to identify small-scale ships.

C. Comparisons in radar images

Figure 12 presents the detection results of YOSMR for ships in marine radar images across different scenarios, with a specific focus on assessing its performance in recognizing ships with various wake features. As shown in the radar images, each real ship spot is accompanied by motion wake characteristics of varying lengths. In contrast, as the islands and coastline remain stationary in the shore-based radar, these targets do not exhibit any pronounced or extended wakes. By evaluating the wake features surrounding the spots, YOSMR can accurately differentiate between ship targets and confounding entities such as islands. Furthermore, during the image annotation process, we have precisely distinguished the true radar reflections from other objects, providing a robust guide for the algorithm training. This enables YOSMR to learn the salient features of the ship targets, ultimately facilitating accurate classification. From Figure 12a, it can be observed that YOSMR accurately identifies ships with long-wake features in various environments, indicating its strong identification capability for targets with distinctive features. This clearly demonstrates the algorithm's ability to effectively address and mitigate the degradation of target localization accuracy caused by interruptions or changes in ship wake patterns. Figure 12b reveals that YOSMR achieves high identification accuracy for ships with short wake features in different radar conditions. It demonstrates robust feature extraction capabilities for small-scale ship targets, avoiding both missed detections and false results. This highlights its ability to effectively mitigate the influence of confounding factors, such as islands and weather patterns that exhibit similar features. Figure 12c demonstrates YOSMR's precise detection of densely packed ships, exhibiting high ship localization accuracy. Moreover, YOSMR performs well even in extreme scenarios such as close-range encounters and crossing trajectories. The analysis suggests that YOSMR's utilization of small-scale feature receptive fields enables it to capture pixel-level features and positional information of ships, even in situations where small targets are densely clustered.

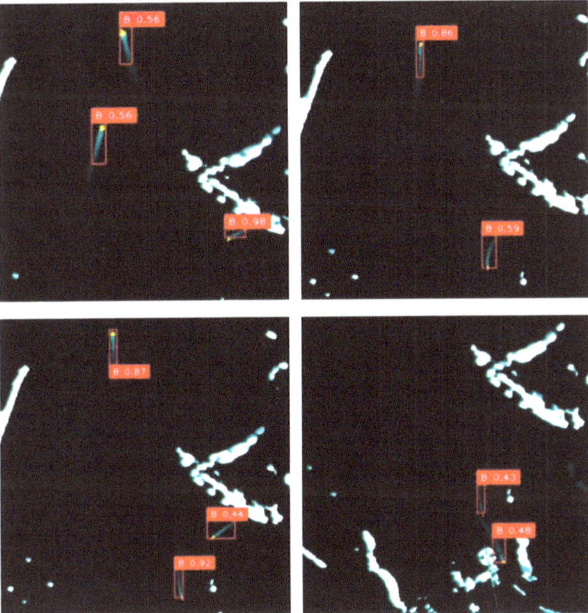

(**a**) Identification of long-wake ships

Figure 12. *Cont.*

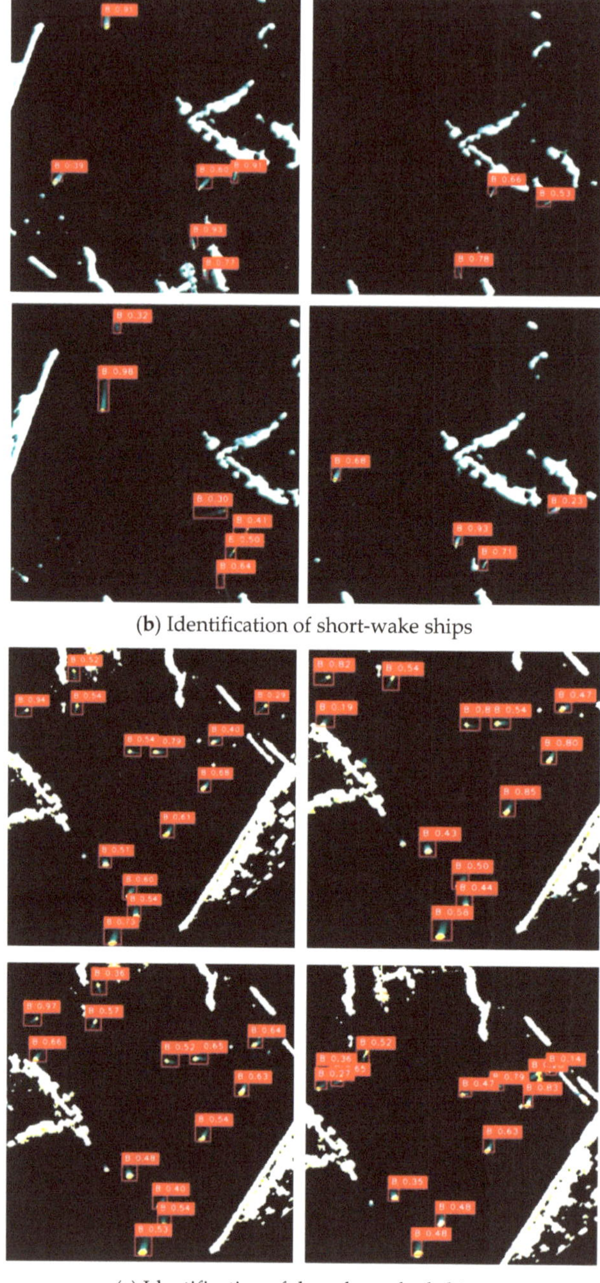

(b) Identification of short-wake ships

(c) Identification of densely packed ships

Figure 12. The identification results of YOSMR for different marine radar images. Regardless of the ship scales, YOSMR exhibits remarkable efficiency in identifying ship spots, effectively capturing target information across various sizes. Moreover, in navigation-intensive environments, this model excels in accurately localizing targets and demonstrates a reduced occurrence of false positives.

Within the 3 nautical miles perceptual range, as depicted in the radar images presented in Figure 10, the proposed model demonstrates the ability to accurately identify ship signatures, as illustrated in Figure 12. It is well-established that as the radar detection range expands, the effective feature points of individual ships undergo a rapid diminution. We have also conducted experiments using radar images of varying scales, which have revealed that without increasing the training sample size, YOSMR can maintain an identification precision of approximately 83% for ship targets within the 3.5 nautical mile range. However, when tasked with detecting ships across larger radar-sensing domains, the model begins to exhibit a higher incidence of omission errors. Therefore, the current iteration of YOSMR has been calibrated to operate within an effective perceptual range of 3 nautical miles.

A salient point to note is that upon scrutinizing the radar images, we have observed that a minority of islands or reefs are accompanied by some short luminous streaks, which could impact the recognition accuracy for certain ship targets. Through experiments, we have found that in 1024×1024 resolution images, when the real wake associated with a ship occupies more than 120 pixels, YOSMR is consistently able to accurately identify the target. However, when the number of pixels occupied by the ship's wake falls below this threshold, or when the luminous streaks surrounding islands approach this value, the model becomes prone to probabilistic misclassification.

D.   Comparisons under noise interference

Radar systems are susceptible to electromagnetic interference (EMI) with similar or adjacent frequencies, leading to anomalies in echo returns, which can subsequently impact radar imaging. The shore-based radar used in this study is equipped with countermeasures against co-frequency asynchronous interference, which can mitigate the effects of EMI at the source. However, to quantify the resilience of the proposed method against varying degrees of interference, we simulated different types and intensities of EMI using salt-and-pepper noise, speckle noise, and Gaussian noise, as outlined in Table 3, to assess their impact on ship detection in radar images.

Table 3. Noise Generation Mechanisms.

Noise Type	Noise Implementation	Noise Configuration
Salt-and-Pepper Noise	Direct addition of noise to the original image	Adheres to a random distribution
Speckle Noise	Multiplication of the original image and noise, then superimposed	Follows a standard normal distribution
Gaussian Noise	Direct addition of noise to the original image	Conforms to a Gaussian distribution

In more detail, the salt-and-pepper noise adheres to a random distribution pattern, and its impact on real-world detection performance is relatively more pronounced. When the proportion of randomly occurring black and white noise pixels occupies less than 30% of the entire image, the proposed YOSMR can still detect the ship targets, though the confidence levels may be affected, and occasional misclassifications may occur. Regarding Gaussian noise, which follows a normal distribution, when the standard deviation is set below 25, the designed model can accurately identify the targets. However, when the standard deviation exceeds 25, more target omissions will arise. In comparison, speckle noise has a relatively less obvious influence on the existing image pixels. This is because the image features a predominately black pixel background surrounding the true ship targets, and this type of noise minimally interferes with the black pixels, thereby preserving the inherent pixel characteristics of the ship targets. Consequently, the performance of YOSMR remains largely unaffected by speckle noise.

Figure 13 presents the real detection results of YOSMR in radar images under various interference signals. The original images contain three types of targets, i.e., ships with short wakes, ships with long wakes, and dense ships, which are common and prevalent in radar images. The empirical investigation conducted on real images has revealed that salt-and-pepper noise significantly impacts radar images, resulting in substantial distortion of pixel characteristics for ships and other objects. In contrast, the influence of speckle noise and Gaussian noise on ship-specific pixel information is comparatively minimal. Empirical observations reveal that YOSMR achieves accurate detection of all ships under the influence of speckle noise and Gaussian noise, demonstrating commendable robustness against these forms of interference. However, when subjected to the interference of salt-and-pepper noise, YOSMR exhibits a slight decrease in confidence for ship identification and introduces one false positive in dense scenarios. Nonetheless, it maintains a satisfactory performance for other ships.

E.   Comparisons of small-scale ship identification

The identification of small targets has always been a focal point and challenge in various computer vision tasks, including ship detection in marine radar images. To evaluate the performance of YOSMR and other typical algorithms in detecting small-scale ships, an experiment was conducted using a dedicated subset of the Radar3000 dataset. This subset included two scenarios, i.e., cross-traveling and dense environment, both of which contained numerous small-scale, and even minuscule, ships with short wakes. The presence of interference objects, such as reefs, also had a noticeable impact on the model's detection. For this experiment, we selected the commonly used YOLOv5(L) and YOLOv8(L) as the comparative algorithms, as they are widely recognized for their effectiveness in detecting small objects. The results, as shown in Table 4, revealed that YOSMR outperformed the comparison algorithms in all aspects, demonstrating its adequate effectiveness in detecting various small-scale ships.

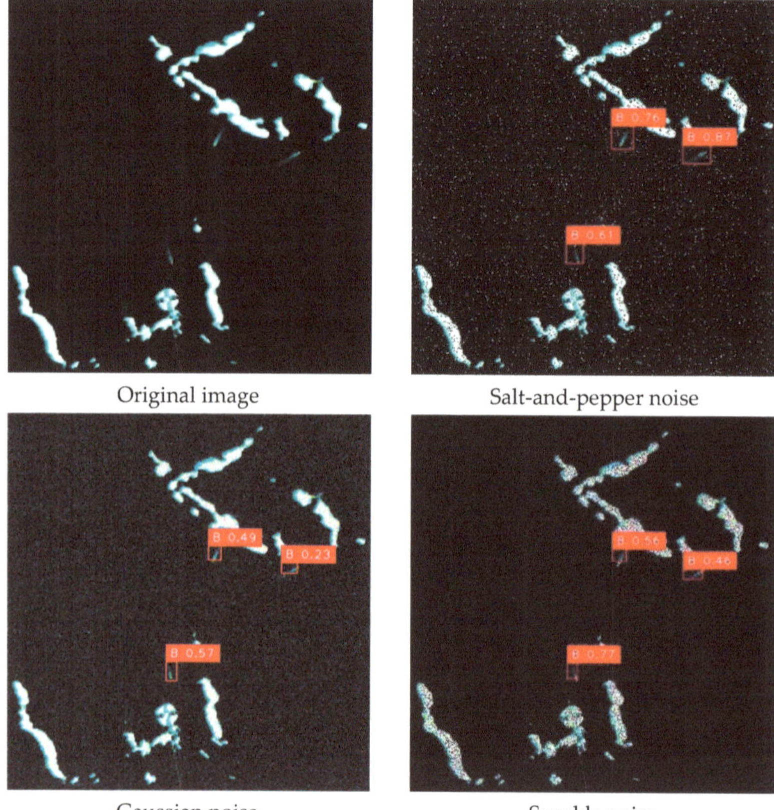

Original image   Salt-and-pepper noise

Gaussian noise   Speckle noise

(**a**) Identification of short-wake ships

Figure 13. *Cont.*

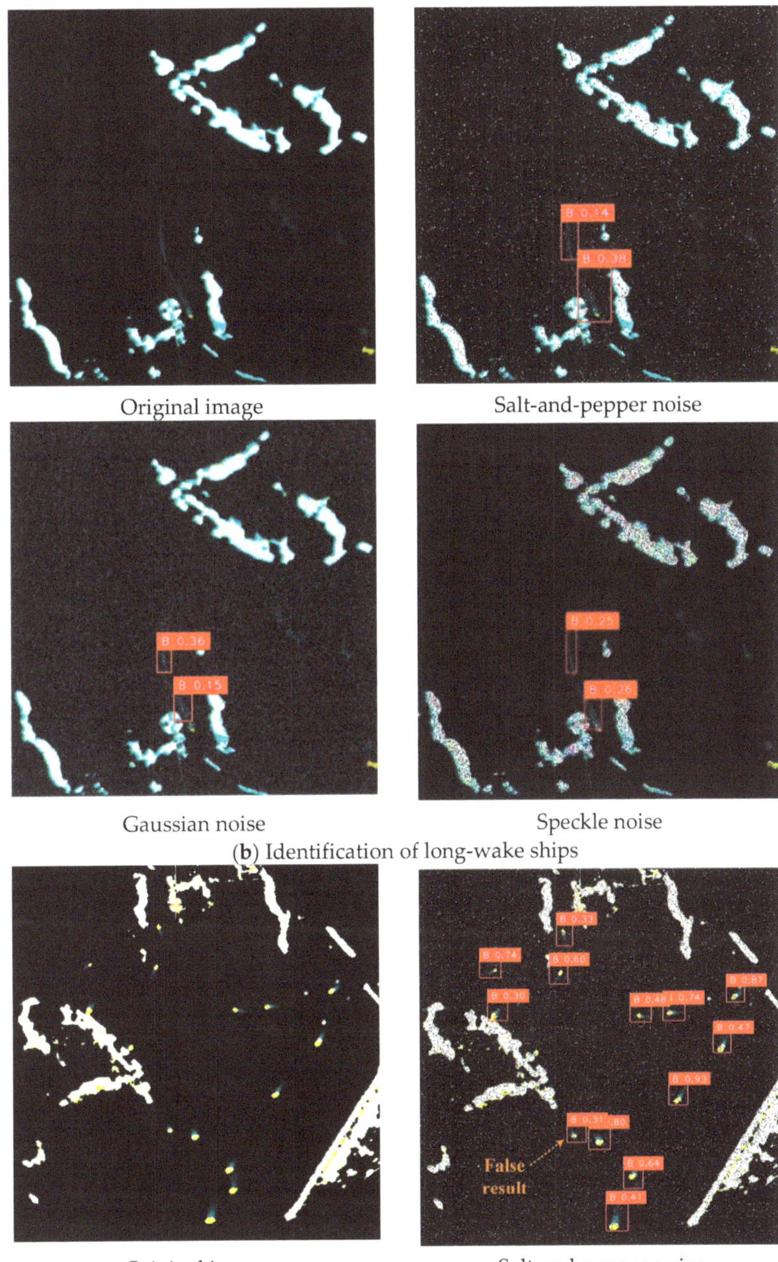

Figure 13. *Cont.*

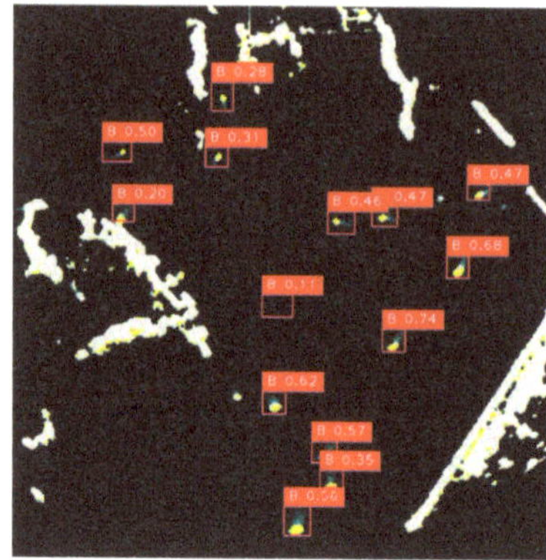 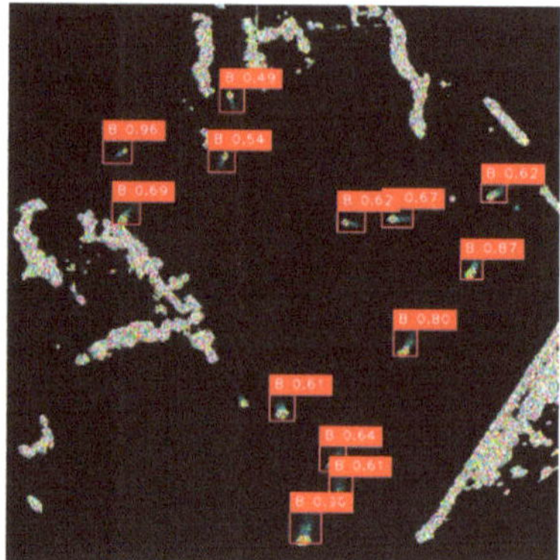

Gaussian noise   Speckle noise

(c) Identification of densely packed ships

Figure 13. Identification of YOSMR for different types of ships under various noises. YOSMR exhibits resilience to a certain degree of interference, as it can effectively discern and accurately locate the majority of authentic ship spots despite the varying impact on target confidence scores caused by different types of disturbances.

Table 4. Comparison of different algorithms in small-scale ship identifications.

Algorithms	Detected Ships	True Ships	False Alarms	Recall	Pr
YOLOv5(L)	1419	1303	116	0.9086	0.9183
YOLOv8(L)	1427	1307	120	0.9114	0.9159
YOSMR	1440	1325	115	0.9240	0.9201

In the context of small-scale ship identification, although YOSMR exhibits competent performance across various evaluation metrics compared to the comparative algorithms, it is still essential to validate its effectiveness on real radar images. To visualize this experiment, we have selected several representative images and presented them in Figure 14. As shown in the experimental results, YOLOv5(L) exhibits susceptibility to misidentification issues when dealing with tiny ship spots, whereas YOLOv8(L) is equally prone to misidentification in dense scenarios. This can be attributed to the utilization of relatively large receptive fields and deep convolutional layers in both methodologies, which may inadvertently diminish their sensitivity towards capturing intricate nuances of small-scale ships. As a consequence, the above models may erroneously categorize background elements or extraneous objects as ships. Conversely, YOSMR outperforms both the YOLOv5(L) and YOLOv8(L) in terms of ship localization precision. It demonstrates lower rates of mistakes and omission, indicating that the lightweight algorithm, through the construction of a rational network structure, can extract finer pixel and texture features, thereby enhancing the detection effect of small targets by the model.

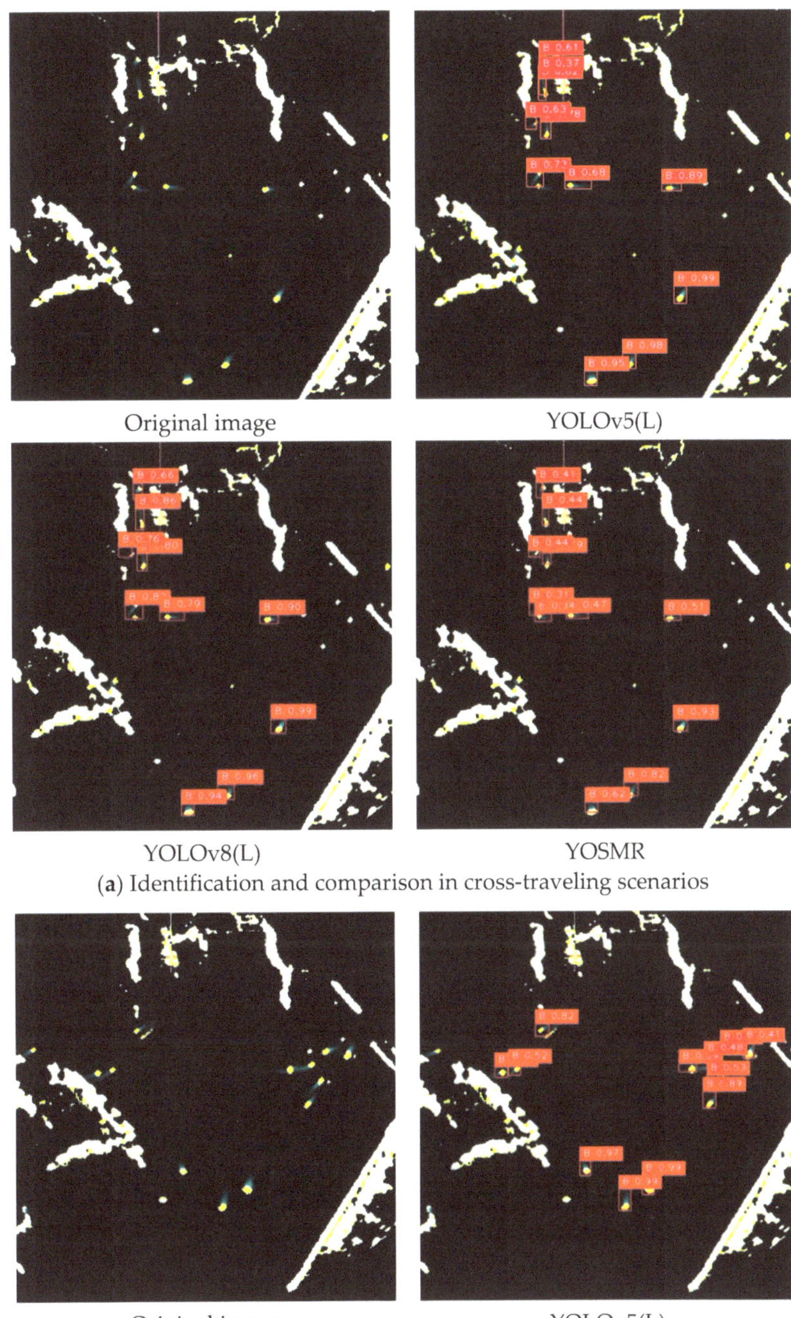

(**a**) Identification and comparison in cross-traveling scenarios

Figure 14. *Cont.*

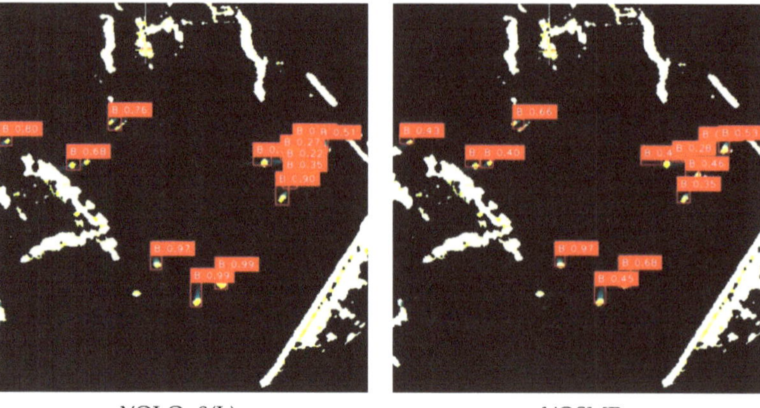

YOLOv8(L)   YOSMR

(**b**) Identification and comparison in dense scenarios

Figure 14. Comparison of different algorithms for small-scale ship identification. It should be noted that the image presented above has been appropriately cropped and magnified from the original radar image to provide a clearer visualization of the detection results of the spots. In the challenging context of identifying small-scale ships, typical models often produce a considerable number of false positive targets in their recognition outcomes. This issue persists even with advanced models such as YOLOv8 and the latest version of YOLOv5, which are widely acknowledged. However, through comparison, it becomes apparent that the proposed YOSMR exhibits remarkable detection performance, particularly in its ability to effectively suppress false positive targets.

F.   Comparative experiments on other datasets

Given the absence of publicly accessible marine radar datasets to assess the practical performance of the proposed YOSMR across alternative scenarios, we have selected a set of representative ship targets within SAR images to conduct corresponding experiments. This decision is informed by the distinctive characteristics of SAR scenes, which feature a high proportion of small targets and a rich diversity of contexts, posing notable challenges for target detection. Employing such a dataset provides a compelling means to evaluate the transferability of our method. The images utilized for model training, validation, and testing were sourced from the RSDD-SAR dataset [37], which encompasses a diverse array of imaging modalities and resolutions, totaling 7000 images and 10,263 ship instances, thereby offering broad representational coverage. The experimental findings reveal that YOSMR achieves recall and precision rates of 0.8612 and 0.8733, respectively, demonstrating considerable adaptability. However, the model also exhibits a comparatively elevated misidentification rate for small targets and densely populated ship scenes, consistent with its performance in radar-based scenarios. Accordingly, our subsequent efforts will focus on optimizing this specific shortcoming.

G.   Comparative Experiments with CFAR Methods

In order to more comprehensively evaluate the capabilities of various methods in suppressing clutter and accurately detecting real targets within the shore-based radar in port waters, we have designed a set of experiments to compare two categories of approaches. The first category comprises target detection algorithms tailored for radar echo signals, which have been extensively deployed in a myriad of practical applications, namely the CFAR method and its diverse improved variants. These methods endeavor to maximize the detection rate under a fixed false alarm probability by meticulously designing dynamic thresholds to filter the radar echo signals. The second category encompasses deep learning-based detection algorithms designed for radar images, such as the constructed YOSMR. As mentioned earlier, the prerequisite for the efficacious implementation of such methods is the pre-processing of radar echo signals to obtain radar images. In the present research, the radar image is generated based on a customized CFAR method for clutter suppression and a Kalman filtering-based target tracking approach, which have eliminated a certain proportion of clutter interference while preserving the majority of true targets.

For the specific experiments, within the traditional methods, we have selected two variants of the CFAR approach, namely CA-CFAR, and OS-CFAR, to be compared against the proposed YOSMR. Since the conventional methods are tailored for radar echo signals, we have chosen the Probability of Detection(PD) and Probability of False Alarm(PFA) as the evaluation metrics [38]. To ensure the uniformity of the evaluation criteria, the YOSMR will also be assessed based on these same metrics. The experimental data is derived from the same shore-based radar located at the Zhoushan passenger terminal, comprising both radar echo and image data collected at the same timestamps.

The experimental results, elucidated in Tables 5 and 6, demonstrate YOSMR's preferable performance in interference suppression compared to two conventional methods. The analysis suggests that while CA-CFAR exhibits commendable performance in scenarios with sparse targets and relatively low interference, it suffers from elevated miss-detection probabilities in ship-dense waters with pronounced clutter variations. Conversely, OS-CFAR showcases enhanced detection rates for multiple targets. As previously expounded, radar image, through efficacious suppression of clutter interference while maintaining satisfying detection rates, lays the groundwork for YOSMR's image-oriented detection capabilities. The empirical data corroborates that YOSMR, by employing identification and localization of ship spots within images, further attenuates interference and mitigates target misidentification instances, consequently diminishing the Probability of False Alarm (PFA).

Table 5. Probability of Detection of various CFAR methods under different PFA settings.

	CA-CFAR	OS-CFAR
PFA(0.1)	0.663	0.837
PFA(0.05)	0.637	0.825
PFA(0.01)	0.557	0.762

Table 6. Probability of Detection and Probability of False Alarm of YOSMR under different PFA settings.

	Detected PD	Detected PFA
PFA(0.1)	0.847	0.044
PFA(0.05)	0.823	0.023
PFA(0.01)	0.753	0.0087

In summary, the detection methods that operate directly on the radar echoes tend to exhibit relatively higher target detection rates, yet also confront correspondingly elevated false alarm rates. Conversely, the detection approaches oriented towards radar images can effectively strike a balance between the detection rate and the false positive rates, with the capability to further reduce the false alarm rate in particular.

4. Conclusions and Discussion

This paper presents YOSMR, an algorithm for ship identification in marine radar images. YOSMR leverages the MobileNetV3(Large) network in its feature extraction module, which demonstrates satisfactory capabilities in extracting ship-specific features from radar images. Additionally, YOSMR employs lightweight design principles in its feature fusion network by replacing certain traditional convolutions with depthwise separable convolutions. This design significantly reduces the model's parameter count and computational complexity while enhancing its ability to capture salient ship features. Moreover, YOSMR incorporates the SPP module after the feature network to enhance the extraction capabilities of deep convolution features. In the prediction structure, YOSMR optimizes

NMS using the Cluster-NMS method and employs the α-DIoU loss function to improve the accuracy of predicted bounding boxes. The experimental results indicate that in the context of marine radar images, YOSMR exhibits a more satisfying performance than conventional approaches. It achieves accurate detection of both long-wake and short-wake ships in various scenarios, particularly excelling in extreme conditions such as recognizing tiny ships under complex navigation scenarios. YOSMR achieves recall, accuracy, and precision values of 0.9308, 0.9204, and 0.9215, respectively. Furthermore, YOSMR significantly reduces convolutional parameter count and real-time computational costs to 12.4 M and 8.63 G, respectively, compared to the standard YOLO series. The proposed YOSMR effectively balances model size and identification accuracy, making it suitable for deployment in embedded monitoring devices.

Considering the limited range of ship classes currently covered in the Radar3000 dataset, we plan to actively pursue the collection of radar images from typical inland waterways and coastal ports. This will enable the inclusion of a more diverse set of ship characteristics, such as cargo ships, container ships, tankers, and tugboats. Particular attention will be given to enhancing the detection of dense ship formations and small-scale ships. Moreover, the next explorations will integrate multi-object tracking algorithms to obtain comprehensive ship trajectory data, which will enable the analysis of ship traffic patterns and enhance safety monitoring in specific waterway scenarios, thereby expanding the practicality of this research.

Author Contributions: Conceptualization, Z.K., F.M. and C.C.; methodology, Z.K. and F.M.; software, Z.K.; validation, Z.K.; formal analysis, F.M. and C.C.; investigation, F.M.; resources, J.S.; data curation, J.S.; writing—original draft preparation, Z.K.; writing—review and editing, Z.K. and F.M.; visualization, Z.K.; supervision, F.M. and C.C. All authors have read and agreed to the published version of the manuscript.

Funding: This research was funded by the National Key R&D Program of China, grant number 2021YFB1600400, and the National Natural Science Foundation of China, grant number 52171352.

Institutional Review Board Statement: Not applicable.

Informed Consent Statement: Not applicable.

Data Availability Statement: We have made the dataset public on the following website: https://github.com/kz258852/dataset_M_Radar (Accessed on 30 July 2024).

Acknowledgments: We are deeply grateful to our colleagues for their exceptional support in developing the dataset and assisting with the experiments.

Conflicts of Interest: Author Jie Sun was employed by the company Nanjing Smart Water Transportation Technology Co., Ltd. The remaining authors declare that the research was conducted in the absence of any commercial or financial relationships that could be construed as a potential conflict of interest.

References

1. Liu, R.W.; Yuan, W.; Chen, X.; Lu, Y. An enhanced CNN-enabled learning method for promoting ship detection in maritime surveillance system. *Ocean Eng.* **2021**, *235*, 109435. [CrossRef]
2. Chen, P.; Li, X.; Zheng, G. Rapid detection to long ship wake in synthetic aperture radar satellite imagery. *J. Oceanol. Limnol.* **2018**, *37*, 1523–1532. [CrossRef]
3. Han, J.; Kim, J.; Son, N. Coastal navigation with marine radar for USV operation in GPS-restricted Situations. *J. Inst. Control Robot. Syst.* **2018**, *24*, 736–741. [CrossRef]
4. Wen, B.; Wei, Y.; Lu, Z. Sea clutter suppression and target detection algorithm of marine radar image sequence based on spatio-temporal domain joint filtering. *Entropy* **2022**, *24*, 250. [CrossRef]
5. Wu, C.; Wu, Q.; Ma, F.; Wang, S. A novel positioning approach for an intelligent vessel based on an improved simultaneous localization and mapping algorithm and marine radar. *Proc. Inst. Mech. Eng. Part M J. Eng. Marit. Environ.* **2018**, *233*, 779–792. [CrossRef]
6. Zhou, Y.; Wang, T.; Hu, R.; Su, H.; Liu, Y.; Liu, X.; Suo, J.; Snoussi, H. Multiple kernelized correlation filters (MKCF) for extended object tracking using x-band marine radar data. *IEEE Trans. Signal Process.* **2019**, *67*, 3676–3688. [CrossRef]

7. Qiao, S.; Fan, Y.; Wang, G.; Mu, D.; He, Z. Radar target tracking for unmanned surface vehicle based on square root sage–husa adaptive robust kalman filter. *Sensors* **2022**, *22*, 2924. [CrossRef] [PubMed]
8. Ma, F.; Chen, Y.; Yan, X.; Chu, X.; Wang, J. Target recognition for coastal surveillance based on radar images and generalised bayesian inference. *IET Intell. Transp. Syst.* **2017**, *12*, 103–112. [CrossRef]
9. Redmon, J.; Farhadi, A. Yolov3: An incremental improvement. *arXiv* **2018**, arXiv:1804.02767.
10. Bochkovskiy, A.; Wang, C.-Y.; Liao, H.-Y.M. Yolov4: Optimal speed and accuracy of object detection. *arXiv* **2020**, arXiv:2004.10934.
11. Wang, C.-Y.; Bochkovskiy, A.; Liao, H.-Y.M. Scaled-yolov4: Scaling cross stage partial network. In Proceedings of the IEEE/CVF Conference on Computer Vision and Pattern Recognition (CVPR), Nashville, TN, USA, 20–25 June 2021; pp. 13024–13033.
12. Jocher, G. YOLOv5 by Ultralytics. Available online: https://github.com/ultralytics/yolov5 (accessed on 21 September 2023).
13. Girshick, R. Fast R-CNN. In Proceedings of the IEEE/CVF International Conference on Computer Vision (ICCV), Santiago, Chile, 7–13 December 2015; pp. 1440–1448.
14. Cai, Z.; Vasconcelos, N. Cascade R-CNN: High quality object detection and instance segmentation. *IEEE Trans. Pattern Anal. Mach. Intell.* **2021**, *43*, 1483–1498. [CrossRef]
15. Zhu, X.; Lyu, S.; Wang, X.; Zhao, Q. TPH-YOLOv5: Improved YOLOv5 based on transformer prediction head for object detection on drone-captured scenarios. In Proceedings of the IEEE/CVF International Conference on Computer Vision (ICCV), Montreal, QC, Canada, 10–17 October 2021; pp. 2778–2788.
16. Chen, S.; Zhan, R.; Wang, W.; Zhang, J. Domain adaptation for semi-supervised ship detection in SAR images. *IEEE Geosci. Remote Sens. Lett.* **2022**, *19*, 1–5. [CrossRef]
17. Miao, T.; Zeng, H.; Yang, W.; Chu, B.; Zou, F.; Ren, W.; Chen, J. An improved lightweight retinanet for ship detection in SAR images. *IEEE J. Sel. Top. Appl. Earth Obs. Remote Sens.* **2022**, *15*, 4667–4679. [CrossRef]
18. Mou, X.; Chen, X.; Guan, J.; Zhou, W.; Liu, N.; Yang, D. Clutter suppression and marine target detection for radar images based on INet. *J. Radars* **2020**, *9*, 640–653. [CrossRef]
19. Chen, X.; Mu, X.; Guan, J.; Liu, N.; Zhou, W. Marine target detection based on marine-faster r-cnn for navigation radar plane position indicator images. *Front. Inf. Technol. Electron. Eng.* **2022**, *23*, 630–643. [CrossRef]
20. Chen, X.; Su, N.; Huang, Y.; Guan, J. False-alarm-controllable radar detection for marine target based on multi features fusion via CNNs. *IEEE Sens. J.* **2021**, *21*, 9099–9111. [CrossRef]
21. Howard, A.; Sandler, M.; Chen, B.; Wang, W.; Chen, L.C.; Tan, M.; Chu, G.; Vasudevan, V.; Zhu, Y.; Pang, R.; et al. Searching for MobileNetV3. In Proceedings of the IEEE/CVF International Conference on Computer Vision (ICCV), Seoul, Republic of Korea, 27 October–2 November 2019; pp. 1314–1324.
22. Han, K.; Wang, Y.; Tian, Q.; Guo, J.; Xu, C.; Xu, C. Ghostnet: More features from cheap operations. In Proceedings of the IEEE/CVF Conference on Computer Vision and Pattern Recognition (CVPR), Seattle, WA, USA, 13–19 June 2020; pp. 1577–1586.
23. Chollet, F. Xception: Deep learning with depthwise separable convolutions. In Proceedings of the IEEE/CVF Conference on Computer Vision and Pattern Recognition (CVPR), Honolulu, HI, USA, 21–26 July 2017; pp. 1800–1807.
24. He, K.; Zhang, X.; Ren, S.; Sun, J. Spatial pyramid pooling in deep convolutional networks for visual recognition. *IEEE Trans. Pattern Anal. Mach. Intell.* **2015**, *37*, 1904–1916. [CrossRef] [PubMed]
25. Zheng, Z.; Wang, P.; Ren, D.; Liu, W.; Ye, R.; Hu, Q.; Zuo, W. Enhancing geometric factors in model learning and inference for object detection and instance segmentation. *IEEE Trans. Cybern.* **2022**, *52*, 8574–8586. [CrossRef]
26. Zheng, Z.; Wang, P.; Liu, W.; Li, J.; Ye, R.; Ren, D. Distance-IoU loss: Faster and better learning for bounding box regression. In Proceedings of the AAAI Conference on Artificial Intelligence (AAAI), New York, NY, USA, 7–12 February 2020; pp. 12993–13000.
27. Lin, T.Y.; Dollár, P.; Girshick, R.; He, K.; Hariharan, B.; Belongie, S. Feature pyramid networks for object detection. In Proceedings of the IEEE/CVF Conference on Computer Vision and Pattern Recognition (CVPR), Honolulu, HI, USA, 21–26 July 2017; pp. 936–944.
28. Liu, S.; Qi, L.; Qin, H.; Shi, J.; Jia, J. Path aggregation network for instance segmentation. In Proceedings of the IEEE/CVF Conference on Computer Vision and Pattern Recognition (CVPR), Salt Lake City, UT, USA, 18–23 June 2018; pp. 8759–8768.
29. Solovyev, R.; Wang, W.; Gabruseva, T. Weighted boxes fusion: Ensembling boxes from different object detection models. *Image Vis. Comput.* **2021**, *107*, 104117. [CrossRef]
30. Cao, S.; Zhao, D.; Liu, X.; Sun, Y. Real-time robust detector for underwater live crabs based on deep learning. *Comput. Electron. Agric.* **2020**, *172*, 105339. [CrossRef]
31. Ma, F.; Wu, Q.; Yan, X.; Chu, X.; Zhang, D. Classification of automatic radar plotting aid targets based on improved fuzzy c-means. *Transp. Res. Part C Emerg. Technol.* **2015**, *51*, 180–195. [CrossRef]
32. Ma, F.; Chen, Y.; Yan, X.; Chu, X.; Wang, J. A novel marine radar targets extraction approach based on sequential images and bayesian network. *Ocean Eng.* **2016**, *120*, 64–77. [CrossRef]
33. Wang, C.-Y.; Bochkovskiy, A.; Liao, H.-Y.M. YOLOv7: Trainable bag-of-freebies sets new state-of-the-art for real-time object detectors. In Proceedings of the IEEE/CVF Conference on Computer Vision and Pattern Recognition (CVPR), Vancouver, BC, Canada, 18–22 June 2023; pp. 7464–7475.
34. Jocher, G. YOLO by Ultralytics. Available online: https://github.com/ultralytics/ultralytics (accessed on 21 February 2024).
35. Lv, J.; Chen, J.; Huang, Z.; Wan, H.; Zhou, C.; Wang, D.; Wu, B.; Sun, L. An anchor-free detection algorithm for SAR ship targets with deep saliency representation. *Remote Sens.* **2022**, *15*, 103. [CrossRef]

36. Wan, H.; Chen, J.; Huang, Z.; Xia, R.; Wu, B.; Sun, L.; Yao, B.; Liu, X.; Xing, M. AFSar: An anchor-free SAR target detection algorithm based on multiscale enhancement representation learning. *IEEE Trans. Geosci. Remote Sens.* **2022**, *60*, 5219514. [CrossRef]
37. Xu, C.; Su, H.; Li, J.; Liu, Y.; Yao, L.; Gao, L.; Yan, W.; Wang, T. RSDD-SAR: Rotated ship detection dataset in SAR images. *J. Radars* **2022**, *11*, 581–599. [CrossRef]
38. Chen, X.; Liu, K.; Zhang, Z. A PointNet-based CFAR detection method for radar target detection in sea clutter. *IEEE Geosci. Remote Sens. Lett.* **2024**, *21*, 3502305. [CrossRef]

Journal of
*Marine Science
and Engineering*

Article

Distributed Swarm Trajectory Planning for Autonomous Surface Vehicles in Complex Sea Environments

Anqing Wang [1], Longwei Li [1], Haoliang Wang [2], Bing Han [1,3] and Zhouhua Peng [1,*]

[1] College of Marine Electrical Engineering, Dalian Maritime University, Dalian 116026, China;
anqingwang@dlmu.edu.cn (A.W.); lilongwei@dlmu.edu.cn (L.L.); han.bing@coscoshipping.com (B.H.)
[2] College of Marine Engineering, Dalian Maritime University, Dalian 116026, China;
haoliang.wang12@dlmu.edu.cn
[3] Shanghai Ship and Shipping Research Institute Co., Ltd., Shanghai 200135, China
*   Correspondence: zhpeng@dlmu.edu.cn

Abstract: In this paper, a swarm trajectory-planning method is proposed for multiple autonomous surface vehicles (ASVs) in an unknown and obstacle-rich environment. Specifically, based on the point cloud information of the surrounding environment obtained from local sensors, a kinodynamic path-searching method is used to generate a series of waypoints in the discretized control space at first. Next, after fitting B-spline curves to the obtained waypoints, a nonlinear optimization problem is formulated to optimize the B-spline curves based on gradient-based local planning. Finally, a numerical optimization method is used to solve the optimization problems in real time to obtain collision-free, smooth and dynamically feasible trajectories relying on a shared network. The simulation results demonstrate the effectiveness and efficiency of the proposed swarm trajectory-planning method for a network of ASVs.

Keywords: autonomous surface vehicles; kinodynamic path searching; uniform B-spline curves; nonlinear optimization

Citation: Wang, A.; Li, L.; Wang, H.; Han, B.; Peng, Z. Distributed Swarm Trajectory Planning for Autonomous Surface Vehicles in Complex Sea Environments. *J. Mar. Sci. Eng.* **2024**, *12*, 298. https://doi.org/10.3390/jmse12020298

Academic Editor: Rafael Morales

Received: 20 December 2023
Revised: 28 January 2024
Accepted: 30 January 2024
Published: 7 February 2024

1. Introduction

The autonomous surface vehicle (ASV), as an intelligent, miniaturized, and versatile unmanned marine transport platform that operates through remote control or autonomous navigation, has a variety of applications in offshore ocean engineering [1–13]. Among the various applications of ASVs, the trajectory planning of ASVs is particularly crucial to provide a safe trajectory. In order to obtain the optimal trajectory, trajectory planning of the ASV is described as a constrained optimization problem by using local sensor and global map information.

The trajectory-planning problem of an ASV has been widely studied in the literature [14–21]. Specifically, in [14], a temporal-logic-based ASV path-planning method is employed, which enables the ASV to pass through heavy harbor traffic to an intended destination in a collision-free manner. In [15], an evolutionary-based path-planning approach is proposed for an ASV to accomplish environmental monitoring tasks. In [16], an extension of the hybrid-A* algorithm is proposed to plan optimal ASV paths under kinodynamic constraints in a leader-following scenario. Another hybrid-A*-based two-stage method is provided in [17] for energy-optimized ASV trajectory planning with experimental validation. In [18], a novel receding horizon multi-objective planner is developed for an ASV performing path planning in complex urban waterways. In [19], an essential visibility graph approach is proposed to generate optimal paths for an ASV with real-time collision avoidance. Furthermore, in [20], the particle swarm optimization algorithm together with the visibility graph is applied to the ASV path-planning problem among complex shorelines and spatiotemporal environmental forces. Recently, in [21], a hybrid artificial potential field method is proposed for an ASV cruising in a dynamic riverine environment. However, it is worth mentioning that the works [14–21] are dedicated to the path planning of a single ASV.

Compared to a single ASV, multiple ASVs are more likely to complete difficult tasks such as exploration and development, territorial sea monitoring and route planning [4,22–32]. Therefore, the trajectory planning of multiple ASVs, as one of the key topics in the field of marine science, has attracted increasing interest in scientific research, and much success has been achieved. Existing path-planning methods for multiple ASVs include model predictive control [33–35], the RRT method [36], satellite maps [37], the priority target assignment method [38], the Voronoi-Visibility roadmap [39], and the multidimensional rapidly exploring random trees star algorithm [40], to list just a few. However, due to the uncertainties of obstacles, and limitations in sensor range and communication bandwidth, the trajectory planning of multiple ASVs in an unknown—an especially an obstacle-rich—marine environment still presents a great challenge. In fact, to the best of the our knowledge, as for the autonomous navigation of multiple ASVs, little research is available that constructs a multi-objective optimization problem with optimal energy consumption, dynamic feasibility, and obstacle avoidance distance in unknown environments based on local sensor information.

In light of the above discussions, this paper focuses on the distributed swarm trajectory-planning problem for multiple ASVs in a complex and obstacle-rich marine environment. Firstly, by utilizing local sensors, the surrounding environment's point cloud data are acquired, and a kinodynamic path-search technique is employed to generate a series of waypoints within a discretized control space. Subsequently, B-spline curves are fitted to these waypoints, and a nonlinear optimization problem is formulated. This problem is optimized using a gradient-based local planning approach, which allows the generation of collision-free, smooth, and dynamically feasible trajectories through a shared network. Compared with the existing trajectory-planning methods, the proposed trajectory-planning method has the following features:

- In contrast to the existing trajectory-planning methods dedicated to a single ASV [14–21] and multiple USVs [36–40], this work constructs a multi-objective optimization problem with optimal energy consumption, dynamic feasibility, curve smoothness, obstacle avoidance distance, and endpoint distance constraints based on local sensor information. By utilizing local onboard sensor information, ASVs can accomplish autonomous navigation requiring no external localization and computation or a pre-built map in unknown complex environments.

- By adopting the proposed planning method, the optimal trajectory generation problem for multiple ASVs is divided into initial trajectory generation and back-end trajectory optimization. A heuristic-based kinodynamic path search is employed to efficiently find a safe, feasible, and minimum-time initial path. The initial path is then optimized by a B-spline optimization incorporating gradient-based local planning. By this means, the generated trajectories are able to meet the dynamic feasibility with enhanced safety.

- By adopting the proposed planning method, ASVs can perceive the surrounding environment in real time, using point cloud information obtained from local sensors to generate real-time local optimal trajectories. Moreover, this work utilizes multiple local optimal trajectories to form a global trajectory, which better meets the requirements of real-time obstacle avoidance and generates shorter sail distance for short-range autonomous navigation.

The rest of this paper is organized as follows. Some preliminaries and the problem formulation are given in Section 2. Section 3 discusses the process of swarm trajectory planning of ASVs. Section 4 provides some simulation results to demonstrate the effectiveness of the proposed method. Section 5 concludes this paper with some concluding remarks and future works.

2. Preliminaries and Problem Formulation

2.1. Preliminaries

$\mathbb{R}^n$ denotes the n-dimensional Euclidean space. The Euclidean norm is denoted by $\|\cdot\|$. $\lambda_{\min}(\cdot)$ and $\lambda_{\max}(\cdot)$ represent the minimum and maximum eigenvalues of a square matrix $(\cdot)$, respectively. $\mathbb{N}^+$ denotes a positive integer. $\mathbb{R}_+$ and $\mathbb{R}_-$ represent the positive and negative real numbers, respectively.

2.2. Problem Formulation

In the trajectory planning, each ASV is governed by the kinematic model expressed as follows:

$$\begin{cases} \dot{x} = u\cos\psi - v\sin\psi \\ \dot{y} = u\sin\psi + v\cos\psi \\ \dot{\psi} = r \end{cases} \tag{1}$$

where $[x, y]$ and ψ denote the position and yaw angle in an earth-fixed frame, respectively. u, v, and r are the surge velocity, sway velocity, and yaw rate in a body-fixed frame, respectively. For the trajectory-planning task, it is assumed that $v = 0$ such that

$$\begin{cases} \dot{x} = u\cos\psi \\ \dot{y} = u\sin\psi \\ \dot{\psi} = r \end{cases} \tag{2}$$

In this paper, a real-time trajectory-planning method is proposed for multiple ASVs to reach the designated target points in a complex and obstacle-rich environment, as shown in Figure 1.

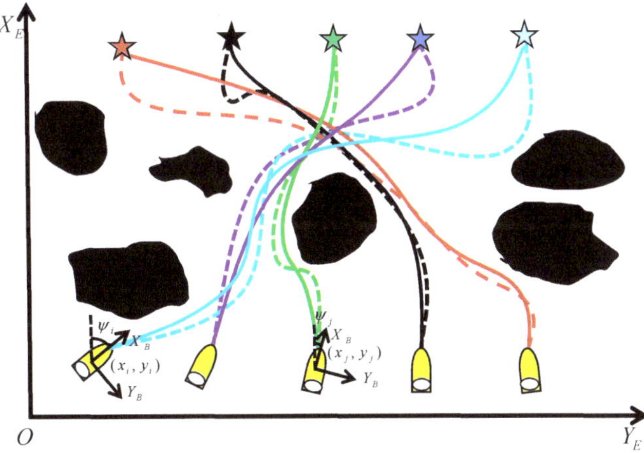

Figure 1. An illustration of trajectory planning for multiple ASVs to reach the designated locations. The dashed lines represent trajectories that have not yet been optimized, the solid lines indicate the optimized trajectories, and the black obstacles represent the complex static environments.

3. Swarm Trajectory Planning

This section gives a general description of the proposed swarm trajectory-planning method, including the construction of occupancy grid maps, the kinodynamic path-searching method, the curve-fitting method based on cubic uniform B-spline, the construction of nonlinear optimization problems under multiple constraint conditions, and the numerical optimization algorithm for solving the formulated optimization problem. The proposed trajectory-planning framework for multiple ASVs is shown in Figure 2.

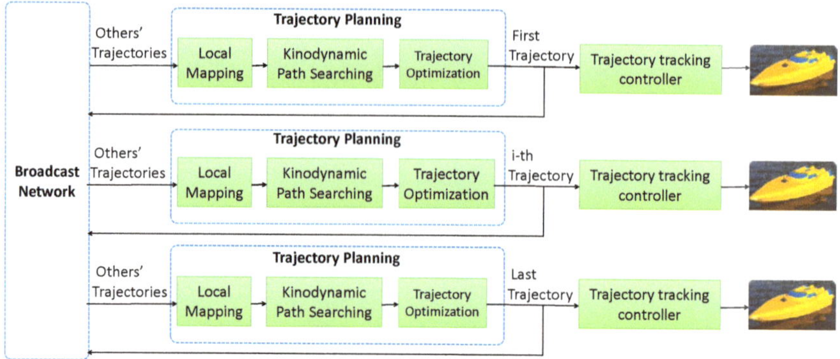

Figure 2. The trajectory-planning framework for multiple ASVs.

3.1. Construction of Occupancy Grid Map

The sensing radius of the sensor is set to r_{id} and takes a circular area for point cloud acquisition. The position of the ASV is set to $p_{id} = [x_{id}, y_{id}, z_{id}] \in \mathbb{R}^3$. The point cloud position coordinates obtained by sampling the surrounding environment are set to $p_{tid} = [x_{tid}, y_{tid}, z_{tid}] \in \mathbb{R}^3$, and the quaternion of the ASV is set to $q_{id} = [w_{qid}, x_{qid}, y_{qid}, z_{qid}] \in \mathbb{R}^4$. In order to obtain the point cloud data within the desired range, the angle value corresponding to the maximum height of point cloud collection is set to $\alpha \in \mathbb{R}$; then, the constraints for point cloud acquisition are given as

$$\begin{cases} \dfrac{z_{tid} - z_{id}}{r_{id}} \leq \tan(\alpha) \\ \begin{bmatrix} x_{tid} - x_{id} \\ y_{tid} - y_{id} \\ z_{tid} - z_{id} \end{bmatrix} \begin{bmatrix} 1 - 2(y_{qid}^2 + z_{qid}^2) \\ 2(x_{qid}y_{qid} + z_{qid}w_{qid}) \\ 2(x_{qid}z_{qid} - y_{qid}w_{qid}) \end{bmatrix} \geq 0 \end{cases} \tag{3}$$

According to (3), a dense point cloud map of the ASVs' forward direction can be obtained. To represent the presence of obstacles, an occupied grid map is constructed by partitioning the point cloud into grids with each being one of two states: filled or empty. For a grid m_i in an occupancy grid map, the probability of the grid state being occupied is denoted as $p(m_i)$ and the probability of the grid state being free is denoted as $p(\overline{m_i})$. The t-th observation is set to z_t. To calculate the posterior probability of the grid state based on existing observations, $p(m_i|z_{1:t})$ and $p(\overline{m_i}|z_{1:t})$ should be calculated. A grid is occupied if $p(m_i|z_{1:t}) \geq \kappa$ with κ being a preset threshold. The probability of the grid state being occupied can be specifically denoted as

$$p(m_i|z_{1:t}) = \frac{p(z_t|z_{1:t-1}, m_i)p(m_i|z_{1:t-1})}{p(z_t|z_{1:t-1})} \tag{4}$$

and the t-th observation is obtained as follows:

$$p(m_i|z_t) = \frac{p(z_t|m_i)p(m_i)}{p(z_t)} \tag{5}$$

According to the Markov assumption that the results of the first $t - 1$ observations are independent of the result of the t-th observation, one has that

$$p(z_t|z_{1:t-1}, m_i) = p(z_t|m_i) \tag{6}$$

Expanding (4) based on the Bayes formula, it follows that

$$p(m_i|z_{1:t}) = \frac{p(m_i|z_t)p(z_t)p(m_i|z_{1:t-1})}{p(m_i)p(z_t|z_{1:t-1})} \tag{7}$$

and

$$p(\overline{m_i}|z_{1:t}) = \frac{p(\overline{m_i}|z_t)p(z_t)p(\overline{m_i}|z_{1:t-1})}{p(\overline{m_i})p(z_t|z_{1:t-1})} \tag{8}$$

Define $l_t(m_i)$ as the t-th update posterior probability for a grid m_i; then, one has that

$$\begin{cases} l_t(m_i) = \log \dfrac{p(m_i|z_{1:t})}{p(\overline{m_i}|z_{1:t})} \\[2mm] l_{t-1}(m_i) = \log \dfrac{p(m_i|z_{1:t-1})}{p(\overline{m_i}|z_{1:t-1})} \end{cases} \tag{9}$$

Further simplifying the above equation using (7), $l_t(m_i)$ can be described as

$$l_t(m_i) = \log \frac{p(m_i|z_t)}{p(\overline{m_i}|z_t)} - \log \frac{p(m_i)}{p(\overline{m_i})} + l_{t-1}(m_i) \tag{10}$$

Substituting (5) into (10) yields

$$l_t(m_i) = \log \frac{p(z_t|m_i)}{p(z_t|\overline{m_i})} + l_{t-1}(m_i) \tag{11}$$

By assuming that the sensor model will not change during the environmental modeling process, the sensor model formulas $p(z_t|m_i)$ and $p(z_t|\overline{m_i})$ are constant. Then, according to (9), one has that

$$p(m_i|z_{1:t}) = \frac{a^{l_t(m_i)}}{1 + a^{l_t(m_i)}} \tag{12}$$

where a is the base of the *log* function. Based on the posterior probability, the occupied grid map can be updated. The occupied grid map constructed by ASVs during autonomous navigation is depicted in Figure 3.

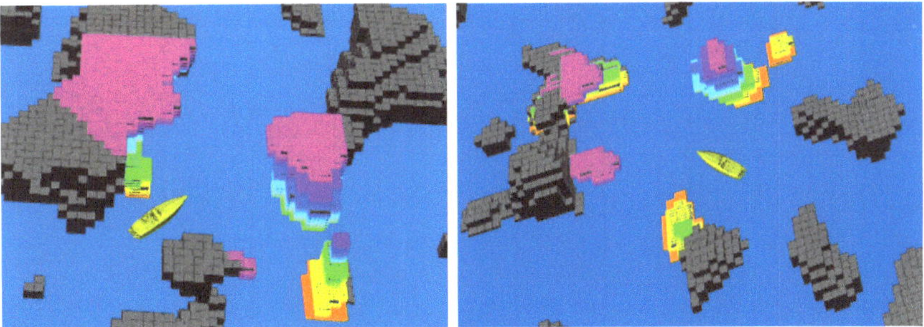

Figure 3. An illustration of the occupied grid map in a complex and obstacle-rich environment. The colored point cloud part represents the environmental information perceived by the ASV, and the gray part represents the unknown obstacle environment.

3.2. Kinodynamic Path Searching

Inspired by the hybrid A* search proposed for autonomous vehicles in [41], the kinodynamic path searching is applied for ASV to obtain a safe and reliable trajectory in an occupancy grid map while minimizing time cost. The mechanism of the kinodynamic path searching is illustrated in Figure 4.

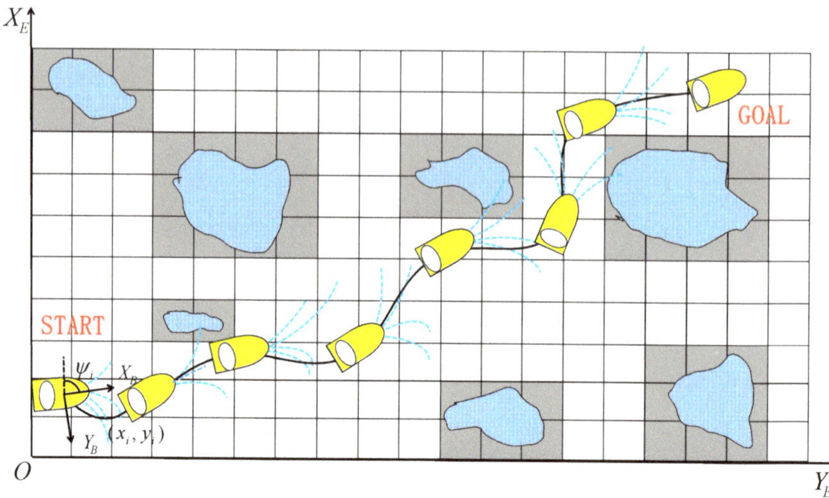

Figure 4. A schematic diagram of the kinodynamic path searching. The blue dashed line represents the motion primitives obtained by expanding the state point by Equation (14), and the black curve represents the optimal curve selected based on the heuristic function.

Similar to the standard A*, we use openlist and closelist to denote the open set and the closed set, respectively. A structure node is used to record the position of the expansion point as well as the g_c and f_c cost (Section 3.2.2). The nodes expand iteratively, and the one with the smallest f_c is saved in openlist. Then, CheckCollision($\cdot$) checks the safety and dynamic feasibility of the nodes. The searching process ends once any node reaches its goal successfully or the AnalyticExpand($\cdot$) succeeds. Details of the kinodynamic path searching are shown in Algorithm 1.

Algorithm 1 Kinodynamic Path Searching

---

while $openlist.$**empty**$()$ **do**
  $openlist$ **pop** n_c
  $closelist$ **insert** n_c
  if $Reachgoal(n_c)$ **or** $AnalyticExpand(n_c)$ **then**
    return $Path()$
  end if
  $expandnodes \leftarrow NodeExpand(n_c)$
  $nodes \leftarrow NodePrune(expandnodes)$
end while
for each $n_i \in nodes$ **do**
  if $closelist$ **contain** n_i **and** $CheckCollision(n_i)$ **then**
    $g_t \leftarrow n_c.g_c + CurrentCost(n_i)$
    if $openlist$ **contain** n_i **then**
      $openlist$ **add** n_i
      if $g_t \geq n_i.g_c$ **then**
        $continue$
      end if
    end if
  end if
end for
$n_i.parent \leftarrow n_c, n_i.g_c \leftarrow g_t$
$n_i.f_c \leftarrow n_i.g_c + h(n_i)$

3.2.1. Primitives Generation

The motion primitives are generated by using NodeExpand($\cdot$) in Algorithm 1. Accordingly, we firstly define the discretization sampling of r_i, T_j, and u_k, $i = 1, 2, \cdots, m$, $j = 1, 2, \cdots, n, k = 1, 2$ with $m, n \in \mathbb{N}^+$ as follows:

$$\begin{cases} r_i \in \{r_0, r_1, \ldots r_m\} \\ T_j \in \{T_0, T_1, \ldots, T_n\} \\ u_k \in \{-u_{max}, u_{max}\} \end{cases} \tag{13}$$

where $r_i, i = 1, 2, \cdots, m$ denotes a set of discretized yaw rates, $T_j, j = 1, 2, \cdots, n$, denotes a set of durations, and $u_k, k = 1, 2$ denotes the control input.

Let $p_0 = [x_0, y_0, \psi_0] \in \mathbb{R}^3$ be the node recording the current pose of the ASV, and let $p_t = [x_t, y_t, \psi_t] \in \mathbb{R}^3$ be the pose of the ASV after sampling. Recalling the ASV kinematic model (2) and applying the control variables u_k and r_i for duration T_j, the pose of the ASV can be calculated by

$$\begin{cases} x_t = x_0 + \displaystyle\int_0^{T_j} u_k \cos(\psi_t) d\tau \\[2mm] y_t = y_0 + \displaystyle\int_0^{T_j} u_k \sin(\psi_t) d\tau \\[2mm] \psi_t = \psi_0 + \displaystyle\int_0^{T_j} r_i d\tau \end{cases} \tag{14}$$

3.2.2. Heuristic Cost

In this subsection, a heuristic function $h(\cdot)$ is designed to speed up the searching in Algorithm 1 as used in A*. Heuristic cost refers to constructing a boundary optimal problem using the Pontryagin extremum principle, considering multiple target points generated by the A* path search method. Computing the cost values of all candidate target points, the one with the minimal cost is selected as the optimal target point. Then, based on the selected optimal target point, the A* algorithm is used to iteratively search toward the endpoint.

In order to find a trajectory that is optimal in time, the Pontryagins minimum principle is applied to design a heuristic cost function $J(T)$ as follows:

$$\begin{cases} J(T) = T + \displaystyle\sum_{\mu \in \{x,y\}} (\frac{1}{3}\alpha_\mu^2 T^3 + \alpha_\mu \beta_\mu T^2 + \beta_\mu^2 T) \\[4mm] \begin{bmatrix} \alpha_\mu \\ \beta_\mu \end{bmatrix} = \begin{bmatrix} -\frac{12}{T^3} & \frac{6}{T^2} \\ \frac{6}{T^2} & -\frac{2}{T} \end{bmatrix} \begin{bmatrix} p_{\mu f} - p_{\mu 0} - v_{\mu 0} T \\ v_{\mu f} - v_{\mu 0} \end{bmatrix} \end{cases} \tag{15}$$

where $p_{\mu 0}, p_{\mu f}$ are the current and goal position, respectively, and $v_{\mu 0}, v_{\mu f}$ are the current and goal velocity, respectively.

In order to obtain the minimum heuristic cost, α_μ and β_μ are substituted into $J(T)$ to obtain the solutions of $\partial J(T)/\partial T = 0$. The solution which minimizes $J(T)$ is denoted as $T_{\min}$, and $J(T_{\min})$ represents the heuristic cost h_c for the current node. Moreover, g_c is used to represent the actual cost of the trajectory from the start position (x_s, y_s) to the current state (x_c, y_c), and it is calculated as $g_c = \sqrt{(x_s - x_c)^2 + (y_s - y_c)^2}$. Thus, the final cost of the current state is obtained by $f_c = g_c + h_c$.

3.2.3. Collision Check

Collisions are checked based on the occupied grid map constructed by point clouds from local sensors. We assume that the entire ASV is included in a maximum rectangular box; then, a series of rectangular boxes is constructed sequentially on the trajectory points along the generated trajectory. All rectangular boxes are detected at a frequency of 20 Hz to determine whether they overlap with the surrounding point clouds or not. If there is overlap, it is determined that a collision has occurred, and multiple ASVs' trajectories need to be replanned. Otherwise, it is determined that the trajectory is safe.

To be specific, approximating an ASV by a rectangle, a set of footprints S representing the area occupied by the ASV is obtained by x_i, y_i and ψ_i along the footprints, as illustrated in Figure 5. The union of footprints S is called the swath along trajectory, which needs to be checked for collisions, as can be found in CheckCollision($\cdot$) in Algorithm 1. With the obtained swath, the occupancy grids of the swath are calculated to see if they overlap with obstacles on the occupancy grid map, as shown in Figure 6.

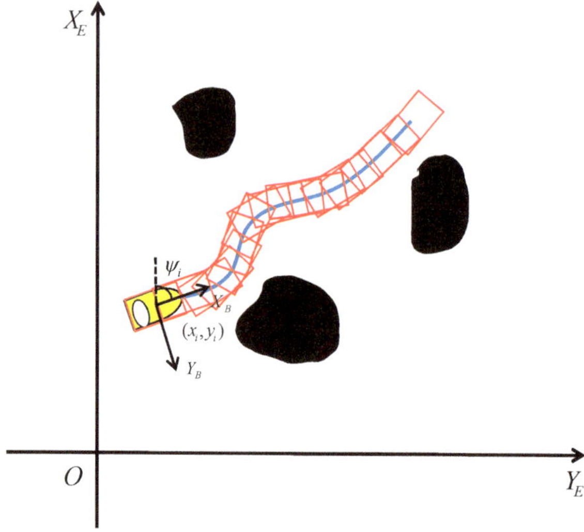

Figure 5. An illustration of the swath along trajectory. The red rectangle represents the area occupied by the ASV. The black area represents complex static obstacles.

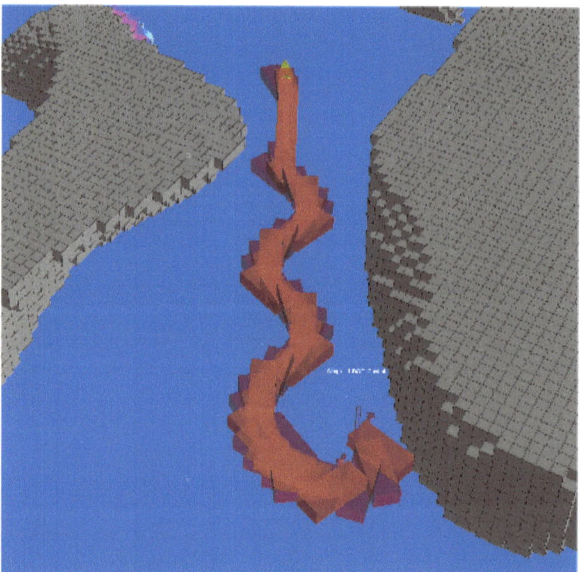

Figure 6. Schematic diagram of collision checking during autonomous navigation of an ASV. The red cube area represents the area occupied by the ASV, while the gray area represents complex static obstacles.

3.2.4. Analytic Expansion

Generally speaking, it is difficult for discretized input to reach the endpoint accurately. To this end, an analytic expansion scheme AnalyticExpand($\cdot$) in Algorithm 1 is induced to speed up the trajectory searching. When the current node $p_{\mu 0}$ is close to the endpoint $p_{\mu f}$, the same approach used in Section 3.2.2 is directly applied to compute a trajectory from $p_{\mu 0}$ to $p_{\mu f}$ without generating primitives. If the trajectory can pass the safety and dynamic feasibility check, the path searching is terminated in advance. This strategy proves to be beneficial in enhancing efficiency, particularly in a sparse environment, as it greatly improves the success rate of the algorithm and terminates the searching earlier.

3.3. Trajectory Smoothing

Theoretically, the safety and dynamic feasibility of the path generated by kinodynamic path searching cannot be strictly guaranteed, as kinodynamic path searching ignores the distance information of obstacles and does not consider curve smoothness. Therefore, a B-spline curve is used in this section to fit the path curve.

3.3.1. Cubic Uniform B-spline

Let $q = \{q_0, q_1, \ldots, q_n\}$ be the control points obtained from kinodynamic path searching and $\Theta = \{\theta_0, \theta_1, \ldots, \theta_m\}$ be the knot vector with $q_k \in \mathbb{R}^2$, $\theta_i \in \mathbb{R}$ and $m = n + p + 1$. A B-spline is a piecewise polynomial determined by its degree p and $n + 1$ control points q and Θ. A cubic B-spline trajectory, used to fit the above control points, is parameterized by θ. For a uniform cubic B-spline trajectory, it is noted that each knot span satisfies $\Delta\theta_k = \theta_{k+1} - \theta_k = \Delta\theta$.

The convex hull property of B-spline curves is illustrated in Figure 7. For $\theta \in [\theta_i, \theta_{i+1})$, $i = 0, 1, \cdots, m - 1$, the four control points of a cubic uniform B-spline trajectory set within the knot vector θ are $q_{k-3}, q_{k-2}, q_{k-1}, q_k$, where $3 \leq k \leq n$. To construct B-spline curves, the B-spline basis function firstly needs to be calculated, of which the degree is set to p. Denoting the k-th B-spline basis function of degree p as $N_{k,p}$, the 0-order and the p_k-order basis function are given as follows:

$$
\begin{cases}
N_{k,0} = \begin{cases} 0 & if\ \theta_i \leq \theta < \theta_{i+1} \\ 1 & otherwise \end{cases} \\
N_{k,p} = \dfrac{\theta - \theta_i}{\theta_{i+p} - \theta_i} N_{k,p-1}(\theta) + \dfrac{\theta_{i+p+1} - \theta}{\theta_{i+p+1} - \theta_{i+1}} N_{k+1,p-1}(\theta)
\end{cases}
\tag{16}
$$

For ease of implementation, p is set to $p = 3$. Correspondingly, the four basis functions are set to $N_{k-3,3}, N_{k-2,3}, N_{k-1,3}$, and $N_{k,3}$.

Defining a normalized variable $s(\theta) = (\theta - \theta_i)/\Delta\theta$, and substituting $s(\theta)$ into (16), one has that

$$
\begin{cases}
N_{k-3,3} = \dfrac{1}{6}(1 - s(\theta))^3 \\
N_{k-2,3} = \dfrac{1}{2}s(\theta)^3 - s(\theta)^2 + \dfrac{2}{3} \\
N_{k-1,3} = -\dfrac{1}{2}s(\theta)^3 + \dfrac{1}{2}s(\theta)^2 + \dfrac{1}{2}s(\theta) + \dfrac{1}{6} \\
N_{k,3} = \dfrac{1}{6}s(\theta)^3
\end{cases}
\tag{17}
$$

Therefore, the following matrix can be calculated to represent the coefficients of cubic uniform B-spline trajectories:

$$
\begin{cases}
p(s(\theta)) = n_k Q_k \\
n_k = \begin{bmatrix} N_{k-3,3} & N_{k-2,3} & N_{k\ 1,3} & N_{k,3} \end{bmatrix} \\
Q_k = \begin{bmatrix} q_{k-3} & q_{k-2} & q_{k-1} & q_k \end{bmatrix}^T
\end{cases}
\tag{18}
$$

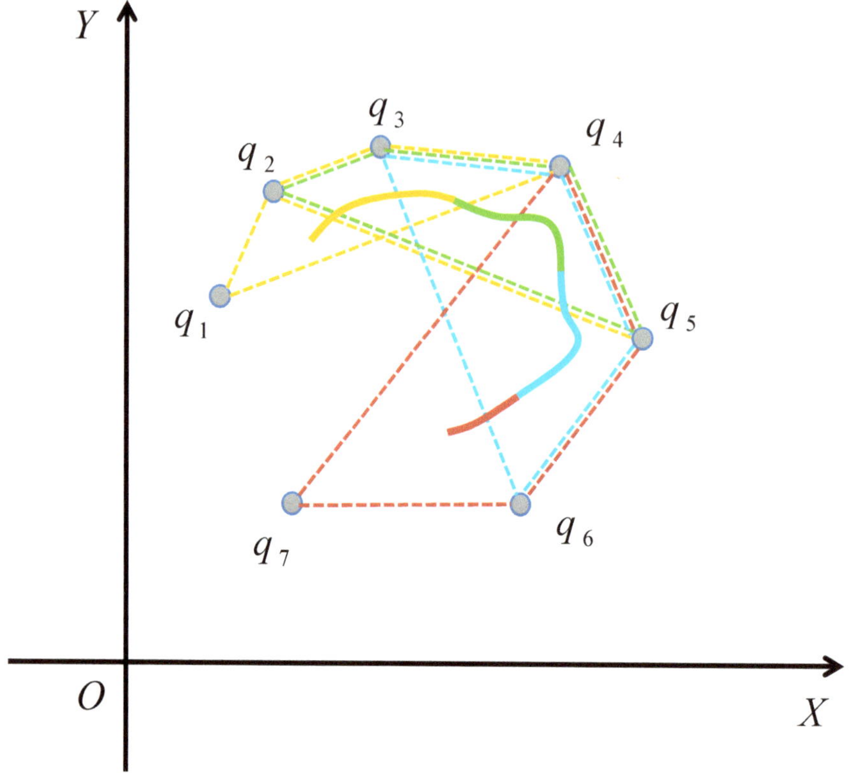

Figure 7. The convex hull property of B-spline curves with gray points representing the control points. Each segment of the curve is included in the convex hull constructed by every four control points.

3.3.2. Convex Hull Property

In order to ensure the dynamic feasibility, it is necessary to construct a nonlinear constrained optimization problem for the first-order and second-order derivatives of B-spline trajectory. For this purpose, it is necessary to prove that the derivative of a B-spline is also a B-spline. With control points $\{q_0, q_1, \ldots, q_n\}$ and basis functions $N_{k,p}$ defined above, a p-degree B-Spline $C(\theta)$ can be obtained as follows:

$$C(\theta) = \sum_{k=0}^{n} N_{k,p}(\theta) q_k \tag{19}$$

with its first-order and second-order derivatives being

$$\begin{cases} \dot{C}(\theta) = \sum_{k=0}^{n-1} N_{k+1,p-1}(\theta) v_k \\ \ddot{C}(\theta) = \sum_{k=0}^{n-2} N_{k+2,p-2}(\theta) a_k \end{cases} \tag{20}$$

Moreover, the first derivative of the basis function is given as follows:

$$\frac{dN_{k,p}(\theta)}{d\theta} = \frac{pN_{k,p-1}(\theta)}{\theta_{i+p} - \theta_i} + \frac{-pN_{k+1,p-1}(\theta)}{\theta_{i+p+1} - \theta_{i+1}} \tag{21}$$

Substituting (21) into (20) yields

$$\dot{C}(\theta) = \sum_{k=0}^{n-1} N_{k+1,p-1}(\theta)\left(\frac{pq_{k+1}}{\theta_{i+p+1} - \theta_{i+1}} + \frac{-pq_k}{\theta_{i+p+1} - \theta_{i+1}}\right) \tag{22}$$

Thus, the control points of the B-spline $C(\theta)$'s first derivative v_k can be computed by

$$v_k = \frac{pq_{k+1}}{\theta_{i+p+1} - \theta_{k+1}} + \frac{-pq_k}{\theta_{i+p+1} - \theta_{i+1}} \tag{23}$$

Since the B-spline trajectory used in Section 3.3.1 is uniform, Equation (23) is further simplified as follows:

$$v_k = \frac{q_{k+1} - q_k}{\Delta\theta} \tag{24}$$

Similarly, the second-order and third-order derivatives of the B-spline trajectory can be further derived as follows:

$$\begin{cases} a_k = \dfrac{v_{k+1} - v_k}{\Delta\theta} = \dfrac{1}{\Delta\theta^2}(q_{k+2} - 2q_{k+1} + q_k) \\ j_k = \dfrac{a_{k+1} - a_k}{\Delta\theta} = \dfrac{1}{\Delta\theta^3}(q_{k+3} - 3q_{k+2} + 3q_{k+1} - q_k) \end{cases} \tag{25}$$

3.4. Nonlinear Optimization

For the B-spline trajectory defined by a set of control points $\{q_0, q_1, \ldots, q_n\}$ in Section 3.3, a nonlinear optimization method is constructed in this subsection to further ensure the safety of the curve. The overall optimization function is defined as follows:

$$f_{total} = \lambda_1 f_{sm} + \lambda_2 f_c + \lambda_3 (f_v + f_a) + \lambda_4 f_{sw} + \lambda_5 f_t, \tag{26}$$

where f_{sm} and f_c are optimization terms for trajectory smoothness and collision distance, respectively. f_v and f_a are constrained optimization terms of velocity and acceleration, respectively. f_{sw} and f_t are optimization terms for the collision distance between ASVs and endpoint arrival, respectively. $\lambda_1, \lambda_2, \lambda_3, \lambda_4$ and λ_5 are the designed weight coefficients.

3.4.1. Smoothness Penalty

The smoothness cost f_{sm} is defined by a function that uses the integral of the squared jerk. According to (25), the jerk (i.e., the 3rd-order derivative of the position) of the trajectory is minimized to obtain a smooth trajectory. f_{sm} is defined as follows:

$$f_{sm} = \sum_{k=2}^{n-3} ||j_k||^2 = \sum_{k=2}^{n-3} ||q_{k+2} - 3q_{k+1} + 3q_k - q_{k-1}||^2 \tag{27}$$

3.4.2. Collision Distance Penalty

Initially, a naive B-spline trajectory is given regardless of whether the control points collide with an obstacle or not, as depicted by the black solid line passing though the obstacle in Figure 8. Therefore, the naive trajectory needs to be optimized by the A* algorithm to obtain a collision-free trajectory Γ.

Specifically, for control points q_k on the colliding segment, points on the obstacle boundary denoted by p_k are assigned with corresponding repulsive direction vectors v_k. The distance between q_k and p_k is denoted by $d_k = (q_k - p_k)^T v_k$, as shown in Figure 8. In order to avoid generating p_k, v_k repeatedly, the B-spline trajectory can only be optimized when $d_k > 0$. By ensuring that d_k is less than the safe distance s_f, the control points can be kept away from the obstacles. In order to optimize the collision distance of the B-spline trajectory, a twice continuously differentiable function F_c is applied as follows:

$$\begin{cases} c_k = s_f - d_k \\ F_c = \begin{cases} 0 & c_k \leq 0 \\ c_k^3 & 0 < c_k \leq s_f \\ 3s_f c_k^2 - 3s_f^2 c_k + s_f^3 & s_f < c_k \end{cases} \end{cases} \tag{28}$$

The collision penalty function is denoted as f_c and its derivative can be obtained as follows:

$$\begin{cases} f_c = \sum_{k=2}^{N-3} F_c(q_k) \\ J_c = \dfrac{\delta f_c}{\delta q_k} = \begin{cases} -3c_k v_k & c_k \leq s_f \\ -6s_f c_k + 3s_f^2 & s_f < c_k \end{cases} \end{cases} \tag{29}$$

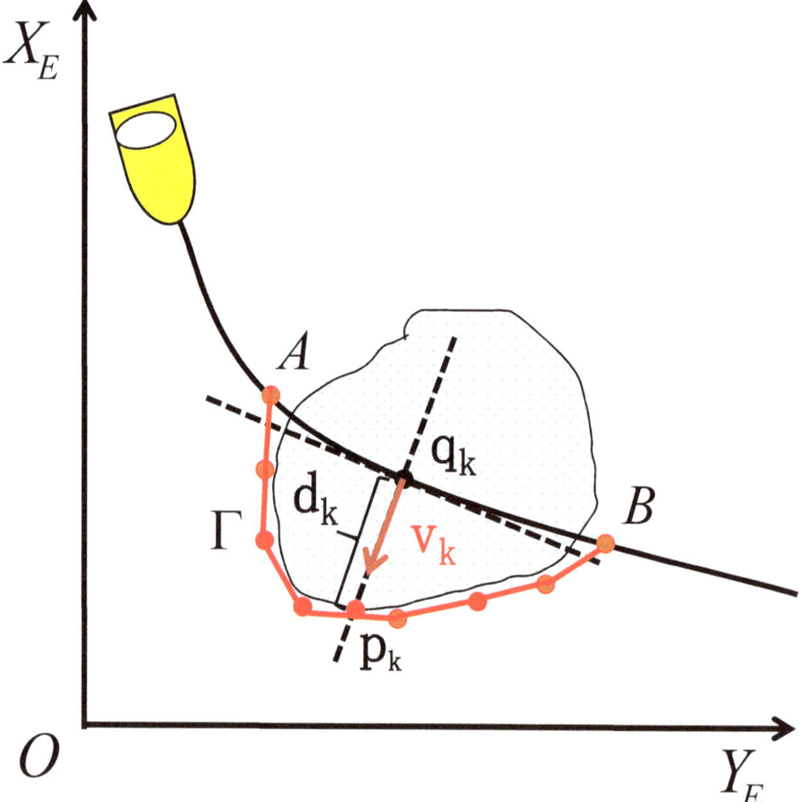

Figure 8. The black solid line passing though the obstacle represents the naive B-spline trajectory to be optimized. The red solid line represents the edge of the obstacle obtained by A* search.

3.4.3. Dynamic Feasibility Penalty

The dynamic feasibility can be ensured by constraining the high-order derivatives of B-spline trajectories at discrete control points $\{q_0, q_1, \ldots, q_n\}$. Specifically, due to the convex hull property, constraining derivatives of the control points is sufficient for constraining the whole B-spline. Therefore, the constrained optimization terms of velocity and acceleration are given, respectively, as follows:

$$
\begin{cases}
f_v = \sum_{k=3}^{N-3} F_v(v_k) \\[4pt]
F_v(v_k) = \begin{cases} 0 \ (0 \leq v_k \leq v_{max}) \\ (v_k - v_{max})^3 \ (v_{max} \leq v_k \leq v_j) \\ a_1 v_k^2 + b_1 v_k + c_1 \ (v_k \geq v_j) \end{cases} \\[18pt]
f_a = \sum_{k=4}^{N-3} F_a(a_k) \\[4pt]
F_a(a_k) = \begin{cases} 0 \ (0 \leq a_k \leq a_{max}) \\ (a_k - a_{max})^3 \ (a_{max} \leq a_k \leq a_j) \\ a_2 a_k^2 + b_2 a_k + c_2 \ (a_k \geq a_j) \end{cases}
\end{cases}
\tag{30}
$$

where $a_1, b_1, c_1, a_2, b_2,$ and c_2 are design parameters to ensure the second-order continuous differentiability of F_v and F_a. v_{max} and a_{max} are the derivative limits, respectively. v_j and a_j are the splitting points of quadratic and cubic curves, respectively. According to (24) and (25), the first-order derivatives corresponding to f_v and f_a are given, respectively, as follows:

$$
\begin{cases}
\dfrac{\delta v_k}{\delta q_k} = -\dfrac{1}{\Delta\theta} \\[10pt]
\dfrac{\delta f_{v_k}}{\delta q_k} = \begin{cases} \sum_{k=3}^{N-3} 3\dfrac{\delta v_k}{\delta q_k}(v_k - v_{max})^2 \ (v_{max} \leq v_k \leq v_j) \\[6pt] \sum_{k=3}^{N-3} 2\dfrac{\delta v_k}{\delta q_k}a_2 v_k + \dfrac{\delta v_k}{\delta q_m}b_2 \ (v_k \geq v_j) \end{cases} \\[18pt]
\dfrac{\delta a_k}{\delta q_k} = \dfrac{1}{\Delta\theta^2} \\[10pt]
\dfrac{\delta f_{a_k}}{\delta q_k} = \begin{cases} \sum_{k=3}^{N-3} 3\dfrac{\delta a_k}{\delta q_k}(a_k - a_{max})^2 \ (a_{max} \leq a_k \leq a_j) \\[6pt] \sum_{k=3}^{N-3} 2\dfrac{\delta a_k}{\delta q_k}a_2 a_k + \dfrac{\delta a_\mu}{\delta q_k}b_2 \ (a_k \geq a_j) \end{cases}
\end{cases}
\tag{31}
$$

3.4.4. Swarm Distance Penalty

An illustration of swarm distance penalty is depicted in Figure 9. Similar to the collision distance penalty and dynamic feasibility penalty, the swarm distance penalty function is formulated as follows:

$$
\begin{cases}
d_{sw} = (q_{i,k} - q_{j,k})^2 \\[4pt]
s_\mu = s_{sw} - d_{sw} \\[4pt]
f_{sw} = \sum_{k=0}^{1/2N} F_s(s_\mu) \\[4pt]
F_{sw}(s_\mu) = \begin{cases} s_\mu^2 \ (s_\mu \geq 0) \\ 0 \ (s_\mu < 0) \end{cases}
\end{cases}
\tag{32}
$$

where d_{sw} and s_{sw} represent the actual distance between swarm trajectories and the preset safety distance, respectively; $q_{i,k}$ and $q_{j,k}$ represent the control point of the i-th and j-th trajectories at time k, respectively. The derivative of the swarm distance penalty function can be obtained as follows:

$$J_{sw} = \frac{\delta f_{sw}}{\delta q_k} = \begin{cases} \sum_{k=0}^{1/2N} -2(q_{i,k} - q_{j,k}) \ (s_\mu \geq 0) \\ 0 \ (s_\mu < 0) \end{cases} \tag{33}$$

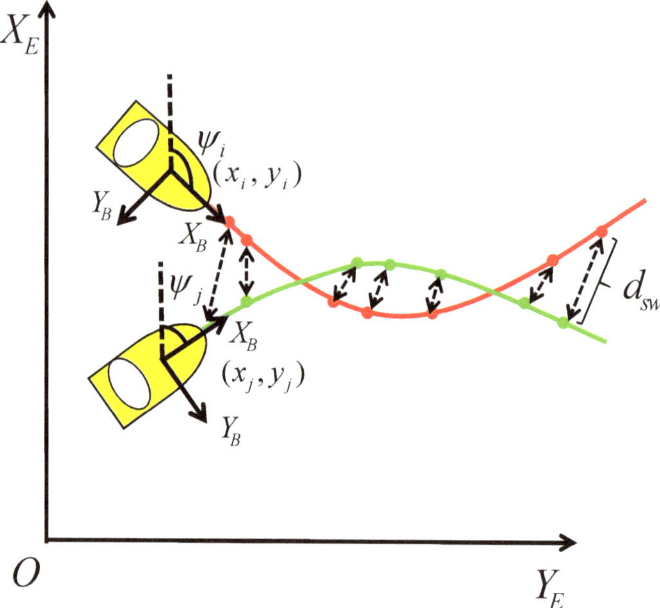

Figure 9. An illustration of swarm distance penalty. The red line represents the i-th trajectory, while the green line represents the j-th trajectory. $q_{i,k}$ and $q_{j,k}$ represent the control points corresponding to the i-th and j-th trajectories, respectively, at time k.

3.4.5. Endpoint Arrival Penalty

To ensure that each ASV can reach the endpoint, the last three control points of the B-spline trajectory are set to q_{k-2}, q_{k-1}, q_k, and $k \in [2, n]$, respectively. Let f_t be the penalty function for reaching the endpoint; then, one has that

$$f_t = (\frac{1}{6}(q_{k-2} + 4q_{k-1} + q_k) - G)^2, \tag{34}$$

where $G \in \mathbb{R}^2$ represents the endpoint. The first derivative of f_t is obtained as

$$J_t = \frac{\delta f_t}{\delta q_k} = \frac{1}{3}(q_{k-2} + 4q_{k-1} + q_k) - 2G \tag{35}$$

3.5. Numerical Optimization

The nonlinear optimization problem has the following two characteristics. Firstly, the total penalty function f_{total} will be updated in real time based on changes in the environment. Secondly, the quadratic optimization term about dynamic feasibility and obstacle avoidance distance will make f_{total} closer to the quadratic form, which means that the utilization of Hessian information can significantly improve the speed of solution. However, in the process of trajectory planning for ASVs, solving the inverse Hessian information is prohibitive in real time. Therefore, the limited memory BFGS (L-BFGS) method is adopted to achieve accurate values through a series of iterations. The details of the optimization process are summarized in Algorithm 2.

Algorithm 2 Numerical Optimization

Initialize x^0, $g^0 \leftarrow \nabla f(x^0)$, $B_0 \leftarrow I$, $k \leftarrow 0$
   while $\|g^k\| > \delta$ **do**
      $d \leftarrow -B^k g^k$
      $t \leftarrow$ *Lewis Overton line search*
      $x^{k+1} \leftarrow x^k + td$
      $g^{k+1} \leftarrow \nabla f(x^{k+1})$
      $B^{k+1} \leftarrow LBFGS(g^{k+1} - g^k, x^{k+1} - x^k)$
      $k \leftarrow k + 1$
   end while

For an unconstrained optimization problem $\min f(x)$, the iterative updating method for x is similar to Newton's method. Specifically, the update of x is given as follows:

$$x^{k+1} = x^k - tB^k g^k, \tag{36}$$

where B^k is updated at every iteration of the LBFGS method as summarized in Algorithm 3, g^k represents the gradient at x^k, and t is the step length obtained through the Lewis–Overton line search method, as summarized in Algorithm 4.

Algorithm 3 The L-BFGS algorithm

Initialize $s^k = x^{k+1} - x^k$, $y^k = g^{k+1} - g^k$, $\rho^k = 1/(y^{k^T} s^k)$, $d \leftarrow g^k$
   for $i = k - 1, k - 2, ..., k - m$ **do**
      $\alpha^i \leftarrow \rho^i s^{i^T} d$
      $d \leftarrow d - \rho^i y^i$
   end for
   $\gamma \leftarrow \rho^{k-1} y^{k-1^T} y^{k-1}$
   $d \leftarrow d/\gamma$
   for $i = k - m, k - m + 1, ..., k - 1$ **do**
      $\beta \leftarrow \rho^i y^{i^T} d$
      $d \leftarrow d + s^i(\alpha^i - \beta)$
   end for

Algorithm 4 Lewis–Overton line search

Initialize $l \leftarrow 0$, $u \leftarrow +\infty$, $\alpha \leftarrow \alpha_0$
   if $S(\alpha) fails$ **then**
      $u \leftarrow \alpha$
   else if $C(\alpha) fails$ **then**
      $l \leftarrow \alpha$
   else
      return α
   end if
   if $u < +\infty$ **then**
      $\alpha \leftarrow (l + u)/2$
   else
      $\alpha \leftarrow 2l$
   end if

It is noted that $S(\alpha)$ and $C(\alpha)$ in Algorithm 4 represent strong Wolfe conditions and weak Wolfe conditions, respectively, which are given as follows:

$$strong\ wolfe\ conditions:\begin{cases} f(x^k) - f(x^k + td) \geq -c_1 * td^T \nabla f(x^k) \\ |d^T \nabla f(x^k + td)| \leq |c_2 d^T \nabla f(x^k)| \end{cases} \tag{37}$$

$$weak\ wolfe\ conditions:\begin{cases} f(x^k) - f(x^k + td) \geq -c_1 * td^T \nabla f(x^k) \\ d^T \nabla f(x^k + td) \geq c_2 d^T \nabla f(x^k) \end{cases} \tag{38}$$

where $c_1 = 10^{-4}, c_2 = 0.9$.

3.6. Broadcast Network

Once an ASV generates a new trajectory in the complex environment, it will publish the trajectory to other ASVs through a broadcast network. Other ASVs will save the trajectory and generate their own safety trajectory when necessary based on the saved trajectory. Meanwhile, each ASV checks the collision condition when the neighbor's trajectory is received from the network. If the received trajectory collides with the trajectories of other ASVs, this strategy can solve the problem. In addition, considering the increasing number of ASVs, each ASV should compare its current position with the trajectories received from neighboring ASVs before conducting trajectory planning.

Remark 1. *The current algorithm proposed in the article aims at obstacle avoidance and collision avoidance of multiple ASVs in complex static obstacle environments. When dynamic obstacles exist and enter the perception range of the ASV, it will stop and continuously replan until a passable path appears. Therefore, it would still be feasible for online application while meeting unknown moving obstacles. Nevertheless, in the case of unknown moving obstacles, it is necessary for the ASV to have strong braking ability when sailing at a fast speed. In our future work, for the case of unknown moving obstacles, we will try to add dynamic point cloud filtering to complete the reconstruction of the surrounding obstacle environment. Moreover, point cloud recognition and/or clustering algorithms will be added to determine the position and speed of unknown dynamic obstacles. With relevant constraint conditions constructed, the safety of autonomous navigation for multiple ASVs can be guaranteed.*

4. Simulation Results

In this section, simulations are provided to illustrate the performance of the proposed distributed trajectory-planning method for multiple ASVs in an unknown environment with lots of static obstacles. Two cases are considered as follows:

Case 1: Swarm trajectory planning for four ASVs. In the simulation, swarm trajectories were planned for four ASVs by various planners in the same scenario. In particular, the proposed method is compared with the enhanced conflict-based search (ECBS) method and the A* + B spline method in terms of sail distance (d_{sail}), sail time (t_{sail}), and collision times per ASV.

Figure 10 shows trajectories planned by various planners in the same scenario with the same initial and final positions. It is observed from Table 1 and Figure 10 that all planners can generate collision-free trajectories, while the proposed one needs shorter sail distance and sail time compared to the other two planners. In particular, compared with the A* + B spline method, the proposed method solves the multi-objective optimization problem and achieves optimization subject to obstacle avoidance distance, curve smoothness, and collision avoidance distance between ASVs, allowing multiple ASVs to reach the specified target points in a shorter time and with lower energy loss. Meanwhile, compared with the ECBS method using global planning, the proposed one utilizes multiple local optimal trajectories to form a global trajectory, which better meets the requirements of real-time obstacle avoidance and generates shorter sail distance for short-range navigation. However, local planning using our method and the A* + B spline method generates a global path using multi-segment local trajectories, which is prone to getting stuck in local optima. This is also what we want to solve in the future.

Table 1. Comparisons between different planners.

Planner	d_{sail} (m)	t_{sail} (s)	Collision
ECBS	12.6	11.6	0
A* + B spline	13.6	9.2	0
Proposed	10.6	8.3	0

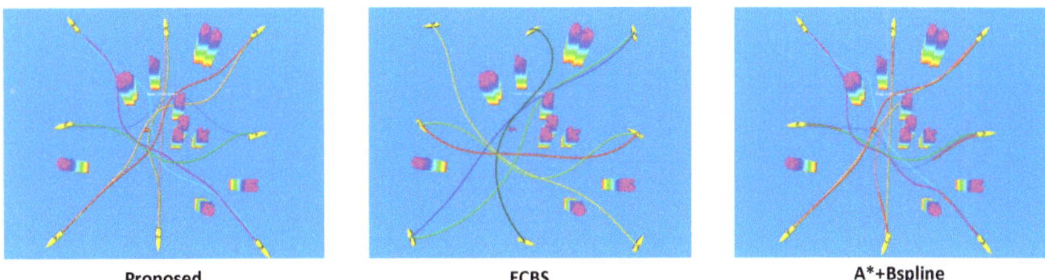

Figure 10. Trajectories planned by various planners in the same scenario.

Case 2: Swarm trajectory planning for seven ASVs. In this case, we conducted simulations on seven ASVs with a random map generated by the Berlin algorithm, where the generated trajectories are shown in Figure 11. The generated trajectories of seven ASVs in the xy-plane are shown in Figure 12. The evolution of the velocity and acceleration of the ASVs are shown in Figures 13 and 14, respectively. Due to the existence of speed and acceleration optimization terms, the navigation speed and acceleration of ASVs do not exceed 2.5 m/s and 3 m/s^2, respectively. The distances of each ASV to the goals, the distances of each ASV to the closest ASV, and the distances of each ASV to the closest obstacle are shown in Figures 15–17, respectively. It is noted that the safety threshold of distance bewteen each ASV and the closest ASV is set to 0.4 m, and the safety threshold of distance between each ASV and the closest obstacle is set to 0.5 m. At the same time, due to the existence of boundary constraints, ASVs can reduce their speed and acceleration to 0 when they reach the goals. Moreover, when the target arrival constraint exists, the ASVs can reach the goals accurately.

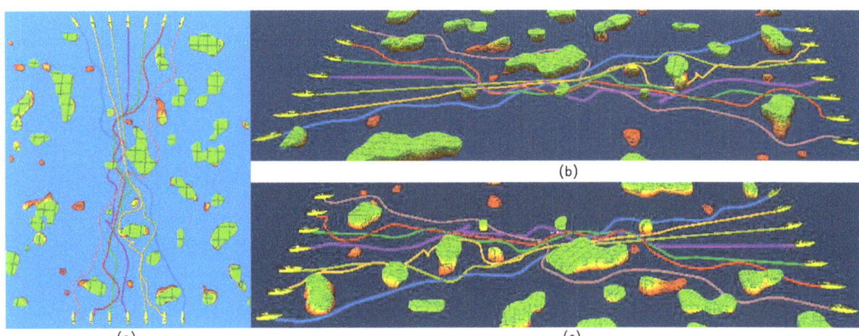

Figure 11. Seven ASVs conduct autonomous navigation in a simulated environment. (**a**–**c**) represent the top view, left-side view, and right-side view, respectively.

To further verify the effectiveness of the proposed method, we conducted multiple simulations based on different numbers of ASVs under different obstacle coverage ranges and different initial conditions. Setting the obstacle coverage rate to 50% and 75%, setting obstacle shape roughness to 10% and 15%, and setting the number of ASVs to 5, 7, and 9, respectively, different simulations were conducted, as shown in Figure 18. As a result, one can see that the proposed method performs well under different obstacle coverage ranges and different initial conditions.

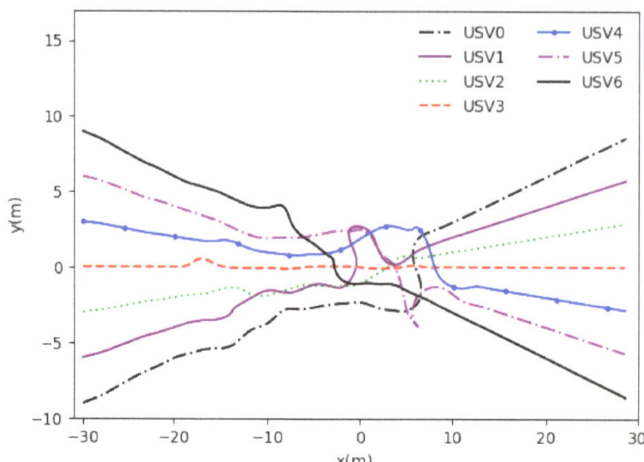

Figure 12. The trajectories of seven ASVs in the xy-plane.

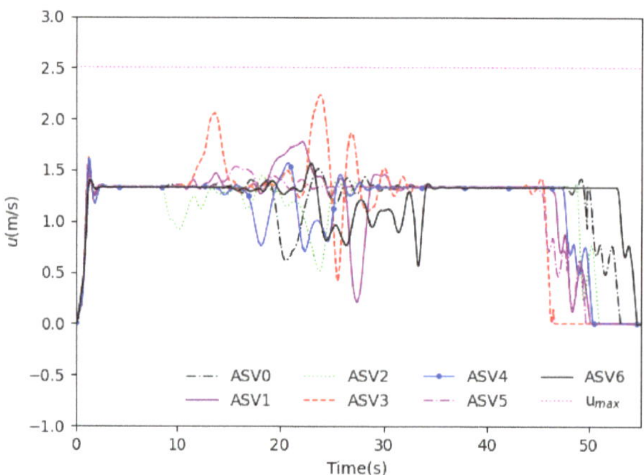

Figure 13. The speed of seven ASVs with each type of line corresponding to one ASV.

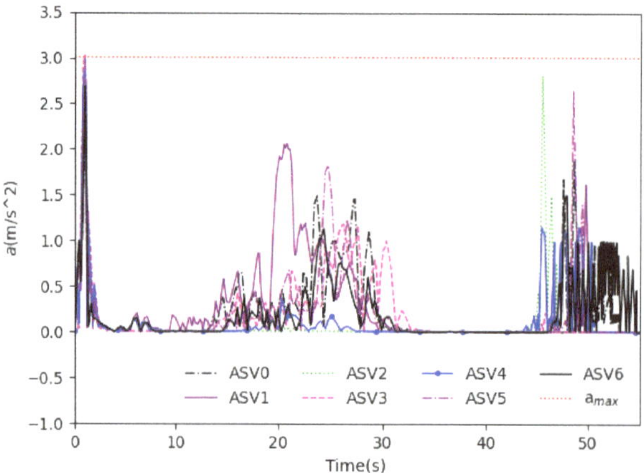

Figure 14. The acceleration of seven ASVs with each type of line corresponding to one ASV.

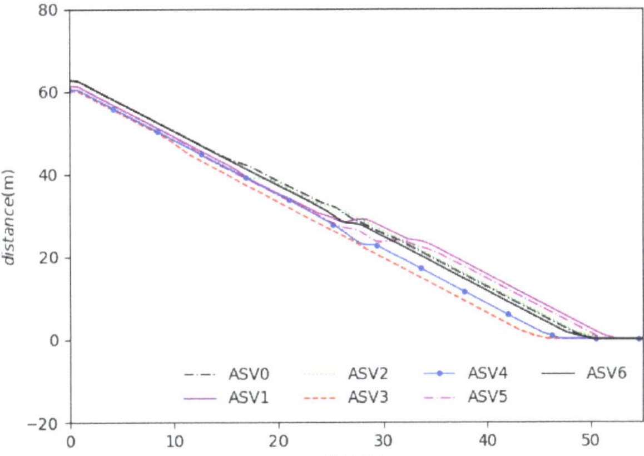

Figure 15. The distances of each ASV to the goals with each type of line corresponding to one ASV.

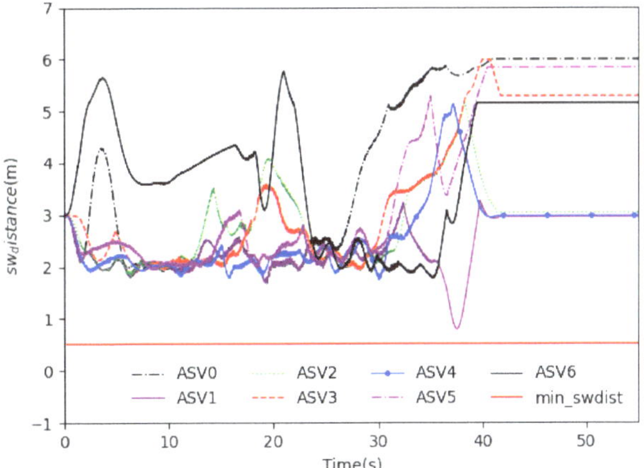

Figure 16. The distances of each ASV to the closest ASV with each type of line corresponding to one ASV.

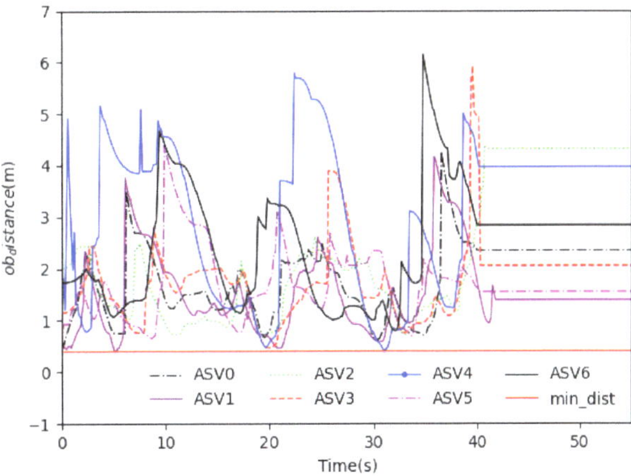

Figure 17. The distances of each ASV to the closest obstacle with each type of line corresponding to one ASV.

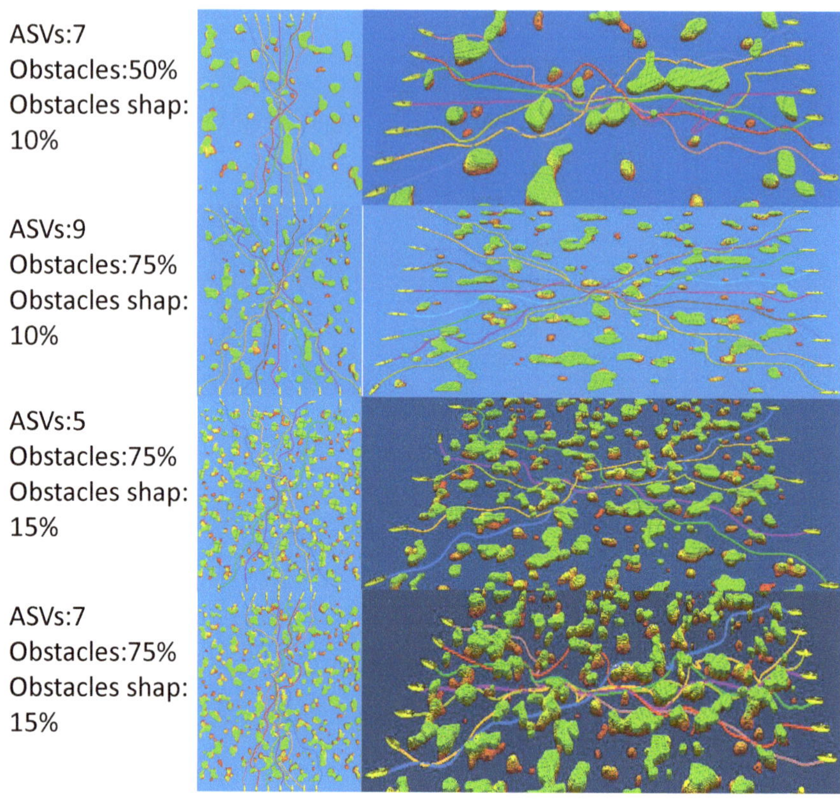

Figure 18. Autonomous navigation of different numbers of ASVs under different obstacle coverage ranges and different initial conditions (**Left**: Top View, **Right**: Left-side View).

5. Conclusions

In this paper, a swarm trajectory-planning method is proposed for multiple ASVs using distributed asynchronous communication. The issues of curve smoothness, dynamic feasibility, collision avoidance between ASVs, and obstacle avoidance are transformed into a non-constrained nonlinear optimization problem. Efficient solutions are proposed for generating smooth and collision-free trajectories that ASVs can track. Since real-time local planning and collision-detection strategies have been adopted, it is effective to reduce the total navigation time and avoid obstacles in a marine environment. In the future, we will further address the issue of formation of ASVs subject to multiple constraints and static obstacles.

Author Contributions: Conceptualization, investigation, methodology, validation, formal analysis, A.W.; resources, data curation, writing—original draft preparation, L.L.; writing—review and editing, H.W.; visualization, supervision, B.H.; project administration, funding acquisition, Z.P. All authors have read and agreed to the published version of the manuscript.

Funding: This research was funded in part by the National Key R&D Program of China under Grant 2022ZD0119902, in part by the Key Basic Research of Dalian under Grant 2023JJ11CG008, in part by the National Natural Science Foundation of China under Grant 51979020, in part by the Top-notch Young Talents Program of China under Grant 36261402, in part by the Doctoral Scientific Research Foundation of Liaoning Province under Grant 2023-BS-155, in part by the Basic Scientific Research Project of Higher Education Department of Liaoning Province under Grant LJKZ0044, in part by the National Natural Science Foundation of China under Grant 62203081, in part by the Postdoctoral

Research Foundation of China under Grant 2022M720628, and in part by Doctoral Science Research Foundation of Liaoning under Grant 2022-BS-097.

Institutional Review Board Statement: Not applicable.

Informed Consent Statement: Not applicable.

Data Availability Statement: Data are contained within the article.

Conflicts of Interest: Author Bing Han was employed by the company Shanghai Ship and Shipping Research Institute Co., Ltd.; The remaining authors declare that the research was conducted in the absence of any commercial or financial relationships that could be construed as a potential conflict of interest.

References

1. Zheng, Z.; Sun, L. Path following control for marine surface vessel with uncertainties and input saturation. *Neurocomputing* **2016**, *177*, 158–167. [CrossRef]
2. Annamalai, A.S.K.; Sutton, R.; Yang, C.; Culverhouse, P.; Sharma, S. Robust adaptive control of an uninhabited surface vehicle. *J. Intell. Robot. Syst.* **2014**, *78*, 319–338. [CrossRef]
3. Peng, Z.; Gu, N.; Zhang, Y.; Liu, Y.; Wang, D.; Liu, L. Path-guided time-varying formation control with collision avoidance and connectivity preservation of under-actuated autonomous surface vehicles subject to unknown input gains. *Ocean. Eng.* **2019**, *191*, 106501. [CrossRef]
4. Peng, Z.; Wang, D.; Li, T.; Han, M. Output-Feedback Cooperative Formation Maneuvering of Autonomous Surface Vehicles With Connectivity Preservation and Collision Avoidance. *IEEE Trans. Cybern.* **2019**, *50*, 2527–2535. [CrossRef] [PubMed]
5. Wang, D.; Huang, J. Adaptive neural network control for a class of uncertain nonlinear systems in pure-feedback form. *Automatica* **2002**, *38*, 1365–1372. [CrossRef]
6. Peng, Z.; Wang, D.; Zhang, H.; Sun, G. Distributed neural network control for adaptive synchronization of uncertain dynamical multiagent systems. *IEEE Trans. Neural Netw. Learn. Syst.* **2014**, *25*, 1508–1519. [CrossRef] [PubMed]
7. Zhang, G.; Han, J.; Li, J.; Zhang, X. Distributed Robust Fast Finite-Time Formation Control of Underactuated ASVs in Presence of Information Interruption. *J. Mar. Sci. Eng.* **2022**, *10*, 1775. [CrossRef]
8. Jing, Q.; Wang, H.; Hu, B.; Liu, X.; Yin, Y. A universal simulation framework of shipborne inertial sensors based on the ship motion model and robot operating system. *J. Mar. Sci. Eng.* **2021**, *9*, 900. [CrossRef]
9. Wang, H.; Yin, Y.; Jing, Q. Comparative Analysis of 3D LiDAR Scan-Matching Methods for State Estimation of Autonomous Surface Vessel. *J. Mar. Sci. Eng.* **2023**, *11*, 840. [CrossRef]
10. Lee, D.; Woo, J. Reactive Collision Avoidance of an Unmanned Surface Vehicle through Gaussian Mixture Model-Based Online Mapping. *J. Mar. Sci. Eng.* **2022**, *10*, 472. [CrossRef]
11. Veitch, E.; Alsos, O.A. Human-centered explainable artificial intelligence for marine autonomous surface vehicles. *J. Mar. Sci. Eng.* **2021**, *9*, 1227. [CrossRef]
12. Chen, L.; Hopman, H.; Negenborn, R.R. Distributed model predictive control for vessel train formations of cooperative multi-vessel systems. *Transp. Res. Part C Emerg. Technol.* **2018**, *92*, 101–118. [CrossRef]
13. Geertsma, R.; Negenborn, R.; Visser, K.; Hopman, J. Design and control of hybrid power and propulsion systems for smart ships: A review of developments. *Appl. Energy* **2017**, *194*, 30–54. [CrossRef]
14. Izzo, P.; Veres, S.M. Intelligent planning with performance assessment for Autonomous Surface Vehicles. In Proceedings of the OCEANS 2015, Genova, Italy, 18–21 May 2015; IEEE: Piscataway, NJ, USA, 2015; pp. 1–6.
15. Arzamendia, M.; Gregor, D.; Reina, D.; Toral, S.; Gregor, R. Evolutionary path planning of an autonomous surface vehicle for water quality monitoring. In Proceedings of the 2016 9th International Conference on Developments in eSystems Engineering (DeSE), Liverpool, UK, 31 August–2 September 2016; IEEE: Piscataway, NJ, USA, 2016; pp. 245–250.
16. Willners, J.S.; Petillot, Y.R.; Patron, P.; Gonzalez-Adell, D. Kinodynamic Path Planning for Following and Tracking Vehicles. In Proceedings of the OCEANS 2018 MTS/IEEE, Charleston, SC, USA, 22–25 October 2018; IEEE: Piscataway, NJ, USA, 2018.
17. Bitar, G.; Martinsen, A.B.; Lekkas, A.M.; Breivik, M. Two-Stage Optimized Trajectory Planning for ASVs Under Polygonal Obstacle Constraints: Theory and Experiments. *IEEE Access* **2020**, *8*, 199953–199969. [CrossRef]
18. Shan, T.; Wang, W.; Englot, B.; Ratti, C.; Rus, D. A Receding Horizon Multi-Objective Planner for Autonomous Surface Vehicles in Urban Waterways. In Proceedings of the 2020 59th IEEE Conference on Decision and Control (CDC), Jeju, Republic of Korea, 14–18 December 2020; pp. 4085–4092.
19. D'Amato, E.; Nardi, V.A.; Notaro, I.; Scordamaglia, V. A Visibility Graph approach for path planning and real-time collision avoidance on maritime unmanned systems. In Proceedings of the 2021 International Workshop on Metrology for the Sea; Learning to Measure Sea Health Parameters (MetroSea), Reggio Calabria, Italy, 4–6 October 2021; IEEE: Piscataway, NJ, USA, 2021; pp. 400–405.
20. Krell, E.; King, S.A.; Garcia Carrillo, L.R. Autonomous Surface Vehicle energy-efficient and reward-based path planning using Particle Swarm Optimization and Visibility Graphs. *Appl. Ocean Res.* **2022**, *122*, 103125. [CrossRef]
21. Hu, S.; Tian, S.; Zhao, J.; Shen, R. Path Planning of an Unmanned Surface Vessel Based on the Improved A-Star and Dynamic Window Method. *J. Mar. Sci. Eng.* **2023**, *11*, 1060. [CrossRef]

22. Liu, Z.; Zhang, Y.; Yu, X.; Yuan, C. Unmanned surface vehicles: An overview of developments and challenges. *Annu. Rev. Control* **2016**, *41*, 71–93. [CrossRef]
23. Peng, Z.; Wang, J.; Wang, D.; Han, Q.L. An overview of recent advances in coordinated control of multiple autonomous surface vehicles. *IEEE Trans. Ind. Inform.* **2020**, *17*, 732–745. [CrossRef]
24. Zhang, G.; Huang, C.; Li, J.; Zhang, X. Constrained coordinated path-following control for underactuated surface vessels with the disturbance rejection mechanism. *Ocean. Eng.* **2020**, *196*, 106725. [CrossRef]
25. Liu, L.; Wang, D.; Peng, Z. Coordinated path following of multiple underacutated marine surface vehicles along one curve. *ISA Trans.* **2016**, *64*, 258–268. [CrossRef]
26. Liu, L.; Wang, D.; Peng, Z.; Li, T.; Chen, C.L.P. Cooperative Path Following Ring-Networked Under-Actuated Autonomous Surface Vehicles: Algorithms and Experimental Results. *IEEE Trans. Cybern.* **2020**, *50*, 1519–1529. [CrossRef]
27. Fossen, T.I.; Strand, J.P. Passive nonlinear observer design for ships using Lyapunov methods: full-scale experiments with a supply vessel. *Automatica* **1999**, *35*, 3–16. [CrossRef]
28. Loria, A.; Fossen, T.I.; Panteley, E. A separation principle for dynamic positioning of ships: Theoretical and experimental results. *IEEE Trans. Control Syst. Technol.* **2000**, *8*, 332–343. [CrossRef]
29. Tan, G.; Wang, Z. Generalized dissipativity state estimation of delayed static neural networks based on a proportional-integral estimator with exponential gain term. *IEEE Trans. Circuits Syst. II Express Briefs* **2020**, *68*, 356–360. [CrossRef]
30. Cui, R.; Yang, C.; Li, Y.; Sharma, S. Adaptive neural network control of AUVs with control input nonlinearities using reinforcement learning. *IEEE Trans. Syst. Man Cybern. Syst.* **2017**, *47*, 1019–1029. [CrossRef]
31. Majid, M.H.A.; Arshad, M.R. Hydrodynamic Effect on V-Shape Pattern Formation of Swarm Autonomous Surface Vehicles (ASVs). *Procedia Comput. Sci.* **2015**, *76*, 186–191. [CrossRef]
32. Wei, H.; Shi, Y. Mpc-based motion planning and control enables smarter and safer autonomous marine vehicles: Perspectives and a tutorial survey. *IEEE/CAA J. Autom. Sin.* **2023**, *10*, 8–24. [CrossRef]
33. Negenborn, R.; De Schutter, B.; Hellendoorn, J. Multi-agent model predictive control for transportation networks: Serial versus parallel schemes. *Eng. Appl. Artif. Intell.* **2008**, *21*, 353–366. [CrossRef]
34. Negenborn, R.; Maestre, J. Distributed Model Predictive Control: An Overview and Roadmap of Future Research Opportunities. *IEEE Control Syst. Mag.* **2014**, *34*, 87–97.
35. Ferranti, L.; Lyons, L.; Negenborn, R.R.; Keviczky, T.; Alonso-Mora, J. Distributed Nonlinear Trajectory Optimization for Multi-Robot Motion Planning. *IEEE Trans. Control Syst. Technol.* **2023**, *31*, 809–824. [CrossRef]
36. Ouyang, Z.; Wang, H.; Huang, Y.; Yang, K.; Yi, H. Path planning technologies for USV formation based on improved RRT. *Chin. J. Ship Res.* **2020**, *15*, 18–24.
37. Yang, J.M.; Tseng, C.M.; Tseng, P. Path planning on satellite images for unmanned surface vehicles. *Int. J. Nav. Archit. Ocean. Eng.* **2015**, *7*, 87–99. [CrossRef]
38. Jin, X.; Er, M.J. Cooperative path planning with priority target assignment and collision avoidance guidance for rescue unmanned surface vehicles in a complex ocean environment. *Adv. Eng. Inform.* **2022**, *52*, 101517. [CrossRef]
39. Niu, H.; Savvaris, A.; Tsourdos, A.; Ji, Z. Voronoi-visibility roadmap-based path planning algorithm for unmanned surface vehicles. *J. Navig.* **2019**, *72*, 850–874. [CrossRef]
40. Cui, R.X.; Li, Y.; Yan, W.S. Mutual information-based multi-AUV path planning for scalar field sampling using multidimensional RRT. *IEEE Trans. Syst. Man Cybern. Syst.* **2016**, *46*, 993–1004. [CrossRef]
41. Dolgov, D.; Thrun, S.; Montemerlo, M.; Diebel, J. Path planning for autonomous vehicles in unknown semi-structured environments. *Int. J. Robot. Res.* **2010**, *29*, 485–501. [CrossRef]

Journal of
Marine Science and Engineering

Article

An Improved VO Method for Collision Avoidance of Ships in Open Sea

Mao Zheng [1,2,*], Kehao Zhang [1,2], Bing Han [3], Bowen Lin [1,2], Haiming Zhou [1,2], Shigan Ding [1,2], Tianyue Zou [1,2] and Yougui Yang [4]

[1] Intelligent Transportation Systems Research Center, Wuhan University of Technology, Wuhan 430070, China; 333151@whut.edu.cn (K.Z.); linbw163@163.com (B.L.); zhouhaiming@whut.edu.cn (H.Z.); dsg1998@126.com (S.D.); zoutianyue1214@163.com (T.Z.)
[2] National Engineering Research Center for Water Transport Safety, Wuhan 430070, China
[3] National Engineering Research Center of Ship & Shipping Control System, Shanghai 200134, China; han.bing@coscoshipping.com
[4] Beibu Gulf Port Qinzhou Terminal Co., Ltd., Qinzhou 535000, China; 13877796966@163.com
* Correspondence: zhengmao@whut.edu.cn

Abstract: In order to effectively deal with collisions in various encounter situations in open water environments, a ship collision avoidance model was established, and multiple constraints were introduced into the velocity obstacle method, a method to determine the ship domain by calculating the safe distance of approach was proposed. At the same time, the ship collision avoidance model based on the ship domain is analyzed, and the relative velocity set of the collision cone is obtained by solving the common tangent line within the ellipse. The timing of starting collision avoidance is determined by calculating the ship collision risk, and a method for ending collision avoidance is proposed. Finally, by comparing the simulation experiments of the improved algorithm with those of the traditional algorithm and the actual ship experiment results of manual ship maneuvering, it is shown that the method can effectively avoid collisions based on safe encounter distances that comply with navigation experience in different encounter situations. At the same time, it has better performance in collision avoidance behavior. It has certain feasibility and practical applicability.

Keywords: velocity obstacle; the ship domain; COLREGs; collision avoidance; collision risk index

Citation: Zheng, M.; Zhang, K.; Han, B.; Lin, B.; Zhou, H.; Ding, S.; Zou, T.; Yang, Y. An Improved VO Method for Collision Avoidance of Ships in Open Sea. *J. Mar. Sci. Eng.* **2024**, *12*, 402. https://doi.org/10.3390/jmse12030402

Academic Editor: Alessandro Ridolfi

Received: 20 December 2023
Revised: 20 January 2024
Accepted: 13 February 2024
Published: 26 February 2024

1. Introduction

Compared with other types of transportation, maritime transportation costs are much lower, and its transportation capacity is huge. With the increasing prosperity of international trade, maritime transportation as an important mode of transportation has achieved great development. At the same time, the industry of ship construction is developing in the direction of intellectualization and enlargement in capacity, which has further improved the transportation efficiency of shipping. However, due to the huge volume of shipping business and a more complex navigation environment, ship navigation safety faces important challenges and impacts [1].

The safety of ship navigation has always been an important area of concern and research by domestic and foreign scholars. According to water accident surveys, most water accidents are caused by human factors, and ship collision accidents account for an important proportion of them [2]. With the continuous development of intelligent and networked technologies, the automation level of ships has been significantly improved. Therefore, research on autonomous ship collision avoidance has also become a focus in the field of safe ship navigation [3]. Continuous research and development in this field provides new opportunities and challenges for preventing accidents and reducing collision risks, which is of great significance to ship navigation safety.

One of the focuses of research on ship collision avoidance behavior is the timing of taking collision avoidance actions and the safe passage distance. The research results in this area mainly focus on the ship domain and the dynamic boundary. The concept of ship domain was first proposed by Fujii [4] and used statistical methods to establish an elliptical ship domain model related to the ship length. Goodwin [5] considered the Convention on the International Regulations for Preventing Collisions at Sea, 1972 (COLREGs) and established a ship domain consisting of three unequal sector regions. However, these ship domains do not take into account the influence of human factors, environmental factors, and ship maneuverability, and their practical application results are not ideal. Smierzchalski [6] determined a hexagonal ship domain model using ship speed and parameters of turning test. Wang et al. [7] fully considered relevant factors such as COLREGs, ship maneuverability, and length and speed of ships, and proposed a quaternary ship domain. These methods determine the size of the ship domain by finding the minimum passage distance of the ship. In order to prevent the ship domain from being infringed, Davi et al. [8] proposed the concept of the dynamic boundary based on the Goodwin model, and defined the dynamic boundary as the domain where the ship begins to take collision avoidance actions. However, because the dynamic boundary cannot take into account the impact of ship speed on collision avoidance timing, the prediction accuracy of the dynamic boundary is not high. Therefore, most researchers use ship collision risk models to predict collision avoidance timing. In the early days, most ships used distance at closest point of approach (DCPA) and time at closest point of approach (TCPA) to estimate the collision risk of the ship to avoid collisions. Although this method is intuitive, simple, and has advantages, factors such as distance, ship speed ratio, and target ship maneuvering may be ignored in the calculation, resulting in inaccurate prediction results [9]. With the development of computer technology, new methods are constantly emerging. Currently, researchers use fuzzy theory, expert systems, and neural network methods [10] to calculate ship collision risk. Chen et al. [11] used neural network and fuzzy reasoning to propose a calculation method for ship collision risk. Zhao et al. [12] combined the ship domain model and the collision risk model, used the quaternary ship domain to determine the safe distance and considered the COLREGs, and proposed a fuzzy evaluation-based calculation method for unmanned surface vessel collision risk. Li et al. [13] proposed an improved Rule-aware Time-varying Collision Risk Model, which considers the estimation of target ship motion and the corresponding risk uncertainty analysis process, and integrates ship maneuverability, COLREGs, and good seamanship. Abebe et al. [14] proposed a new method for calculating the collision risk index (CRI) by combining machine learning with D-S theory to increase the efficiency of the computations while preserving the prediction accuracy of the CRI.

Research on ship collision avoidance mainly focuses on two aspects: ship collision avoidance models and path planning methods. Ship collision avoidance models can be divided into ship domain model and collision risk model. In terms of path planning and collision avoidance, safe navigation of ships requires efficient collision avoidance algorithms to deal with dynamic environments. Commonly used algorithms include artificial potential field method [15], A* algorithm [16], particle swarm algorithm [17], etc. These traditional ship collision avoidance algorithms are simple and easy to understand, and have strong search capabilities in global planning. However, they are mostly used to avoid collisions with static obstacles. They cannot respond quickly in the dynamic environment during navigation and may fall into local optimal solution [18].

When dealing with dynamic obstacles or complex encounter situations, the velocity obstacle (VO) algorithm can more flexibly adapt to situation changes and provide a more accurate collision avoidance strategy. At the same time, it is easier to generate a global optimal solution than traditional methods. Therefore, the VO algorithm is widely used by researchers. Hong et al. [19] constructed an obstacle model and a collision risk model, and completed the dynamic obstacle avoidance of unmanned surface vessels by introducing the VO algorithm with multiple constraints. Zhang et al. [20] combined the dynamic window method and the VO algorithm, and proposed a virtual obstacle method; Ma et al. [21] introduced the artificial potential field method and inland river collision avoidance rule constraints into the VO algorithm to achieve collision avoidance of stand-on ships and static obstacles; Zhang et al. [22] proposed a collision avoidance decision-making system for inland river ships based on the VO algorithm, and improved the ship motion model (MMG) suitable for inland rivers; Wang et al. [23] proposed a new collision avoidance decision-making system specifically designed for autonomous ships, where the VO algorithm is combined with a finite state machine (FSM) and considers various constraints such as ship maneuverability, COLREGs, and navigation technology.

Nevertheless, there are some problems in applying the traditional VO algorithm for ship collision avoidance. When ships are in different encounter situations, they will use different safe distance of approach to pass by, and this algorithm often simplifies obstacles into circular targets, resulting in the same encounter distance between two ships in different situations, which is obviously not in line with navigation reality. COLREGs has detailed regulations on ship collision avoidance. The traditional VO algorithm may have certain limitations in complying with these rules, making it difficult to achieve strict compliance with COLREGs [24]. At the same time, the traditional VO algorithm applied to ship collision avoidance can only provide a collision avoidance strategy based on the current status, but it is obviously difficult to determine how to avoid collision at the right time; the method of ending collision avoidance mostly uses $TCPA < 0$, while under this condition, when collision avoidance ends and returns to the original course or track, $TCPA$ will experience a trend of first increasing and then decreasing, which means that the own ship will create a collision risk with the target ship again when returning to the original course or track [25].

Therefore, in order to make up for the shortcomings of the traditional algorithm, this paper combines COLREGs and draws on the suggestions of experienced captains, introduces these experiences into the VO algorithm, and proposes a ship domain of the same proportion as the Fujii model, so that the give-way ship can comply with COLREGs. Based on the rules, the collision avoidance action of the two ships is completed with the optimal collision avoidance course angle. Finally, we also propose a method for determining when to start and end collision avoidance. These improvements are designed to improve the safety and efficiency of ship collision avoidance and make it more in line with the needs of actual navigation situations. The specific flow chart is shown in Figure 1.

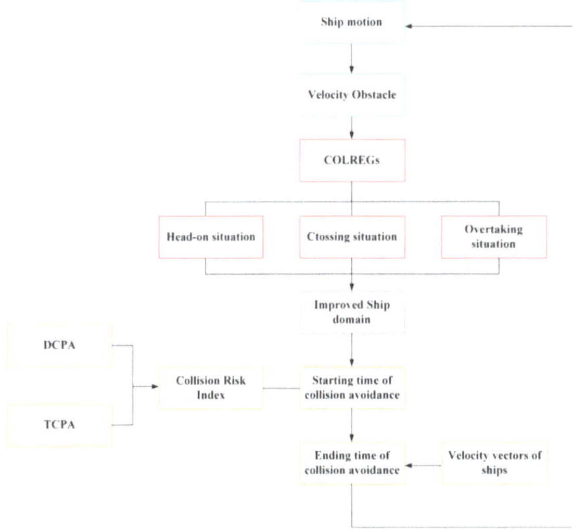

Figure 1. The flowchart of algorithm implementation.

2. The Velocity Obstacle Algorithm

2.1. Methods and Principles

Assume that the position of ship A is P_A, the velocity is $\vec{V_A}$, and the radius is R_A; obstacle B is located at P_B, the velocity is $\vec{V_B}$, and the radius is R_B. According to the radius of the robot, the obstacle is expanded into a circle with radius R, $R = R_A + R_B$. Then the mobile robot can be simplified into a particle [26].

As shown in Figure 2, define the relative velocity of A and B as $\vec{V_{AB}} = \vec{V_A} - \vec{V_B}$. When the robot moves at the relative velocity V_{AB}, the obstacle can be regarded as a static obstacle; a ray $\lambda(P_A, V_{AB})$ denotes a line starting from point A (denoted as P_A) in the direction of vector V_{AB} (representing the vector from point A to point B), assuming that the relative velocity V_{AB} remains unchanged, when the ray $\lambda(P_A, V_{AB})$ intersects the obstacle B, it is considered that A and B will collide at some time in the future. The set of V_{AB} that cause A and B to collide is defined as Relative Collision Cone (RCC), that is, the relative collision area in the relative velocity space of A and B, which can be expressed as:

$$RCC = \{V_{AB} | \lambda(P_A, V_{AB} \cap B \neq \varnothing)\} \tag{1}$$

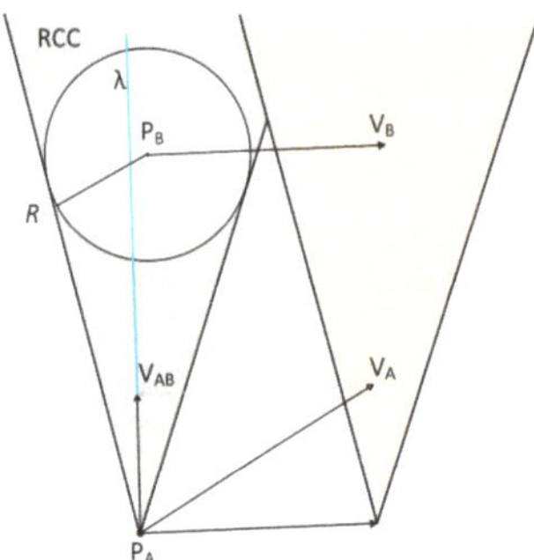

Figure 2. The geometric definition of the VO algorithm.

Translate the relative collision cone along the direction of V_{AB} to the end of the velocity vector to form a new cone-shaped area, the absolute collision cone. If V_A falls into this area, A and B will collide at some point in the future, causing A and B to collide. The set of V_A is called Velocity Obstacle:

$$VO = RCC \oplus V_A \tag{2}$$

where $\oplus$ represents the Minkowski sum. If V_A is adjusted so that it does not belong to the velocity vector in the VO set, the robot can avoid obstacles and prevent collisions and continue to move.

2.2. Application in Ship

The VO algorithm is mainly used in collision avoidance of robots, using the expanded volume of the robot as a collision area to avoid collisions. This method allows two robots to complete collision avoidance behaviors at a relatively close distance. Later, the VO algorithm was gradually applied to ship collision avoidance. Due to the ship-to-ship effect, if ships also encounter each other at a close distance during collision avoidance, the risk of ship collision will be greatly increased [27]. Therefore, ships need to ensure that they pass at a safe distance during the encounter. The safe distance can be defined as safe distance of approach (S_{DA}) [28]. As shown in Figure 3, the VO algorithm is used to establish a ship collision avoidance model. Turn the own ship into a particle, represented by O, and the velocity vector is $\vec{V_O}$; set the target ship as a circle with a radius of $R = S_{DA}$, so that the two ships maintain a safe distance, represented by P, and the velocity vector is $\vec{V_P}$, then the relative velocity of the two ships is $\vec{V_{OP}} = \vec{V_O} - \vec{V_P}$.

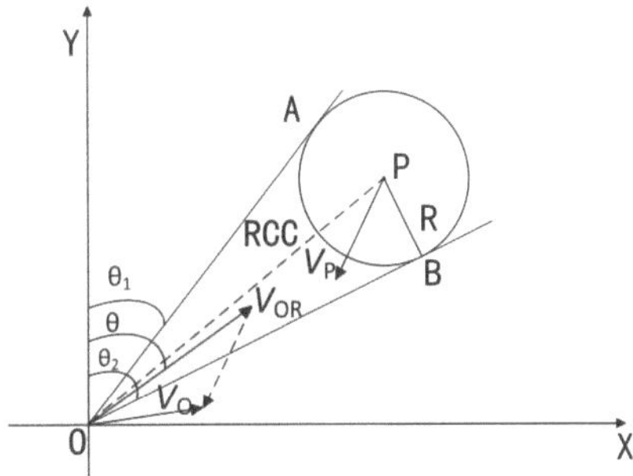

Figure 3. Ship velocity obstacle model.

At this time, the target ship can be regarded as stationary when the own ship is sailing at velocity $\vec{V_{op}}$. If the relative velocity $\vec{V_{op}}$ falls within the cone RCC in the figure, then the two ships will collide at some point. That is, $\vec{V_{op}} \in RCC$, then the own ship and the target ship are in danger of collision; if $\vec{V_{op}} \notin RCC$, then the own ship and the target ship are safe. To sum up, the core of the VO algorithm is to determine the collision cone, as shown with grey area in Figure 3.

In Figure 3, the relative coordinates (ship mounted coordinate system) were built, in which the y-axis points north, and the x-axis points east. The clockwise rotation angle of the positive y-axis is the course angle. Assuming that the own ship's routes OA and OB are tangent to the radius of the target ship, the course angle of OA is θ_1, the course angle of OB is θ_2, the relative velocity of the own ship is $\vec{V_R}$, and the course angle is θ. In order to avoid collision, θ should satisfy $\theta \notin [\theta_1, \theta_2]$.

Assume the current position of the own ship (X_O, Y_O) and the position of the target ship (X_P, Y_P), then θ_1 can be obtained by the following formula:

$$\theta_1 = \theta_{OP} - \angle AOP \tag{3}$$

where θ_{OP} is the angle between the line segment OP and the y-axis. The calculation formula of θ_{OP} and $\angle AOP$ is as follows:

$$\theta_{OP} = arctan(\frac{X_P - X_O}{Y_P - Y_O}) \tag{4}$$

$$\angle AOP = arcsin(\frac{r}{\sqrt{(Y_P - Y_O)^2 + (X_P - X_O)^2}}) \tag{5}$$

Substitute Formulas (4) and (5) into Formula (3) to get:

$$\theta_1 = arctan(\frac{X_P - X_O}{Y_P - Y_O}) - arcsin(\frac{r}{\sqrt{(Y_P - Y_O)^2 + (X_P - X_O)^2}}) \tag{6}$$

In the same way, θ_R can be calculated:

$$\theta_2 = arctan(\frac{X_P - X_O}{Y_P - Y_O}) + arcsin(\frac{r}{\sqrt{(Y_P - Y_O)^2 + (X_P - X_O)^2}}) \tag{7}$$

In summary, if $\theta \in [\theta_1, \theta_2]$, then the two ships are in danger of collision; otherwise, the two ships are safe.

3. Improved Velocity Obstacle

3.1. Velocity Obstacle Based on the Ship Domain

3.1.1. Determine the Ship Domain

In general, the value of safe distance of approach (S_{DA}) is vaguely defined based on the interpretation of rules by captains, human experts, and the practical experience of ship pilots [29]. However, the same value of safe distance of approach often appears in different encounter situations, resulting in a decrease in collision avoidance accuracy. Therefore, this paper establishes a mathematical model to quantify S_{DA} based on the calculating method of safe distance of approach proposed by Li [30] and Chen [31], COLREGs, and navigation practices.

The quantification of S_{DA} is divided into two parts, collision area and operating margin. The collision area requires a circular area that takes the size of the two ships, the estimated relative position error, and the error caused by the ship's navigational yaw motion into consideration. First, Kalman filter [32] is used to calculate the track error of the ship when sailing. Assuming that the filter variance is P, the $2P$ is regarded as the systematic error caused by environmental interference such as wind, waves, and currents.

During the navigation process, the ship's trajectory is not a line, but a track belt due to course deviation. Generally, the turning center of a ship is about one-third of the width of the ship position at the bow. As shown in Figure 4, assuming the length of the ship is L and the course deviation is ε, the width of the ship's track band is:

$$\begin{cases} W_O = 2 \times \frac{2}{3}L_O \cdot sin\varepsilon_O \\ W_T = 2 \times \frac{2}{3}L_T \cdot sin\varepsilon_T \end{cases} \tag{8}$$

where L_O, W_O, ε_O represent the length of own ship, the width of the trajectory belt and course deviation, and L_T, W_T, ε_T represent the length of target ship, the width of the trajectory belt and course deviation. Therefore, the error in collision between the two ships due to course deviation is $\frac{1}{2}(W_O + W_T)$.

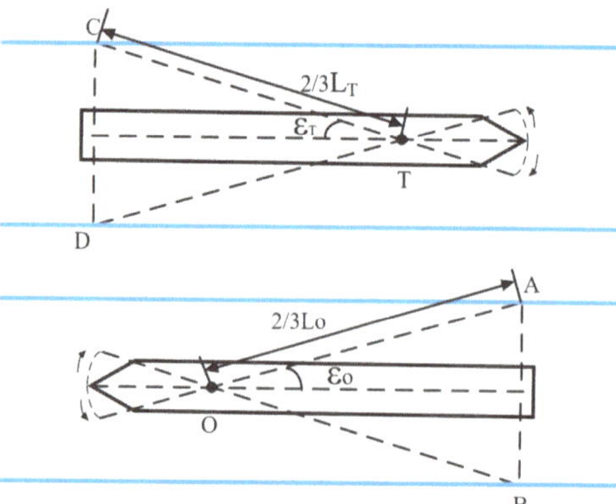

Figure 4. The trajectory belt of ships.

And based on the size of the two ships, the radius R_c of the ship collision area is obtained by

$$R_c = (L_o + L_T) + 2P + \frac{1}{2}(W_O + W_T) \tag{9}$$

where L_o and L_T represent the length of the own ship and the target ship, respectively, and P is the filter variance.

The operating margin is the distance between the own ship and the target ship after the own ship turns $90°$ at full speed and full rudder angle. Based on the above analysis, the value of S_{DA} in each encounter situation can be obtained:

- Head-on situation and crossing situation: $S_{DA} = R_f$;
- Own ship being overtaken by the target ship: $S_{DA} = R_f - A_d$;
- Own ship overtaking the target ship: $S_{DA} = R_f - V_s \times T_{90}$;
- Other situations: $S_{DA} = R_f - D_T/2$.

where $R_f = R_c + A_d + (V_s \times T_{90})$, T_{90}, A_d, D_T, and V_s are, respectively, the time for the ship to rotate 90° at full speed and full rudder angle, the advance, the tactical diameter of the turning, and the relative speed. Figure 5 shows the various parameters of the ship turning test.

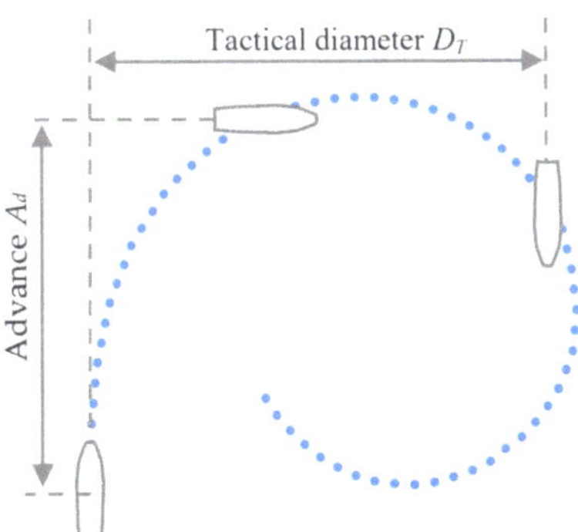

Figure 5. Diagram of ship turning test.

Generally, the critical safe distance of approach is largest in crossing situations, lower in overtaking situations, and smallest in head-on situations [31]. However, using the above formula to evaluate S_{DA} in an overtaking situation is much smaller than the actual navigation requirements.

Therefore, this paper makes S_{DA} in the overtaking situation equal to the value in the head-on situation. In maritime traffic, it is believed that to ensure ships sail safely and effectively, a certain safety distance must be kept between ships, and this safety distance can be called as the ship domain radius [33]. Therefore, the ship domain can be determined by the safety distance of approach. To improve the accuracy and safety of ship collision avoidance operation, an elliptical-type ship domain model with the same long and short axis ratio as the Fujii model for collision avoidance is adopted.

The Fujii model takes into account the International Regulations for Preventing Collisions at Sea, ensuring that collision avoidance decisions comply with international regulations, contributing to the compliant autonomous collision avoidance of vessels. As an elliptical model compared to a circular model, it better aligns with the length-to-width ratio of vessels and maintains the minimum distance between two vessels, especially in head-on and crossing encounter situations. However, since the Fujii model is mainly applied to large ships, the length calculated by this model is much smaller than S_{DA} when applied to small- and medium-size ships. To solve this problem, in Figure 6a, a scaling factor f is introduced to enlarge the original semidiameters in Fujii model, shown as black solid ellipse. The dotted ellipse is a schematic diagram of the ship domain, in which the major axis semidiameter A_a and minor axis semidiameter B_b are enlarged by f. B_b is exactly equal to S_{DA}. Meanwhile, the scaling factor can be defined as $f = B_b/b = A_a/a$, while, a,b are the major and minor axis semidiameters of Fujii model ellipse, equal to 4 times and 1.6 times of the ship's length, respectively, to ensure the encounter distance between ships greater than (at least equal to) S_{DA} to prevent the target ship invading the own ship domain.

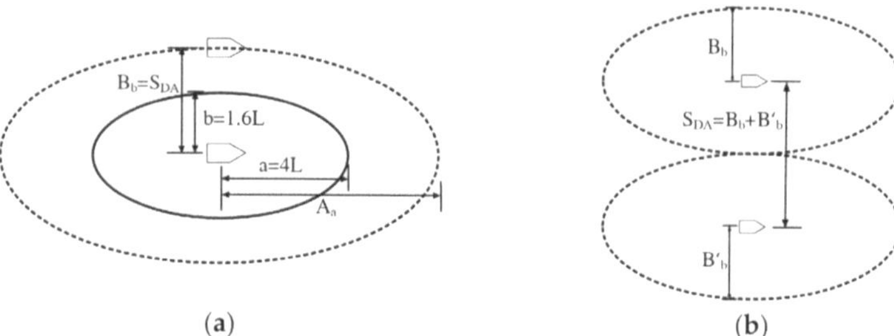

(**a**)   (**b**)

Figure 6. The ship domain: (**a**) Fujii model enlarged by scaling factor; (**b**) The state where the distance between two ships is minimum.

In addition, it should be emphasized that when the target ship is always located outside the boundary of the own ship's domain, it can be considered that the own ship's domain is not infringed; at the same time, it cannot be considered the target ship's domain is not infringed by the own ship, as the sizes of these two ships' domains may not be the same. Therefore, this paper adopts the mutual non-intrusion principle between these two ships, the S_{DA} can be defined as the sum of minor axis semidiameters of these two ships' domains. As a result, the minimum distance between two ships in overtaking encounters is always larger than S_{DA}, as shown in Figure 6b.

3.1.2. Collision Avoidance Analysis Based on Elliptical Ship Domain

The traditional VO algorithm uses the "expanding circle" method to simply superimpose the collision area. If the shapes of both moving objects are irregular, this method would introduce larger errors. This paper adopts the method of finding common tangent lines of two ships' ellipse domains to avoid errors.

Draw two inner common tangent lines in the ship domains of the own ship and the target ship, as shown in Figure 7. Draw parallel lines OM and ON of these two inner common tangent lines through the origin O to form the relative collision cone RCC, the shadow area. In order to avoid collision, the relative velocity $\vec{V_p}$ needs to fall outside the RCC. Assume the common tangent equation $Ax + By + C = 0$, the ship domain equation of the own ship is $y = f_o(x)$, and the ship domain equation of the target ship is $y = f_p(x)$. The four common tangents (including two inner common tangents and two outer common tangents) can be found by the joint equation, after which the two internal tangents are identified by the relationship between the size of the angle between the common tangent and the positive direction of the y-axis to obtain the range of relative collision area.

Let $\theta_i (i = 1, 2, 3, 4; 0 \leq \theta_i < 2\pi)$ be the angles between these four common tangents and the positive direction of the y-axis. Sort θ_i according to size: $0 \leq \theta_1 < \theta_2 \leq \theta_3 < \theta_4 < 2\pi$. Under normal circumstances, the two internal common tangents are the minimum and maximum values, respectively, but it is necessary to additionally consider the situation where the common tangents are on both sides of the y-axis. θ represents the set of relative velocity angles that cause the own ship and the target ship to collide:

$$\begin{cases} \theta \in [\theta_1, \theta_4] & (\theta_4 - \theta_1) < \pi \\ \theta \in [0, \theta_1] \cup [\theta_2, 2\pi) & (\theta_2 - \theta_1) > \pi \\ \theta \in [0, \theta_2] \cup [\theta_3, 2\pi) & (\theta_3 - \theta_2) > \pi \\ \theta \in [0, \theta_3] \cup [\theta_4, 2\pi) & (\theta_4 - \theta_3) > \pi \end{cases} \tag{10}$$

The own ship will collide with the target ship if the relative velocity of own ship and target ship satisfies θ at a certain moment during sailing. In order to avoid the collision, it is necessary to select the desired relative velocity angle $RV(0 \leq RV < 2\pi)$ that can avoid the relative collision cone, and in the ideal situation, making RV equal to the boundary angle of the relative collision cone can ensure that the ships go through at a safe distance during collision avoidance operation. However, in the collision avoidance process, to avoid the influence of small perturbations on the collision avoidance system due to environmental interference and other uncertainty factors, resulting in the mutual infringement of the ship domains of the two ships, a correction factor θ_r is introduced to modify the $RV_{\theta r} = RV - \theta_r$, so that the $RV_{\theta r}$ is always outside the relative collision cone RCC.

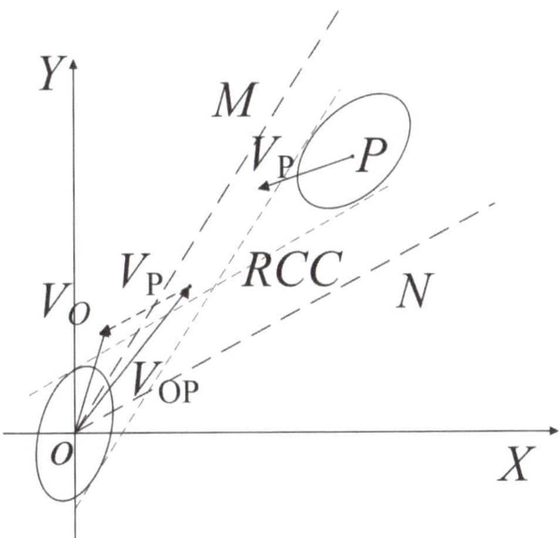

Figure 7. Collision avoidance analysis.

3.2. Application to the Convention on the International Regulations for Preventing Collisions at Sea

When selecting the desired relative velocity angle $RV_{\theta r}$, the traditional VO algorithm generally uses an angle closer to the original velocity vector to update the $RV_{\theta r}$. However, to avoid collisions, ships must conduct collision avoidance operations according to COLREGs. Therefore, when updating RV, each type of encounter situation should be checked according to COLREGs. This paper put forward a method to classify encounter situations [34] referring to the COLREGs, as shown in Figure 8.

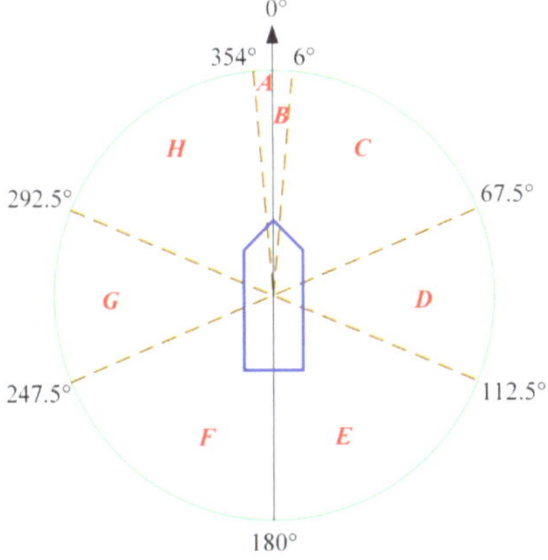

Figure 8. The relative position of the encounter between own ship and target ship.

The initial encounter situation of ships is divided into three situations: head-on, crossing, and overtaking situation. As in different types of encounter situations, different actions should be adopted. Based on the target ship's initial position, cross situations could be divided into starboard crossing and port crossing. Ulteriorly, the course difference between these two ships could be a useful item to classify the overtaking, crossing, and head-on situations. As shown in Table 1, when the target ship appears in area C, while the course difference between these two ships falls in 180–270°, this kind of crossing encounter situations could be called small course difference crossing from starboard side, for own ship, turning right and pass on the stern of target ship is suitable. When the target ship appears in area D, while the course difference between these two ships falls in 270–360°, this kind of crossing encounter situations could be called as large course difference crossing form starboard side, for own ship, turning right and pass on the stern of target ship could be dangerous, turning left may be the only suitable action. Similarly, descriptions of other kinds of encounter situations and the suitable actions for the own ship are put forward in Table 1 based on multiple rounds of discussion with seven experienced captains, maritime experts, and autopilot system developers.

Table 1. Classification of ship encounter.

Types of Encounter	The Area Where the Target Ship Is Located	Course Difference	Encounter Situation	Suitable Action for Own Ship
Head-on	A, B	175–185°	Head-on	Turning right
Starboard side crossing	D	270–360°	Large course difference crossing form starboard side	Turning right
	C	180–270°	Small course difference crossing form starboard side	Turning right
Port side crossing	G	0–90°	port large angle crossing	Keeping course and speed
	H	90–180°	port large angle crossing	Keeping course and speed
Overtaking	A, H	0–67.5°	Own ship pass on the starboard side of the target ship	Turning right
	B, C	292.5–360°	Own ship pass on the port side of the target ship	Turning left
Being overtaken	E	0–67.5° 292.5–360°	Target ship pass on the starboard side of own ship	Keeping course and speed
	F		Target ship pass on the port side of own ship	Keeping course and speed

At the same time, according to the analysis of encounter situations by Zhong [35], the initial distance between the own ship and target ship could be divided into collision danger, close-quarters situation, and imminent danger according to the avoidance stage, as shown in Table 2. Significantly, imminent danger means the minimum safe turning space of the own ship, equals to S_{DA}.

Table 2. Classification of ship situations.

Type	Minimum Distance (*n* Mile)	Maximum Distance (*n* Mile)
Collision risk	1	6
Close-quarters situation	0.25	1
Imminent danger	0	0.25

Select the expected relative velocity RV when the ship is in collision danger. The selected rules could be classified as follows:

- In head-on and starboard crossing, the ship is the give-way ship, and could alter its course according to the description in Table 1 to avoid collision. That is:

$$RV = \begin{cases} \theta_4 + \theta_r & (\theta_4 - \theta_1) < \pi, \theta \in [\theta_1, \theta_4] \\ \theta_1 + \theta_r \, (\theta_2 - \theta_1) > \pi, \theta \in [0, \theta_1] \cup [\theta_2, 2\pi) \\ \theta_2 + \theta_r \, (\theta_3 - \theta_2) > \pi, \theta \in [0, \theta_2] \cup [\theta_3, 2\pi) \\ \theta_3 + \theta_r \, (\theta_4 - \theta_3) > \pi, \theta \in [0, \theta_3] \cup [\theta_4, 2\pi) \end{cases} \tag{11}$$

- For port side crossing encounter situation where the target ship overtakes the own ship, the own ship is the stand-on ship and should keep its course and speed, and there is no need to calculate *RV*;

- When the own ship is overtaking the target ship and the own ship's velocity $\left| \vec{V_O} \right|$ module is greater than the target ship, the own ship is the give-way ship and should alter course to avoid collision. According to the 7–8th row of Table 1, the RV can be concluded as follows:

$$RV = \begin{cases} \theta_1 - \theta_r & (\theta_4 - \theta_1) < \pi, \theta \in [\theta_1, \theta_4] \\ \theta_2 - \theta_r \, (\theta_2 - \theta_1) > \pi, \theta \in [0, \theta_1] \cup [\theta_2, 2\pi) \\ \theta_3 - \theta_r \, (\theta_3 - \theta_2) > \pi, \theta \in [0, \theta_2] \cup [\theta_3, 2\pi) \\ \theta_4 - \theta_r \, (\theta_4 - \theta_3) > \pi, \theta \in [0, \theta_3] \cup [\theta_4, 2\pi) \end{cases} \tag{12}$$

After determining the vector *RV* through the above method, the course or velocity of the own ship should be controlled by automatic systems making the relative velocity angle of the own ship and the target ship approaching the *RV* to conduct dynamic collision avoidance. This method could operate well in simulation systems; however, the ship's speed adjustment frequency is too high for the actual engine system, and in this research, the ship's speed is not changeable to prevent engine damage. For actual ship's collision avoidance operation at sea, we mainly adjust course. The expected course C_m that the ship can avoid the relative collision cone RCC is obtained by the following formula:

$$tanRV = \frac{V_o sinC_m - V_{Px}}{V_o cosC_m - V_{Py}} \tag{13}$$

where V_{Px} is the component of the target ship's speed along the *x*-axis, and V_{Py} is the component of the target ship's speed along the *y*-axis. If C_m has two values, choose the desired course which is closer to the current course.

3.3. The Determination of the Start and End Time of Collision Avoidance
3.3.1. Collision Risk Determines the Start Time of Collision Avoidance

In actual situations, there are many factors that affect Collision Risk Index (CRI). Subjective factors include sea state, visibility, etc. Objective factors include DCPA, TCPA, etc. This paper introduced the weighted sum of the corresponding membership functions of DCPA and TCPA to generate the collision risk, thereby determining the start time of collision avoidance [36].

- Membership function of DCPA could be written as a piecewise function $\mu(D_{CPA})$:

$$\mu(D_{CPA}) = \begin{cases} 1 & D_{CPA} \leq d_1 \\ \frac{1}{2} - \frac{1}{2}sin\left[\frac{\pi}{d_2 - d_1}\left(D_{CPA} - \frac{d_1 + d_2}{2}\right)\right] & d_1 < D_{CPA} < d_2 \\ 0 & d_2 \leq D_{CPA} \end{cases} \tag{14}$$

where d_1 indicates the minimum distance between two ships, in this paper d_1 is the minimum distance at which the ship domains of the two ships are tangent; d_2 indicates the minimum distance at which the two ships can safely pass, $d_2 = ?d_1$

- Membership function of TCPA could also be written as a piecewise function $\mu(T_{CPA})$:

$$\mu(T_{CPA}) = \begin{cases} 1 & T_{CPA} \leq t_1 \\ \left(\frac{t_2 - T_{CPA}}{t_2 - t_1}\right)^2 & t_1 < T_{CPA} < t_2 \\ 0 & T_{CPA} \geq t_2 \end{cases} \tag{15}$$

where,

$$t_1 = \begin{cases} \frac{\sqrt{D_1^2 - D_{CPA}^2}}{V_r} & D_{CPA} \leq D_1 \\ \frac{DCPA - D_1}{V_r} & D_{CPA} > D_1 \end{cases} \tag{16}$$

$$t_2 = \frac{\sqrt{D_2^2 - D_{CPA}^2}}{V_r} \tag{17}$$

where D_1 is the distance between these two ships at the last opportunity to use helm, which is generally 12 times the length of the own ship; D_2 is the dynamic boundary, that is, the distance at which the own ship starts turning, and is obtained as follows:

$$D_2 = 1.7\cos(B_t - 19°) + \sqrt{4.4 + 2.89\cos^2(B_t - 19°)} \tag{18}$$

where B_t is the true direction of the target ship relative to own ship.

By weighting the membership functions of DCPA and TCPA, the collision risk of the ship is obtained as:

$$\mu_r = \alpha_d \cdot \mu(D_{CPA}) + \alpha_t \cdot \mu(T_{CPA}) \tag{19}$$

where α_d and α_t are the weighting coefficients of their respective functions.

The ship starts to avoid collision when it meets the following two conditions:

- The relative velocity vectors of the two ships are in the relative collision cone RCC;
- Based on the collision avoidance test results of multiple encounter situations and the collision avoidance suggestions of the experienced captain, the ship collision risk assessment μ_r is larger than 0.72.

3.3.2. Conditions for Ending Collision Avoidance

Theoretically, $TCPA < 0$ means the collision risk is low enough, and returning to original course or track should be conducted; however, in our simulation, the desired course C_m oscillation is caused by the algorithm. The reason is that returning to the original course makes T_{CPA} greater than zero. To avoid this phenomenon, the collision avoidance could be terminated when the ship satisfies Equation (20).

$$V_{LOS} \notin RCC, V_{orig} \notin RCC(V_{des} \notin RCC) \tag{20}$$

where V_{LOS} is the velocity vector of trajectory tracking planning, V_{orig} is the velocity vector when the own ship moves along the original course direction at the current speed, and V_{des} is the velocity vector when the own ship moves along the original track direction at the current speed.

4. Experimental Results and Analysis

4.1. Actual Ship Experiment

The experiment is divided into two parts: actual ship experiment and simulation experiment. The actual ship experiment uses a built-in intelligent ship navigation test system to record the movement data during ship collision avoidance, and the test evaluation system evaluates the entire process of collision avoidance. This actual ship experiment was operated by an experienced captain. The simulation experiment uses the Matlab simulation platform to simulate ship collision avoidance using the improved VO algorithm and the traditional VO algorithm and record movement data during the collision avoidance process.

Finally, the established evaluation system evaluates manual collision avoidance, traditional algorithm collision avoidance, and improved algorithm collision avoidance.

For the actual ship experiment, a sea-going fishing vessel was updated as the test platform, which is equipped with a course keeping autopilot, achieving the course keeping error between 1 and 2° during the test. To collect the vessel's movement and turning data, a real time kinematic (RTK) module based on GPS and a gyrocompass. The test was conducted in Zhanjiang Bay area, as shown in Figure 9.

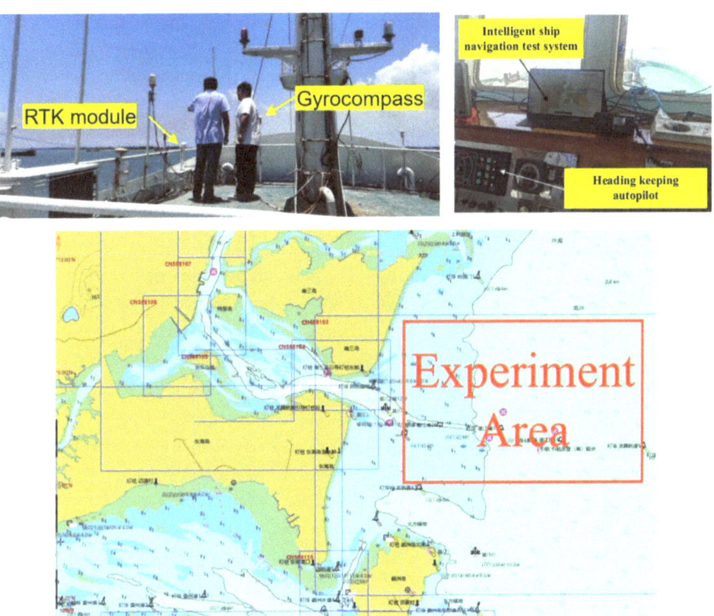

Figure 9. The updating of own ship and the experimental area in Zhanjiang city of China.

The specific parameters of the experimental platform are shown in Table 3.

Table 3. Parameters of the ship.

Length (m)	Width (m)	Total Tonnage (t)	Speed (kn)
43.9	7.3	353	3–12

To verify the S_{DA} value in OV model, this paper conducted an actual ship turning test and obtained specific parameters as shown in Table 4.

Table 4. The results of the turning test.

	D_T/(m)	D_O/(m)	A_d/(m)	T_r/(m)	T_{90}/(s)
Full rudder 30°	148.1256	146.4241	109.3487	76.8136	44

where D_T is Tactical diameter; D_O is Steady diameter; A_d is Advance; T_r is Transfer; T_{90} is the time for the ship's course to turn 90°. The average speed during turning test is about 3 m/s, the length of own ship and target ship could be simply regarded as the same. The safe distance of approach (S_{DA}) could be calculated as 0.251 NM.

The full rudder turning test of the test ship is shown in Figure 10:

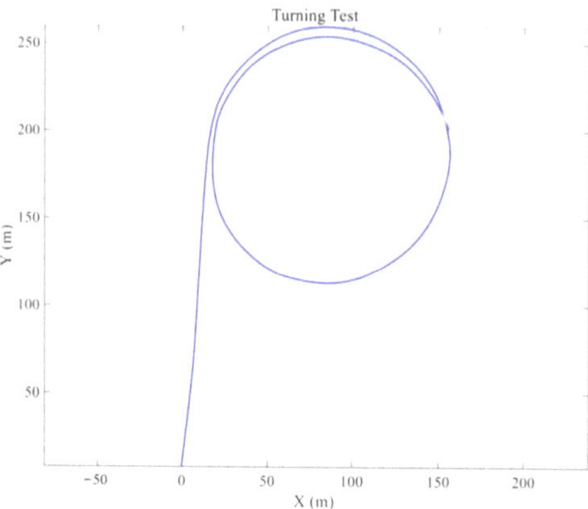

Figure 10. Turning test of 30°.

The following is a brief introduction to the evaluation index system of the ship's intelligent navigation test system: the collision avoidance process is evaluated from three aspects: safety, economy, and practicality. The safety evaluation index is mainly used to judge whether the collision avoidance process meets the regulatory requirements and ensures the safety of the ship, including whether it enters imminent danger, the minimum encounter distance, etc. The economic evaluation index is used to evaluate whether the collision avoidance process is economical, whether it can shorten the operation time, and shorten the distance, including deviation factor, maximum alteration of course, total collision avoidance time, and other indicators. Practicality evaluation indicators are used to judge whether it meets the requirements of simple operation and whether it is suitable for actual ship operations, including indicators such as steering frequency. The analytic hierarchy process (AHP) is used to study the weights of the above indicators to form a final complete evaluation system.

Actual ship experiment collision avoidance test method process: The test ship maintains a constant speed and travels in the designated test waters, coordinates are converted to the preset encounter situation according to the longitude, latitude, and course of the test ships, and ships' courses are generated by the AIS simulator. The distance between the two ships is about 1–6 NM (nautical mile). The own ship is operated by the captain to avoid collisions. In order to improve the efficiency of the actual ship experiment, when there is no collision risk between these two test ships and the own ship returns ±2° to the original course, the collision avoidance is ended. During this process, the experiment evaluation system records the navigation data of the collision avoidance process and generates an evaluation report.

4.2. The Evaluation of Experimental Data

In accordance with the international rules for avoiding collisions at sea, this article simulates encounter situations in three situations: head-on, crossing, and overtaking, in which the ship has the duty to avoid. In order to make the experiment results of the improved VO algorithm, the traditional VO algorithm, and the actual ship experiment comparable in these three situations, it is necessary to maintain situation consistency between the actual ship experiment and the simulation experiment. The initial parameter settings of the actual ship experiment are shown in Table 5.

Table 5. Initial parameters for collision avoidance experiments.

Situation	Own Ship			Target Ship		
	Initial Coordinates (N, E)	Course (°)	Speed (kn)	Initial Coordinates (N, E)	Course (°)	Speed (kn)
Head-on	(20.756941°, 110.625902°)	324.4	9.1	(20.812550°, 110.583753°)	147	8.6
Starboard side crossing	(20.784380°, 110.616869°)	117.2	7.9	(20.747173°, 110.611780°)	44	12.5
Overtaking	(20.689691°, 110.630284°)	356.7	8.6	(20.710497°, 110.626862°)	356	4.5

- Head-on situation.

Figures 11–13 are, respectively, the collision avoidance trajectories of the improved VO, the traditional VO, and the actual ship experiment in the head-on situation. In the improved VO, the red line is the navigation trajectory of the target ship, and the blue line is the navigation trajectory of the own ship. The closed graphics around the two ships represent the ship domain, with the long axis being 0.4 NM and the short axis being 0.16 NM. In the traditional VO, the red line is the navigation trajectory of the target ship, and the blue line is the navigation trajectory of the own ship, and the closed figure around the target ship represents the size of the collision circle of the traditional algorithm, with a radius of 0.25 NM. In the actual ship experiment, MMSI: 111111111 represents the navigation trajectory of the target ship (In order to facilitate the identification of AIS simulation ships, the MMSI is specially set to 111111111), MMSI: 412382898 represents the navigation trajectory of the own ship, and the green enclosed shape around the own ship represents the 1 NM marked circle, while the red enclosed shape represents the 0.25 NM marked circle. When collision avoidance begins, the improved algorithm calculates the ship collision risk in real time. When $\mu_r > 0.72$, the ship starts to avoid collision. The collision avoidance timing for the traditional algorithm is when the distance between two ships (D_{OT}) is less than 3600 m. The collision avoidance timing in the actual ship experiment is determined by the captain's discretion. The ship will turn right to avoid the situation in accordance with the COLREGs rules and return to the original course after the danger is eliminated.

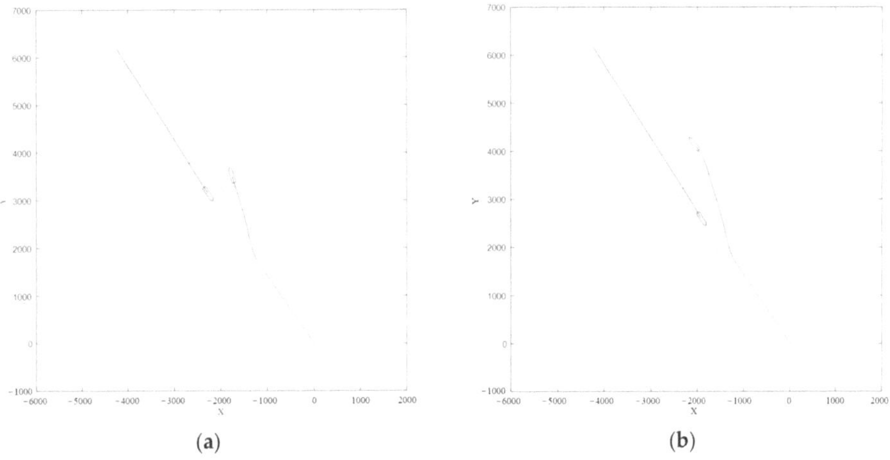

(a) (b)

Figure 11. Improved Velocity Obstacle in the head-on situation: (**a**) the trajectory of the two ships when they are at the closest distance; (**b**) the trajectory when collision avoidance is completed.

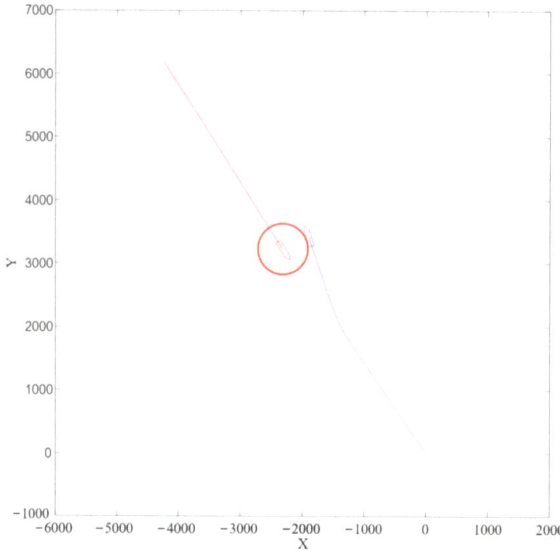

Figure 12. Traditional Velocity Obstacle in the head-on situation.

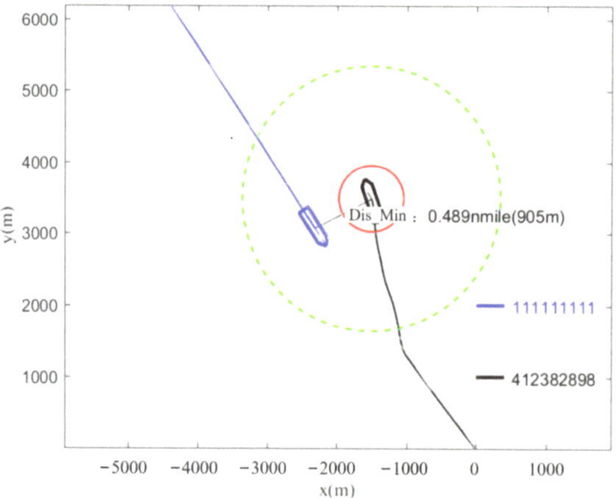

Figure 13. The actual ship experiment in the head-on situation.

Figure 14a,b and Figure 15 show the parameter changes of the improved VO, the traditional VO, and the actual ship experiment in the head-on situation. The rudder angle, course, D_{OT} and DCPA were analyzed under three experiments. In the improved VO, the own ship started to avoid collision at 207 s, 3900 m away from the target ship, turned right to the maximum course of 347.4°, and started returning to the original course at 426 s. The minimum of D_{OT} was 622 m. In the traditional VO algorithm, the own ship started to avoid collision at 223 s, 3600 m away from the target ship, turned right to the maximum course of 341.8°, and started returning to the original course at 434 s. The minimum of D_{OT} was 482 m. In the actual ship experiment, the own ship started to avoid collision at 179 s, 4200 m away from the target ship, turned right to the maximum course of 355°, and started returning to the original course at 398 s. The minimum of D_{OT} was 905 m.

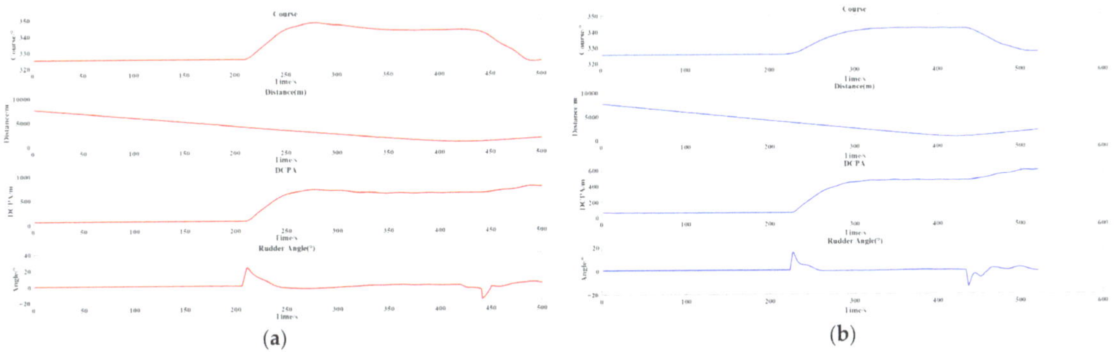

Figure 14. Parameter change diagram of the algorithm in the head-on situation: (**a**) Improved Velocity Obstacle; (**b**) Traditional Velocity Obstacle.

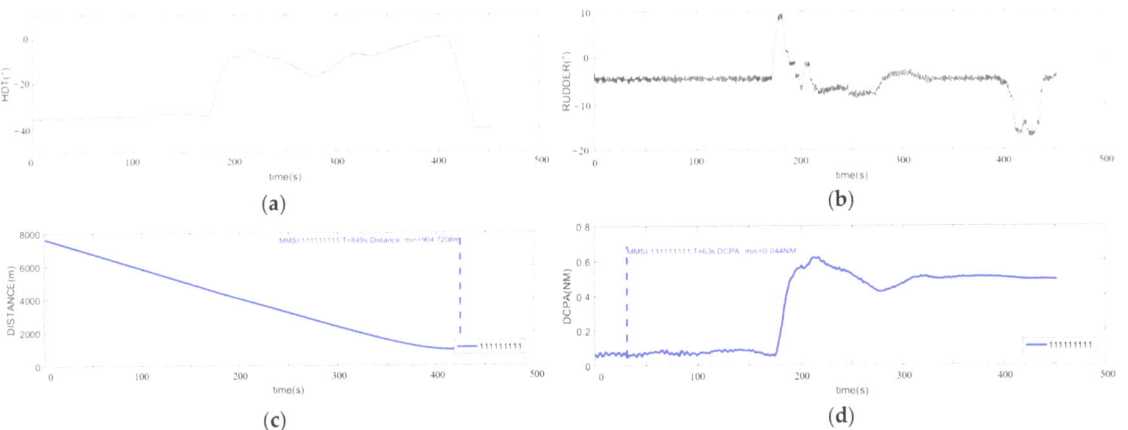

Figure 15. Parameter change diagram of the actual ship experiment in the head-on situation: (**a**) course graph; (**b**) rudder angle graph; (**c**) distance graph; (**d**) DCPA graph.

As shown in Table 6, it is the evaluation system's ranking of each index score of the three collision avoidance methods.

Table 6. Evaluation conclusion in the head-on situation.

Evaluation Index	Improved VO	Traditional VO	Manual Ship Handling
Own ship is in imminent danger	①	③	①
Minimum encounter distance	②	③	①
Deviation factor	②	①	③
Maximum alteration of course	②	③	①
Collision avoidance time	②	①	②
Steering frequency	①	①	③
Total	①	③	②

where ① indicates the highest ranking in the score of this indicator; ② indicates the second highest ranking; ③ indicates the lowest ranking.

- Starboard side crossing situation.

Figures 16–18 show the collision avoidance trajectories of the improved VO, traditional VO, and actual ship experiment in crossing situation. In the improved VO, the red line is the navigation trajectory of the target ship, and the blue line is the navigation trajectory of the own ship. The closed graphics around the two ships represent the ship domain, with the long axis being 0.35 NM and the short axis being 0.14 NM.

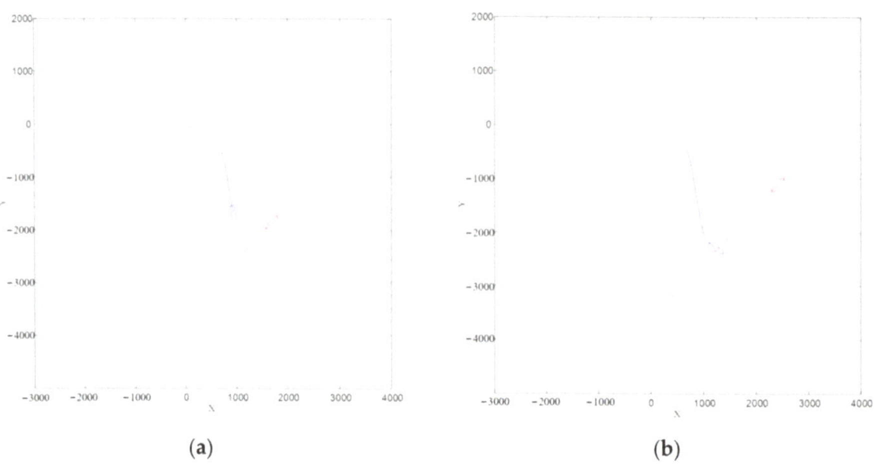

(a)   (b)

Figure 16. Improved Velocity Obstacle in the starboard side crossing situation: (**a**) the trajectory of the two ships when they are at the closest distance; (**b**) the trajectory when collision avoidance is completed.

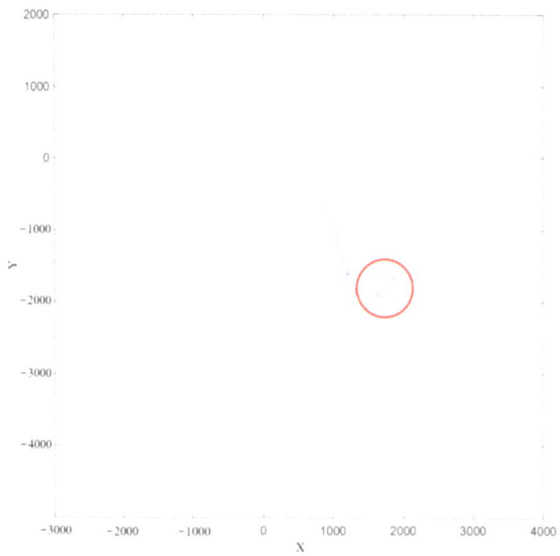

Figure 17. Traditional Velocity Obstacle in the starboard side crossing situation.

Figures 19a,b and 20 show the parameter changes of the improved VO, the traditional VO, and the actual ship experiment in the crossing situation. The rudder angle, course, D_{OT} and DCPA were analyzed under three experiments. In the improved VO, the own ship started to avoid collision at 44 s, 3600 m away from the target ship, turned right to the maximum course of 171.4°, and started returning to the original course at 213 s. The minimum of D_{OT} was 762 m. In the traditional VO, the own ship started to avoid collision at 44 s, 3600 m away from the target ship, turned right to the maximum course of 163.6°, and started returning to the original course at 220 s. The minimum of D_{OT} was 489 m. In the actual ship experiment, the own ship started to avoid collision at 39 s, 3800 m away from the target ship, turned right to the maximum course of 188°, and started returning to the original course at 208 s. The minimum of D_{OT} was 987 m.

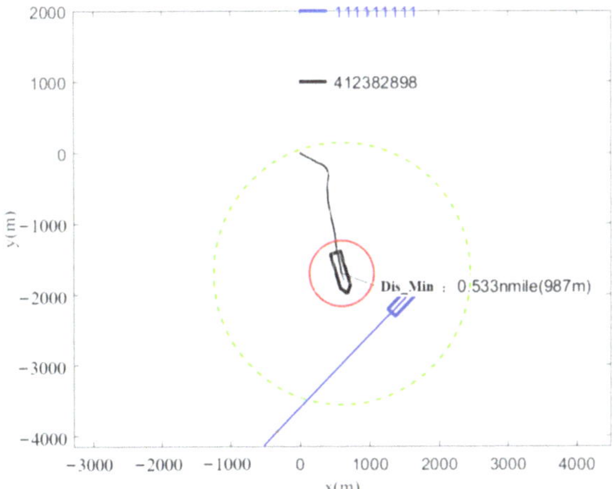

Figure 18. The actual ship experiment in the starboard side crossing situation.

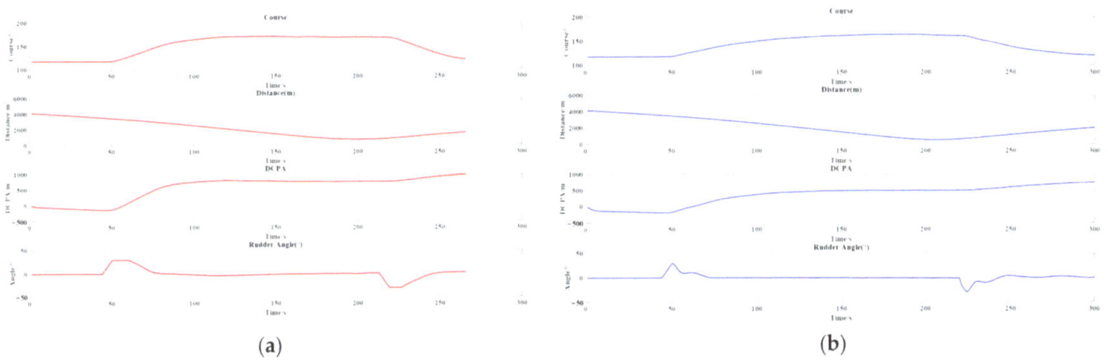

(**a**)   (**b**)

Figure 19. Parameter change diagram of the algorithm in the starboard side crossing situation: (**a**) Improved Velocity Obstacle; (**b**) Traditional Velocity Obstacle.

As shown in Table 7, it is the evaluation system's ranking of each index score of the three collision avoidance methods.

Table 7. Evaluation conclusion in the starboard side crossing situation.

Evaluation Index	Improved VO	Traditional VO	Manual Ship Handling
Own ship is in imminent danger	①	③	①
Minimum encounter distance	②	③	①
Deviation factor	②	①	③
Maximum alteration of course	②	③	①
Collision avoidance time	①	③	①
Steering frequency	①	①	③
Total	①	③	②

where ① indicates the highest ranking in the score of this indicator; ② indicates the second highest ranking; ③ indicates the lowest ranking.

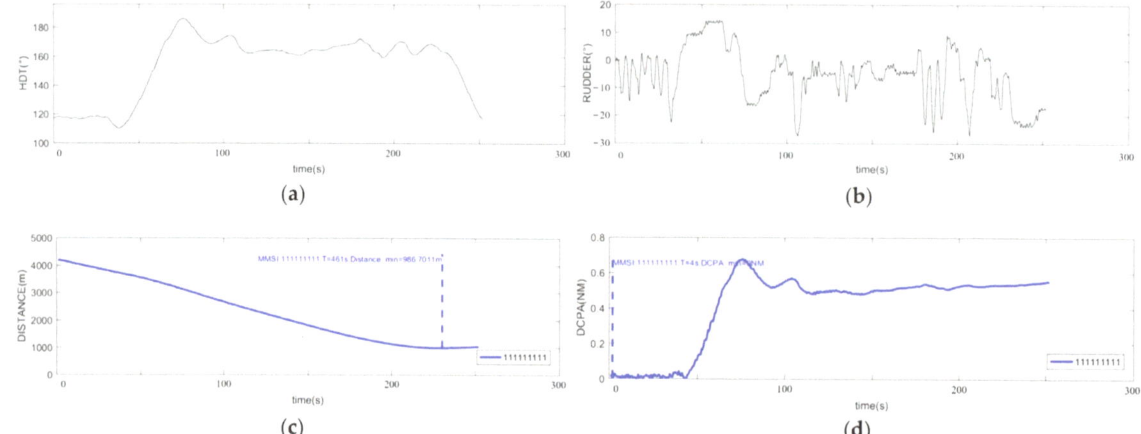

Figure 20. Parameter change diagram of the actual ship experiment in the starboard side crossing situation: (**a**) course graph; (**b**) rudder angle graph; (**c**) distance graph; (**d**) DCPA graph.

- Port side crossing situation.

Figure 21 shows the collision avoidance trajectories of the improved VO in the port side crossing situation. In the improved VO, the red line is the navigation trajectory of the target ship, and the blue line is the navigation trajectory of the own ship. The closed graphics around the two ships represent the ship domain, with the long axis being 0.35 NM and the short axis being 0.14 NM. When the target ship is on the port side of the own ship, this ship has no responsibility for avoidance. If the target ship does not take avoidance measures, this ship will start emergency collision avoidance and return to its original course after the collision danger is eliminated.

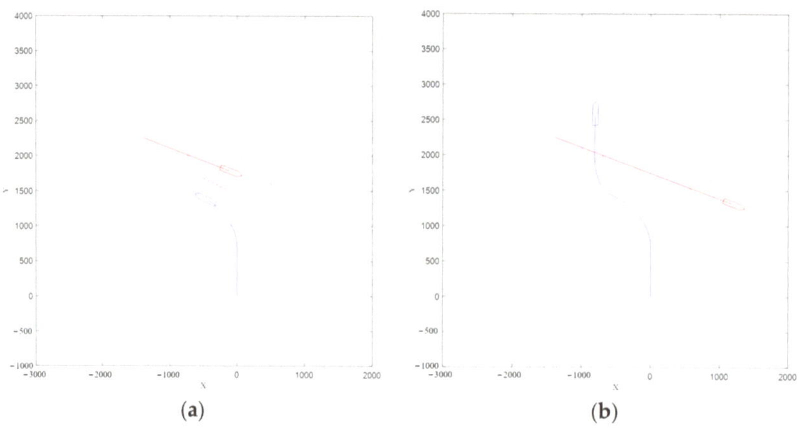

Figure 21. Improved Velocity Obstacle in the port side crossing situation: (**a**) the trajectory of the two ships when they are at the closest distance; (**b**) the trajectory when collision avoidance is completed.

Figure 22 shows the parameter changes of the improved VO in the port side crossing situation. The rudder angle, course, D_{OT} and DCPA were analyzed under the experiment. In the improved VO, the own ship started to avoid collision at 55 s, 1850 m away from the target ship, turned right to the minimum course of $-64.4°$, and started returning to the original course at 130 s. The minimum of D_{OT} was 515 m.

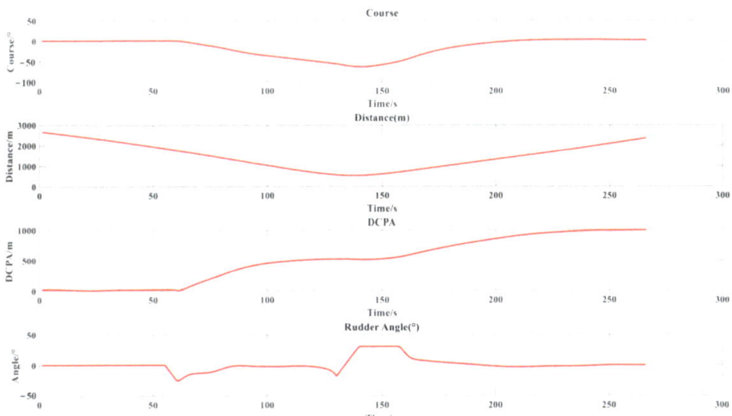

Figure 22. Parameter change diagram of the Improved Velocity Obstacle in the port side crossing situation.

- Overtaking situation.

Figures 23–25 show the collision avoidance trajectories of the improved VO, the traditional VO, and the actual ship collision avoidance experiment in the overtaking situation. Among them, the red line in the improved VO is the navigation trajectory of the target ship, and the blue line is the navigation trajectory of the own ship. The closed graphics around the two ships represent the ship domain, with the long axis being 0.4 NM and the short axis being 0.16 NM.

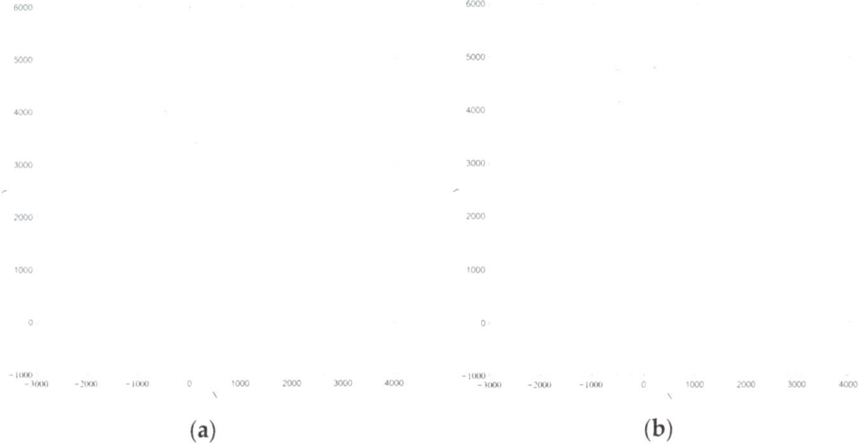

(a)   (b)

Figure 23. Improved Velocity Obstacle in the overtaking situation: (**a**) the trajectory of the two ships when they are at the closest distance; (**b**) the trajectory when collision avoidance is completed.

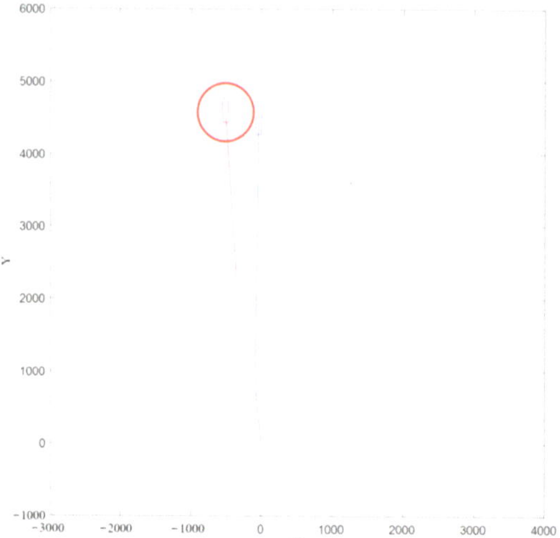

Figure 24. Traditional Velocity Obstacle in the overtaking situation.

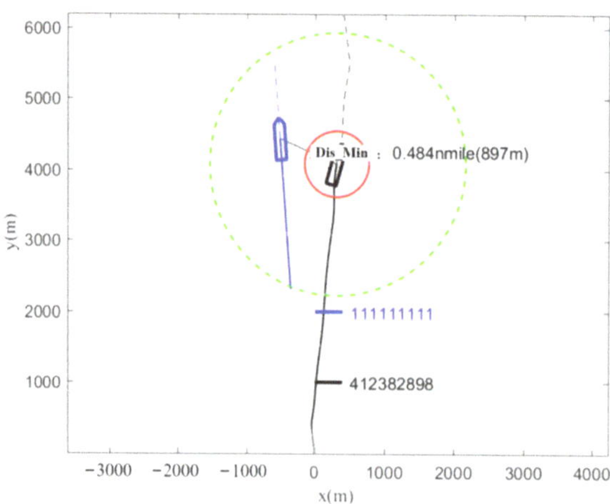

Figure 25. The actual ship experiment in the overtaking situation.

Figure 26a,b and Figure 27 show the parameter changes of the improved VO, the traditional VO, and the actual ship experiment in the overtaking situation. The rudder angle, course, D_{OT} and DCPA were analyzed under three experiments. In the improved VO, the own ship started to avoid collision at 37 s, 2200 m away from the target ship, turned right to the maximum course of 3.6°, and started returning to the original course at 562 s. The minimum of D_{OT} was 717 m. In the traditional VO, the own ship started to avoid collision at 1 s, 2343 m away from the target ship, turned right to the maximum course of 0.2°, and started returning to the original course at 514 s. The minimum of D_{OT} was 480 m. In the actual ship experiment, the own ship started to avoid collision at 38 s, 2200 m away from the target ship, turned right to the maximum course of 18°, and started returning to the original course at 589 s. The minimum of D_{OT} was 897 m.

As shown in Table 8, it is the evaluation system's ranking of each index score of the three collision avoidance methods.

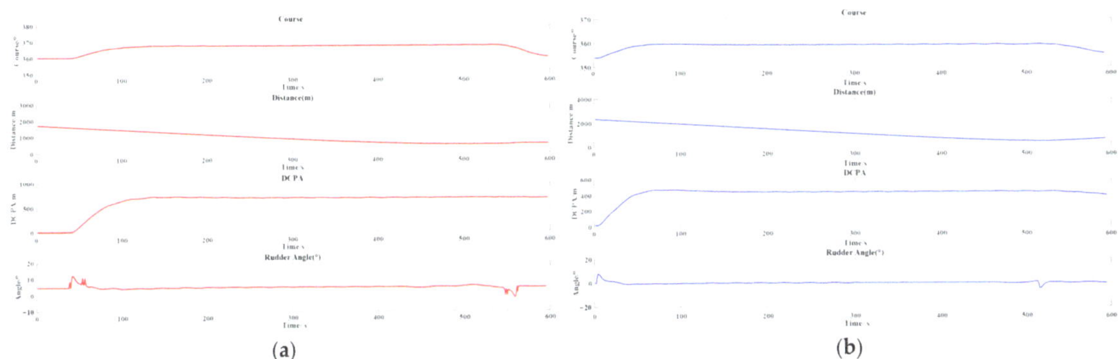

Figure 26. Parameter change diagram of the algorithm in the overtaking situation: (**a**) Improved Velocity Obstacle; (**b**) Traditional Velocity Obstacle.

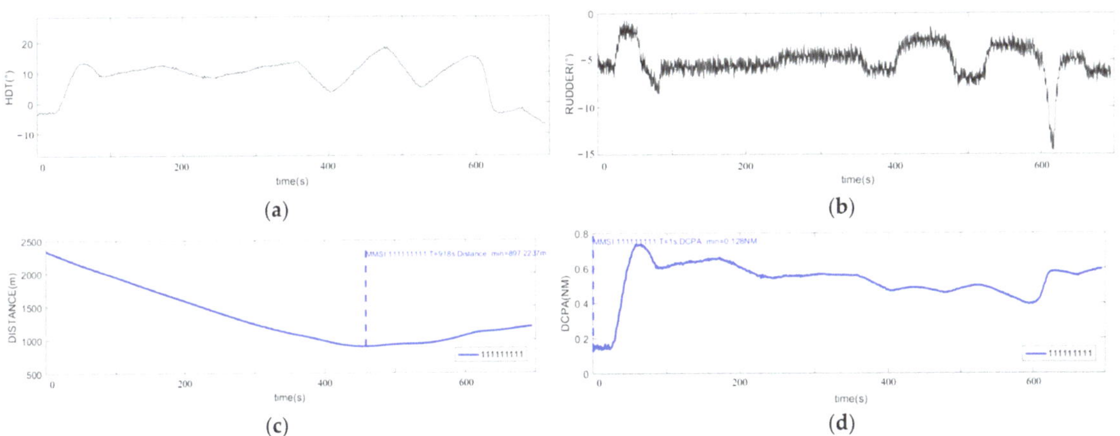

Figure 27. Parameter change diagram of the actual ship experiment in the overtaking situation: (**a**) course graph; (**b**) rudder angle graph; (**c**) distance graph; (**d**) DCPA graph.

Table 8. Evaluation conclusion in the overtaking situation.

Evaluation Index	Improved VO	Traditional VO	Manual Ship Handling
Own ship is in imminent danger	①	③	①
Minimum encounter distance	②	③	①
Deviation factor	②	①	③
Maximum alteration of course	②	③	①
Collision avoidance time	②	①	③
Steering frequency	①	①	③
Total	①	③	②

where ① indicates the highest ranking in the score of this indicator; ② indicates the second highest ranking; ③ indicates the lowest ranking.

From the comparison of collision avoidance results in four encounter situations (head-on, starboard side crossing, port side crossing, and overtaking), it can be seen that the improved VO algorithm can strictly ensure that the ship domains of the own ship and target ship do not invade each other during collision avoidance. The minimum distance between the two ships is always greater than 0.25 NM, and the timing of starting and ending collision avoidance is relatively close to the actual ship experiment. This method can use a smaller deviation factor to complete collision avoidance at a safe distance, while taking into account safety, economy, and practicality. In the evaluation index system of collision avoidance behavior, it is significantly better than other collision avoidance methods.

5. Conclusions

This paper determines the ship domain by calculating the safe distance of approach in different encounter situations. It employs an approach of finding the tangent of ellipses to ensure non-intersection of the ship domains of the own ship and target ship. Additionally, it combines the COLREGs and the VO algorithm to further determine the desired course. Finally, it determines the appropriate initiation timing for collision avoidance based on the risk of ship collision. By comparing the simulation experiments of the improved VO algorithm with traditional algorithms and actual ship experimental results, it can be seen that the improved VO algorithm can strictly abide by COLREGs to achieve collision avoidance when facing head-on, crossing, and overtaking encounter situations, and ensure that the minimum distance between two ships during collision avoidance is always greater than the imminent danger distance (0.25 NM) defined in the actual ship experiment. At the same time, the improved algorithm fits well with the captain's collision avoidance behavior in the actual ship experiment, which shows that this method can effectively deal with the collision situations of various encounter situations in open water environments. And it has certain feasibility and practical applicability in terms of collision avoidance behavior.

Author Contributions: Conceptualization, K.Z. and B.L.; methodology, M.Z., K.Z., B.H., B.L. and S.D.; software, K.Z., H.Z. and S.D.; formal analysis, M.Z., K.Z., B.H., B.L. and H.Z.; investigation, K.Z., H.Z., S.D., T.Z. and Y.Y.; resources, M.Z., B.H. and B.L.; data curation, K.Z., B.H., B.L., H.Z. and S.D.; writing—original draft preparation, K.Z.; writing—review and editing, M.Z., B.L. and T.Z.; visualization, M.Z. and B.H.; supervision, M.Z.; project administration, T.Z.; funding acquisition, M.Z. All authors have read and agreed to the published version of the manuscript.

Funding: This research was funded by the National Key R&D Program of China (2022YFB4301405), the National Natural Science Foundation of China (No. 52001243), the Fund of Guangxi Science and Technology Program (AB23026132).

Institutional Review Board Statement: Not applicable.

Informed Consent Statement: Not applicable.

Data Availability Statement: Data are contained within the article.

Conflicts of Interest: Author Yougui Yang was employed by the company Beibu Gulf Port Qinzhou Terminal Co., Ltd. The remaining authors declare that the research was conducted in the absence of any commercial or financial relationships that could be construed as a potential conflict of interest.

References

1. Lin, B.; Zheng, M.; Chu, X.; Mao, W.; Zhang, D.; Zhang, M. An overview of scholarly literature on navigation hazards in Arctic shipping routes. *Environ. Sci. Pollut. Res.* **2023**, 1–17. [CrossRef]
2. Shi, X.; Zhuang, H.; Xu, D. Structured survey of human factor-related maritime accident research. *Ocean Eng.* **2021**, *237*, 109561. [CrossRef]
3. Weng, J.; Liao, S.; Wu, B.; Yang, D. Exploring effects of ship traffic characteristics and environmental conditions on ship collision frequency. *Marit. Policy Manag.* **2020**, *47*, 523–543. [CrossRef]
4. Toyoda, S.; Fujii, Y. Marine traffic engineering. *J. Navig.* **1971**, *24*, 24–34. [CrossRef]
5. Goodwin, E.M. A statistical study of ship domains. *J. Navig.* **1975**, *28*, 328–344. [CrossRef]
6. Smierzchalski, R.; Michalewicz, Z. Modelling of a ship trajectory in collision situations at sea by evolutionary algorithm. *IEEE Trans. Evol. Comput.* **2000**, *4*, 227–241. [CrossRef]
7. Wang, N. An intelligent spatial collision risk based on the quaternion ship domain. *J. Navig.* **2010**, *63*, 733–749. [CrossRef]
8. Davis, P.V.; Dove, M.J.; Stockel, C.T. A computer simulation of marine traffic using domains and arenas. *J. Navig.* **1980**, *33*, 215–222. [CrossRef]
9. Bukhari, A.C.; Tusseyeva, I.; Kim, Y.G. An intelligent real-time multi-vessel collision risk assessment system from VTS view point based on fuzzy inference system. *Expert Syst. Appl.* **2013**, *40*, 1220–1230. [CrossRef]

10. Ma, D.; Chen, X.; Ma, W.; Zheng, H.; Qu, F. Neural Network Model-Based Reinforcement Learning Control for AUV 3-D Path Following. *IEEE Trans. Intell. Veh.* **2023**, 1–13. [CrossRef]

11. Chen, J.; Chen, H.; Liu, K. A method of estimating ship collision risk based on fuzzy neural network. *Ship Sci. Technol.* **2008**, 135–138.

12. Zhao, G.; Wang, C.; Zhou, J.; Li, Y. Collision risk calculation of unmanned surface vehicle on improved fuzzy evaluation method. *Syst. Eng. Electron.* **2023**, 1–9.

13. Li, M.; Mou, J.; Chen, P.; Rong, H.; Chen, L.; van Gelder, P.H.A.J.M. Towards real-time ship collision risk analysis: An improved R-TCR model considering target ship motion uncertainty. *Reliab. Eng. Syst. Saf.* **2022**, *226*, 108650. [CrossRef]

14. Abebe, M.; Noh, Y.; Seo, C.; Kim, D.; Lee, I. Developing a ship collision risk Index estimation model based on Dempster-Shafer theory. *Appl. Ocean Res.* **2021**, *113*, 102735. [CrossRef]

15. Li, L.; Wu, D.; Huang, Y.; Yuan, Z.M. A path planning strategy unified with a COLREGS collision avoidance function based on deep reinforcement learning and artificial potential field. *Appl. Ocean Res.* **2021**, *113*, 102759. [CrossRef]

16. He, Z.; Liu, C.; Chu, X.; Negenborn, R.R.; Wu, Q. Dynamic anti-collision A-star algorithm for multi-ship encounter situations. *Appl. Ocean Res.* **2022**, *118*, 102995. [CrossRef]

17. Zheng, Y.; Zhang, X.; Shang, Z.; Guo, S.; Du, Y. A decision-making method for ship collision avoidance based on improved cultural particle swarm. *J. Adv. Transp.* **2021**, *2021*, 8898507. [CrossRef]

18. Singh, Y.; Sharma, S.; Sutton, R.; Hatton, D.; Khan, A. A constrained A* approach towards optimal path planning for an unmanned surface vehicle in a maritime environment containing dynamic obstacles and ocean currents. *Ocean Eng.* **2018**, *169*, 187–201. [CrossRef]

19. Hong, X.; Xu, Z.; Wei, X.; Zhu, K.; Chen, Y. Dynamic obstacle avoidance of unmanned surface vehicle based on improved speed obstacle method. *Opt. Precis. Eng.* **2021**, *29*, 2126–2139. [CrossRef]

20. Zhang, Y.Y.; Qu, D.; Ke, J.; Li, X. Dynamic obstacle avoidance for USV based on velocity obstacle and dynamic window method. *J. Shanghai Univ. Nat. Sci. Ed.* **2017**, *23*, 1–16.

21. Ma, J.; Su, Y.; Xiong, Y.; Zhang, Y.; Yang, X. Decision-making method for collision avoidance of ships in confined waters based on velocity obstacle and artificial potential field. *Chin. J. Saf. Sci.* **2020**, *30*, 60–66.

22. Zhang, G.; Wang, Y.; Liu, J.; Cai, W.; Wang, H. Collision-avoidance decision system for inland ships based on velocity obstacle algorithms. *J. Mar. Sci. Eng.* **2022**, *10*, 814. [CrossRef]

23. Wang, S.; Zhang, Y.; Li, L. A collision avoidance decision-making system for autonomous ship based on modified velocity obstacle method. *Ocean Eng.* **2020**, *215*, 107910.

24. Kuwata, Y.; Wolf, M.T.; Zarzhitsky, D.; Huntsberger, T.L. Safe maritime autonomous navigation with COLREGS, using velocity obstacles. *IEEE J. Ocean. Eng.* **2013**, *39*, 110–119. [CrossRef]

25. Zhao, L.; Fu, X. A novel index for real-time ship collision risk assessment based on velocity obstacle considering dimension data from AIS. *Ocean Eng.* **2021**, *240*, 109913. [CrossRef]

26. Fiorini, P.; Shiller, Z. Motion planning in dynamic environments using velocity obstacles. *Int. J. Robot. Res.* **1998**, *17*, 760–772. [CrossRef]

27. Du, P.; Ouahsine, A.; Tran, K.T.; Sergent, P. Simulation of the overtaking maneuver between two ships using the non-linear maneuvering model. *J. Hydrodyn.* **2018**, *30*, 791–802. [CrossRef]

28. Sun, F.; Cai, Y.; Ma, J. A Study of Test Methods and Indicators for Marine Intelligent Anti-collision strategy. *Traffic Inf. Saf.* **2019**, *37*, 84–93.

29. Zhang, L.; Wang, H.; Meng, Q. Big data–based estimation for ship safety distance distribution in port waters. *Transp. Res. Rec.* **2015**, *2479*, 16–24. [CrossRef]

30. Li, L. Determination of elements such as the safe distances of approach in automatic collision avoidance studies of ships. *J. Dalian Marit. Univ.* **2002**, 23–26. [CrossRef]

31. Chen, C.; Li, G.; Li, F.; Li, L.; Chen, G. The Risk Threshold of Ship Collision in Different Waters. *China Navig.* **2020**, *43*, 27–32.

32. Perera, L.P.; Soares, C.G. Ocean vessel trajectory estimation and prediction based on extended Kalman filter. In Proceedings of the Second International Conference on Adaptive and Self-Adaptive Systems and Applications, Lisbon, Portugal, 21–26 November 2010; pp. 14–20.

33. Zhao, J. *Principles of Ship Collision Avoidance*; Dalian Maritime University Press: Dalian, China, 1998.

34. Naeem, W.; Irwin, G.W.; Yang, A. COLREGs-based collision avoidance strategies for unmanned surface vehicles. *Mechatronics* **2012**, *22*, 669–678. [CrossRef]

35. Zhong, J. Analysis of the risk of collision, close quarters situation and immediate Danger. *J. Shanghai Marit. Univ.* **1999**, *20*, 77–79.

36. Hu, Y.; Zhang, A.; Tian, W.; Zhang, J.; Hou, Z. Multi-ship collision avoidance decision-making based on collision risk index. *J. Mar. Sci. Eng.* **2020**, *8*, 640. [CrossRef]

Journal of
*Marive Science
and Engineering*

Article

Dynamic Projection Method of Electronic Navigational Charts for Polar Navigation

Chenchen Jiao [1], Xiaoxia Wan [1,*], Houpu Li [2] and Shaofeng Bian [2]

[1] School of Geodesy and Geomatics, Wuhan University, Wuhan 430079, China; jiaocchen@whu.edu.cn
[2] College of Electrical Engineering, Naval University of Engineering, Wuhan 430033, China;
     lihoupu1985@126.com (H.L.); 2016301610356@whu.edu.cn (S.B.)
*   Correspondence: wan@whu.edu.cn

Abstract: Electronic navigational charts (ENCs) are geospatial databases compiled in strict accordance with the technical specifications of the International Hydrographic Organization (IHO). Electronic Chart Display and Information System (ECDIS) is a Geographic Information System (GIS) operated by ENCs for real-time navigation at sea, which is one of the key technologies for intelligent ships to realize autonomous navigation, intelligent decision-making, and other functions. Facing the urgent demand for high-precision and real-time nautical chart products for polar navigation under the new situation, the projection of ENCs for polar navigation is systematically analyzed in this paper. Based on the theory of complex functions, we derive direct transformations of Mercator projection, polar Gauss-Krüger projection, and polar stereographic projection. A rational set of dynamic projection options oriented towards polar navigation is proposed with reference to existing specifications for the compilation of the ENCs. From the perspective of nautical users, rather than the GIS expert or professional cartographer, an ENCs visualization idea based on multithread-double buffering is integrated into Polar Region Electronic Navigational Charts software, which effectively solves the problem of large projection distortion in polar navigation applications. Taking the CGCS2000 reference ellipsoid as an example, the numerical analysis shows that the length distortion of the Mercator projection is less than 10% in the region up to 74°, but it is more than 80% at very high latitudes. The maximum distortion of the polar Gauss-Krüger projection does not exceed 10%. The degree of distortion of the polar stereographic projection is less than 1% above 79°. In addition, the computational errors of the direct conversion formulas do not exceed 10^{-9} m throughout the Arctic range. From the point of view of the computational efficiency of the direct conversion model, it takes no more than 0.1 s to compute nearly 8 million points at $1' \times 1'$ resolution, which fully meets the demand for real-time nautical chart products under information technology conditions.

Keywords: polar navigation; ENCs; complex function; dynamic projection; multithread-double buffer

Citation: Jiao, C.; Wan, X.; Li, H.; Bian, S. Dynamic Projection Method of Electronic Navigational Charts for Polar Navigation. *J. Mar. Sci. Eng.* **2024**, *12*, 577. https://doi.org/10.3390/jmse12040577

Academic Editor: Xinqiang Chen

Received: 6 March 2024
Revised: 27 March 2024
Accepted: 27 March 2024
Published: 28 March 2024

1. Introduction

With global warming causing the rapid melting of polar sea ice, the significance of the Arctic region in shipping, energy, and security has been increasingly highlighted. The Arctic route is poised to become the new major maritime artery [1–4]. Currently, there are several challenges in polar navigation, including rapidly changing marine environments, limited communication capabilities, and inadequate navigation safety measures [5–7]. There are potential risks associated with ships exploring the Arctic, such as the occurrence of extreme events, such as ship collisions leading to oil spills [8,9]. Hence, to ensure safe navigation or operations in polar regions, vessels must rely on high-precision charts as essential information support, which makes polar charts indispensable, serving as a key prerequisite for the realization of polar navigation and resource development. However, when encountering the unique environment of polar regions, the common Electronic Chart Display and Information System (ECDIS) based on true heading reference is not fully applicable, which could cause systemic problems in navigation parameter definitions, piloting, positioning, and calculations [10,11]. Among them, choosing the projection method for charts is a fundamental issue that urgently needs to be solved [12,13].

The projection of ENCs forms the cornerstone of research in navigation applications. The presentation of navigational information, the characteristics of navigational errors, and the implementation of navigational methodologies are determined. Since the middle of the 20th century, extensive research has been carried out on polar nautical charting by map cartographers such as Beresford, Snyder, Pearson, Smith, and others [14–18]. The availability of the modified Lambert projection, Gauss-Krüger projection, gnomonic projection, polar spherical projection, and azimuthal equidistant projection has been analyzed. With the development of ECDIS, the study of polar chart projection has entered a new stage. The projection to be used for Electronic Navigational Charts (ENCs) was the focus of [19,20]. However, the current international standards do not provide specific requirements for map projection to be used in ENCs and ECDIS. The choice of projection is still up to the manufacturer, which leads to different systems using different methods and creating different problems. Based on the existing literature [15,16], it is considered that the development of a computer-based method for the projection of ENCs for polar navigation should satisfy the following conditions:

- Conformal projection, in order to facilitate angle measurements and pilotage.
- Length and area distortions are kept as small as possible, in order to facilitate accurate measurement of distance and area measurements.
- The grid lines of latitude and longitude should be simple, in order to facilitate the construction of grids and the measurement of headings.
- Great circle routes should be as straight as possible, in order to facilitate navigation along them.

Therefore, Mercator projection, polar Gauss-Krüger projection, and polar stereographic projection are selected as projections that can be used for ENCs in the polar region. The Mercator projection is commonly used in low- and mid-latitude charts [21]. Furthermore, a mature set of marine navigation techniques based on the Mercator projection has been developed, which is in line with the charting habits of mariners, but the projection has a large deformation in the polar region [22]. The polar Gauss-Krüger projection is divided into 3° or 6° zones [23], making it difficult to be fully represented. The existing studies related to polar stereographic projection are based on the sphere [24]. However, the high-precision earth model is a rotating ellipsoid, and inherent principle errors are inevitably present in the projections based on the sphere model when navigating in the polar region.

Aiming at the problem of large distortion of the Mercator projection in the polar region, the selection of an appropriate reference latitude or mapping area in order to reasonably control the degree of distortion has been proposed by scholars [25], which makes it possible to follow the Mercator projection within a certain range of distortion. To address the problem of the poor availability of the Gauss-Krüger projection in the polar region, Bowring, using the inherent connection between complex (variable) functions and conformal mapping, derived the formula for the transverse Mercator projection without zones [26]. With the help of a computer algebra system, Shaofeng Bian provided formulas for the Gauss-Krüger projection complex function without band splitting, and the complex function expressions of scale ratio and meridian convergence angle were derived [27,28]. The non-iterative formulas of forward and inverse solutions of the Gauss-Krüger projection based on Lee's formula were derived by Jiachun Guo [29]. To some extent, the difficulty of applying the traditional Gauss-Krüger projection has been solved by the introduction of the complex function. In most cases, polar stereographic projection has been studied based on the sphere. Drawing on the method established by the ellipsoidal sundial projection formula, the double polar stereographic projection was proposed by Chaojiang Wen [30]. That is, the ellipsoidal surface is first conformally projected onto a suitable transition sphere, and then that sphere is projected onto the plane in the manner of polar stereographic projection.

All of the above solutions effectively address the shortcomings of the three projections in practice to a certain extent. Nevertheless, in navigation, nature entails a process of constant position change, especially in marine navigation. The use of a single projection makes it often difficult to meet the accuracy requirements of map representation, and it is necessary to design an adaptive map projection according to the specific characteristics of the navigation to satisfy the needs of high-precision and high-reliability applications in polar navigation [31–34]. In addition to this, how to make the map visualization system better meet the needs of the users and ensure its more effective utilization is also an issue that needs to be put into focus.

Therefore, in order to improve the accuracy of map representation in navigation applications, this paper introduces complex functions, considering their unique role and obvious advantages in conformal transformation. On the basis of existing research, complex function expressions for three projections are given, and direct conversion formulas based on the complex functions of these three projections are derived. In addition, a dynamic projection of the ENCs oriented towards polar navigation is proposed, the conversion accuracy and efficiency of the three projections are analyzed, and a visualization algorithm based on multithread-double buffer is designed. The results show that the constructed dynamic projection method of the ENCs for polar navigation can improve the accuracy of map expression in polar navigation applications and can provide a reference for the compilation and application of high-precision polar electronic nautical charts.

This article is organized as follows. Section 2 includes the complex function expression of the conformal projection of a polar chart based on the ellipsoid. In Section 3, a design approach to the dynamic projection of the ENCs is presented. In Section 4, the distortions and their availability within each latitude band are elaborated, the most suitable navigation-oriented projections within that range are given, and the transformations between the different projections are analyzed. Section 5 presents a visualization idea for an ENC-based multithread-double buffer, and Section 6 discusses the results of the analysis, followed by the conclusions.

2. Complex Function Expression of Conformal Projection of Polar Charts Based on the Ellipsoid

2.1. Mercator Projection Coordinates in Complex Form

According to the general formula for normal cylindrical projection, combined with the properties of complex functions, the complex function expression for the forward solution of the Mercator projection is as follows:

$$z_1 = x_1 + iy_1 = r_0 w \tag{1}$$

In Formula (1), z_1 represents the Mercator projection complex coordinates; x_1 is the ordinate, which is the southbound coordinate; and y_1 represents the abscissa, that is, the east coordinate. $w = q + il$ is the expression for isometric latitude in the field of complex functions, where q is the isometric latitude, corresponding to the expression of geodetic latitude B, and l is the longitude difference. In addition, $r_0 = \dfrac{a\cos B_0}{\sqrt{1-e^2\sin^2 B_0}}$ represents the cylindrical radius, a represents the semi-major axis of the ellipsoid, and B_0 refers to the reference latitude.

The simple shape of the latitude and longitude gridlines of the Mercator projection makes it easy to map and calculate. In addition, rhumb lines are projected as straight lines, which allows marine navigation users to visualize the shape and direction of the routes when planning the routes, so that accurate navigation can be carried out. Hence, the Mercator projection is the most commonly used method in nautical charts. However, the large distortion at high latitudes becomes a major factor limiting the application of Mercator projection in the polar regions.

The expression for the complex function of the inverse solution of the Mercator projection is obtained by a slight change in Equation (1):

$$w = q + il = \frac{x_1 + iy_1}{r_0} \tag{2}$$

The Mercator projection is a conformal projection, meaning that the angles are not distorted. The scale distortion u_1 of the secant Mercator projection at the reference latitude B_0 and latitude B can be expressed as follows, where e represents the first eccentricity:

$$u_1 = \frac{\cos B_0}{\cos B} \sqrt{\frac{1 - e^2 \sin^2 B}{1 - e^2 \sin^2 B_0}} - 1 \tag{3}$$

It is clear that the secant projection is a similar variation to the tangent projection in the theoretical study of map projections. The scale factor of the Mercator projection can be determined by the reference latitude. According to a previous report [35], the reference latitude is generally determined by minimizing the distortion at the maximum deformation, with the purpose of reducing the distortion and making the deformation uniformly distributed in the region. B_0 is given by $\bar{B} = \frac{B_N + B_S}{2}$ and $\Delta B = \frac{B_N - B_S}{2}$, with B_N and B_S being the southern and northern latitudes of the region, respectively:

$$B_0 = \bar{B} + \frac{\left(12 - (4 - 7e^2)\cos \bar{B} - e^2\left(\cos 2\bar{B} - \cos 3\bar{B} - 2\cos 4\bar{B}\right)\right)}{8\sin 2\bar{B}} \Delta B^2 \tag{4}$$

2.2. Polar Gauss-Krüger Projection Coordinates in Complex Form

According to the literature [36], the expression of the complex function of the non-singular polar Gauss-Krüger projection for the forward solution in the polar region is given below:

$$z_2 = x_2 + iy_2 = a\alpha_0\theta + a\sum_{k=1}^{5} (-1)^{k-1}\alpha_{2k}\sin 2k\theta \tag{5}$$

Similarly, a sketch of the Gauss-Krüger projection is drawn in Geocart, as shown in Figure 1.

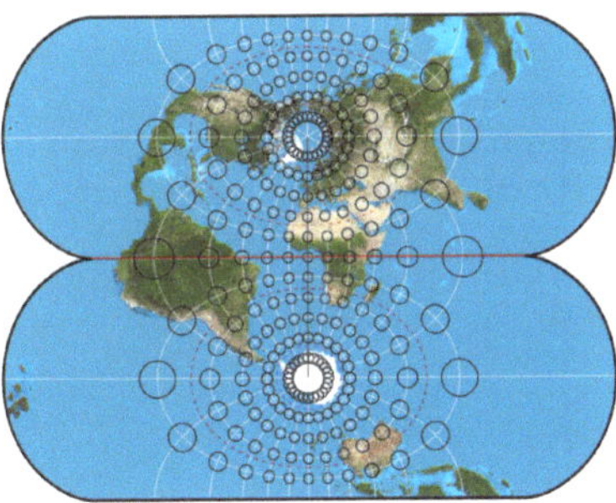

Figure 1. Sketch of Gauss-Krüger projection.

Given that power series expressions based on the third flattening n have a more compact form and better convergence [37,38], the coefficients can be expressed in terms of n as follows:

$$
\begin{cases}
\alpha_0 = -1 + n - \frac{5}{4}n^2 + \frac{5}{4}n^3 - \frac{81}{64}n^4 + \frac{81}{64}n^5 \\
\alpha_2 = \frac{1}{2}n - \frac{7}{6}n^2 + \frac{77}{48}n^3 - \frac{1111}{720}n^4 + \frac{2281}{1920}n^5 \\
\alpha_4 = \frac{13}{48}n^2 - \frac{209}{240}n^3 + \frac{3817}{2880}n^4 - \frac{6917}{6720}n^5 \\
\alpha_6 = \frac{61}{240}n^3 - \frac{1663}{1680}n^4 + \frac{14459}{8960}n^5 \\
\alpha_8 = \frac{49561}{161280}n^4 - \frac{221401}{161280}n^5 \\
\alpha_{10} = \frac{34729}{80640}n^5
\end{cases}
$$

In the polar Gauss-Krüger projection plane, z_2 represents the polar Gauss-Krüger projection complex coordinates and x_2, y_2 are the ordinate and abscissa, respectively. In order to eliminate singularities near the poles, the conformal co-latitude $\theta = 2\arctan[\exp(-w)]$ is introduced. Two points need to be noted. One is that the coefficients of the expression in this paper are slightly different from those in the literature [36], which is due to the fact that the origin of the expression is moved to the pole for ease of mapping in the Arctic. The other is that $\mathrm{Re}(\sin kz) = \sin kx\cosh ky$ and $\mathrm{Im}(\sin kz) = \cos kx\sinh ky$. These equalities hold for any complex number $z = x + iy$ and any natural number $k \geq 1$ [39].

We obtain the expression for the complex function of the inverse solution by using the symbolic iteration method:

$$
\begin{cases}
w = \frac{z_2}{a\alpha_0} = \frac{x_2 + iy_2}{a\alpha_0} \\
\theta = w + \sum_{k=1}^{5} b_{2k}\sin 2kw
\end{cases}
\tag{6}
$$

The coefficients in Equation (6) are expended in terms of n up to n^5:

$$
\begin{cases}
b_2 = \frac{1}{2}n - \frac{2}{3}n^2 + \frac{37}{96}n^3 - \frac{1}{360}n^4 - \frac{81}{512}n^5 \\
b_4 = -\frac{1}{48}n^2 - \frac{1}{15}n^3 + \frac{437}{1440}n^4 - \frac{46}{105}n^5 \\
b_6 = \frac{17}{480}n^3 - \frac{37}{840}n^4 - \frac{209}{4480}n^5 \\
b_8 = -\frac{4397}{161280}n^4 + \frac{11}{504}n^5 \\
b_{10} = \frac{4583}{161280}n^5
\end{cases}
$$

A clerical error in the inverse solution coefficients b_2 and b_8, *as presented in* the literature [36], is corrected here.

Additionally, the angle of the polar Gauss-Krüger projection is not distorted. The length distortion u_2 can be calculated as the derivative of the coordinate z_2 at a specific point.

$$
u_2 = |z_2{}'| - 1 = \left| \frac{(1 - e^2\sin^2 B)^{1/2}\sin\theta\left(-\alpha_0 - \sum_{k=1}^{5} 2k\alpha_{2k}\cos 2k\theta\right)}{\cos B} \right| - 1
\tag{7}
$$

The Universal Transverse Mercator Projection (UTM) is a Gauss-Krüger projection with a central meridian projection length ratio of 0.9996.

2.3. Polar Stereographic Projection Coordinates in Complex Form

According to the general formula for azimuthal conformal projection, the complex function expression for the forward solution of the double polar stereographic projection is as follows:

$$
u_2 = |z_2{}'| - 1 = \left| \frac{(1 - e^2\sin^2 B)^{1/2}\sin\theta\left(-\alpha_0 - \sum_{k=1}^{5} 2k\alpha_{2k}\cos 2k\theta\right)}{\cos B} \right| - 1
\tag{8}
$$

In Formula (8), z_3 is defined as stereographic projection complex coordinates, x_3, y_3 are the ordinate and abscissa, respectively, φ_0 stands for equiangular reference latitude. Normally, the conformal spherical radius is considered as $R_\varphi\left(\frac{\pi}{2}\right) = \frac{a}{\sqrt{1-e^2}}\left(\frac{1-e}{1+e}\right)^{e/2}$. The Universal Polar Stereographic Projection (UPS) is a polar stereographic projection for $\varphi_0 = 81°06'25''.3$.

Moreover, a sketch of the polar stereographic projection is drawn in Geocartv3.2.0, as shown in Figure 2.

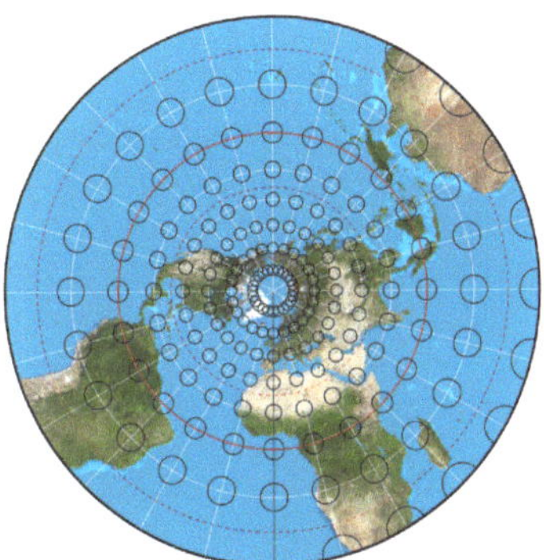

Figure 2. Sketch of polar stereographic projection.

We obtain the complex function expression for the inverse solution of the double polar stereographic projection from Equation (8):

$$exp(-w) = -\frac{\sec^2\left(\frac{\pi}{4} - \frac{\varphi_0}{2}\right)}{2R_\varphi}(x_3 + iy_3) \tag{9}$$

Since the double polar stereographic projection is a conformal projection, the angular distortion is 0. The length distortion u_3 is related to the geodetic latitude B and the conformal reference latitude φ_0, independent of the longitude difference l, which can be expressed as follows [25]:

$$u_3 = \frac{2\cos^2\left(\frac{\pi}{4} - \frac{\varphi_0}{2}\right)\sqrt{1 - e^2\sin^2 B}}{\sqrt{1-e^2}(1 + \sin B)}\left(\frac{1-e}{1+e} \cdot \frac{1 + e\sin B}{1 - e\sin B}\right)^{\frac{e}{2}} - 1 \tag{10}$$

3. Method Design for Dynamic Chart Projection

The core objective of polar dynamic chart projection is to accurately represent maps in navigation applications by adopting suitable projection methods. Therefore, the design of the projection is key. In addition, solving problems based on trajectory points in real time is crucial. The specific method of dynamic chart projection designed in this paper is as follows: firstly, the inverse operation is carried out to calculate the latitude and longitude according to the inverse solution expressions of the three projections, and then the latitude and longitude are converted to plane coordinates according to the forward solution expressions. The essence of the dynamic projection is the transformation between different projections. At present, the numerical transformation method and indirect transformation method are mainly used for the translation between different projections under the ellipsoid model [40]. However, these methods are more complicated and inefficient in the calculation process, and both of them fail to establish a direct transformation between the projected coordinates. Therefore, the formulas for the direct conversion of the projections in this section are derived for fast visualization in the ENCs of the polar region.

3.1. Transformation between Mercator Projection and Polar Gauss-Krüger Projection

The Mercator projection coordinates (x_1, y_1) are substituted into the complex function expression for the inverse solution of the Mercator projection (Equation (2)) to obtain the latitude and longitude coordinates. Subsequently, it is substituted into the complex function expression of the forward solution of polar Gauss-Krüger projection (Equation (5)), to obtain the complex function expression of the Mercator projection transformed to polar Gauss-Krüger projection as follows:

$$
\begin{cases}
\theta = 2\arctan\exp\left(-\frac{x_1+iy_1}{r_0}\right) \\
z_2 = x_2 + iy_2 = a\alpha_0\theta + a\sum_{k=1}^{5}(-1)^{k-1}\alpha_{2k}\sin 2k\theta
\end{cases}
\tag{11}
$$

The polar Gauss-Krüger projection coordinates (x_2, y_2) are known. The latitude and longitude coordinates are obtained by substituting into Formula (6). Then, by substituting these coordinates into Formula (1), we obtain the expression of the complex function of the polar Gauss-Krüger projection transformed to the Mercator projection:

$$
\begin{cases}
\theta = \frac{x_2+iy_2}{a\alpha_0} + \sum_{k=1}^{5} b_{2k}\sin\frac{2k(x_2+iy_2)}{a\alpha_0} \\
z_1 = x_1 + iy_1 = -r_0\ln\tan\frac{\theta}{2}
\end{cases}
\tag{12}
$$

It should be noted that the central meridian of the two projections derived from Equations (11) and (12) should be the same, and if not, it should be corrected.

3.2. Transformation between Polar Gauss-Krüger Projection and Polar Stereographic Projection

Similarly, by substituting the polar Gauss-Krüger projection coordinates (x_2, y_2) into Equation (6) to obtain the latitude and longitude, and later substituting into the expression of the complex function of the forward solution of the polar stereographic projection (Equation (8)), we obtain the expression of the complex function of the Gauss-Krüger projection, transformed to the polar stereographic projection:

$$
\begin{cases}
\theta = \frac{x_2+iy_2}{a\alpha_0} + \sum_{k=1}^{5} b_{2k}\sin\frac{2k(x_2+iy_2)}{a\alpha_0} \\
z_3 = x_3 + iy_3 = -2R_\varphi\cos^2\left(\frac{\pi}{4} - \frac{\varphi_0}{2}\right)\tan\frac{\theta}{2}
\end{cases}
\tag{13}
$$

Given the polar stereographic projection coordinates (x_3, y_3), we obtain the latitude and longitude by inverse solution, and then substitute into Formula (5) to obtain the expression of the complex function of polar stereographic projection, transformed to polar Gauss-Krüger projection:

$$
\begin{cases}
\theta = -2\arctan\left[\frac{\sec^2\left(\frac{\pi}{4} - \frac{\varphi_0}{2}\right)}{2R_\varphi}(x_3 + iy_3)\right] \\
z_2 = x_2 + iy_2 = a\alpha_0\theta + a\sum_{k=1}^{5}(-1)^{k-1}\alpha_{2k}\sin 2k\theta
\end{cases}
\tag{14}
$$

3.3. Transformation between Polar Gauss-Krüger Projection and Polar Stereographic Projection

We obtain the expression for the complex function of the Mercator projection transformed to polar stereographic projection by substituting (x_1, y_1) into Equation (2) and then into Equation (8):

$$
z_3 = x_3 + iy_3 = -2R_\varphi\cos^2\left(\frac{\pi}{4} - \frac{\varphi_0}{2}\right)\exp\left(-\frac{x_1+iy_1}{r_0}\right)
\tag{15}
$$

Substituting (x_3, y_3) into Formula (9) to get the latitude and longitude and then substituting into Formula (1), we can get the expression of the complex function of polar stereographic projection transformed to Mercator projection:

$$
z_1 = x_1 + iy_1 = -r_0\ln\left[-\frac{\sec^2\left(\frac{\pi}{4} - \frac{\varphi_0}{2}\right)}{2R_\varphi}(x_3 + iy_3)\right]
\tag{16}
$$

4. Evaluation of Dynamic Projection

With the help of Mathematica and MATLAB, the China Geodetic Coordinate System 2000 (CGCS2000) and the Arctic boundary data from the National Oceanic and Atmospheric Administration (NOAA, https://www.noaa.gov/, accessed on 26 March 2024) are taken as examples for the numerical analysis in this paper. We discuss and analyze the distortion of polar navigation projection from two aspects: length distortion and longitude and latitude grid line distortion, focusing on the distortion in each latitude segment of $66.5°{\sim}70°, 70°{\sim}75°, 75°{\sim}80°, 80°{\sim}85°, 85°{\sim}90°$. Availability is analyzed and the most suitable navigation-oriented projection type for that projection range is given. The transition between different projections is translated using the direct transformation formulas derived in Section 3, and the translation accuracy and efficiency are analyzed. The reference ellipsoid constants are $a = 6\ 378\ 137$ m and $f = 1/298.257\ 222\ 101$.

4.1. Length Distortion

Length distortion is a primary factor in determining which projection is used, and controlling length distortion is crucial in cartography. The variations of u_1, u_2, u_3, with longitude difference l at different circles of latitude, ranging from the Arctic Circle $66.5°$ to the North Pole $90°$ are shown in Figure 3. To provide a more visual representation of the distortion within the polar regions, the length distortions calculated based on Equations (3), (7), and (10) are listed in Table 1.

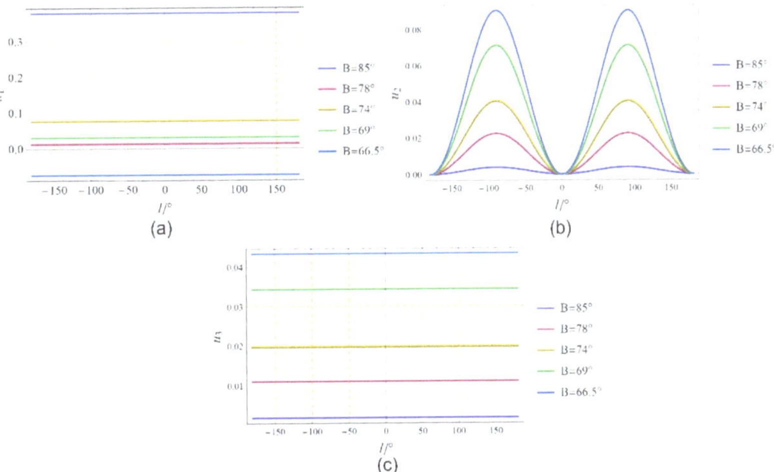

Figure 3. Variation of length distortion with longitude difference *l* ((**a**) Mercator projection; (**b**) polar Gauss-Krüger projection; (**c**) polar stereographic projection).

The results in Figure 4 and Table 1 show the following:

(1)   In the Mercator projection, there is no length distortion at the reference latitude, while the length distortion is greater than 0 beyond the reference latitude and less than 0 within it. The length distortions are all less than 10% in the region up to $74°$, whereas in polar regions with very high latitudes, the maximum distortion can exceed 80%. This suggests that it is possible to control the degree of distortion of the Mercator projection by adjusting the reference latitude, but the Mercator projection is still significantly distorted at high latitudes.

(2)   The length distortions for $l = 60°$ are listed in Table 1. In the polar Gauss-Krüger projection, the farther away from the standard meridian, the larger the length distortion is for a fixed longitude difference. At the same latitude, the length distortion increases and then decreases with the longitude difference, reaching a maximum at $l = \pm90°$. The maximum distortion is calculated to be no more than 10%. This shows

that the bandwidth can be broadened by using a complex function to represent the Gauss-Krüger projection, which facilitates uniform representation of the land and nautical charts and provides an important reference for scientific research and nautical charting in the polar region.

(3) The overall distortion of the double polar stereographic projection in the polar region is relatively small, especially above 79°, where the distortion is less than 1%. It has been able to satisfy the compilation of large and medium scale marine charts, which is very important for the application of high-precision polar navigation.

Table 1. Length distortions at significant nodes within 66.5° to 90° of three projections.

$B/°$	Type of Projection		
	Mercator Projection	Polar Gauss-Krüger Projection ($l = 60°$)	Polar Stereographic Projection
66.5	−0.074787	0.065517	0.043240
69	0.029360	0.051923	0.034335
70	0.078520	0.046964	0.031079
71	−0.089466	0.042272	0.027993
74	0.075366	0.029762	0.019747
75	0.145209	0.026102	0.017328
76	−0.130658	0.022693	0.015073
78	0.011497	0.016613	0.011045
79	0.102135	0.013938	0.009270
80	0.211028	0.011502	0.007653
81	−0.232651	0.009304	0.006193
83	−0.015044	0.005616	0.003741
85	0.377225	0.002861	0.001906
86	−0.838213	0.001830	0.001219
88	−0.676627	0.000457	0.000305
90	∞	0	0

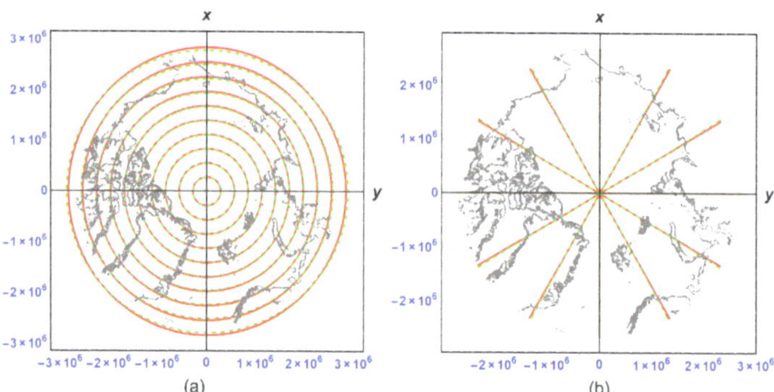

Figure 4. Comparison of the two projections. (**a**) Schematic of the latitude lines at the Arctic; (**b**) schematic of the meridians at the Arctic, where the red solid curves represent the polar stereographic projection; the green dashed curves represent the polar Gauss-Krüger projection; and the gray line shows the boundary line of the Arctic.

4.2. Distortions of Longitude and Latitude Grid Lines

The establishment of latitude and longitude grids on the plane is the basis for determining the precise position of a ship during polar navigation. The grid of latitude and longitude lines for the Mercator projection is widely known, so it will not be discussed in this paper. Based on the polar Gauss-Krüger projection and polar stereographic projection formulas, the latitude and longitude are plotted as shown in Figure 4. In this case, the meridian starts at 0° longitude, the interval longitude difference is 30°, and the latitude circle starts at 90° latitude. The red solid curves represent the polar stereographic projection; the green dashed curves represent the polar Gauss-Krüger projection; and the gray line shows the boundary line of the Arctic. From Figure 4, it can be observed that the meridians in the polar Gauss projection are radial straight lines centered on the poles, and the latitudinal lines are depicted as concentric circles centered on the poles, whereas the meridians in the polar Gauss-Krüger projection resemble an inverse hyperbola, and the latitudinal lines resemble an ellipse.

To analyze specifically the differences between the latitude and longitude grids of these two projections at high latitudes, the distribution of the difference Δx and Δy between the x and y coordinates in the interval of latitude 87° to 90° and longitude difference −90° to 90° are shown in Figures 5 and 6. Figures 5 and 6 show that Δx decreases with increasing longitude difference and Δy increases with increasing longitude difference when latitude is fixed. In addition, Δx and Δy decrease with increasing latitude when the longitude difference is constant. Normally, in 1:1,500,000 nautical charts, the difference of coordinates on the charts within the region is negligible within 1 mm. Correspondingly, the actual coordinate differences should be less than 1500 m. It is clear that the two projections only have coordinate differences near the poles simultaneously less than 1500 m under the ellipsoid model. Therefore, it is easier to use the dynamic projection to match other auxiliary navigation.

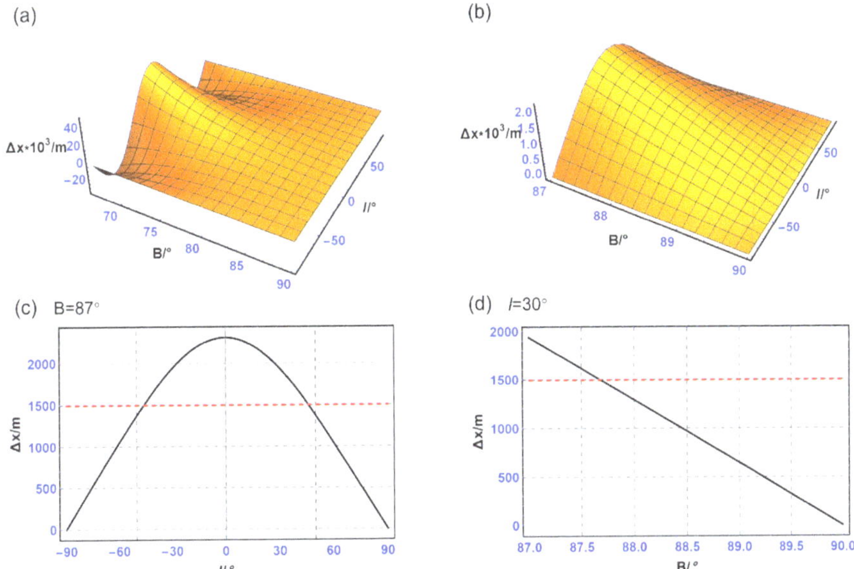

Figure 5. Distribution of Δx. (a) Distribution of Δx in the latitude range of 66.5°~90°, longitude difference −90° ~ 90°; (b) distribution of Δx in the latitude range of 87°~90°, longitude difference −90° ~ 90°; (c) distribution of Δx as a function of longitude difference l for $B = 87°$; (d) distribution of Δx with latitude B at $l = 30°$. The black solid line shows the curve of Δx as a function of l and B. The red dotted line indicates the maximum allowable distance (1500 meters) for Δx in a 1:1.5 million chart.

Referring to the existing recommendations for the compilation of ENCs [41,42], the length distortion of berthing and harbor at scales of 1:22,000 and above should be limited to within 5%. The length distortions in the 1:22,000 to 1:90,000 approach are within 10%. The range of distortions in the general scale, ranging from 1:350,000 to 1:1,500,000, is controlled to 40% or less. For overviews at 1:1,500,000 and smaller scales, the range of length distortion can be limited to about 50%. By this convention, the adoption of dynamic projection in the polar region is suggested in this paper. Based on the analysis of the length distortions and the longitude and latitude grid distortions, the projections oriented towards polar navigation are listed in Table 2. It is recommended that the compilation scales for ENCs be based upon standard radar ranges in nautical miles. The smaller the selectable range, the larger the scale of the ENCs, and the more detailed the geographic information displayed on the ECDIS screen. Maritime users can flexibly manage and browse chart information by adjusting selectable distances to suit different navigational needs and scenarios. Here, for the purpose of uniformity of units in the paper, we standardize nautical miles to meters, 1 n mile = 1852 m.

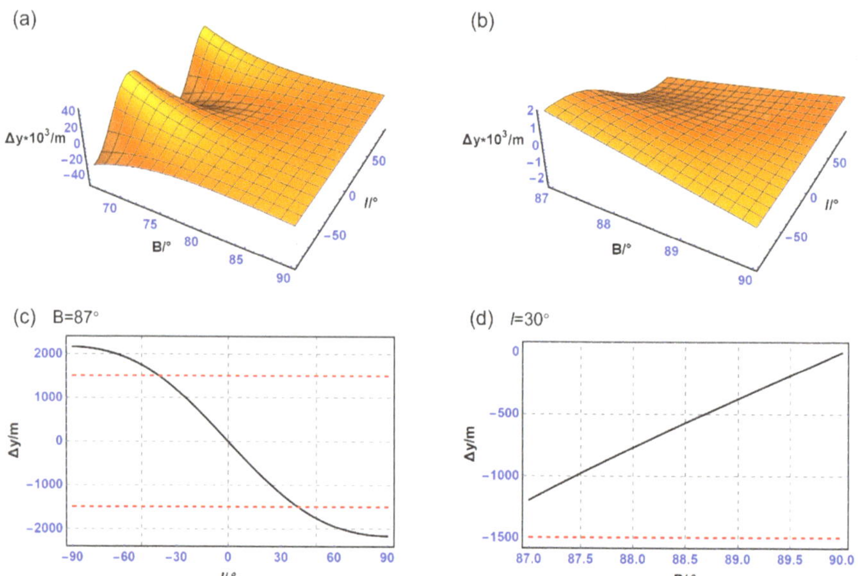

Figure 6. Distribution of Δy (**a**) Distribution of Δy in the range of latitude $66.5°{\sim}90°$, longitude difference $-90° \sim 90°$; (**b**) distribution of Δy in the latitude range of $87°{\sim}90°$, longitude difference $-90° \sim 90°$; (**c**) distribution of Δy as a function of longitude difference l for $B = 87°$; (**d**) distribution of Δy with latitude B at $l = 30°$. The black solid line shows the curve of Δy as a function of l and B. The red dotted line indicates the maximum allowable distance (1500 meters) for Δy in a 1:1.5 million chart.

Table 2. Projections facing polar navigation at different scales.

Name	Scale Range	Selectable Range (m)	Deformation	Range of Latitudes	Projection
Berthing	>1:4000	463	<5%	$66.5° \sim 69°$	Mercator Projection
				$69° \sim 79°$	Polar Gauss-Krüger Projection
				$79° \sim 90°$	Polar Stereographic Projection
Harbor	1:4000~1:22,000	2778	<5%	$66.5° \sim 69°$	Mercator Projection
				$69° \sim 79°$	Polar Gauss-Krüger Projection
				$79° \sim 90°$	Polar Stereographic Projection
Approach	1:1:22,000~1:90,000	11,112	<10%	$66.5° \sim 74°$	Mercator Projection
				$74° \sim 79°$	Polar Gauss-Krüger Projection
				$79° \sim 90°$	Polar Stereographic Projection
Coastal	1:90,000~1:350,000	44,448	<30%	$66.5° \sim 83°$	Mercator Projection
				$83° \sim 90°$	Polar Stereographic Projection
General	1:350,000~1:1,500,000	177,792	<40%	$66.5° \sim 85°$	Mercator Projection
				$85° \sim 90°$	Polar Stereographic Projection
Overview	<1:1,500,000	407,440	<50%	$66.5° \sim 85°$	Mercator Projection
				$85° \sim 90°$	Polar Stereographic Projection

4.3. Accuracy of Direct Transformation between Projections

The aim of this paper is to carry out an accuracy analysis of the direct transformation between the established projections, as follows: Firstly, $B \in [66.5°, 90°]$ and $l \in [-90°, 90°]$ are substituted into Equations (1), (5), and (8) to get the true value of the projected coordinates. Secondly, the true values of the coordinates are substituted into the projected conversion formulas (Equations (11)~(16)) derived in this paper to get the computed values. Lastly, the computed values are subtracted from the true values to get the computational errors of the conversion formulas. The computational errors of the transformation between the polar Gauss-Krüger projection and the polar stereographic projection are listed in this section, limited by the length of the article. The computational errors of the direct transformation of the polar Gauss-Krüger projection to the polar spherical projection are recorded as Δx_{23}, Δy_{23}, while the computational errors of the direct transformation of the polar spherical projection to the polar Gauss projection are noted as Δx_{32} and Δy_{32}. The variation of the computational errors is shown in Figure 7. As can be seen from Figure 7, the computational errors of the derived direct transformation formulas for both polar Gauss-Krüger projection and polar stereographic projection are less than 10^{-9} m in this paper. The computational errors of the other direct transformation formulas were calculated to be less than 10^{-9} m. The correctness of the derived direct transformation formulas can be proved numerically by taking into account the computational errors inherent in Mathematica.

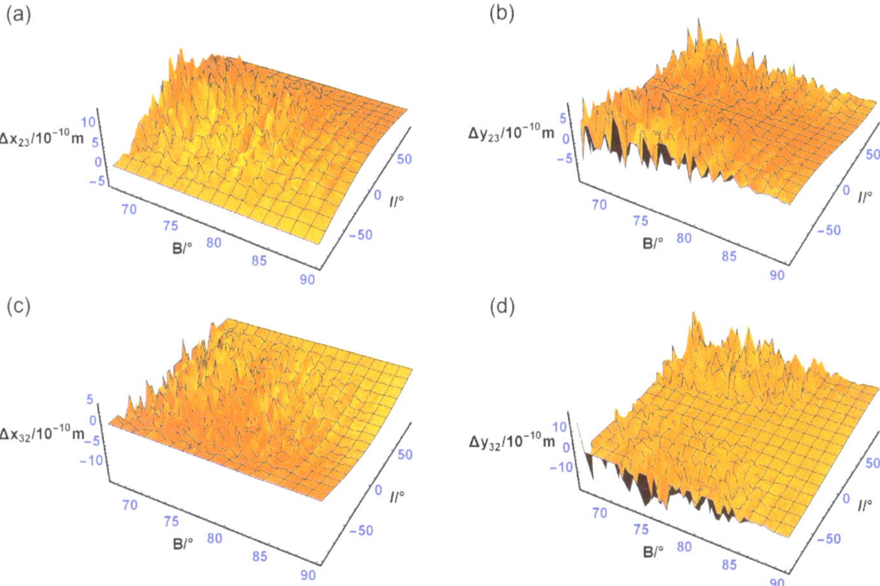

Figure 7. Calculation error of the transformation between polar Gauss-Krüger projection and polar stereographic projection. (**a**) Calculation error of Δx_{23}; (**b**) calculation error of Δy_{23}; (**c**) calculation error of Δx_{23}; (**d**) calculation error of Δy_{23}, where the direct transformation of the polar Gauss-Krüger projection to the polar spherical projection are recorded as Δx_{23}, Δy_{23}, and the computational errors of the direct transformation of the polar spherical projection to the polar Gauss projection are noted as Δx_{23}, Δy_{23}.

4.4. Calculation Efficiency Analysis

In order to verify the efficiency of the established direct transformation model among the three projections, $B \in [66.5°, 90°]$, $l \in [0°, 90°]$ are selected as the study area, and the transformation between the Gauss-Krüger projection of the polar region and the polar stereographic projection is used as an example. The computational time used for the three resolutions of $1° \times 1°$, $1' \times 1'$, $0.1' \times 0.1'$ is measured and is shown in Table 3. The resolution of $1° \times 1°$ requires the computation of 2 115 points; the resolution of $1' \times 1'$ includes 7.614 million points; and the resolution of $0.1' \times 0.1'$ requires the calculation of 761.4 million points. Here, the $0.1' \times 0.1'$ resolution is chosen because of the format of XX°XX.XXXX′ in which nautical users enter the coordinates of points on ECDIS. In the table, t_1 is defined as the computational time used for the direct conversion of the polar Gauss-Krüger projection to polar stereographic projection, and t_2 represents the computational time used for the direct conversion of polar stereographic projection to polar Gauss-Krüger projection. The algorithm is tested in the following environment:

Table 3. Calculational time (unit: s).

Form	Time (s)	Resolution		
		$1° \times 1°$	$1' \times 1'$	$0.1' \times 0.1'$
$(x_2, y_2) \rightarrow (x_3, y_3)$	t_1	0.0018389	0.081655	197.8277
$(x_3, y_3) \rightarrow (x_2, y_2)$	t_2	0.0019954	0.067192	203.3265

Hardware environment: the processor is AMD Ryzen 7 5800H with Radeon Graphics 3.20 GHz, RAM is 16.0 GB; the graphics card is AMD Radeon (TM) Graphics.

Software environment: Windows 11, 64-bit, MATLAB R2019av9.6.0.

According to Table 3, the time taken for the direct transformation between the polar Gauss-Krüger projection and the polar stereographic projection also does not exceed 0.1 s for the calculation of nearly 8 million points at $1' \times 1'$ resolution, which can better satisfy the higher requirements of high-precision ENCs in terms of resolution and projection transformation efficiency.

5. Visualization of Dynamic Chart Projection

The real-time capability of the dynamic projection can meet the needs of route changing at any time when the ship navigation path is determined. In this paper, a method based on buffer analysis is proposed. The projection area is determined according to the real-time position of the object, and adaptive visualization is achieved using a multithread-double buffer dynamic scheduling algorithm. The specific process of dynamic chart projection implementation is shown in Figure 8. The radius r of the buffer is determined by the relationship between the screen size and the current map scale. The radius r of the buffer is determined by the relationship between the screen size and the current map scale. Screen width and height are represented by w and h, respectively, and scale is the current scale.

$$r = \frac{\sqrt{w^2 + h^2}}{scale} \tag{17}$$

The current point of the ship's navigation, which is placed at the center of the screen, is set as the initial point for the projection. A circular buffer is formed with the center of the projected datum and a radius of r, and all objects within the buffer are projected. The next projection reference point is judged by the real-time position of the ship's movement. Specifically, this is achieved by dividing the screen into four parts, as shown in Figure 9, and if the point is located in area i, then D_i is used as the next projection reference point. For fast transitions in navigation scenarios, a double buffering technique can be used. The double buffer technique means that two buffers (the front buffer and the back buffer) are used to store and display images alternately. In general, the front buffer is used to display the current image, and the back buffer is used to render the new image. While the vessel is moving, the back buffer is exchanged with the chart information displayed on the screen of the ECDIS, so that the ENCs are ensured to display updates smoothly and quickly. Dynamic chart projection is required to visualize, while completing the projection transformation based on the next projection datum, which can be made more efficient using multithreaded techniques. The main thread is dedicated to loading the chart data centered on the current projection reference point. It also focuses on providing the navigation-related details that the user is interested in during the navigation process. The sub-thread is applied to load the chart data for the next projection datum. Because the projection area corresponding to the current projection datum covers a larger area than the current screen, fast visualization of the projection area can be achieved by the technique of double buffering.

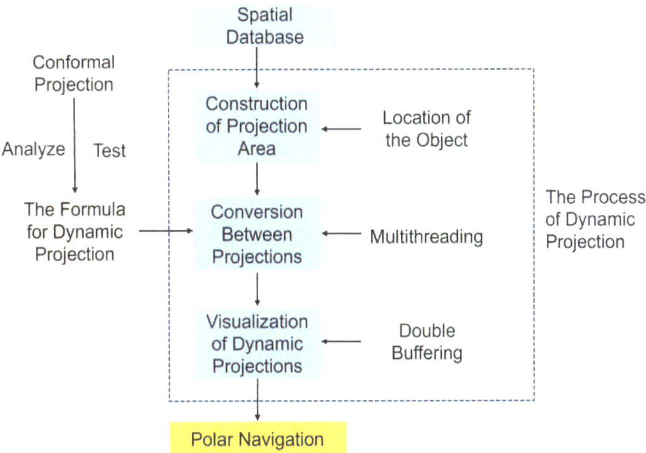

Figure 8. Schematic diagram of the realization process of dynamic chart projection.

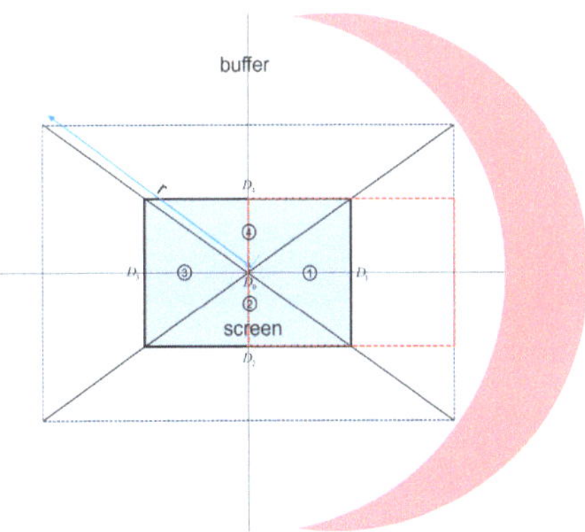

Figure 9. Selection of the projection reference point and the corresponding projection (the dark blue portion of the figure in the thick black rectangular box is the front buffer, which displays the chart information from the ECDIS screen. The light blue portion and the red portion are the back buffers. Assuming that the next position of the ship is located in zone 1, the next projection datum is D_1. The red part is the buffer with D_1 as the projection datum buffer).

6. Conclusions

In order to solve the problem that the projection used in ENCs is not fully applicable to polar navigation, an ENC projection suitable for polar navigation, based on the theory of complex function, is analyzed in detail. Direct transformations of Mercator projection, polar Gauss-Krüger projection, and polar stereographic projection are derived, dynamic projections oriented to polar navigation are designed, and an ENCs visualization idea based on multithread-double buffering is developed. Taking the CGCS2000 reference ellipsoid as an example for calculation, the Mercator projection has less than 10% distortion up to 74° latitude, but distortion exceeds 80% at extreme high latitudes. The maximum distortion of the polar Gauss-Krüger projection does not exceed 10%. Polar stereographic projection is shown to be less than 1% above 79°. From the perspective of longitude and latitude grid lines, the polar Gauss-Krüger projection and the polar stereographic projection differ only to a small extent near the poles. By combining the existing specifications for the compilation and mapping of ENCs, recommendations for the projections oriented towards polar navigation at different scales and in different latitude bands for different applications are given. Taking the transformation between the Gauss-Krüger projection and the polar stereographic projection in the polar region as an example, the computational error of the direct transformation formula is less than 10^{-9} m, and the time does not exceed 0.1 s for the calculation of nearly 8 million points at $1' \times 1'$ resolution, which fully meets the demand of high-precision ENCs for resolution and projection transformation efficiency.

In conclusion, the adverse effects on navigation caused by projection errors in ENCs can be effectively eliminated by the established dynamic projection and visualization methods for polar navigation. By providing high-precision spatial and temporal information services and improving the visualization of navigation software for smart ships, it can better serve polar scientific research, ocean shipping, and other related fields, while also reducing navigation risks.

Author Contributions: Conceptualization, C.J. and X.W.; formal analysis, C.J. and S.B.; funding acquisition, X.W.; investigation, C.J. and H.L.; methodology, C.J., X.W., H.L. and S.B.; resources, C.J. and X.W.; software, C.J. and H.L.; supervision, S.B.; validation, H.L.; visualization, C.J. and S.B.; writing—original draft, C.J.; writing—review and editing, S.B. and X.W. All authors have read and agreed to the published version of the manuscript.

Funding: This work was supported by The National Science Foundation for Outstanding Young Scholars of China (42122025); The National Natural Science Foundation of China (42074010).

Institutional Review Board Statement: Not applicable.

Informed Consent Statement: Informed consent was obtained from all subjects involved in the study.

Data Availability Statement: Data are contained within the article.

Acknowledgments: The earth parameters are from CGCS2000 and the arctic boundary data in this paper come from the National Oceanic and Atmospheric Administration (NOAA), which are public data. Geocartv3.2.0, MATLAB, and Mathematica were used for image presentation and data analysis. We express our sincere gratitude to these organizations and software providers.

Conflicts of Interest: The authors declare no conflicts of interest.

References

1. Yastrebova, A.; Höyhtyä, M.; Boumard, S. Positioning in the Arctic Region: State-of-the-Art and Future Perspectives. *IEEE Access* **2021**, *9*, 53964–53978. [CrossRef]
2. Li, X.; Otsuka, N.; Brigham, L. Spatial and temporal variations of recent shipping along the Northern Sea Route. *Polar Sci.* **2021**, *27*, 100569. [CrossRef]
3. Screen, J.; Simmonds, I. The central role of diminishing sea ice in recent Arctic temperature amplification. *Nature* **2010**, *464*, 1334–1337. [CrossRef] [PubMed]
4. Beveridge, L.; Fournier, M.; Lasserre, F.; Huang, L. Interest of Asian Shipping Companies in Navigating the Arctic. *Polar Sci.* **2016**, *10*, 404–414. [CrossRef]
5. Liang, S.; Zeng, J.; Li, Z. Spatio-temporal analysis of the melt onset dates over Arctic sea ice from 1979 to 2017. *Acta Oceanol. Sin.* **2022**, *41*, 146–156. [CrossRef]
6. Cheng, J.; Liu, J.; Zhao, L. Survey on polar marine navigation and positioning system. *Chin. J. Ship Res.* **2021**, *16*, 16–29.
7. Plass, S.; Clazzer, F.; Bekkadal, F. Current situation and future innovations in Arctic communications. In Proceedings of the Vehicular Technology Conference, Boston, MA, USA, 6–9 September 2015; pp. 1–7.
8. Liu, Z.; Zhang, B.; Zhang, M. A quantitative method for the analysis of ship collision risk using AIS data. *Ocean Eng.* **2023**, *272*, 113906. [CrossRef]
9. Chen, X.; Liu, S.; Liu, R.W. Quantifying Arctic oil spilling event risk by integrating an analytic network process and a fuzzy comprehensive evaluation model. *Ocean Coast. Manag.* **2022**, *228*, 106326. [CrossRef]
10. Yao, Y.; Xu, X.; Li, Y. Transverse Navigation under the Ellipsoidal Earth Model and its Performance in both Polar and Non-polar areas. *J. Nav.* **2016**, *1*, 1–18. [CrossRef]

11. Naumann, J. Grid navigation with polar stereographic charts. *Eur. J. Navig.* **2011**, *9*, 4–8.

12. Lapon, L.; Ooms, K.; Maeyer, P. The Influence of Map Projections on People's Global-Scale Cognitive Map: A Worldwide Study. *Int. J. Geo Inf.* **2020**, *9*, 196. [CrossRef]

13. Baselga, S. Two Conformal Projections for Constant-Height Surface to Plane Mapping. *J. Surv. Eng.* **2021**, *147*, 06020004. [CrossRef]

14. Beresford, P. Map Projection Used in Polar Regions. *J. Nav.* **1953**, *6*, 29–37. [CrossRef]

15. Snyder, J. *Map Projections-A Working Manual*; United States Government Printing Office: Washington, DC, USA, 1987.

16. Grafarend, E.; You, R.; Syffus, R. *Map Projections*; Springer: Berlin/Heidelberg, Germany, 2014.

17. Pearson, F. *Map Projections: Theory and Applications*; CRC Press: Boca Raton, FL, USA, 1990.

18. Smith, L.; Stephenson, S. New Trans-Arctic shipping routes navigable by midcentury. *Proc. Natl. Acad. Sci. USA* **2013**, *110*, E1191–E1195. [CrossRef] [PubMed]

19. Pallikaris, A.; Tsoulos, L. Map projections and visualization of navigational paths in electronic chart systems. In Proceedings of the 3rd International Conference on Cartography and GIS, Nessebar, Bulgaria, 15–20 June 2010; pp. 15–20.

20. Skopeliti, A.; Tsoulos, L. Choosing a Suitable Projection for Navigation in the Arctic. *Mar. Geod.* **2013**, *36*, 234–259. [CrossRef]

21. Peter, O. *The Mercator Projections*; Edinburgh University Press: Edinburgh, UK, 2008; pp. 23–29.

22. Pallikaris, A. Choosing Suitable Map Projections for World-wide Depiction of Electronic Charts in ECDIS. *Coordinates* **2014**, *10*, 21–28.

23. Xiong, J. *Ellipsoid Geodesy*; Chinese People's Liberation Army Publishing House: Beijing, China, 1988.

24. Wang, H.; Zhang, W.; Zhang, P. Polar Navigation Method Based on Polar Stereographic Projection. *Command Control Simul.* **2016**, *38*, 106–109.

25. Wen, C.; Bian, H.; Gao, X. Availability Analysis of Mercator Projection in Nautical Charts' Compilation in Polar Regions. *Appl. Mech. Mater.* **2014**, *535*, 556–561. [CrossRef]

26. Bowring, B. The Transverse Mercator Projection—A Solution by Complex Numbers. *Surv. Rev.* **1978**, *30*, 325–342. [CrossRef]

27. Bian, S.; Li, Z.; Li, H. The Non-singular Formula of Gauss Projection in Polar Regions by Complex Numbers. *Acta Geod. Cartogr. Sin.* **2014**, *43*, 348–352+359.

28. Li, Z.; Bian, S.; Jin, L. Forward and Inverse Expressions of Polar Gauss Projection without Zoning Limitations. *Acta Geod. Cartogr. Sin.* **2017**, *46*, 780–788.

29. Guo, J.; Shen, W.; Ning, J. Development of Lee's exact method for Gauss-Krüger projection. *J. Geod.* **2020**, *94*, 58. [CrossRef]

30. Wen, C.; Bian, H. Polar grid rumble route based on polar stereographic projection. *Acta Geod. Cartogr. Sin.* **2019**, *48*, 18–23.

31. Jenny, B. Adaptive composite map projections. *IEEE Trans. Vis. Comput. Graph.* **2012**, *18*, 2575–2582. [CrossRef]

32. Jenny, B.; Šavrič, B.; Patterson, T. A compromise aspect-adaptive cylindrical projection for world maps. *Int. J. Geogr. Inf. Sci.* **2015**, *29*, 935–952. [CrossRef]

33. Palikaris, A.; Mavraeidopoulos, A. Electronic navigational charts: International standards and map projections. *J. Mar. Sci. Eng.* **2020**, *8*, 248. [CrossRef]

34. Gosling, P.; Symeonakis, E. Automated map projection selection for GIS. *Cartogr. Geogr. Inf. Sci.* **2020**, *47*, 261–276. [CrossRef]

35. Jiao, C. Computer Algebraic Analysis and Optimization of Common Map Projection Deformation. Master's Thesis, China University of Geosciences (Wuhan), Wuhan, China, 2022.

36. Li, X.; Li, H.; Liu, G.; Bian, S. Optimization of Complex Function Expansions for Gauss-Krüger Projections. *ISPRS Int. J. Geo Inf.* **2022**, *11*, 566. [CrossRef]

37. Li, X.; Li, H.; Liu, G.; Bian, S.; Jiao, C. Simplified Expansions of Common Latitudes with Geodetic Latitude and Geocentric Latitude as Variables. *Appl. Sci.* **2022**, *12*, 7818. [CrossRef]

38. Karney, C.F.F. On auxiliary latitudes. *Surv. Rev.* **2023**, *56*, 165–180. [CrossRef]

39. Bermejo, M.; Otero, J. Simple and highly accurate formulas for the computation of Transverse Mercator coordinates from longitude and isometric latitude. *J. Geod.* **2009**, *83*, 1–12. [CrossRef]

40. Wang, K.; Ye, S.; Gao, P.; Yao, X.; Zhao, Z. Optimization of Numerical Methods for Transforming UTM Plane Coordinates to Lambert Plane Coordinates. *Remote Sens.* **2022**, *14*, 2056. [CrossRef]

41. *GB 12320-2022*; Compilation Specifications for Chinese Nautical Charts. Standards Press of China: Beijing, China, 2022.

42. International Hydrographic Organization. *Electronic Navigational Charts (ENC)–Production, Maintenance and Distribution Guidance*; Edition 2.1.0; Publication S-65; IHO: Monaco, France, 2017.

Journal of
Marine Science and Engineering

Article

The Analysis of Intelligent Functions Required for Inland Ships

Guozhu Hao [1,2], Wenhui Xiao [1], Liwen Huang [1,2,*], Jiahao Chen [1], Ke Zhang [2,3] and Yaojie Chen [4]

[1] School of Navigation, Wuhan University of Technology, Wuhan 430063, China; whuthgz@whut.edu.cn (G.H.); xiaowh143@gmail.com (W.X.); cjh255242@whut.edu.cn (J.C.)
[2] Hubei Key Laboratory of Inland Shipping Technology, Wuhan 430063, China; zhangk@wti.ac.cn
[3] China Waterborne Transport Research Institute, Beijing 100088, China
[4] School of Computer Science and Technology, Wuhan University of Science and Technology, Wuhan 430081, China; chenyaojie@wust.edu.cn
* Correspondence: chieryuzhu@126.com

Abstract: Sorting out the requirements for intelligent functions is the prerequisite and foundation of the top-level design for the development of intelligent ships. In light of the development of inland intelligent ships for 2030, 2035, and 2050, based on the analysis of the division of intelligent ship functional modules by international representative classification societies and relevant research institutions, eight necessary functional modules have been proposed: intelligent navigation, intelligent hull, intelligent engine room, intelligent energy efficiency management, intelligent cargo management, intelligent integration platform, remote control, and autonomous operation. Taking the technical realization of each functional module as the goal, this paper analyzes the status quo and development trend of related intelligent technologies and their feasibility and applicability when applied to each functional module. At the same time, it clarifies the composition of specific functional elements of each functional module, puts forward the stage goals of China's inland intelligent ship development and the specific functional requirements of different modules under each stage, and provides reference for the Chinese government to subsequently formulate the top-level design development planning and implementation path of inland waterway intelligent ships.

Keywords: inland intelligent ships; functional module; intelligent technologies; functional requirements

Citation: Hao, G.; Xiao, W.; Huang, L.; Chen, J.; Zhang, K.; Chen, Y. The Analysis of Intelligent Functions Required for Inland Ships. *J. Mar. Sci. Eng.* **2024**, *12*, 836. https://doi.org/10.3390/jmse12050836

Academic Editor: Mihalis Golias

Received: 15 March 2024
Revised: 5 May 2024
Accepted: 15 May 2024
Published: 17 May 2024

1. Introduction

Compared with traditional ships, intelligent ships possess numerous advantages, such as safety, reliability, energy conservation, environmental friendliness, and economic efficiency. With the rapid development and widespread application of technologies such as artificial intelligence, the Internet of Things, cloud computing, and big data, intelligent ships based on digitization and aiming for autonomy have become a new focus in the shipbuilding industry, international shipping, and maritime circles [1]. In recent years, the development of intelligent ships has achieved remarkable results. In January 2022, the Japanese container ship Mikage completed a fully autonomous navigation test from Tsuruga Port in Fukui Prefecture to Sakaiminato Port in Tottori Prefecture over a total distance of about 270 km. In addition to the Automatic Identification System (AIS) of the ship and the radar, the ship was equipped with visual cameras, infrared cameras for nighttime, and an artificial intelligence (AI) learning system for detecting other ships. In April 2022, China's first self-developed autonomous 300 TEU container ship, "Zhi Fei," made its maiden navigation at Qingdao Port, which has three driving modes: manual, remote control, and unmanned autonomous navigation. It is capable of realizing intelligent perception and cognition of the navigation environment, autonomous route planning, intelligent collision avoidance, automatic berthing and unberthing, and remote control navigation. By the end of 2023, the ship had sailed over 20,000 nautical miles, with its intelligent navigation system consistently operating safely. In April 2022, the Norwegian ship "Yara Birkeland" entered commercial operation as the world's first fully electric, unmanned container ship equipped with remote control and autonomous navigation systems. In June 2022, the South Korean super-large natural gas (LNG) carrier "Prism Courage" completed an oceanic intelligent navigation experiment. The same month, the unmanned electric ship "The May Flower" conducted intelligent perception and decision-making with its AI captain and edge computing system during its first fully autonomous transatlantic navigation. In January 2023, the world's first scientific research ship with remote control and autonomous navigation in open waters, "Zhuhai Yun," was delivered for use in Guangzhou, China. It is expected that the International Maritime Organization (IMO) will issue the "Maritime Autonomous Surface Ship Code (MASS Code)" by the end of 2024 and implement it on 1 January 2025. This code is a comprehensive set of regulations tailored for MASS to address issues that existing maritime organization documents cannot adequately address or have not yet addressed for MASS. Reviewing the current research status of major international research institutions in the field of intelligent ships, the development of intelligent ships primarily focuses on ocean-going ships, with relatively few applications in inland ships. There are several reasons for this. Firstly, at the international level, intelligent ship development is still in the early stages of system development and testing, and further refinement, integration, and reliability verification of intelligent ship technology are needed. Secondly, compared to sea navigation environments, inland waterway navigation environments are more complex. From a safety perspective, current inland ship operations still heavily rely on subjective judgments, decisions, and responses based on human experience. Thirdly, compared with the scale of sea ships and marine transportation, the scale of manufacturing and operating bodies of inland ships is small, and there is still a lack of economic capacity and development consciousness in the introduction of intelligent technology.

At present, there are relatively few international studies on intelligent inland ships. This article summarizes the functional classification or grading of intelligent ships by international representatives of classification societies and shipbuilding companies, systematically organizes the technology, and takes into account the specific aspects of inland waterway navigation. It extracts functional modules that meet the development needs of inland intelligent ships and combines them with the current development status and technological forecasts of ship intelligence technology. It proposes functional requirements for the development of China's inland intelligent ships by 2030, 2035, and 2050.

The remaining part of this paper is organized as follows. Section 2 briefly outlines the international classification of smart ship functional modules and analyzes the necessity of functional modules. Section 3 describes in detail the technologies related to smart ships, including intelligent perception technology, intelligent communication technology, intelligent evaluation technology, intelligent decision-making technology, and intelligent control technology. Section 4 gives a prediction of the functional demand for inland waterway smart ships under different stages. Section 5 summarizes the conclusion and describes the future research direction of inland intelligent ships.

2. Analysis of Functional Modules for Inland Intelligent Ships

2.1. Current Classification Status of Functional Modules for Intelligent Ships

Currently, there is no universal consensus among international research institutions regarding the classification of intelligent ship functions or standards for intelligent grades. In December 2015, the China Classification Society (CCS), taking into account both domestic and international experiences in intelligent ship applications and the future direction of ship intelligence, developed and issued the world's first "Rules for Intelligent Ships". Subsequently, it underwent multiple iterations, was updated, and reissued as the "Rules for Intelligent Ships (2024)" in December 2023, which takes safety, economy, high efficiency, and environmental protection as the starting points and introduces the new concept of artificial intelligence. It divides intelligent ship functional modules into eight categories: intelligent navigation, intelligent hull, intelligent engine room, intelligent energy efficiency management, intelligent cargo management, intelligent integration platform, remote control, and autonomous operation [2]. In February 2017, Lloyd's Register issued the "Code for Unmanned Marine Systems," which adopts a system similar to traditional ship regulations. Its chapters are highly consistent with traditional ship regulations and are divided into sections such as structure, stability, control, electrical, navigation, propulsion systems, and firefighting. From the perspective of unmanned operation, the regulation provides corresponding discussions on the scope, purpose, functional objectives, and performance requirements of unmanned systems [3]. In October 2018, Det Norske Veritas (DNV) proposed in its "Class Guideline Smartship" that intelligent ships should possess a total of five intelligent features: enhanced foundation, operational enhancement, performance enhancement, safety and reliability enhancement, and enhanced condition monitoring [4]. In June 2022, the American Bureau of Shipping (ABS) introduced the "Smart Functions for Marine Vessels and Offshore Units," proposing that intelligent ships should include five aspects of intelligent functions: structural health monitoring, machinery health monitoring, asset efficiency monitoring, operational performance management, crew assistance, and functional enhancement [5]. In January 2020, the Japan Ship Classification Society (JSCS) outlined two functional goals for intelligent ships from the perspective of supporting crew operations in its "Guidelines for Automated/Autonomous Operation of ships." These goals include designing and developing unmanned ships and short-distance small ships with the aim of reducing the number of crew members and designing and developing partial automation or remote support for onboard tasks. This guideline does not directly classify the autonomous level of ships but categorizes automation operation systems and remote operation systems from the perspectives of system design, development, installation, and operation [6]. In 2019, the European Union launched the Autonomous Ship Research and Development Program. In this program, the functions of smart ships are typically categorized as autonomous navigation systems, intelligent energy management, intelligent ship operations, communication and remote monitoring, and autonomous safety systems to meet the needs and challenges of autonomous ship navigation [7]. In November 2021, the Netherlands Forum Smart Shipping (SMASH) published the Smart Shipping Roadmap, which sets out a vision for the development of smart shipping in the Netherlands towards 2030. Its short-term goal is to reduce the number of ship drivers through ship automation and intelligent technology and to realize "autonomous human assistance" on ships on a small scale [8]. Furthermore, relevant international shipbuilding enterprises and research institutions have also put forward their respective research focuses in the development of intelligent ships, as shown in Table 1.

Table 1. Summary of intelligent ship specifications of internationally relevant agencies.

Organization	Date	Related Documents	Primary Content
China Classification Society (CCS)	December 2015 (Updated as of December 2023)	Rules for Intelligent Ships 2024	Intelligent Navigation, Intelligent Hull, Intelligent Engine Room, Intelligent Energy Efficiency Management, Intelligent Cargo Management, Intelligent Integration Platform, Remote Control, Autonomous Operation
Lloyd's Register of Shipping (LR)	February 2017	Code for Unmanned Marine Systems	Structure, Stability, Control, Electrical, Navigation, Propulsion System, Firefighting
Det Norske Veritas (DNV GL)	October 2018	Class Guideline Smartship	Enhanced Foundation, Operational Enhancement, Performance Enhancement, Safety and Reliability Enhancement, Enhanced Condition Monitoring

Table 1. *Cont.*

Organization	Date	Related Documents	Primary Content
American Bureau of Shipping (ABS)	June 2022	Smart Functions for Marine Vessels and Offshore Units	Structural Health Monitoring, Machinery Health Monitoring, Asset Efficiency Monitoring, Operational Performance Management, Crew Assistance and Functional Enhancement
Nippon Kaiji Kyokai (NK)	January 2020	Guidelines for Automated/Autonomous Operation of ships	Streamlining and unmanned crewing of small, short-distance ships; automation or remote operation of part of the ship's operations, mainly in support of the crew
European Union (EU)	2019	Autonomous Ship Research and Development Program	Autonomous Navigation Systems, Intelligent Energy Management, Intelli-gent Ship Operations, Communication and Remote Monitoring, Autonomous Safety Systems
Netherlands Forum Smart Shipping (SMASH)	November 2021	Smart Shipping Roadmap	In the short term, focus on reducing the number of ship drivers through ship automation and intelligent technology, and realize "autonomous human assistance" for ships on a small scale
Rolls-Royce	2014	Advanced Autonomous Waterborne Applications (AAWA)	Focusing the functional research and development of intelligent ships on two aspects: firstly, intelligent subsystems, and secondly, realizing the intelligence of the whole ship's platform [9]
Hai Lanxin	2016	Intelligent Ship 1.0 Specialization	Focusing on the development of ship intelligent assisted autopilot system, completed the ship assisted autopilot system with sensing, decision-making and execution functions [10]
Hyundai Heavy Industries Group	2017	Intelligent Ship Program	Focusing on the research and development of intelligent navigation, intelligent berthing and other auxiliary systems for ships [11]

By sorting out the functional module division of intelligent ships of the above eight institutions or organizations, it can be seen that the current mainstream classification of ship autonomy level mainly targets specific functions or operations and elaborates in detail what functions or operations can be realized by the system when it is at different levels of autonomy, covering from manual operation to full autonomy.

2.2. Necessity Analysis of Functional Modules

CCS "Rules for Intelligent Ships (2024)" has a fairly complete framework of intelligent ship specifications and corresponding functional and technical requirements, which is more suitable for guiding the development of intelligent ships on inland waterways in China. As illustrated in Figure 1, the regulations propose eight intelligent modules from a technical perspective, forming a complete system of functional and technical requirements for intelligent ships. However, in the context of inland ship development towards 2030, 2035, and 2050, the actual development status of inland ships needs to be taken into account. For the intelligent hull module, the ship structure serves as the most fundamental system unit and one of the most stable and reliable components of a ship. The current level of modern manufacturing is sufficient to ensure the stable and reliable operation of the hull throughout its lifecycle. There is scarce evidence of inland maritime accidents caused by excessive damage to the ship's hull structure. Therefore, investing excessive research and development costs in the already stable and reliable hull structure in the short term may not be economically justified. As for the remote control and autonomous operation module, the realization of remote control and autonomous operation of the ship not only requires the ship itself to have a high level of intelligence but also must rely on reliable, stable, low-latency means of communication and an intelligent remote control platform, which is the integrated embodiment of the ship end-ship and shore communication-shore control center. The integration of the ship's intelligence and external intelligent technology, not only by the intelligent ship itself, can be realized independently. Moreover, the inland navigation environment has the characteristics of narrow and long water bodies, which are more limited and complex compared to the marine environment. The safety risk of remote control of inland ships is greater when other intelligent functions are not yet mature. At the same time, there are still many problems with the large number of inland ships, complex ship types, generally low level of advanced system equipment, and the overall quality of crew that still needs to be further improved. In addition, compared to ocean-going vessels, inland ships can always maintain short-distance contact with onshore bases during navigation, and the demand for remote control is weaker than that of ocean-going vessels. Therefore, remote control of ships does not have significant economic advantages in the short term [12].

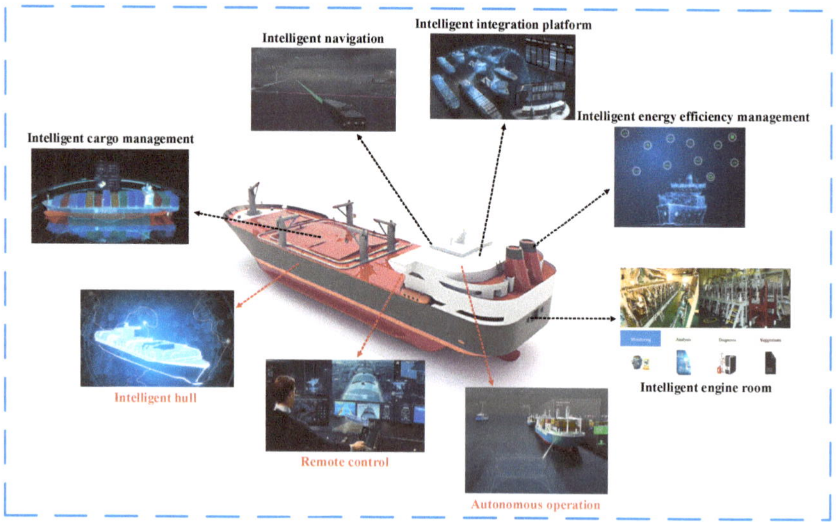

Figure 1. Analysis of functional modules for inland intelligent ships.

3. Intelligent Ship Technology

From the dimension of science and technology, an intelligent ship requires a high degree of integration of information and control, which needs to be supported by a complete technical system. Taking the equivalent replacement of manpower with intelligent functions as the criterion and analyzing it from the perspective of the technical realization path of functional modules, the intelligent ship technology system specifically includes the complete technology chain composed of intelligent perception technology, intelligent communication technology, intelligent evaluation technology, intelligent decision-making technology and intelligent control technology. The details are as follows:

3.1. Ship Intelligent Perception Technology

Perception technology is the basis for realizing ship intelligence, which mainly includes navigation environment information perception, ship state monitoring, information analysis and processing, as shown in Figure 2. Intelligent perception technology utilizes various onboard sensor devices to perceive and monitor information regarding the ship's own state (including dynamic information, hull structural condition, energy efficiency and energy consumption, mechanical equipment and system operation conditions), external environment, cargo and cargo hold condition, etc., and realize the intelligent collection and processing of ship state information data.

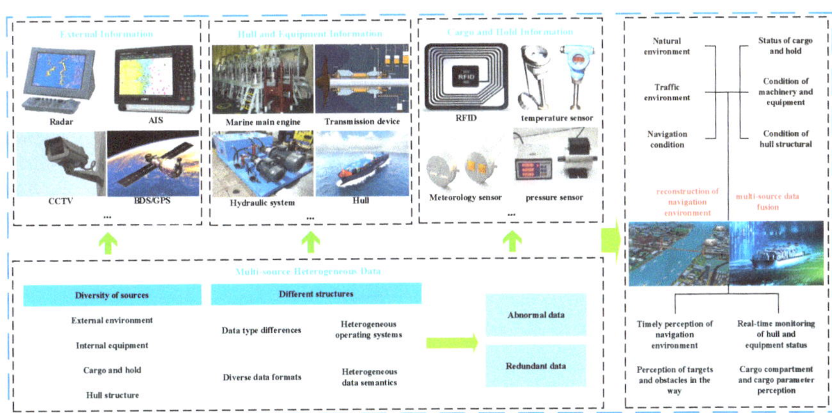

Figure 2. Ship intelligent perception technology diagram.

3.1.1. Navigation Environment Information Perception

Navigation environment information sensing technology mainly utilizes GPS, AIS, radar, depth sounder, motion sensor, wind speed, direction meter, and other sensing devices to obtain various environmental information outside the ship, which provides more reliable data support for subsequent key technologies such as intelligent assessment and intelligent decision-making.

Thompson et al. [13] proposed an efficient LiDAR-based target segmentation method for the marine environment that utilizes 3D occupancy grid segmentation to effectively map large areas. Xu et al. [14] proposed a novel network architecture for small SAR ship target feature extraction and multi-field feature fusion combined with dual-feature mobile processing based on bridge node and feature assumptions, which solved the problem of misdetections and false detections in the detection of small SAR ship targets. Ye et al. [15] proposed an EA-YOLOv4 algorithm with an augmented attention mechanism, which utilizes a convolutional block attention module (CBAM) to search for features in the channel dimension and spatial dimension, respectively, to improve the feature perception ability of the model for ship targets. Hu et al. [16] propose to add the natural image quality evaluation (NIQE) index in the generative adversarial network (GAN) to make the generated image have a better effect than the real image set in the existing dataset, which effectively solves the problems of underwater image distortion, low visibility, low contrast, and other problems.

3.1.2. Ship State Monitoring

Ship condition monitoring involves monitoring the condition of the ship's structure, equipment, and loaded cargo, which can be achieved by collecting parameters related to the ship's hull, engine room, cargo, and energy consumption systems.

Wang et al. [17] introduced machine learning algorithms and proposed a ship's machinery room equipment's condition monitoring method that combines manifold learning and Isolation Forest. By reducing the complexity of the data through dimensionality reduction in raw data, intelligent monitoring of ship machinery room equipment's condition is achieved. Zhuang et al. [18] designed a ship's electromechanical equipment's vibration signal data acquisition system based on wireless sensor networks. This system can collect multi-channel monitoring data in real-time and accurately, improving the accuracy of ship electromechanical equipment's vibration signal detection and analysis capabilities. Hover et al. [19] developed and applied a hull monitoring planning algorithm, dividing the hull into an open hull and complex regions. The open hull part is mapped using integrated acoustic and visual methods, while the complex part achieves high-resolution full imaging coverage of all structures through large-scale planning programs. Zhan [20] developed a ship energy consumption data acquisition and transmission system based on the virtual instrument Labview platform by optimizing the traditional ship energy consumption data acquisition system, which improved the integration degree of the system as well as the accuracy of the energy consumption data. Tong et al. [21] realized the automatic acquisition of parameters such as humidity and gas content of the cabin by installing explosion-proof automatic acquisition and analysis equipment on the dome of the forward part of the cargo hold and achieved the automatic acquisition of parameters such as moisture and gas content of the cabin to reduce the danger of cargo management for LNG carriers. Jiang [22] designed a ship's dangerous goods detection system based on IoT technology, RFID imaging, and other technologies. This system has excellent cargo information acquisition capabilities and can accurately read label information for different categories of dangerous goods for detection. Hu et al. [23] proposed a ship's cargo hold environmental monitoring and control system based on the SSM framework. By integrating Socket monitoring interfaces into the SSM framework, communication between local clients and cloud servers is established, enabling effective monitoring of cargo hold environmental parameters.

3.1.3. Information Analysis and Processing

Information analysis and processing technology involves collecting, organizing, analyzing, and mining data collected through perception using information technologies such as machine learning. This technology helps improve the operational efficiency and safety of ships.

Liu [24] designed and implemented a data conversion algorithm that transforms relational models into XML Schema. Based on this algorithm, a multi-source heterogeneous data integration platform was designed to address the challenges of integrating multiple heterogeneous data sources and dynamic information service patterns in shipping. Chen et al. [25] analyzed the IoT data mining technology of the ship big data platform, reasonably set the content of different functional layers, formed an efficient operation and management chain, and solved the problems of wide data sources and an unstable network of the ship monitoring platform. Liu et al. [26] developed a fuzzy logic-based multi-sensor data fusion algorithm and proposed a two-stage fuzzy logic association method. By integrating it with the Kalman filter, the system's data calculation performance was effectively optimized.

Currently, the application of relevant perception technologies on ships is already widespread. However, in the development process of inland intelligent ships towards 2030, 2035, and even 2050, challenges persist in their application. These challenges include overcoming adverse weather conditions, numerous obstacles in inland waterways, limited detection range, and insufficient accuracy in perception, all of which affect the ability to obtain reliable data. With the enhancement of the ship's sensing ability, the amount of data collected by the ship's sensors will also increase [27], so communication technology needs to be continuously improved with the development of sensing technology.

3.2. Ship Intelligent Communication Technology

Communication technology can further integrate and share the information collected and processed by perception technology, open the channel of information between various systems of the ship, and exchange and communicate the corresponding state information of its own ship with the outside world to realize reliable, stable. and low-latency intelligent data exchange, as shown in Figure 3.

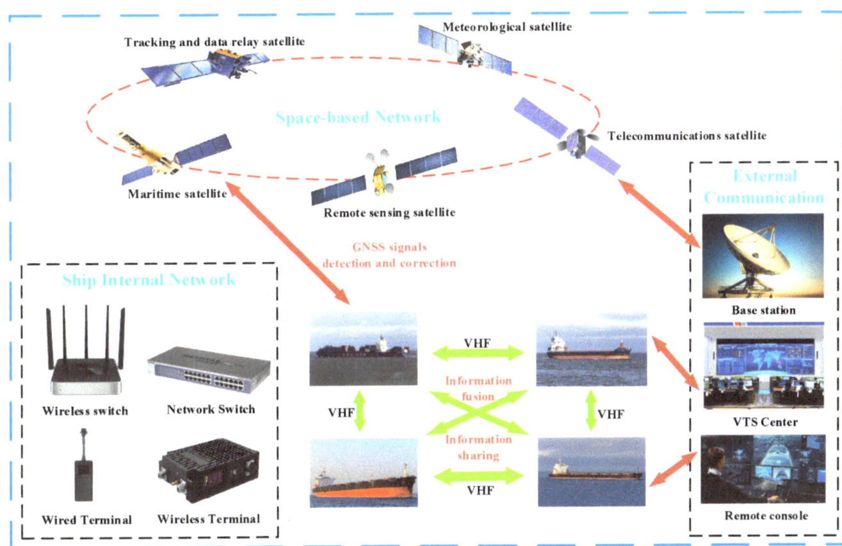

Figure 3. Ship intelligent communication technology diagram.

Zhang [28] proposed an intelligent power allocation algorithm based on deep reinforcement learning (DNN) within the framework of the ship's Internet of Things short packet communication network structure, under the constraint of data transmission security capacity. This algorithm demonstrates good stability in data transmission. Yang et al. [29] introduced a ship communication information transmission channel control algorithm based on an improved bandwidth estimation algorithm. By calculating bandwidth sample values and updating the filtering bandwidth sample value thresholds, intelligent control of the large data transmission channel was achieved. Yoo et al. [30] proposed a distributed state quantization formation design method under a directed network for a low-complexity designated performance control scheme, realizing quantization communication. Cai et al. [31] proposed a marine IoRT system based on deep reinforcement learning and a GEO/LEO heterogeneous network for IoRT data collection and transmission. Data are forwarded to the ground data center through satellite links, achieving seamless coverage and capacity expansion.

Based on the various information data collected by perception technology, communication technology needs to break through the barrier of mutual independence of the data of various systems and equipment and centralize and integrate the information on the external environment, the ship, and the condition of the cargo in order to comprehensively improve the operating efficiency of the ship. In addition, on the inland ships, the fusion of information data and other ships or shore facilities for data exchange, can be more effective coordination between ships in a variety of navigational environments, collision avoidance operations, and ship route planning, berthing and unberthing, loading and unloading of goods and other operations. However, considering the high dynamic changes in the navigation environment of intelligent ships and the large amount of sensor data, etc., the performance of the 5G mobile communication system, which is being vigorously deployed by various countries, is still difficult to satisfy the requirements of intelligent communication for ships; therefore, the research on intelligent communication technology should also focus on new methods and new technologies [32].

3.3. Ship Intelligent Evaluation Technology

Evaluation technology can realize the use of computers to simulate human perception, analysis, thinking, decision-making, and other processes to cognitively calculate the relevant data of each system and the uncertainty, imprecision, and partially real problems, and then make evaluations and give relevant suggestions. It mainly includes navigational posture assessment, hull structure condition assessment, energy consumption and energy efficiency condition assessment, cargo and cargo hold condition assessment, fault diagnosis, etc., as shown in Figure 4.

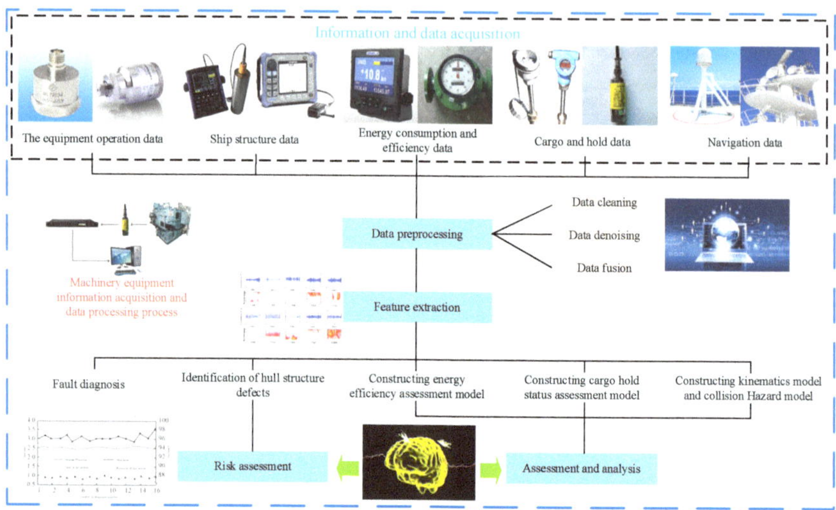

Figure 4. Ship intelligent assessment technology diagram.

3.3.1. Navigational Posture Assessment

Realizing real-time assessment of ship navigation posture is one of the keys to enhancing the navigation safety of intelligent inland ships. Through the equipment information collected by perception technology or the system data obtained by communication technology fusion, using deep learning and other intelligent assessment technology to carry out cognitive computation on the navigation environment in the waters and the attitude of the ship, to make judgments on the real-time encounter situation of the ship, and to make predictions and assessments of the encounter situation in a short period of time in the future through the training of adopting different collision avoidance measures in the current situation and to give appropriate maneuvering suggestions.

Cheng et al. [33] proposed a fuzzy logic model to estimate the adaptability risk of crossing gaps and conducted a collision risk assessment of conflict points in the scenario. The model takes into account the relationship between the adaptive risk of the selected crossing gap and the collision risk of the conflict point, which can reduce the ship operation risk from the source of risk development. Bi et al. [34], using the alpha-shape algorithm and Voronoi diagram, categorized safety assessment indicators for coastal waters into five risk levels: very low, low, moderate, high, and very high. They then applied entropy weight theory to calculate the weights of evaluation indicators, establishing a model for assessing safety risks in coastal waters that fully considers the impact of objective factors and the uncertainty of safety assessment indicators. Chen et al. [35] introduced fuzzy theory into the ship risk assessment model, which can adapt well to changes in heading, speed, and position, to some extent addressing the problem of poor comprehensive evaluation and collision avoidance effectiveness of ships. Xu et al. [36] proposed an intelligent hybrid collision avoidance algorithm based on deep reinforcement learning that can accurately judge the collision situation, give reasonable collision avoidance actions, and realize effective collision avoidance in the complex environment of dynamic and static obstacles. Xin et al. [37] expanded the application of complex network theory and node deletion methods, quantifying the interactions and dependencies between multiple ships in collision scenarios and enabling collision risk assessment at any spatial scale.

3.3.2. Hull Structure Condition Assessment

Ship structural condition assessment involves real-time monitoring of dynamic parameters of the ship's structure to identify key information, such as stress distribution and fatigue damage, and to perform condition assessment and prediction. This technology can effectively enhance ship safety, navigation efficiency, and operational efficiency.

Akpan et al. [38] modeled the corrosion growth as a time-varying stochastic function of the hull structural component thickness reduction over time, used the second-order reliability method (SORM) to calculate the instantaneous reliability of the main hull structure, and put forward a time-varying reliability calculation method for the corrosive structure of the ship based on the hazardous rate function. Wang [39] took the structural condition monitoring system of a new type of polar ship as the research object and gave a set of reasonable measurement point arrangements, load inversion, and strength assessment scheme through theoretical research and computational analysis. Liu [40] based on the design technical indexes of ultra-large ships and the operational characteristics of the actual ship, and combined with all kinds of norms and standards, established an assessment system of the impact of the effects of torsion, thumping vibration, and other effects on the structural safety of the ship's hull, which fills in the blanks of the monitoring norms and assessment standards of the ultra-large ships. Lang et al. [41] established a fatigue assessment model for 2800 TEU container ship based on measured data using machine learning technology, which can accurately capture the nonlinear increase in fatigue. Compared with the traditional spectral method, this method can realize more accurate monitoring of ship fatigue damage.

3.3.3. Energy Consumption and Energy Efficiency Condition Assessment

Ship Energy Consumption and Energy Efficiency Condition Assessment is to realize real-time analysis and assessment of ship energy consumption and energy efficiency by integrating sensors, data acquisition, processing, and intelligent analysis technologies to provide decision-making support for ship management. This technology is of great significance for reducing operation cost, saving energy, and reducing emissions.

Wang [42] adopted the fuzzy set analysis method to form a set pair for ship energy consumption and ideal operating conditions. Comprehensive evaluation results were obtained from the analysis of proximity, uncertainty, and trend levels, addressing the fuzziness and uncertainty issues of factors affecting ship energy consumption evaluation. Fan et al. [43] considered the stochastic nature of environmental parameters and established a new ship energy efficiency model based on the Monte Carlo simulation method, which was applied to ship performance simulation. This facilitated ship managers in evaluating maritime ship energy efficiency, thereby promoting energy conservation and emission reduction in the shipping industry. Wang et al. [44] used Long Short-Term Memory (LSTM) neural network with better prediction performance for a sequential dataset to establish a ship energy consumption prediction model and used a genetic algorithm to optimize the network structure and hyper-parameters, which greatly improved the prediction accuracy of the energy consumption model, which is of great significance for the optimization and improvement of the energy efficiency of ships.

3.3.4. Cargo and Cargo Hold Condition Evaluation

The assessment of cargo and cargo hold conditions primarily relies on the coordinated operation of various sensors and sensing systems. Through real-time monitoring, data processing, predictive assessment, and other means, it achieves a comprehensive understanding and effective management of the condition of cargo and cargo holds.

Gao et al. [45], based on embedded development, collect real-time data on humidity, temperature, oxygen concentration, smoke concentration, and cold well liquid level inside the cargo hold, detect the condition of hatch closure, and capture real-time video information inside the cargo hold, designing a comprehensive cargo hold monitoring system. Lan et al. [46], based on the risk transmission model of the cargo transportation process chain, constructed a model system for risk analysis and quantitative assessment of cargo transportation based on real-time dynamic big data fusion technology, thereby achieving informatization and modern intelligent supervision of the entire process and all aspects of cargo. He [47] proposed a weighted average combined assessment method for the security of critical information in ship cargo hold surveillance video, which uses BP neural network, support vector machine, and extreme learning machine to assess the security of critical information in ship cargo hold surveillance video and improves the results of the assessment of the security of critical information in ship cargo hold surveillance video.

3.3.5. Fault Diagnosis

Fault diagnosis involves analyzing relevant data from various systems to determine if they are in a stable condition. If the equipment is found to be in poor condition, corresponding evaluations are made based on the type and severity of the fault, and alerts or warnings are issued accordingly.

Jiang et al. [48] divided the intelligent fault diagnosis of ship power units into three key stages: data signal acquisition, data feature extraction, and fault identification and prediction. They proposed that the goal should be to achieve condition-based maintenance and health management of ship power units and advocated for establishing a cloud-based data monitoring system. Cheng [49] applied the artificial neural network algorithm from the artificial immune algorithm to diagnose faults in ship electronic equipment, addressing the inefficiency of manual fault diagnosis methods. Ozturk et al. [50] proposed an intelligent fault diagnosis system for ship mechanical systems using a classification tool based on support vector machine principles. Liu et al. [51] designed a convolutional neural network intelligent fault detection optimization algorithm based on frequency domain information features. The algorithm detects a small number of false alarms, but the detection effect is significantly improved compared with the previous one, which can provide a valuable reference for robust fusion of sensors on surface ships. Tang et al. [52] developed a ship engine room remote fault diagnosis system based on a hybrid B/S and C/S architecture for ship power unit fault diagnosis. This system is stable, reliable, and accurate in fault diagnosis, providing a promising solution for the development of intelligent ships.

In the actual operation of inland ships, the ships are less manned, and the assessment of ship intelligence will be beneficial to improve the operational efficiency of the equipment and the navigation safety of the ship, discover the negligence that the manpower fails to discover in time, reduce the maintenance cost, and guarantee the safe operation of the ship. In the subsequent development of ship intelligence, the comprehensive use of intelligent databases and intelligent machine computing, fully expanding the application of artificial intelligence technology in intelligent ships, integrating the advantages of more advanced computer technology [53], more sophisticated instruments, and more efficient expert systems [54], thus enhancing the intelligence level of ship assessment technology.

3.4. Ship Intelligent Decision-Making Technology

Decision-making technology can combine the recommendations made by the assessment technology, comprehensively apply artificial intelligence technologies such as expert databases, neural networks [55], genetic algorithms, machine learning [56], etc., and retrieve databases from the ship or the shore at the same time, intelligently correlate the reasoning and analysis of the ship's historical data, automatically calculate the optimal solution of each solution, and provide the function of visualization and human–machine interaction so as to assist the decision making of the crew or autonomous decision making. It also provides visualization and human–computer interaction functions to assist the crew in decision-making or autonomous decision-making, and realizes the goal of keeping the related systems and equipment in efficient operation. Ship intelligent decision-making technology includes navigation assistance decision-making, hull and equipment maintenance assistance decision-making, energy efficiency management assistance decision-making and intelligent loading assistance decision-making. Among them, the hull structure and equipment are more stable and reliable, so the current research on intelligent decision-making technology mainly focuses on navigation-related decision-making, i.e., route planning and intelligent collision avoidance, as shown in Figure 5.

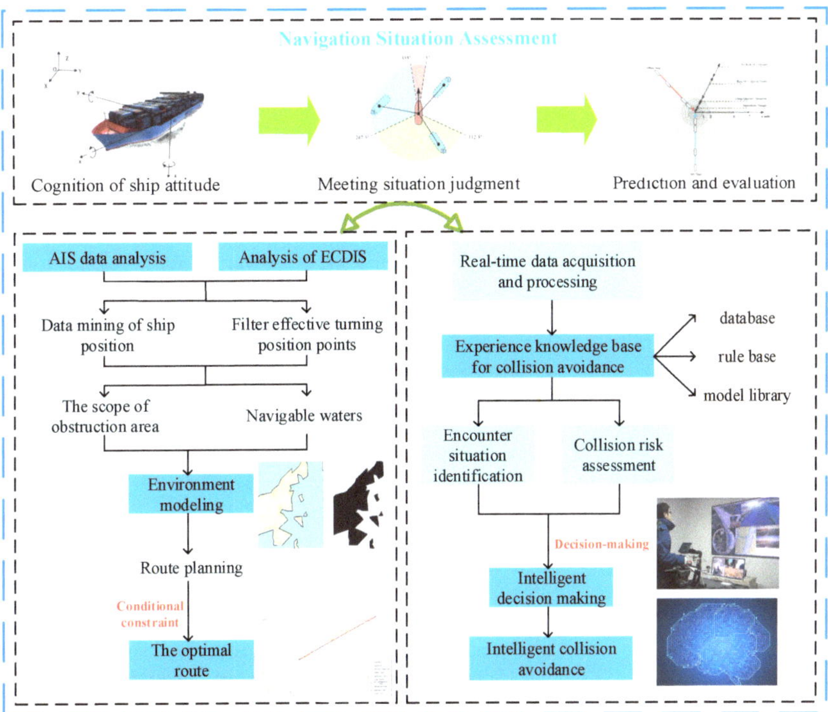

Figure 5. Ship intelligent decision-making technology diagram.

3.4.1. Route Planning

Route Intelligent Planning is a navigation method that obtains real-time traffic and environmental information about the water ahead through various sensors, such as AIS and radar, during navigation and then intelligently selects the ship's position and course within the waterway to optimize the route for safe, efficient, and environmentally friendly navigation.

Liu et al. [57] proposed a hybrid heuristic approach integrating genetic algorithms and particle swarm optimization algorithms to enhance the accuracy and robustness of route planning in constrained waterways. Pan et al. [58] utilized the Delaunay triangulation algorithm to devise a method for adjusting the weights of navigable networks under environmental disturbances, enabling the planning of suitable routes for ships of different scales in various environments. Liang et al. [59] introduced an efficient, robust, adaptive, and implementable route planning algorithm based on leader-node ant colony optimization and a trajectory maintenance control algorithm using nonlinear feedback, enhancing both the efficiency and safety of ship navigation. Ma et al. [60] employed a hierarchical mapping method to separate the decision layer from the weather information layer, directly obtaining route and speed decision schemes conforming to ship maneuverability and crew habits. They proposed a new strategy that simultaneously optimizes ship routes and speeds, significantly reducing the cost of route generation. Zhou et al. [61] proposed a ship path planning method based on historical trajectory data and the SARIMA model, effectively addressing ship collision issues caused by buoy displacement in the navigation of large, slow autonomous ships.

3.4.2. Intelligent Collision Avoidance

Intelligent collision avoidance is the core issue of intelligent ship navigation. Based on the historical trajectory and position, the ship predicts the future course of the ship, evaluates the risk of collision according to the actual situation, and chooses the best time to carry out the automatic collision avoidance operation in order to complete the safe avoidance.

Zhang et al. [62–64] proposed an autonomous decision-making model for complex encounter situations of multi-ship based on the deduction of ship maneuvering process, model predictive control (MPC), modified velocity obstacle (VO) algorithm, and grey cloud model. Wang et al. [65] employed the Twin Delayed Deep Deterministic Policy Gradient (TD3) reinforcement learning algorithm to address the coordinated control problem of unmanned surface ships (USVs) regarding speed and heading. The TD3-IC controller exhibits outstanding robustness and adaptability even in the presence of disturbances and variations in USV parameters. Li [66] combined deep reinforcement learning algorithms with autonomous navigation decision-making technology, proposing an autonomous navigation decision-making algorithm based on an improved Deep Q-Network (DQN) model. By establishing a reasonable state space, a discrete action space, and a comprehensive reward function, this algorithm enhances autonomous navigation decision-making. Xu et al. [67] tackled the intelligent collision avoidance decision-making problem in multi-ship encounters, presenting an improved Sparrow Search Optimization algorithm based on Gaussian mutation and Tent chaotic mapping. This algorithm efficiently finds collision avoidance paths that are safe and economical, providing collision avoidance decision-making references for ship navigators. Zhao et al. [68] devised an autonomous collision avoidance algorithm suitable for intelligent ships based on navigation experience. By constructing a dynamic collision avoidance knowledge base, this algorithm automatically acquires dynamic collision avoidance knowledge, estimates real-time danger assessment thresholds, generates, verifies, and optimizes collision avoidance decision implementation schemes, thus achieving autonomous collision avoidance in intelligent ship navigation. Wang et al. [69] introduced an unmanned ship collision avoidance method based on deep reinforcement learning, incorporating the MMG model to account for ship maneuvering characteristics. This method ensures autonomous collision avoidance in complex environments while complying with COLREG regulations. Wang et al. [70] proposed an autonomous collision avoidance sequential decision-making chain construction method based on humanoid thinking. The construction process involves situational awareness, collision risk identification, collision avoidance rule library and strategy set construction, humanoid thinking sequence collision avoidance strategy generation, collision avoidance process monitoring and strategy adjustment, and restoration of navigational conditions, thereby enhancing the collision avoidance decision-making capability of unmanned ships in multi-ship encounter scenarios.

In the process of intelligent development of inland ships, intelligent decision-making technology can effectively reduce the labor intensity of crew members, mitigate various risks caused by human operational errors, and enable crew members to make the most accurate judgments and achieve the most reliable decisions in the shortest possible time. In order to realize the change in decision-making mode from driver to human–machine integration, the main body of ship control should fully understand and master the information from various sources and make efforts to improve the reliability of the navigational environment situation and the assessment of the operation condition of the system and equipment. Through in-depth research on quantitative techniques of collision avoidance rules and good seamanship, the safety and reliability of intelligent decision-making on ships can be improved.

3.5. Ship Intelligent Control Technology

The control technology can realize intelligent control such as route speed optimization, system and equipment maintenance, energy efficiency control, and automatic loading/unloading based on the optimized loading/unloading scheme under different navigation scenarios and complex environmental conditions through the corresponding decision-making scheme, as shown in Figure 6.

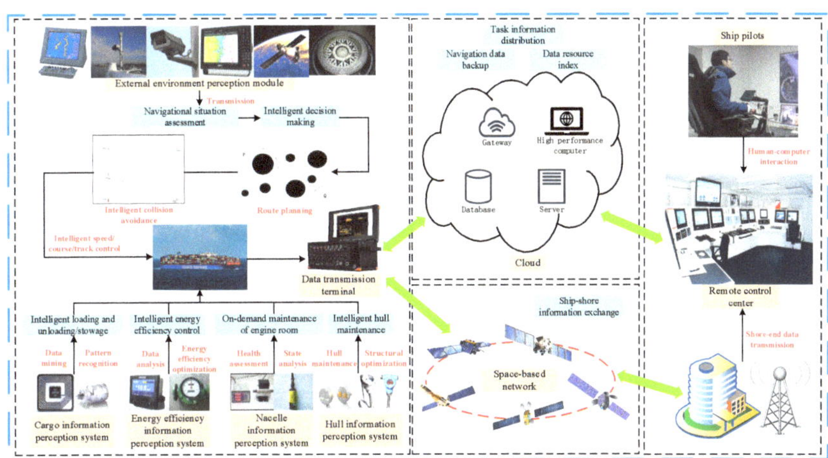

Figure 6. Ship intelligent control technology diagram.

3.5.1. Motion Control

Intelligent ship motion control refers to the use of intelligent technology and electronic technology to control the ship's motion, including heading, speed, attitude angle, and other parameters, in order to improve the ship's automation and intelligence level.

Wu et al. [71] employed the MMG separation model to establish a mathematical model for twin rudder twin propeller interference ships under marine navigation conditions. They proposed an improved practical tool for nonlinear control systems called CMAC-VPID, achieving trajectory control for twin-rudder twin-propeller ships. Teng et al. [72] proposed a hierarchical model predictive control (H-MPC) tracking method combining model predictive control (MPC) and hierarchical control, which effectively achieves intelligent ship tracking of target ships. Considering wind and wave disturbances, Liu et al. [73] transformed them into equivalent rudder angles generated by the ship and proposed an ITCA algorithm combining the Nomoto model and sliding mode control, capable of automatically stabilizing berthing according to the ideal heading angle in undisturbed conditions. Li [74] addressed the design problem of the guidance subsystem under time-varying environmental disturbances and proposed an integral parameter adaptive ILOS guidance law, providing a rational and effective theoretical optimization method for the practical application of intelligent ships in engineering. Pham et al. [75] combined neural networks with fuzzy logic control to design an autonomous ship steering system operating in disturbed environments. They selected an ANFIS controller, significantly enhancing system stability and trajectory accuracy.

3.5.2. Remote Control

Remote control refers to the use of technologies such as positioning navigation, auxiliary control, and beyond visual line of sight operation to manipulate ships from an onshore control center or other remote locations.

Yoshida et al. [76] improved the methodology for developing a regulatory framework for RO capability by applying a typical case to a remote control system based on the previous work. It identifies the trend of ship-perceived failures and provides the required information and additional requirements. Basnet et al. [77] integrated models such as Noisy-OR gates, Parent-divorcing, etc., and proposed a new risk analysis method by combining the improved STPA with BN, which can provide reliable support for real-time decision-making by remote pilotage. Chen et al. [78] proposed a delay-compensated state estimation method for remotely controlled ships with uncertain delay navigation measurements. This method effectively improves the stability and effectiveness of remote control.

3.5.3. Energy Efficiency Control

Energy efficiency control refers to the optimal management of energy consumption and emissions of ships through intelligent technical means, which is an important means to realize energy savings and emission reduction in ships and improve operational efficiency.

Perera et al. [79] identified potential energy-saving scenarios during ship operation based on proposed energy flow pathways. They also discussed the use of appropriate navigation strategies within designated ECAs to reduce exhaust emissions. Wang et al. [80] established dynamic optimization models for ship energy efficiency, considering time-varying environmental factors and non-linear ship energy efficiency systems, and designing control algorithms and controllers for dynamic optimization of ship energy efficiency. This method effectively enhances ship energy efficiency and reduces CO_2 emissions. Guo [81] aimed to reduce the energy efficiency ratio of ship energy efficiency control systems effectively. He established a distributed ship energy efficiency data collector using distributed data collection technology to optimize the distribution of energy efficiency data collection resources. Chen et al. [82] developed a hybrid optimization algorithm combining the chaos algorithm with GWO to design a nonlinear model predictive control energy management strategy. This strategy maximizes optimality to achieve rational energy distribution.

3.5.4. Automatic Loading and Unloading and Intelligent Stowage

Automatic loading and unloading along with intelligent cargo stowage refer to the idea and method of simulating the dock dispatcher through artificial intelligence algorithms, etc., and realizing the automation and intelligence of ship loading, unloading, and dispensing by taking into account the situation of the equipment, the state of the stacks, etc.

Qin et al. [83] proposed a pseudo-gradient estimation algorithm based on the multi-innovation theory and a model-free control law based on multi-innovation. They developed specifications for open-loop control system based on the characteristics of liquid cargo loading and unloading systems and constructed the software and hardware framework OCSSLA for liquid cargo loading and unloading systems, which partially addressed key issues in intelligent control systems for liquid cargo loading and unloading. Liu [84] considering both ship safety and economy, established a multi-objective constrained optimization mathematical model for bulk carriers with ship trim control as a constraint. The intelligent loading of bulk carriers fully utilizes the computational power of computers, compensating for the deficiencies of traditional manual loading. The loading plans provided effectively improve the longitudinal force situation of the ship. Wang [85] based on practical work experience, proposed a method to inspect ship loading conditions using NAPAMANAGER, achieving automatic inspection of ship loading conditions.

3.5.5. Hull and Equipment Maintenance as Needed

Maintenance according to the condition of the hull and equipment, that is, based on the assessment results, to develop a reasonable safety management maintenance plan and optimization program, and through the maintenance management system to achieve effective maintenance of the hull and equipment. It is of great significance in reducing maintenance costs, improving navigation safety, and extending the life of equipment.

Hou et al. [86] utilized graph theory to establish a mathematical model of diesel engine systems. Building upon intelligent damage assessment and reconstruction algorithms, they designed an intelligent damage assessment and reconstruction system for damaged diesel engine systems based on UML and implemented it in the VB programming environment. This significantly enhances the usability, survivability, and safety of ships. Yan [87], through the analysis of vibration signals from diesel generators, constructed a probabilistic neural network model for identifying faults in diesel generators. Based on the health assessment and fault identification of ship diesel generators, intelligent operational maintenance management strategies were formulated, establishing a modern ship health management and intelligent operation and maintenance system. Hong et al. [88], leveraging digital twin technology and big data analytics, implemented a remote operation and maintenance system for ship intelligence. They established a comprehensive analysis of real-time operating condition, fault prediction, and situational maintenance virtual interaction models for key ship equipment, forming an effective equipment health management system.

For the action program after decision-making, the control technology will be able to execute specific operations to practice. The research on inland intelligent ships is by no means an overnight success but requires long-term R&D investment and experience accumulation, and the degree of intelligence of the ship will be gradually and progressively enhanced. In the recent development process, the crew in the ship to support decision-making and control will still be the main way, intelligent control technology or will rely on robots, drones, and other advanced technology [89], by being able to replace some of the manpower as the goal and then realize autonomous decision-making and control.

Currently, with the continuous development of artificial intelligence, big data, the Internet of Things, and other high-tech scientific and technological fields, the application of intelligent ship technology needs to follow the trend of modern science and technology development, combined with the existing advanced technology to carry out a wider range of in-depth use and integration, and continue to create a more practical and efficient new technologies to achieve the full coverage of the intelligent ship functional requirements of the technology chain, this can be specified by the illustration shown in Figure 7.

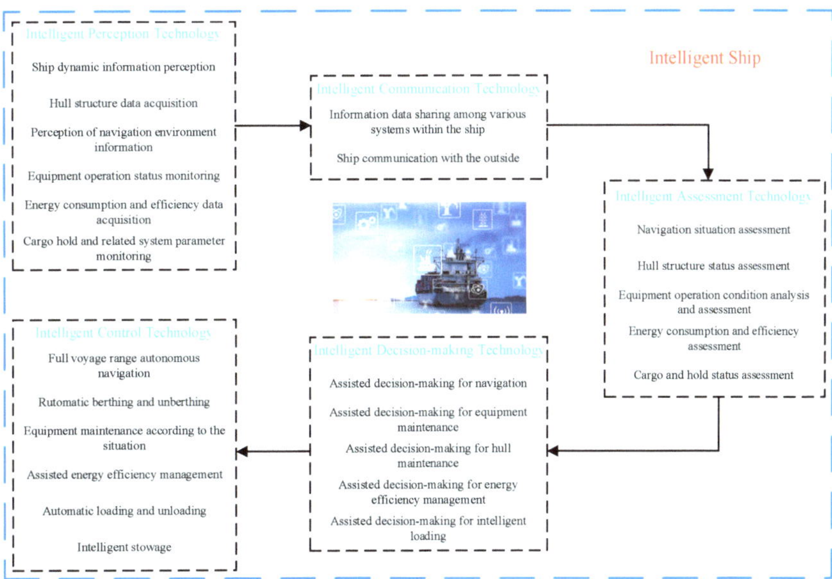

Figure 7. Intelligent ship technology chain.

4. Inland Intelligent Ship Functional Requirements Forecast

4.1. Association between Inland Intelligent Ship Functions and Intelligent Technologies

The intelligent function of inland ships is the application of intelligent ship technology. The development of intelligent technology requires the accumulation of time, and the development of intelligent ships must also follow the objective law of gradual and phased development. The development of inland intelligent ships must be based on the objective development law of intelligent technology: the degree of intelligence from assisted decision-making gradually transitions to full autonomy, and the scope of intelligence follows the direction of development from individual systems to parts of the system and then to the whole ship.

4.2. Overall Development Goals of Inland Intelligent Ships

China generally formulates a development strategy for a period of five years. Currently, China is in the final stage of the 14th Five-Year Plan (2020–2025), so we chose the next Five-Year Plan (2026–2030) as the first stage of the development of inland intelligent ships. At the same time, the Chinese government has issued development planning documents such as the "Outline for the Development of Inland Transportation [90]" and the "Implementing Opinions on Accelerating the Green and Intelligent Development of Inland Ships [91]", which sets strategic development goals related to inland shipping and intelligent inland ships for 2030, 2035, and 2050. Therefore, we chose 2030 as the short-term development planning node for inland intelligent ships and 2035 and 2050 as the mid-term and long-term planning nodes, respectively.

Based on the current situation of China's inland ships, the development plan for inland shipping and intelligent ships formulated by the Chinese government, as well as domestic and international trends in intelligent technology. Based on normative technical documents such as the "MASS Code "and the "Rules for Intelligent Ships (2024)", we make functional predictions for intelligent inland ships in China at different stages of development.

The development of inland intelligent ships in the near future (to 2030) will still be dominated by modularized intelligent aids and focus on the five key functional modules, namely, intelligent navigation, intelligent engine room, intelligent energy-efficiency management, intelligent cargo management, and intelligent integration platform, in order to ensure efficient utilization of scientific and technological resources. In the medium term (towards 2035), the functions of each module should be more perfect, stable, and reliable, and progress has been made in the intelligent hull module, which can realize the linkage of multiple intelligent modules of the ship and the formation of a new shipping industry characterized by full intelligence. In the long-term development plan facing 2050, remote control and fully autonomous control functions will be realized, and the eight functional modules mentioned in the "Rules for Intelligent Ships (2024)" will be comprehensively covered so as to form the global intelligence and high degree of intelligence of inland ships, as shown in Table 2.

Table 2. Overall prediction of functions for inland intelligent ships.

Development Stage	Planning Year	Overall Prediction of Intelligent Functional Goals	Functional Module
Near Term	2030	Modular intelligence, primarily at an initial level with assistance	Intelligent Navigation, Intelligent Engine Room, Intelligent Energy Efficiency Management, Intelligent Cargo Management, Intelligent Integration Platform
Mid Term	2035	Multi-intelligent module linkage, intelligent level advancement	Addition of intelligent hull module
Long Term	2050	Global intelligence, highly intelligent	Addition of remote control and autonomous operation modules

4.3. Specific Development Goals and Predictions for Inland Intelligent Ships at Different Stages

In the near-term development of inland intelligent ships towards 2030, inland intelligent ships should be able to realize local modularized intelligence, and the required functional modules mainly focus on navigational safety and green and efficient operation, which specifically include five functional modules: intelligent navigation, intelligent engine room, intelligent energy efficiency management, intelligent cargo management, and intelligent integration platform, as shown in Table 3.

Table 3. Forecast of inland intelligent ship functionality requirements for 2030.

Functional Module	Goal of Intelligence	Division of Intelligent Functional Stages
Intelligent Navigation	Achieving reliable ship situational awareness and environmental information perception, equipped with route planning functionality	(1) Intelligent perception function (2) Monitoring function for ship's floating state and dynamic motion (3) Design optimization function for route and speed
Intelligent Engine Room	Achieving monitoring, diagnosis, and evaluation of ship engine room condition and equipment operation, and providing intelligent decision support and maintenance plans based on problem types	(1) Engine room condition monitoring function (2) Engine room equipment health assessment function (3) Engine room auxiliary decision-making function (4) Engine room condition-based maintenance function (5) Remote control of the main propulsion device from the wheelhouse and periodic unmanned watchkeeping capability
Intelligent Energy Efficiency Management	Assessing the ship's energy efficiency, navigation, and loading condition to provide evaluation results and solutions such as speed optimization and optimal loading based on longitudinal trim optimization	(1) Ship energy efficiency online intelligent monitoring function (2) Ship speed intelligent optimization function (3) Optimal loading function based on trim optimization
Intelligent Cargo Management	Monitoring of cargo condition onboard and related systems, combined with the ship's cargo condition and port terminal condition, to achieve formulation and optimization of loading/unloading plans, as well as process risk alerting and decision-making	(1) Function of sensing the condition of cargo, cargo holds, and related systems (2) Function of formulating and optimizing cargo loading/unloading plans (3) Function of alarm for abnormal states, analysis of causes, and formulation of assisted decision-making
Intelligent Integration Platform	Complete the standardization of interface types for various intelligent modules within inland ships, enabling the integration of existing module information of intelligent ships, with the integration platform being open-ended	(1) Integration of local area network systems within the ship (2) Formation of a unified digital twin system by various intelligent modules (3) Preliminary data processing function (4) Integration of information and data between existing modules

In the mid-term development phase of inland intelligent ships towards 2035, it is predicted that it will mainly realize the perfection of single-module intelligent function and multi-modular intelligent function linkage. The functional modules at this stage mainly include intelligent navigation, intelligent hull, intelligent engine room, intelligent energy efficiency management, intelligent cargo management, and intelligent integration platform, as shown in Table 4.

Table 4. Forecast of inland intelligent ship functionality requirements for 2035.

Functional Module	Goal of Intelligence	Division of Intelligent Functional Stages
Intelligent Navigation	Achieve the intelligent motion requirements of various ships for safe and efficient navigation, anchoring, and berthing/departing in various navigation scenarios	(1) Integration of multiple functions of the 2030 intelligent navigation module (2) Autonomous navigation function for regular routes (3) Fully autonomous navigation function for the entire navigation (4) Automatic berthing and unberthing
Intelligent Hull	Achieve three-dimensional modeling and maintenance of the hull, providing auxiliary decision-making for the maintenance and replacement of hull and deck machinery during the operational phase of the ship	(1) Hull structure and deck machinery monitoring function (2) Formulation of hull structure and deck machinery maintenance plans (3) Record and evaluation of hull structure condition (4) Formulation of structure replacement plans
Intelligent Engine Room	Achieve fully autonomous operation and realize the goal of a fully intelligent engine room system	(1) Integration of multiple functions of the 2030 intelligent engine room module, adapting to the development of engine rooms in new energy-powered ships (LNG, electric, etc.) (2) Continuous normal operation of engine room equipment within unmanned duty cycles
Intelligent Energy Efficiency Management	Achieve real-time monitoring, evaluation, and optimization of ship energy efficiency, realizing the goal of complete intelligence	(1) Integration of multiple functions of the 2030 intelligent energy efficiency management module (2) Fully automated energy efficiency management
Intelligent Cargo Management	Implement fully intelligent cargo management, including automatic generation and optimization of cargo stowage plans, as well as autonomous loading and unloading	(1) Integration of multiple functions of the 2030 intelligent cargo module (2) Automatic generation and optimization of cargo loading plans (3) Automatic loading and unloading functions (4) Intelligent ballast water management functions
Intelligent Integration Platform	Integrate the newly added information management system with the capability of data exchange among multiple modules	(1) Integration of multiple functions of the 2030 intelligent integration platform module (2) Ship-shore information data communication (3) Information data communication among multiple modules

In the long-term development phase of inland intelligent ships towards 2050, according to the current pace of development in artificial intelligence, the Internet of Things, big data, and other technologies, it is expected that significant progress will have been made in inland intelligent ships. It is anticipated that the ultimate functions of remote control and autonomous operation will be achieved. Remote control of ships refers to the ability of a ship to be controlled by a remote control station or position outside the ship, enabling the ship's operation. Autonomous operation of ships refers to the ability to achieve fully autonomous operation in open waters or throughout the entire navigation without the need for onboard crew operation. Both functionalities require the foundation of the aforementioned intelligent modules to fulfill their roles effectively. The functional modules at this stage mainly include the intelligent hull, intelligent integration platform, remote control, and autonomous operation of ships, as shown in Table 5. In this stage, the intelligent navigation, intelligent engine room, intelligent energy efficiency management, and intelligent cargo management modules should have already been fully implemented and integrated into the intelligent integration platform, thus no longer requiring separate discussion.

Table 5. Forecast of inland intelligent ship functionality requirements for 2050.

Functional Module	Goal of Intelligence	Division of Intelligent Functional Stages
Intelligent Hull	Achieve the goal of fully intelligent ship hull, including self-diagnosis, and autonomous handling capabilities	(1) Integration of multiple functions of the 2035 intelligent hull module (2) Local strength monitoring of the hull, real-time monitoring of overall longitudinal strength, and stability calculation (3) Intelligent adjustment of ballast water, heading, and speed to ensure the ship is always in a safe state (4) Fully autonomous hull maintenance and upkeep
Intelligent Integration Platform	Realize comprehensive monitoring and intelligent management of various ships, including engineering and research ships, and achieve real-time two-way data exchange with shore-based systems	(1) Integration of multiple functions of the 2035 intelligent integration platform module (2) Information sharing and presentation function (3) Providing support for other intelligent applications on the basis of meeting its own information display and data diagnosis functions
Remote Control	Capable of being controlled by a remote control station or control position outside the ship, enabling unmanned operation of the ship	(1) Stable and applicable wireless communication equipment for ships with sufficient bandwidth (2) Beyond-line-of-sight control, scene perception, and real-time sharing of video information (3) Intelligent detection, alarm, and control processing functions
Autonomous Operation	Fully autonomous operation throughout the entire navigation	(1) Achieve fully autonomous navigation and comprehensive analysis decision-making from berth to berth (2) Real-time monitoring, evaluation, decision-making, and intelligent control of all ship systems

5. Conclusions

This paper systematically examines the current status of the division of intelligent ship function modules by international major classification societies, shipbuilding companies, and organizations such as the European Union and analyzes the required function modules for intelligent inland ships. From the perspective of the technological implementation of intelligent functions, a complete intelligent ship technology system is constructed, including intelligent perception technology, intelligent communication technology, intelligent evaluation technology, intelligent decision-making technology, and intelligent control technology. Through in-depth analysis of the technological connotation, it can be concluded that these five modules are all necessary key technologies and critical for realizing the intelligence of inland ships and are of great significance for the development of intelligent inland vessels and inland navigation. Taking into account the current situation of China's inland ships, the development plans on inland shipping and intelligent ships issued by government departments, and the development trend of intelligent technology, the functional requirements of intelligent inland ships in the near, medium, and long term are predicted. This article can provide suggestions for the development of intelligent inland ship implementation strategies in China and other regions of the world.

Ship intelligence is an inevitable trend in the development of the shipping industry. At present, the application of inland navigation is uncommon, and the intelligent development of inland ships based on intelligent technology is still in the primary stage. Sorting out the demand for intelligent functions and the many bottlenecks faced in the process of future intelligent development is conducive to clarifying the development direction of inland intelligent ships and promoting the development of inland intelligent ships in an orderly manner.

Realizing the ultimate goal of fully autonomous navigation of intelligent ships not only requires the ships themselves to have a high level of intelligence but also requires the joint progress of multiple supporting technologies such as shore-based platforms, network communications, intelligent waterways, and intelligent ports. Therefore, multiple technologies should integrate and promote each other, and then jointly promote the formation of high-quality shipping systems.

In addition, in terms of the development of intelligent ships on inland waterways in terms of laws and regulations, management mechanisms, production benefits, and other aspects of the same, there are many problems to be further standardized and breakthroughs. The next step also needs to be coordinated by some parties to coordinate the conflict of interest between the application of technology and market demand and steadily promote the development of ship intelligence.

Author Contributions: Conceptualization, G.H. and L.H.; Methodology, G.H. and W.X.; Software, W.X. and Y.C.; Validation, G.H., L.H. and K.Z.; Formal analysis, W.X. and K.Z.; Data curation, W.X. and J.C.; Writing—original draft, G.H. and W.X.; Writing—review and editing, G.H., W.X. and L.H.; Visualization, W.X., J.C. and Y.C.; Supervision, G.H., L.H. and K.Z.; Project Administration, G.H. and L.H.; Funding Acquisition, L.H. All authors have read and agreed to the published version of the manuscript.

Funding: This research was funded by the National Natural Science Foundation of China, grant number 52071248, the Research Program of Hubei Key Laboratory of Inland Shipping Technology, grant number NHHY2023004.

Institutional Review Board Statement: Not applicable.

Informed Consent Statement: Not applicable.

Data Availability Statement: The original contributions presented in the study are included in the article, further inquiries can be directed to the corresponding author.

Conflicts of Interest: The authors declare no conflicts of interests.

References

1. Wu, Z. Ship Automatic Navigation and Navigation Intelligence. *China Marit. Saf.* **2017**, 16–19. [CrossRef]
2. China Classification Society. *Rules for Intelligent Ships 2024*; CCS: Beijing, China, 2023.
3. Systems Lloyd's Register of Shipping. *Code for Unmanned Marine Systems*; Lloyd's Register: London, UK, 2017.
4. Det Norske Veritas. *Class Guideline Smartship*; DNV: Oslo, Norway, 2018.
5. American Bureau of Shipping. *Smart Functions for Marine Vessels and Offshore Units*; ABS: Houston, TX, USA, 2022.
6. Nippon Kaiji Kyokai. *Guidelines for Automated/Autonomous Operation of Ships*; ClassNK: Tokyo, Japan, 2020.
7. European Union. *Autonomous Ship Research and Development Program*; European Union: Maastricht, The Netherlands, 2019.
8. Netherlands Forum Smart Shipping. *Smart Shipping Roadmap*; Netherlands Forum Smart Shipping: Rotterdam, The Netherlands, 2021.
9. Weng, Y. Kongsberg's unmanned ship "ambitions". *China Ship Surv.* **2019**, 42–44.
10. Blogchina. Hailanxin, the Leading "Unicorn" in the Field of Intelligent Ships and Intelligent Oceans! Available online: https://net.blogchina.com/blog/article/956325013 (accessed on 17 July 2019).
11. International Ship Network. World's First! Hyundai Heavy Industries Builds First "Unmanned" Bulk Carrier. Available online: https://www.eworldship.com/html/2020/Shipyards_0414/158625.html (accessed on 14 April 2020).
12. Yu, Q. Development and Prospects of Inland Waterway Transport Infrastructure and Ship Technology. *Ship Boat* **2024**, *35*, 96–108. [CrossRef]
13. Thompson, D.; Coyle, E.; Brown, J. Efficient LiDAR-Based Object Segmentation and Mapping for Maritime Environments. *IEEE J. Ocean. Eng.* **2019**, *44*, 352–362. [CrossRef]
14. Xu, Z.; Zhai, J.; Huang, K.; Liu, K. DSF-Net: A Dual Feature Shuffle Guided Multi-Field Fusion Network for SAR Small Ship Target Detection. *Remote Sens.* **2023**, *15*, 4546. [CrossRef]
15. Ye, Y.; Zhen, R.; Shao, Z.; Pan, J.; Lin, Y. A Novel Intelligent Ship Detection Method Based on Attention Mechanism Feature Enhancement. *J. Mar. Sci. Eng.* **2023**, *11*, 625. [CrossRef]
16. Hu, K.; Zhang, Y.; Weng, C.; Wang, P.; Deng, Z.; Liu, Y. An Underwater Image Enhancement Algorithm Based on Generative Adversarial Network and Natural Image Quality Evaluation Index. *J. Mar. Sci. Eng.* **2021**, *9*, 691. [CrossRef]
17. Wang, R.; Chen, H.; Guan, C. Condition Monitoring Method for Marine Engine Room Equipment Based on Machine Learning. *Chin. J. Ship Res.* **2021**, *16*, 158–167. [CrossRef]
18. Zhuang, L. Design of Vibration Signal Data Acquisition System for Ship Mechanical and Electrical Equipment. *J. Coast. Res.* **2019**, *97*, 254–260. [CrossRef]

19. Hover, F.S.; Eustice, R.M.; Kim, A.; Englot, B.; Johannsson, H.; Kaess, M.; Leonard, J.J. Advanced Perception, Navigation and Planning for Autonomous In-Water Ship Hull Inspection. *Int. J. Robot. Res.* **2012**, *31*, 1445–1464. [CrossRef]
20. Zhan, H. Research on Energy Consumption Data Collection and Transmission Technology Based on Labview. *Ship Sci. Technol.* **2020**, *42*, 100–102.
21. Tong, J.; Yang, X.; Liu, H. Analysis on the Current Status of Smart Cargo Management Technologies for LNG Carriers. *Navigation* **2023**, 27–31. [CrossRef]
22. Jiang, J. Ship Dangerous Goods Monitoring System Based on Internet of Things Technology. *Ship Sci. Technol.* **2023**, *45*, 171–174.
23. Hu, W.; Ren, H.; Wang, M. Ship Cargo Compartment Environment Measurement and Control System Based on SSM Framework. In Proceedings of the 4th International Conference on Environmental Science and Material Application (ESMA), Xi'an, China, 15–16 December 2018.
24. Liu, P. Research on Multi Source and Dynamic Information Service Technology of Intelligent Ship. Master's Thesis, Harbin Engineering University, Harbin, China, 2017.
25. Chen, X.; Wei, D.; Pan, H.; Wang, Z.; Wang, G. Research on Internet of Things Data Mining Based on Ship Big Data Platform. *China Sci. Technol. Overv.* **2023**, 58–60.
26. Liu, W.; Liu, Y.; Gunawan, B.A.; Bucknall, R. Practical Moving Target Detection in Maritime Environments Using Fuzzy Multi-sensor Data Fusion. *Int. J. Fuzzy Syst.* **2021**, *23*, 1860–1878. [CrossRef]
27. Zhang, J.; Letaief, K.B. Mobile Edge Intelligence and Computing for the Internet of Vehicles. *Proc. IEEE* **2020**, *108*, 246–261. [CrossRef]
28. Zhang, Y. Intelligent Allocation of Ship Short Packet Communication Resources in the Internet of Things Environment. *Ship Sci. Technol.* **2023**, *45*, 170–173.
29. Yang, H.; Zhang, C.; Chen, G.; Zhao, B. Big Data-Oriented Intelligent Control Algorithm for Marine Communication Transmission Channel. *J. Coast. Res.* **2019**, *93*, 741–746. [CrossRef]
30. Yoo, S.J.; Park, B.S. Approximation-Free Design for Distributed Formation Tracking of Networked Uncertain Underactuated Surface Vessels under Fully Quantized Environment. *Nonlinear Dyn.* **2023**, *111*, 6411–6430. [CrossRef]
31. Cai, Y.; Wu, S.; Luo, J.; Jiao, J.; Zhang, N.; Zhang, Q. Age-Oriented Access Control in GEO/LEO Heterogeneous Network for Marine IoRT. In Proceedings of the IEEE Global Communications Conference (GLOBECOM), Rio de Janeiro, Brazil, 4–8 December 2022; IEEE: Piscataway, NJ, USA, 2020; pp. 662–667.
32. Wei, Z.; Ma, H.; Zhang, Q.; Feng, Z. Challenge and Trend of Sensing, Communication and Computing Integrated Intelligent Internet of Vehicles. *ZTE Technol. J.* **2020**, *26*, 45–49. [CrossRef]
33. Cheng, Z.; Zhang, Y.; Wu, B.; Soares, G. Traffic-Conflict and Fuzzy-Logic-Based Collision Risk Assessment for Constrained Crossing Scenarios of a Ship. *Ocean Eng.* **2023**, *274*, 114004. [CrossRef]
34. Bi, J.; Gao, M.; Zhang, W.; Zhang, X.; Bao, K.; Xin, Q. Research on Navigation Safety Evaluation of Coastal Waters Based on Dynamic Irregular Grid. *J. Mar. Sci. Eng.* **2022**, *10*, 733. [CrossRef]
35. Chen, X.; Shi, F.; Wang, Z.; Yang, Y.; Liu, W.; Sun, Y. Safety Evaluation of Multi-Ship Encountering Situation Via Fuzzy Logic Model. *J. Guangxi Univ.* **2022**, *47*, 1327–1336. [CrossRef]
36. Xu, X.; Lu, Y.; Liu, G.; Cai, P.; Zhang, W. COLREGs-Abiding Hybrid Collision Avoidance Algorithm Based on Deep Reinforcement Learning for USVs. *Ocean Eng.* **2022**, *247*, 110749. [CrossRef]
37. Xin, X.; Liu, K.; Loughney, S.; Wang, J.; Li, H.; Ekere, N.; Yang, Z. Multi-Scale Collision Risk Estimation for Maritime Traffic in Complex Port Waters. *Reliab. Eng. Syst. Saf.* **2023**, *240*, 109554. [CrossRef]
38. Akpan, U.O.; Koko, T.S.; Ayyub, B.; Dunbar, T.E. Reliability Assessment of Corroding Ship Hull Structure. *Nav. Eng. J.* **2003**, *115*, 37–48. [CrossRef]
39. Wang, Y. Research on Ship Structure Monitoring and Assessment System for the Polar Ship. Master's Thesis, Harbin Engineering University, Harbin, China, 2018.
40. Liu, H. Research on Structural Sate Assessment and Navigation Safety Decision of Ultra Large Ships Based on Monitoring Data. Master's Thesis, Harbin Engineering University, Harbin, China, 2020.
41. Lang, X.; Wu, D.; Tian, W.; Zhang, C.; Ringsberg, J.W.; Mao, W. Fatigue Assessment Comparison between a Ship Motion-Based Data-Driven Model and a Direct Fatigue Calculation Method. *J. Mar. Sci. Eng.* **2023**, *11*, 2269. [CrossRef]
42. Wang, Y. Application of Fuzzy Set Pair Analysis Method in Ship Energy Consumption Assessment. In Proceedings of the IEEE International Conference on Information, Communication and Engineering (ICICE), Fujian University of Technology, Xiamen, China, 17–20 November 2017; pp. 569–572.
43. Fan, A.; Yan, X.; Bucknall, R.; Yin, Q.; Ji, S.; Liu, Y.; Song, R.; Chen, X. A Novel Ship Energy Efficiency Model Considering Random Environmental Parameters. *J. Mar. Eng. Technol.* **2020**, *19*, 215–228. [CrossRef]
44. Wang, K.; Hua, Y.; Huang, L.; Guo, X.; Liu, X.; Ma, Z.; Ma, R.; Jiang, X. A Novel GA-LSTM-Based Prediction Method of Ship Energy Usage Based on the Characteristics Analysis of Operational Data. *Energy* **2023**, *282*, 128910. [CrossRef]
45. Gao, R.-j.; Wang, D.-l.; Zou, C.-m.; Xu, Y.-m.; Chen, M.; Jin, C. The Design of Bulk Carrier Cargo Holds State Integrated Monitoring System. In Proceedings of the International Seminar on Applied Physics, Optoelectronics and Photonics (APOP), Shanghai, China, 28–29 May 2016.

46. Lan, R.; Chang, W.; Shen, W.; Jia, Z. Research on Quantitative Risk Assessment Method of Packaged Cargoes Carried by Ship Based on Online Dynamic Big Data Fusion Technology. In Proceedings of the International Symposium on Power Electronics and Control Engineering (ISPECE), Xi'an University of Technology, Xi'an, China, 28–30 December 2018.
47. He, W. Security Evaluation of Key Information of Ship Warehouse Monitoring Video. *Ship Sci. Technol.* **2020**, *42*, 187–189.
48. Jiang, J.; Hu, Y.; Fang, Y.; Li, F. Application and Prospects of Intelligent Fault Diagnosis Technology for Marine Power System. *Chin. J. Ship Res.* **2020**, *15*, 56–67. [CrossRef]
49. Cheng, L. Research on Fault Intelligent Diagnosis of Ship Electronic Equipment Based on Artificial Immune Algorithm. *Ship Sci. Technol.* **2020**, *42*, 184–186.
50. Ozturk, U.; Cicek, K.; Celik, M. An Intelligent Fault Diagnosis System on Ship Machinery Systems Based on Support Vector Machine Principles. In Proceedings of the 26th Conference on European Safety and Reliability (ESREL), Glasgow, Scotland, 25–29 September 2016; pp. 1949–1953.
51. Liu, J.; Wu, Z.; Cao, Y.; He, Q.; Guo, M. An Fault Detection Algorithm for Intelligent Ship Integrated Navigation System. *Ship Sci. Technol.* **2023**, *45*, 155–158. [CrossRef]
52. Tang, G.; Fu, X.; Chen, N. Ship Power Plant Remote Fault Diagnosis System Based on B/S and C/S Architecture. *Ship Eng.* **2018**, *40*, 66–71. [CrossRef]
53. Xu, J.; Ramos, S.; Vazquez, D.; Lopez, A.M. Hierarchical Adaptive Structural SVM for Domain Adaptation. *Int. J. Comput. Vis.* **2016**, *119*, 159–178. [CrossRef]
54. Shi, Y. Research on Engine Fault Diagnosis of Gasoline Vehicle Based on Artificial Intelligence. Master's Thesis, Jilin University, Changchun, China, 2019.
55. Shang, L.; Cheng, Y. Integrated Power System Health Assessment of Large-Scale Unmanned Surface Ships Based on Convolutional Neural Network Algorithm. In *IOP Conference Series: Earth and Environmental Science*; Purpose-Led Publishing: Bristol, UK, 2018; p. 042052.
56. Guo, S.; Zhang, X.; Zheng, Y.; Du, Y. An Autonomous Path Planning Model for Unmanned Ships Based on Deep Reinforcement Learning. *Sensors* **2020**, *20*, 426. [CrossRef]
57. Liu, Z.; Liu, J.; Zhou, F.; Liu, R.W.; Xiong, N. A Robust GA/PSO-Hybrid Algorithm in Intelligent Shipping Route Planning Systems for Maritime Traffic Networks. *J. Internet Technol.* **2018**, *19*, 1635–1644. [CrossRef]
58. Pan, W.; Dong, Q.; Xu, X.; He, P.; Xie, X. Route Planning of Intelligent Ships Considering Navigation Enviroment Factors. *J. Dalian Marit. Univ.* **2021**, *42*, 76–84. [CrossRef]
59. Liang, C.; Zhang, X.; Han, X. Route Planning and Track Keeping Control for Ships Based on the Leader-Vertex Ant Colony and Nonlinear Feedback Algorithms. *Appl. Ocean Res.* **2020**, *101*, 102239. [CrossRef]
60. Ma, W.; Han, Y.; Tang, H.; Ma, D.; Zheng, H.; Zhang, Y. Ship Route Planning Based on Intelligent Mapping Swarm Optimization. *Comput. Ind. Eng.* **2023**, *176*, 108920. [CrossRef]
61. Zhou, S.; Wu, Z.; Ren, L. Ship Path Planning Based on Buoy Offset Historical Trajectory Data. *J. Mar. Sci. Eng.* **2022**, *10*, 674. [CrossRef]
62. Zhang, K.; Huang, L.; Liu, X.; Chen, J.; Zhao, X.; Huang, W.; He, Y. A Novel Decision Support Methodology for Autonomous Collision Avoidance Based on Deduction of Manoeuvring Process. *J. Mar. Sci. Eng.* **2022**, *10*, 765. [CrossRef]
63. Zhang, K.; Huang, L.; He, Y.; Zhang, L.; Huang, W.; Xie, C.; Hao, G. Collision Avoidance Method for Autonomous Ships Based on Modified Velocity Obstacle and Collision Risk Index. *J. Adv. Transp.* **2022**, *2022*, 1534815. [CrossRef]
64. Zhang, K.; Huang, L.; He, Y.; Wang, B.; Chen, J.; Tian, Y.; Zhao, X. A real-time multi-ship collision avoidance decision-making system for autonomous ships considering ship motion uncertainty. *Ocean Eng.* **2023**, *278*, 114205. [CrossRef]
65. Wang, Y.; Zhao, S.; Wang, Q. Cooperative Control of Velocity and Heading for Unmanned Surface Vessel Based on Twin Delayed Deep Deterministic Policy Gradient with an Integral Compensator. *Ocean Eng.* **2023**, *288*, 115943. [CrossRef]
66. Li, Y. Research on Autonomous Navigation Decision-Making Technology Based on Deep Reinforcement Learning. Master's Thesis, China Ship Research and Development Academy, Wuxi, China, 2023.
67. Xu, Y.; Lv, J.; Liu, J.; Li, L.; Guan, H. Multi-Vessel Intelligent Collision Avoidance Decision-Making Based on CSSOA. *Chin. J. Ship Res.* **2023**, *18*, 88–96. [CrossRef]
68. Zhao, Y.; Yuan, R.; Liu, S.; Zhang, J. The Algorithm Research on the Autonomous Collision Prevention with Intelligence Ship Based on Navigational Experience. *Navig. Tianjin* **2023**, 68–74.
69. Wang, W.; Huang, L.; Liu, K.; Wu, X.; Wang, J. A COLREGs-Compliant Collision Avoidance Decision Approach Based on Deep Reinforcement Learning. *J. Mar. Sci. Eng.* **2022**, *10*, 944. [CrossRef]
70. Wang, X.; Wang, G.; Wang, Q.; Han, J.; Chen, L.; Wang, B.; Shi, H.; Hirdaris, S. A Construction Method of a Sequential Decision Chain for Unmanned-Ship Autonomous Collision Avoidance Based on Human-Like Thinking. *J. Mar. Sci. Eng.* **2023**, *11*, 2218. [CrossRef]
71. Wu, H.; Guo, C.; Yang, F. Intelligent Control Adopted CMAC for Twin-Rudder Twin-Propeller Ship Course Tracking. In Proceedings of the 36th Chinese Control Conference (CCC), Dalian, China, 26–28 July 2017; pp. 568–573.
72. Teng, Y.; Liu, Z.; Zhang, L.; Liu, C.; Wei, N. On Intelligent Ship Tracking Control Method Based on Hierarchical MPC. In Proceedings of the 40th Chinese Control Conference (CCC), Shanghai, China, 26–28 July 2021; pp. 2814–2819.

73. Liu, Z.; Wang, Q.; Sun, J.; Yang, Y.; Society, I.C. Intelligent Tracking Control Algorithm for Under-Actuated Ships through Automatic Berthing. In Proceedings of the 6th IEEE Cyber Science and Technology Congress (CyberSciTech), Electr Network, Virtual, 25–28 October 2021; pp. 232–237.
74. Li, C. Research on Optimal Tracking Control Method of Intelligent Ship with Time-varying Environment Disturbance. Master's Thesis, Wuhan University of Technology, Wuhan, China, 2021.
75. Pham, D.-A.; Han, S.-H. Designing a Ship Autopilot System for Operation in a Disturbed Environment Using the Adaptive Neural Fuzzy Inference System. *J. Mar. Sci. Eng.* **2023**, *11*, 1262. [CrossRef]
76. Yoshida, M.; Shimizu, E.; Sugomori, M.; Umeda, A. Regulatory Requirements on the Competence of Remote Operator in Maritime Autonomous Surface Ship: Situation Awareness, Ship Sense and Goal-Based Gap Analysis. *Appl. Sci.* **2020**, *10*, 8751. [CrossRef]
77. Basnet, S.; BahooToroody, A.; Chaal, M.; Lahtinen, J.; Bolbot, V.; Banda, O.A.V. Risk Analysis Methodology Using STPA-Based Bayesian Network- Applied to Remote Pilotage Operation. *Ocean Eng.* **2023**, *270*, 113569. [CrossRef]
78. Chen, S.; Xiong, X.; Wen, Y.; Jian, J.; Huang, Y. State Compensation for Maritime Autonomous Surface Ships' Remote Control. *J. Mar. Sci. Eng.* **2023**, *11*, 450. [CrossRef]
79. Perera, L.P.; Mo, B. Emission Control Based Energy Efficiency Measures in Ship Operations. *Appl. Ocean Res.* **2016**, *60*, 29–46. [CrossRef]
80. Wang, K.; Yan, X.; Yuan, Y.; Jiang, X.; Lin, X.; Negenborn, R.R. Dynamic Optimization of Ship Energy Efficiency Considering Time-Varying Environmental Factors. *Transp. Res. Part D-Transp. Environ.* **2018**, *62*, 685–698. [CrossRef]
81. Guo, H. Design of Ship Energy Efficiency Control System with Distributed Data Acquisition Technology. *Ship Sci. Technol.* **2021**, *43*, 118–120.
82. Chen, L.; Gao, D.; Xue, Q. Energy Management Strategy of Hybrid Ships Using Nonlinear Model Predictive Control via a Chaotic Grey Wolf Optimization Algorithm. *J. Mar. Sci. Eng.* **2023**, *11*, 1834. [CrossRef]
83. Qin, P.; Lin, Y.; Chen, M. Ship Liquid Cargo Handling System Based on Decoupling Control and PSO Algorithm. In Proceedings of the 7th International Symposium on Test Measurement, Beijing, China, 5–8 August 2007; pp. 5224–5228.
84. Liu, C. Intelligent Stowage of Bulk Carrier. Ph.D. Thesis, Dalian Maritime University, Dalian, China, 2017.
85. Wang, Q. Development of Automatic Inspection Procedure of Loading Condition Based on NAPA MANAGER. Master's Thesis, Dalian University of Technology, Dalian, China, 2018.
86. Hou, Y.; Pu, J.-Y. The Intelligent Damage Assessment and Intelligent Reconfiguration System of Ship Diesel System. In Proceedings of the 8th ACIS International Conference on Software Engineering, Artificial Intelligence, Networking and Parallel/Distributed Computing, Qingdao, China, 30 July–1 August 2007; p. 56.
87. Yan, B. Research on Intelligent Operation and Maintenance Management Method of Marine Diesel Generator. Master's Thesis, Southwest Jiaotong University, Chengdu, China, 2021.
88. Hong, X.; Li, J.; Zeng, J.; Zhan, Y.; Yang, M. Digital Twin-Based Analytics for Remote Ship Operation. *Mar. Equip. Mater. Mark.* **2021**, *29*, 17–20. [CrossRef]
89. Hetherington, S. Unmanned Ships. *Ausmarine* **2016**, *38*, 29.
90. PRC Ministry of Transport. *Outline for the Development of Inland Transportation*; Ministry of Transport of the People's Republic of China: Beijing, China, 2020.
91. PRC Ministry of Transport. *Implementing Opinions on Accelerating the Green and Intelligent Development of Inland Ships*; Ministry of Transport of the People's Republic of China: Beijing, China, 2022.

Journal of
*Marine Science
and Engineering*

MDPI

Article

Integrating Computational Fluid Dynamics for Maneuverability Prediction in Dual Full Rotary Propulsion Ships: A 4-DOF Mathematical Model Approach

Qiaochan Yu [1,2], Yuan Yang [3], Xiongfei Geng [1], Yuhan Jiang [4], Yabin Li [5] and Yougang Tang [3,*]

[1]   China Waterborne Transport Research Institute, Beijing 100088, China; yuqiaochan@wti.ac.cn (Q.Y.)
[2]   School of Electrical and Information Engineering, Tianjin University, Tianjin 300072, China
[3]   School of Civil Engineering, Tianjin University, Tianjin 300354, China
[4]   Intelligent Navigation (Qingdao) Technology Co., Ltd., Qingdao 266200, China
[5]   Qingdao Institute of Shipping Development Innovation, Qingdao 266200, China
*   Correspondence: tangyougang_td@163.com

Abstract: To predict the maneuverability of a dual full rotary propulsion ship quickly and accurately, the integrated computational fluid dynamics (CFD) and mathematical model approach is performed to simulate the ship turning and zigzag tests, which are then compared and validated against a full-scale trial carried out under actual sea conditions. Initially, the RANS equations are solved, employing the Volume of Fluid (VOF) method to capture the free water surface, while a numerical simulation of the captive model test is conducted using the rigid body motion module. Secondly, hydrodynamic derivatives for the MMG model are obtained from the CFD simulations and empirical formula. Lastly, a four-degree-of-freedom mathematical model group (MMG) maneuvering model is proposed for the dual full rotary propulsion ship, incorporating full-scale simulations of turning and zigzag tests followed by a full-scale trial for comparative validation. The results indicate that the proposed method has a high accuracy in predicting the maneuverability of dual full-rotary propulsion ships, with an average error of less than 10% from the full-scale trial data (and within 5% for the tactical diameters in particular) in spite of the influence of environmental factors such as wind and waves. It provides experience in predicting the maneuverability of a full-scale ship during the ship design stage.

Keywords: ship maneuvering; CFD; fully rotary propulsion; MMG mathematical model; full-scale trial

Citation: Yu, Q.; Yang, Y.; Geng, X.; Jiang, Y.; Li, Y.; Tang, Y. Integrating Computational Fluid Dynamics for Maneuverability Prediction in Dual Full Rotary Propulsion Ships: A 4-DOF Mathematical Model Approach. *J. Mar. Sci. Eng.* **2024**, *12*, 762. https://doi.org/10.3390/jmse12050762

Academic Editor: María Isabel Lamas Galdo

Received: 29 March 2024
Revised: 26 April 2024
Accepted: 29 April 2024
Published: 30 April 2024

1. Introduction

Ship maneuverability, a pivotal aspect of maritime performance research, is essential for the assurance of navigational safety. Methods for predicting ship maneuvering performance are categorized into three principal types [1]: no simulation, system-based simulation and computational fluid dynamics (CFD)-based simulation. No simulation methods encompass database approaches, full-scale trials and free-running model tests. The former primarily utilizes regression analysis for swift evaluation, while the latter employs targeted experiments to directly ascertain maneuverability performance. The system-based simulation approach integrates hydrodynamic coefficients with equations of motion for ship maneuverability, facilitating the calculation of the ship's trajectory and associated motion parameters to predict its maneuverability.

Broadly, the assessment of ship maneuverability favors methods that are straightforward, efficient and cost-effective. These include approaches grounded in regression formulas based on characteristic parameters, database methods, free-running model test and numerical analyses employing mathematical models [2]. Mathematical models for ship maneuvering are principally divided into response models and hydrodynamic models. Response models establish a direct link between the ship's state of motion and rudder actions via navigational tests, enabling the analysis and resolution of maneuvering characteristics. Regarding hydrodynamic models, two primary mathematical models are predominantly utilized. (See Figure 1).

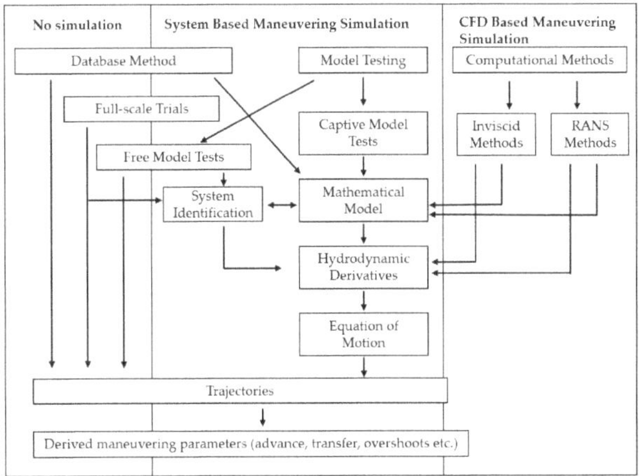

Figure 1. Maneuvering prediction methods [1].

The first model is the integrated structure model delineated by Abkowitz [3], which examines the hull, propeller and rudder collectively along with the cumulative force exerted. The second model, the Ship Maneuvering Mathematical Model Group (MMG model [4]), introduced by the Japanese Towing Tank Committee (JTTC), conducts separate hydrodynamic calculations for the hull, propeller and rudder, taking into account their interferences. Meng et al. [5] developed a response mathematical model for the vessel YUKUN, integrating Support Vector Regression with a modified grey wolf optimizer to obtain reference values for the model's parameters. Fossen [6] performed the theory and practice research on nonlinear ship control in response to the mathematical model. Svilicic et al. [7] assessed collision risks for the KVLCC2 ship through accurate modelling of ship maneuverability using non-linear FEM (NFEM). The Abkowitz maneuvering model is implemented in the LS-Dyna software code and is therefore coupled with FEM calculations. Shin et al. [8] investigated the maneuverability of a KCS equipped with energy-saving devices utilizing the MMG model. Reichel [9] introduced a novel characterization of forces on azimuth thrusters within the motion mathematical model, employing MMG methodology and experimentally validating its accuracy.

Both methods necessitate precise hydrodynamic derivative calculations to develop an accurate model of ship maneuverability. The extensive application of CFD techniques facilitates more precise outcomes in ship hydrodynamics analysis, design and maneuverability forecasting. Given this backdrop, numerous researchers have advanced static and dynamic simulations to study ship maneuvering movements. Sun et al. [10] utilized the STAR-CCM+ software to model the Planar Motion Mechanism (PMM) of the hull through the overlapping grid technique, deriving hydrodynamic derivatives subsequently integrated into the MMG model for twin waterjet propulsion vessels. Ahmad et al. [11] performed Oblique Towing Tests (OTT) and dynamic Planar Motion Mechanism (PMM) analyses on the vessel DTMB 5512 using CFD simulations to determine the hydrodynamic derivatives.

In ship maneuverability modeling, the four-Degree-of-Freedom (4-DOF) MMG model is broadly embraced for its clear-cut derivatives and efficient computation, particularly emphasizing the impact of ship roll. R. Rajita et al. [12] addressed the calculation of linear, nonlinear and roll-coupled hydrodynamic derivatives for a container ship through CFD-based numerical simulations of static and dynamic tests across various roll angles. Li et al. [13] executed Oblique Towing Tests (OTT) and dynamic Circular Motion Tests (CMT) to collect essential data for MMG model identification, and they simulated the free-running maneuverability test using a body force propeller approach that obviates the need for detailed flow field construction around the propeller. Guo et al. [14] investigated the 4-DOF ship maneuvering motion in calm water for the ONR tumblehome model by a system-based method. The result indicates the validity of the CFD-based modelling method for the hull−propeller−rudder interaction of twin-screw ships. Okuda et al. [15] applied the 4-DOF MMG method as a practical simulation method that includes the roll-coupling effect to predict the maneuvering of a KCS at fast speeds. Dash et al. [16] developed a 4-DOF simulation method for the maneuvering motion of a ship with a twin propeller and twin rudder system. The hydrodynamic derivatives and parameters were determined by the PMM tests, and roll-induced bifurcation in maneuvering was discussed by the simulations.

Likewise, the discretized propeller approach can simulate free-running maneuverability tests. Shen et al. [17] implement the dynamic overset grid technique into naoe-FOAM-SJTU solver to simulate standard 10/10 zig-zag maneuver and modified 15/1 zig-zag maneuver of KCS, which showed good agreement with the experiment data. Wang et al. [18] studied the free running test of the ONR Tumblehome ship model under course keeping control with twin actual rotating propellers and moving rudders, and a new course keeping control module was developed using a feedback controller based on the CFD solver. Carrica et al. [19] conducted a study on a KCS container ship performing a zigzag maneuver in shallow water experimentally and numerically using direct discretization of a moving rudder and propeller. The zigzag maneuver at the nominal rudder rate uses grids of up to 71.3 million points. Sanada et al. [20] performed research on the hull–propeller–rudder interaction at the Korea Research Institute of Ships using a combined experimental fluid dynamic and CFD method, with an innovative approach being employed for the analysis of steady state circular motions. Nonetheless, both techniques demand significant computational resources and time costs to precisely model ship maneuvering movements, especially for a full rotary ship. Consequently, integrating mathematical modeling with CFD simulations of captive model tests has proven to enhance forecasting speed while maintaining a balance between rapidity and accuracy [21].

Full rotary propellers, as opposed to conventional rudder and propeller setups, possess enhanced maneuvering capabilities, adeptly dealing with intricate scenarios like stationary rotation and sideways motion [22]. It is shown that a ship's stability can be jeopardized in terms of excessive heeling in calm water or parametric rolling in extreme waves due to low GM and damping characteristics. The effect of GM or loading conditions due to the accommodation of full rotary propellers have been more apparent between runs in design draught and scantling draught conditions [23]. Currently, limited calculations and simulations fully account for the impact of dual full rotary propellers on ship maneuverability, with reliance primarily being on free-running tests or full-scale trials. Neatby et al. [24] performed comprehensive full-scale trials, encompassing turning circles, effective turning tests and crash stops, on a vessel equipped with dual Z-drive thrusters. Reichel [9] introduced a 3-DOF mathematical model grounded in MMG methodology, conducting both numerical simulations and experimental validations on a pod-driven coastal tanker. This approach verified the model's capability to discern performance trends, even in vessels with unstable trajectories. Piaggio et al. [25] showcased the findings of a comparative analysis between spade and flap rudder configurations versus pod-driven systems for a select fleet, demonstrating that appropriately designed pod units do not compromise yaw control capabilities.

Conducting full rotary propulsion ship free-running tests via direct CFD simulations necessitates a finer grid mesh, thereby increasing the demand for computational resources and extending the time required for analysis. Meanwhile, the current research on the maneuverability of dual full-rotary propulsion ships lacks consideration of rolling conditions, and the high DOF motion of the propeller makes direct CFD simulation more difficult. Viewed comprehensively, research on fast motion prediction of dual full-rotary propulsion ships is still relatively scant, and related theoretical studies and practical problems still need to be examined.

This paper introduces a fast-time prediction technique for predicting ship maneuverability. Utilizing CFD methods, this study simulates the captive model test of the "Zhifei" [26] ship to derive its hydrodynamic derivatives. Furthermore, it presents a 4-DOF MMG mathematical model for dual full rotary propulsion ships. Numerical simulations on full-scale tests were performed on the "Zhifei" ship for turning and zigzag maneuvers in calm waters, with the results being compared against experimental full-scale trial data to confirm the viability of the proposed maneuvering model for dual full-rotary propulsion ships. This approach offers a reliable solution for the precise prediction of maneuverability during the ship's design stage.

2. Mathematical Model and Method

2.1. Coordinate and MMG Model

The development of mathematical models for maneuvering motions necessitates establishing earth-fixed and ship-fixed coordinate systems to delineate the pertinent motion variables. Within the earth-fixed coordinate framework, ship movement is characterized by the spatial coordinates $[x, y, z]^\mathrm{T}$ and the orientation angles $[\varphi, \theta, \psi]^\mathrm{T}$. $O_0 - X_0Y_0Z_0$ is fixed to a specific point on the earth's surface, with the Z_0 axis being oriented vertically downwards. $G - xyz$ is attached to the ship's center of gravity, with the x axis being directed towards the bow and the y axis towards the starboard side, while the z axis extends vertically downwards. Generally, analyzing such problems requires facilitating the mutual conversion between these two coordinate systems, which are shown in Figure 2.

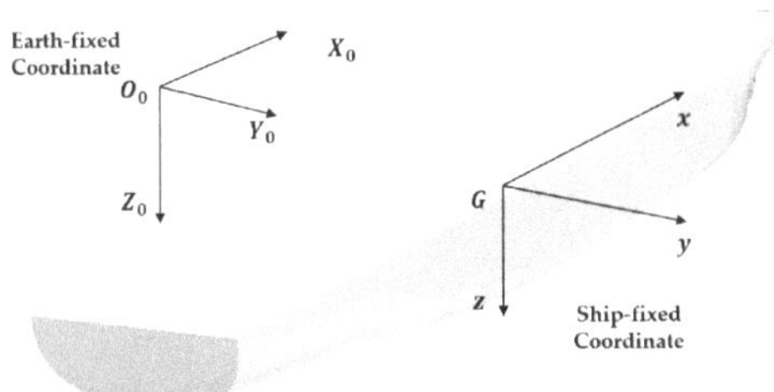

Figure 2. Coordinate systems.

Following the concept of segregated ship-motion mathematical modeling, the array of forces and moments exerted on the ship is categorically allocated to the bare hull and the propeller for computational purposes. The four-degree-of-freedom motion equations for a dual full rotary propelled ship within the designated coordinate system are derived, accounting for the ship's rolling state, as follows [14]:

$$
\begin{aligned}
(m + m_x)\dot{u} - (m + m_y)vr &= X_\mathrm{H} + X_\mathrm{P} \\
(m + m_y)\dot{v} + (m + m_x)ur &= Y_\mathrm{H} + Y_\mathrm{P} \\
(I_{xx} + J_{xx})\dot{p} &= K_\mathrm{H} + K_\mathrm{P} \\
(I_{zz} + J_{zz})\dot{r} &= N_\mathrm{H} + N_\mathrm{P}
\end{aligned}
\tag{1}
$$

where X_P, Y_P, K_P and N_P are the longitudinal thrust force, lateral thrust force, rolling moment and yawing moment acted on the ship by the full rotary propeller, respectively, while X_H, Y_H, K_H and N_H are the hydrodynamic forces (moments) acting on different degrees of freedom of the hull by all other external forces except the propellers. Additionally, m is the mass of the ship and m_x and m_y are the additional mass of the ship in the x axis and y axis, respectively. It is caused by the co-motion of the water around the hull of the ship. I_{zz} and J_{zz} are the inertia of the ship around the z axis and the additional inertia, respectively.

2.2. Hull Hydrodynamic and Propeller Thrust Model

The hydrodynamic forces can be divided into two categories according to their causes: one is fluid inertial forces and the other is fluid viscous forces. On the basis of the Kijimas research [27], a model for estimating the hydrodynamic forces of the hull is summarized based on the consideration of the rolling moment caused by the ship's motion:

$$
\begin{aligned}
X_H &= X(u) + X_{vv}v^2 + X_{vr}vr + X_{rr}r^2 \\
Y_H &= Y_v v + Y_r r + Y_{v|v|}v|v| + Y_{r|r|}r|r| + Y_{vvr}v^2 r + Y_{vrr}vr^2 + Y_{H1}(v, r, \varphi) \\
K_H &= -K_1(\dot{\varphi}) - K_2(\varphi) - Y_H z_H \\
N_H &= N_v v + N_r r + N_{v|v|}v|v| + N_{r|r|}r|r| + N_{vvr}v^2 r + N_{vrr}vr^2 + N_{H1}(v, r, \varphi)
\end{aligned}
\tag{2}
$$

where $X(u)$ is the ship resistance when sailing straight, $K_1(\dot{\varphi})$ is the rolling damping moment, $K_2(\varphi)$ is the rolling restoring moment, $Y_H z_H$ is the rolling moment of the hull hydrodynamic force Y_H on the x axis, while z_H is the z axis coordinate of the point where Y_H acts.

This study focuses on a ship equipped with dual full rotary propeller propulsion, as illustrated in Figure 3. The propellers are symmetrically positioned, with a longitudinal distance of L_{op} from the ship's center of gravity and a lateral separation of L_{ps} between them.

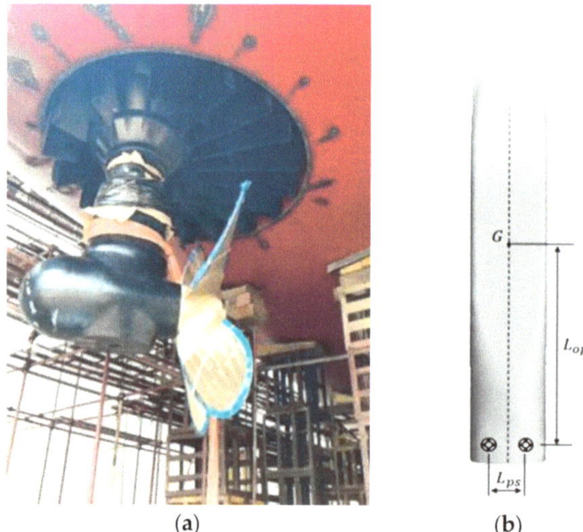

Figure 3. Schematic. (**a**) Full Rotary Propeller of Ship "ZhiFei". (**b**) Propellers Position.

In the propulsion system of ships with full rotary twin-propellers, both forward movement and steering capabilities are achieved by manipulating the propellers' orientation or exploiting the differential in their rotational speeds. This technique allows for the generation of axial thrust by the propellers in static water, as follows [28]:

$$
\begin{aligned}
T_p &= (1 - t_p)\rho n_p^2 D^4 k_{T(p)} \\
T_s &= (1 - t_p)\rho n_s^2 D^4 k_{T(s)}
\end{aligned}
\tag{3}
$$

where t_p is the thrust deduction coefficient, the subscript p and s represent the portside and starboard propellers, respectively, ρ is the density of seawater, D is the diameter of the propeller disk, n_p and n_s are the rotation speed of the left and right propellers, respectively, $k_{T(p)}$ and $k_{T(s)}$ are the coefficients of the left and right propeller thrusts open water characteristic, respectively, and calculated as follows:

$$
K_T = a_0 + a_1 J + a_2 J^2
\tag{4}
$$

where a_0, a_1 and a_2 are the propeller coefficients, J is the propeller advanced ratio, which is calculated as $J = u(1 - w_p)/nD$, where w_p is the wake fraction at propeller position.

When the left and right propellers work simultaneously at rotation angles δ_p and δ_s, respectively, the resulting axial thrust can be decomposed along the attached coordinate system GX and GY. Based on previous studies, we proposed the following calculation method for dual full rotary propulsion ships:

$$
\begin{aligned}
X_P &= \left(T_p \cos \delta_p + T_s \cos \delta_s\right) \\
Y_P &= \left(T_p \sin \delta_p + T_s \sin \delta_s\right) \\
K_P &= -\cos \varphi \left(T_p \sin \delta_p + T_s \sin \delta_s\right) z_P + \tfrac{1}{2} \sin \varphi \left(T_p \sin \delta_p - T_s \sin \delta_s\right) L_{ps} \\
N_P &= \tfrac{1}{2}\left(T_p \cos \delta_p - T_s \cos \delta_s\right) L_{ps} - Y_P L_{op}
\end{aligned}
\tag{5}
$$

where z_P is the distance of the horizontal center axis of propeller and the center of gravity.

2.3. Governing Equations

This study employs a CFD method to simulate the hydrodynamic forces acting on a ship. The viscous flow is approximated using the Reynolds-averaged Navier–Stokes (RANS) equations. This approach converts a transient problem into a steady-state problem by averaging over time the random fluctuation terms in the viscous flow, thereby facilitating problem resolution. The continuity equation applicable to viscous flow is presented as follows:

$$
\frac{\partial(u_i)}{\partial x_i} = 0
\tag{6}
$$

The Reynolds-averaged Navier–Stokes equation is:

$$
\rho \frac{\partial u_i}{\partial t} + \rho u_j \frac{\partial u_i}{\partial x_j} = -\frac{\partial P}{\partial x_i} + \frac{\partial}{\partial x_j}\left(\mu \frac{\partial u_i}{\partial x_j} - \rho \overline{u_i' u_j'}\right)
\tag{7}
$$

where $\mu_{i,j} (i,j = 1,2,3)$ are the mean velocity vectors, P is the time-averaged value of pressure, $\rho \overline{u_i' u_j'}$ is the Reynolds stress, while u_i and u_j are the time-averaged values of the velocity component.

2.4. Full-Scale Test

As shown in Figure 4, a full-scale ship maneuverability test was conducted for the "Zhifei" in the China Nǚdao sea area under the conditions of a northeast wind of level 3–4 and sea state 3, with an average draft of 3.317 m. The test concluded with the ship's 10° turning motion, $\pm 10°$ zigzag motion and ship resistance test when sailing straight. Motion data was collected using the SPS351 DGPS receiver, with a maximum dynamic error not exceeding 5 m.

Figure 4. Full-scale ship maneuverability test.

The database controlled ship−shore data synchronization based on the network status, where the ship-side network could establish a connection with the shore-side and the line status could support data communication. Typically, the signal data was stored during the experiment and uniformly transferred to the ground control PC at the shore terminal at the end of each day's experiment. The signal sampling frequency was 1 Hz.

3. Numerical Simulation of Captive Model Test

3.1. Computational Settings and Convergence Analysis

This study focuses on the "Zhifei" 300TEU smart container ship, China's inaugural coastal intelligent navigation container ship. The principal parameters concerning its hull, propeller and the numerical computations for the scaled model are detailed in Table 1. To derive the hydrodynamic derivatives of the hull, the OTT and CMT are simulated by using the RANS solver platform STAR-CCM+ (CD-adapco Company, German)with the specific calculation conditions outlined in Table 2.

Table 1. Main parameters of hull and propeller.

Name	Symbol	Unit	Ship	Model
Scale factor	λ		1	28.5
Length overall	L_{oa}	m	117.15	4.111
Length between perpendiculars	L_{pp}	m	111.3	3.905
Breadth	B	m	17.32	0.607
Draft	d	m	4.8	0.158
Displacement	Δ	m^3	4800	0.207
Block coefficient	C_b		0.7797	0.7797
Metacentric height	GM	m	5.1	0.179
Vertical center of gravity (from keel)	KG	m	6.16	0.216
Propeller diameter	D	m	2.7	0.947
Wake fraction at propeller position	ω_p		0.183	0.183
Number of blades	Z		4	4

Table 2. Computational case conditions.

Test	Fr	β(degree)	r'
Oblique Tow Test (OTT)	0.226	$\pm 11, \pm 9, \pm 6, \pm 2, 0$	0
Circular Motion Test (CMT)	0.226	0	$\pm 0.2, \pm 0.4, \pm 0.6$
Circular Motion Test (CMT)	0.226	$\pm 11, \pm 9, \pm 6, \pm 2$	$-0.2, -0.4, -0.6$

The computational region is defined as a cuboid, as depicted in Figure 5. The inlet, sides, top and bottom of the flow field domain are set as velocity inlet, and the outlet is set as pressure outlet. The hull surface is defined as no-slip walls to model the interface accurately. The dimensional size of the flow field domain is $-4L_{pp} < x < 2L_{pp}$, $-2.5L_{pp} < y < 2.5L_{pp}, -2.5L_{pp} < z < 1.0L_{pp}$. The velocity field function is used to simulate the velocity of each boundary. The simulation region is discretized using a trimmed mesher approach. Mesh refinement is applied to regions surrounding the hull and the free water surface to precisely capture the flow dynamics during the vessel's movement. Additionally, mesh refinement is employed at the bow and stern to enhance the resolution of the flow field captured. In the boundary layer, a four-layer prism is employed to maintain y+ around 30 for the majority of the region. The k-ε turbulence model is selected and integrated with the two-layer all y+ Wall Treatment to accurately represent the free water surface via the volume of the fluid method. Additionally, the dynamic fluid−body interaction (DFBI) module is utilized to numerically simulate the captive movement of the ship model. The computational region comprises approximately 2.55×10^6 grids in total. The grid count varies slightly under different operating conditions, with the hull surface and the grids of the computational domain being partitioned, as illustrated in Figure 6.

The convergence analysis for longitudinal forces on the hull under the condition of $Fr = 0.226$ was conducted during the direct flight test of the ship model, employing the methodology advocated by the International Towing Tank Conference. The mesh size and time step were scaled by a constant factor of $\sqrt{2}$. The stability of the calculation results is judged by the convergence parameter R_G, which is defined below [29]. These three cases show monotonically converge consistently and satisfy the computational requirements when $0 < R_G < 1$. (See Table 3).

$$R_G = \frac{S_2 - S_1}{S_3 - S_2} \tag{8}$$

where: S_1, S_2, S_3 are calculated results for fine, moderate and rough levels, respectively.

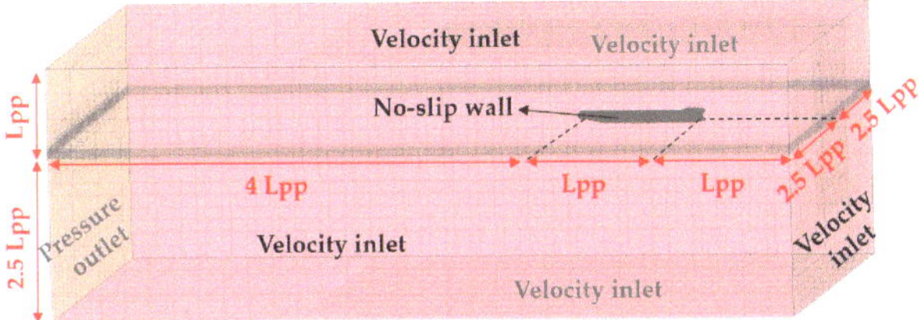

Figure 5. Computational region of captive model tests.

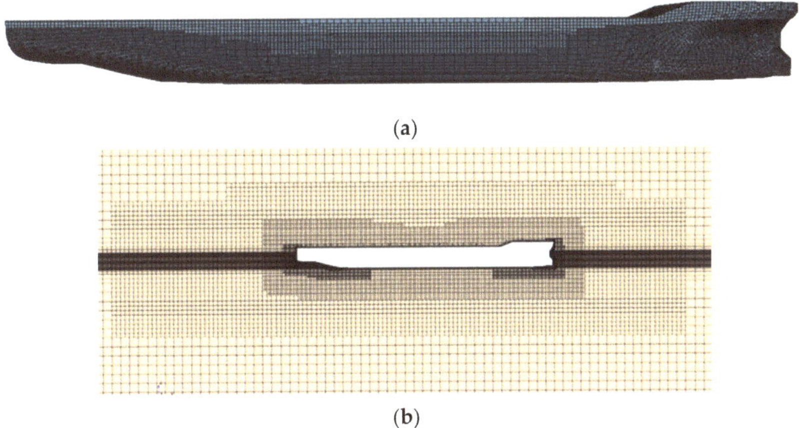

(**a**)

(**b**)

Figure 6. The grids in computational region: (**a**) Hull surface; (**b**) Computational region.

Table 3. Convergence analysis of mesh sizes and time steps.

Case	Name	Value	Solution Time [1]	Longitudinal Resistance	R_G
G_1		0.05 (m)	49,800 (s)	12.1556 (N)	
G_2	Mesh Size [2]	0.07 (m)	14,065 (s)	12.2694 (N)	0.341
G_3		0.10 (m)	8665 (s)	12.6028 (N)	
T_1		0.007 (s)	30,343 (s)	12.2495 (N)	
T_2	Time Step	0.010 (s)	14,065 (s)	12.2694 (N)	0.279
T_3		0.014 (s)	9586 (s)	12.3405 (N)	

[1] Time required to simulate 45 s in solver. [2] Mesh size G_1, G_2, G_3 contains the total number of grids 5.22×10^6, 2.55×10^6, 1.26×10^6, respectively.

The findings indicate that both grid size and time step exhibit convergence. Diminishing either the grid size or the time step further minimizes the errors in the computational outcomes while leading to a significant increase in the solution time. Upon verifying the precision of the calculations and considering time efficiency, the simulation method using a medium grid size G_2 and a medium time step T_2 was chosen in this study.

3.2. Resistance Validation of Straight Sailing

Conducting a numerical simulation of the resistance during straight sailing allows for additional validation of the numerical model's calculation accuracy. Furthermore, fitting the resistance function of the ship model at varying speeds enables the derivation of X(u). The specific calculation scenarios and their outcomes are detailed in Table 4. A comparison of the modeled ship's resistance with actual ship test values, post-Froude-resistance transformation, reveals an error margin of approximately 5%, with a consistent overall trend being observed. The error is within an acceptable range, taking into account factors such as scale effects and experimental error. Figure 7 illustrates the free surface wave patterns at various speeds, accurately depicting the symmetrical Kelvin waves generated by the bow and stern. The clarity of peak and trough contours further signifies the simulation's effectiveness.

Table 4. Comparison of calculation conditions and results of direct flight resistance.

$Fr = U/\sqrt{gL}$	U (m/s)	Numerical Modelled Values (N)	Froude Transform Value (N)	Test Value (N)	Error (%)
0.129	0.8	3.934	73,298.84	76,159.31	3.76
0.178	1.1	7.625	125,651.22	135,501.75	7.27
0.226	1.4	12.269	242,683.36	250,553.62	3.14
0.275	1.7	19.853	396,594.89	421,314.91	5.87
0.323	2	40.871	695,527.64	647,785.64	7.37

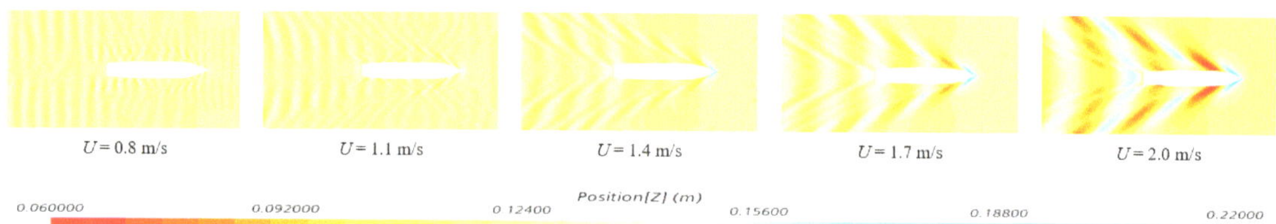

Figure 7. Free surface wave patterns at various speeds.

The calculated resistance values in the table are plotted as resistance curves, as shown in Figure 8, and the resistance function of the ship "Zhifei" is fitted as $X(u) = 420666.76u^2 - 672733.19u + 348350.78$.

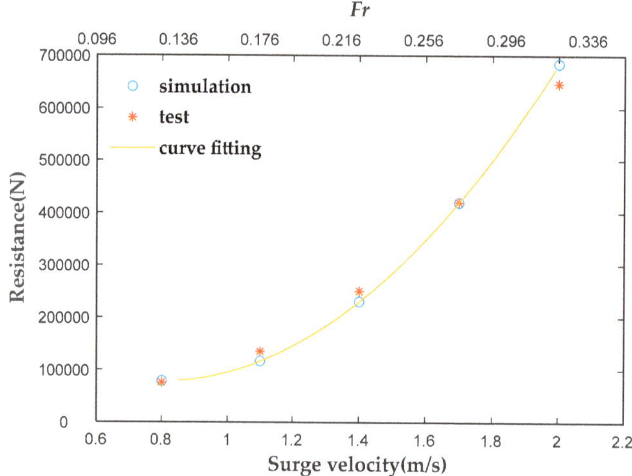

Figure 8. Resistance value and fitted curve.

3.3. Captive Model Test Simulation

The free surface wave patterns under varying test conditions are shown as follows: Figure 9 shows the simulation result of wave pattern in an oblique towing test at $\beta = 0°, 2°, 9°$. When $\beta = 0°$, symmetrical kelvin waves are formed at the bow and stern of the ship, and the height of the rising waves on both sides is basically identical. In contrast, as β increases, asymmetrical waves emerge on either side of the hull, becoming more pronounced. The waves at the bow and stern on the windward side converge, whereas those on the leeward side diverge, which becomes more apparent as β increases.

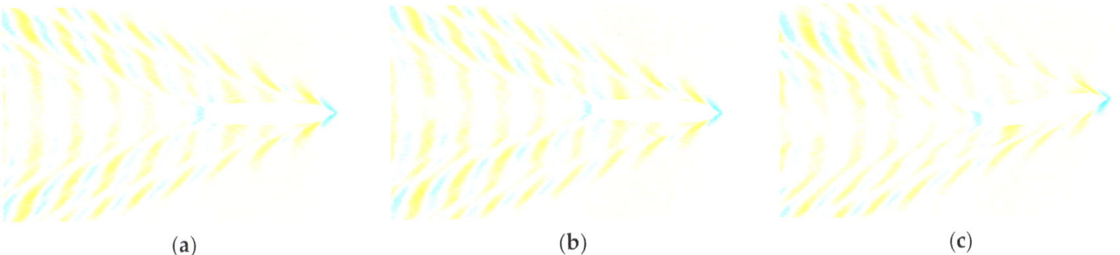

(a)   (b)   (c)

Figure 9. Wave pattern in OTT test: (**a**) $\beta = 0°$; (**b**) $\beta = -2°$; (**c**) $\beta = 9°$.

Figures 10–12 depicts the circular motion test simulation with $\beta = 0°, -2°, 9°$. In Figure 10, with $\beta = 0°$, the waves generated at the bow and stern of the ship continuously intersect behind the ship while sailing. The bow wave crest progressively moved towards the port side as the rate of turn r' increased, while the stern wave also curves towards the port side in response to the ship's movement. Meanwhile, the intersection becomes more obvious, and the wave area on both sides of the ship is more compact and shifted to the port side.

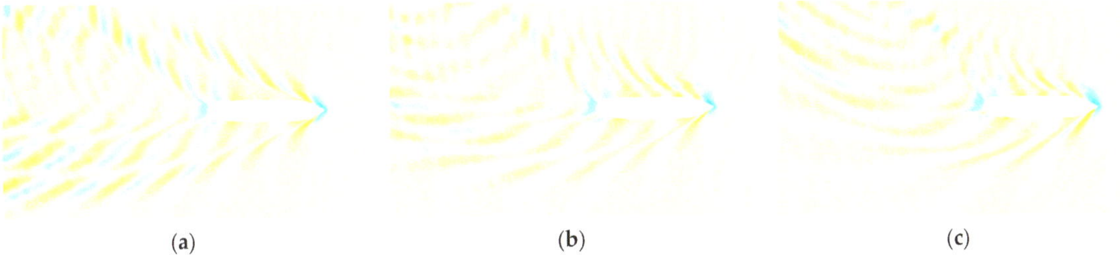

(a)   (b)   (c)

Figure 10. Wave pattern in CMT test with $\beta = 0°$: (**a**) $r' = -0.2$; (**b**) $r' = -0.4$; (**c**) $r' = -0.6$.

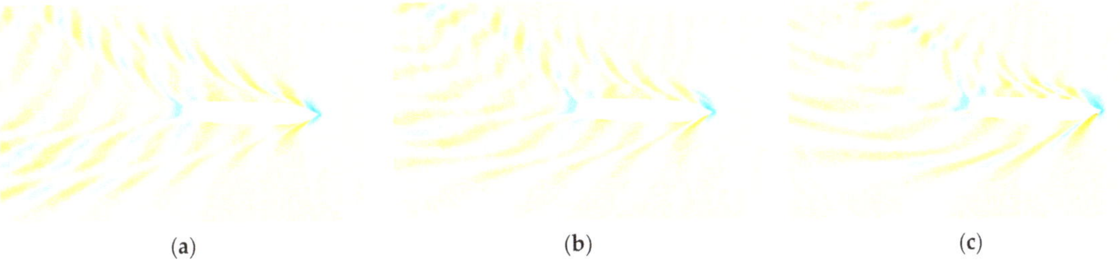

(a)   (b)   (c)

Figure 11. Wave pattern in CMT test with $\beta = -2°$: (**a**) $r' = -0.2$; (**b**) $r' = -0.4$; (**c**) $r' = -0.6$.

In Figure 11, with $\beta < 0°$, the wave amplitude along the hull significantly increases, with the predominant wave distribution shifting to the starboard side, indicating that the bow and stern waves gradually converge from the port side towards and spread to the starboard side. Meanwhile, comparing with the same r', the height of the bow rising wave increases obviously. With $\beta > 0°$, the amplitude of the waves along the hull is reduced, and the primary area of wave distribution for both bow and stern is on the port side, as shown in Figure 12.

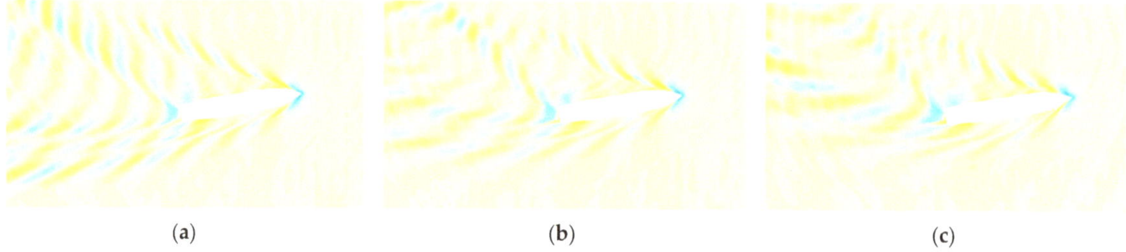

(a)   (b)   (c)

Figure 12. Wave pattern in CMT test with $\beta = 9°$: (**a**) $r' = -0.2$; (**b**) $r' = -0.4$; (**c**) $r' = -0.6$.

Figure 13 illustrates how the waterline at the bow changes with the drift angle, highlighting that the drift angle induces an asymmetric wave pattern on both sides of the bow. This asymmetry becomes increasingly pronounced with larger drift angles.

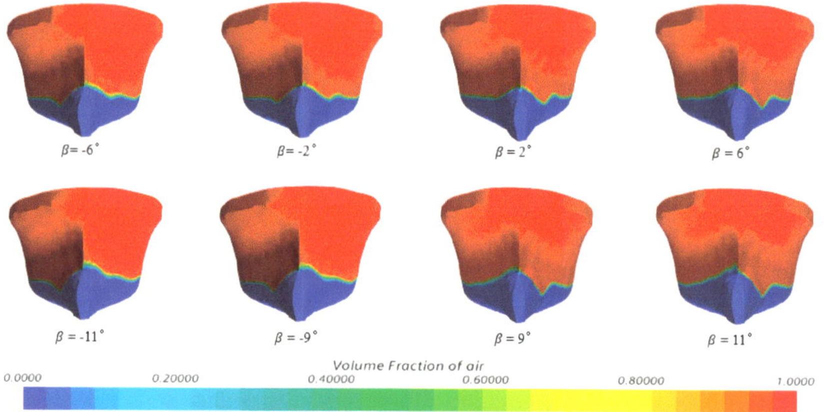

Figure 13. Bow waterline at different drift angles.

Figure 14 depicts the distribution of the pressure coefficient along the hull's bottom, providing insights into the hydrodynamic pressures exerted on the ship's underbody during various test conditions. With increasing turn rate r', the pressure on the starboard side of the ship intensifies. At a drift angle of $\beta < 0°$, the pressure concentration at the starboard side of the bow escalates, expanding the high-pressure zone as r' rises, with a continuous increase in pressure amplitude. Conversely, at $\beta > 0°$, the port side experiences more concentrated pressure, with the high-pressure region progressively moving to the starboard side as r' increases. A gradual increase in pressure on the starboard bow leads to the emergence of negative pressure on both the port side and the starboard side at the stern.

Figures 15–17 display the forces and moments on the hull measured and fitted during the oblique towing test and circular motion test. It is observed that all three parameters tend to increase as the drift angle rises. Within the drift angle range specified by the oblique towing test, the lateral force and the yaw moment demonstrate an approximate linear response. However, the longitudinal force shows minimal sensitivity to drift angle variations. When $\beta = 0°$, the longitudinal force acting on the hull can be represented by $X = X(u)$. Similarly, at smaller β values, the longitudinal force on the hull remains essentially constant. As the drift angle β increases, the lateral force correspondingly rises, with its rate of increase accelerating alongside β. In circular motion tests, changes in drift angle and the yaw velocity significantly influence both the lateral force and the yaw moment. When $\beta = 0°$, the behavior of longitudinal force mirrors that observed in the oblique towing test. However, the lateral force and yaw moment exhibit heightened sensitivity to variations in the yaw velocity. As yaw velocity increases, the lateral force on the hull progressively rises, with each increment being larger than the last. Meanwhile, the yaw moment diminishes, with its overall magnitude slightly decreasing.

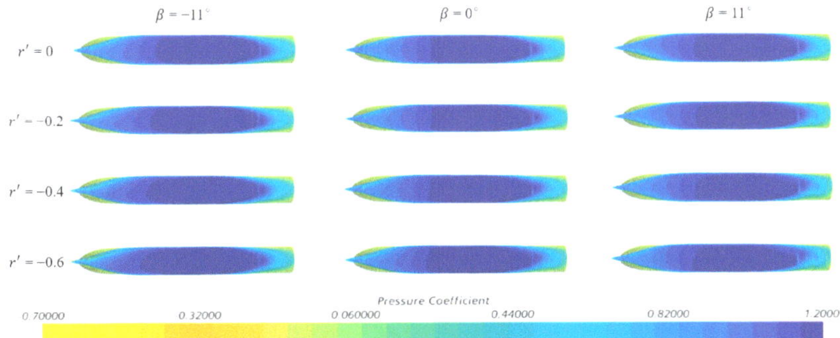

Figure 14. Hull's bottom pressure distribution in varies test conditions.

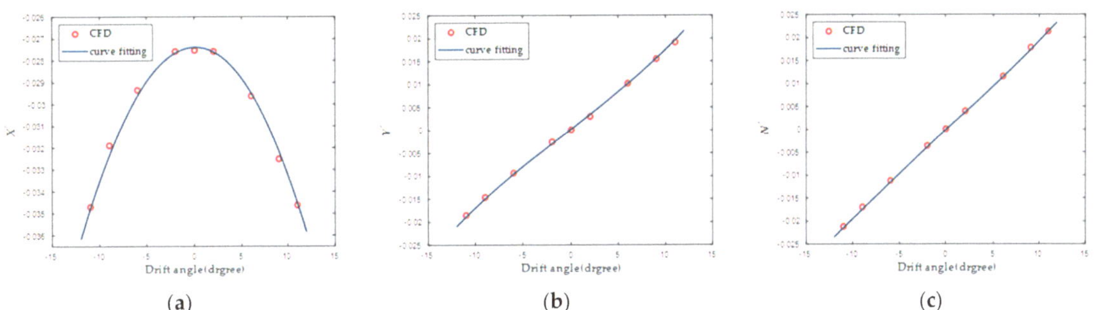

Figure 15. Simulation results of OTT: (**a**) longitudinal force; (**b**) lateral force; (**c**) yaw moment.

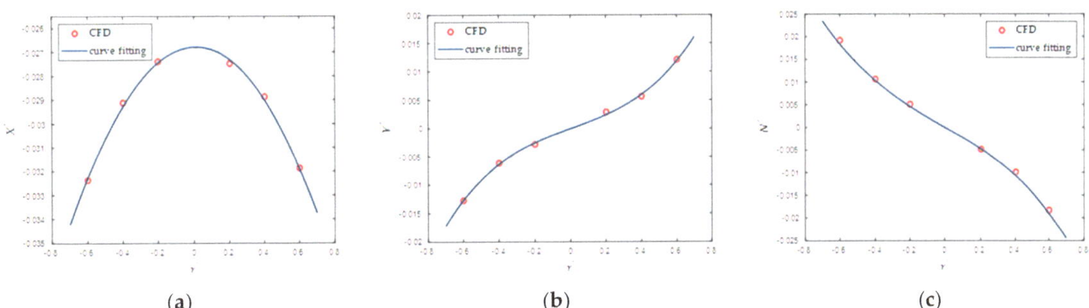

Figure 16. Simulation results of CMT (when $\beta = 0°$): (**a**) longitudinal force; (**b**) lateral force; (**c**) yaw moment.

Through simulations at varying yaw velocities, it is observed that lower yaw velocity results in smoother transitions in the three curves, indicating a lesser impact from drift angle actions. Additionally, at drift angle $\beta > 0°$, the longitudinal force on the hull attains a minimum value at some point.

A least squares regression analysis of the simulation outcomes yielded the hydrodynamic derivatives presented in Table 5. The empirical formulas are obtained from reference [27,30] and enable direct calculation of the hydrodynamic derivatives from parameters such as the ship length and breadth. When these results are compared to empirical formulas, some discrepancies are shown in the CFD findings. The calculated values of the linear hydrodynamic derivatives align with those from empirical formulas, maintaining a similar order of magnitude. However, greater differences are observed in some nonlinear hydrodynamic derivatives, potentially because the numerical simulations fail to precisely forecast hydrodynamic forces at higher drift angles. Furthermore, all lateral hydrodynamic derivatives appear to be underestimated, potentially because of challenging flow separations occurring at the hull's curvature under conditions of significant drift angles and high bow angular velocities.

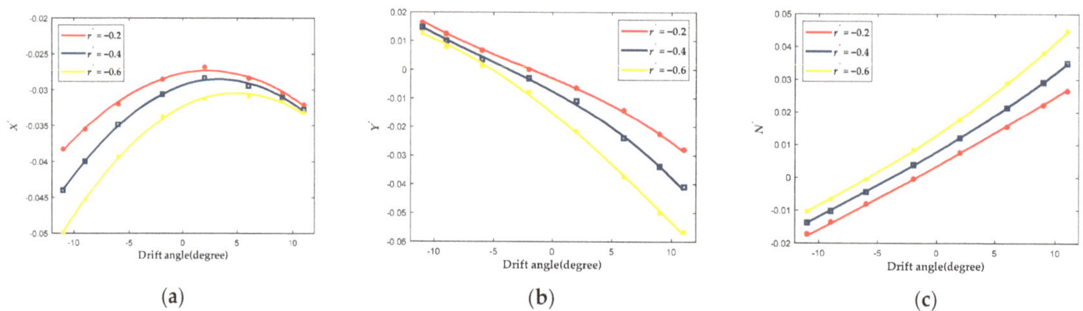

(a)　　　　　(b)　　　　　(c)

Figure 17. Simulation results of CMT (when $r' = -0.2\sim-0.6$): (**a**) longitudinal force; (**b**) lateral force; (**c**) yaw moment.

Table 5. Results of hydrodynamic derivative calculations and comparisons.

Hydrodynamic Derivative	CFD	Empirical Formula	Difference (%)	Hydrodynamic Derivative	CFD	Empirical Formula	Difference (%)
X_{uu}	−0.1015	−0.0837	21.29	Y_{vvr}	0.05243	0.0405	29.46
X_{vv}	−0.0474	−0.1391	65.91	Y_{vrr}	1.0639	0.2341	554.46
X_{vr}	−0.0019	−0.0007	164.15	N_v	−0.0784	−0.0467	67.88
X_{rr}	0.0780	0.1010	22.77	N_r	−0.0184	−0.0331	44.41
Y_v	−0.1295	−0.3082	57.98	N_{vv}	−0.0161	−0.0114	41.23
Y_r	0.0072	−0.1633	104.42	N_{rr}	−0.0209	−0.0227	7.92
Y_{vv}	−0.5809	−2.0374	71.49	N_{vvr}	−0.1012	−0.1845	45.15
Y_{rr}	−0.0219	−0.0080	174.24	N_{vrr}	0.0840	0.0257	226.85

Overall, the hydrodynamic derivatives derived from CFD calculations show acceptable differences from those calculated using empirical formulas, with large differences in some of the higher order and cross-coupled hydrodynamic derivatives.

4. Maneuverability Simulation and Verification

Utilizing the four-degree-of-freedom MMG equations for a fully rotary propelled ship, this study calculates hydrodynamic derivatives through empirical formulae and CFD simulations. Subsequently, time-domain differential equations are solved to facilitate computer simulations of the ship maneuvering dynamics, enabling the determination of its motion trajectory and maneuvering characteristics. The study conducts numerical simulations of the ship's 10° turning motion and ±10° zigzag motion under the assumption that both left and right propellers maintain constant rotational speeds and receive identical motion commands throughout the simulation process.

Concurrently, a full-scale trial is conducted in calm waters using the same commands, allowing for a direct comparison between the simulated tests and actual ship performance, as depicted in Figure 18. Additionally, a comparison of characteristic parameters for the turning and zigzag motions is presented in Tables 6 and 7.

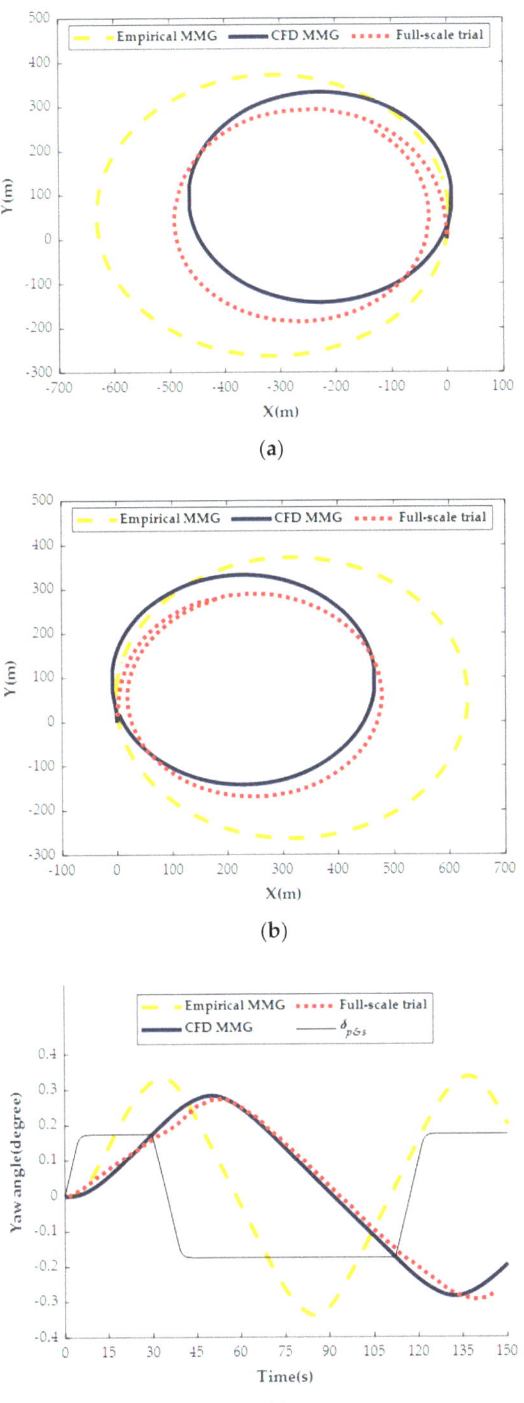

Figure 18. Simulation and test curves: (**a**) left turning at 10°; (**b**) right turning at 10°; (**c**) Zigzag test motion at ±10°.

Table 6. Comparison of characteristic parameters of turning test.

Main Parameters	Test Case	Full-Scale Trial	Empirical MMG	Error (%)	CFD MMG	Error (%)
Tactical Diameter (m)	Left turning at 10°	487.1	632.26	29.80	465.59	4.42
	Right turning at 10°	474.5	629.56	32.68	464.54	2.10
Advance (m)	Left turning at 10°	289.8	372.42	28.51	332.89	14.87
	Right turning at 10°	287.2	372.54	29.71	332.75	15.86
Transfer (m)	Left turning at 10°	195.9	329.03	67.96	220.98	12.80
	Right turning at 10°	205.8	323.69	57.28	217.19	5.53
Stabilized rolling angle (deg)	Left turning at 10°	1.99	1.64	18.00	1.72	13.57
	Right turning at 10°	1.98	1.66	17.01	1.71	13.64

Table 7. Comparison of characteristic parameters of zigzag test.

Main Parameters	Test Case	Full-Scale Trial	Empirical MMG	Error (%)	CFD MMG	Error (%)
1st overshoot (deg)	Zigzag test at ±10°	6.20	9.16	47.74	6.28	1.03
2nd overshoot (deg)		7.00	9.44	34.85	6.42	8.28

The graphical data illustrates that the ship's maneuvering parameters simulated during the CFD MMG tests align with the parameters observed in the full-scale trial. Furthermore, the ship's turning capabilities and directional stability meet the IMO's standards for maneuverability. The simulation accurately reproduced the tactical diameters observed during the turning motion to within a 5% error of the real ship test data. Nevertheless, the overall motion trajectory deviates slightly from the full-scale trial, which is caused by the interference of wind and wave factors present in the full-scale test, leading to a lateral shift in the ship's trajectory. While the current simulation only considers calm water conditions, as a result, the simulation errors for both advance and transfer are significantly larger and add to the uncertainty of rolling angle simulation; however, the transfer error is reduced when sailing upwind in the right turning at 10°. The error for the overall parameter characteristics is kept within 15%, which is an acceptable threshold, although there are also errors arising from scale effects inherent in the simulation. The CFD MMG method exhibits higher accuracy compared to the empirical MMG model.

During the zigzag maneuvering motion test, the simulated yaw direction and test curve largely align, exhibiting minimal error in the first overshoot angle. However, as calculation iterations progress, the cumulative time error incrementally escalates. Nevertheless, the overall deviation of the zigzag maneuvering motion parameters derived from the simulation remains below 10%, closely mirroring the full-scale trial data.

The methodology employed in this study markedly diminishes simulation errors across all maneuvering motion characteristic parameters compared to empirical MMG method. It can calculate more accurate hydrodynamic parameters in advance and obtain the ship's motion response through a rapid mathematical model calculation, and so its computational cost is significantly reduced compared with the CFD direct simulation method and experiments. This enhances the precision of maneuverability predictions at an acceptable computational expense, rendering it highly conducive to validating and optimizing ship maneuverability during the design phase.

5. Conclusions

This study offers a comprehensive prediction of the "Zhifei" ship's maneuvering motion, utilizing CFD technology and empirical formulas. This approach presents a viable method for accurately and rapidly forecasting maneuvering performance at the design stage of contemporary ships. Initially, the study conducts CFD numerical simulations on the captive motion model of "Zhifei", deriving all necessary hydrodynamic derivatives for the ship's maneuvering through both regression analysis and empirical formulas. Subsequently, based on the MMG mathematical model, a maneuvering motion mathematical model suitable for dual full-rotary propulsion ships is formulated. The model's turning and zigzag maneuvering motions is then numerically simulated. Finally, a full-scale trial maneuverability test is conducted, and the data from this test is compared and analyzed alongside the simulation outcomes, leading to the following key conclusions:

(1) The RANS-based numerical simulation method effectively predicts the hydrodynamic characteristics of the "Zhifei" hull. The hydrodynamic curves derived from the captive model tests align with the ship's behavior trend. However, there is a notable deviation in the hydrodynamic characteristics obtained post-regression analysis when compared to empirical formulas.

(2) This study introduces a four-degree-of-freedom maneuverability prediction technique for dual full-rotary propulsion ships. It evaluates the turning and zigzag maneuvering motion simulation results—derived from empirical formulas and CFD simulations—against full-scale trial test data, validating the mathematical model's efficacy for dual full-rotary propulsion ships. The results show that all maneuverability parameters have an error of less than 15% from the full-scale trial data. The error of the tactical diameter in the turning test is less than 5%, and the rest of the parameters may be affected by the wind and waves with an error of about 10%. The errors in the zigzag test are all within 10%.

(3) Currently, the research presented in this paper is limited to the four-degree-of-freedom maneuvering motion of ships in calm water. However, in real-world conditions, the interaction of wind, waves, currents and other environmental factors introduces some degree of error in the comparative results. Consequently, incorporating these factors into ship maneuvering studies represents the next focal point of future work.

Author Contributions: Conceptualization, Y.Y. and Q.Y.; methodology, Y.Y. and Y.T.; software, Y.Y., X.G. and Y.L.; validation, Y.T., Y.L. and X.G.; formal analysis, X.G. and Y.L.; investigation, X.G. and Y.L.; resources, Y.T., Y.L. and Y.J.; data curation, Q.Y. and Y.Y.; writing—original draft preparation, Y.Y. and Y.T.; writing—review and editing, Y.Y. and Y.J.; visualization, Y.T., Q.Y. and Y.J.; supervision, Q.Y. and Y.J.; project administration, Q.Y. and Y.J.; funding acquisition, X.G., Q.Y. and Y.J. All authors have read and agreed to the published version of the manuscript.

Funding: This research was funded by National Key Research and Development Program of China, grant number 2022YFB4301401.

Institutional Review Board Statement: Not applicable.

Informed Consent Statement: Informed consent was obtained from all subjects involved in the study.

Data Availability Statement: Data are contained within the article.

Conflicts of Interest: Author Yuhan Jiang was employed by the company Intelligent Navigation (Qingdao) Technology Co., Ltd. The remaining authors declare that the research was conducted in the absence of any commercial or financial relationships that could be construed as a potential conflict of interest.

Nomenclature

φ	roll angle	m_x	additional masses of the hull in the x axis
θ	pitch angle	m_y	additional masses of the hull in the y axis
ψ	yaw angle	X_H	hydrodynamic forces (moments)
u	surge velocity	X_P	longitudinal thrust force
v	sway velocity	Y_P	lateral thrust force
r	yaw velocity	K_P	rolling moment by propeller
m	hull mass	N_P	yawing moment by propeller
T	propeller thrust	I_{zz}	moment of inertia of the hull mass around the z axis
D	diameter of the propeller disk	J_{zz}	moment of additional inertia of the hull mass around the z axis

ρ	density of water	I_{xx}	moment of inertia of the hull mass around the x axis
n	rotation speed of propellers	J_{xx}	moment of additional inertia of the hull mass around the x axis
J	propeller advanced ratio	z_H	z axis coordinate of the point where Y_H acts
δ	rotation angle of propellers	L_{op}	longitudinal distance between propellers and ship's center of gravity
β	drift angle	L_{ps}	lateral distance between propellers
B	breadth	t_p	thrust deduction coefficient
λ	scale factor	ω_p	wake fraction at propeller position.
d	draft	R_G	convergence parameter
Δ	displacement	C_b	block coefficient
Z	number of blades	D_T	tactical diameter
Fr	Froude number	$X(u)$	resistance of the hull during straight sailing
		L_{pp}	length between perpendiculars
		L_{oa}	length overall

References

1. The Maneuvering Committee Final Report and Recommendations to the 24th ITTC. *Res. Propos.* **2005**, *1*, 137–140.
2. Yu, L.W.; Wang, S.Q.; Ma, N. Study on wave-induced motions of a turning ship in regular and long-crest irregular waves. *Ocean Eng.* **2021**, *195*, 108807. [CrossRef]
3. Abkowitz, M.A.; Liu, G. Measurement of ship resistance, powering and manoeuvring coefficients from simple trials during a regular voyage. *Soc. Nav. Archit. Mar. Eng.* **1988**, *96*, 97–128.
4. Ogawa, A.; Koyama, T.; Kijima, K. MMG report-I, on the mathematical model of ship manoeuvring. *Bull. Soc. Nav. Archit.* **1977**, *575*, 22–28.
5. Meng, Y.; Zhang, X.F.; Zhu, J.X. Parameter identification of ship motion mathematical model based on full-scale trial data. *Int. J. Nav. Archit. Ocean Eng.* **2022**, *14*, 100437. [CrossRef]
6. Fossen, T.I. A survey on Nonlinear Ship Control: From Theory to Practice. *IFAC Proc.* **2000**, *33*, 1–16. [CrossRef]
7. Sviličić, Š.; Rudan, S. Modelling Manoeuvrability in the Context of Ship Collision Analysis Using Non-Linear FEM. *J. Mar. Sci. Eng.* **2023**, *11*, 497. [CrossRef]
8. Shin, Y.J.; Kim, M.C.; Lee, K.W.; Jin, W.S.; Kim, J.W. Maneuverability Performance of a KRISO Container Ship (KCS) with a Bulb-Type Wavy Twisted Rudder and Asymmetric Pre-Swirl Stator. *J. Mar. Sci. Eng.* **2023**, *11*, 2011. [CrossRef]
9. Maciej, R. Prediction of manoeuvring abilities of 10,000 DWT pod-driven coastal tanker. *Ocean Eng.* **2017**, *136*, 201–208.
10. Sun, H.W.; Yang, J.L.; Liu, B.; Li, H.W.; Xiao, J.F.; Sun, H.B. Research on Maneuverability Prediction of Double Waterjet Propulsion High Speed Planing Craft. *J. Mar. Sci. Eng.* **2022**, *10*, 1978. [CrossRef]
11. Hajivand, A.; Mousavizadegan, S.H. Virtual simulation of maneuvering captive tests for a surface vessel. *Int. J. Nav. Archit. Ocean Eng.* **2015**, *7*, 848–872. [CrossRef]
12. Shenoi, R.R.; Krishnankutty, P.; Selvam, R.P. Study of manoeuvrability of container ship by static and dynamic simulations using a RANSE-based solver. *Ships Offshore Struct.* **2014**, *11*, 3:316–334. [CrossRef]
13. Li, S.L.; Liu, C.G.; Chu, X.M.; Zheng, M.; Wang, Z.P.; Kan, J.Y. Ship maneuverability modeling and numerical prediction using CFD with body force propeller. *Ocean Eng.* **2022**, *264*, 112454. [CrossRef]
14. Guo, H.P.; Zou, Z.J. System-based investigation on 4-DOF ship maneuvering with hydrodynamic derivatives determined by RANS simulation of captive model tests. *Appl. Ocean. Res.* **2017**, *68*, 11–25. [CrossRef]
15. Okuda, R.; Yasukawa, H.; Matsuda, A. Validation of maneuvering simulations for a KCS at different forward speeds using the 4-DOF MMG method. *Ocean Eng.* **2023**, *284*, 115174. [CrossRef]
16. Dash, A.K.; Chandran, P.P.; Khan, M.K.; Nagarajan, V.; Sha, O.P. Roll-induced bifurcation in ship maneuvering under model uncertainty. *J. Mar. Sci. Technol.* **2016**, *21*, 689–708. [CrossRef]
17. Shen, Z.R.; Wan, D.C.; Carrica, P.M. Dynamic overset grids in openfoam with application to KCS self-propulsion and maneuvering. *Ocean Eng.* **2015**, *108*, 287–306. [CrossRef]
18. Wang, J.H.; Zou, L.; Wan, D.C. CFD simulations of free running ship under course keeping control. *Ocean Eng.* **2017**, *141*, 450–464. [CrossRef]
19. Carrica, P.M.; Mofidi, A.; Eloot, K.; Delefortrie, G. Direct simulation and experimental study of zigzag maneuver of KCS in shallow water. *Ocean Eng.* **2016**, *112*, 117–133. [CrossRef]
20. Sanada, Y.; Park, S.; Kim, D.H.; Wang, Z.Y.; Stern, F.; Yasukawa, H. Experimental and computational study of hull–propeller–rudder interaction for steady turning circles. *Phys. Fluids* **2021**, *33*, 127117. [CrossRef]
21. Yasukawa, H.; Yoshimura, Y. Introduction of MMG standard method for ship maneuvering predictions. *J. Mar. Sci. Technol.* **2015**, *20*, 37–52. [CrossRef]
22. Gierusz, W. Simulation model of the LNG carrier with podded propulsion Part 1: Forces generated by pods. *Ocean Eng.* **2015**, *108*, 105–114. [CrossRef]

23. Turan, O.; Ayaz, Z.; Aksu, S.; Kanar, J.; Bednarek, A. Parametric rolling behaviour of azimuthing propulsion-driven ships. *Ocean Eng.* **2008**, *35*, 1339–1356. [CrossRef]
24. Neatby, H.C.; Thornhill, E. Turning and stopping of a ship with twin Z-drive thrusters. *Ocean Eng.* **2024**, *293*, 116641. [CrossRef]
25. Piaggio, B.; Sommariva, G.; Franceschi, A.; Villa, D.; Viviani, M. Twin-screw vessel manoeuvrability: The traditional twin-rudder configuration vs. pod-drives. *Ocean Eng.* **2023**, *271*, 113725. [CrossRef]
26. China Launches Autonomous Container Ship Zhi Fei. Available online: https://marineindustrynews.co.uk/china-launches-autonomous-container-ship-zhi-fei/#:~:text=China%20is%20reported%20to%20have%20put%20the%20world%E2%80%99s,and%20a%20normal%20operating%20speed%20of%208%20knots (accessed on 4 May 2022).
27. Kijima, K.; Katsuno, T.; Nakiri, Y.; Furukawa, Y. On the manoeuvring performance of a ship with the parameter of loading condition. In Proceedings of the Autumn Meeting of the Society of Naval Architects of Japan, 27–28 November 1990; pp. 141–148.
28. Kim, D.J.; Yun, K.; Yeo, D.J.; Kim, Y.G. Initial and steady turning characteristics of KCS in regular waves. *Appl. Ocean. Res.* **2020**, *105*, 102421. [CrossRef]
29. He, Y.Y.; Mou, J.M.; Chen, L.Y.; Zeng, Q.S.; Huang, Y.M.; Chen, P.F.; Zhang, S. Will sailing in formation reduce energy consumption? Numerical prediction of resistance for ships in different formation configurations. *Appl. Energy* **2022**, *312*, 118695. [CrossRef]
30. Grim, O.; Hachmann, D.; Hansen, H.-J.; Isensee, J.; Kastner, S.; Keil, H.; Laudan, J.; Lee, K.-Y.; Misra, P.N.; Ple, E.; et al. Ship structure loads and stresses. *Ocean Eng.* **1980**, *7*, 571–658. [CrossRef]

Journal of
*Marine Science
and Engineering*

MDPI

Review

State-of-the-Art Review and Future Perspectives on Maneuvering Modeling for Automatic Ship Berthing

Song Zhang [1,2], Qing Wu [1,2], Jialun Liu [1,3,4,*], Yangying He [5] and Shijie Li [1,2]

1   State Key Laboratory of Maritime Technology and Safety, Wuhan University of Technology, Wuhan 430063, China; songzhang@whut.edu.cn (S.Z.)
2   School of Transportation and Logistics Engineering, Wuhan University of Technology, Wuhan 430063, China
3   Intelligent Transportation Systems Research Center, Wuhan University of Technology, Wuhan 430063, China
4   National Engineering Research Center for Water Transport Safety, Wuhan 430063, China
5   School of Navigation, Wuhan University of Technology, Wuhan 430063, China
*   Correspondence: jialunliu@whut.edu.cn

Abstract: Automatic berthing is at the top level of ship autonomy; it is unwise and hasty to hand over the control initiative to the controller and the algorithm without the foundation of the maneuvering model. The berthing maneuver model predicts the ship responses to the steerage and external disturbances, and provides a foundation for the control algorithm. The modular MMG model is widely adopted in ship maneuverability studies. However, there are two ambiguous questions on berthing maneuver modeling: What are the similarities and differences between the conventional MMG maneuvering model and automatic berthing maneuvering model? How can an accurate automatic berthing maneuvering model be established? To answer these two questions, this paper firstly performs bibliometric analysis on automatic berthing, to discover the hot issues and emphasize the significance of maneuver modeling. It then demonstrates the similarities and differences between the conventional MMG maneuvering model and the automatic berthing maneuvering model. Furthermore, the berthing maneuver specifications and modeling procedures are explained in terms of the hydrodynamic forces on the hull, four-quadrant propulsion and steerage performances, external disturbances, and auxiliary devices. The conclusions of this work provide references for ship berthing mathematical modeling, auxiliary device utilization, berthing aid system improvement, and automatic berthing control studies.

Keywords: automatic berthing; maneuver modeling; bibliometric analysis; motion specifications; hydrodynamic characteristics

Citation: Zhang, S.; Wu, Q.; Liu, J.; He, Y.; Li, S. State-of-the-Art Review and Future Perspectives on Maneuvering Modeling for Automatic Ship Berthing. *J. Mar. Sci. Eng.* **2023**, *11*, 1824. https://doi.org/10.3390/jmse11091824

Academic Editor: Diego Villa

Received: 18 August 2023
Revised: 14 September 2023
Accepted: 16 September 2023
Published: 19 September 2023

1. Introduction

1.1. Background

According to statistics, 89%~96% of collision accidents [1] are caused by human error, and nearly 70% of accidents are related to the bad ship skills of the operators in the port [2]. To reduce or even eliminate the collisions caused by human errors, the maritime sector is moving rapidly towards autonomous shipping.

With the proposal of E-Navigation, the China Classification Society (CCS), Det Norske Veritas (DNV), Germanischer Lloyd (GL), Lloyd's Register of Shipping (LR), and other authority institutions successively published corresponding regulations and standards on autonomous ships. Regarded as the last-mile issue in ship operation, the berthing maneuver is the most complicated and dangerous part of the mission, with comprehensive consideration of restricted and busy waterways, off-design ship performances, and strong external disturbances. In sum, automatic berthing is at the top level of autonomy [3–5].

1.2. Application of Automatic Berthing

Shipping and technology institutions and enterprises in Asia and Europe have conducted automatic berthing studies and experiments and made remarkable progress; the applications of the automatic berthing systems are shown in Figures 1 and 2. Between 2018 and 2019, Mitsui E&S Shipbuilding Co., Ltd. (MES-S) (Tokyo, Japan), Mitsui O.S.K. Lines, Ltd. (MOL) (Tokyo, Japan), Tokyo University of Marine Science and Technology (TUMST), Akishima Laboratories (Mitsui Zosen), and MOL Ferry conducted a total of 54 auto berthing operations using a virtual pier in open water, with the training ship 'Shioji Maru' [6]. And in 2022, the project team announced the success of an actual demonstration test of their jointly developed auto berthing and un-berthing system, equipped on the large-sized car ferry 'Sunflower Shiretoko' [7]. In 2022, MOL (Mitsui O.S.K. Lines, Ltd.) completed the world's first containership sea trial for unmanned docking and undocking, with a 1870 DWT containership 'Mikage' [8]. In 2020, KASS (Korea Autonomous Surface Ship) Project [9] brought together KRISO, KAIST, Korea Maritime and Ocean University, and other institutions, to investigate autonomous ships and to release the study objective on berthing aid systems and automatic berthing prototypes. In China, Navigation Brilliance performed a series of autonomous ship tests, with the application of a berthing control system on a training ship 'ZhiTeng' [10], and achieved assisted berthing and automatic berthing on a 117 m 300 TEU container ship 'ZhiFei' [11].

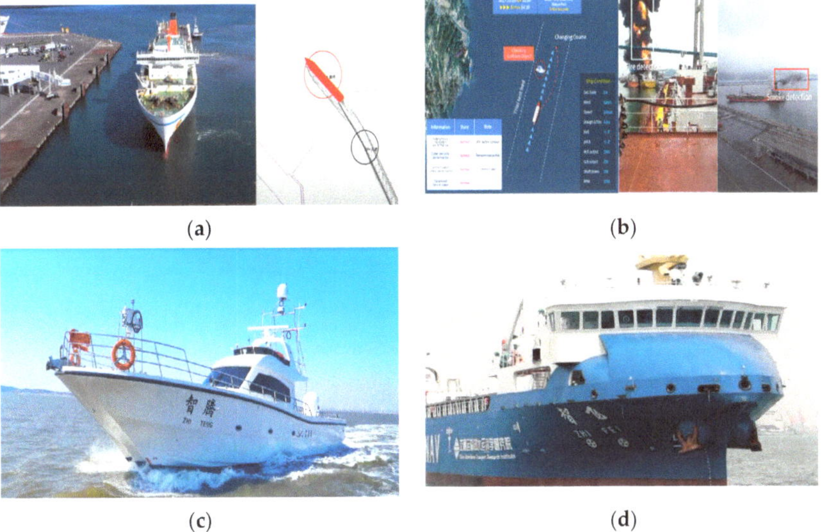

Figure 1. Application of the automatic berthing systems in Asia. (**a**) Car ferry 'Sunflower Shiretoko' [7]. (**b**) KASS berthing aid system [9]. (**c**) Training ship 'ZhiTeng' [10]. (**d**) Container ship 'ZhiFei' [11].

In 2018, Rolls-Royce and Wartsilia successively announced their achievement with the automatic berthing control system. Rolls-Royce and the Finnish state-owned ferry operator Finferries [12] successfully demonstrated automatic berthing with a developed autonomous navigation system, without any intervention from the crew, with a fully autonomous 53.8 m double-ended car ferry 'Falco'. Wartsilia [13] successfully carried out a world-first autodocking test on an 83m-long ferry 'Folgefonn'. The test covers the full ship docking procedure, performs a gradual slowing of speed, and activates the line-up and docking maneuver fully automatically until the ship is secured at the berth. In 2021, Kongsberg and Yara [14] used the world's first fully electric and autonomous container ship 'YARA Birkeland', an 80 m 120TEU open-top container ship, preliminarily achieving automatic berthing with the assistance of a Macgregor intelligent mooring system. In the same year, Volvo Penta [15] released their assisted docking system for boat docking, to remove the dynamics of wind and current and to improve the control for maneuvering in tight spaces: this is the first commercial application of an integrated berthing assistant system. An overview of the automatic berthing applications is presented in Table 1.

(a)

(b)

(c)

(d)

Figure 2. Application of the automatic berthing systems in Europe. (**a**) Car ferry 'Falco' [12]. (**b**) Ferry 'Folgefonn' [13]. (**c**) Container ship 'YARA Birkeland' [14]. (**d**) Volvo Penta docking assistance system [15].

Table 1. Application of automatic berthing.

	Name	Type	Date	Affiliations	Overview for Development
Asia	Shioji Maru	Training ship (49 m)	2018	MES-S, MOL, TUMST, etc.	▪ Carrying out the tests using a virtual pier.
	Sunflower Shiretoko	Car ferry (190 m)	2022	MOL Ferry	▪ Tests carried out on the service routes and an actual pier.
	Mikage	Container ship (95.4 m, 194TEU)	2022	MOL Ferry	▪ Calculating and visually displaying gaps and angles.
	/	System	2020	KASS	▪ Berthing aid system.
	ZhiTeng	Training ship (21 m)	2019	China waterborne transport research institute, etc.	▪ Intelligent situation awareness system; ▪ Autonomous navigation decision-making system; ▪ Autonomous control system.
	ZhiFei	Container ship (117 m, 300TEU)	2021	China waterborne transport research institute, etc.	▪ Three driving modes: manual driving, remote control, and autonomous navigation; ▪ Independent route planning, intelligent collision avoidance, automatic berthing, and disembarking.

Table 1. *Cont.*

	Name	Type	Date	Affiliations	Overview for Development
Europe	Falco	Car ferry (53.8 m)	2018	Rolls-Royce	■ Real-time, detailed pictures of surroundings; ■ 50 km remote control.
	Folgefonn	Ferry (83 m)	2018	Wartsilia	■ Hybrid propulsion; ■ Automatic wireless charging; ■ Automatic vacuum mooring; ■ Automated docking.
	YARA Birkeland	Container ship (117 m, 300TEU)	2021	Kongsberg and Yara	■ Fully electric container feeder; ■ Remote and unmanned operations.
	PENTA	Fully integrated assisted docking system	2021	Volvo Penta	■ Dynamic variable compensating; ■ Straight line movement without manual compensation; ■ Stop, slow maneuver functionality; ■ Rotation around a fixed point; ■ Micro repositioning and alignment and lateral thrust for lateral docking; ■ Human–machine interaction.

1.3. Contributions

The berthing maneuver in the harbor area is one of the key problems of ship manipulation, as the course stability and helm response of the ship is rather different from that in open-water conditions. This paper aims to explore the hot issues in automatic berthing maneuver modeling, demonstrate the similarities and differences of the conventional maneuvering modeling group (MMG) [16] model and the berthing maneuver MMG model, and emphasize the significance of berthing maneuver modeling. The main contributions of this paper are as follows:

(1) Conducts bibliometric and statistical analysis on existing automatic berthing research, and extracts the hot issues of automatic berthing.
(2) Demonstrates the similarities and differences between the conventional MMG model and berthing maneuver MMG model.
(3) Summarizes the motion specifications and hydrodynamic performances of the berthing maneuver, and provides proper mathematical expressions.

1.4. Outline

The outline of this paper is organized as follows: Section 1 introduces the technical background and application status of automatic berthing, illustrating the contributions and outline of the present paper. Section 2 performs bibliometric analysis on automatic berthing, generalizes six main topics of automatic berthing study, and indicates the similarities and differences of the conventional MMG model and the berthing maneuver MMG model. Section 3 introduces the advantages and disadvantages of three common mathematical modeling methods, and provides suggestions on the utilization of the berthing maneuver modeling method. Section 4 concludes with four motion specifications and hydrodynamic characteristics, and gives a specific mathematical modeling procedure. Conclusions and perspectives on berthing maneuver modeling are provided in Section 5. The workflow of this paper is illustrated in Figure 3.

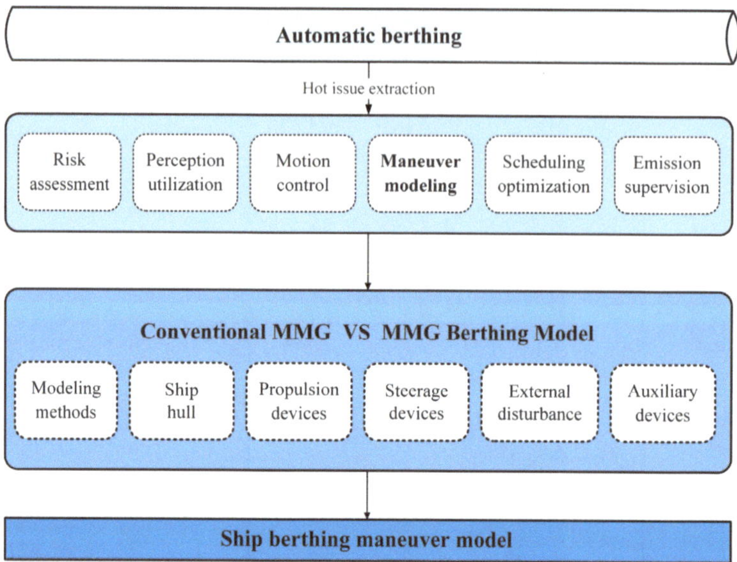

Figure 3. Workflow of the present paper.

2. Bibliometric Analysis

Scientometric analysis [17,18] presents high-level insights into the research domain; the field tendencies, important issues, study contents and methods can be readily visualized, conveniently identified and interpreted. In this section, bibliometric analysis is conducted to show the timeline and source distribution of automatic ship berthing within the collected literature database. A global correlation analysis and research focus of each study subject are discussed via research density in the following subsections.

2.1. Literature Search and Visualization

In the present work, the reference and citation database Web of Science (WoS), and bibliometric software VOS-viewer are adopted to collect references, and analyze the important issues and correlation of current references related to automatic berthing. The method and process of a literature index [19] and visualization are as follows:

(1) The first step is to search the literature in the WoS database and the KCI-Korean journal database, with the following index keywords in the theme, abstract, and keywords: ("berthing*" OR "docking*") AND ("ASV" OR "unmanned surface vessel" OR "unmanned surface vehicle" OR "autonomous surface vessel" OR "autonomous surface vehicle" OR "ship") NOT ("underwater" OR "ROV" OR "UUV" OR "AUV" OR "aircraft" OR "drone" OR "car" OR "truck" OR "launch" OR "recovery" OR "cell" OR "actuator");

(2) The second step is to go through the collected literature and remove the research that is out of this work's scope; 115 papers are retained;

(3) The third step is to supplement studies and papers that are the source of certain research or cited in the selected papers but not included in the database; finally, 134 papers are added. With the literature collection and filter, a total of 249 articles consistent with the research scope are collected.

(4) The fourth step is to extract the research objects, methods, contents, and publication time from the titles, abstracts, and keyword section of the collected references, and establish a bibliometric database.

(5)   The fifth step is to set up the threshold for the occurrence number in the extraction database, and then plot the network illustration on automatic berthing studies and density diagrams of the detailed research methods and techniques.

2.2. Global Analysis

The overall research objective dependency statistic and timeline distribution on the automatic berthing study are illustrated in Figures 4 and 5. The global network contains four highly correlated clusters, where the red band relates to the Berthing Maneuver, the blue group is associated with Control Method, and the collections of green and yellow knots are in connection with Mathematical Modeling and Safety Factors, respectively. To some extent, the four aspects with strong relevance indicate that the docking operation itself is a complicated maneuvering procedure and that the study of automatic berthing is not isolated, but correlated with other research interests, such as study constraints or objectives.

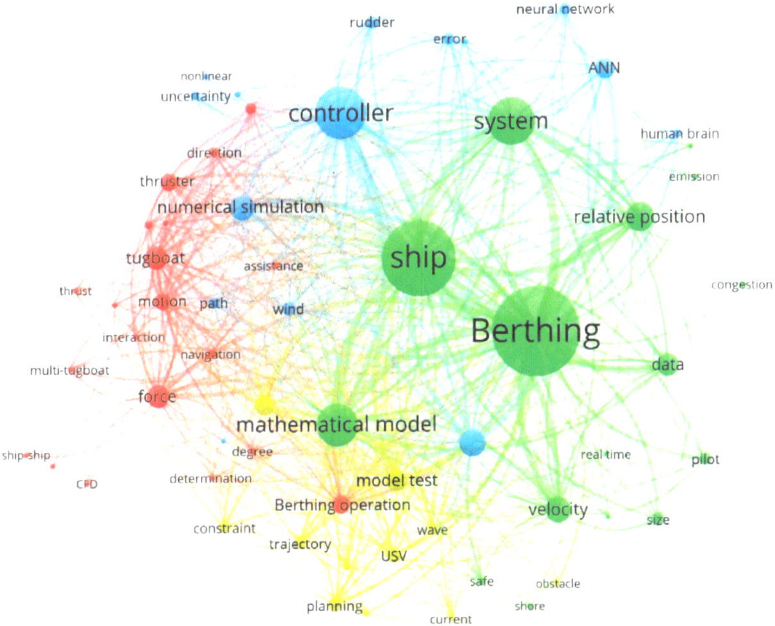

Figure 4. Overall research objective dependency statistics of the automatic berthing study.

In the database of this work, the earliest study [20] on automatic berthing could date back to the late 1980s, followed by the work of Kouichi Shouji [21] and Hiroyuki Yamato [22]. As Takeo Koyama stated, the automatic berthing system is a knowledge-based system, involving the production rules that are mostly acquired from the shipmasters, pilots, and circumstance-dependent parameters. The subsequent works on automatic berthing adopt the knowledge-based or expert-based framework as the study constraints.

Systematic studies on automatic berthing started in the 2000s, focusing on the hydrodynamic forces and ship motions in the berthing procedure. Then the control method and system came into sight, followed by neural network and other intelligence algorithms, and the low-speed low-frequency, and high-accuracy berthing control problem were realized upon the simulation level. As test methods and measurement accuracy improved by leaps and bounds, the maneuverability model on ship docking was introduced to describe the non-linear ship motion responses and mechanism and to elevate the ship motion control effectiveness. Up until now, multi-factor (trajectory, external disturbance, maneuvering velocity error, and emission, etc.)-coupled motion control on self-berthing and multi-tug assistant berthing have been important issues, among which pilot, navigation, and trajectory refer to the berthing plan, and velocity corresponds to the approaching angle and velocity in the berthing maneuver. In addition, with the development of learning and identification algorithms and data processing, the clusters of data are found to play an important role in motion control and statistical analysis study.

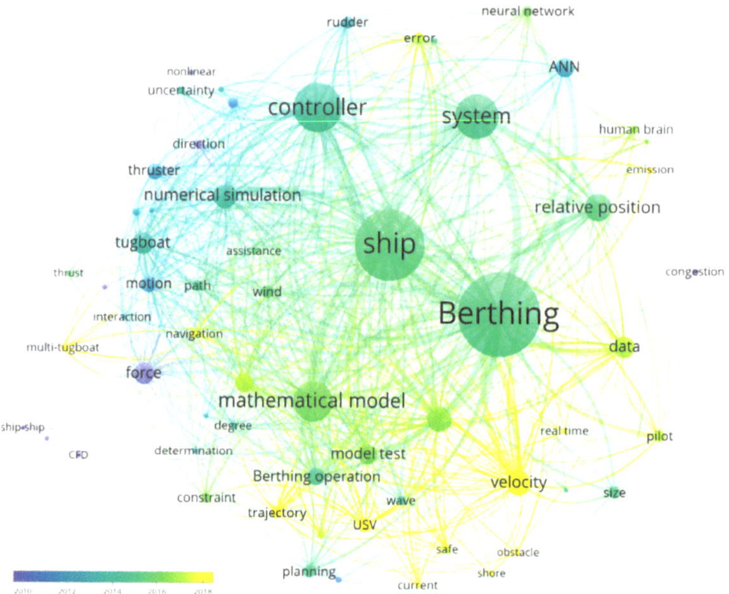

Figure 5. Timeline distribution of the automatic berthing study.

2.3. Correlation Analysis

In conclusion, the study keywords of automatic berthing are Risk Assessment, Scheduling Optimization, Emission Supervision, Perception Utilization, Motion Control, and Maneuverability Modeling. Scheduling optimization [23] and emission supervision are studied to relieve the pressure on traffic flow, improve the operation efficiency of the harbor area, and reduce air pollution. However, these two aspects make little contribution to the automatic berthing technology, and hence are not discussed here. Detailed statistical analyses of the other branches are conducted in the following parts.

2.3.1. Risk Assessment, Perception Utilization and Motion Control

Risk assessment extracts the manipulating principles and concerns of berthing operation, and quantifies the automatic control indices; it is the insurance for automatic berthing. As illustrated in Figure 6a, the safety factor could be categorized by order of importance into 'ship skills', 'quay layouts', 'external disturbances', 'ship characteristics', 'traffic flow', and 'port regulation'. Each aspect is interrelated and constrained. Ship skills, wall distance, approaching angle, and lateral speed, are essential to determine the ship's safety berthing principles [24,25], which are affected by ship actuation level and external disturbances [26]. As for under-actuated ships, traditional large ships, or unexpected weather conditions, generally it is suggested or required that the ship berth with the assistance of tugs. Moreover, the quay layouts of water depth, berth orientation, and position could also hold up the berthing process.

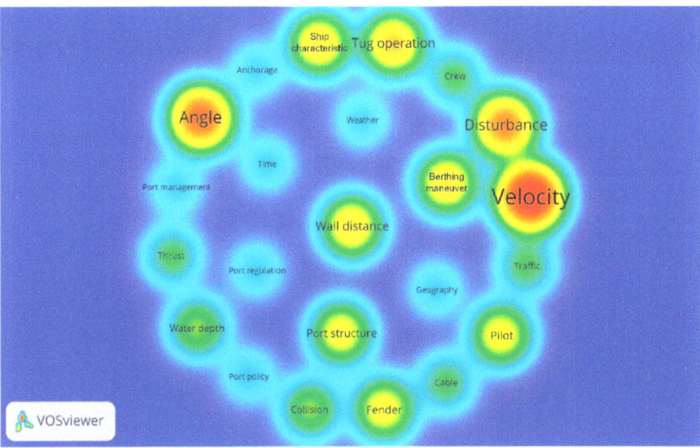

(**a**) Safety factor.

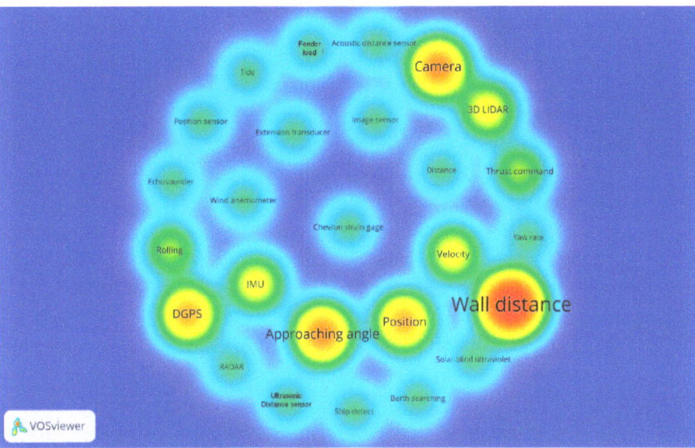

(**b**) Perception element and technique.

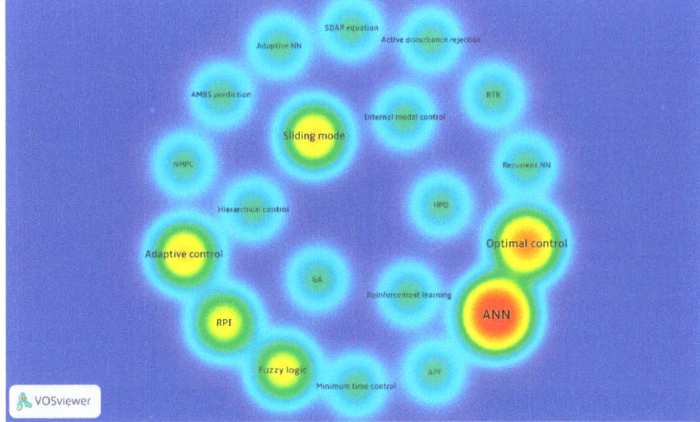

(**c**) Motion control methods.

Figure 6. Density of safety factor, perception element and technique, and motion control methods in berthing procedure.

Perception utilization acts as a pilot in automatic berthing, supports berthing strategy elaboration, environmental and state perception, and is the extra eye [27] and premise of automatic berthing. As illustrated in Figure 6b, the perception element and acquisition methods could be categorized by order of importance into 'berth and obstacle perception', 'orientation and position perception', 'own-ship state perception', 'environmental perception', and 'target-ship perception'. Each aspect is indispensable. In correspondence with safety berthing principles, approaching angle, lateral speed, and wall distance are the most significant indices. Among these, the approaching angle, lateral speed, and other own-ship states are monitored through DGPS (Differential Global Positioning System) and IMU (inertial measurement unit), the berth and bollard location and target-ship detection are determined by the camera [28], millimeter wave RADAR [29], and other position sensors, while in severe weather conditions 3D LIDAR [30], ultrasonic sensor, solar-blind ultraviolet and other measurement gages are employed to make up the deficiencies.

According to statistics, 70% of accidents are related to the bad ship skills of the drivers in the port [2,31], and thus ship motion control is of vital importance to the berthing operation. In the traditional berthing process, ship control is a complicated system with multi-input sources and multi-output terminals. It requires the officers in charge to collect and resolve massive data expressing the external conditions and own-ship states, the shipmaster and pilot to make up a berthing plan and alternative plan, and the chief officer and chief engineer to convert and supervise the instruction execution conditions of the shipmaster.

The automatic berthing process [32] is described as the following: move the ship with low speed from pose A in the proximity of the harbor to pose B lying right next to it, while simultaneously avoiding all static and dynamic obstacles. The hidden scientific control problems are to position the target ship to the final pose B with real-time feedback of the perception elements (wall distance, lateral speed, and approaching angle) in the restricted water area and with strong environmental disturbances [33]. In the berthing process, path planning [34,35], trajectory tracking [36,37], stabilization and robustness control are essential control targets. Furthermore, with regard to multi-tug-assistance berthing control, it is necessary to exert constant control on the thrust allocation induced by the assistant tugboats, and to monitor the status of the target ship [38–40]. Whether for the self-berthing or tug-assisted berthing, wall distance, approaching angle, ship speed, and yaw rate are the control indices. Accordingly, ship berthing control is a low-speed low-frequency, and high-accuracy berthing control problem.

The density of motion control methods is illustrated in Figure 6c: model predictive control [41], fuzzy logic control [42], adaptive control [43], sliding mode control [44], optimal control [45], and artificial neural network-based control [46,47] are the most common and proven control technologies. Most of the above control methods are only effective on a specified scene, and once the external conditions change the parameters of the control system are ineffective. To resolve such deficiencies, artificial neural networks and other learning algorithms [48,49] are adopted. However, the learning algorithms and intelligent algorithms are fed on massive data, which represents a high cost; moreover, once the imported berthing condition is not involved in the training database, the system makes incorrect decisions, even breaking down. Furthermore, it is reported that most marine accidents are caused by ship–ship collision and ship–shore collision [50]. In order to reduce the risk of collision accidents, a collision avoidance algorithm [51] is embedded into the control system to determine and implement the required safety margin distance between the moored ships, moving ships, and the obstacles, which increases the system load to some extent. Thus, it is essential to improve the robustness of the control system.

2.3.2. Maneuverability Modeling

The ship maneuvering model is grounded in the mechanical characteristics, in order to denote ship responses under different internal and external inputs, and is the foundation of the automatic berthing control system. There exist two main effects and applications:

one is to predict the maneuvering characteristics and help ship designers and operators know about the handling performance and the other is to provide a kinematic and dynamic foundation for the control system. As illustrated in Figure 7, the modeling methods could be categorized by intention into the 'mechanism model' and 'control model'. The 'mechanism model' involves the Abkowitz model and MMG model. In the Abkowitz model [52], the ship is considered as a whole, and the hydrodynamic forces acting on the system are expressed as the function of ship motion, rudder steerage, and external disturbances, while the MMG model [16] treats the ship as an organism composed of a ship hull, propeller, rudder, bow thruster, wind, wave, and current. Additionally, auxiliary devices like tugs, cables, anchors, and waterway constraints like shallow-water effect, bank effect, and ship–ship interaction could also be represented in the MMG model. The 'control model' treats the ship as a multiple-input and multiple-output (MIMO) system, mainly containing the dynamic model, Nomoto model, and model-free model. The dynamic model [53] solves ship motion control issues with matrix formation obtained from the rigid-body kinetics. The Nomoto model [54], the so-called ship response model, is introduced to indicate the relationship between ship turning ability and rudder steerage, and is commonly adopted in automated rudder exploitation. And the model-free model [55,56] is a new form of resolving ship motion control in the black box model.

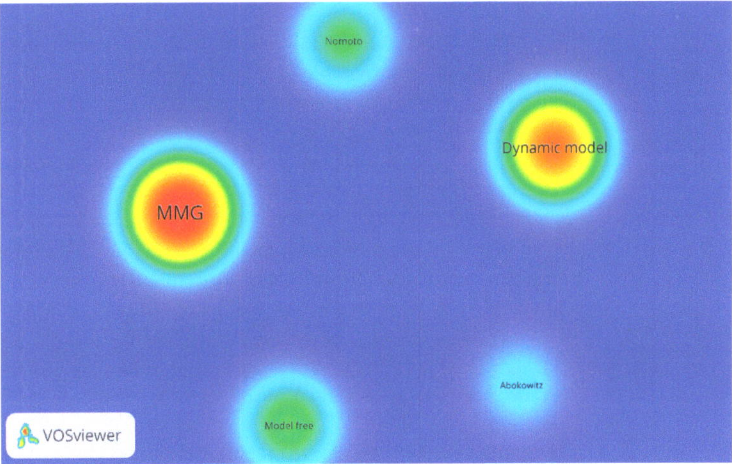

Figure 7. Density of modeling method in a berthing procedure.

In the Abkowitz model, the ship hull, propeller, rudder and external disturbances are treated as a whole: the number of hydrodynamic derivatives exceeds 60, the physical meaning of some remains unclear, and it takes a lot more captive model tests to obtain the hydrodynamic derivatives. With regard to the dynamic model and the Nomoto model, the whole ship is considered as a MIMO control system. These methods are widely used in ship motion control; however, the hydrodynamic performance is eliminated to a certain extent. In comparison, the modular MMG model independently describes the hydrodynamic forces on the ship hull, propeller, thruster, rudder, and external conditions, with few interaction coefficients forging a bond with each other. Moreover, each coefficient has a distinct physical meaning, and could be obtained with fewer captive model tests.

In practice, small ships and actuated or over-actuated ships usually perform self-berthing. Large ships often berth with the assistance of tugs, and, when conditions permit, could also conduct independent berthing. Normally, the control actuators such as the propeller and rudder are designed for relatively high speed (design speed, economic speed, or constant speed). However, during the berthing process the ship undergoes much more complicated external conditions, such as extreme low speed, high drifting, propeller reversal, shallow-water effect, bank effect, and heavy traffic flow, which eventually lead to distinct changes in the hydrodynamic forces acting on the ship hull, propeller thrust, and rudder steerage force. It is of great practical significance to study the hydrodynamic and maneuvering characteristics of the berthing maneuver. Considering the model form, influencing factors, and manipulating features, the MMG model now is utilized in berthing maneuvering modeling, and will be discussed in detail in the following sections.

2.4. Discussion of MMG Model

Some studies [57–61] on ship berthing control adopt the conventional MMG mathematical model framework as the foundation of the control algorithm. In these research studies, a number of assumptions are proposed [16]:

- Hydrodynamic forces acting on the ship are treated quasi-steadily.
- The lateral velocity component is small compared with the longitudinal velocity component.

Automatic berthing control studies [62–66] considering the berthing maneuver characteristics have shown satisfactory results in comparison of model tests. In these research studies, the berthing maneuver specifications are discussed:

- Hydrodynamic forces acting on the ship have strong non-linearity; the ship longitudinal velocity is small, and is of the same order as the lateral velocity and yaw moment.
- Thrust and steerage forces have four-quadrant characteristics.
- The ship is vulnerable to external disturbances.
- Ship motion is assisted by auxiliary devices like side thrusters, tugs, cables, and anchors.

In sum, automatic berthing simulation results based on both methods are acceptable. However, there exist two ambiguous questions on the maneuver modeling framework for the berthing maneuver motion control:

(1) What are the similarities and differences between the conventional MMG maneuvering model and the automatic berthing maneuvering model?
(2) How can an accurate automatic berthing maneuvering model be established?

To answer the first question, the similarities and differences between the conventional MMG model and the berthing maneuver model are summarized in Table 2. With regard to the modeling methods, uniform methods (including the data-based method, system-based method, and CFD-based method) are adopted to obtain the ship hydrodynamic performances for both the conventional model and the berthing maneuver model. The main differences are found in the hull motion characteristics, propulsion and steerage device performances, external disturbances, and auxiliary devices. In the conventional MMG model a moderate speed is concerned, the hydrodynamic forces are treated as linear, and the should be smaller than 20 degrees; the resultant inflow angle to the thruster and drift-angle rudder is small, and the ship motion is relatively insensitive to the external disturbances. However, in the berthing maneuver process [67,68] the ship undergoes conditions like low advance speed, large drifting, four-quadrant thrust and low rudder effect, and the ship is vulnerable and sensitive to external disturbances.

In brief, there exist distinct differences between the conventional moderate speed MMG model and the berthing maneuver MMG model, and it is essential to build a proper and accurate maneuvering model for automatic ship berthing. The kind of effects the differences lead to and how to establish an accurate model will be answered in the following sections.

Table 2. Comparison between conventional MMG model and berthing maneuver model.

<table>
<tr><th colspan="3">Indices</th><th>Conventional MMG Model</th><th>MMG Berthing Maneuver Model</th></tr>
<tr><td rowspan="1">Similarities</td><td colspan="2">Modeling methods</td><td>Data-based
System-based
CFD-based</td><td>Data-based
System-based
CFD-based</td></tr>
<tr><td rowspan="7">Differences</td><td rowspan="3">Hull</td><td>Ship speed</td><td>$0.1 < Fr < 0.3$</td><td>$Fr < 0.1$</td></tr>
<tr><td>Drifting</td><td rowspan="2">$|\beta| = [0, 20°]$
$|r'| = [0, 0.6]$</td><td rowspan="2">$|\beta| = [0, 180°]$
$|r'| > 0.6$</td></tr>
<tr><td>Rotation rate</td></tr>
<tr><td rowspan="2">Propulsion,
Steerage devices</td><td>Thruster</td><td>First-quadrant inflow angle</td><td>Four-quadrant inflow angle</td></tr>
<tr><td>Rudder</td><td>Small resultant inflow angle
High rudder effect</td><td>Large resultant inflow angle
Low rudder effect</td></tr>
<tr><td colspan="2">External disturbance</td><td>Insensitive</td><td>Vulnerable and sensitive</td></tr>
<tr><td colspan="2">Auxiliary devices</td><td>None</td><td>Side thruster, tug, and cable</td></tr>
</table>

2.5. Remarks

Taken together, automatic berthing is a coordination of perception utilization, motion control, and mathematical modeling technologies. Perception is the premise, motion control is the key, and the maneuverability model is the foundation. A precise mathematical model is required to make clear ship responses to the internal operations and external disturbances. The modular MMG mathematical model is now widely adopted in the study of maneuver modeling, due to its accessible hydrodynamic forces, clear physical meaning, and explicit and flexible structure. In light of the maneuver specifications, a comparison between the traditional model and the berthing maneuver MMG model is performed, and the traditional MMG model is found to differ from the traditional model in the hydrodynamic forces on the hull, propulsion and steerage devices, external disturbances, and auxiliary devices. As for the automatic berthing, it is essential to establish a berthing maneuver mathematical model.

3. Berthing Modeling Methods

3.1. Mathematical Modeling Methods

Grounded in the mechanical characteristics, the ship maneuvering mathematical model denotes ship responses under different internal and external inputs, and is the foundation of the automatic berthing control system. There exist two main functions, one being to assess and predict ship maneuverability, helping ship designers and operators to know about the handling performances, and the other to provide the kinematic and dynamic foundation for the control system. Based on the experience of the 25th ITTC maneuvering committee [69] and insights obtained from the SIMMAN 2008 workshop, the maneuvering prediction methods are organized into three main parts, the data-based method, the system-based method, and the CFD-based method. The overview of the maneuverability prediction methods is shown in Figure 8.

3.2. Data-Based Methods

The data-based method covers the experimental method and empirical method:

(1) The experimental method mainly contains full/model-scale free-running tests, and captive model tests. The former method establishes a database involving the ship maneuverability indices of advance, transfer, overshoot, track reach, etc., to characterize the turning, yaw-checking, and stopping abilities, and to evaluate the ship's inherent dynamic and course-keeping stabilities (shown in Figure 9a). Meanwhile, the captive model tests measure the hydrodynamic performance of ships to indicate the ship response under various external conditions.

(2) The empirical method makes a quick estimation of the resistance [70] and hydrodynamic derivatives [71,72] of a target ship. This method conducts regression analysis on massive captive model test results of ships within a hull-type set; the hydrodynamic derivatives are expressed as functions of the ship's main dimensions, and form parameters (shown in Figure 9b).

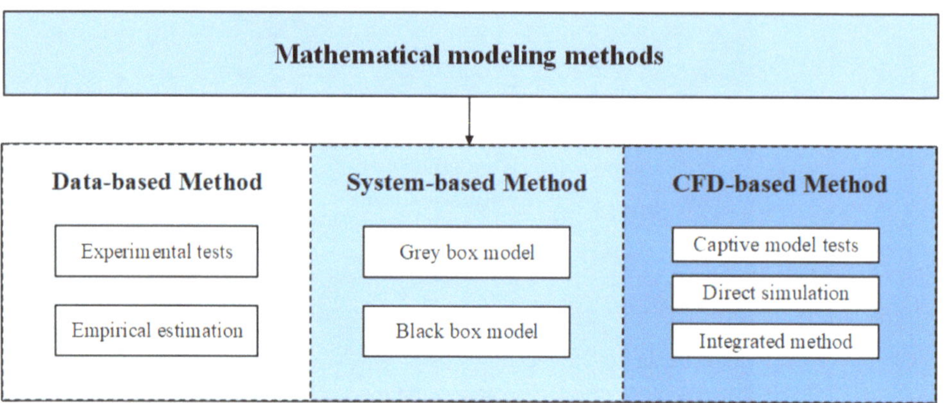

Figure 8. Overview of maneuverability modeling and prediction methods.

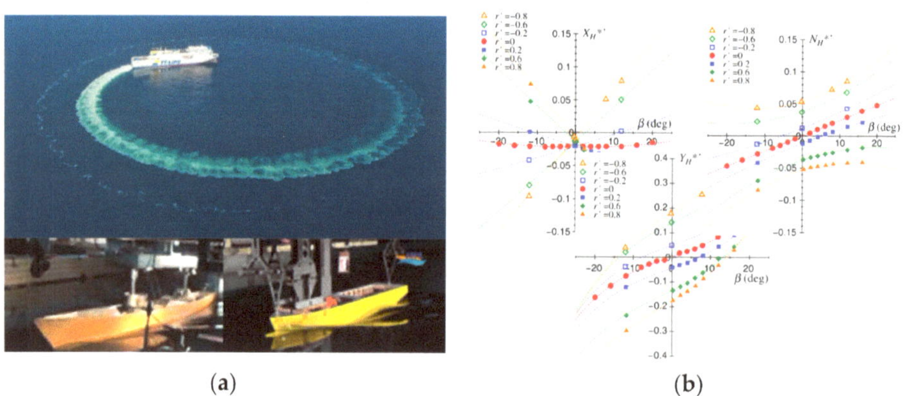

(a)   **(b)**

Figure 9. Data-based method. (**a**) Sea trails [73] and Model tests. (**b**) Regression analysis [16].

The data-based method predicts the resistance and maneuverability of ships with flexibility. However, there exist some defects: the full-scale trails and model-scale tests normally come with a high cost; the database and the regression formulas rely heavily on the ship-type spectrum, and once the studied ship is not involved in the spectrum, or shows obvious differences in the hull profiles, this method cannot predict the maneuverability with certain precession. Additionally, the accuracy of this method is also influenced by scale effects.

3.3. System-Based Methods

The system-based method, also known as the system identification (SI) method, mainly includes the grey box model and black box model. It is a control strategy that establishes a mathematical model equivalent to the measuring system based on the system's input and output data [74]:

(1) The grey box model sets a prior model structure [75]; some identification algorithms, such as maximum likelihood (ML) [76], Kalman filtering (KF) [77], the least

squares method (LS) [78] or the improved algorithm are used to identify the experiment/simulation parameters like ship speed, yaw rate, propeller revolution, rudder angle, and trajectories. The maneuverability of the target ship is then obtained (shown in Figure 10a). However, these methods have some inherent disadvantages: the accuracy is sensitive to signal noise and initial estimations, and simultaneous drift is another critical issue.

(2)   To cope with these defects, the black box model is proposed. No prior information is needed other than the datasets to gain the mapping relationship between system input variables and output variables [79] (shown in Figure 10b). Machine learning and deep learning techniques have been successfully applied as tools to establish the identified model, for example, the least-squares support-vector machine (LS-SVM) [80], the fully connected neural network [81], the deep neural network [55,82], etc.

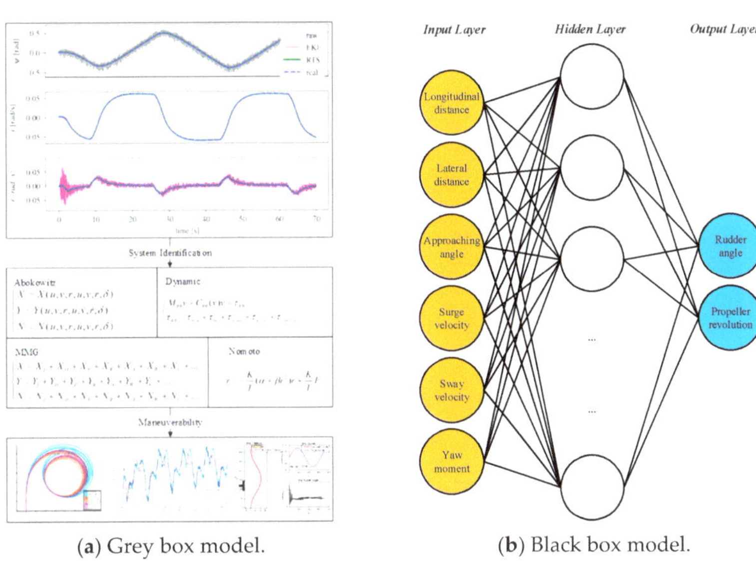

(**a**) Grey box model.   (**b**) Black box model.

Figure 10. System-based methods.

In general, the system identification method is a data-driven modeling method, utilizing the training data collected from simulation results, free-running model tests or full-scale trials. The accuracy is highly dependent on the training data, and the qualified and continuous-excitation input signal ensures better identification performance and generalization ability. However, once the training data are insufficient, or the excitation is weak, the accuracy and efficiency of identification reduces dramatically. Moreover, the specialty of the SI method in coping with the collected data indicates that this method could not predict the ship maneuverability at the design stage or in unknown conditions, but that it could improve the mechanical model accuracy. As for ship berthing maneuver modeling, an integrated SI-CFD method could be utilized to optimize the mechanical model accuracy and to provide a basis for the control algorithm.

3.4. CFD-Based Methods

The CFD-based method is the mapping of physical captive model tests and maneuvering model tests. The hydrodynamic forces are solved with the potential flow method or viscous flow method, and the method itself consists of virtual captive model tests, direct simulation, and the integrated method:

(1)   The virtual captive model tests method conducts specific maneuvering test simulations based on the maneuvering model [83,84] of the target ship, the hydrodynamic characteristics of which are obtained via the CFD method (shown in Figure 11a,

where upper left is the dimensionless longitudinal force X', upper right the propeller thrust coefficient K_T, torque coefficient K_Q, and thrust efficiency η_0, lower left the comparisons of the turning maneuver trajectory, and lower right the comparisons of the heading angle ψ and rudder angle δ time histories). Moreover, the propeller–rudder interaction, ship–ship interaction, ship–bank interaction, shallow-water effects, and detailed ship amplitudes and flow field development could also be obtained. This is the most effective, economical, and widely used method for studying the maneuverability of a ship.

(2)　The direct simulation method indicates that the maneuvering model tests are performed with the CFD method directly (shown in Figure 11b, where the upper part is the turning maneuver, and lower part the zig-zag maneuver). This method can assess the maneuverability of a ship under various external conditions and working conditions (calm water, regular and irregular waves, constraint water, wind, propeller reversal, etc.), and observe the response of the target ship to rudder/propeller operations. However, this method demands longer research periods and stronger computing power, and it is not the best option for the maneuverability study at present.

(3)　In consideration of the research requirements, efficiency, and rapidity, a hybrid method that integrates the empirical method and CFD method is constructed, based on the study experience and solid foundation [85–87]. The hydrodynamic performances of the ship hull, propeller, and rudder are obtained from the CFD method, and the hull–propeller–rudder interaction factors are solved by empirical methods. The accuracy of this method is affected by the ship-type diversity involved in the empirical formula.

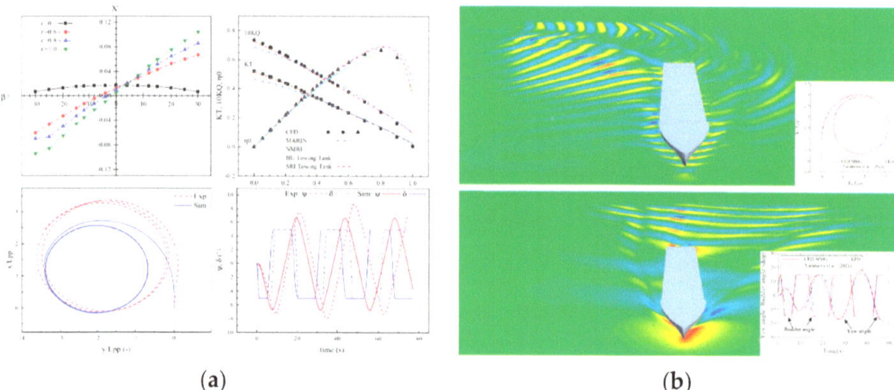

(a)　　　　　　　　　　　　　　　(b)

Figure 11. CFD-based methods. (**a**) Conventional CFD simulation [87]. (**b**) Direct CFD simulation [88].

3.5. Remarks

The berthing maneuver modeling methods include data-based, system-based and CFD-base methods (shown in Table 3). Although the data-based methods are the true reflection of navigation practice, an enormous amount of time and money are injected into experiment design and data collection. The system-based methods are very much data-dependent; however, in the berthing process, the ship hydrodynamic forces, propeller thrust, and rudder steerage force are strongly non-linear and vulnerable to external disturbances, so that it is hard to obtain valid data. Moreover, the physical meaning is unclear and the model structure is unknown in the black box model. The CFD method could intentionally control and change the study circumstances and the output hydrodynamic forces with acceptable accuracy. Hence, the CFD-based method could be used to establish the berthing maneuvering model. Under certain circumstances, a hybrid method integrates the empirical method, and the CFD method is introduced to further improve the calculation efficiency.

Table 3. Specifications of various maneuverability prediction methods.

	Prediction Methods	Advantages	Disadvantages
Data-based	Full/model-scale free-running tests	▪ Reflection of reality ▪ Solid and repeatable ▪ Custom external conditions	▪ Large basin ▪ Physical insight loss ▪ High cost
	Captive model tests	▪ High accuracy ▪ Solid and repeatable ▪ Physical insights ▪ Custom external conditions	▪ Massive tests ▪ High sensor precision ▪ High cost
	Empirical methods	▪ Quick prediction ▪ High adaptability ▪ Low cost	▪ Insufficient for uncovered hull forms ▪ Physical insight loss ▪ Reliance on mathematical model type
System-based	Grey box	▪ Full/model-scale model ▪ Adaptable to various model types ▪ Low cost	▪ Reliance on an algorithm and data accuracy ▪ Susceptible to noise and external disturbance
	Black box	▪ Full/model-scale model ▪ Model-free ▪ Low cost	
CFD-based	Virtual captive model tests	▪ Full/model-scale model ▪ Physical insight ▪ Sufficient accuracy ▪ Custom external conditions	▪ Reliance on simulation solver and algorithm accuracy ▪ Requirement of high grid density and computing power
	Direct simulation		
	Integrated method	▪ Quick prediction ▪ Sufficient accuracy ▪ Low cost	▪ Reliance on existing database and algorithm accuracy ▪ High computing power

4. Berthing Maneuver Modeling

In the berthing process, ship motion is determined by the hydrodynamic forces acting on the ship hull, thrust force induced by the propeller and thruster, steerage force generated by the rudder, external forces like water-cushion effects, wind, and current, and additional forces provided by auxiliary devices like the side thruster, tugs, and cables. In conclusion, the forces could be classified into four hydrodynamic features, within which the most important factors are as follows:

(1) Low-speed effect, large drifting and yaw rate. Compared with the service speed, in the berthing process the longitudinal velocity is very low, the lateral speed and yaw rate are of the same magnitude, the hydrodynamic forces and moments acting on the ship hull present strong non-linearity, and the ship motion covers the full drifting conditions.

(2) Four-quadrant propulsion and steerage. Ships in berthing operation need to operate the main engine frequently to adjust the ship amplitude and maintain the rudder steerage. Under such circumstances, the propeller and rudder work in four-quadrant conditions, and their performances are explicitly different from the designed capabilities.

(3) External disturbances. Under berthing maneuver, thrust due to the propeller cannot counteract the disturbances induced by external environments. To maintain the steerage of the ship, the external disturbances, including the water-cushion effect (shallow-water effect, bank effect, ship–ship interaction), wind, and current, should be considered.

(4)   Auxiliary-device-induced forces. Due to the small velocity and low propeller revolution, the rudder is affected by the wake, and the crabbing motion and turning motion of the ship usually count on the assistance of auxiliary devices like side thrusters, tugs, and anchors.

4.1. Hydrodynamic Forces Acting on the Hull

4.1.1. Ship Speed

In accordance with the ship maneuvering velocity, surface ship motions could be divided into three categories: low-speed motion, moderate-speed motion, and high-speed motion [89]. In consideration of ship safety, marine structure safety, and personnel safety, whether in the wharf design guidelines, or the measured actual operation, the berthing ships are asked to impact with the dock at a low or extremely low speed and with a parallel attitude. Namely, the ship berthing maneuver is a typical variable-speed period, ranging from service speed to low speed, and covering harbor entry to ship docking.

Apart for the speed variation, there exist multiple factors affecting the ship's maneuverability in the berthing process. For instance, ship dimension and displacement determine the difficulty of changing/maintaining kinetic states; restricting waters would raise the squat effect, bank effect, and bank-cushion effect; winds and tidal currents could lead to drifting and shifting of a ship; and under the assistance of tugs, ship maneuverability could be greatly improved. To better understand the motion response of ships under different steerage and internal and external conditions, it is important to perform further studies to reveal the hydrodynamic mechanism and to characterize the maneuvering indices.

4.1.2. Drifting and Rotation Rate

In the berthing process, ships undergo larger drifting ($|\beta| = [0, 180°]$) and greater turning angular velocity ($|r'| > 0.6$); the ship longitudinal velocity component is of the same dimension as the lateral component velocity and yaw rate, or even far smaller. Thus, concerns about the hydrodynamic forces move from the friction-dominant longitudinal force to the pressure-dominant lateral force and yaw moment. Under such states, the hydrodynamic forces present strong non-linearity, and the traditional linear mathematical model cannot describe the hydrodynamic forces and moments accurately.

Within the framework of the modular mathematical model, three resolutions are introduced to express the hydrodynamic forces acting on the ship hull:

(1)   The piecewise model [89] involving the small drifting model, moderate drifting model, and large drifting model. It should be noted that the moderate drifting model is the interpolation of the small drifting model and the large drifting model.
(2)   The unified model [68,90], based on the cross-flow drag theory, to express the ocean and harbor-area maneuvering.
(3)   The table model [91,92], with direct application of the hydrodynamic forces.

The small-drift-angle model [93] is expressed as:

$$\begin{cases} X_H = R_0 + X_{vv}v^2 + X_{vr}vr + X_{rr}r^2 \\ Y_H = Y_v v + Y_r r + Y_{|v|v}|v|v + Y_{|v|r}|v|r + Y_{|r|r}|r|r \\ N_H = N_v v + N_r r + N_{|v|v}|v|v + N_{vvr}v^2 r + N_{vrr}vr^2 \end{cases}, \tag{1}$$

where X_H and Y_H are the hydrodynamic forces, and N_H is the hydrodynamic moment acting on the ship hull, R_0 is the resistance under the straight moving condition, v the lateral component of ship velocity, r the yaw rate, and X_{vv}, Y_r, N_{vvr}, et al. the hydrodynamic derivatives.

Based on cross-flow drag theory, the large-drift-angle model [94] is expressed as:

$$\left\{ \begin{array}{l} X_H = X_H(r=0) + X_{vr}vr + X_{rr}r^2 \\ Y_H = Y_H(r=0) + Y_r|u|r + \frac{1}{2}\rho dC_d\left\{ Lv|v| - \int_{-L/2}^{L/2} |v + C_{rY}xr|(v + C_{rY}xr)dx \right\} \\ N_H = N_H(r=0) + N_r|u|r - \frac{1}{2}\rho LdC_d\left\{ Lv|v| - \int_{-L/2}^{L/2} |v + C_{rN}xr|(v + C_{rN}xr)xdx \right\} \end{array} \right. , \tag{2}$$

where $X_H(r=0)$, $Y_H(r=0)$, $N_H(r=0)$ represent the hydrodynamic derivatives related to the lateral speed, u is the longitudinal speed, v the lateral speed, r the yaw rate, C_d the cross-flow resistance coefficient with drift angle $\beta = 90°$, C_{rY} and C_{rN} the correction factor, L the ship length, d the ship draft, and x the longitudinal distance from the mid-ship point. Based on cross-flow drag theory, the unified model [68] is defined as:

$$\left\{ \begin{array}{l} X_H = \frac{1}{2}\rho Ld\left\{ \left[X'_{0(F)} + (X'_{0(A)} - X'_{0(F)})(|\beta|/\pi) \right]uU + \left((m'_y + X'_{vr})Lvr \right) \right\} \\ Y_H = \frac{1}{2}\rho Ld\left[Y'_v v|u| + (Y'_r - m'_x)Lru - (\frac{C_d}{L})\int_{-L/2}^{L/2} |v + C_{rY}xr|(v + C_{rY}xr)\cdot dx \right] , \\ N_H = \frac{1}{2}\rho L^2 d\left[N'_v vu + N'_r Lr|u| - (\frac{C_d}{L^2})\int_{-L/2}^{L/2} |v + C_{rN}xr|(v + C_{rN}xr)x\cdot dx \right] \end{array} \right. \tag{3}$$

where $X'_{0(F)}$ and $X'_{0(A)}$ are the straight forward and astern resistance coefficients, and m_x and m_y are the added masses of x and y axis directions, respectively.

And the table model [91] is defined as:

$$\left\{ \begin{array}{l} X_H = \frac{1}{2}\rho Ld\left\{ \left[U^2 + (Lr)^2 \right]C_{HX}(\beta, \alpha_r) - U^2 R'_0\cos\beta \right\} \\ Y_H = \frac{1}{2}\rho Ld\left[U^2 + (Lr)^2 \right]C_{HY}(\beta, \alpha_r) , \\ N_H = \frac{1}{2}\rho L^2 d\left[U^2 + (Lr)^2 \right]C_{HN}(\beta, \alpha_r) \end{array} \right. \tag{4}$$

C_{HX}, C_{HY}, and C_{HN} are the hydrodynamic force coefficients represented as functions of the ship drift angle β and the yaw rate angle α_r, $\beta = \tan^{-1}(-v/u)$, $\alpha_r = \tan^{-1}(rL/U)$. The hydrodynamic forces X_H, Y_H and N_H are non-dimensionalized by $0.5\rho Ld[U^2 + (Lr)^2]$ and $0.5\rho L^2 d[U^2 + (Lr)^2]$.

4.2. Propulsion and Steerage Devices

The propeller and rudder are the key elements in ship maneuvering [95]. In the berthing process, the main engine and rudder are frequently operated to achieve the turning, braking, and reversing of the ship. Under such operations, the propeller and rudder work with off-design conditions. With respect to the safety and efficiency concerns of ship docking/undocking, it is important to fully understand the performances of the propeller and rudder. The off-design performance refers to the propeller and rudder characteristics under the following telegraph conditions: ahead ship ahead, ahead ship astern, astern ship astern, and astern ship ahead. The four-quadrant performance is implied by the relationship between thrust and inflow angle (K_T-β_p), or the thrust and propeller hydrodynamic pitch angle (K_T-θ_p); the diagrams of the four-quadrant propulsion and steerage are summarized in Figure 12, and the correspondence of ship velocity U and propeller resolution n_p is shown in Table 4:

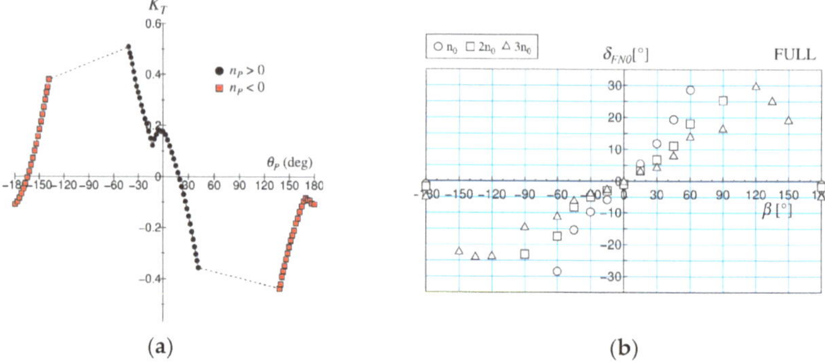

Figure 12. Four-quadrant steerage of propeller and rudder. (**a**) Propeller four-quadrant performance [67]. (**b**) Rudder four-quadrant performance [96].

Table 4. Correspondence of U and n_p to four-quadrant motion and θ_p.

Motion	Quadrant	θ_p	U	n_p
ahead ship ahead telegraph	I	0–90°	ahead	normal
ahead ship astern telegraph	II	90°–180°	ahead	reverse
astern ship astern telegraph	III	−180°–−90°	astern	reverse
astern ship ahead telegraph	IV	−90°–0	astern	normal

At present, the modeling of the four-quadrant propeller and rudder is mature and practical, the expression of the four-quadrant propeller mainly relies on the findings of Van Lammeren [97], and the rudder force is depicted by the work of Yoshimura [98] and Yasukawa [96]. The four-quadrant propeller performance [97] could be expressed as:

$$\begin{cases} X_P = (\rho/2)S_P V_r^2 \left[(1 - t_p)K_T(\theta_P)C_T(\theta_P)\right] \\ X_P = (\rho/2)S_P V_r^2 C_{PY}(\theta_P) \\ N_P = (\rho/2)S_P V_r^2 C_{PN}(\theta_P) \end{cases} \tag{5}$$

where X_p, Y_p, N_p are the thrust force on the longitudinal, lateral and yaw directions, respectively, S_p is the area of propeller span, V_r the resultant inflow velocity to the propeller, K_T the thrust coefficient, C_T the effect thrust coefficient, θ_p the propeller pitch angle, and C_{PY} and C_{PN} are the lateral force and torque moment coefficient, respectively. And the propeller pitch angle and resultant inflow velocity to the propeller is defined as:

$$\begin{cases} \theta_P = \tan^{-1}\left(\frac{V_A}{0.7\pi n_p D_p}\right) \\ V_r = \sqrt{V_A^2 + (0.7\pi n_p D_p)^2} \end{cases} , \tag{6}$$

where V_A is the inflow velocity, n_p the propeller revolution, and D_p the propeller diameter.

The rudder forces [16] on the longitudinal, lateral and yaw directions X_R, Y_R, and N_R are expressed as:

$$\begin{cases} X_R = -(1 - t_R)F_N \sin\delta \\ Y_R = -(1 + a_H)F_N \cos\delta \\ N_R = -(x_R + a_H x_H)F_N \cos\delta \\ F_N = (1/2)\rho A_R U_R^2 f_\alpha \sin\alpha_R \end{cases} , \tag{7}$$

where F_N is the rudder normal force, t_R the rudder deduction factor, δ the rudder angle, α_H the rudder force increase factor, x_R the longitudinal coordinate of rudder position, x_H the longitudinal coordinate of the acting point of the additional lateral force, A_R the profile area of the rudder, U_R the resultant inflow velocity to the rudder, f_α the rudder lift gradient coefficient, and α_R the effective inflow angle to the rudder. Under small drifting conditions, the longitudinal inflow velocity component u_R to the rudder is expressed as:

$$u_R = \varepsilon u(1 - w_P)\sqrt{\eta\left\{1 + \kappa\left(\sqrt{1 + \frac{8K_F}{\pi J_P^2}} - 1\right)\right\}^2 + (1 - \eta)}, \tag{8}$$

where J_P is the advance ratio. Under large drifting conditions [96], the longitudinal inflow velocity component to the rudder is expressed as:

$$u_R = \begin{cases} u_R^{**} & u = 0 \\ u_R^{*} & u \neq 0, (u_R^{**} - u_R^{*})\mathrm{sgn}(u) < 0, \\ u_R^{**} & u \neq 0, (u_R^{**} - u_R^{*})\mathrm{sgn}(u) > 0 \end{cases} \tag{9}$$

$$\begin{cases} u_R^{**} = 0.7\pi n_p D_p C_{UR} \\ u_R^{*} = u_p\varepsilon\left\{\eta\kappa\left(\mathrm{sgn}(u)\sqrt{1 + \frac{8K_T}{\pi J_p^2}} - 1\right) + 1\right\} \end{cases}, \tag{10}$$

where ε is the ratio of wake fraction at propeller and rudder positions, u the ship longitudinal velocity component, ω_p the rudder wake fraction ratio, η the propeller-diameter-to-rudder-span ratio, κ and C_{UR} are experimental constants, and u_p is the longitudinal inflow velocity component to the propeller.

The essential issues in the four-quadrant performance are the inflow angle and relative position of the propeller and rudder. For instance, in straight-ahead conditions, the propeller is affected by the ship hull wake, and the rudder is affected by the hull wake and propeller slipstream. In maneuvering conditions, the propeller thrust is impacted by the hull wake and current, while the rudder steerage force is impacted by the superposition of hull wake, propeller slipstream, and current. In reverse navigation, the propeller and rudder are only influenced by the current flow, but the induced forces meet a decrease due to the relative position of the ship.

In practice, it is very hard for a ship to achieve self-berthing only with the adoption of an internal combustion engine and bow thrusters. Thus, to improve the thrust efficiency and ship maneuverability, several new types of propulsion systems are introduced, and the comparisons of the common propulsion systems are listed in Table 5. With the investigation [27] of ship masters, pilots, and port managers, the three combined propulsion systems are considered the most effective methods for self-berthing: an all-electric ship with azimuth thrusters, a ship with conventional propelling system and jet thrusters, and a ship with jet propulsion and thrusters. The mathematical descriptions [99–102] of such propulsion systems differ from the traditional propeller–rudder system, and further studies are needed to reveal the thrust performances. All things considered, precise expression of the propeller thrust and the hull–propeller–rudder interaction factors are essential for accurate maneuverability prediction.

4.3. External Disturbance

In the ship berthing process, the external disturbances mainly refer to the water-cushion effect (shallow-water effect, bank effect, and ship–ship interaction) (shown in Figure 13), wind, and current. All these phenomena could be attributed to the pressure distribution variation upon the ship hull, and finally lead to changes in ship resistance and attitude. The water-cushion effects occur in confined waters [103], the flow velocity difference is observed between the bow and stern section, or between the port side and starboard side, and the asymmetric flow causes pressure difference on a ship and makes additional resistance and amplitude variation. With regard to the wind, it acts on the superstructure of the ship, as the larger the windward area, the stronger the wind effect. And the current normally influences the resultant inflow velocity and angle to the thruster and rudder.

Table 5. Overview of propulsion systems.

Propulsion	Advantages	Disadvantages
Conventional propulsion	■ Inexpensive fuel ■ Low-cost installation ■ Long-lasting	■ Heavy ■ Valuable space ■ Pollutant
Azimuth electric diesel	■ Effective design ■ Reduced noise and vibration ■ Redundancy ■ Efficiency ■ Maneuverability	■ Very Expensive ■ Difficult maintenance ■ High-quality distributing network ■ Significant safety level
Mechanically azimuth thruster	■ Maneuverability ■ Hardly needs tugs ■ No need for rudders	■ Gearbox needed ■ Expensive ■ Less efficient than conventional propulsion
Stern-bow thrusters	■ Assistance with ship turning ■ Docking without tugs	■ Only effective under 3 knots of sailing speed ■ Resistance increase
Water-jet propulsion	■ Maneuverability ■ Improved shallow water operation ■ Reduced noise	■ Expensive ■ Less efficient than a propeller at low speed ■ Risk of intake grill clog.
Water-jet thrusters	■ Smaller hull penetration ■ More efficient than bow thrusters at advancing	■ In need of powerful pumps

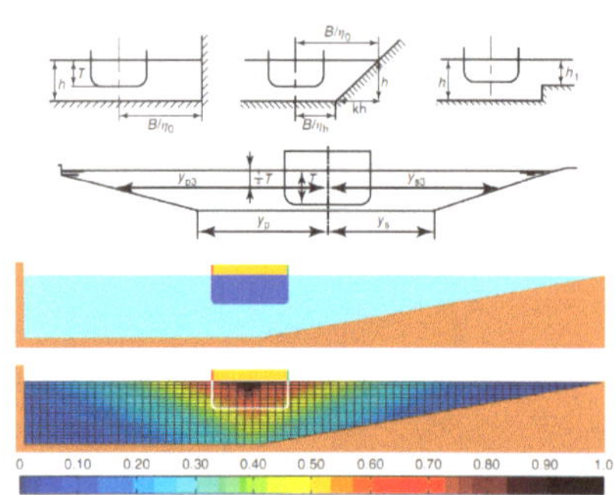

(**a**) Influence factors of water-cushion effects.

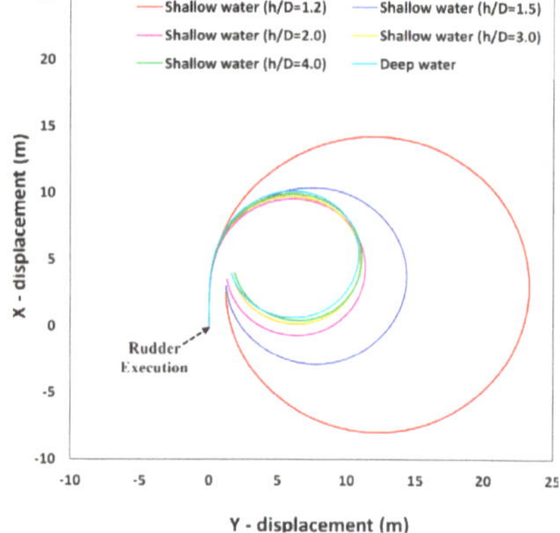

(**b**) Shallow-water effects on maneuverability.

Figure 13. Water-cushion effects [104].

4.3.1. Shallow-Water Effect

In the harbor area, the water depth is relatively shallower, and the ship may be affected by the shallow-water effect. The shallow-water effect brings ship squat and trim and is affected by the under-keel clearance, ship speed, and seabed topography. It is commonly accepted that the shallow-water effect emerges when the water-depth-to-draft ratio (h/T) is smaller than 4; PIANC [105] made an arbitrary distinction among deep ($h/T > 3.0$ or UKC > 200%), medium deep ($1.5 < h/T < 3.0$), shallow ($1.2 < h/T < 1.5$), and very shallow water ($h/T < 1.2$). Moreover, to indicate the gravity influence on fluid motion or the characteristics of the wave-making resistance in shallow water, the Froude depth number F_{rh} is induced [106]: $F_{rh} < 1$ is called a subcritical flow, $F_{rh} \approx 1$ is denoted as a critical flow, and $F_{rh} > 1$ is characterized as a supercritical flow. Generally, the shallower the water depth, the greater the shallow-water effect, namely the added resistance, ship squat, and trim amplitude increase with the decrease in water depth [104,107–110]. The decrease in water depth boosts the damping moment on the ship hull and leads to smaller turning angular velocity and drift angles, and in turn, the relatively small drift angle reduces the ship's turning rate. Accordingly, the hull–propeller–rudder interaction factors are significantly affected [94,111] with the reduction in water depth; the thrust reduction factor t_p [112], effective wake coefficient ω_p [113], and rudder force increase factor α_H present an increasing tendency, the acting point of the rudder-induced additional lateral force (x_H) moves towards the bow slightly, and the flow straightening coefficient γ_R drops by a specific point and reverses to growth [112], namely, follows the cubic parabola trend. As for the steering resistance deduction factor t_R [112], it is assumed to be constant, like the deep-water conditions. As a result, in shallow-water maneuvering conditions, ships obtain better sailing and heading stability, and worse turning ability.

4.3.2. Bank Effect

The ship berthing maneuver is the process of approaching and stopping near the quay wall, and the ship is influenced by the bank effect under such conditions. The bank effect arouses the bow cushion, bank suction, and heel, and is affected by ship–bank distance, ship speed, water depth, propeller action, and bank geometry [114]. Most of these parameters and their influence on bank effects are not independent of each other. Generally, no significant bank influence is observed when the bank proximity distance is greater than three times the ship breadth [115]; within the ship-to-bank distance spectrum of $[0.25B, B]$, the bank effects are most significant, and the interaction effects increase dramatically as the lateral distance decreases [116]. Lataire [117] conducted over 10,000 captive model tests to investigate the influence of bank characteristics on ship–bank interaction: for instance, the distance of significant influence of a ship to a vertical piercing bank is introduced as a function of ship breadth and Froude depth number, and the lateral force and yaw moment induced by the submerged slop bank is expressed as an exponential function of that induced by the surface piercing bank [118]. In brief, the closer the ship–bank distance, the faster the ship speed, and the shallower the water depth, the severer the bank effects. Furthermore, under extreme shallow-water conditions ($1.1 < h/T < 1.25$), the bank repulsion effect is observed on the bare hull; however, due to the propeller revolution, the bank repulsion changes into bank attraction. With regard to the hydrodynamic derivatives, it is found to make a relatively small difference, compared to shallow-water conditions [119]. Hence, for safe navigation of self-berthing ships, it is necessary to maintain larger rudder angles than in the conventional operation [120] and to take the helm to direct the bow towards the closet bank, to compensate for the bank-induced yaw moment [104].

4.3.3. Ship–Ship Interaction

The ship–ship interaction induces ship attraction, repulsion, and heel, and is affected by ship status (overtaking, head-on, parallel, and moored-passing), ship–ship distance, ship speed, water depth, and ship dimension and profile, and by secondary influences from the propeller and rudder [121]. The hydrodynamic interactions vanish when the longitudinal distance between adjacent ships is larger than twice the ship length [122]. Generally, for safety berthing concerns, such a close interval is not supposed to be allowed or observed in the self-berthing process. As for the tug-assistant berthing, with over four times the scale, tugs would not exert any significant influence on the own-ship either. Consequently, the ship–ship interaction could be ignored in the studies on ship berthing, hydrodynamic performance and maneuverability modeling.

4.3.4. Wind and Current

The external disturbances in the berthing period mainly refer to the wind and current effects. As shown in Figure 14, ship floating is observed along the wind or current-flow direction. The wind acts upon the ship's superstructure, and the wind load influence on ship maneuverability becomes much more apparent when the ship's speed is lower than the wind speed. Wind forces the ship to drift off course and decreases ship stability. Knowing the wind load characteristics could help avoid collisions from happening and improve the feasibility under certain conditions. Refs. [123–125] summarize the wind effect on ship maneuverability as a function of ship speed, wind speed, wind direction, and windward area; such a functional relation could be utilized to estimate the wind damping force. Moreover, the wind has a significant effect on ship speed loss under the scope of head-to-beam wind direction, and, compared with this, ship stability gradually degrades within the range from beam wind to quartering wind [126–128]. Furthermore, with the increase in wind velocity, the rudder becomes less effective [129]. The wind effect on the ship hull could be expressed as:

$$\begin{cases} X_{wind} = 0.5\rho A_o U_r^2 C_{wx}(\alpha_r) \\ Y_{wind} = 0.5\rho A_l U_r^2 C_{wy}(\alpha_r) \\ N_{wind} = 0.5\rho A_l L_{oa} U_r^2 C_{wn}(\alpha_r) \end{cases}, \tag{11}$$

where A_o is the orthographic projection area above the waterline, A_l the lateral projection area above the waterline, U_r the relative wind speed, C_{wx}, and C_{wy} are the wind pressure force coefficients on the x and y axes, C_{wn} is the wind pressure moment coefficient around the z axis, and α_r is the relative wind angle.

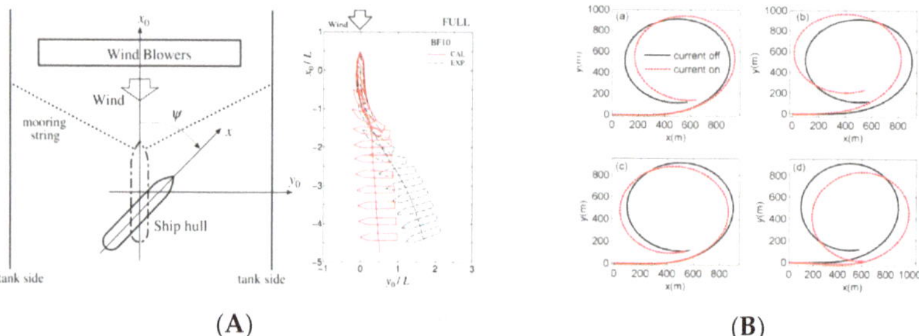

Figure 14. External disturbances on ship maneuverability. (**A**) Wind effect [92]. (**B**) Current effect [130].

Affected by the wind, wave, tide, seabed topography, and obstacles, currents behave non-uniformly in both the horizontal and vertical direction. However, studies such as [131] indicated that it is acceptable to concern the horizontal current flow, due to the uncertainty of which it is hard to directly express the complex and random current flow; hence, present works are performed on the uniform and steady current. The current flow mainly affects the flow around the ship, and it is essential to contain the current effect for small ships, as its impact on speed over the ground outweighs the wind effect [132]. The modeling of steady current is relatively simple: the ship speed U over ground is resolved into the current velocity U_c and the ship speed U_0 relative to the current, and the hydrodynamic characteristics and the interactions among the hull, propeller, and propeller are solved based on the ship's speed relative to the current. The ship's longitudinal and lateral speed relative to the current are expressed as:

$$\begin{cases} u_r = u - u_c = u - U_c \cos(\psi_c - \psi) \\ v_r = v - v_c = v - U_c \sin(\psi_c - \psi) \end{cases}, \tag{12}$$

where, u_r and v_r are the ship's longitudinal and lateral speed relative to the current, u and v are the ship's longitudinal and lateral speed relative to the ground, u_c and v_c are the current speed, ψ is the heading angle, and ψ_c is the current angle.

4.4. Auxiliary Devices

The berthing process requires the ship to maintain a low or extremely low speed, and under such working conditions a larger rudder steerage is utilized to rectify the heading deviation and keep the course; otherwise, the ship would lose its rudder effects. Resolution is carried out with the introduction of auxiliary devices such as side thrusters, tugs, and anchors. The diagram is shown in Figure 15.

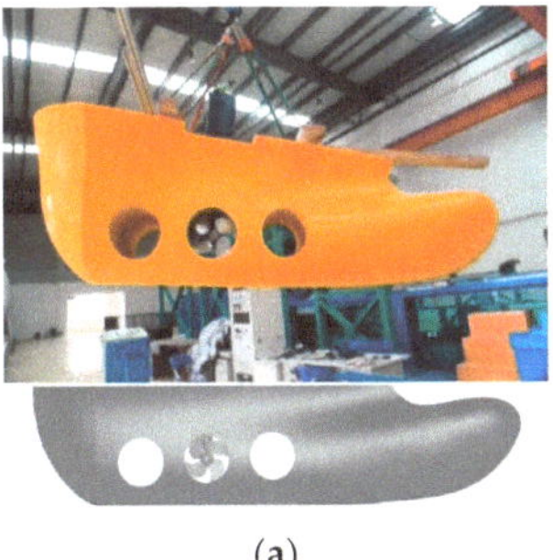

(a)

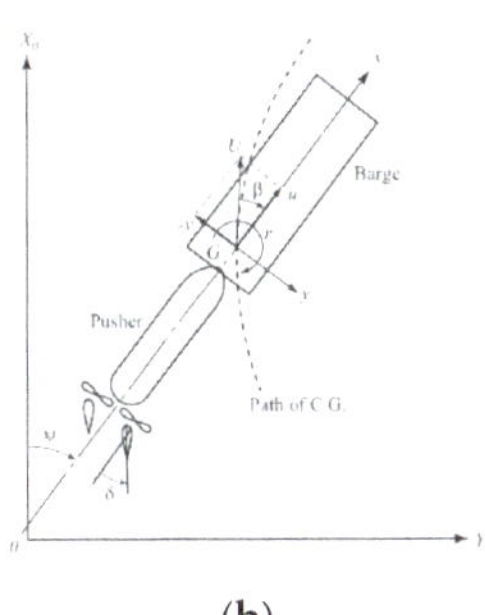

(b)

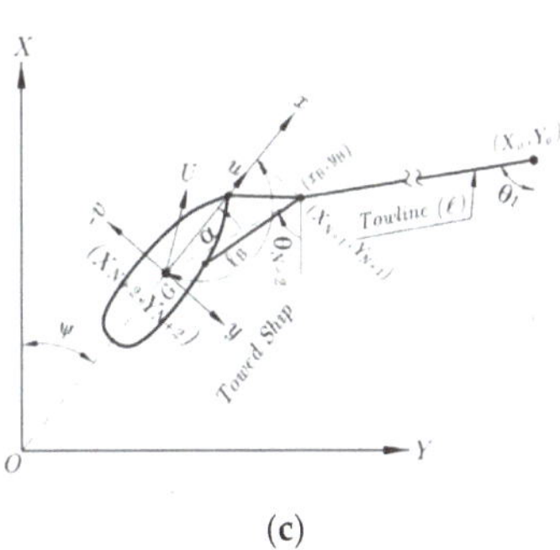

(c)

Figure 15. Auxiliary devices. (**a**) Side thruster [133]. (**b**) Pushing of tug [134]. (**c**) Towing of tug [135].

4.4.1. Side Thruster

The side thruster, especially the bow thruster, is used to provide lateral force and yaw moment for slewing motions. Normally, for best-turning effects, bow thrusters are assembled as far forward as possible or under the bottom of the keel. Studies [136,137] indicated that a ship's forward speed has a great influence on the effectiveness of the bow thruster, and that the generated force decreases remarkably with the increase in ship speed. As for the lateral force induced by the bow thruster, the drifting effect of the ship hull should be considered when the drift angle is over $10°$, while the yaw moment seems to be immune to the ship's drifting.

4.4.2. Tug Assistance

In the berthing process, whether it is for ships with small displacement, or ships with larger dimensions, the essential indices for ship berthing are the same, namely the approaching angle, lateral speed, and wall distance. Tugs are utilized to release the berthing risk and improve efficiency. Based on the survey of 15 ports in China, 120 m of the ship length is normally regarded as a boundary for tug usage [138], and it is acceptable for ships with shorter ship lengths to conduct self-berthing. As for ships with greater dimensions, it is mandatory to change direction, turn around, pull up, and parallel berth with the assistance of tugboats [139]. However, when it comes to liquid cargo ships, engineering ships, ships with damage, extreme weather, and other conditions, ships are obliged to berth with tugs.

Tugs assist the target ship in a direct (pushing) or indirect (towing) way, to lead the ship to the quay/berthing place, and maintain proper speed and attitude. There are two methods expressing the tug direct tug pushing in berthing maneuver modeling: one treats the tug-induced force as an azimuth external force, and the motion and specifications of the tug itself are eliminated. The other takes the pusher–barge system for reference and regards the pusher and the barge as integral. Yang [140,141] simplified the tug as a force and obtained acceptable accuracy on berthing operation mathematical modeling and harbor area navigation simulations. Series studies [134,142–144] were carried out on an inland pusher–barge-system maneuvering modeling, which analyzed the effects of barge ship formation, the profile, numbers, and pusher locations on the system, discussed the resistance variation and thrust efficiency, and established the mathematical model to predict the turning and course keeping ability of the system. In conclusion, the modeling method of the pusher–barge system is similar to the individual ship, and this method could be expanded to the modeling study on tug-pushing modeling in the berthing process.

The modeling of indirect towing via cables is complicated. The tug-towed system involves three closely related parts, the tug, the towed ship, and the cable line, which are connected to each other. From the point of view of mathematical modeling, researchers [126,135,145–148] conducted massive work on the force characteristics of the system and the individuals in linear and non-linear conditions, indicating that the focuses of the system are the cable length, towing point locations, cable angles, and cable status. Considering that the slack towing line evokes impulse forces, which adds risk to the towing operation, it is suggested that the cables are kept under strain.

The anchor functions through the cables, and the resolution is similar to the tug-towed system.

4.5. Remarks

As the ship conducts self-berthing or berthing with tug assistance, the force characteristics are completely different from the traditional maneuvers. The hydrodynamic performance, four-quadrant propulsion and steerage devices, external disturbances, and auxiliary devices, are the four main aspects. The hydrodynamic performance indicates the hydrodynamic forces acting on the hull under low-speed, high-drifting and high-yaw-rate conditions, and the cross-flow drag is the main concern. The four-quadrant propulsion and steerage performances focus on the propeller thrust and rudder steerage force under arbitrary resultant inflow angles. Ships in the berthing process are sensitive to external disturbances, especially the shallow-water and bank effects caused by the water-cushion effect, wind pressure provoked by the wind, and flow speed variation induced by the tidal current. Hydrodynamic forces, four-quadrant propulsion and steerage devices are essential to the berthing maneuver modeling; the external disturbances and auxiliary devices are supplementary to the model's scope and accuracy.

5. Conclusions

An autonomous ship is encouraged by the increasing shipping demand and technology, and automatic berthing control is at the top level of ship autonomy, due to the complicated and dangerous low-speed operation. A precise berthing maneuver model is the foundation for the automatic control system, providing a reference for the utilization of auxiliary devices and offering the responses of the ship's motion within the berthing aid system to the steerage operation and external disturbances.

In the present work, a bibliometric study on automatic ship berthing is performed to search for the important issues, and six specific study fields are obtained: risk assessment, perception utilization, motion control, maneuverability modeling, scheduling optimization, and emission supervision. With regard to automatic berthing, risk assessment is the guarantee, perception utilization is the premise, motion control is the key, and maneuver modeling is the foundation.

The modular MMG mathematical model is found to better describe ship motion and responses to internal operation and external disturbances. Both conventional MMG and berthing maneuver MMG models are adopted in the automatic berthing control studies. To make clear the similarities and differences, the following difficult questions are discussed:

(1)　What are the similarities and differences between the conventional MMG maneuvering model and the automatic berthing maneuvering model?
(2)　How can an accurate automatic berthing maneuvering model be established?

Uniform mathematical modeling methods (data-based, system-based, and CFD-based methods) could be used to establish the conventional MMG model and the berthing maneuver MMG model. Moreover, four specified characteristics of berthing maneuver are concluded to exist: low-speed-, high-drifting- and high-yaw-rate-induced hydrodynamic forces, the arbitrary resultant inflow angle causing four-quadrant thrust and steerage performances, external disturbances provoked by the water-cushion effects, wind, and tidal currents, and the additional forces provided by auxiliary devices.

With the aim of practical use, future work is put on the agenda for establishing the mathematical model expressing the berthing maneuver and comparing and validating the applicability and accuracy of the conventional MMG model and berthing maneuver MMG model.

Author Contributions: Conceptualization, J.L.; methodology, S.Z., J.L. and Y.H.; software, S.Z. and Y.H.; validation, S.Z.; formal analysis, S.Z. and Y.H.; investigation, S.Z., J.L. and Y.H.; data curation, S.Z., J.L. and Y.H.; writing—original draft preparation, S.Z., and Y.H.; writing—review and editing, Q.W., J.L. and S.L.; visualization, S.Z., and Y.H.; supervision, Q.W. and J.L.; project administration, J.L.; funding acquisition, J.L. and S.L. All authors have read and agreed to the published version of the manuscript.

Funding: This research was funded by the National Key R&D Program of China, grant number 2022YFB4301402, the National Natural Science Foundation of China, grant number 52272425 and 62003250.

Institutional Review Board Statement: Not applicable.

Informed Consent Statement: Not applicable.

Data Availability Statement: Not applicable.

Conflicts of Interest: The authors declare no conflict of interest.

References

1. Rothblum, A.M. Human error and marine safety. In Proceedings of the National Safety Council Congress and Expo, Orlando, FL, USA, 16–18 October 2000.
2. Ahmed, Y.A.; Hasegawa, K. Implementation of automatic ship berthing using artificial neural network for free running experiment. *IFAC Proc. Vol.* **2013**, *46*, 25–30. [CrossRef]
3. China Classification Society. *Rules for Intelligent Ships*; China Classification Society: Ningbo, China, 2015.
4. China Classification Society. *Rules for Intelligent Ships*; China Classification Society: Ningbo, China, 2020.
5. China Classification Society. *Rules for Intelligent Ships*; China Classification Society: Ningbo, China, 2023.
6. Executive, T.M. Video: Japanese Demonstration of Autonomous Docking System. Available online: https://maritime-executive.com/article/video-japanese-demonstration-of-autonomous-docking-system (accessed on 24 May 2021).
7. SAFETY4SEA. Watch: World's First Successful Trail of Auto Berthing and un-Berthing System. Available online: https://safety4sea.com/watch-worlds-first-successful-trial-of-auto-berthing-and-un-berthing-system/ (accessed on 24 May 2021).
8. Executive, T.M. First Autonomous Navigation and Berthing Test on a Containership. Available online: https://maritime-executive.com/article/first-autonomous-navigation-and-berthing-test-on-a-containership (accessed on 25 January 2022).
9. Project, K.A.S.S. Detailed Task. Available online: https://kassproject.org/en/task/task.php (accessed on 24 May 2021).
10. Qingdaonews. Water Test of the First Domestic Autonomous Navigation System Experimental Ship. Available online: https://news.qingdaonews.com/qingdao/2019-05/17/content_20362302.htm (accessed on 17 May 2019).
11. Brilliance, N. The First Unmanned Container Ship in China Successfully Undocked. Available online: http://www.brinav.com/pages/News/NewsDetails.aspx?ID=2022 (accessed on 1 July 2021).
12. Rolls-Royce. Rolls-Royce and Finferries Demonstrate World's First Fully Autonomous Ferry. Available online: https://www.rolls-royce.com/media/press-releases/2018/03-12-2018-rr-and-finferries-demonstrate-worlds-first-fully-autonomous-ferry.aspx (accessed on 3 December 2018).
13. Corporation, W. World's first Autodocking Installation Successfully Tested by Wärtsilä. Available online: https://www.wartsila.com/media/news/26-04-2018-world-s-first-autodocking-installation-successfully-tested-by-wartsila-2169290 (accessed on 26 April 2018).
14. MARITIME, K. Autonomous Ship Project, Key Facts About Yara Birkeland: The Zero Emission, Autonomous Container Feeder. Available online: https://www.kongsberg.com/zh-hans/maritime/support/themes/autonomous-ship-project-key-facts-about-yara-birkeland/ (accessed on 24 May 2021).
15. PENTA, V. Launching the First Fully Integrated Assisted Docking System. Available online: https://www.volvopenta.com/about-us/news-page/2021/jan/launching-the-first-fully-integrated-assisted-docking-system/ (accessed on 24 May 2021).
16. Yasukawa, H.; Yoshimura, Y. Introduction of MMG standard method for ship maneuvering predictions. *J. Mar. Sci. Technol.* **2015**, *20*, 37–52. [CrossRef]
17. Li, J.; Goerlandt, F.; Reniers, G. Mapping process safety: A retrospective scientometric analysis of three process safety related journals (1999–2018). *J. Loss Prev. Process Ind.* **2020**, *65*, 104141. [CrossRef]

18. Li, J.; Goerlandt, F.; Reniers, G. An overview of scientometric mapping for the safety science community: Methods, tools, and framework. *Saf. Sci.* **2021**, *134*, 105093. [CrossRef]

19. He, Y.; Mou, J.; Chen, L.; Zeng, Q.; Chen, P.; Zhang, S. Survey on hydrodynamic effects on cooperative control of Maritime Autonomous Surface Ships. *Ocean. Eng.* **2021**, *235*, 109300. [CrossRef]

20. Koyama, T.; Yan, J.; Huan, J.K. A systematic study on automatic berthing control (1st report). *J. Soc. Nav. Archit. Jpn.* **1987**, *1987*, 201–210. [CrossRef]

21. Shouji, K.; Ohtsu, K.; Mizoguchi, S. An automatic berthing study by optimal control techniques. *IFAC Proc. Vol.* **1992**, *25*, 185–194. [CrossRef]

22. Yamato, H.; Koyama, T.; Nakagawa, T. Automatic berthing system using expert system. *J. Soc. Nav. Archit. Jpn.* **1993**, *1993*, 327–337. [CrossRef]

23. Meng, W.; Li, G.; Hua, L. A goal-programming based optimal port docking scheme under COVID-19. *Ocean. Coast. Manag.* **2022**, *225*, 106222.

24. Han, S.; Wang, L.; Wang, Y. A potential field-based trajectory planning and tracking approach for automatic berthing and COLREGs-compliant collision avoidance. *Ocean. Eng.* **2022**, *266*, 112877. [CrossRef]

25. Lee, H.-T.; Lee, J.-S.; Son, W.-J.; Cho, I.-S. Development of machine learning strategy for predicting the risk range of ship's berthing velocity. *J. Mar. Sci. Eng.* **2020**, *8*, 376. [CrossRef]

26. Liao, Y.; Jia, Z.; Zhang, W.; Jia, Q.; Li, Y. Layered berthing method and experiment of unmanned surface vehicle based on multiple constraints analysis. *Appl. Ocean. Res.* **2019**, *86*, 47–60. [CrossRef]

27. Gravendeel, A.; Noorland, J.; Jacobs, S. *Automatic Berthing*; Mainport Rotterdam University: Rotterdam, Netherlands, 2017.

28. Digerud, L.; Volden, Ø.; Christensen, K.A.; Kohtala, S.; Steinert, M. Vision-based positioning of Unmanned Surface Vehicles using Fiducial Markers for automatic docking. *IFAC-Pap. Online* **2022**, *55*, 78–84. [CrossRef]

29. Wang, Z.; Zhang, Y. Estimation of ship berthing parameters based on Multi-LiDAR and MMW radar data fusion. *Ocean. Eng.* **2022**, *266*, 113155. [CrossRef]

30. Hu, B.; Liu, X.; Jing, Q.; Lyu, H.; Yin, Y. Estimation of berthing state of maritime autonomous surface ships based on 3D LiDAR. *Ocean. Eng.* **2022**, *251*, 111131. [CrossRef]

31. Chen, L.; Negenborn, R.; Huang, Y.; Hopman, H. Survey on cooperative control for waterborne transport. *IEEE Intell. Transp. Syst. Mag.* **2020**, *13*, 71–90. [CrossRef]

32. Bitar, G.I. *Towards the Development of Autonomous Ferries*; NTNU: Trondheim, Norway, 2017.

33. Zhang, Q.; Im, N.-K.; Ding, Z.; Zhang, M. Review on the research of ship automatic berthing control. In *Offshore Robotics Volume I Issue 1, 2021*; Springer: Singapore, 2022; pp. 87–109.

34. de Kruif, B.J. Applied trajectory generation to dock a feeder vessel. *IFAC-Pap. Online* **2022**, *55*, 172–177. [CrossRef]

35. Wang, X.; Deng, Z.; Peng, H.; Wang, L.; Wang, Y.; Tao, L.; Lu, C.; Peng, Z. Autonomous docking trajectory optimization for unmanned surface vehicle: A hierarchical method. *Ocean. Eng.* **2023**, *279*, 114156. [CrossRef]

36. Bitar, G.; Martinsen, A.B.; Lekkas, A.M.; Breivik, M. Trajectory planning and control for automatic docking of ASVs with full-scale experiments. *IFAC-Pap. Online* **2020**, *53*, 14488–14494. [CrossRef]

37. Rachman, D.M.; Maki, A.; Miyauchi, Y.; Umeda, N. Warm-started semionline trajectory planner for ship's automatic docking (berthing). *Ocean. Eng.* **2022**, *252*, 111127. [CrossRef]

38. Bidikli, B.; Tatlicioglu, E.; Zergeroglu, E. Robust dynamic positioning of surface vessels via multiple unidirectional tugboats. *Ocean. Eng.* **2016**, *113*, 237–245. [CrossRef]

39. Du, Z.; Reppa, V.; Negenborn, R.R. Cooperative control of autonomous tugs for ship towing. *IFAC-Pap. Online* **2020**, *53*, 14470–14475. [CrossRef]

40. Wu, G.; Zhao, X.; Sun, Y.; Wang, L. Cooperative maneuvering mathematical modeling for multi-tugs towing a ship in the port environment. *J. Mar. Sci. Eng.* **2021**, *9*, 384. [CrossRef]

41. Wang, S.; Sun, Z.; Yuan, Q.; Sun, Z.; Wu, Z.; Hsieh, T.-H. Autonomous piloting and berthing based on Long Short Time Memory neural networks and nonlinear model predictive control algorithm. *Ocean. Eng.* **2022**, *264*, 112269. [CrossRef]

42. Nguyen, V.S. Investigation on a novel support system for automatic ship berthing in marine practice. *J. Mar. Sci. Eng.* **2019**, *7*, 114. [CrossRef]

43. Baek, S.; Woo, J. Model reference adaptive control-based autonomous berthing of an unmanned surface vehicle under environmental disturbance. *Machines* **2022**, *10*, 244. [CrossRef]

44. Xu, Z.; Galeazzi, R. Guidance and Motion Control for Automated Berthing of Twin-waterjet Propelled Vessels. *IFAC-Pap. Online* **2022**, *55*, 58–63. [CrossRef]

45. Martinsen, A.B.; Bitar, G.; Lekkas, A.M.; Gros, S. Optimization-based automatic docking and berthing of ASVs using exteroceptive sensors: Theory and experiments. *IEEE Access* **2020**, *8*, 204974–204986. [CrossRef]

46. Mizuno, N.; Kuboshima, R. Implementation and evaluation of non-linear optimal feedback control for ship's automatic berthing by recurrent neural network. *IFAC-Pap. Online* **2019**, *52*, 91–96. [CrossRef]

47. Nguyen, V.S. Investigation of a multitasking system for automatic ship berthing in marine practice based on an integrated neural controller. *Mathematics* **2020**, *8*, 1167. [CrossRef]

48. Mizuno, N.; Koide, T. Application of Reinforcement Learning to Generate Non-linear Optimal Feedback Controller for Ship's Automatic Berthing System. *IFAC-Pap. Online* **2023**, *56*, 162–168. [CrossRef]

49. Shimizu, S.; Nishihara, K.; Miyauchi, Y.; Wakita, K.; Suyama, R.; Maki, A.; Shirakawa, S. Automatic berthing using supervised learning and reinforcement learning. *Ocean. Eng.* **2022**, *265*, 112553. [CrossRef]
50. Chopra, K. A Detailed Explanation of How a Ship Is Manoeuvered to a Port. Available online: https://www.marineinsight.com/guidelines/a-detailed-explanation-of-how-a-ship-is-manoeuvered-to-a-port/ (accessed on 4 March 2021).
51. Wakita, K.; Akimoto, Y.; Rachman, D.M.; Miyauchi, Y.; Naoya, U.; Maki, A. Collision probability reduction method for tracking control in automatic docking/berthing using reinforcement learning. *arXiv* **2022**, arXiv:2212.06415.
52. Abkowitz, M.A. *Lectures on Ship Hydrodynamics-Steering and Manoeuvrability*; Hydro & Aerodynamic Laboratory: Lyngby, Denmark, 1964.
53. Fossen, T.I. *Handbook of Marine Craft Hydrodynamics and Motion Control*; John Wiley & Sons: Hoboken, NJ, USA, 2011.
54. Nomoto, K.; Taguchi, K.; Honda, K.; Hirano, S. On the steering qualities of ships. *J. Zosen Kiokai* **1956**, *1956*, 75–82. [CrossRef]
55. Wang, Z.; Kim, J.; Im, N. Non-parameterized ship maneuvering model of Deep Neural Networks based on real voyage data-driven. *Ocean. Eng.* **2023**, *284*, 115162. [CrossRef]
56. Ouyang, Z.; Liu, S.; Zou, Z. Nonparametric modeling of ship maneuvering motion in waves based on Gaussian process regression. *Ocean. Eng.* **2022**, *264*, 112100. [CrossRef]
57. Im, N. A study on ship automatic berthing with assistance of auxiliary devices. *Int. J. Nav. Archit. Ocean. Eng.* **2012**, *4*, 199–210.
58. Li, S.; Liu, J.; Negenborn, R.R.; Wu, Q. Automatic docking for underactuated ships based on multi-objective nonlinear model predictive control. *IEEE Access* **2020**, *8*, 70044–70057. [CrossRef]
59. Piao, Z.; Guo, C.; Sun, S. Research into the automatic berthing of underactuated unmanned ships under wind loads based on experiment and numerical analysis. *J. Mar. Sci. Eng.* **2019**, *7*, 300. [CrossRef]
60. Zhang, Q.; Zhu, G.; Hu, X.; Yang, R. Adaptive neural network auto-berthing control of marine ships. *Ocean. Eng.* **2019**, *177*, 40–48. [CrossRef]
61. Wu, G.; Zhao, X.; Wang, L. Modeling and Simulation of Automatic Berthing based on Bow and Stern Thruster Assist for Unmanned Surface Vehicle. *J. Mar. Sci. Eng.* **2021**, *3*, 16–21. [CrossRef]
62. Ahmed, Y.A.; Hannan, M.A.; Siang, K.H. Artificial Neural Network controller for automatic ship berthing: Challenges and opportunities. *Mar. Syst. Ocean. Technol.* **2020**, *15*, 217–242. [CrossRef]
63. Ahmed, Y.A.; Hasegawa, K. Automatic ship berthing using artificial neural network trained by consistent teaching data using nonlinear programming method. *Eng. Appl. Artif. Intell.* **2013**, *26*, 2287–2304. [CrossRef]
64. Hasegawa, K.; Kitera, K. Mathematical model of manoeuvrability at low advance speed and its application to berthing control. In Proceedings of the 2nd Japan-Korea Joint Workshop of Ship and Marine Hydrodynamics, Osaka, Japan, 28–30 June 1993; pp. 144–153.
65. Miyauchi, Y.; Maki, A.; Umeda, N.; Rachman, D.M.; Akimoto, Y. System parameter exploration of ship maneuvering model for automatic docking/berthing using CMA-ES. *J. Mar. Sci. Technol.* **2022**, *27*, 1065–1083. [CrossRef]
66. Sawada, R.; Hirata, K.; Kitagawa, Y. Automatic berthing control under wind disturbances and its implementation in an embedded system. *J. Mar. Sci. Technol.* **2023**, 452–470. [CrossRef]
67. Okuda, R.; Yasukawa, H.; Yamashita, T.; Matsuda, A. Maneuvering simulations at large drift angles of a ship with a flapped rudder. *Appl. Ocean. Res.* **2023**, *135*, 103567. [CrossRef]
68. Yoshimura, Y.; Nakao, I.; Ishibashi, A. Unified mathematical model for ocean and harbour manoeuvring. In Proceedings of the MARSIM2009, Panama City, FL, USA, 17–20 August 2009.
69. ITTC. The Manoeuvring Committee Final Report and Recommendations. In Proceedings of the 25th International Towing Tank Conference (ITTC), Fukuoka, Japan, 14–20 September 2008.
70. Holtrop, J.; Mennen, G. A statistical power prediction method. *Int. Shipbuild. Prog.* **1978**, *25*, 290. [CrossRef]
71. Kijima, K.; Katsuno, T.; Nakiri, Y.; Furukawa, Y. On the manoeuvring performance of a ship with theparameter of loading condition. *J. Soc. Nav. Archit. Jpn.* **1990**, *1990*, 141–148. [CrossRef]
72. Yoshimura, Y.; Masumoto, Y. Hydrodynamic database and manoeuvring prediction method with medium high-speed merchant ships and fishing vessels. In Proceedings of the International MARSIM Conference, Singapore, 23–27 April 2012; pp. 1–9.
73. Pekic, S. TT-Line Starts Sea Trials of Its 1st Green Ship. Available online: https://www.offshore-energy.biz/tt-line-starts-sea-trials-of-their-1st-green-ship/ (accessed on 19 November 2021).
74. Zhang, Z.; Zhang, Y.; Wang, J.; Wang, H. Parameter identification and application of ship maneuvering model based on TO-CSA. *Ocean Eng.* **2022**, *266*, 113128. [CrossRef]
75. Liu, S.; Ouyang, Z.; Chen, G.; Zhou, X.; Zou, Z. Black-box modeling of ship maneuvering motion based on Gaussian process regression with wavelet threshold denoising. *Ocean Eng.* **2023**, *271*, 113765. [CrossRef]
76. Astrom, K.J.; Kallstrom, C.; Norrbin, N.; Bystrom, L. The identification of linear ship steering dynamics using maximum likelihood parameter estimation. *Medd. Från Statens Skeppsprovningsanst.* **1975**, *75*, 105.
77. Banazadeh, A.; Ghorbani, M. Frequency domain identification of the Nomoto model to facilitate Kalman filter estimation and PID heading control of a patrol vessel. *Ocean Eng.* **2013**, *72*, 344–355. [CrossRef]
78. Zhang, G.; Zhang, X.; Pang, H. Multi-innovation auto-constructed least squares identification for 4 DOF ship manoeuvring modelling with full-scale trial data. *ISA Trans.* **2015**, *58*, 186–195. [CrossRef] [PubMed]
79. Zhang, Y.; Wang, Z.; Zou, Z. Black-box modeling of ship maneuvering motion based on multi-output nu-support vector regression with random excitation signal. *Ocean Eng.* **2022**, *257*, 111279. [CrossRef]

80. Xu, H.; Soares, C.G. Hydrodynamic coefficient estimation for ship manoeuvring in shallow water using an optimal truncated LS-SVM. *Ocean Eng.* **2019**, *191*, 106488. [CrossRef]
81. He, H.; Wang, Z.; Zou, Z.; Liu, Y. Nonparametric modeling of ship maneuvering motion based on self-designed fully connected neural network. *Ocean Eng.* **2022**, *251*, 111113. [CrossRef]
82. Jiang, Y.; Hou, X.; Wang, X.; Wang, Z.; Yang, Z.; Zou, Z. Identification modeling and prediction of ship maneuvering motion based on LSTM deep neural network. *J. Mar. Sci. Technol.* **2022**, *27*, 125–137. [CrossRef]
83. Guo, H.; Zou, Z.; Liu, Y.; Wang, F. Investigation on hull-propeller-rudder interaction by RANS simulation of captive model tests for a twin-screw ship. *Ocean Eng.* **2018**, *162*, 259–273. [CrossRef]
84. Liu, Y.; Zou, L.; Zou, Z.; Guo, H. Predictions of ship maneuverability based on virtual captive model tests. *Eng. Appl. Comput. Fluid Mech.* **2018**, *12*, 334–353. [CrossRef]
85. Liu, J.; Hekkenberg, R.; Quadvlieg, F.; Hopman, H.; Zhao, B. An integrated empirical manoeuvring model for inland vessels. *Ocean Eng.* **2017**, *137*, 287–308. [CrossRef]
86. Lu, S.; Cheng, X.; Liu, J.; Li, S.; Yasukawa, H. Maneuvering modeling of a twin-propeller twin-rudder inland container vessel based on integrated CFD and empirical methods. *Appl. Ocean. Res.* **2022**, *126*, 103261. [CrossRef]
87. Zhang, S.; Wu, Q.; Liu, J.; He, Y.; Li, S. Twin-screw ASD tug maneuvering prediction based on integrated CFD and empirical methods. *Ocean Eng.* **2023**, *269*, 113489. [CrossRef]
88. Li, S.; Liu, C.; Chu, X.; Zheng, M.; Wang, Z.; Kan, J. Ship maneuverability modeling and numerical prediction using CFD with body force propeller. *Ocean Eng.* **2022**, *264*, 112454. [CrossRef]
89. Jia, X.; Yang, Y. *Ship motion mathematical model: Mechanism modeling and identification modeling*; Dalian Maritime University Press: Dalian, China, 1999.
90. Sutulo, S. Extension of polynomial ship mathematical models to arbitrary manoeuvres. In Proceedings of the The International Shipbuilding Conference in Commemoration of the Centenary of the Krylov Ship Research Institute (ISC), St. Petersburg, Russia, 8–12 October 1994; pp. 291–298.
91. Sutulo, S.; Soares, C.G. Development of a core mathematical model for arbitrary manoeuvres of a shuttle tanker. *Appl. Ocean. Res.* **2015**, *51*, 293–308. [CrossRef]
92. Yasukawa, H.; Hirata, N.; Nakayama, Y.; Matsuda, A. Drifting of a dead ship in wind. *Ship Technol. Res.* **2023**, *70*, 26–45. [CrossRef]
93. Inoue, S.; Hirano, M.; Kijima, K. Hydrodynamic derivatives on ship manoeuvring. *Int. Shipbuild. Prog.* **1981**, *28*, 112–125. [CrossRef]
94. Yoshimura, Y. Mathematical model for the manoeuvring ship motion in shallow water (2nd Report)-Mathematical model at slow forward speed. *J. Kansai Soc. Nav. Archit. Jpn.* **1988**, *210*, 77–84.
95. Oltmann, P.; Sharma, S. *Simulation of Combined Engine and Rudder Maneuvers Using an Improved Model of Hull-Propeller-Rudder Interactions*; Hamburg University of Technology: Hamburg, Germany, 1984.
96. Yasukawa, H.; Ishikawa, T.; Yoshimura, Y. Investigation on the rudder force of a ship in large drifting conditions with the MMG model. *J. Mar. Sci. Technol.* **2021**, *26*, 1078–1095. [CrossRef]
97. Van Lammeren, W.; van Manen, J.v.; Oosterveld, M. *The Wageningen B-Screw Series*; National Academy of Sciences: Washington, DC, USA, 1969; Volume 77, p. 43.
98. Yoshimura, Y.; Sasaki, N.; Takekawa, M. Prediction of ship manoeuvrability with a flapped rudder. *J. Soc. Nav. Archit. Jpn.* **1997**, *1997*, 191–196. [CrossRef]
99. Huang, G.; Li, X.; Zhu, Y.; Nie, S. Experimental Study on Reaction Thrust Characteristics of Water Jet for Conical Nozzle. *China Ocean Eng.* **2009**, *23*, 669–678.
100. Jiang, F.; Li, Y.; Gong, J. Study on the manoeuvre characteristics of a trimaran under different layouts by water-jet self-propulsion model test. *Appl. Ocean. Res.* **2021**, *108*, 102550. [CrossRef]
101. Reichel, M. Prediction of manoeuvring abilities of 10000 DWT pod-driven coastal tanker. *Ocean Eng.* **2017**, *136*, 201–208. [CrossRef]
102. Yasukawa, H.; Hirata, N.; Tanaka, S.; Kose, K. Study on Maneuverability of a Ship with Azimuthing Propellers. *J. Jpn. Soc. Nav. Archit. Ocean. Eng.* **2009**, *9*, 155–165.
103. ITTC. ITTC Recommended Procedures and Guidelines. In *Free Running Model Tests*; No. 7.5-02-06-01; ITTC: Lawrence, KS, USA, 2008.
104. Vantorre, M.; Eloot, K.; Delefortrie, G.; Lataire, E.; Candries, M.; Verwilligen, J. Maneuvering in shallow and confined water. In *The Encyclopedia of Marine Offshore Engineering*; Wiley Online Library: Hoboken, NJ, USA, 2017; pp. 1–17.
105. Pianc. Capability of ship manoeuvring simulation models for approach channels and fairways in harbours. In *Report of Working Group no. 20 of Permanent Technical Committee II. Supplement to Bulletin no. 77*; Pianc: Brussels, Belgium, 1992.
106. Havelock, T.H. The propagation of groups of waves in dispersive media, with application to waves on water produced by a travelling disturbance. *Proc. R. Soc. London. Ser. A Contain. Pap. A Math. Phys. Character* **1908**, *81*, 398–430.
107. Delefortrie, G.; Eloot, K.; Lataire, E.; Van Hoydonk, W.; Vantorre, M. Captive model tests based 6 DOF shallow water manoeuvring model. In Proceedings of the 4th MASHCON-International Conference on Ship Manoeuvring in Shallow and Confined Water with Special Focus on Ship Bottom Interaction, Hamburg, Germany, 23–25 May 2016; pp. 273–286.
108. Eloot, K. *Selection, Experimental Determination and Evaluation of a Mathematical Model for Ship Manoeuvring in Shallow Water*; Ghent University: Ghent, Belgium, 2006.

109. Liu, Y.; Zou, Z.; Zou, L.; Fan, S. CFD-based numerical simulation of pure sway tests in shallow water towing tank. *Ocean Eng.* **2019**, *189*, 106311. [CrossRef]
110. Zeng, Q.; Hekkenberg, R.; Thill, C. On the viscous resistance of ships sailing in shallow water. *Ocean Eng.* **2019**, *190*, 106434. [CrossRef]
111. Yoshimura, Y.; Sakurai, H. Mathematical Model for the Manoeuvring Ship Motion in Shallow Water (3rd Report: Manoeuvrability of a Twin-propeller Twinrudder ship). *J. Kansai Soc. Nav. Archit. Jpn.* **1989**, *211*, 115–126.
112. Yoshimura, Y. Mathematical model for the manoeuvring ship motion in shallow water-Application of MMG mathematical model to shallow water. *J. Kansai Soc. Nav. Archit. Jpn.* **1986**, *200*, 42–51.
113. Kijima, K.; Yoshimura, Y.; Takashina, J. Report of MSS. 2: Mathematical model for ship maneuverability in shallow water. *Bull. Soc. Nav. Archit. Jpn.* **1989**, *718*, 207–220.
114. ITTC. The Manoeuvring Committee–Final Report and Recommendations to the 23rd ITTC. In *Proceedings of the Proceedings of the 23rd ITTC, Venice, Italy, 8–14 September 2002*; pp. 153–234.
115. Liu, H.; Ma, N.; Gu, X. Ship–bank interaction of a VLCC ship model and related course-keeping control. *Ships Offshore Struct.* **2017**, *12*, S305–S316. [CrossRef]
116. Duan, W.; Liu, J.; Chen, J.; Wang, L. Comparison research of ship-to-ship hydrodynamic interaction in restricted water between TEBEM and other computational method. *Ocean Eng.* **2020**, *202*, 107168. [CrossRef]
117. Lataire, E.; Vantorre, M.; Laforce, E.; Eloot, K.; Delefortrie, G. Navigation in confined waters: Influence of bank characteristics on ship-bank interaction. In Proceedings of the International Conference on Marine Research and Transportation, ICMRT, Naples, Italy, 19–21 September 2007; pp. 1–9.
118. Norrbin, N.H. Bank clearance and optimal section shape for ship canals. In Proceedings of the 26th PIANC international navigation congress, Brussels, Belgium, 16–28 June 1985; pp. 167–178.
119. Kijima, K. Calculation of the Derivatives on Ship Manoeuvring in Narrow Waterways by Rectangular Plate. *J. Soc. Nav. Archit. Jpn.* **1972**, *1972*, 155–163. [CrossRef]
120. Yasukawa, H. Bank effect on ship maneuverability in a channel with varying width. *Proc. Trans. West-Jpn. Soc. Nav. Archit.* **1991**, *81*, 85–100.
121. DeMarco Muscat-Fenech, C.; Sant, T.; Zheku, V.V.; Villa, D.; Martelli, M. A Review of Ship-to-Ship Interactions in Calm Waters. *J. Mar. Sci. Eng.* **2022**, *10*, 1856. [CrossRef]
122. Vantorre, M.; Verzhbitskaya, E.; Laforce, E. Model test based formulations of ship-ship interaction forces. *Ship Technol. Res.* **2002**, *49*, 124–141.
123. Fujiwara, T.; Ueno, M.; Nimura, T. Estimation of Wind Forces and Moments acting on Ships. *J. Soc. Nav. Archit. Jpn.* **1998**, *1998*, 77–90. [CrossRef]
124. Isherwood, R.M. Wind resistance of merchant ships. *Trans. RINA* **1973**, *115*, 327–338.
125. Yoshimura, Y.; Nagashima, J. Estimation of the Manoeuvring Behaviour of Ship in Uniform Wind. *J. Soc. Nav. Archit. Jpn.* **1985**, *1985*, 125–136. [CrossRef]
126. Fitriadhy, A.; Yasukawa, H. Course stability of a ship towing system. *Ship Technol. Res.* **2011**, *58*, 4–23. [CrossRef]
127. Kijima, K.; Varyani, K. Wind effect on course stability of two towed vessels. *J. Soc. Nav. Archit. Jpn.* **1985**, *1985*, 137–148. [CrossRef]
128. Kijima, K.; Wada, Y. Course stability of towed vessel with wind effect. *J. Soc. Nav. Archit. Jpn.* **1983**, *1983*, 117–126. [CrossRef]
129. Yasukawa, H.; Hirono, T.; Nakayama, Y.; Koh, K.K. Course stability and yaw motion of a ship in steady wind. *J. Mar. Sci. Technol.* **2012**, *17*, 291–304. [CrossRef]
130. Wang, Y.; Du, W.; Li, G.; Li, Z.; Hou, J.; Hu, H. *Efficient Ship Maneuvering Prediction with Wind, Wave and Current Effects*; IOS Press: Clifton, VA, USA, 2022.
131. Chislett, M.S.; Wied, S. A note on the mathematical modelling of ship manoeuvring in relation to a nautical environment with particular reference to currents. In Proceedings of the International Conference on Numerical and Hydraulic Modelling of Ports and Harbours, Birmingham, UK, 23–25 April 1985; pp. 1–11.
132. Zhou, Y.; Daamen, W.; Vellinga, T.; Hoogendoorn, S.P. Impacts of wind and current on ship behavior in ports and waterways: A quantitative analysis based on AIS data. *Ocean Eng.* **2020**, *213*, 107774. [CrossRef]
133. Feng, Y.; Chen, Z.; Dai, Y.; Zhang, Z.; Wang, P. An experimental and numerical investigation on hydrodynamic characteristics of the bow thruster. *Ocean Eng.* **2020**, *209*, 107348. [CrossRef]
134. Koh, K.K.; Yasukawa, H.; Hirata, N.; Kose, K. Maneuvering simulations of pusher-barge systems. *J. Mar. Sci. Technol.* **2008**, *13*, 117–126.
135. Fitriadhy, A.; Yasukawa, H.; Maimun, A. Theoretical and experimental analysis of a slack towline motion on tug-towed ship during turning. *Ocean. Eng.* **2015**, *99*, 95–106. [CrossRef]
136. Fujino, M.; Saruta, T.; IDA, T. Experimental studies on the effectiveness of the side thruster. *J. Kansai Soc. Nav. Archit. Jpn.* **1978**, *168*, 35–44.
137. Kijima, K. The Effect of Drift Angle on Side Thruster Performance. *Trans. West-Jpn. Soc. Nav. Archit.* **1977**, *54*, 179–192.
138. Zhang, S.; Liu, J.; Li, S.; Ye, J. System design and key technology of ship automatic berthing and unberthing. *China Ship Surv.* **2022**, *10*, 42–47.
139. Yang, Y. Study on ship manoeuvring mathematical model in shiphandling simulator. In *Marine Simulation and Ship Manoeuvrability*; Routledge: London, UK, 1996; pp. 607–615.

140. Yang, Y. Study on mathematical model for simulating ship berthing or unberthing (in Chinese). *J. Dalian Marit. Univ.* **1996**, *22*, 11–15.
141. Yang, Y.; Fang, X. Study on Ship Maneuvering Mathematical Model with Assistance of Tugs in the Harbor. *Navig. China* **1997**, *2*, 19–24. (In Chinese)
142. Koh, K.K.; Yasukawa, H. Comparison study of a pusher–barge system in shallow water, medium shallow water and deep water conditions. *Ocean Eng.* **2012**, *46*, 9–17. [CrossRef]
143. Koh, K.K.; Yasukawa, H.; Hirata, N. Hydrodynamic derivatives investigation of unconventionally arranged pusher-barge systems. *J. Mar. Sci. Technol.* **2008**, *13*, 256–268. [CrossRef]
144. Yasukawa, H.; Hirata, N.; Koh, K.K.; Krisana, P.; Kose, K. Hydrodynamic Force Chracteristics on Maneuvering of Pusher-Barge Systems. *J. Jpn. Soc. Nav. Archit. Ocean. Eng.* **2007**, *5*, 133–142. [CrossRef]
145. Fitriadhy, A.; Yasukawa, H. Turning ability of a ship towing system. *Ship Technol. Res.* **2011**, *58*, 112–124. [CrossRef]
146. Kishimoto, T.; Kijima, K. The manoeuvring characteristics on tug-towed ship systems. *IFAC Proc. Vol.* **2001**, *34*, 173–178. [CrossRef]
147. Yasukawa, H.; Hirata, N.; Nakamura, N.; Matsumoto, Y. Simulations of Slewing Motion of a Towed Ship. *J. Jpn. Soc. Nav. Archit. Ocean. Eng.* **2006**, *4*, 137–146. [CrossRef]
148. Yasukawa, H.; Hirata, N.; Yokoo, R.; Fitriadhy, A. Slack Towline of Tow and Towed Ships during Turning. *J. Jpn. Soc. Nav. Archit. Ocean. Eng.* **2012**, *16*, 41–48.

Journal of
*Marine Science
and Engineering*

Review

A Review on Motion Prediction for Intelligent Ship Navigation

Daiyong Zhang [1,2,3,4], Xiumin Chu [1,2,3], Chenguang Liu [1,2,3,*], Zhibo He [1,2,3,5], Pulin Zhang [1,2,3,5] and Wenxiang Wu [1,2,3,5]

1   State Key Laboratory of Maritime Technology and Safety, Wuhan University of Technology, Wuhan 430063, China; daiyong.whut@gmail.com (D.Z.); chuxm@whut.edu.cn (X.C.); he_zhi_bo@whut.edu.cn (Z.H.); zhangpulin@whut.edu.cn (P.Z.); wuwenxiang@whut.edu.cn (W.W.)
2   National Engineering Research Center for Water Transport Safety, Wuhan University of Technology, Wuhan 430063, China
3   Intelligent Transportation Systems Research Center, Wuhan University of Technology, Wuhan 430063, China
4   School of Naval Architecture, Ocean and Energy Power Engineering, Wuhan University of Technology, Wuhan 430063, China
5   School of Transportation and Logistics Engineering, Wuhan University of Technology, Wuhan 430063, China
*   Correspondence: liuchenguang@whut.edu.cn

Abstract: In recent years, as intelligent ship-navigation technology has advanced, the challenge of accurately modeling and predicting the dynamic environment and motion status of ships has emerged as a prominent area of research. In response to the diverse time scales required for the prediction of ship motion, various methods for modeling ship navigation environments, ship motion, and ship traffic flow have been explored and analyzed. Additionally, these motion-prediction methods are applied for motion control, collision-avoidance planning, and route optimization. Key issues are summarized regarding ship-motion prediction, including online modeling of motion models, real ship validation, and consistency in modeling, optimization, and control. Future technology trends are predicted in mechanism-data fusion modeling, large-scale model, multi-objective motion prediction, etc.

Keywords: intelligent ships; motion prediction; navigation environment modeling; route optimization; motion control

Citation: Zhang, D.; Chu, X.; Liu, C.; He, Z.; Zhang, P.; Wu, W. A Review of Motion Prediction for Intelligent Ship Navigation. *J. Mar. Sci. Eng.* **2024**, *12*, 107. https://doi.org/10.3390/jmse12010107

Academic Editor: Diego Villa

Received: 26 November 2023
Revised: 28 December 2023
Accepted: 3 January 2024
Published: 5 January 2024

1. Introduction

As the size and number of ships increase, the maritime navigation environment and scenarios become increasingly complex, which poses new challenges to maritime safety. Empowered by the integration of cutting-edge technologies such as artificial intelligence (AI), machine learning, and big data, the modern ship is poised for a transformative leap in operational efficiency and decision-making capabilities [1,2]. Autonomous ships represent the pinnacle of maritime automation, operating with minimal or no human intervention. The coexistence and mutual influence of intelligent ships and traditional ships are expected to persist over the long term. Abnormal ship behaviors will increase the risks of collisions, groundings, and reef encounters. Predicting ship behavior through the prediction of the ship-navigation trajectory will effectively mitigate these navigation risks [3–6]. According to statistics from the International Maritime Organization (IMO), approximately 80–85% of accidents that occurred in the past two decades were attributed to human faults [7,8]. By implementing ship automation or assisted navigation systems, the probability of ship accidents can be effectively reduced. Accurate prediction of ship motion not only enhances the intelligence of maritime navigation but also decreases risks during navigation and energy consumption, thereby reducing pollution and greenhouse gas emissions during voyages [9–11].

The prediction of ship motion represents the temporal aspect of a ship Guidance, Navigation, and Control (GNC) system [12]. Ship motion prediction can be divided into different time scales during specific applications in the maritime domain. The long-term prediction is used for traffic-flow management or long-voyage trajectory prediction, short-term prediction is used for collision avoidance or path planning, and extreme short-term prediction is used for ship motion control. Ship motion is influenced by both the maritime environment and ship maneuvering. ship-motion prediction models at different time scales can be established based on either mechanics or data. These models provide the foundation for collision-avoidance path planning, path tracking control, traffic flow simulation, and the optimization of long-distance ship routes. They are integral to the realization of intelligent navigation. Currently, various researchers are conducting studies on the short-term prediction of ship motion, collision-avoidance trajectory prediction and planning, and long-term trajectory prediction [13–15].

To study motion prediction systematically for intelligent navigation, we provide an overview and analysis of the methods and applications of motion prediction across different time scales. In previous studies [16–18], motion prediction is taken only as a part of route planning, motion control, or traffic simulation individually. It is essential to consider the overall motion-prediction models, algorithms, and systems for different time scales. Many people are trying to solve this problem in terms of different aspects, and many models and applications have been presented for the motion prediction and control [19,20]. Since different types of the weather conditions will affect the ship's navigating state and safety, the influence of the environment must be considered, and the high-precision weather forecast can provide effective data support [21–23]. Meanwhile, with the development of new technologies, data collection with high-precision sensors, the state of ship motion, and environmental condition can be measured with high frequency, which will promote motion prediction [24–27]. More and more new technologies will be used in the shipping industry, and it is important to analyze the new direction for trends in technological development.

This review investigates the landscape of ship-motion prediction algorithms and their application in intelligent navigation. This paper is organized as follows: Section 2 systematically analyzes existing algorithms and research advancements through a thorough literature review. Section 3 delves into the environmental factors influencing motion prediction, exploring diverse methods for their modeling and forecasting. Section 4 introduces a range of motion models and prediction algorithms suitable for various time scales. Section 5 demonstrates the practical potential of motion prediction through representative applications. In Section 6, we critically analyze key challenges and emerging research trends identified in the reviewed literature, and Section 7 provides insights for future development. Section 8 concludes this paper.

2. Research Progress on Motion Prediction Based on the Literature Review

In order to analyze the current status and trends in research on ship-motion prediction, this paper conducts a search in the Web of Science (WOS) for literature on methods for ship-motion prediction from the past decade. Keywords from the relevant literature are analyzed using visualization tools to identify research hotpots and trends. Figure 1 depicts a keyword cluster map for the prediction of ship motion. The upper part of the figure illustrates keyword hotspots, while the lower part displays the evolution of research topics over time. Through this literature review, it is evident that earlier research tended to utilize standard ship models, employing methods such as Computational Fluid Dynamics (CFD) or empirical formulas. These methods use data from towing tank experiments with scaled ship models to describe mathematical models of ship motion. Mathematical models are used to characterize the nonlinear motion of ships, taking ship maneuverability and hydrodynamics into account. Research on ship maneuverability, navigation resistance, and hydrodynamics is often combined with the ship's own control and seaworthiness. Over time, as computational resources advance, and with the accumulation of experimental data and developments in research methods, recent studies have increasingly incorporated ship Automatic Identification System (AIS) data. Artificial intelligence, machine learning, and deep learning techniques are utilized to predict ship motion and trajectories, which contribute to enhancing ship navigation safety and collision avoidance applications.

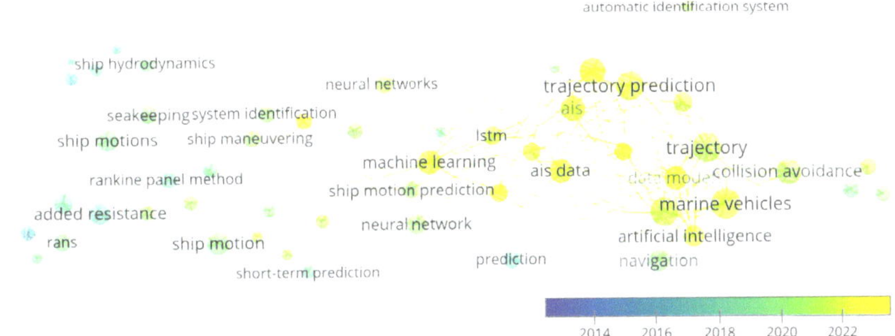

Figure 1. Knowledge graph for the prediction of ship motion.

As shown in Figure 2, the ship's navigation and motion control system represent a complex framework. This includes the ship-motion model and control algorithms. Moreover, it requires the utilization of perception, positioning, and observation systems to determine the ship's real position relative to other ships. Based on the navigational environment of the ship's destination port and sailing area, a safe and rational sailing route is charted. Then, this route, in turn, integrates with the ship's power control system for the ship-motion control by a seafarer or an autonomous system [28].

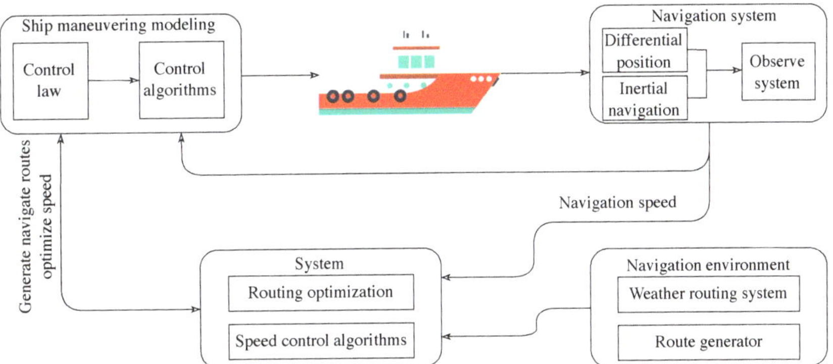

Figure 2. Intelligent ship navigation environment perception, motion models, and control systems.

Figure 3 depicts the process of ship-motion prediction. Before the ship embarks on a voyage, route planners use weather forecasts for the ship's navigational area, environmental constraints such as navigational facilities within the waterways, and the ship state model to establish the ship's navigation trajectory. During the voyage, the ship receives information from other ships through AIS, including their navigation trajectories and departure ports. Ship operators, relying on their navigation experience or traffic-flow-prediction algorithms, predict the ship's long-distance voyage, taking into account the ship's current status, and then plan a new route if necessary.

In situations where navigation trajectories are insufficient or when encountering complex traffic scenarios, ship operators traditionally communicate through Very High-Frequency (VHF) communication to determine navigation priorities and ship routes during these encounters. For unmanned autonomous ships, however, obtaining navigation priorities and predefined routes directly through VHF is challenging. Instead, these ships rely on onboard perception systems to sense the navigation environment. They consider navigation collision-avoidance rules and the predicted trajectories of other ships to determine the optimal route and ensure a safe and reasonable collision-avoidance process. Thus, for unmanned autonomous ships, short-term predictions (approximately 10–30 s) become especially critical.

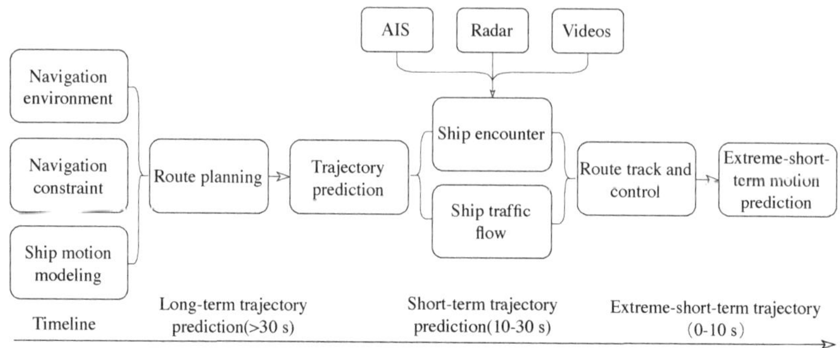

Figure 3. Flowchart showing the process of ship-motion prediction.

During the motion control (trajectory tracking and path following) for the defined route, an unmanned ship needs to predict the ship's motion in an extremely short time, obtain good motion control performance, and guarantee navigation safety, especially for the model-based control methods, such as model predictive control (MPC) and optimal control [29,30]. This motion prediction is defined as extremely short-term prediction (approximately 0–10 s). Extremely short-term prediction should take into account the ship's motion states and environmental disturbances with sensors or observers and employ suitable control algorithms to calculate the steering commands with time delays.

3. Ship-Navigation Environment Modeling and Prediction Methods

3.1. Characterizing Navigational Environmental Factors

The ship-navigation environment typically refers to the hydrological, meteorological, topographical, and traffic conditions around ships' navigational waters. For intelligent ships, it is essential to obtain reliable information about the navigational environment that can be exported to the navigation control system. This information can be acquired through the use of AIS, radar, cameras, sonar, depthometer, pitometer, anemometer, and other sensors and analyzed with the data processing and fusion [31]. Environmental factors affecting ship navigation include both static and dynamic aspects. Specifically, the static factors primarily refer to navigational water depth, shorelines, islands, reefs, and obstacles like sunken ships, which are typically marked on paper or electronic charts and are used to plan routes before a voyage begins. During the voyage, the ship continuously updates its position, calculates distances from static obstacles, and ensures that it avoids accidents like running aground or colliding with obstacles. The dynamic factors mainly involve wind, waves, currents, visibility, traffic flows, and ship-behavior characteristics in the navigation area. Dynamic factors are subject to change over time. They can be predicted dynamically using ocean and weather forecasts, traffic flow predictions, ship behavior detection, or real-time data collection from onboard sensors. These predictions are commonly used for dynamic route optimization, collision avoidance, and navigation control.

3.2. Static Environmental Factors

Electronic charts (nautical charts) provide precise information about static obstacles and can overlay dynamic data from sensors such as the Global Position System (GPS), radar, and AIS onto the electronic chart. Due to the presence of measurement errors in radar data, existing electronic charts and radar systems can introduce radar projection distortions when radar data are directly overlaid on electronic charts. To address this issue, Naus et al.

associated and matched electronic chart information with radar data, projecting radar echo vector data into the coordinate system of the electronic chart [32]. This eliminates radar measurement errors, enhances the accuracy of obstacle identification by radar, and allows ship operators to better monitor the navigation situation.

During autonomous ship navigation, ships heavily rely on real-time data from sensors and electronic charts. While electronic charts contain these data, there are relatively few open-source interfaces for research and development. Blindheim et al. use Python and Electronic Navigation Chart (ENC), an open-source electronic chart, to create interfaces displaying water depths, islands, reefs, and shallows within a ship's navigational area, facilitating ship-route planning and optimization [33]. Zhang et al. employed machine learning algorithms to analyze ship AIS data and quantified the risk of ship grounding using proximity prediction and depth information [34]. This approach is validated on a Ro-Pax ship in the Gulf of Finland, helping to prevent ship grounding and improving ship navigation safety.

To study the impact of offshore wind farms on ship-navigation risks, Xue et al. applied adaptive brainstorming with variable disturbances to collision-avoidance decisions in encounters near wind farms [35]. This approach considers the safety of nearby ships when following traditional collision-avoidance rules, and its effectiveness and reliability have been verified through theoretical calculations and simulations for different encounter scenarios near wind farms.

Addressing the issue of limited environmental perception for ships under restricted sensing conditions, Shi proposes a method for modeling the navigation area map of unmanned ships based on high-resolution satellite imagery [36]. This method employed high-resolution maps for land-sea segmentation, obstacle detail enhancement, and morphological transformation through image processing. It identifies obstacle areas and edge points, constructing a navigation map for unmanned ships to aid in route planning. This approach demonstrated effective recognition of static obstacles in unknown areas but requires frequent updates of satellite imagery.

Current research has made significant progress in studying natural static obstacles that ships encounter during navigation and those that may affect normal ship operations. Various algorithms can reconsider the impact of obstacles and generate reasonable routes. However, challenges arise when dealing with dynamic obstacles that change over time, such as sunken ships and fishing nets. Some obstacles can be updated in electronic charts, but others, like fishing nets, change dynamically with the operating areas of fishing ships. During ship navigation, mariners often need to combine electronic charts with visual observations to adaptively plan routes and avoid challenging situations that may affect normal ship operations.

Let us take an example to illustrate the above issue. Our team conducted a task of autonomous navigation journey in the East China Sea from July 2023 to August 2023. Since all the functions of the ship are in the testing stage, some experienced captains were on the ship during the test, and in emergency situations, the crew has a higher prioritization than the algorithms and the auto navigation model. In Figure 4, the red route represents the static route planned for a ship traveling from Zhoushan to Shanwei Port, taking into account the ship draft and economical cruising speed. The black trajectory represents the actual navigation route during the voyage, which encountered newly established wind farms, navigation announcements from the maritime management groups, a cluster of operational fishing vessels, and dynamic fishing nets during the fishing season. In response to these real-time conditions, the ship crew adjusted the navigation route based on the ship's current status.

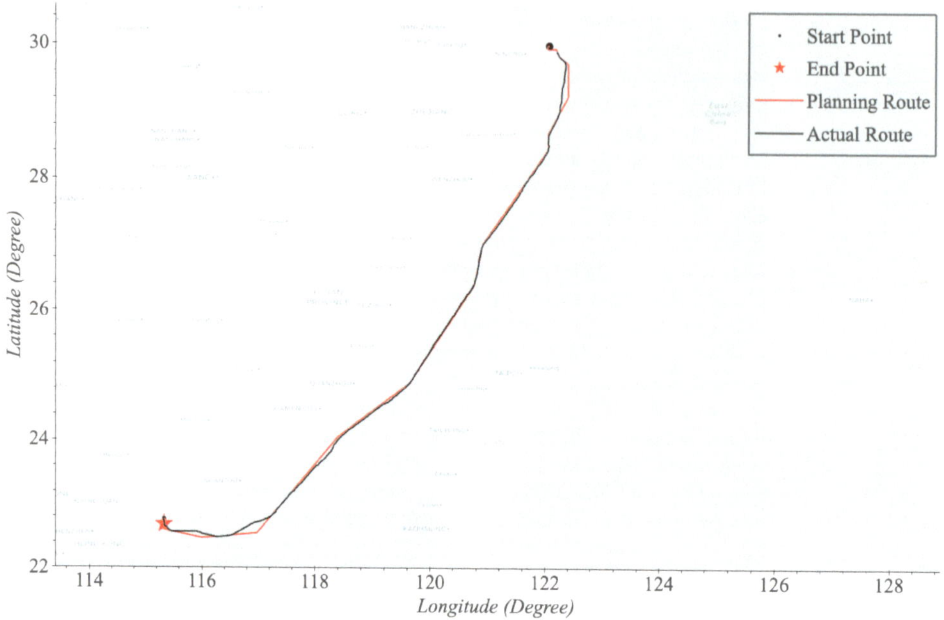

Figure 4. Ship's static planned route and actual navigation route.

3.3. Dynamic Environmental Factors

Dynamic navigational environmental factors primarily refer to disturbances within the navigational area, such as wind, waves, and currents. Wind and current data can be obtained directly through long-term observations and onboard sensors. However, the accurate modeling and prediction of sea waves is challenging due to their stochastic and complex nature. The navigational environment not only affects the safe operation of ships but also introduces uncertainty in fuel consumption, impacting route planning and optimization for long-distance ship voyages. Vettor et al. analyze different weather forecast models and compare them with the navigation routes of a container ship traveling in the North Atlantic [37]. The results show that the fuel consumption error caused by environmental model variations is approximately 10%.

In high-sea wind conditions, the interaction between waves and currents can significantly impact ship navigation. Chen et al. use third-generation wave models to study the effects of waves and currents on ship navigation [38]. Zwolan et al. incorporate wave models into ship navigation simulators to simulate ships' attitude and motion in waves, providing valuable references for crew training and similar purposes [39]. Bingham et al. used Gazebo simulation in ROS2 (Robot Operating System 2nd) to model ship navigation in complex sea conditions, offering mathematical models for different wind, wave, and current conditions to create a simulation test environment for marine robotics [40]. Daisuke et al., through the study of AIS data from ship navigation in areas with currents, observed significant deviations between the ship heading and bow direction [41], effectively predicting currents within the ship's navigational area. Yu employed wavelet transformation to decompose time series related to ocean wave factors and used the transformation results for training and predicting significant wave heights using residual neural networks [42]. The results indicate that wavelet transformation can improve the predictability of significant wave heights. Remya et al. used genetic algorithms to predict tidal currents, which proved to be more effective than harmonic analysis and fluid-dynamics-based methods [43]. Kavousi-Fard et al. decomposed current data into harmonic components and used these components in different Support Vector Regression (SVR) models for current prediction [44]. They tested this approach using data from a bay and found that it could achieve satisfactory prediction accuracy. This method, based on wavelet transformation and SVR, demonstrates effective current prediction.

Chen et al. established a numerical simulation model for different weather conditions combined with the Manoeuvering Modeling Group (MMG) model of the ship by using a weather model for the sea area in Japan, and they studied the impact of different models on the ship's response to weather conditions [45]. This research is of great significance for the environmental perception and modeling of ship. Additionally, machine learning and numerical simulation methods have also been used for the modeling and numerical prediction of ocean wind, waves, and currents [46–49]. Deep neural network models can automatically learn and extract feature relationships in complex data. Xie extracted the spatiotemporal coupled features of ocean currents, captured correlations and dependencies between adjacent sea areas using the Spatial Channel Attention Module (SCAM), and used the Gated Recurrent Unit (GRU) to model the temporal relationships of ocean currents [50]. They developed a deep network model called Spatiotemporal Coupled Attention Network (STCANet), which outperforms traditional models such as History Average (HA) and Autoregressive Integrated Moving Average (ARIMA). Traditional ocean current prediction methods have difficulty considering both the temporal and spatial effects of the prediction of ocean current. Thongniran et al. designed a prediction method that combines spatial and temporal features [51]. This method uses Convolutional Neural Networks (CNNs) for spatial prediction of ocean currents and Gated Recurrent Units (GRUs) for modeling the temporal features of ocean currents. It is compared with traditional models like ARIMA and K-nearest neighbors (KNN) using data from the Gulf of Thailand. The results show that the method combining spatial and temporal features had better prediction accuracy. To study the predictive performance of different machine learning algorithms on oceanographic parameters, Balogun [52] combined ARIMA, Support Vector Regression (SVR), and Long short-term Memory (LSTM) neural network models to model and predict different oceanographic elements. They used data from the Malaysian Peninsula region for validation. The results indicated that different prediction models had varying levels of accuracy for different elements. The accurate prediction of specific ocean parameters requires a reasonable choice of modeling methods and parameters. For the weather factors, Figure 5 shows the ocean weather conditions, including the bathymetry, current and significant sea height for the southeast coast of China, the different weather aspects affect the ship navigation from the motion, fuel consumption, and even safety.

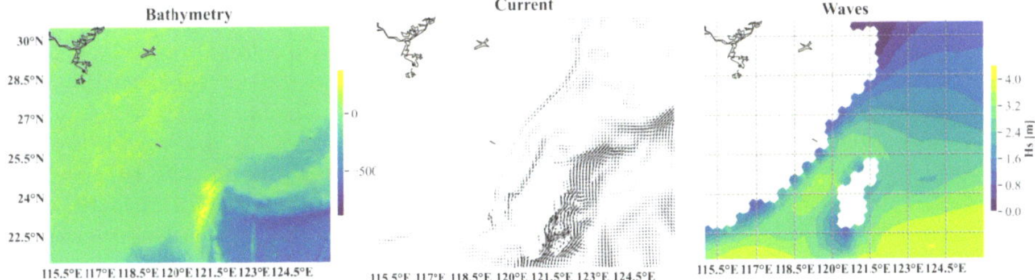

Figure 5. Metocean data for the southeast China Sea.

Different prediction algorithms used in the dynamic environmental factors are compared in Table 1.

Table 1. Summarizing the prediction algorithms used in the dynamic environmental factors.

Algorithms	Advantages	Disadvantages
LSTM [53]	A model that inherits characteristics from RNNs while incorporating gating mechanisms can effectively learn and retain long-term dependencies, addressing the limitation of traditional RNNs in capturing lengthy sequences. These mechanisms assist in gradient preservation, mitigating the issue of vanishing gradients during backpropagation, which results from a stepwise reduction in gradient magnitude.	The gradient issues of traditional RNNs have been partially addressed in LSTM and its variants, but challenges persist. While LSTM is capable of handling sequences on the order of 100 time steps, dealing with sequences of 1000 time steps or more remains a formidable task. Each LSTM cell inherently involves four fully connected layers (Multi-Layer Perceptrons or MLP). When LSTMs cover a substantial time span and are deep in terms of network architecture, the computational load becomes substantial, resulting in longer processing times.
GRU [54]	Thanks to the gating mechanisms that allow for selective information retention and forgetting, models like LSTM excel in capturing long-term dependencies compared to traditional RNNs. They typically require less training time than other types of recurrent neural networks. With fewer parameters than LSTM, they offer quicker training speeds and are less prone to overfitting.	When modeling complex sequential dependencies, it may not perform as well as LSTM. Explaining the gating mechanisms and information flow within the network can be more challenging compared to traditional RNNs. Some hyperparameter tuning may be required to achieve optimal performance. When dealing with extremely long sequences, it may encounter issues similar to other types of recurrent neural networks, such as the problem of vanishing gradients.
ARIMA [55]	ARIMA treats the data sequence generated by the predictive subject over time as a stochastic sequence. It utilizes a specific mathematical model to provide an approximation of this sequence. Once the model is identified, it facilitates the prediction of future values based on past and present values within the time series. This approach is fundamental in time series analysis and forecasting.	The ARIMA model indeed requires data to exhibit stationarity. As such, data need to undergo differencing to achieve stationarity before modeling. In essence, ARIMA models primarily capture linear data patterns and may not perform well in predicting non-linear data. Oceanic factors often encompass non-linear elements, which can pose challenges for ARIMA models in effectively modeling and forecasting such data.
STCANet [50,56]	STCANet, through the integration of spatial and temporal attention mechanisms, excels in capturing the interactions between variables such as wind, waves, currents, and tides. This results in higher predictive accuracy compared to traditional models. Additionally, STCANet demonstrates remarkable performance in modeling dependencies in the context of ocean prediction, which often relies on diverse data sources. It effectively handles the integration of multi-modal data, a crucial aspect of ocean forecasting.	Ocean prediction tasks typically involve handling large-scale spatiotemporal data, which may require substantial computational resources. While STCAN can offer insights into which features are important for prediction through its attention mechanisms, the inherent complexity of deep learning models may make it challenging to fully explain the model's decision-making process.

3.4. Ship Navigation Behavior, Traffic Flow Modeling and Prediction

Maritime traffic flow is a manifestation of ship behaviors, and it contains a wealth of information. It is the result of interactions among various elements involved in maritime traffic according to certain rules. Modeling and analyzing maritime traffic flow can guide ship-navigation decisions and control. The modeling of traffic flow is primarily based on analyzing the characteristics and distribution patterns of historical data. It is then combined with simulation methods to replicate traffic flow. Key aspects of modeling and simulation involve data collection, processing, and traffic-flow-generation methods [57].

Maritime traffic flow, as the macroscopic representation of ship behavior, characterizes the navigational situation of ships in specific areas. Research on modeling maritime traffic flow and ship behavior is beneficial for safety supervision and traffic flow organization and planning.

AIS data are commonly used as a data source for maritime traffic flow. They are utilized for extracting waterway traffic elements, clustering ship behaviors, and predicting ship behaviors. AIS data have various applications in maritime traffic services and collision risk assessments [58]. With the continuous accumulation of AIS data and data mining, there are still many unexplored possibilities [59]. AIS data contain crucial information about the ship-navigation status. In busy waterways, AIS data can be incomplete or missing. To improve the accurate monitoring of ship speeds in such areas, Zhao et al. used UAV (Unmanned Aerial Vehicle) onboard video for ship speed extraction [60]. They employed a simple linear tracking method with depth correlation metrics to extract ship speeds. Furthermore, satellite-based AIS stations enable the monitoring of global ship data. Yan et al. combined support vector machines and random forest methods to classify different types of ships and identify abnormal behaviors in satellite-based AIS data [61]. Considering ship behavior features, the ship identification accuracy reached as high as 92.7%.

Due to the presence of anomalies and errors in ship AIS data, it cannot be directly used for modeling and predicting traffic flows. Guo et al. collect knowledge about ship movements from the raw AIS data [62]. They use interpolation to estimate potential errors in the original trajectories. An improved K-means clustering method is then applied to assign weights to datapoint errors. In another study, the authors analyzed parameters affecting frequencies of ship encounters [63]. They introduced a method for predicting traffic-flow behaviors and ship-encounter frequencies with a time constraint, which finds applications in areas like offshore wind farms and fisheries. Liu et al. proposed a method that incorporates an attention mechanism into GRU and optimizes the GRU parameters using intelligent optimization algorithms for the detection of ship anomalies [6]. This approach trains on ship AIS trajectory data using the TensorFlow framework and reduces the time required for ship anomaly identification. Han et al. introduced a density-based spatial clustering method called DBSCAN (Density-Based Spatial Clustering of Applications with Noise) [64]. They adjusted clustering-algorithm parameters using a data-driven approach to model ship behavior based on ship AIS trajectory points, identifying abnormal ship behaviors. The effectiveness of behavior extraction is validated using AIS data from the Gulf of Mexico and the Great Lakes.

Zhang et al. use an adaptive particle swarm optimization algorithm to adjust the structure parameters of a BP neural network [65]. They developed an improved Particle Swarm Optimization-Back Propagation (PSO-BP) prediction model to predict the total ship traffic flow in a harbor area, which was validated in the Port of Los Angeles and demonstrated good results. To enhance the accuracy of prediction of the ship's traffic flow, Ye et al. leveraged the advantages of the encoder-decoder structure to capture long-term dependencies in time series data [66], this improvement aims to address the issue of cumulative errors that traditional iterative methods often face. They proposed a multi-step prediction method for ships' traffic flow based on an LSTM encoder, which involves the statistical analysis of ships' traffic flow and is validated to be effective. Shi et al. based their modeling of ships' abnormal behavior on trajectory data, approaching it from a spatial information perspective [3]. They modeled the rules of collective ship behavior and use a graph-based approach to determine abnormal ship behavior from spatial information. Then, they apply the Isolation Random Forest algorithm to detect abnormal ship behaviors.

For AIS data processing, the data cannot be used directly, as shown in Figure 6, and we have taken an example of using different algorithms include Bi-directional Long short-term Memory Recurrent Neural Networks (BLSTM-RNNs), Artificial Neural Network (ANN) and Piecewise Cubic Hermite Interpolating Polynomial (PCHIP) to restore the lost data; the BLSTM-RNNs has a good result for the AIS data restoration for the curve and straight trajectories in Hubei Wuhan, where the RMSE is near about 25 m, which is acceptable for the ship trajectories in inland area. Our analysis utilizes approximately one month of AIS data from August 2022, focused on the Zhoushan region of Zhejiang province. This area exhibits a highly complex traffic flow, encompassing numerous ferry routes between stops, cargo ships from diverse ports, and anchorage activities. To analyze these historical AIS trajectories, we employed a distance-based clustering method. This approach successfully identified nine distinct groups, with each subgroup characterized by similar trajectory patterns. Moving forward, we intend to leverage these insights to model individual ship-navigation behavior, enabling future research endeavors in trajectory prediction, behavioral discrimination, anomaly detection, and data mining.

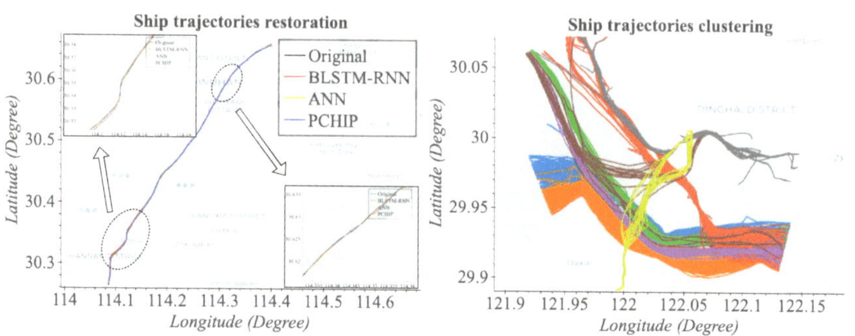

Figure 6. Ship trajectory restoration using different algorithms and trajectories clustering using a distance-based algorithm.

Navigation-environment modeling is relevant to ship route planning and optimization, ship-collision-avoidance route design, and the best timing for switching the collision-avoidance routes in real-time navigation control. The comprehensive modeling of the navigation environment takes into account factors affecting ship navigation control, trajectory prediction, and route optimization across different time dimensions. The ship-navigation environment modeling and prediction methods are summarized in Table 2.

Table 2. Methods for characterizing the navigation environment.

Type	Data Source	Impact Mode	Scope of Application	Perception Modeling Methods
Static factors	ENC	Navigational static objects, including water depth in the navigation area, islands and reefs, bridges, shipwrecks, navigation rules, and non-navigable areas	Applied to global optimization, local collision avoidance, and navigation control.	Combine ship GNSS, electronic charts, and perception sensors to sense the static factors in the navigation area, and use methods like artificial potential fields, image recognition, etc., in combination with electronic charts to model them [67,68].
Dynamic factors	Weather forecasting and prediction	Weather factors affecting ship navigation, including wind, waves, and currents.	Meteorological forecasts are used for ship route optimization, while short-term, high-precision weather forecasts are applied to ship motion control.	Incorporate weather forecasts and onboard sensors for prediction, utilizing LSTM, CNN, ARMA, hydrodynamic simulation, and more [43,44,69].
Traffic flow	Ship radar, AIS, vessel traffic services (VTS) system.	Applied to global optimization, local collision avoidance, and navigation control.	At the level of maritime navigation organization, modeling traffic flows will impact ships	Combining historical data with algorithms such as random number generation, probability space modeling, spatial clustering, CNN, DBSCAN, and LSTM to model traffic flow. In the short term, onboard sensors predict and anticipate ship trajectories, which serve as inputs for navigation decisions [41,45,63,64,70].

4. Ship Motion Modeling and Prediction Methods

4.1. Ship Motion Model

The motion of a ship in the water involves six degrees of freedom, with a typical consideration of sway, heave, and yaw as the primary three degrees of freedom. The propulsion methods for ships mainly involve propellers and rudders. The number of independent control variables is usually less than the number of motion degrees of freedom, resulting in underactuated characteristics in ship motion. Additionally, since ships are affected by external factors like wind, waves, and currents during navigation, predicting unmanned ship motion is a complex task. The ship coordinate system and motion models are illustrated in Figure 7.

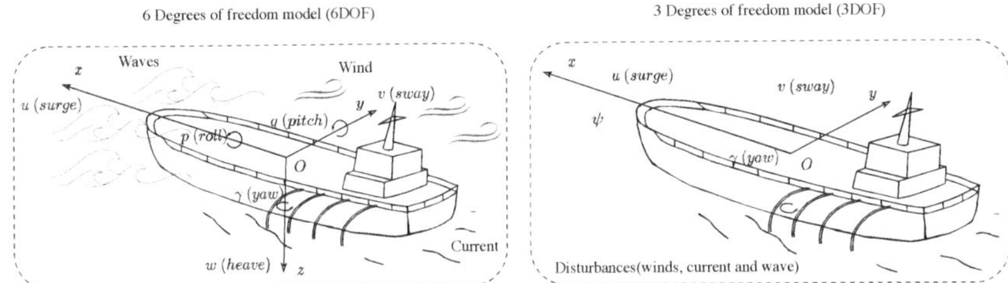

Figure 7. Six-DOF and three-DOF ship motion models.

ship-motion models are generally classified into hydrodynamic models and responding models [71]. Hydrodynamic models include multiple linear and nonlinear hydrodynamic parameters and disturbance coefficients and can be further divided into Abkowitz models and Maneuvering Modeling Group (MMG) models [12,72]. Abkowitz models analyze and solve the forces acting on the ship hull, propellers, and rudders as a whole, while MMG models create separate hydrodynamic models for the ship hull, propellers, and rudders to analyze the individual hydrodynamic effects.

Responding models represent another form of mathematical ship-motion model. Starting from an engineering control perspective, Nomoto views a ship as a dynamic system, with the rudder angle as the system input and the heading angle or yaw rate as the system output [73]. These models can be classified into first-order and second-order linear models and nonlinear models. Depending on the level of simplification of the mathematical model, they can be categorized as single-degree-of-freedom, three-degree-of-freedom, four-degree-of-freedom, and six-degree-of-freedom models [73–76]. Different mathematical motion models contain various dynamic parameters, which are estimated through maneuvering experiments or derived from model-scale tests of hydrodynamic parameters; the CFD can also be used for parameter calculations [12]. A three-degree-of-freedom (DOF) Abkowitz motion model is shown in Equation (1).

$$\begin{cases} m(\dot{u} - ru - x_G r^2) = X \\ m(\dot{v} + rv + x_G \dot{r}) = Y \\ I_{zz}\dot{r} + mx_G(\dot{v} + rv) = N \end{cases} , \qquad (1)$$

where m represents the actual mass of the ship; x_G is the longitudinal coordinate of the ship's center of gravity along the x axis; and X, Y, and N represent the components of hydrodynamic forces and moments acting on the ship in the three degrees of freedom u, v, and r, respectively. I_z is the moment of inertia of the ship around the vertical axis through its center of gravity, which governs its resistance to rotation around that axis. In the context of ship navigation, when considering the impact of external factors such as wind, waves, and currents on a ship's motion, a coupled superposition approach is applied. The model (1) involves combining the forces acting on the ship in these environmental conditions and, guided by empirical formulas, integrating the ship's dynamic state. This process is used to assess how the interference from these environmental factors affects the navigational state of the ship.

Accordingly, a three-degree-of-freedom (DOF) MMG motion model is shown in Equation (2).

$$\begin{cases} m(\dot{u} - rv - x_G r^2) = X_H + X_P + X_R \\ m(\dot{v} + rv + x_G \dot{r}) = Y_H + Y_P + Y_R \\ I_{zz}\dot{r} + mx_G(\dot{v} + rv) = N_H + N_P + N_R \end{cases}, \tag{2}$$

where the ship's inertia force term is included on the left side of the equation. On the right side of the equation, X_H, X_P, and X_R represent the hydrodynamic forces in the forward direction caused by the viscous fluid, the propulsion force from the propeller, and the force from the rudder, respectively. Y_P and Y_R represent the hydrodynamic forces in the lateral direction caused by the viscous fluid, the propulsion force from the propeller, and the force from the rudder, respectively. N_H, N_P, and N_R represent the hydrodynamic moments (causing rotation) in the yawing direction (yaw) caused by the viscous fluid, propulsion torque from the propeller, and the torque from the rudder, respectively. When a ship is affected by wind and waves, it is necessary to analyze the forces exerted on the ship by these environmental factors. These forces are then decomposed into the ship's direction of motion, allowing for the quantification of the interference experienced by the ship.

Accordingly, a responding motion model is shown in Equation (3).

$$T_1 T_2 \ddot{r} + (T_1 + T_2)\dot{r} + \alpha r + \beta r^3 = K(\delta + \delta_r) + KT_3\dot{\delta}, \tag{3}$$

where δ represents the rudder angle or steering angle. r represents the yaw rate, which is the change rate of the ship heading. K, T_1, T_2, and T_3 are control parameters related to the control system. δ_r represents the rudder deflection angle. α and β are constant coefficients or parameters. As the parameters of responsive models are typically derived from ship-maneuvering experiments to describe the influence of steering on a ship heading, it becomes challenging to directly superimpose external disturbances from the maritime environment onto the ship-motion model. When considering the effects of wind and waves on a ship heading, it is common to simulate them using stochastic noise. As for the influence of ocean currents, it is addressed by vectorially superimposing the current's velocity direction directly onto the ship's navigational state.

Alongside the accumulation of ship maneuvering control and navigation data, apart from the model (1)–(3) mentioned earlier, black-box models based on neural networks, data, and similar techniques are gradually finding applications in ship motion prediction and navigation control [77–80]. Black-box models can take into account a wider range of inputs, leading to higher predictive accuracy. However, these models are typically trained using complex mathematical algorithms, and their internal structure is typically composed of a large number of parameters, making them difficult to understand and explain, so these models are less amenable to mathematical representation and heavily rely on the accuracy of input data.

Three types of ship handling models are commonly used for ship motion control. Model (1) is relatively complex and provides high accuracy. It is based on Taylor's expansions of the forces acting on the ship in different directions. Each hydrodynamic parameter in this model has a clear mathematical interpretation. However, this model has many hydrodynamic derivatives that lack direct physical meaning, making them challenging to measure directly. Parameter-identification methods are often used to determine these non-measurable hydrodynamic parameters, allowing for the accurate simulation of ship handling. While it provides high accuracy, the modeling and parameter identification can be demanding in simulation environment.

The model (2) differentiates the effects of the hull, rudder, and propeller on ship handling performance. Model parameters can often be obtained directly through towing tank experiments, free-running model tests, or CFD. However, high requirements are placed on the hydrodynamic derivatives and the disturbance coefficients between the hull and the rudder/propeller. Models (1) and (2) are frequently used with model-scale ship tests, and their parameters correlate strongly with model accuracy. However, it may not fully capture environmental disturbances to the ship's motion.

Model (3) only considers the relationship between control inputs and state outputs. The parameters for this model can be obtained directly through full-scale ship maneuvering experiments, and it does not suffer from scale effects. Model (3) is commonly employed in automatic steering systems and heading controllers in the operations of real ships.

Due to the need for extensive data accumulation and model training, black-box models are currently less commonly applied in practical ship state and navigation control. Most algorithms are still in the research and testing phase. However, as ships become more intelligent and accumulate a growing amount of navigation data, this trend will likely promote the application of data-driven black-box models in ship motion prediction and navigation control.

The choice of model depends on the specific application, the level of precision required, and the available data and experiments. Different mathematical models are suitable for different scenarios. It is essential to choose the appropriate model based on the specific research requirements. Ship motion models are valuable for studying ship handling and seaworthiness. They are used to investigate ship control maneuvers and navigation attitude control, often in combination with the impact of marine environments on ship motion and disturbances. These models find applications in simulation studies, where they can help to analyze and understand various aspects of ship behavior under different conditions.

4.2. Ship Extreme Short-Term Motion Prediction

Ship motion models encompass the ship motion states and control inputs. Precise motion models can accurately describe the ship's state and predict its motion trends. high-precision modeling and parameter identification of ship motion models are especially important. The identification of ship maneuvering motion relies on mathematical models of ships, parameter-identification methods, and maneuvering motion data. Mathematical models of ships' maneuvering motion include the models mentioned above. Common parameter identification methods include least squares [81,82], Kalman filtering [83,84], support vector machines [85,86], neural networks [87,88], least squares support vector machine methods [89–91], particle swarm optimization algorithms [92,93], and Bayesian methods [94,95], among others. Several scholars have conducted in-depth research on this [96–98]. Maneuvering motion data are primarily obtained through zigzag tests and turning trials.

In extremely short-term prediction periods, due to the brief prediction duration, variations in the ship navigation environment can be ignored. These methods only focus on the immediate impact of the current environment on the ship's motion to predict its state. Methods using mathematical models for the prediction of ship motion mainly consist of linear and nonlinear predictive methods. Prediction methods based on mathematical models of ship motion can predict certain aspects of a ship's motion state. However, due to the nonlinear and time-varying nature of ship-motion models, complex motion states require more sophisticated and accurate motion models. For example, the ship-heading prediction, which represents a single-degree-of-freedom motion prediction, often employs the responsiveness model proposed by Nomoto [99]. Large ships on long-distance ocean voyages use different heading directions planned along the route for long-distance course-keeping. Early automatic steering is used to control the heading.

For the speed control in unmanned ships, a data-driven model-free elastic speed control method is presented in [100]. This method learns the rules of elastic speed control from a neural network predictor, determining the optimal input for motor control. The effectiveness of the algorithm is validated through simulation testing. Xu et al. applied a physics-informed neural network for parameter identification of a three-degree-of-freedom motion model in unmanned ships [101]. By combining data-driven and physical model advantages, they constructed a loss function for predicting the motion attitude of unmanned ships based on velocity and steering models. Compared to traditional neural network models, this approach shows better predictive performance.

In [102], a black-box motion model for unmanned ships is established. They employed a weighted Least-squares support vector machine based on the Sparrow search algorithm for parameter identification of unmanned ship motion models. To enhance the algorithm's stability and robustness, they introduced weighted least squares and the Sparrow Search Algorithm (SSA) for parameter optimization. The results show that the proposed black-box motion model has good generalization ability and could effectively predict the motion state of unmanned ships. Wang et al. collected data on the changes in ship steering angle input and ship heading change during ship operations [103]. They compared seven different regression algorithms and selected the best one to establish a ship-heading prediction model. They introduced an Antlion Optimizer (ALO) algorithm to search for the optimal weights for the prediction model, which was used in ship-course-keeping control.

The navigation attitude of unmanned ships is a complex, time-varying, and nonlinear system. Traditional algorithms for predicting unmanned ship attitudes could suffer from low accuracy, poor robustness, and limited practical application. Zhang et al. combined CNN with LSTM to build a data-driven neural network model for predicting the roll attitude changes in unmanned ships [104]. They used CNN to extract time series features and LSTM to predict the attitude at the next time step, which showed good prediction accuracy when validated on a real dataset of unmanned ship motion.

The extreme short-term motion-prediction performance with different methods is compared in [105], which can be seen in Figure 8. Some algorithms used for the extremely short-term ship-motion prediction were summarized in Table 3.

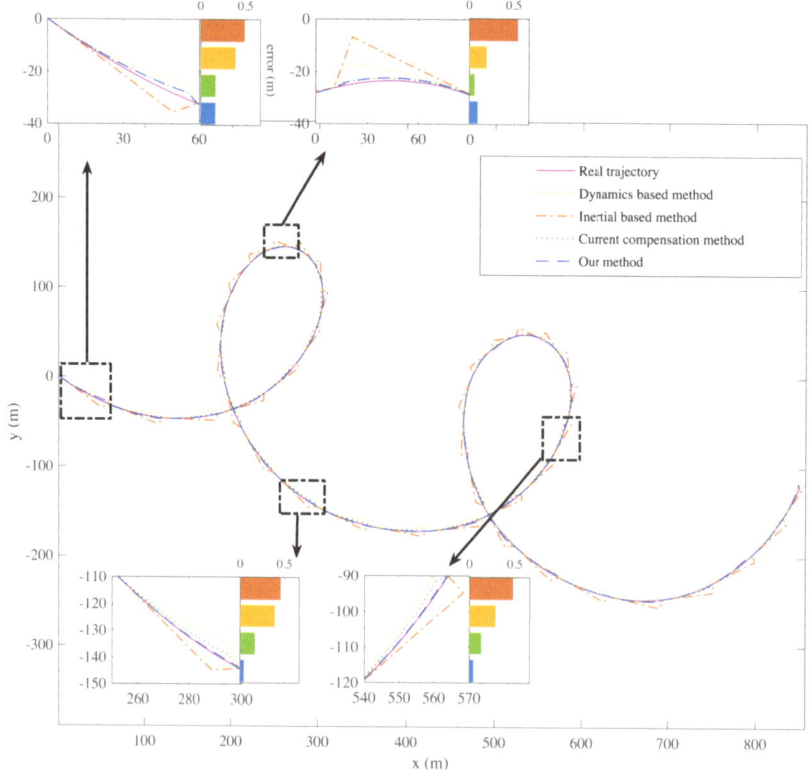

Figure 8. Extremely short-term trajectory prediction [105].

Table 3. Summarizing the algorithms used for extremely short-term ship-motion prediction.

Algorithms	Description	Advantages	Disadvantages
CNN [53]	Extracts spatial features from data like wave patterns.	Effective for capturing wave-induced motions and good for short-term prediction.	May require additional data preprocessing, black-box nature can hinder interpretability.
SSA [102]	Uses a black-box model and a nature-inspired metaheuristic.	Simple and easy to implement.	May not be as efficient as other metaheuristics for complex problems.
Kalman filtering [83,84]	Estimates system states with noise and uncertainties.	Robust to noise, can handle non-linear systems.	Requires accurate system model, and computational cost increases with model complexity.
LSTM [104]	Captures temporal dependencies and learns complex dynamics.	High accuracy, handles nonlinearities.	Requires large datasets, computationally expensive.

4.3. Short-Term Ship-Motion Prediction

During a ship's voyage, it needs to perceive dynamic and static obstacles using various sensors and, based on its navigation status, avoid these obstacles. For intelligent autonomous ships, the generation of collision-avoidance paths often depends on collision-avoidance planning algorithms, especially for short-time and short-distance navigation re-planning. For the planning of collision-avoidance trajectories, time sensitivity is crucial. When dealing with the problem of collision avoidance among multiple autonomous ships, a distributed multi-unmanned collaborative ship-planning algorithm based on deep reinforcement learning has been proposed in [106]. This approach uses an improved reciprocal velocity obstacle as the reinforcement learning reward function and plans collision-avoidance routes for different obstacles based on gated recurrent unit neural networks. Xie et al. combined the Model Predictive Control (MPC) algorithm with a three-degree-of-freedom ship-motion model [107]. The improved Turn-Neyman algorithm is integrated with maritime collision-avoidance rules to solve the ship predictive collision-avoidance problem. The algorithm's reliability was validated through simulations using the KRISO Very Large Crude Carrier no. 2 (KVLCC2) standard ship model.

Existing ship-collision-avoidance planning methods are primarily based on the Time to Closest Point of Approach (TCPA) and Distance to Closest Point of Approach (DCPA) between one's ship and the target ship. They often do not consider the uncertainty of ship positions and velocities. In [108], a ship-prediction probability collision0avoidance method is proposed based on the Kalman filter, which combines an Unscented Kalman Filter to predict the ship state and obtain the probability of ship positions. This information is then used to plan the optimal collision-avoidance route. Zhang et al. represent a ship's time and path during collision avoidance using space reconstruction and describe the ship's nonlinear motion with the MMG model [109]. It combines Nonlinear Model Predictive Control (NMPC) with an Extended Kalman Filter to address the ship motion prediction problem during the collision avoidance process, demonstrating the timeliness and reliability of the algorithm through simulations. The popular algorithms employed for short-term motion prediction are summarized in the Table 4.

Table 4. Summarizing the algorithms used for the ship's short-term motion prediction.

Algorithms	Description	Advantages	Disadvantages
MPC [106,107]	Optimizes future trajectory based on predicted motions and constraints.	Accounts for control limitations and environmental disturbances.	High computational cost, requires accurate model and prediction.
NMPC [108,109]	Extends MPC with non-linear models for improved accuracy.	Handles complex dynamics and uncertainties.	Increases computational cost and complexity.

He et al. transformed the ship collision avoidance motion planning problem into a multi-constrained nonlinear optimization problem with controllability, navigation environment, and navigation rules [110]. A novel Model Predictive Artificial Potential Field (MPAPF) motion-planning method is proposed to generate the ship-collision-avoidance path. The different collision avoidance methods include the Dynamic Anti-collision A-star(DAA*), COLREGS-RRT, and Asexual Reproduction Optimization-Artificial Potential Field (ARO-APF) based on short-term prediction are compared in Figure 9.

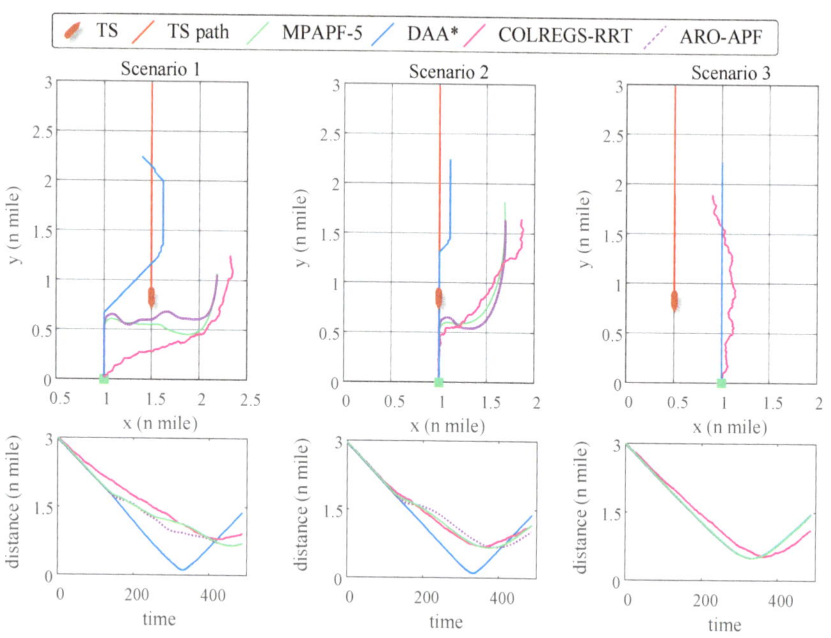

Figure 9. Different collision-avoidance methods based on motion prediction.

4.4. Ship Traffic-Flow Modeling and Long-Term Trajectory Prediction

Predicting long-distance ship trajectories based on estimated ship traffic flow, historical trajectories, and time series data can contribute to enhancing the safety of maritime environments, organizing ship traffic, and optimizing resource allocation in special water areas. It is beneficial for ship-route planning and optimization, ultimately improving a ship's navigational capabilities. Ships' navigation trajectories contain various characteristics of traffic flow. The accurate prediction of ship navigation trajectories can facilitate the statistical analysis and modeling of traffic flow.

To fully exploit the traffic flow information contained in ship AIS data and accurately predict ship trajectories, ref. [111] employs a prediction model called the MHA-BiGRU model, which is based on multi-head attention mechanisms and bidirectional recurrent units (BiGRUs). This model filters and modifies historical ship data, retaining more information and making predictions on time series data. The model correlates information from both historical and future ship trajectories, providing high precision, reliability, and ease of implementation.

Ref. [70] uses various machine learning and deep learning algorithms to predict and analyze the historical movement trajectories of ships in different navigational environments. This analysis supports decision making for intelligent ship route planning, collision avoidance, and traffic management operations. Ref. [112] utilizes historical AIS data from the Świnoujście port in Poland to establish ship traffic spatial distribution and probability models. These models analyze factors influencing the safe distance between ships, aiming to enhance maritime safety and reduce collision risks.

Ref. [113] proposes three trajectory-similarity search models for historical trajectory points, trajectory databases, and historical trajectories, which are used for short-term and long-term ship-trajectory prediction and are tested for algorithm reliability using ships' AIS data. Ref. [66] constructs a Long short-term Memory Encoder (LSTM-ED) for training ship AIS data to develop and validate ship traffic-flow models. This approach shows better performance than traditional traffic-flow statistical baseline methods. Considering the presence of abnormal jump values in long-distance trajectory data, ref. [114] integrates the LSTM structure into the deep learning Transformer framework, which leverages the strengths of Transformer and LSTM in dealing with long-distance dependencies in temporal and spatial features. The method employs time-window shifting and smoothing filtering to maintain trajectory smoothness, enabling the model to predict long-distance trajectories. Ref. [115] utilizes a hierarchical clustering method to extract behavioral features from AIS data. This approach is more efficient than traditional Douglas-Peucker (DP) and Least-squares methods. The LSTM algorithm is then used to predict ship trajectories, resulting in lower prediction errors compared to traditional RNN algorithms. The Temporal Convolutional Network (TCN) has strong time memory capabilities and performs well in time series prediction. Ref. [116] combines the attention mechanism with ship-trajectory time series (TTCN). They utilize one-dimensional convolutional units to extract high-dimensional features from the data and introduce mechanisms to enhance the learning of important features. This enables the training and learning of ultra-long time series data for ship trajectories. They use AIS data to construct a ship traffic flow dataset and compare it with traditional prediction methods, achieving better accuracy.

In order to investigate the efficiency of the trajectory prediction with different mentioned algorithms, we take some AIS data to implement the prediction algorithms. In Figure 10, we show different algorithms contains the LSTM, attention-based LSTM(ATT-LSTM), transformer with deep embedded clustering(TRFM DEC), CNN-LSTM, and Bi LSTM that we tested for long-term and short-term trajectory prediction; for the long and curved trajectories, the LSTM method has good performance in prediction, while for the short-term prediction, the CNN-based LSTM has good results for prediction, and the average error is 0.5 m for short-term prediction.

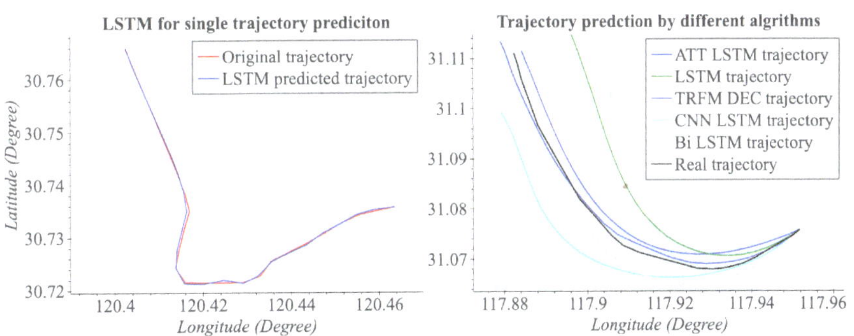

Figure 10. Long- term and short-term trajectory prediction based on different algorithms.

To summarize, the ship-motion-prediction methods can be broadly categorized into three types based on different prediction lead times, prediction intervals, and prediction accuracy.

(1) Extremely short-term motion prediction: This type of prediction is used for ship motion control, where high demands are placed on the prediction lead time, accuracy, and update frequency. It is employed throughout the entire voyage, repeatedly, until the journey is complete.

(2) short-term motion prediction: This type of prediction is applied to avoidance of ship collision, planning, and decision making. When ships encounter each other, it is necessary for them to react accordingly until the specific encounter scenario ends. After ensuring there is no risk of collision, a short-term motion prediction cycle is completed.

(3) long-term trajectory prediction: This type of prediction is used in ship traffic organization and voyage-route optimization. It primarily serves voyage planning, taking into account the origin of the voyage task and the dynamic characteristics of the route. An optimal route is established for safe navigation. If the navigation conditions change during the voyage, the long-term motion predictions may need to be adjusted accordingly.

Different prediction types use different data sources, methods, and time frames. The choice of the most suitable prediction method for ship motion depends on factors like the actual ship situation, the sailing environment, and specific characteristics.

5. Applications of Ship Motion Prediction

5.1. Motion Control

Automated ship control takes into account the maneuverability of ships in complex environments and the influences of wind, waves, and currents on ship handling. Besides these disturbances, there might be other unknown disturbances to be considered, and observers need to be used to estimate these unknown disturbances. In trajectory-tracking control, it is essential to compensate for the effects of environmental disturbances. In order to improve the accuracy of navigation control, many methods are gradually applied in navigation tests and simulations.

MPC is a method that balances modeling, prediction, and control. It can use historical data to create a simple model of a ship's maneuverability for trajectory prediction. By incorporating an extended observer, it can perceive the actual sailing conditions of the ship. For example, ref. [117] proposes a guidance strategy based on NMPC for path following in unmanned ships. This strategy overcomes the limitations of Line-Of-Sight (LOS) navigation and uses an established state observer for tracking control of heading and speed.

To address the shortcomings of a ship motion control framework, which typically plans before tracking, ref. [118] proposes an improved artificial potential field method combined with the MPC algorithm. This approach generates control trajectories for ship-collision avoidance. In complex navigational environments, it allows ships to navigate around obstacles safely, serving as the algorithmic foundation for controlling unmanned ships in complex sea areas.

The sailing environment can impact the ship's motion model and control. For uncertain wave disturbances on ships, ref. [119] introduces a parameter identification method using Least-squares support-vector Machines (LS-SVMs) to identify parameters for a four-degree-of-freedom MMG model. They also designed an online sliding-window modeling method to predict ship motion under wave interference.

To achieve real-time high-precision and anti-interference motion control for unmanned ships in complex scenarios, a study [120] designed a NMPC trajectory tracking controller based on a three-degree-of-freedom hydrodynamic model for unmanned ships. This controller considers disturbances like wind, waves, and currents and uses a nonlinear disturbance observer to provide online compensation for the control model. The study's findings confirm the effectiveness of the control algorithm.

5.2. Collision Avoidance Planning

Predicting collision-avoidance trajectories for ships requires considering the sailing status of both the ship and other ships. The ship's sailing status can be directly obtained through positioning or inertial navigation. In contrast, information about other ships' sailing statuses must be perceived using sources like AIS (Automatic Identification System), radar, video, and electronic charts to determine the motion of target obstacles [121]. The acquired information regarding other ships' motions often contains errors and uncertainties. Hence, it needs the use of short-range prediction algorithms to predict their movement trends and calculate minimum encounter distance and closest encounter time. These algorithms must meet high standards for short-term ship-state prediction and encounter-situation assessment.

To address the challenge of accurately estimating the DCPA and TCPA of encountering ships, ref. [122] decomposes ship AIS trajectory data into linear and nonlinear components. They then employed ARIMA and LSTM methods to model these two components. This modeling is utilized for trajectory prediction and collision-avoidance route planning.

Refs. [123,124] combined the principles of MPC with maritime collision avoidance rules (COLREG) to design a trajectory-based collision-avoidance planning method. They constructed window functions to reduce tracking errors and cost acceleration, which improved the algorithm's reliability. This method predicts the sailing trajectories of encountering ships, uses convex free sets and maritime collision avoidance rules to develop collision-avoidance strategies, and generates planned routes.

To mitigate dangerous situations caused by abrupt changes in ship speed during trajectory prediction, ref. [125] proposed an improved ant colony algorithm using an artificial potential field. This modification enhances the convergence speed and global optimization of the algorithm, resulting in smoother collision-avoidance planning trajectories and reducing the probability of accidents.

Zhu and Ding [126] introduced a method for determining the optimal collision-avoidance point, which helps in defining the most effective collision-avoidance route based on the relative velocity and kinematic parameters between unmanned ships and obstacles. It calculates the ship's best collision point, using the speed and heading-angle increments that can be reached by the ship to constrain dynamic window sampling. Simulation testing demonstrates that the proposed optimal collision-avoidance point method is more efficient and robust.

In a real-world navigation scenario, ships have to contend with the motion of other ships and various uncertainties caused by interference. To address these challenges, we implemented the virtual potential field and dynamic ship field methods for planning collision-avoidance paths. In October 2022, we specifically selected ships encountered in sheltered waters for testing, as shown in Figure 11, and we built a software platform and implemented the algorithms with Qt (verison 5.2), C++ (version 11), Python (version 3.10), and ENC (s57 standard) map for the automatic control for a 40 m length USV. We tested the collision avoidance and control algorithms by using an intelligent ship equipped with radar, AIS, and an optoelectronic system. In the open sea environment, we designed a predetermined route for the ship, which subsequently navigated along this planned trajectory. Meanwhile, the target ship, located ahead and on a directly opposing heading, maintained its course throughout the encounter. To avoid collisions, the ship dynamically replanned its route based on the relative velocity between the vessels. Aiming to prevent entering the target ship's domain, the ship adjusted its speed to approximately 7 knots, maintaining a horizontal separation of approximately 120 m-roughly four times the ship's length.

5.3. Ship Voyage Optimization

In ship route optimization, various factors need to be considered, including sailing speed, water depth in the sailing area, disturbances like wind and currents, ship loading, fuel prices, and more. The goal is to achieve the lowest fuel consumption, minimize operations, reduce sailing time, maximize fuel efficiency, and minimize leasing costs [127–130]. Ships at sea can optimize their routes based on the influence of wind and currents, aiming to save fuel and ensure safe navigation. When sailing near the coast, factors such as islands and tides also need to be considered. The sailing environment significantly affects a ship's movement. By combining a ship's current status with historical traffic flow data and meteorological information specific to the sailing area, long-distance trajectories can be predicted and optimized, thus enhancing navigation safety and reducing the impact of the environment on navigation.

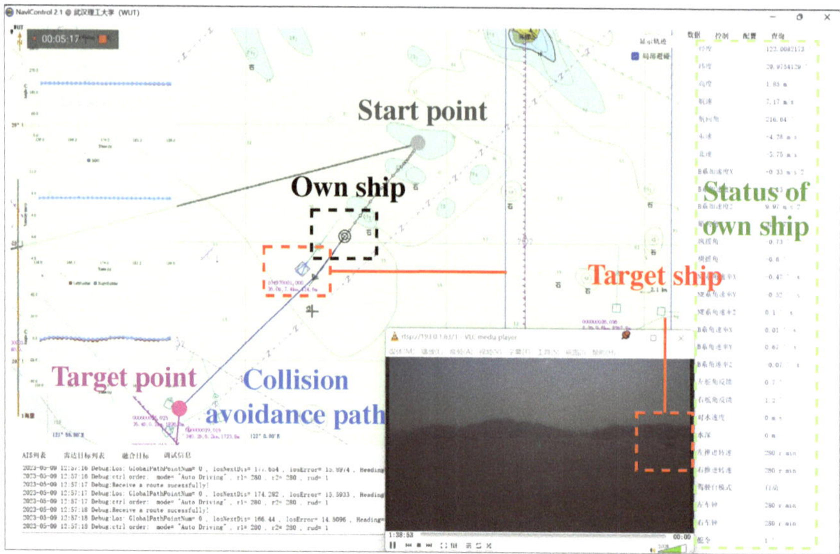

Figure 11. Dynamic domain-based collision avoidance sea trial in Zhoushan.

Ship-route optimization mainly consists of environmental modeling and optimization searching. To reduce the complexity of environmental modeling, the planning space is typically structured and transformed into a state-space search problem, which is further converted into a state-space path-search problem. These problems are addressed using search algorithms to obtain optimal solutions. Common ship-route-optimization algorithms include A*, Dijkstra, artificial potential fields, Rapidly Exploring Random Tree (RRT, RRT*), Isochrone, Dynamic Programming, 3D Dynamic Programming, genetic algorithms, simulated annealing, particle swarm optimization, neural networks, and more [131–141].

Taking into account the influence of ocean currents and marine environments, ref. [142] uses reinforcement learning in combination with ship motion models and map information to optimize ship routes. For unmanned ships operating in complex and dynamic environments, ref. [143] combines the accelerated A* algorithm with the visibility graph algorithm, using a quadtree to perform fast searches on the visibility graph and improving trajectory update efficiency. Many optimization and planning algorithms apply to optimizing ship routes in nearshore waters, considering aspects like maritime safety, avoiding grounding or collisions, and fuel savings [144].

In traditional RRT algorithms, the planned trajectories may suffer from issues like lack of smoothness and excessive travel distances. Guo et al. [145] employed quadratic Bézier curves and the Dijkstra algorithm for trajectory pruning within the RRT algorithm, making the generated trajectories smoother. They also used Morphin for global trajectory pruning, and simulations confirmed its suitability for route planning for unmanned ships in complex navigation environments.

Route-optimization algorithms each have their pros and cons. In the application of route optimization for unmanned ships, it is also necessary to incorporate maritime collision-avoidance rules, consider the maneuverability of the ship, and take into account the navigation environment. To address the shortcomings of optimization algorithms, it is possible to improve them by combining and using different algorithms, all in alignment with the ship's objective functions.

Route-optimization algorithms can be used in voyage planning for long-distance routes based on objective optimization functions. In actual ship navigation, when encountering other ships, it is essential to follow maritime collision-avoidance rules according to the current collision situation. Different collision scenarios require considering maritime collision avoidance rules, the ship's collision responsibility, and the necessary actions to be taken. When a collision-avoidance maneuver is needed, it involves adjusting the ship route based on the ship's current state, considering factors such as ship maneuverability, target ship length, width, DCPA, TCPA, the ship domain, and navigation conditions. This new route is followed for a certain period while implementing maneuvers such as altering the course, reducing the speed, or a combination of both [146].

In dynamic encounter scenarios, it is necessary to consider the motion state of other ships. To ensure safe navigation, path-planning algorithms are applied to dynamically re-plan routes in local areas, generating new collision-avoidance trajectories that are both safe and efficient. Common collision-avoidance planning algorithms include A*, the artificial potential field method, velocity obstacle method, and RRT, among others [147–150].

The ship's navigation environment significantly affects the accurate prediction of ship motion. The traditional modeling of motion models and parameter identification can adequately describe a ship's motion status through mathematical modeling. However, achieving the real-time prediction of a ship's motion status in complex navigation environments requires the rational selection of motion modeling and state prediction based on the specific focus of ship-motion control. By incorporating motion models into the modeling process, short-term and long-term trajectories of the ship can be predicted, providing insights into the ship's motion response in the navigation environment. The accurate prediction of ship-motion status and practical application necessitates selecting appropriate prediction methods based on the application scenarios and ensuring the proper application of the prediction results.

6. Analysis of the Key Issues

Ship navigation research has tackled factors affecting sailing, traffic flow, motion models, and route optimization as separate modules. The focus now shifts toward integrating these for the better modeling and prediction of ship-navigation environments. In current research, simulation models are widely used to investigate existing problems and explore methods in depth. However, in practical engineering applications, the time-varying and nonlinear nature of the navigation environment and motion models present limitations in simulation-based methods. Unmanned ships highlight the need for the integrated modeling of navigation environments, dynamic traffic flow, and context-aware motion models, paving the way for automatic route planning and control decisions. The challenge lies in seamlessly integrating the navigation environment, traffic flow, and ship motion models for autonomous ships, enabling intelligent route planning and control in complex situations. These areas have seen relatively limited research. The specific limitations are as follows.

6.1. Online Modeling of Ship Motion

Due to the influence of the navigation environment, the motion models and model parameters of ships exhibit time-varying and uncertain characteristics. The accuracy of mathematical models of ship motion, based on the CFD simulations, empirical formulas, and experimental ship handling, is no longer suitable for precise ship control. Leveraging data from diverse sources like environment, traffic flow, and sensor readings can enhance the observation and modeling of disturbances impacting ship motion. By incorporating these disturbances into the model-building and parameter-identification process, intelligent ship-control systems can achieve higher accuracy and minimize errors. Combining environmental models within the navigational area with ship-motion-control models for online model establishment and parameter identification is essential for path optimization and control decision-making in intelligent ships. This approach effectively addresses navigation, decision-making, and control issues in intelligent ships and is a key indicator of the autonomous driving level and degree of automation in intelligent ships.

Figure 12 shows the prediction of ship speed during navigation using the xG-Boost (Extreme Gradient Boosting) machine learning algorithm. In the upper part of the figure, the blue curve represents the prediction of ship speed based solely on the ship propeller RPM (revolutions per minute). The red curve, on the other hand, takes into account additional factors such as the wind encountered during navigation and ocean current data specific to the navigational area. By comparing these two curves, it becomes evident that increasing the number of input parameters in the model significantly enhances the accuracy of ship speed prediction.

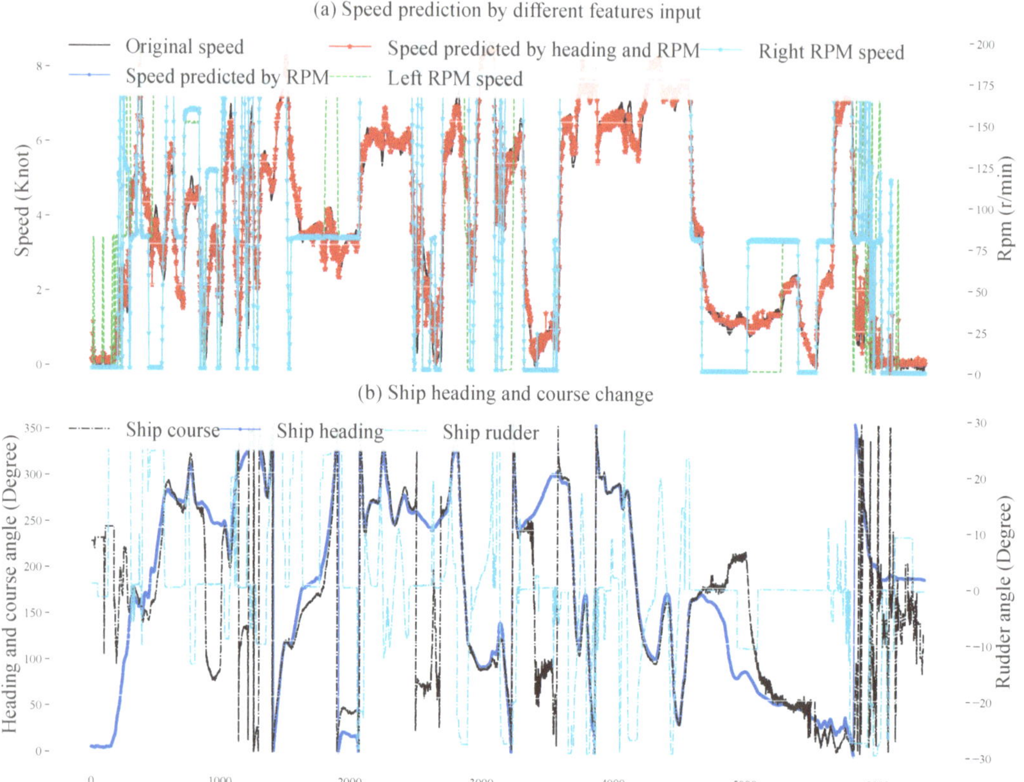

Figure 12. Ship navigation speed prediction with different input characters and parameters based on the xG-Boost method in complex sea conditions.

6.2. Limitations of Simulation Validation

Based on the existing literature, most scholars primarily use simulation to study the modeling of ship motion in response to the effects of navigation environments, parameter identification, model development, and control. Environmental influences are often generated using random methods, while real navigation environments are complex, dynamic, and nonlinear, making it challenging to directly describe key factors affecting ship-motion modeling and control with mathematical equations. It can be challenging for simulated environments to replicate the actual navigation state of ships accurately. Additionally, conducting real ship testing for large ships is costly. Current research often relies on scale models for experiments, which can reflect essential issues to some extent. However, due to scale effects, the results may vary for medium and large ships. There is a considerable gap between ship-simulation research and engineering applications. Precise data collection, historical playback, and specific ship-modeling and testing methods tailored to particular ships are required to meet the needs of motion prediction and control. For the implementation of these planning, collision avoidance, and control algorithms, many steps need to be tested before the final application in real conditions, these algorithms will not perform as well as the simulation tests.

As shown in Figure 13, prior to the final sea trials of our collision-avoidance algorithms for unmanned surface vessels (USVs), we conducted extensive simulations to validate their performance under safe and controlled conditions. These simulations comprised four key stages:

(1) Virtual collision avoidance with geographic information system (GIS) data in Figure 13a: We established a virtual environment using GIS data, where a virtual ship navigated directly toward a virtual USV in the open sea. This initial step assessed the robustness of the algorithm under predictable conditions.

(2) Virtual collision avoidance with static obstacles in Figure 13b: This stage introduced virtual static obstacles alongside sea trials, further challenging the algorithm's ability to navigate complex environments.

(3) Virtual dynamic collision avoidance in Figure 13c: Building on the previous stages, we implemented virtual dynamic obstacles within sea trial simulations, simulating encounters with moving objects. This provided a more realistic test of the algorithm's adaptability.

(4) Real ship encounter in Figure 13d: The final stage involved a real ship navigating toward the USV, allowing us to observe the algorithm's performance in a live setting and verify its ability to generate optimal collision-avoidance paths.

It is important to acknowledge that significant effort is required to bridge the gap between simulation and real-world testing. Sea trials often present unforeseen circumstances that cannot be fully replicated in simulated environments. Consequently, the transition from simulation to sea trials necessitates iterative refinement and validation. In essence, while simulations offer valuable insights, their limitations necessitate real-world testing for comprehensive performance evaluation.

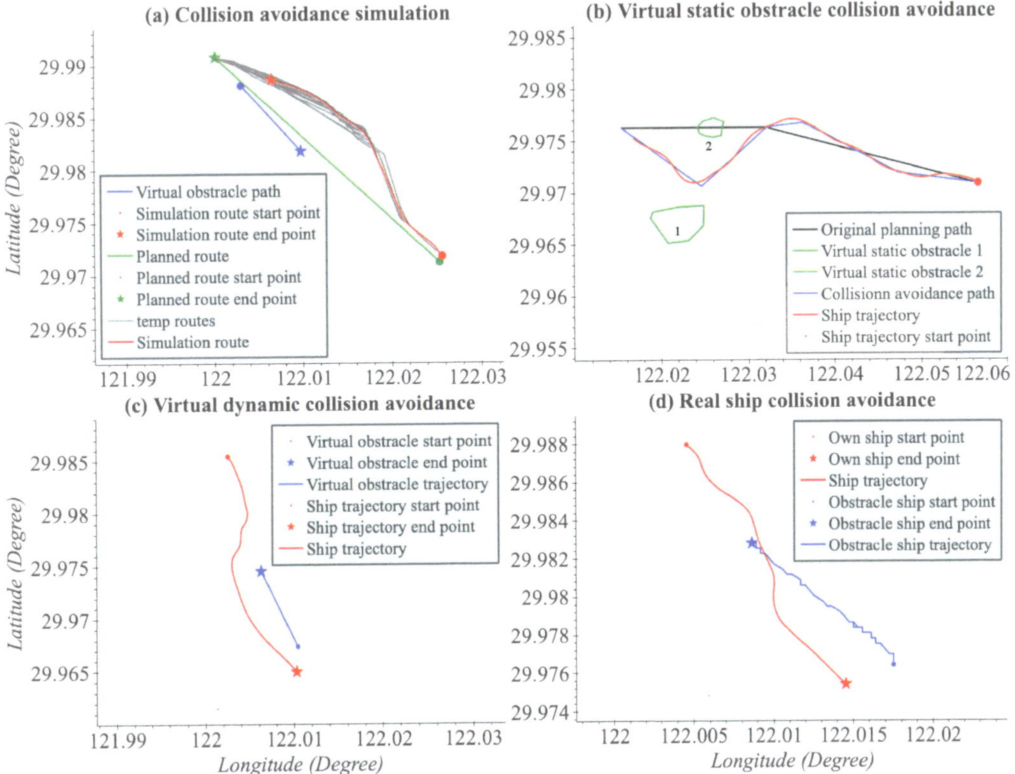

Figure 13. Ship-trajectory tracking and collision-avoidance control simulation, virtual test, and sea trial in a real condition.

6.3. Consistency in Modeling, Optimization, and Control

In the existing body of literature, the prevailing trend in current research leans towards the independent development of ship motion models, optimization algorithms, and various other aspects. A notable drawback of this approach is the absence of a unified standard interface for both input and output data across these models. This deficiency gives rise to a lack of general and consistent models, particularly in the realms of ship-route optimization, collision-avoidance planning, and navigation-motion control.

From the vantage point of intelligent ship navigation and control, there emerges a critical need to amalgamate diverse systems. This integration should aim to simplify the complex processes involved in motion modeling, motion prediction, and navigation control. The ultimate goal is to achieve a higher degree of automation and intelligence in the domain of ship navigation control. The absence of a standardized interface impedes the seamless exchange of information among different modules, hindering the establishment of a cohesive and interoperable framework for ship navigation.

In the shipping industry, there are many aspects that need to be considered between the planning and the final navigation or ship control. As shown in Figure 14, before the ship leaves the ports, the planners will draw up a route according to the departure, destination, and navigation task. Sometimes, the route will not be the best option for every ship, and the researchers will optimize the route based on the forecasted weather and the ship-performance model, while the performance model is summarized by the empirical formula or the historical data. For the ship's operation, once the route is designed or optimized, the ship operators will drive it based on their experience or the maneuverability of the ship, wile the maneuverability is changed based on the different navigate status. For intelligent navigation, all the aspects need to be considered using the performance model, maneuverability, and control algorithms.

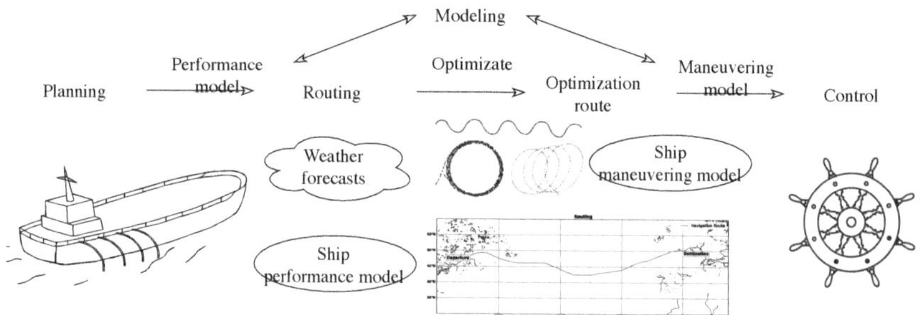

Figure 14. Ship route planning, optimization, and control system.

To address this challenge, future research efforts should concentrate on the development of a standardized interface that can serve as a common language for communication among various ship-navigation systems. This unified interface would facilitate the integration of motion models, optimization algorithms, and other components, fostering a more cohesive and interoperable approach to ship route optimization, collision-avoidance planning, and navigation motion control. Such an integrated and standardized framework holds the potential to streamline the development and implementation of intelligent navigation and control systems for ships, paving the way for enhanced safety, efficiency, and automation in maritime operations.

7. Trends in Technological Development

With the development of computer technology, artificial intelligence technology, marine environmental observation technology, and ship motion prediction technology, great progress has been made. In the future, ship-motion-prediction technology will develop in the following directions.

7.1. Trajectory Prediction Based on the Fusion of Multi-Source Sensor Data

In complex navigation scenarios requiring collision avoidance, a single sensor has limitations on target recognition. Ship operators need to combine the information from different sensors manually or according to their experience, making it difficult to perceive obstacles effectively. This necessitates data fusion from multiple sensors for trajectory prediction to provide more accurate assessments of ship-navigation encounters. Existing perceptual environment modeling primarily focuses on the recognition and tracking of static and dynamic obstacles in collision-avoidance decision-making scenarios. This research combines the ship's navigation status with the recognition and tracking of obstacles. With the advancement of ship intelligence and unmanned navigation, unmanned ships adhere to maritime collision avoidance rules designed for manned ships. Nevertheless, unmanned ships typically exhibit superior maneuverability and increased levels of automation. In the context of autonomous navigation, it becomes crucial to autonomously and accurately perceive the surrounding navigation environment. Combining navigation-environment modeling with route planning for unmanned ships can enhance the reliability of route selection and navigation control, harnessing the advantages of intelligent navigation decisions. The fusion of multi-source shipboard data with route planning plays a crucial role in optimizing navigation routes and navigation control. The fusion of AIS, radar, and GPS data is applied for different navigation trajectories, which is used in Figure 15.

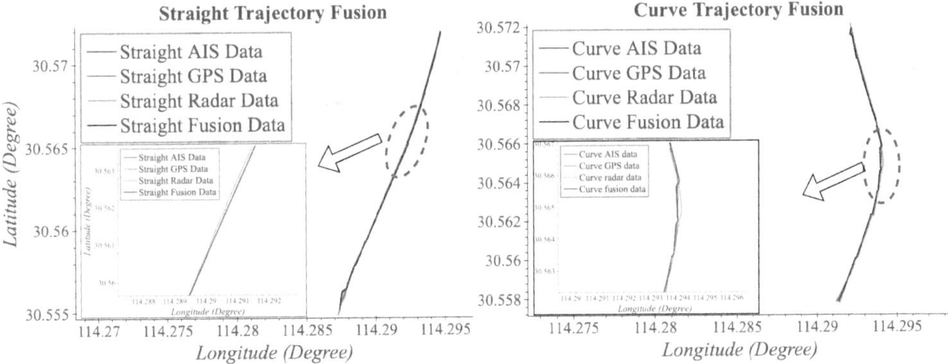

Figure 15. Trajectory fusion from AIS, radar and GPS based on the straight and curved conditions.

In adverse weather conditions, it is essential to consider factors like wind, waves, and currents that can significantly impact the navigation of unmanned ships, particularly in high-sea conditions. In such scenarios, it is about not only planning collision-avoidance routes but also optimizing navigation-control strategies based on the current sea conditions. This is crucial for reducing the impact of waves on the ship, thereby enhancing both the efficiency and safety of ship navigation.

7.2. Integration of Environmental Modeling, Route Optimization, and Motion Control

Traditional navigation and control systems have three relatively independent modules: guidance, navigation, and control. With the development of intelligent navigation and unmanned technologies in ships, these modules become more closely intertwined. From a planning perspective, static and dynamic environmental models of navigation will significantly influence route planning for ships. In actual navigation, weather conditions and environmental modeling can impact route optimization and control for local ship encounters. Starting with the establishment of environmental models and extending to route optimization and control decision making, integrated system models should be created. The integration spans from perception systems to decision systems, utilizing optimization algorithms to select and establish the best models that are suitable for decision making and control in intelligent and unmanned ships.

In the context of intelligent and unmanned ships, route optimization is focused on long-distance, extended-time motion planning, where factors like distance, energy consumption, and voyage duration are considered. On the other hand, ship motion control emphasizes short-distance, short-term motion tracking control, accounting for navigation risks and collision avoidance decisions during the voyage. Both aspects need to be effectively combined. It is necessary to choose appropriate routes while also considering the current encounter scenarios and selecting the right collision avoidance routes and control decision-making methods to ensure the economic efficiency and safety of navigation.

The issue of transitioning between global, long-distance route optimization and local, short-distance collision-avoidance decision making will be a focal point of research for intelligent ships and unmanned ships in autonomous navigation control. This transition impacts not only the economic effectiveness of ships but also their navigational safety. As illustrated in Figure 16, we have employed both 2D Dynamic Dijkstra and 3D Dynamic Dijkstra by considering the variable speed of each sub-route, to optimize the path from America to the English Channel across the North Atlantic Ocean. The optimized route not only saves more distance but also yields a significantly lower wave height compared to the original route.

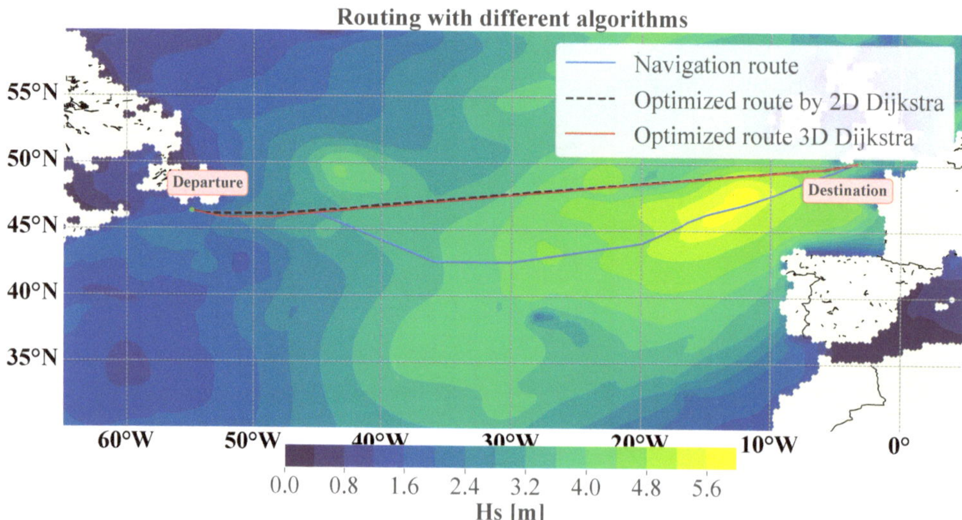

Figure 16. Route planning with different weather models and optimization methods.

7.3. Mechanism and Data-Fusion-Driven Modeling

As ships evolve toward increased automation and intelligence, traditional ships are adopting advanced sensors to support navigational decision making and control. In open sea areas, these technologies serve as valuable assistants for ship operators, yet in complex navigation scenarios, the precision of models and algorithms may fall short of expectations. Shipping companies, armed with extensive operational and navigational data, possess valuable experience for decision making and operations. Leveraging this wealth of data and experience is pivotal for enhancing navigation safety and precision while reducing the labor intensity for operators.

The volume of data related to navigational environments and decision-making control is anticipated to surge as accumulation continues. high-precision, large-scale operational data play a pivotal role in propelling the advancement of intelligent and unmanned ships, as well as contributing to the digitization of the maritime industry. Extracting valuable insights from this extensive data pool, especially regarding key factors influencing navigational decision making and control for intelligent and unmanned ships, promises substantial advancements.

The increasing utilization of machine learning and deep learning algorithms is poised to significantly augment perception and decision-making capabilities during ship navigation. This shift toward advanced algorithms is expected to result in enhanced autonomy for intelligent ships, leading to reduced human intervention, heightened efficiency and safety, and lowered operating costs. The continuous evolution through data-driven modeling, accumulation, iteration, and updates will contribute to the increased accuracy of ships' navigational environment and motion-control models. The principle of the mechanism and the data fusion-driven modeling method is shown in Figure 17.

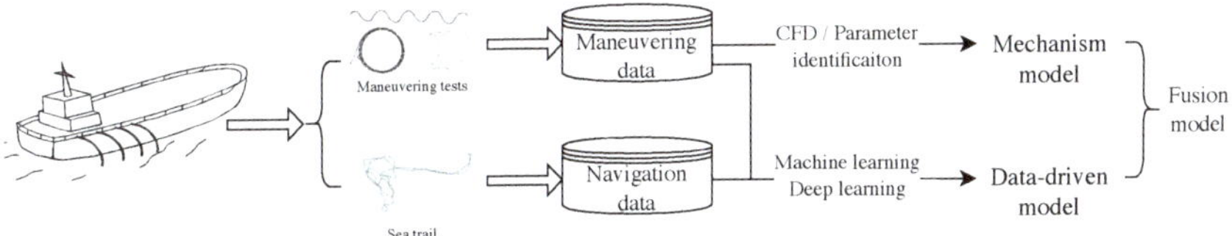

Figure 17. Mechanism and data-driven model fusion.

7.4. Time Series Modeling and Multi-Objective Prediction

Conventional predictions of a ship's motion state heavily rely on precise mathematical models dedicated to a ship's motion. These models are intricately integrated with a ship's responses in diverse settings, utilizing model-prediction methods to anticipate a ship's motion. The intricate challenge lies in the time-varying nature of ship motion and navigational environments, rendering the use of mathematical models complex. Hence, the establishment of time series models derived from high-precision data emerges as a critical necessity to seamlessly integrate data-driven ship motion prediction and control.

Moreover, deploying sensitivity-analysis methods allows for a comprehensive examination of key factors influencing ship motion control and their corresponding impacts. By identifying these critical factors, one can effectively model and predict navigational environment data. This process not only facilitates ship-route optimization but also contributes to the enhanced control of navigational environments, further emphasizing the importance of a data-driven approach in refining ship-motion prediction and control mechanisms. The performance of time series modeling and multi-objective prediction is shown in Figure 18. In Figure 18a the correlation of each state is presented with different shades of color, and Figure 18b,c shows the speed and roll angle prediction based on the input time series with the weather and control command information. The speed and roll can be predicted very precisely.

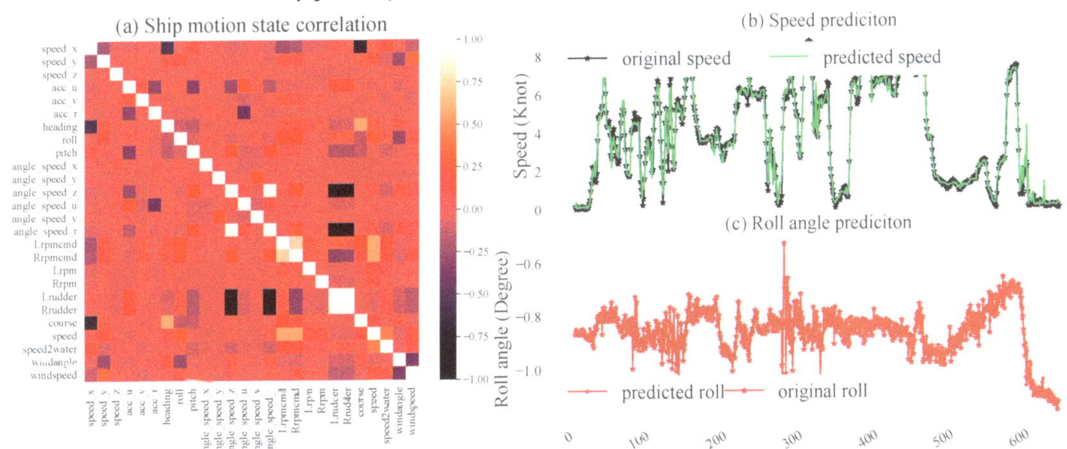

Figure 18. Time series modeling and multi-objective prediction.

8. Conclusions

The vast reach of the oceans has, throughout history, ignited a flame of discovery and advancement. This dedication to progress extends to the field of ship navigation, which is currently experiencing a significant evolution in the wake of technological advancements. This transformation is fueled by several key trends and active research areas that are reshaping the future of intelligent ship navigation.

(1) Dawn of the autonomous age: Intelligence takes the helm. The most conspicuous transformation lies in the burgeoning field of ship autonomy and intelligence. Cutting-edge technologies like artificial intelligence are no longer futuristic fantasies; they are being strategically implemented in unmanned ships. Sophisticated navigational environment sensing and modeling prediction methods act as catalysts, paving the way for ever-more intelligent vessels. This synergistic integration is not just about automation; it is about reshaping the maritime industry, ushering in an era of unparalleled efficiency and safety. Imagine unmanned ships seamlessly navigating complex waterways, optimizing routes in real time, and evading dangers with preternatural precision. This is the future that intelligent ship navigation promises.

(2) Precision takes center stage: Modeling the seas with mathematical elegance. At the heart of intelligent navigation lies the power of precise mathematical modeling. Accurate models of the navigational environment and ship motion are the bedrock upon which robust algorithms are built. The future holds exciting advancements in simulation, virtual testing, and digital twinning-revolutionary methodologies that will redefine the validation and testing of design and control algorithms for unmanned ships. Think of it as pushing a virtual ship through a simulated ocean, testing its responses to every wave and current, all before it ever sets sail. This iterative refinement process will lead to more robust and reliable autonomous ships that are ready to conquer the real-world seas.

(3) Conquering time: A holistic approach to navigational prediction. The future of intelligent ship navigation isn't confined to a single time horizon. A holistic approach that encompasses long-term, mid-term, and short-term ship motion prediction methods is rapidly emerging. This multi-temporal perspective addresses the multifaceted challenges faced by ships, from voyage planning and route optimization to real-time collision avoidance and dynamic decision making. Imagine a ship that can predict not only its immediate path but also anticipate potential hazards and weather patterns hours and even days in advance. This level of foresight will revolutionize maritime safety and operational efficiency, paving the way for a truly intelligent and autonomous future at sea.

Author Contributions: Conceptualization, X.C., C.L. and D.Z.; methodology, D.Z. and C.L.; software, D.Z., Z.H. and P.Z.; validation, D.Z., Z.H. and W.W.; writing—review and editing, D.Z., C.L. and X.C.; supervision, X.C. and C.L.; funding acquisition, X.C. All authors have read and agreed to the published version of the manuscript.

Funding: This research was funded by the National Key R&D Program of China (2022YFB2602305), National Natural Science Foundation of China (52001240), Key Research and Development Program of Guangxi Zhuang autonomous region (GuikeAA23062052), the Natural Science Foundation of Chongqing, China (cstc2021jcyj-msxmX1220), the Open Project Program of Science and Technology on Hydrodynamics Laboratory (6142203210204).

Conflicts of Interest: The authors declare no conflicts of interest.

Abbreviations

The following abbreviations are used in this manuscript:

AI	Artificial Intelligence
AIS	Automatic Identification System
ALO	Antlion Optimizer
ANN	Artificial Neural Network
ARIMA	Autoregressive Integrated Moving Average
ARMA	Autoregressive Moving Average

ARO-APF	Asexual Reproduction Optimization-Artificial Potential Field
ATT-LSTM	Attention-based LSTM
BiGRU	Bidirectional gate recurrent unit
BLSTM-RNNs	Bi-directional Long short-term Memory Recurrent Neural Networks
BP	Back Propagation
CDF	Computational Fluid Dynamic
CNN	Convolutional neural network
DAA*	Dynamic Anti-collision A-star
DBSCAN	Density-Based Spatial Clustering of Applications with Noise
DCPA	Distance to Closest Point of Approach
DOF	Degrees of Freedom
DP	Douglas-Peucker
ENC	Electronic Navigation Chart
GIS	Geographic Information System
GNC	Guidance, Navigation and Control
GNSS	Global Navigation Satellite System
GPS	Global Position System
GRU	Gated Recurrent Unit
HA	History Average
IMO	International Maritime Organization
KNN	K-nearest neighbors
LOS	Line of Sight
LSTM	Long short-term Memory
LSTM-ED	Long short-term Memory Encoder
LS-SVM	Least-squares support-vector machine
MHA-BiGRU	multi-head attention mechanism and bidirectional gate recurrent unit
MMG	Maneuvering Modeling Group
MLP	Multilayer Perceptron
MPC	Model Predictive Control
MPAPF	Model Predictive Artificial Potential Field
NMPC	Nonlinear Model Predictive Control
PCHIP	Piecewise Cubic Hermite Interpolating Polynomial
PSO	Particle Swarm Optimization
PSO-BP	Particle Swarm Optimization-Back Propagation
ROS2	Robot Operation System 2nd
RPM	Revolutions Per Minute
RRT	Rapidly exploring Random Tree
SCAM	Spatial Channel Attention Module
SSA	Sparrow Search Algorithm
STCANet	Spatiotemporal Coupled Attention Network
SVR	Support Vetor Regression
TCN	Temporal Convolutional Network
TCPA	Time to Closest Point of Approach
TRFM DEC	Transformer with Deep Embedded Clustering
TTCN	Tiered Temporal Convolutional Network
UAV	Unmanned Aerial Vehicle
VHF	Very High Frequency
KVLCC2	KRISO Very Large Crude Carrier no. 2
USV	Unmmanned Surface Vessels
VTS	Vessel Traffic Service
WOS	Web of Science
xG-Boost	eXtreme Gradient Boosting

References

1. Munim, Z.H.; Haralambides, H. Advances in maritime autonomous surface ships (MASS) in merchant shipping. *Marit. Econ. Logist.* **2022**, *24*, 181–188 [CrossRef]
2. Forti, N.; d'Afflisio, E.; Braca, P.; Millefiori, L.M.; Carniel, S.; Willett, P. Next-Gen Intelligent Situational Awareness Systems for Maritime Surveillance and Autonomous Navigation [Point of View]. *Proc. IEEE* **2022**, *110*, 1532–1537. [CrossRef]

3. Shi, Y.; Long, C.; Yang, X.; Deng, M. Abnormal Ship Behavior Detection Based on AIS Data. *Appl. Sci.* **2022**, *12*, 4635. [CrossRef]
4. Yao, L. Research Status and Development Trend of Intelligent Ships. *Int. Core J. Eng.* **2019**, *5*, 49–57 . [CrossRef]
5. Chen, Q.; Lau, Y.Y.; Zhang, P.; Dulebenets, M.A.; Wang, N.; Wang, T.N. From concept to practicality: Unmanned vessel research in China. *Heliyon* **2023**, *9*, e15182. [CrossRef]
6. Liu, H.; Liu, Y.; Li, B.; Qi, Z. Ship Abnormal Behavior Detection Method Based on Optimized GRU Network. *J. Mar. Sci. Eng.* **2022**, *10*, 249. [CrossRef]
7. Hasanspahić, N.; Vujičić, S.; Frančić, V.; Čampara, L. The Role of the Human Factor in Marine Accidents. *J. Mar. Sci. Eng.* **2021**, *9*, 261. [CrossRef]
8. Eliopoulou, E.; Alissafaki, A.; Papanikolaou, A. Statistical Analysis of Accidents and Review of Safety Level of Passenger Ships. *J. Mar. Sci. Eng.* **2023**, *11*, 410. [CrossRef]
9. Pedrozo, R. Advent of a New Era in Naval Warfare: Autonomous and Unmanned Systems. In *Autonomous Vessels in Maritime Affairs: Law and Governance Implications*; Johansson, T.M., Fernández, J.E., Dalaklis, D., Pastra, A., Skinner, J.A., Eds.; Springer International Publishing: Cham, Switzerland, 2023; pp. 63–80. [CrossRef]
10. Wang, J.; Xiao, Y.; Li, T.; Chen, C.L.P. A Survey of Technologies for Unmanned Merchant Ships. *IEEE Access* **2020**, *8*, 224461–224486. [CrossRef]
11. Liu, J.; Yan, X.; Liu, C.; Fan, A.; Ma, F. Developments and Applications of Green and Intelligent Inland Vessels in China. *J. Mar. Sci. Eng.* **2023**, *11*, 318. [CrossRef]
12. Fossen, T. *Handbook of Marine Craft Hydrodynamics and Motion Control*; Wiley: Hoboken, NJ, USA, 2021.
13. He, Z.; Liu, C.; Chu, X.; Negenborn, R.R.; Wu, Q. Dynamic anti-collision A-star algorithm for multi-ship encounter situations. *Appl. Ocean Res.* **2022**, *118*, 102995. [CrossRef]
14. Geng, X.; Li, Y.; Sun, Q. A Novel Short-Term Ship Motion Prediction Algorithm Based on EMD and Adaptive PSO–LSTM with the Sliding Window Approach. *J. Mar. Sci. Eng.* **2023**, *11*, 466. [CrossRef]
15. Zhang, M.; Taimuri, G.; Zhang, J.; Hirdaris, S. A deep learning method for the prediction of 6-DoF ship motions in real conditions. *Proc. Inst. Mech. Eng. Part M J. Eng. Marit. Environ.* **2023**, *237*, 147509022311578. [CrossRef]
16. Lyu, H.; Hao, Z.; Li, J.; Li, G.; Sun, X.; Zhang, G.; Yin, Y.; Zhao, Y.; Zhang, L. Ship Autonomous Collision-Avoidance Strategies—A Comprehensive Review. *J. Mar. Sci. Eng.* **2023**, *11*, 830. [CrossRef]
17. Zaccone, R.; Martelli, M. A collision avoidance algorithm for ship guidance applications. *J. Mar. Eng. Technol.* **2020**, *19*, 62–75. [CrossRef]
18. Zhou, S.; Wu, Z.; Ren, L. Ship Path Planning Based on Buoy Offset Historical Trajectory Data. *J. Mar. Sci. Eng.* **2022**, *10*, 674. [CrossRef]
19. Sørensen, M.E.N.; Breivik, M.; Skjetne, R. Comparing Combinations of Linear and Nonlinear Feedback Terms for Ship Motion Control. *IEEE Access* **2020**, *8*, 193813–193826. [CrossRef]
20. Skulstad, R.; Li, G.; Fossen, T.I.; Vik, B.; Zhang, H. A Hybrid Approach to Motion Prediction for Ship Docking—Integration of a Neural Network Model Into the Ship Dynamic Model. *IEEE Trans. Instrum. Meas.* **2021**, *70*, 1–11. [CrossRef]
21. Cheng, W.; Yan, Y.; Xia, J.; Liu, Q.; Qu, C.; Wang, Z. The Compatibility between the Pangu Weather Forecasting Model and Meteorological Operational Data. *arXiv* **2023**, arXiv:2308.04460.
22. Bi, K.; Xie, L.; Zhang, H.; Chen, X.; Gu, X.; Tian, Q. Accurate medium-range global weather forecasting with 3D neural networks. *Nature* **2023**, *619*, 533–538. [CrossRef]
23. Bi, K.; Xie, L.; Zhang, H.; Chen, X.; Gu, X.; Tian, Q. Pangu-Weather: A 3D High-Resolution Model for Fast and Accurate Global Weather Forecast. *arXiv* **2022**, arXiv:2211.02556.
24. Rodger, M.; Guida, R. Classification-Aided SAR and AIS Data Fusion for Space-Based Maritime Surveillance. *Remote Sens.* **2021**, *13*, 104. [CrossRef]
25. Rong, H.; Teixeira, A.; Guedes Soares, C. Maritime traffic probabilistic prediction based on ship motion pattern extraction. *Reliab. Eng. Syst. Saf.* **2022**, *217*, 108061. [CrossRef]
26. Silva, K.M.; Maki, K.J. Data-Driven system identification of 6-DoF ship motion in waves with neural networks. *Appl. Ocean Res.* **2022**, *125*, 103222. [CrossRef]
27. Schirmann, M.L.; Collette, M.D.; Gose, J.W. Data-driven models for vessel motion prediction and the benefits of physics-based information. *Appl. Ocean Res.* **2022**, *120*, 102916. [CrossRef]
28. Wang, S.; Zhang, Y.; Zhang, X.; Gao, Z. A novel maritime autonomous navigation decision-making system: Modeling, integration, and real ship trial. *Expert Syst. Appl.* **2023**, *222*, 119825. [CrossRef]
29. Model predictive ship trajectory tracking system based on line of sight method. *Bull. Pol. Acad. Sci. Tech. Sci.* **2023**, *71*, e145763. [CrossRef]
30. Øveraas, H.; Halvorsen, H.S.; Landstad, O.; Smines, V.; Johansen, T.A. Dynamic Positioning Using Model Predictive Control With Short-Term Wave Prediction. *IEEE J. Ocean. Eng.* **2023**, *48*, 1065–1077. [CrossRef]
31. Thombre, S.; Zhao, Z.; Ramm-Schmidt, H.; Vallet Garcia, J.M.; Malkamaki, T.; Nikolskiy, S.; Hammarberg, T.; Nuortie, H.; H. Bhuiyan, M.Z.; Sarkka, S.; et al. Sensors and AI Techniques for Situational Awareness in Autonomous Ships: A Review. *IEEE Trans. Intell. Transport. Syst.* **2022**, *23*, 64–83. [CrossRef]
32. Naus, K.; Wąż, M.; Szymak, P.; Gucma, L.; Gucma, M. Assessment of ship position estimation accuracy based on radar navigation mark echoes identified in an Electronic Navigational Chart. *Measurement* **2021**, *169*, 108630. [CrossRef]

33. Blindheim, S.; Johansen, T.A. Electronic Navigational Charts for Visualization, Simulation, and Autonomous Ship Control. *IEEE Access* **2022**, *10*, 3716–3737. [CrossRef]
34. Zhang, M.; Kujala, P.; Hirdaris, S. A machine learning method for the evaluation of ship grounding risk in real operational conditions. *Reliab. Eng. Syst. Saf.* **2022**, *226*, 108697. [CrossRef]
35. Xue, H.; Qian, K. Ship collision avoidance based on brain storm optimization near offshore wind farm. *Ocean Eng.* **2023**, *268*, 113433. [CrossRef]
36. Shi, B. Obstacles modeling method in cluttered environments using satellite images and its application to path planning for USV. *Int. J. Nav. Archit. Ocean Eng.* **2019**, *11*, 202–210. [CrossRef]
37. Vettor, R.; Guedes Soares, C. Reflecting the uncertainties of ensemble weather forecasts on the predictions of ship fuel consumption. *Ocean Eng.* **2022**, *250*, 111009. [CrossRef]
38. Chen, C. Case study on wave-current interaction and its effects on ship navigation. *J. Hydrodyn.* **2018**, *30*, 411–419. [CrossRef]
39. Zwolan, P.; Czaplewski, K. Sea waves models used in maritime simulators. *Zesz. Nauk. Akad. Morska W Szczecinie* **2012**, *32*, 186–190.
40. Bingham, B.; Agüero, C.; McCarrin, M.; Klamo, J.; Malia, J.; Allen, K.; Lum, T.; Rawson, M.; Waqar, R. Toward Maritime Robotic Simulation in Gazebo. In Proceedings of the OCEANS 2019 MTS/IEEE SEATTLE, Seattle, WA, USA, 27–31 October 2019; pp. 1–10. [CrossRef]
41. Inazu, D.; Ikeya, T.; Waseda, T.; Hibiya, T.; Shigihara, Y. Measuring offshore tsunami currents using ship navigation records. *Prog. Earth Planet. Sci.* **2018**, *5*, 38. [CrossRef]
42. Yu, X.; Liu, Y.; Sun, Z.; Qin, P. Wavelet-Based ResNet: A Deep-Learning Model for Prediction of Significant Wave Height. *IEEE Access* **2022**, *10*, 110026–110033. [CrossRef]
43. Remya, P.; Kumar, R.; Basu, S. Forecasting tidal currents from tidal levels using genetic algorithm. *Ocean Eng.* **2012**, *40*, 62–68. [CrossRef]
44. Kavousi-Fard, A.; Su, W. A Combined Prognostic Model Based on Machine Learning for Tidal Current Prediction. *IEEE Trans. Geosci. Remote Sens.* **2017**, *55*, 3108–3114. [CrossRef]
45. Chen, C.; Shiotani, S.; Sasa, K. Numerical ship navigation based on weather and ocean simulation. *Ocean Eng.* **2013**, *69*, 44–53. [CrossRef]
46. Lou, R.; Lv, Z.; Dang, S.; Su, T.; Li, X. Application of machine learning in ocean data. In *Multimedia Systems*; Springer: Berlin/Heidelberg, Germany, 2021; pp. 1–10.
47. Isozaki, I.; Uji, T. Numerical prediction of ocean wind waves. *Pap. Met. Geophys.* **1973**, *24*, 207–231. [CrossRef] [PubMed]
48. Saruwatari, A.; Yoneko, Y.; Tajima, Y. Effects of wave, tidal current and ocean current coexistence on the wave and current predictions in the tsugaru strait. *Coast. Eng. Proc.* **2014**, *34*, 42. [CrossRef]
49. Group, T.W. The WAM model—A third generation ocean wave prediction model. *J. Phys. Oceanogr.* **1988**, *18*, 1775–1810. [CrossRef]
50. Xie, C.; Chen, P.; Man, T.; Dong, J. STCANet: Spatiotemporal Coupled Attention Network for Ocean Surface Current Prediction. *J. Ocean Univ. China* **2023**, *22*, 441–451. [CrossRef]
51. Thongniran, N.; Vateekul, P.; Jitkajornwanich, K.; Lawawirojwong, S.; Srestasathiern, P. Spatio-Temporal Deep Learning for Ocean Current Prediction Based on HF Radar Data. In Proceedings of the 2019 16th International Joint Conference on Computer Science and Software Engineering (JCSSE), Chonburi, Thailand, 10–12 July 2019; pp. 254–259. [CrossRef]
52. Balogun, A.L.; Adebisi, N. Sea level prediction using ARIMA, SVR and LSTM neural network: Assessing the impact of ensemble Ocean-Atmospheric processes on models' accuracy. *Geomat. Nat. Hazards Risk* **2021**, *12*, 653–674. [CrossRef]
53. Karevan, Z.; Suykens, J.A. Transductive LSTM for time-series prediction: An application to weather forecasting. *Neural Netw.* **2020**, *125*, 1–9. [CrossRef]
54. Chhetri, M.; Kumar, S.; Pratim Roy, P.; Kim, B.G. Deep BLSTM-GRU Model for Monthly Rainfall Prediction: A Case Study of Simtokha, Bhutan. *Remote Sens.* **2020**, *12*, 3174. [CrossRef]
55. Jain, G. A Study of Time Series Models ARIMA and ETS. Available online: https://papers.ssrn.com/sol3/papers.cfm?abstract_id=2898968 (accessed on 22 September 2023)
56. Duan, J.; Chang, M.; Chen, X.; Wang, W.; Zuo, H.; Bai, Y.; Chen, B. A combined short-term wind speed forecasting model based on CNN–RNN and linear regression optimization considering error. *Renew. Energy* **2022**, *200*, 788–808. [CrossRef]
57. Xu, W.; Wu, S. Ship Agent model for traffic flow simulation in inland waterway. *IOP Conf. Ser. Mater. Sci. Eng.* **2020**, *768*, 072104. [CrossRef]
58. Goerlandt, F.; Kujala, P. Traffic simulation based ship collision probability modeling. *Reliab. Eng. Syst. Saf.* **2011**, *96*, 91–107. [CrossRef]
59. Wolsing, K.; Roepert, L.; Bauer, J.; Wehrle, K. Anomaly Detection in Maritime AIS Tracks: A Review of Recent Approaches. *J. Mar. Sci. Eng.* **2022**, *10*, 112. [CrossRef]
60. Zhao, J.; Chen, Y.; Zhou, Z.; Zhao, J.; Wang, S.; Chen, X. Extracting Vessel Speed Based on Machine Learning and Drone Images during Ship Traffic Flow Prediction. *J. Adv. Transp.* **2022**, *2022*, 3048611. [CrossRef]
61. Yan, Z.; Song, X.; Zhong, H.; Yang, L.; Wang, Y. Ship Classification and Anomaly Detection Based on Spaceborne AIS Data Considering Behavior Characteristics. *Sensors* **2022**, *22*, 7713. [CrossRef] [PubMed]

62. Li, K.; Guo, J.; Li, R.; Wang, Y.; Li, Z.; Miu, K.; Chen, H. The Abnormal Detection Method of Ship Trajectory with Adaptive Transformer Model Based on Migration Learning. In Proceedings of the Spatial Data and Intelligence, Nanchang, China, 13–15 April 2023; Meng, X., Li, X., Xu, J., Zhang, X., Fang, Y., Zheng, B., Li, Y., Eds.; Springer Nature: Cham, Switzerland, 2023; pp. 204–220.

63. Itoh, H. Method for prediction of ship traffic behaviour and encounter frequency. *J. Navig.* **2022**, *75*, 106–123. [CrossRef]

64. Han, X.; Armenakis, C.; Jadidi, M. Modeling Vessel Behaviours by Clustering AIS Data Using Optimized DBSCAN. *Sustainability* **2021**, *13*, 8162. [CrossRef]

65. Zhang, Z.G.; Yin, J.C.; Wang, N.N.; Hui, Z.G. Vessel traffic flow analysis and prediction by an improved PSO-BP mechanism based on AIS data. *Evol. Syst.* **2019**, *10*, 397–407. [CrossRef]

66. Li, Y.; Ren, H. Vessel Traffic Flow Prediction Using LSTM Encoder-Decoder. In Proceedings of the 2022 5th International Conference on Signal Processing and Machine Learning, Dalian China, 4–6 August 2022; pp. 1–7. [CrossRef]

67. Li, M.; Li, B.; Qi, Z.; Li, J.; Wu, J. Optimized APF-ACO Algorithm for Ship Collision Avoidance and Path Planning. *J. Mar. Sci. Eng.* **2023**, *11*, 1177. [CrossRef]

68. He, Q.; Hou, Z.; Zhu, X. A Novel Algorithm for Ship Route Planning Considering Motion Characteristics and ENC Vector Maps. *J. Mar. Sci. Eng.* **2023**, *11*, 1102. [CrossRef]

69. He, G.; Dong, G.; Yao, C.; Sun, X. Real-Time Deterministic Prediction of Ship Motion Based on Multi-Layer LSTM. In Proceedings of the International Conference on Offshore Mechanics and Arctic Engineering, Melbourne, Australia, 11–16 June 2023; Volume 7: CFD & FSI. Available online: https://asmedigitalcollection.asme.org/OMAE/proceedings-pdf/OMAE2023/86892/V007T08A0 23/7041468/v007t08a023-omae2023-102195.pdf (accessed on 22 September 2023).

70. Li, H.; Jiao, H.; Yang, Z. AIS data-driven ship trajectory prediction modelling and analysis based on machine learning and deep learning methods. *Transp. Res. Part E Logist. Transp. Rev.* **2023**, *175*, 103152. [CrossRef]

71. Tzeng, C.Y.; Chen, J.F. Fundamental properties of linear ship steering dynamic models. *J. Mar. Sci. Technol.* **2009**, *7*, 2. [CrossRef]

72. Abkowitz, M.A. Measurement of Hydrodynamic Characteristics from Ship Maneuvering Trials by System Identification; Technical Report; 1980. Available online: https://trid.trb.org/view/157366 (accessed on 22 September 2023).

73. Nomoto, K.; Taguchi, K. On steering qualities of ships (2). *J. Zosen Kiokai* **1957**, *1957*, 57–66. [CrossRef] [PubMed]

74. Yoshimura, Y.; Nomoto, K. Modeling of manoeuvring behaviour of ships with a propeller idling, boosting and reversing. *J. Soc. Nav. Archit. Jpn.* **1978**, *1978*, 57–69. [CrossRef] [PubMed]

75. Nomoto, K.; Taguchi, K.; Honda, K.; Hirano, S. On the steering qualities of ships. *J. Zosen Kiokai* **1956**, *1956*, 75–82. [CrossRef] [PubMed]

76. Khattab, O.M.; Nomoto, K. Steering Control of a Ship in a Canal (Part I). In *Journal of the Kansai Society of Naval Architects, Japan 169*; The Japan Society of Naval Architects and Ocean Engineers: Tokyo, Japan, 1978; pp. 41–55.

77. Lang, X.; Wu, D.; Mao, W. Comparison of supervised machine learning methods to predict ship propulsion power at sea. *Ocean Eng.* **2022**, *245*, 110387. [CrossRef]

78. Karagiannidis, P.; Themelis, N. Data-driven modelling of ship propulsion and the effect of data pre-processing on the prediction of ship fuel consumption and speed loss. *Ocean Eng.* **2021**, *222*, 108616. [CrossRef]

79. Cheliotis, M.; Lazakis, I.; Theotokatos, G. Machine learning and data-driven fault detection for ship systems operations. *Ocean Eng.* **2020**, *216*, 107968. [CrossRef]

80. Wang, T.; Li, G.; Hatledal, L.I.; Skulstad, R.; AEsoy, V.; Zhang, H. Incorporating Approximate Dynamics Into Data-Driven Calibrator: A Representative Model for Ship Maneuvering Prediction. *IEEE Trans. Ind. Inf.* **2022**, *18*, 1781–1789. [CrossRef]

81. Hu, J.H.; Hu, D.B.; Xiao, J.B. Model parameter identification and simulation of ship power system based on recursive least squares method. In Proceedings of the Applied Mechanics and Materials, Sydney, NSW, Australia, 11–14 December 2012; Trans Tech Publications: Wollerau, Switzerland, 2012; Volume 220, pp. 482–486.

82. Jian-Chuan, Y.; Zao-Jian, Z.; Feng, X. Parametric identification of Abkowitz model for ship maneuvering motion by using partial least squares regression. *J. Offshore Mech. Arct. Eng.* **2015**, *137*, 031301. [CrossRef]

83. Shi, C.; Zhao, D.; Peng, J.; Shen, C. Identification of ship maneuvering model using extended Kalman filters. In *Marine Navigation and Safety of Sea Transportation*; CRC Press: Boca Raton, FL, USA, 2009; pp. 355–360.

84. Dong, Z.; Yang, X.; Zheng, M.; Song, L.; Mao, Y. Parameter identification of unmanned marine vehicle manoeuvring model based on extended Kalman filter and support vector machine. *Int. J. Adv. Robot. Syst.* **2019**, *16*, 1729881418825095. [CrossRef]

85. Wang, T.; Li, G.; Wu, B.; Æsøy, V.; Zhang, H. Parameter identification of ship manoeuvring model under disturbance using support vector machine method. *Ships Offshore Struct.* **2021**, *16*, 13–21. [CrossRef]

86. Luo, W.; Zou, Z. Parametric identification of ship maneuvering models by using support vector machines. *J. Ship Res.* **2009**, *53*, 19–30. [CrossRef]

87. Luo, W.; Zhang, Z. Modeling of ship maneuvering motion using neural networks. *J. Mar. Sci. Appl.* **2016**, *15*, 426–432. [CrossRef]

88. Xing, Z.; McCue, L. Parameter identification for two nonlinear models of ship rolling using neural networks. In Proceedings of the 10th International Conference on Stability of Ships and Ocean Vehicles, St. Petersburg, Russia, 21–26 June 2009; pp. 421–428.

89. Alexandersson, M.; Zhang, D.; Mao, W.; Ringsberg, J.W. A comparison of ship manoeuvrability models to approximate ship navigation trajectories. *Ships Offshore Struct.* **2022**, *18*, 550–557. [CrossRef]

90. Chen, L.; Yang, P.; Li, S.; Tian, Y.; Liu, G.; Hao, G. Grey-box identification modeling of ship maneuvering motion based on LS-SVM. *Ocean Eng.* **2022**, *266*, 112957. [CrossRef]

91. Xu, H.; Hinostroza, M.; Hassani, V.; Guedes Soares, C. Real-time parameter estimation of a nonlinear vessel steering model using a support vector machine. *J. Offshore Mech. Arct. Eng.* **2019**, *141*, 061606. [CrossRef]

92. Luo, W.; Guedes Soares, C.; Zou, Z. Parameter identification of ship maneuvering model based on support vector machines and particle swarm optimization. *J. Offshore Mech. Arct. Eng.* **2016**, *138*, 031101. [CrossRef]

93. Chen, Y.; Song, Y.; Chen, M. Parameters identification for ship motion model based on particle swarm optimization. *Kybernetes* **2010**, *39*, 871–880. [CrossRef]

94. Ding, F. Several multi-innovation identification methods. *Digit. Signal Process.* **2010**, *20*, 1027–1039. [CrossRef]

95. Xie, S.; Chu, X.; Liu, C.; Liu, J.; Mou, J. Parameter identification of ship motion model based on multi-innovation methods. *J. Mar. Sci. Technol.* **2020**, *25*, 162–184. [CrossRef]

96. Wang, S.; Wang, L.; Im, N.; Zhang, W.; Li, X. Real-time parameter identification of ship maneuvering response model based on nonlinear Gaussian Filter. *Ocean Eng.* **2022**, *247*, 110471. [CrossRef]

97. Luo, W.; Guedes Soares, C.; Zou, Z. Parameter identification of ship manoeuvring model based on particle swarm optimization and support vector machines. In Proceedings of the International Conference on Offshore Mechanics and Arctic Engineering, Nantes, France, 9–14 June 2013; American Society of Mechanical Engineers: New York, NY, USA, 2013; Volume 55393, p. V005T06A071.

98. Liu, S.; Song, J.; Li, B.; Li, G. Investigation of steering dynamics ship model identification based on pso-lssvr. In Proceedings of the 2008 2nd International Symposium on Systems and Control in Aerospace and Astronautics, Shenzhen, China, 10–12 December 2008; IEEE: Piscataway, NJ, USA, 2008; pp. 1–6.

99. Li, Z.; Sun, J. Disturbance Compensating Model Predictive Control With Application to Ship Heading Control. *IEEE Trans. Contr. Syst. Technol.* **2011**, *20*, 5713831. [CrossRef]

100. Gao, S.; Liu, L.; Wang, H.; Wang, A. Data-driven model-free resilient speed control of an autonomous surface vehicle in the presence of actuator anomalies. *ISA Trans.* **2022**, *127*, 251–258. [CrossRef] [PubMed]

101. Xu, P.F.; Han, C.B.; Cheng, H.X.; Cheng, C.; Ge, T. A Physics-Informed Neural Network for the Prediction of Unmanned Surface Vehicle Dynamics. *J. Mar. Sci. Eng.* **2022**, *10*, 148. [CrossRef]

102. Song, L.; Hao, L.; Tao, H.; Xu, C.; Guo, R.; Li, Y.; Yao, J. Research on Black-Box Modeling Prediction of USV Maneuvering Based on SSA-WLS-SVM. *J. Mar. Sci. Eng.* **2023**, *11*, 324. [CrossRef]

103. Wang, L.; Li, S.; Liu, J.; Wu, Q. Data-driven model identification and predictive control for path-following of underactuated ships with unknown dynamics. *Int. J. Nav. Archit. Ocean Eng.* **2022**, *14*, 100445. [CrossRef]

104. Zhang, W.; Wu, P.; Peng, Y.; Liu, D. Roll Motion Prediction of Unmanned Surface Vehicle Based on Coupled CNN and LSTM. *Future Internet* **2019**, *11*, 243. [CrossRef]

105. Zhang, D.; Chu, X.; Wu, W.; He, Z.; Wang, Z.; Liu, C. Model identification of ship turning maneuver and extreme short-term trajectory prediction under the influence of sea currents. *Ocean Eng.* **2023**, *278*, 114367. [CrossRef]

106. Xue, D.; Wu, D.; Yamashita, A.S.; Li, Z. Proximal policy optimization with reciprocal velocity obstacle based collision avoidance path planning for multi-unmanned surface vehicles. *Ocean Eng.* **2023**, *273*, 114005. [CrossRef]

107. Xie, S.; Chu, X.; Zheng, M.; Liu, C. Ship predictive collision avoidance method based on an improved beetle antennae search algorithm. *Ocean Eng.* **2019**, *192*, 106542. [CrossRef]

108. Kim, J.H.; Lee, S.; Jin, E.S. Collision avoidance based on predictive probability using Kalman filter. *Int. J. Nav. Archit. Ocean Eng.* **2022**, *14*, 100438. [CrossRef]

109. Zhang, M.; Hao, S.; Wu, D.; Chen, M.L.; Yuan, Z.M. Time-optimal obstacle avoidance of autonomous ship based on nonlinear model predictive control. *Ocean Eng.* **2022**, *266*, 112591. [CrossRef]

110. He, Z.; Chu, X.; Liu, C.; Wu, W. A novel model predictive artificial potential field based ship motion planning method considering COLREGs for complex encounter scenarios. *ISA Trans.* **2023**, *134*, 58–73. [CrossRef] [PubMed]

111. Bao, K.; Bi, J.; Gao, M.; Sun, Y.; Zhang, X.; Zhang, W. An Improved Ship Trajectory Prediction Based on AIS Data Using MHA-BiGRU. *J. Mar. Sci. Eng.* **2022**, *10*, 804. [CrossRef]

112. Nowy, A.; Łazuga, K.; Gucma, L.; Androjna, A.; Perkovič, M.; Sŕše, J. Modeling of Vessel Traffic Flow for Waterway Design–Port of Świnoujście Case Study. *Appl. Sci.* **2021**, *11*, 8126. [CrossRef]

113. Alizadeh, D.; Alesheikh, A.A.; Sharif, M. Vessel Trajectory Prediction Using Historical Automatic Identification System Data. *J. Navig.* **2021**, *74*, 156–174. [CrossRef]

114. Jiang, D.; Shi, G.; Li, N.; Ma, L.; Li, W.; Shi, J. TRFM-LS: Transformer-Based Deep Learning Method for Vessel Trajectory Prediction. *J. Mar. Sci. Eng.* **2023**, *11*, 880. [CrossRef]

115. Ma, H.; Zuo, Y.; Li, T. Vessel Navigation Behavior Analysis and Multiple-Trajectory Prediction Model Based on AIS Data. *J. Adv. Transp.* **2022**, *2022*, 6622862. [CrossRef]

116. Lin, Z.; Yue, W.; Huang, J.; Wan, J. Ship Trajectory Prediction Based on the TTCN-Attention-GRU Model. *Electronics* **2023**, *12*, 2556. [CrossRef]

117. Bejarano, G.; Manzano, J.M.; Salvador, J.R.; Limon, D. Nonlinear model predictive control-based guidance law for path following of unmanned surface vehicles. *Ocean Eng.* **2022**, *258*, 111764. [CrossRef]

118. Wang, X.; Liu, J.; Peng, H.; Qie, X.; Zhao, X.; Lu, C. A Simultaneous Planning and Control Method Integrating APF and MPC to Solve Autonomous Navigation for USVs in Unknown Environments. *J. Intell. Robot. Syst.* **2022**, *105*, 36. [CrossRef]

119. Chen, L.; Yang, P.; Li, S.; Liu, K.; Wang, K.; Zhou, X. Online modeling and prediction of maritime autonomous surface ship maneuvering motion under ocean waves. *Ocean Eng.* **2023**, *276*, 114183. [CrossRef]

120. Asfihani, T.; Chotimah, K.; Fitria, I.; Subchan. Ship Heading Control Using Nonlinear Model Predictive Control. In Proceedings of the 2020 3rd International Seminar on Research of Information Technology and Intelligent Systems (ISRITI), Yogyakarta, Indonesia, 10 December 2020; pp. 306–309. [CrossRef]

121. Lazarowska, A. Review of Collision Avoidance and Path Planning Methods for Ships Utilizing Radar Remote Sensing. *Remote Sens.* **2021**, *13*, 3265. [CrossRef]

122. Abebe, M.; Noh, Y.; Kang, Y.J.; Seo, C.; Kim, D.; Seo, J. Ship trajectory planning for collision avoidance using hybrid ARIMA-LSTM models. *Ocean Eng.* **2022**, *256*, 111527. [CrossRef]

123. Thyri, E.H.; Breivik, M. Collision avoidance for ASVs through trajectory planning: MPC with COLREGs-compliant nonlinear constraints. *MIC* **2022**, *43*, 55–77. [CrossRef]

124. Akdağ, M.; Fossen, T.I.; Johansen, T.A. Collaborative Collision Avoidance for Autonomous Ships Using Informed Scenario-Based Model Predictive Control. *IFAC-PapersOnLine* **2022**, *55*, 249–256. [CrossRef]

125. Gao, P.; Zhou, L.; Zhao, X.; Shao, B. Research on ship collision avoidance path planning based on modified potential field ant colony algorithm. *Ocean. Coast. Manag.* **2023**, *235*, 106482. [CrossRef]

126. Zhu, H.; Ding, Y. Optimized Dynamic Collision Avoidance Algorithm for USV Path Planning. *Sensors* **2023**, *23*, 4567. [CrossRef]

127. Du, Y.; Meng, Q.; Wang, S.; Kuang, H. Two-phase optimal solutions for ship speed and trim optimization over a voyage using voyage report data. *Transp. Res. Part B Methodol.* **2019**, *122*, 88–114. [CrossRef]

128. Yu, H.; Fang, Z.; Fu, X.; Liu, J.; Chen, J. Literature review on emission control-based ship voyage optimization. *Transp. Res. Part D Transp. Environ.* **2021**, *93*, 102768. [CrossRef]

129. Wang, Y.; Meng, Q.; Kuang, H. Jointly optimizing ship sailing speed and bunker purchase in liner shipping with distribution-free stochastic bunker prices. *Transp. Res. Part C Emerg. Technol.* **2018**, *89*, 35–52. [CrossRef]

130. Li, X.; Sun, B.; Guo, C.; Du, W.; Li, Y. Speed optimization of a container ship on a given route considering voluntary speed loss and emissions. *Appl. Ocean Res.* **2020**, *94*, 101995. [CrossRef]

131. James, R.W. *Application of Wave Forecasts to Marine Navigation*; New York University: New York, NY, USA, 1957.

132. Wang, H.; Mao, W.; Eriksson, L. A Three-Dimensional Dijkstra's algorithm for multi-objective ship voyage optimization. *Ocean Eng.* **2019**, *186*, 106131. [CrossRef]

133. Zis, T.P.; Psaraftis, H.N.; Ding, L. Ship weather routing: A taxonomy and survey. *Ocean Eng.* **2020**, *213*, 107697. [CrossRef]

134. Zhu, X.; Wang, H.; Shen, Z.; Lv, H. Ship weather routing based on modified Dijkstra algorithm. In Proceedings of the 2016 6th International Conference on Machinery, Materials, Environment, Biotechnology and Computer, Tianjin, China, 11–12 June 2016; Atlantis Press: Amsterdam, The Netherlands, 2016; pp. 696–699.

135. Ma, W.; Han, Y.; Tang, H.; Ma, D.; Zheng, H.; Zhang, Y. Ship route planning based on intelligent mapping swarm optimization. *Comput. Ind. Eng.* **2023**, *176*, 108920. [CrossRef]

136. Liu, Z.; Liu, J.; Zhou, F.; Liu, R.W.; Xiong, N. A robust GA/PSO-hybrid algorithm in intelligent shipping route planning systems for maritime traffic networks. *J. Internet Technol.* **2018**, *19*, 1635–1644.

137. Jang, D.u.; Kim, J.s. Development of Ship Route-Planning Algorithm Based on Rapidly-Exploring Random Tree (RRT*) Using Designated Space. *J. Mar. Sci. Eng.* **2022**, *10*, 1800. [CrossRef]

138. Wang, H.; Lang, X.; Mao, W. Voyage optimization combining genetic algorithm and dynamic programming for fuel/emissions reduction. *Transp. Res. Part D Transp. Environ.* **2021**, *90*, 102670. [CrossRef]

139. Zhao, W.; Wang, Y.; Zhang, Z.; Wang, H. Multicriteria ship route planning method based on improved particle swarm optimization–genetic algorithm. *J. Mar. Sci. Eng.* **2021**, *9*, 357. [CrossRef]

140. Moreira, L.; Vettor, R.; Guedes Soares, C. Neural network approach for predicting ship speed and fuel consumption. *J. Mar. Sci. Eng.* **2021**, *9*, 119. [CrossRef]

141. Zhu, Z.; Li, L.; Wu, W.; Jiao, Y. Application of improved Dijkstra algorithm in intelligent ship path planning. In Proceedings of the 2021 33rd Chinese Control and Decision Conference (CCDC), Kunming, China, 22–24 May 2021; IEEE: Piscataway, NJ, USA, 2021; pp. 4926–4931.

142. Yoo, B.; Kim, J. Path optimization for marine vehicles in ocean currents using reinforcement learning. *J. Mar. Sci. Technol.* **2016**, *21*, 334–343. [CrossRef]

143. Shah, B.C.; Gupta, S.K. Long-Distance Path Planning for Unmanned Surface Vehicles in Complex Marine Environment. *IEEE J. Ocean Eng.* **2020**, *45*, 813–830. [CrossRef]

144. Vagale, A.; Oucheikh, R.; Bye, R.T.; Osen, O.L.; Fossen, T.I. Path planning and collision avoidance for autonomous surface vehicles I: A review. *J. Mar. Sci. Technol.* **2021**, *26*, 1292–1306. [CrossRef]

145. Guo, W.; Tang, G.; Zhao, F.; Wang, Q. Global Dynamic Path Planning Algorithm for USV Based on Improved Bidirectional RRT. In Proceedings of the 32nd International Ocean and Polar Engineering Conference, Shanghai, China, 5 June 2022; p. ISOPE–I–22–547.

146. Wang, L.; Wu, Q.; Liu, J.; Li, S.; Negenborn, R.R. State-of-the-art research on motion control of maritime autonomous surface ships. *J. Mar. Sci. Eng.* **2019**, *7*, 438. [CrossRef]

147. Kim, D.; Kim, J.S.; Kim, J.H.; Im, N.K. Development of ship collision avoidance system and sea trial test for autonomous ship. *Ocean Eng.* **2022**, *266*, 113120. [CrossRef]

148. Tengesdal, T.; Johansen, T.A.; Brekke, E.F. Ship Collision Avoidance Utilizing the Cross-Entropy Method for Collision Risk Assessment. *IEEE Trans. Intell. Transport. Syst.* **2022**, *23*, 11148–11161. [CrossRef]

149. Lee, M.C.; Nieh, C.Y.; Kuo, H.C.; Huang, J.C. A collision avoidance method for multi-ship encounter situations. *J. Mar. Sci. Technol.* **2020**, *25*, 925–942. [CrossRef]
150. Seo, C.; Noh, Y.; Abebe, M.; Kang, Y.J.; Park, S.; Kwon, C. Ship collision avoidance route planning using CRI-based A* algorithm. *Int. J. Nav. Archit. Ocean Eng.* **2023**, *15*, 100551. [CrossRef]

Journal of
Marine Science and Engineering

MDPI

Article

PSO-Based Predictive PID-Backstepping Controller Design for the Course-Keeping of Ships

Bowen Lin [1,2], Mao Zheng [1,2,*], Bing Han [3], Xiumin Chu [1,2], Mingyang Zhang [4], Haiming Zhou [1,2], Shigan Ding [1,2], Hao Wu [1,2] and Kehao Zhang [1,2]

[1] Intelligent Transportation Systems Research Center, Wuhan University of Technology, Wuhan 430062, China; linbw163@163.com (B.L.); zhouhaiming@whut.edu.cn (H.Z.)
[2] National Engineering Research Center for Water Transport Safety, Wuhan 430070, China
[3] National Engineering Research Center of Ship & Shipping Control System, Shanghai 200135, China
[4] School of Engineering, Aalto University, 02150 Espoo, Finland
* Correspondence: zhengmao@whut.edu.cn

Abstract: Ship course-keeping control is of great significance to both navigation efficiency and safety. Nevertheless, the complex navigational conditions, unknown time-varying environmental disturbances, and complex dynamic characteristics of ships pose great difficulties for ship course-keeping. Thus, a PSO-based predictive PID-backstepping (P-PB) controller is proposed in this paper to realize the efficient and rapid course-keeping of ships. The proposed controller takes the ship's target course, current course, yawing speed, as well as predictive motion parameters into consideration. In the design of the proposed controller, the PID controller is improved by introducing predictive control. Then, the improved controller is combined with a backstepping controller to balance the efficiency and stability of the control. Subsequently, the parameters in the proposed course-keeping controller are optimized by utilizing Particle Swarm Optimization (PSO), which can adaptively adjust the value of parameters in various scenarios, and thus further increase its efficiency. Finally, the improved controller is validated by carrying out simulation tests in various scenarios. The results show that it improves the course-keeping error and time-response specification by 4.19% and 9.71% on average, respectively, which can efficiently achieve the course-keeping of ships under various scenarios.

Keywords: ship course-keeping; MMG; PID control; predictive control; backstepping control; particle swarm optimization

Citation: Lin, B.; Zheng, M.; Han, B.; Chu, X.; Zhang, M.; Zhou, H.; Ding, S.; Wu, H.; Zhang, K. PSO-Based Predictive PID-Backstepping Controller Design for the Course-Keeping of Ships. *J. Mar. Sci. Eng.* **2024**, *12*, 202. https://doi.org/10.3390/jmse12020202

Academic Editor: Diego Villa

Received: 20 December 2023
Revised: 14 January 2024
Accepted: 18 January 2024
Published: 23 January 2024

1. Introduction

Maritime transportation holds a pivotal share of international trade [1,2]. As the most economical and effective tool of marine transport, it is essential that shipping increases constantly with the development of the national economy and international trade [3–5]. However, complex navigational conditions and unknown time-varying environmental disturbances pose a significant difficulty in the operation of ships [6–8]. Meanwhile, the complex dynamic characteristics (e.g., multiple degrees-of-freedom (DOF), nonlinearity, limitation in rudder angle) further increase the uncertainty of the ship's motion [9–14]. As a result, it is difficult to lead ships moving along a target course efficiently and accurately, especially under harsh environmental conditions [9,15–17]. Thus, it becomes indispensable to carry out research related to ship course-keeping.

To date, research on the course-keeping of ships mainly focuses on (a) improved traditional controllers and (b) data-driven controllers.

Improved traditional controllers are considered an effective approach to realize the course-keeping of ships. Some common methods include improved sliding mode controllers [18–21], improved PID controllers [20,22–24], improved backstepping controllers [9,25–27], and improved bipolar sigmoid functions [28,29]. These methods mainly improve traditional controllers by introducing adaptive control, synergetic control,

control formula improvement, and parameter optimization. However, these methods contain limitations such as a lack of course-keeping accuracy under environmental disturbances and inefficiency in reaching the target course, which may lead to additional course-keeping errors in some scenarios. At the same time, some controllers are complex to build and therefore less available.

Data-driven controllers become feasible for ship course-keeping control with the development of data acquisition and processing technologies. Methods adopted are known as expert knowledge controllers [30], artificial neural networks [31–35], neuro-fuzzy systems [36], and multi-agent systems [37]. These controllers are established based on empirical knowledge or navigation data. Researchers fused ship motion and operation data and processed it using various statistical and intelligent modeling methods to establish automatic ship course-keeping controllers. To improve the accuracy of course-keeping, these controllers have high requirements for the amount and quality of data in particular scenarios. As a result, such controllers are more effective in scenarios with a large amount of data on similar ships. Conversely, it is hard to realize accurate ship course-keeping in scenarios with less navigation data. Meanwhile, the selection of empirical knowledge or navigation data has a substantial impact on the effectiveness of ship course-keeping control, which further increases the uncertainty of the controller.

The above-mentioned controllers for the course-keeping of ships have been applied and validated. Nevertheless, difficulties such as obtaining the data required to train, and the high complexity to establish, limit the accuracy of the course-keeping controllers mentioned above. Additionally, some of the studies lack accuracy in establishing ship motion and environmental disturbance models for simulation tests, which prevents the effectiveness of those course-keeping controllers from being effectively verified [24,27].

Given these research gaps, in this paper, a PSO-based predictive PID-backstepping (P-PB) controller is introduced for the course-keeping of ships. The P-PB controller is designed on the basis of PID and backstepping controllers, thus retaining the simplicity and interpretability of traditional controllers. At the same time, course-overshoot of the controller is avoided by introducing a predictive PID control, which improves the accuracy of course-keeping. Subsequently, the parameters in the P-PB controller are optimized via PSO, which is characterized by its efficiency, and is widely used in the field of ship control to improve its applicability in various scenarios [38–40].

In Section 2, a nonlinear ship model is first introduced, which is adapted for ship motion prediction and simulation tests. Then, the improved PID controller and the backstepping controller are combined to design the P-PB controller. Further, PSO is introduced to optimize parameters in the proposed controller. Section 3 provides comparison tests with other controllers, demonstrating the effectiveness of our approach in various scenarios by using a case ship called KVLCC2. Section 4 serves as the conclusion, which engages in a discussion concerning the distinctive features and advantages of our proposed method, as applied to the field of ship course-keeping.

2. Methodology

The framework of the proposed P-PB controller for ship course-keeping (Figure 1) comprises three steps:

Step (i): Nonlinear ship model. A nonlinear ship model is established based on the MMG model. Consequently, an environmental disturbance model is introduced to simulate ship motion under time-varying disturbance conditions.

Step (ii): Course-keeping controller design. First, the PID control is improved based on a predictive control method. Then, the ship course-keeping controller is established by combining the improved PID and backstepping controllers, thus combining the advantages of both.

Step (iii): Parameter optimization of the ship course-keeping controller. The parameters in the proposed controller are adopted as input, and the minimization of the cumulative course-keeping error is used as a fitness index. Then, the optimal control parameters for course-keeping in a particular scenario are obtained based on PSO.

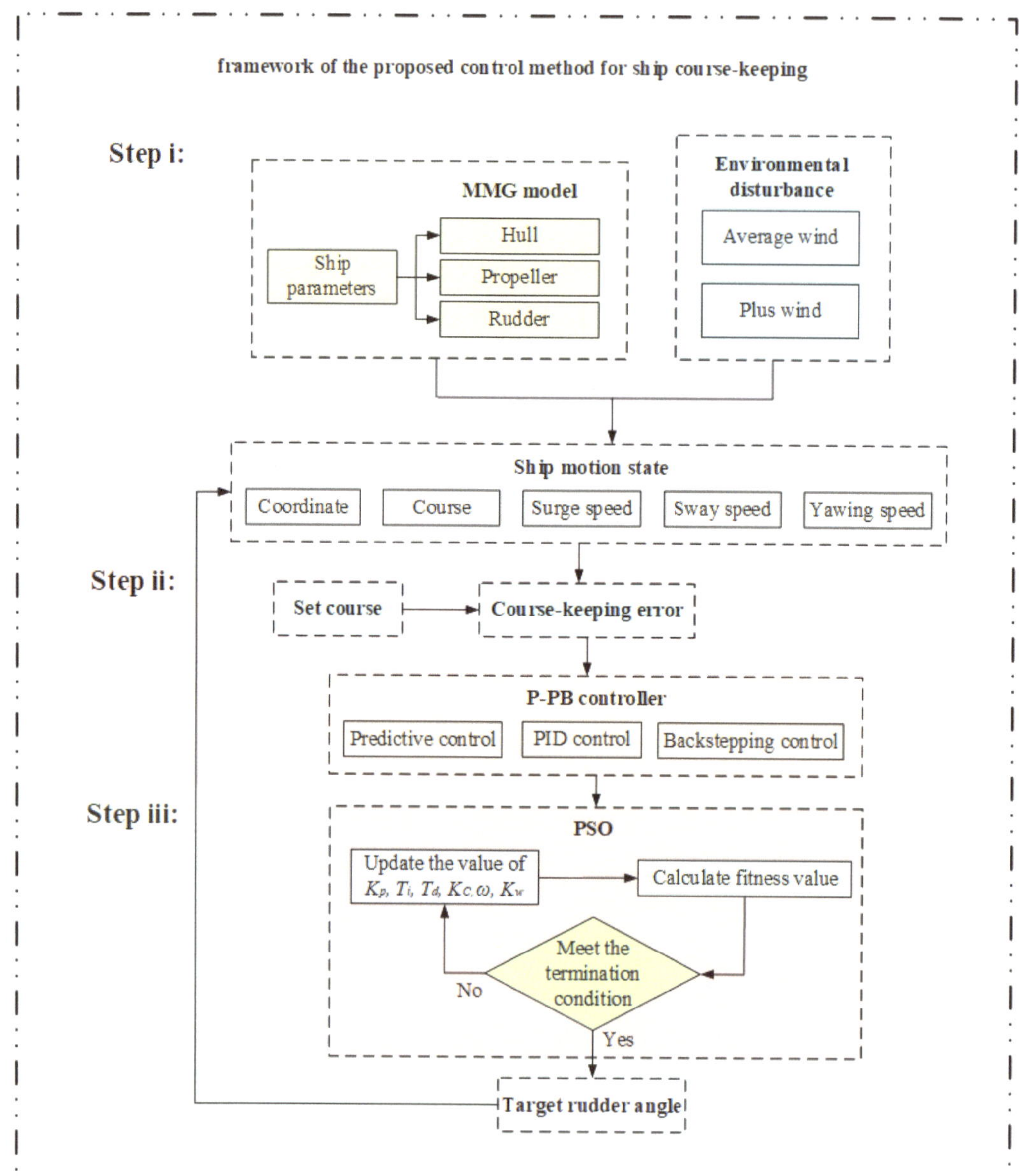

Figure 1. The framework of the proposed P-PB control method for ship course-keeping.

2.1. Nonlinear Ship Model

To simulate ship motion accurately, the ship motion model is established based on the MMG model. Then, the environmental disturbance model is introduced to simulate ship motion under various scenarios. A ship is typically considered a rigid body with six degrees of freedom (DOF) in motion. However, a three-degrees of freedom (3-DOF) ship dynamic model can be used when it comes to the control of ship motion along the horizontal plane [41–43]. The course-keeping controller mainly changes the ship's motion along the horizontal plane; therefore, this paper is based on a 3-DOF nonlinear ship model.

The establishment of a nonlinear ship model contains three steps: the establishment of a ship motion coordinate system, kinematic modeling, and environmental disturbance modeling.

2.1.1. Ship Motion Coordinate System

The space-fixed coordinate system $O_0 - x_0 y_0$ and the ship-fixed coordinate system $O - xy$ are established, respectively, where the x_0 axis points directly north and the y_0 axis points directly east. In terms of the ship-fixed system, the x and y axes point towards the ship's bow and starboard, respectively.

u, v, and r are the ship's surge speed, sway speed, and yawing speed, respectively. ψ is the ship's course, which is defined as the angle between the x_0 and x axes. ψ_T is the wind direction. Finally, the ship-motion coordinate system is set up as in Figure 2.

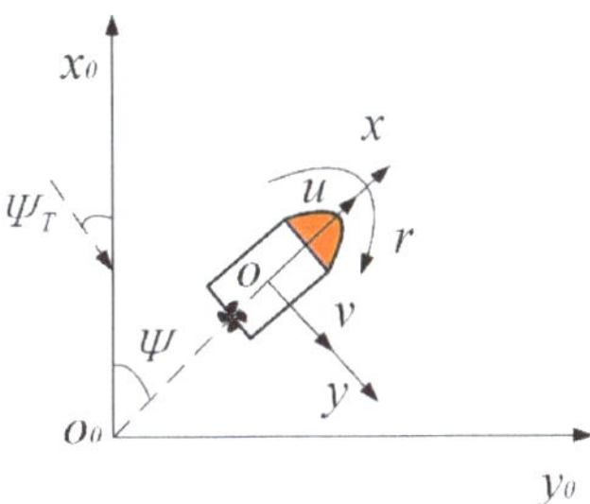

Figure 2. Coordinate system of ship motion.

2.1.2. Kinematic Model

The MMG model divides a ship into hull, propeller, and rudder. In addition, the effect of environmental disturbances on the ship's motion is also considered. Thus, the motion of a ship can be expressed as Equation (1):

$$\begin{cases} dx_t/dt = u\cos(\psi) - v\sin(\psi) \\ dy_t/dt = u\sin(\psi) + v\cos(\psi) \\ d\psi/dt = r \\ (m + m_x)\dot{u} - (m + m_y)vr = X_H + X_P + X_R + X_{wind} \\ (m + m_y)\dot{v} + (m + m_x)ur = Y_H + Y_P + Y_R + Y_{wind} \\ (I_{ZZ} + J_{ZZ})\dot{r} = N_H + N_P + N_R + N_{wind} \end{cases} \tag{1}$$

where x_t and y_t are the position coordinates of origin in time t in the earth-fixed coordinate system. m is ship's mass. m_x and m_y are the added masses of the x and y axis directions. I_{ZZ} is the inertia moment of the ship, J_{ZZ} is the added moment of inertia. X_H, Y_H, and N_H are the surge force, sway force, and yaw moment acting on the ship's hull, respectively. X_P, Y_P, and N_P are the surge force, sway force, and yaw moment generated by the propeller. X_R, Y_R, and N_R are the surge force, sway force, and yaw moment generated by the rudder. X_{wind}, Y_{wind}, and N_{wind} are the wind load in the surge, sway, and yaw direction.

The calculation method of hull fluid force is shown in Equation (2).

$$\begin{cases} X_H = X(u) + X_{vv}v^2 + X_{vr}vr + X_{rr}r^2 \\ Y_H = Y_v v + Y_r r + Y_{|v|v}|v|v + Y_{|v|r}|v|r + Y_{|r|r}|r|r \\ N_H = N_v v + N_r r + N_{|v|v}|v|v + N_{vvr}v^2 r + N_{vrr}vr^2 \end{cases} \tag{2}$$

$X(u)$, X_{vv}, X_{vr}, X_{rr}, Y_v, Y_r, $Y_{|v|v}$, $Y_{|v|r}$, $Y_{|r|r}$, N_v, N_r, $N_{|v|v}$, N_{vvr}, and N_{vrr} are the hydrodynamic factors, which are determined by the empirical formulas proposed by Kijima [44].

By combining the methods proposed by Jia, Yang, and Brogliacan [45,46], the propeller force is calculated using Equation (3):

$$\begin{cases} X_P = \rho n_p^2 D_P^4 (1 - t_P) K_T(J_P) \\ Y_P = \rho n_p^2 D_P^4 K_T \sin(\arccos(u/v))/3 \\ N_P = 0.083 Y_P \end{cases} \tag{3}$$

where ρ is the density of water, t_P is the thrust deduction factor, D_P is the propeller diameter, J_P is the propeller advanced ratio, n_p is the propeller revolution, and K_T is the thrust coefficient of the propeller.

Subsequently, the rudder force is determined using Equation (4) [45]:

$$\begin{cases} X_R = (1 - t_R) F_N \sin \delta \\ Y_R = (1 + \alpha_H) F_N \cos \delta \\ N_R = (x_R + \alpha_H x_H) F_N \cos \delta \end{cases} \tag{4}$$

where t_R is the steering resistance deduction factor, α_H is the rudder force increase factor, x_R is the longitudinal coordinate of rudder position, x_H is the longitudinal coordinate of the acting point of the additional lateral force, F_N is the rudder normal force, and δ is the current rudder angle.

Additionally, due to the large resistance to movement, the turning speed of the rudder is limited [47]. Thus, the ship's rudder movement is characterized in Equation (5).

$$T_E \dot{\delta} = \delta_E - \delta \tag{5}$$

where T_E is the time constant, $\dot{\delta}$ is the rudder turning speed, and δ_E is the command rudder angle.

2.1.3. Environmental Disturbance Model

The operational performance of a ship is significantly vulnerable to external disturbances induced by wind, waves, and currents. However, a disturbance by currents mainly changes a ship's surge and sway speed and has less effect on its yawing moment, thus it can be ignored when it comes to control of a ship's course. In terms of waves, their height and frequency are closely related to the interference of wind. Therefore, the performance of a ship's course-keeping controller under wind disturbance represents a control effect under waves, to a certain extent. Moreover, it is difficult to accurately simulate the effect of wave disturbance on a ship's motion [48]. Consequently, the environmental disturbance model is established based on wind disturbance.

Wind disturbance can be divided into average wind and pulse wind. Between them, average wind is calculated according to the empirical formula in Equation (6) [49],

$$\begin{cases} \alpha_R = -\arctan(\frac{-v - V_R \sin(\psi_T - \psi)}{-u - V_R \cos(\psi_T - \psi)}) - v \\ F_{Xwind} = 0.5 C_x(\alpha_R) \rho_a V_R^2 A_F \\ F_{Ywind} = 0.5 C_y(\alpha_R) \rho_a V_R^2 A_L \\ N_{wind} = 0.5 C_m(\alpha_R) \rho_a V_R^2 A_L L \end{cases} \tag{6}$$

where α_R is the angle between the ship's course and the wind direction; v is the compensation angle of wind; $C_x(\alpha_R)$, $C_y(\alpha_R)$, $C_m(\alpha_R)$ is the wind load factor in surge, sway, and yaw direction; ρ_a is the density of air; V_R is the wind speed; A_F, A_L is the area of the ship exposed to wind in surge and sway direction; and L is the length of ship.

Then, white noise is introduced to calculate the pulse wind, which is calculated in Equation (7) [50]:

$$H(s) = 0.4198s/(s^2 + 0.3638s + 0.3675) \tag{7}$$

where s is the Laplace operator.

2.2. P-PB Course-Keeping Controller Design

The PID controller and the backstepping controller have their own advantages in various scenarios. In this paper, both the PID controller and the backstepping controller are taken into account when building the P-PB controller; thus, the advantages of the two controllers can be combined. The framework for the course-keeping controller is shown in Figure 3.

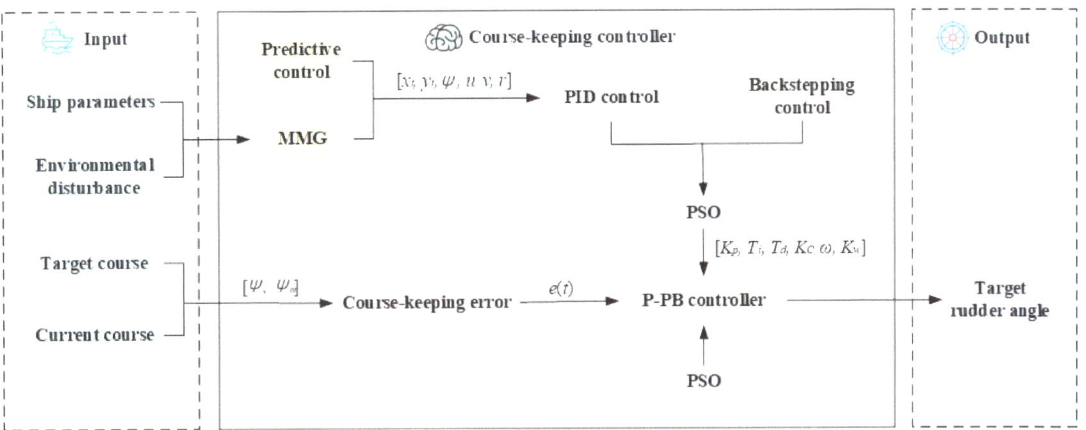

Figure 3. Framework of the P-PB course-keeping controller.

The inputs and outputs are defined first in order to design the course-keeping controller. The main principle of a course-keeping controller is to minimize course-keeping error by adjusting the rudder angle. Therefore, course-keeping error and target rudder angle are selected as the input and output of the proposed controller, respectively.

The course-keeping error is expressed in Equation (8),

$$e(t) = \psi_m - \psi \tag{8}$$

where $e(t)$ is the course-keeping error in time t, and ψ_m is the target course.

Then, the improved PID controller and the backstepping controller are combined to design the P-PB controller.

2.2.1. Improved PID Controller

PID control is a simple and reliable method that is widely adopted in the motion control of ships [22]. The basic formulation of the PID controller can be expressed as Equation (9),

$$u(t) = K_P[e(t) + \int_0^t e(t)\mathrm{dt}/T_i + T_d(\mathrm{d}e(t)/\mathrm{dt})] \tag{9}$$

where $u(t)$ is the output of the PID controller. K_P, T_i, and T_d are the proportional parameter, integral parameter, and derivative parameter, respectively.

To integrate the PID controller with the ship's rudder control, Equation (9) is modified as Equation (10):

$$\delta_E = \delta + K_P[e(t) + \int_0^t e(t)\mathrm{dt}/T_i + T_d(\mathrm{d}e(t)/\mathrm{dt})] \tag{10}$$

Then, to prevent control overshooting caused by the delayed ship motion, the PID controller is improved by introducing predictive control based on the MMG model [5,48].

Synthesizing the simplicity and interpretation of the controller, the improved PID controller is expressed as Equation (11),

$$
\begin{cases}
\begin{aligned}
\delta_{EP} = {} & \delta + K_P[e(t_{x3}) + \int_0^t e(t_{x3})\mathrm{dt}/T_i + T_d(\mathrm{de}(t_{x3})/\mathrm{dt})] \\
& -K_P[e(t_{x2}) + \int_0^t e(t_{x2})\mathrm{dt}/T_i + T_d(\mathrm{de}(t_{x2})/\mathrm{dt})] \\
& +K_P[e(t_{x1}) + \int_0^t e(t_{x1})\mathrm{dt}/T_i + T_d(\mathrm{de}(t_{x1})/\mathrm{dt})]
\end{aligned} & e(t_{x3}) > c_{f2}e(t_0) \\[6pt]
\begin{aligned}
\delta_{EP} = {} & \delta + c_{f1}K_P[e(t_{x3}) + \int_0^t e(t_{x3})\mathrm{dt}/T_i + T_d(\mathrm{de}(t_{x3})/\mathrm{dt})] \\
& -c_{f1}K_P[e(t_{x2}) + \int_0^t e(t_{x2})\mathrm{dt}/T_i + T_d(\mathrm{de}(t_{x2})/\mathrm{dt})] \\
& +c_{f1}K_P[e(t_{x1}) + \int_0^t e(t_{x1})\mathrm{dt}/T_i + T_d(\mathrm{de}(t_{x1})/\mathrm{dt})]
\end{aligned} & c_{f2}e(t_0) > e(t_{x3}) > c_{f3}e(t_0) \\[6pt]
\delta_{EP} = 0 & e(t_{x3}) < c_{f3}e(t_0)
\end{cases}
\tag{11}
$$

where δ_{EP} is the target rudder angle calculated by the improved PID controller, $e(t_x)$ is the course-keeping error predicted by the MMG model after x seconds, and c_f is the control factor with a value between 0 and 1.

2.2.2. Backstepping Controller

The formula of the backstepping controller is expressed as Equation (12),

$$
\begin{cases}
u_{BS} = m_{BS}\ddot{x}_{BS} + c_{BS}\dot{x}_{BS} + d_{BS} \\
y_{BS} = x_{BS}
\end{cases}
\tag{12}
$$

where m_{BS}, c_{BS}, and d_{BS} are the variable parameters, u_{BS} is the input of the controller, and y_{BS} is the output of the backstepping controller.

Using the set $x_{1,BS} = x_{BS}$, $x_{2,BS} = \dot{x}_{BS}$, Equation (12) can be changed to Equation (13).

$$
\begin{cases}
\dot{x}_{1,BS} = x_{2,BS} \\
\dot{x}_{2,BS} = 1/m_{BS}(u_{BS} - c_{BS}x_{2,BS} - d_{BS}) \\
y_{BS} = x_{1,BS}
\end{cases}
\tag{13}
$$

Next, the systematic error $z_{1,BS}$ is calculated.

$$
z_{1,BS} = y_{BS} - y_{d,BS}
\tag{14}
$$

where $y_{d,BS}$ is the desired output.

Subsequently, the Lyapunov function $V_{1,BS}$ is defined in Equation (15).

$$
V_{1,BS} = z_{1,BS}^2/2
\tag{15}
$$

Then, the first order derivative of $V_{1,BS}$ can be expressed as Equation (16).

$$
\dot{V}_{1,BS} = z_{1,BS}\dot{z}_{1,BS} = z_{1,BS}(\dot{y}_{BS} - \dot{y}_{d,BS})
\tag{16}
$$

Consequently, the virtual control volume $a_{1,BS}$ is introduced to make $\dot{V}_{1,BS} \leq 0$.

$$
a_{1,BS} = -\lambda_{1,BS}z_{1,BS} + \dot{y}_{d,BS}
\tag{17}
$$

where $\lambda_{1,BS} \geq 0$ is the constant.

After that, the error variable is defined.

$$
z_{2,BS} = \dot{y}_{BS} - a_{1,BS} = \dot{x}_{1,BS} - a_{1,BS} = x_{2,BS} - a_{1,BS}
\tag{18}
$$

Substituting Equations (17) and (18) into Equation (16), the value of $\dot{V}_{1,BS}$ is expressed as Equation (19).

$$
\dot{V}_{1,BS} = z_{1,BS}\dot{z}_{1,BS} = z_{1,BS}(z_{2,BS} + a_{1,BS} - \dot{y}_{d,BS}) = z_{1,BS}z_{2,BS} - \lambda_{1,BS}z_{1,BS}^2
\tag{19}
$$

Then, set the Lyapunov function $V_{2,BS}$ as Equation (20).

$$\dot{V}_{2,BS} = \dot{V}_{1,BS} + z_{1,BS}z_{2,BS} = z_{1,BS}z_{2,BS} \\ -\lambda_{1,BS}z_{1,BS}^2 + z_{2,BS}[(u_{BS} - c_{BS}x_{2,BS} - d_{BS})/m - \dot{a}_{1,BS}] \tag{20}$$

Thus, the first order derivative of $V_{2,BS}$ can be calculated as Equation (21).

$$\dot{V}_{2,BS} = z_{1,BS}z_{2,BS} - \lambda_{1,BS}z_{1,BS}^2 + z_{2,BS}[(u_{BS} - c_{BS}x_{2,BS} - d_{BS})/m - \dot{a}_{1,BS}] \\ = -(\lambda_{1,BS}z_{1,BS}^2 + \lambda_{2,BS}z_{2,BS}^2) \tag{21}$$

Finally, the backstepping control law u_{BS} is determined using Equation (22).

$$u_{BS} = m_{BS}(\dot{a}_{1,BS} - \lambda_{2,BS}z_{2,BS} - z_{1,BS}) + c_{BS}x_{2,BS} + d_{BS} \tag{22}$$

By combining the backstepping controller with the rudder control, the improved backstepping controller for the course-keeping of the ship [9] is shown in Equation (23),

$$\begin{cases} H(r) = (\alpha + \beta)r \\ b = K/T \\ \delta_{EB} = 1/b[-bH(r) + K_C \sin(\omega e(t))] \end{cases} \tag{23}$$

where α, β, K, and T are the ship's maneuvering indexes, which can be determined in Ref [51]. K_C and ω are the variable parameters of the controller. δ_{EB} is the target rudder angle calculated by the improved backstepping controller.

The backstepping controller has the advantage of a shorter time required to approach the target course, but its course-keeping stability is relatively poor [25]. Meanwhile, it is less accurate under harsh environmental disturbances affected by the dependence of r.

2.2.3. Design of the P-PB Controller

The improved PID and backstepping controller are combined to establish the P-PB controller. The main control law of the P-PB controller is designed using Equation (24),

$$\delta_{PPB} = K_w \delta_{EP} + (1 - K_w)\delta_{EB} \tag{24}$$

where δ_{PPB} is the target rudder angle calculated by the proposed P-PB controller. K_w is the control factor, with a value between 0 and 1.

By changing the value of K_w, the weights of the improved PID and backstepping controllers can be adjusted adaptively. Therefore, the P-PB controller combines the features of both controllers to achieve better control efficiency in various scenarios.

2.3. Parameter Optimization of the Ship Course-Keeping Controller

PSO was presented by Kennedy and R. Eberhart in 1994 and it is adopted as an effective approach to solving dynamic and multi-objective optimizing problems [52]. Therefore, PSO is adopted to optimize parameters in the proposed P-PB controller. The flowchart for optimizing the parameters of the proposed P-PB course-keeping controller is shown in Figure 4.

The values of K_P, T_i, T_d, K_C, ω, and K_w are defined as the key parameters to be optimized. To generate the particle swarm for each parameter, the population size of PSO is set as N, the learning factors are set as $c1$ and $c2$. The inertia weight value is set as w, and the maximum iteration number is set as T.

After initializing the population, the initial position (value) of each parameter $x_{PSO} = [K_P, T_i, T_d, K_C, \omega, K_w]$ and velocity v_{PSO} of the particle group is generated, which makes up the initial particle.

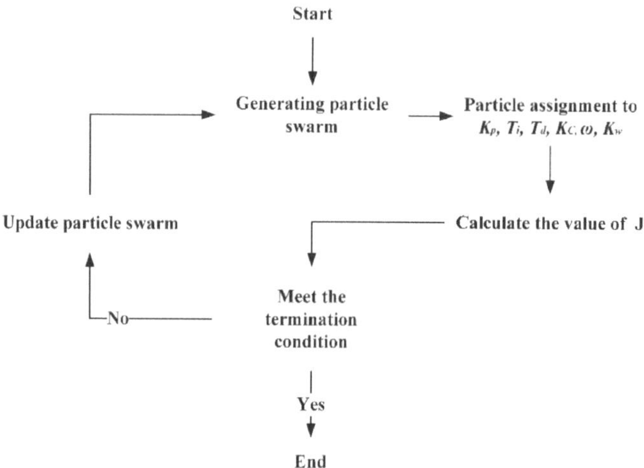

Figure 4. Flowchart of course-keeping controller parameter optimization.

Subsequently, the fitness function is defined to calculate the fitness value of each particle using Equation (25):

$$J = 1/\sum_{s=1}^{step} (\psi_m - \psi) \tag{25}$$

where J is the fitness value, and $step$ is the time duration of the optimization.

By using the MMG model to predict the ship's motion state, the value of J is determined when adopting the current parameters.

Subsequently, more particles are generated and compared to find the maximum fitness value. The position and velocity of each particle which contains different values of the key parameters are updated as Equation (26). Then, the fitness value of each particle is compared with the historical optimal fitness value. Then, the global optimal position g and global optimal fitness value g_{best} are obtained:

$$\begin{cases} v_i^{k+1} = wv_i^k + r_1c_1(g - x_i^k) + r_2c_2(g_{best} - x_i^k) \\ x_i^{k+1} = x_i^k + v_i^{k+1} \end{cases} \tag{26}$$

where r_1, r_2 is a random value between 0 and 1, and i is the current particle number.

Finally, the value of each parameter can be continuously optimized until it meets the termination condition, and the value of key parameters with the max fitness value can be achieved.

3. Application of the P-PB Ship Course-Keeping Controller

3.1. Simulation Preliminaries

A ship named KVLCC2 is selected as the case ship to verify the effectiveness of the P-PB controller. The main parameters of KVLCC2 are shown in Table 1, other parameters are shown in Ref [53].

After establishing the nonlinear ship model of KVLCC2 in Section 2.1, the simulation results of the turning test, with an initial ship speed of 15.5 kn/h, rudder angle of ±35°, and initial course of 0°, are compared with the results of Yasukawa and Yoshimura [53]. A comparison of ship's trajectory is shown in Figure 5.

Table 1. Parameters of KVLCC2.

Parameter	Value	Unit
Length	320	m
Breadth	58	m
Draft	20.8	m
Displacement	312,622	m^3
Open water speed	15.5	kn/h
Initial course	0	deg
Max steering speed of rudder	2.34	deg/s
Max rudder angle	35	deg

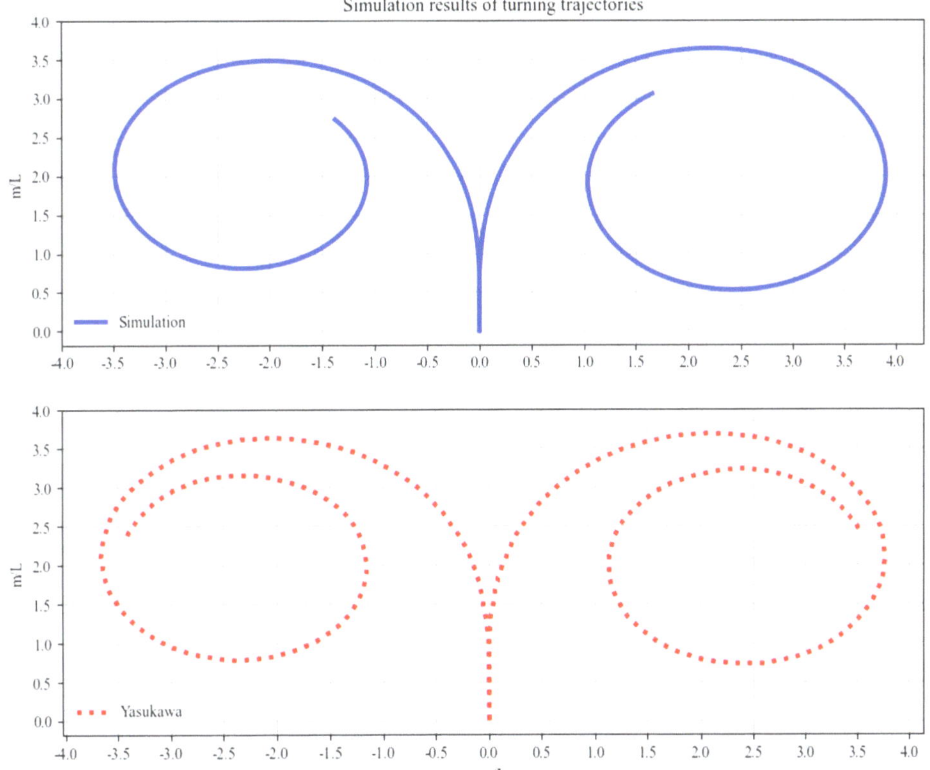

Figure 5. Comparison of ship trajectory; Yasukawa & Yoshimura, 2015 [53].

The accuracy of the ship's turning test can be verified using the formula below [54]:

$$C_\mathrm{M} = \frac{\min(S_\mathrm{D}, R_\mathrm{D})}{\max(S_\mathrm{D}, R_\mathrm{D})} 100\%$$

(27)

where S_D is the simulation result, R_D is the experimental result, and C_M is the consistency evaluation indicator.

Finally, a comparison of the ship's turning test is shown in Table 2. The consistency between simulation results in this study and Yasukawa [53] is 97.52%. The simulation results are in line with the valid experimental results.

Table 2. Comparison of simulation results.

Turning Test	$A_d/L(\delta = 35°)$	$T_d/L(\delta = 35°)$	$A_d/L(\delta = -35°)$	$T_d/L(\delta = -35°)$
Yasukawa & Yoshimura (2015) [53]	3.67	3.71	3.56	3.59
Simulation	3.64	3.90	3.49	3.49
C_M	99.18%	95.13%	98.03%	97.72%
$\overline{C}_M$		97.52%		

A_d and T_d are the advance and tactical diameter of the turning test, respectively.

To verify the efficiency of the P-PB controller under various environmental disturbances, simulation scenarios were established, as described in Table 3.

Table 3. Simulation scenarios for the course-keeping test.

Simulation Scenario	Wind Speed (m/s)	Wind Direction (°)	Target Course (°)			
			0–900 s	900–1800 s	1800–2700 s	2700–3600 s
No wind	0	/	30	10	−5	20
Low speed wind	10	30	30	10	−5	20
High speed wind	20	30	30	10	−5	20

Then, to balance the efficiency and accuracy of PSO when optimizing the P-PB controller, the main parameters of PSO were set, as described in Table 4.

Table 4. Main parameters of PSO.

Parameter	Value
N	20
T	40
$c1$	2
$c2$	2
w	0.5

The change curve of the fitness value for iterations under each stage is shown in Figure 6. The fitness value increased with each iteration and eventually stabilized in each stage, which ensured the effectiveness and completeness of the optimization.

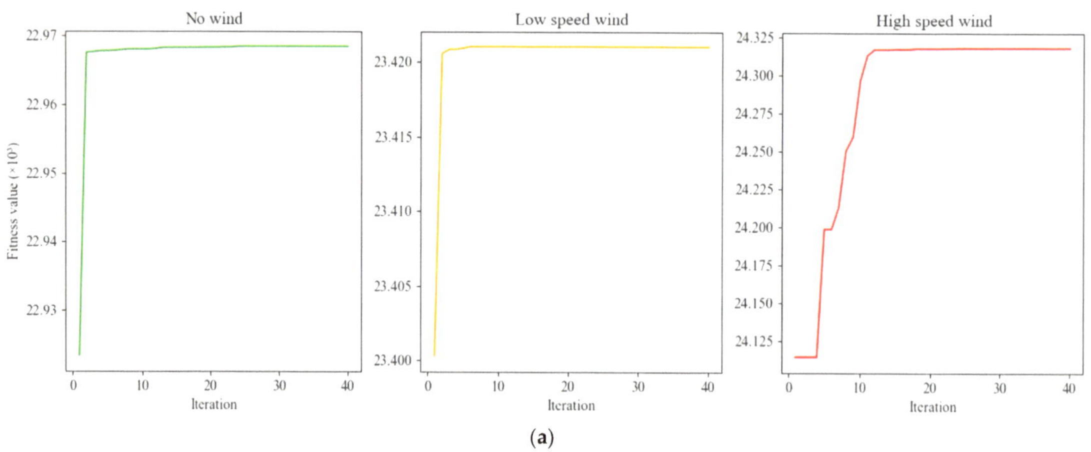

Figure 6. *Cont.*

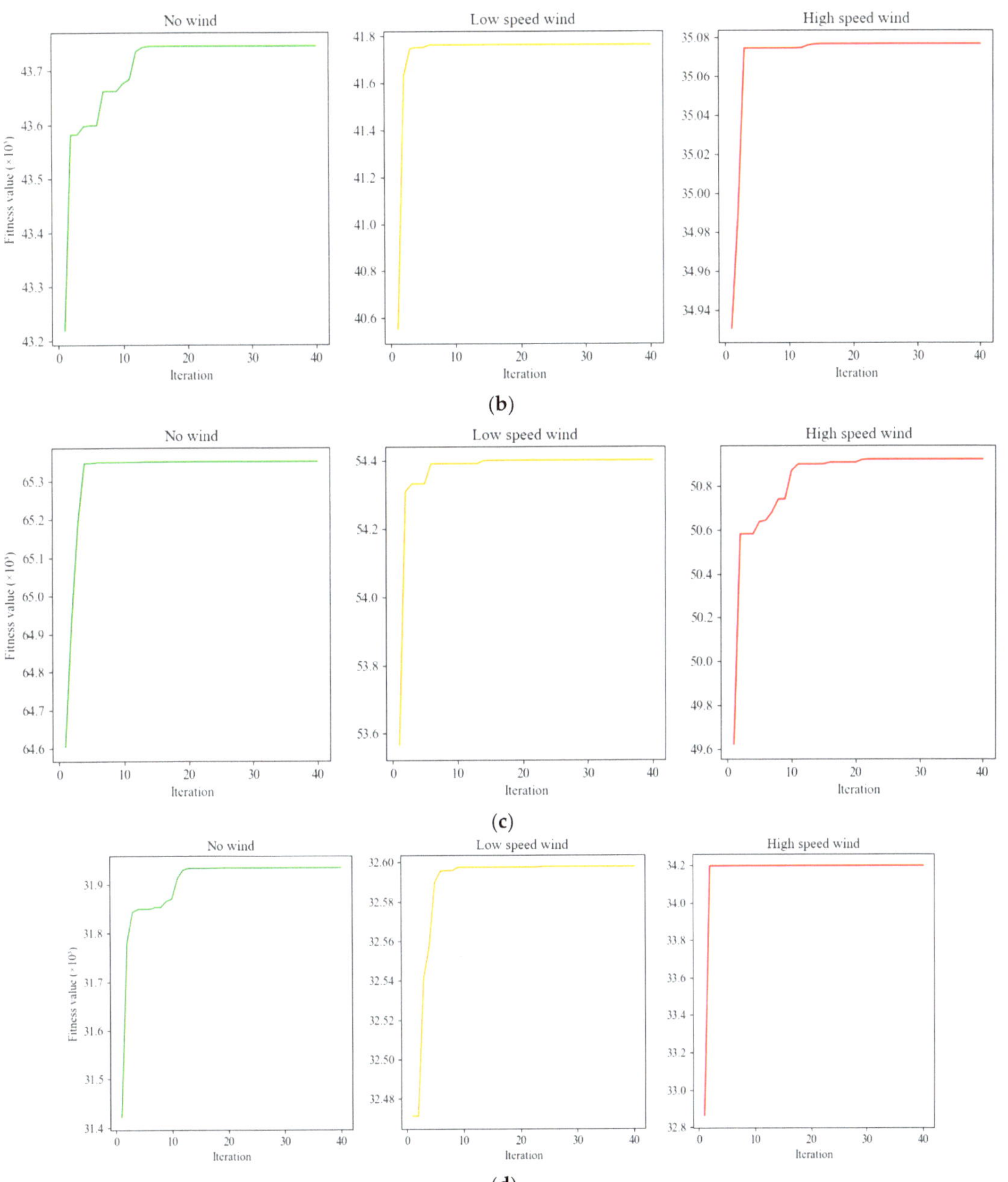

Figure 6. Change curves of fitness values for iterations in (**a**) 0–900, (**b**) 900–1800, (**c**) 1800–2700, and (**d**) 2700–3600 s.

3.2. Comparison and Analysis of Simulation Results

The P-PB controller was valid when compared to the improved PID controllers proposed by Diabac and He [22,23] and the controller based on backstepping control proposed by Zhang [9]. The change in ship's course under these three models and the proposed P-PB controller is compared in various scenarios, as shown in Figure 7.

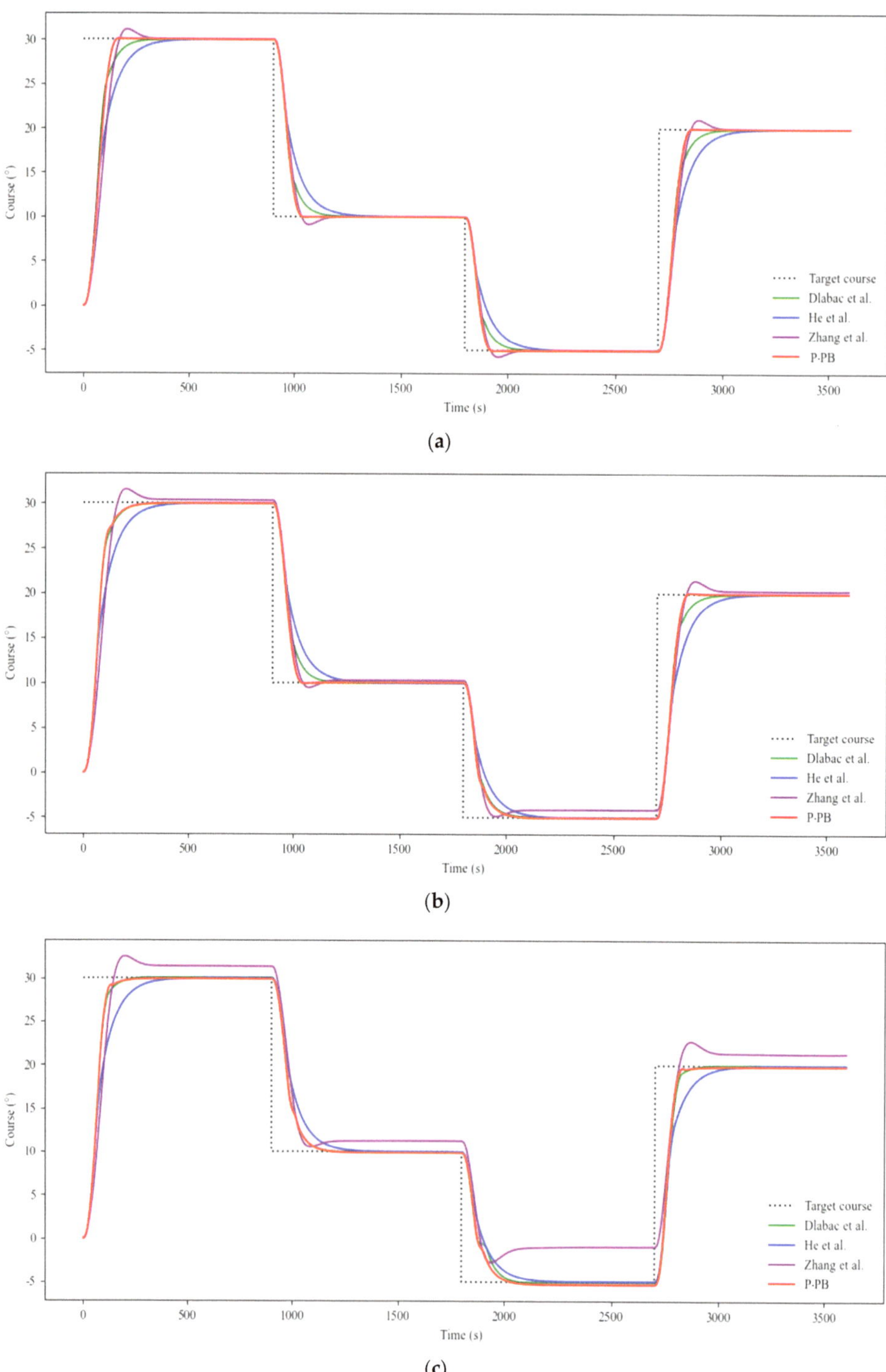

Figure 7. Comparison of the ship's course under (**a**) no wind, (**b**) low-speed wind, and (**c**) high-speed wind scenarios; Dlabac et al. (2019) [22], He et al. (2020) [23], Zhang et al. (2020) [24].

As seen in Figure 7, all the controllers kept the ship on course under various environmental disturbances. When comparing the four controllers, it is obvious that the P-PB controller and the controller proposed by Dlabac [22] and Zhang [24] responded more quickly and tracked the target course satisfactorily in the initial stage when the target course was changing.

However, the overshoot of the backstepping controller proposed by Zhang [24] increased rapidly as the environmental disturbances became harsher. This is because the yawing speed of the ship is considered one of the inputs of the improved backstepping controller proposed by Zhang [24]. When an environmental disturbance is lower, consideration of the ship's yawing speed is an effective way to improve the stability and accuracy of the course-keeping controller. Conversely, when an environmental disturbance is larger, external factors significantly interfere with the ship's yawing speed, which causes the ship's course-keeping error to increase rapidly.

The proposed P-PB controller also considers the ship's yawing speed to achieve efficient course-keeping control when environmental disturbances are lower. Furthermore, when environmental disturbances gradually increase, an overshoot of ship course-keeping is avoided by adjusting the value of K_w using PSO. Therefore, a smaller course-keeping error can be obtained under various scenarios.

The accumulated course-keeping error is compared in Figure 8.

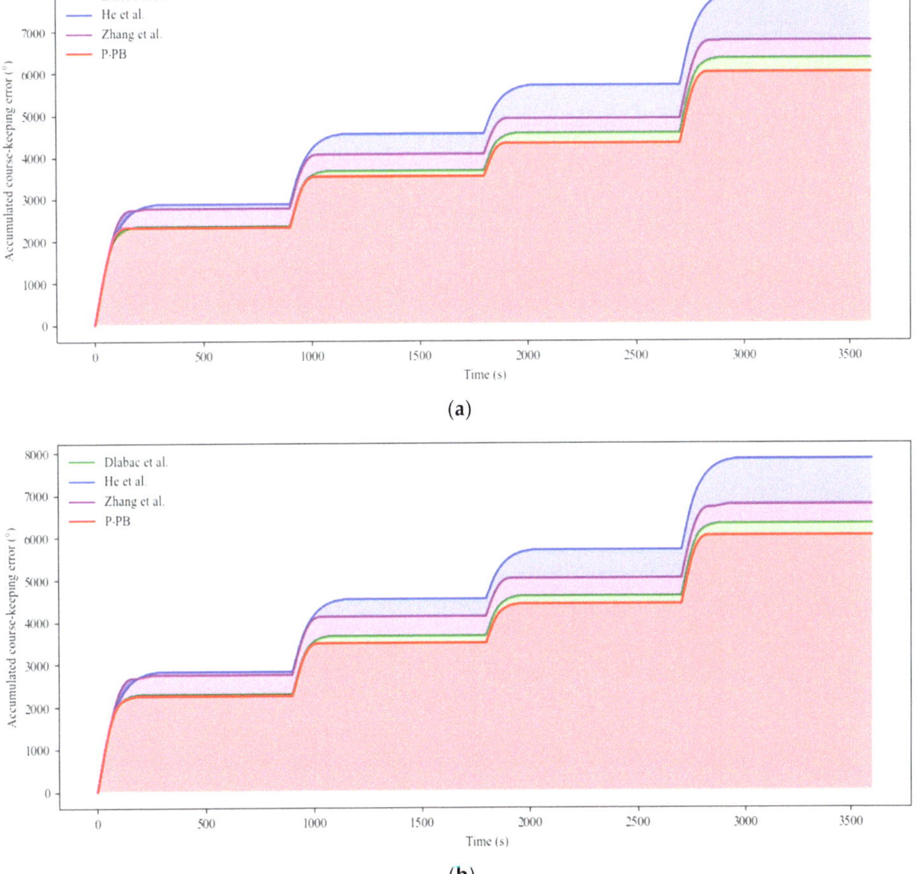

Figure 8. *Cont.*

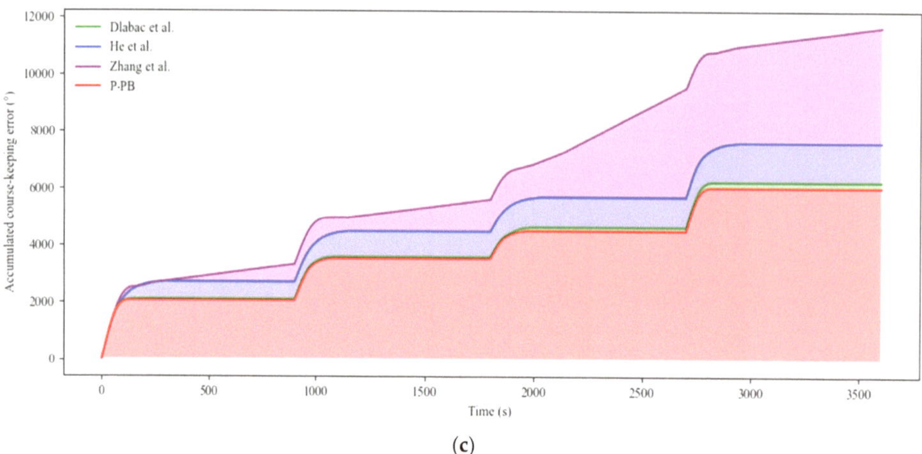

(**c**)

Figure 8. Comparison of the accumulated course-keeping error under (**a**) no wind, (**b**) low-speed wind, and (**c**) high-speed wind scenarios; Dlabac et al. (2019) [22], He et al. (2020) [23], Zhang et al. (2020) [24].

The mean course-keeping error, which is calculated by dividing the accumulated course-keeping error by time, is shown in Figure 9. It illustrates the mean course-keeping error for each controller as a percentage of the accumulated mean course-keeping error for the four controllers, and each circle represents one percent. For example, when under a high-speed wind scenario, the accumulated mean course-keeping error of the four controllers is 8.77°, and the mean course-keeping error of the P-PB controller is 1.67°, which accounts for 19.04% of the mean course-keeping error and therefore occupies 19 circles.

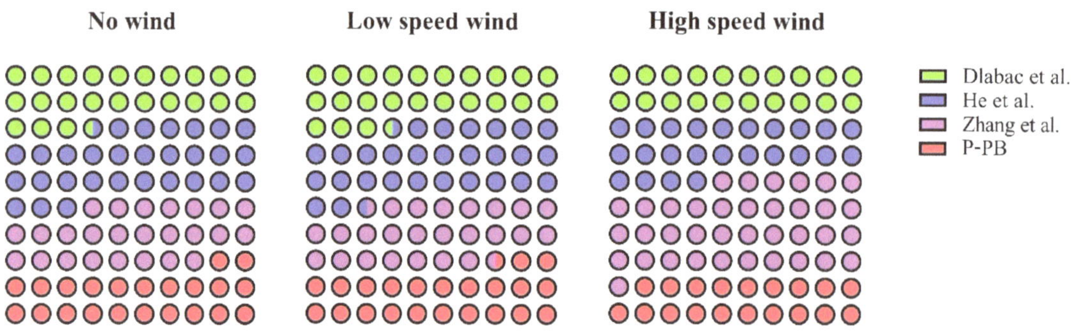

Figure 9. Comparison of the mean course-keeping error; Dlabac et al. (2019) [22], He et al. (2020) [23], Zhang et al. (2020) [24].

The smaller the number of circles corresponding to the controllers, the smaller the average error of the controllers compared to the other controllers, and therefore the more efficient the controller is.

As indicated by Figures 7–9, it is clear that the P-PB controller and the controller proposed by Dlabac [22] achieved course-keeping with the lowest and second-lowest accumulated course-keeping error, respectively, throughout the simulation tests under the three scenarios. Meanwhile, compared to the controller proposed by Dlabac [22], the P-PB controller had a larger advantage in terms of accumulated course-keeping error when environmental disturbances were lower.

The accumulated course-keeping error of the controller proposed by Zhang [24] performed better under lower environmental disturbances. However, its accumulated course-keeping error increased significantly under extreme environmental disturbances. The improved PID controller proposed by He performed stably in various scenarios, but was slightly slower to reach the target course.

The course-keeping error when the ship's course is stabilized is shown in Figure 10.

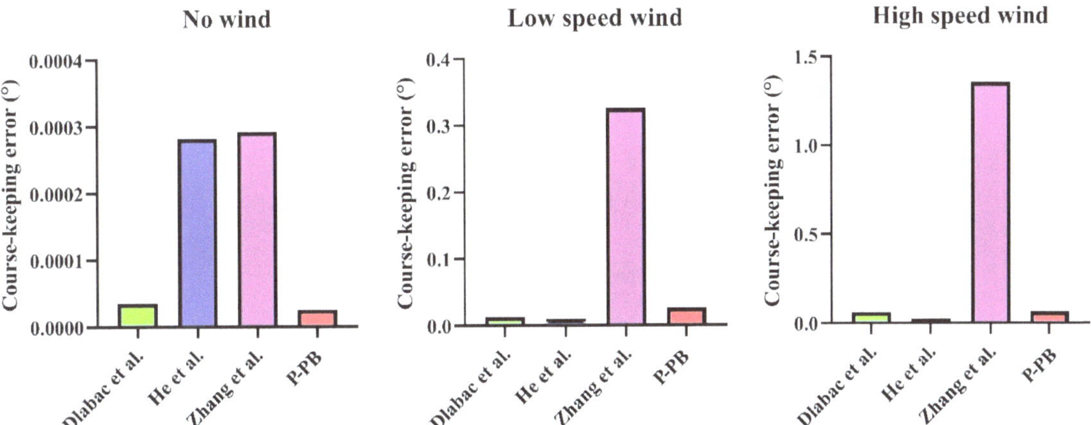

Figure 10. Comparison of the stabilized course-keeping error; Dlabac et al. (2019) [22], He et al. (2020) [23], Zhang et al. (2020) [24].

When there was no environmental disturbance, the stabilized course-keeping error of all four controllers was extremely small. Among them, the controller proposed by Dlabac [22] and the P-PB controller had the lowest stabilized course-keeping error. However, with the gradual increase in environmental disturbances, the course-keeping error of Zhang's [24] proposed controller increased rapidly, and the errors of the P-PB and Dlabac's [22] proposed controllers also increased. He's proposed controller maintained a lower error than the other controllers under environmental disturbances.

The average rudder angle in the various scenarios is shown in Figure 11. The average rudder angles of all ship course-keeping controllers increased when the wind disturbance became harsher.

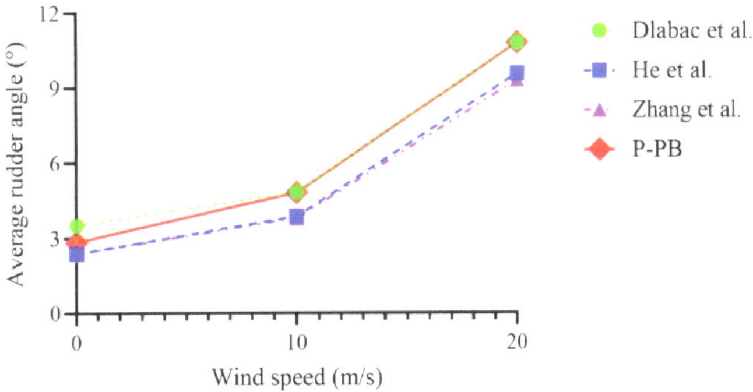

Figure 11. Comparison of average rudder angle; Dlabac et al. (2019) [22], He et al. (2020) [23], Zhang et al. (2020) [24].

When comparing the four controllers, the controllers proposed by Zhang [24] and He [23] achieved course-keeping with a smaller rudder angle under all scenarios. Conversely, the P-PB controller and the controller proposed by Dlabac [22] had a larger average rudder angle and therefore required relatively more energy for control. Furthermore, when comparing the above-two controllers with lower course-keeping errors, the P-PB controller had a relatively small average rudder angle, especially under lower environmental disturbances.

Finally, the mean course-keeping error for the four controllers is shown in Table 5. Compared to other controllers, the P-PB controller improved course-keeping error by 4.19% on average.

Table 5. Course-keeping controller performance.

Course-Keeping Controller	Mean Course-Keeping Error (°)		
	No Wind	Low Speed Wind	High Speed Wind
Dlabac et al. [22]	1.76	1.75	1.73
He et al. [23]	2.20	2.18	2.12
Zhang et al. [24]	1.88	1.88	3.23
P-PB	1.67	1.67	1.68
Improvement	5.11%	4.57%	2.89%
Average improvement		4.19%	

When environmental disturbances increased, the reduction in the ship course-keeping error of some controllers was due the fact that the disturbance matched the direction of the ship's course, and therefore the ship achieved the target course in a shorter period of time.

To provide a comprehensive analysis of the control effectiveness of the P-PB controller, time-response specifications have been compared. Among them, rise time is the time required to adjust the ship's course to 90% of the target course, which represents the rapidity of the controller's response. Overshoot characterizes the maximum deviation of the controller, and the smaller its value the better the stability of the controller. Settling time refers to the time it takes to maintain a course-keeping error within 2% and is a key factor of control stability. The mean value of each specification in the various scenarios is shown in Table 6.

Table 6. Comparison of time response specifications.

Course-Keeping Controller	Time-Response Specification		
	Rise Time (s)	Settling Time (s)	Overshoot (%)
Dlabac et al. [22]	134.83	209.92	0.01
He et al. [23]	191.17	312.17	0.01
Zhang et al. [24]	186.67	195.77	5.35
P-PB	117.42	158.17	0.16
Improvement	12.91%	19.21%	−2.99%
Average improvement		9.71%	

As seen in Table 6, the proposed P-PB controller saw a 9.71% improvement on average in terms of time-response specification. Analyzing the time-response specification, the proposed controller has a faster response speed and better control stability compared to the other controllers, providing further confirmation of the controller's effectiveness. However, some overshoot remains. This is attributed to the fact that the PSO-optimized controller retains a partial dependence on r to reduce the course-keeping error.

In summary, the P-PB controller proposed in this paper can realize a stabilized course-keeping of ships with lower error and better time-response performance, providing a new approach for the automatic control of ships.

4. Conclusions

This study proposes a PSO-based predictive PID-backstepping controller for the course-keeping of ships, taking target course, current course, yawing speed, and predictive motion parameters into consideration. The proposed controller is designed based on the predictive PID controller and backstepping controller. The parameters in the proposed controller are optimized via PSO. Finally, the proposed controller's efficacy was demonstrated by comparing it with other ship controllers in various scenarios. Comparison results illustrate that the proposed controller can achieve the target course more quickly and more precisely under various environmental disturbances, which provides a new approach for the course-keeping of ships. However, the proposed method has a larger average rudder angle, which may lead to higher energy consumption. In future research, improvements can be made to improve energy consumption in ship course-keeping.

Although the research in this study revealed some important findings, there are still some limitations that need to be further researched in the future. Firstly, the hydrodynamic coefficients in this paper were mainly calculated using empirical formulas, which could be further optimized to improve the accuracy of the ship motion model. Second, the current study does not consider the effect of obstacles on course-keeping during navigation.

Author Contributions: All authors contributed to this study. B.L.: Conceptualization, Methodology, Software, Visualization, Writing—original draft, Writing—review & editing; M.Z. (Mao Zheng): Conceptualization, Data curation, Funding acquisition, Investigation, Supervision, Validation, Writing—review & editing; B.H.: Investigation, Conceptualization, Project administration, Supervision, Validation, Writing—original draft, Writing—review & editing; X.C.: Conceptualization, Data curation, Formal analysis, Funding acquisition, Investigation, Project administration, Writing—review & editing; M.Z. (Mingyang Zhang): Supervision, Validation, Writing—original draft, Writing—review & editing; H.Z. and S.D.: Software, Writing—review & editing; H.W. and K.Z.: Writing—review & editing. All authors have read and agreed to the published version of the manuscript.

Funding: This research was funded by the National Key Research and Development Program of China under Grant (2022YFB2602301), Fund of Guangxi Science and Technology Program (AB23026132), the National Natural Science Foundation of China (52001240, 52001243).

Institutional Review Board Statement: Not applicable.

Informed Consent Statement: Not applicable.

Data Availability Statement: Data are contained within the article.

Conflicts of Interest: The authors declare no conflict of interest.

References

1. Sun, Y.; Lu, Z.; Lian, F.; Yang, Z. Study of Channel Upgrades and Ship Choices of River-Shipping of Port Access-Transportation. *Transp. Res. Part Transp. Environ.* **2023**, *119*, 103733. [CrossRef]
2. Liu, Z.; Gao, H.; Zhang, M.; Yan, R.; Liu, J. A Data Mining Method to Extract Traffic Network for Maritime Transport Management. *Ocean Coast. Manag.* **2023**, *239*, 106622. [CrossRef]
3. Clark, X.; Dollar, D.; Micco, A. Port Efficiency, Maritime Transport Costs, and Bilateral Trade. *J. Dev. Econ.* **2004**, *75*, 417–450. [CrossRef]
4. Grewal, D.; Haugstetter, H. Capturing and Sharing Knowledge in Supply Chains in the Maritime Transport Sector: Critical Issues. *Marit. Policy Manag.* **2007**, *34*, 169–183. [CrossRef]
5. Ma, D.; Chen, X.; Ma, W.; Zheng, H.; Qu, F. Neural Network Model-Based Reinforcement Learning Control for AUV 3-D Path Following. *IEEE Trans. Intell. Veh.* **2023**, 1–13. [CrossRef]
6. He, Z.; Liu, C.; Chu, X.; Negenborn, R.R.; Wu, Q. Dynamic Anti-Collision A-Star Algorithm for Multi-Ship Encounter Situa-tions. *Appl. Ocean Res.* **2022**, *118*, 102995. [CrossRef]
7. Chu, Z.; Yan, R.; Wang, S. Evaluation and Prediction of Punctuality of Vessel Arrival at Port: A Case Study of Hong Kong. *Marit. Policy Manag.* **2023**, 1–29. [CrossRef]
8. Lin, B.; Zheng, M.; Chu, X.; Zhang, M.; Mao, W.; Wu, D. A Novel Method for the Evaluation of Ship Berthing Risk Using AIS Data. *Ocean Eng.* **2024**, *293*, 116595. [CrossRef]
9. Zhang, Q.; Zhang, M.; Hu, Y.; Zhu, G. Error-Driven-Based Adaptive Nonlinear Feedback Control of Course-Keeping for Ships. *J. Mar. Sci. Technol.* **2021**, *26*, 357–367. [CrossRef]
10. Zhang, M.; Conti, F.; Le Sourne, H.; Vassalos, D.; Kujala, P.; Lindroth, D.; Hirdaris, S. A Method for the Direct Assessment of Ship Collision Damage and Flooding Risk in Real Conditions. *Ocean Eng.* **2021**, *237*, 109605. [CrossRef]
11. Zhang, M.; Taimuri, G.; Zhang, J.; Hirdaris, S. A Deep Learning Method for the Prediction of 6-DoF Ship Motions in Real Conditions. *Proc. Inst. Mech. Eng. Part M J. Eng. Marit. Environ.* **2023**, *237*, 887–905. [CrossRef]
12. Liu, Z.; Zhang, B.; Zhang, M.; Wang, H.; Fu, X. A Quantitative Method for the Analysis of Ship Collision Risk Using AIS Data. *Ocean Eng.* **2023**, *272*, 113906. [CrossRef]

13. Zhang, M.; Kujala, P.; Musharraf, M.; Zhang, J.; Hirdaris, S. A Machine Learning Method for the Prediction of Ship Motion Trajectories in Real Operational Conditions. *Ocean Eng.* **2023**, *283*, 114905. [CrossRef]

14. Lin, B.; Zheng, M.; Chu, X.; Mao, W.; Zhang, D.; Zhang, M. An Overview of Scholarly Literature on Navigation Hazards in Arctic Shipping Routes. *Environ. Sci. Pollut. Res.* **2023**, 1–17. [CrossRef] [PubMed]

15. Min, B.; Zhang, X. Concise Robust Fuzzy Nonlinear Feedback Track Keeping Control for Ships Using Multi-Technique Improved LOS Guidance. *Ocean Eng.* **2021**, *224*, 108734. [CrossRef]

16. Kim, D.; Song, S.; Jeong, B.; Tezdogan, T.; Incecik, A. Unsteady RANS CFD Simulations of Ship Manoeuvrability and Course Keeping Control under Various Wave Height Conditions. *Appl. Ocean Res.* **2021**, *117*, 102940. [CrossRef]

17. Kim, D.; Tezdogan, T. CFD-Based Hydrodynamic Analyses of Ship Course Keeping Control and Turning Performance in Irregular Waves. *Ocean Eng.* **2022**, *248*, 110808. [CrossRef]

18. Liu, Z. Ship Adaptive Course Keeping Control With Nonlinear Disturbance Observer. *IEEE Access* **2017**, *5*, 17567–17575. [CrossRef]

19. Liu, Z. Ship Course Keeping Using Different Sliding Mode Controllers. *Trans. Famena* **2019**, *43*, 49–60. [CrossRef]

20. Liangqi, L.; Renxiang, B.; Wuchen, S.; Xinyu, L. Ship Track-Keeping Control Based on Sliding Mode Variable Structure PID Controller and Particle Swarm Optimization. In Proceedings of the 2019 6th International Conference on Information Science and Control Engineering (ICISCE), Shanghai, China, 20–22 December 2019; pp. 887–891.

21. Islam, M.M.; Siffat, S.A.; Ahmad, I.; Liaquat, M. Supertwisting and Terminal Sliding Mode Control of Course Keeping for Ships by Using Particle Swarm Optimization. *Ocean Eng.* **2022**, *266*, 112942. [CrossRef]

22. Dlabač, T.; Ćalasan, M.; Krčum, M.; Marvučić, N. PSO-Based PID Controller Design for Ship Course-Keeping Autopilot. *Brodogradnja* **2019**, *70*, 1–15. [CrossRef]

23. He, Y.; Zhang, X.; Yu, Y.; Li, M.; Gong, S.; Jin, Y.; Mou, J. Ship Dynamic Collision Avoidance Mechanism Based on Course Control System. *Ournal Southwest Jiaotong Univ.* **2020**, *55*, 988–993+1027.

24. Zhang, Q.; Ding, Z.; Zhang, M. Adaptive Self-Regulation PID Control of Course-Keeping for Ships. *Pol. Marit. Res.* **2020**, *27*, 39–45. [CrossRef]

25. Wang, C.; Yan, C.; Liu, Z.; Cao, F. An Inverse Optimal Approach to Ship Course-Keeping Control. *IMA J. Math. Control Inf.* **2020**, *37*, 1192–1217. [CrossRef]

26. Islam, M.M.; Siffat, S.A.; Ahmad, I.; Liaquat, M. Robust Integral Backstepping and Terminal Synergetic Control of Course Keeping for Ships. *Ocean Eng.* **2021**, *221*, 108532. [CrossRef]

27. Hu, Y.; Su, W.; Zhang, Q.; Zhang, Y.; Wang, C. A Nonlinear Power Feedback Improvement of the Ship Course-Keeping Controller. *Math. Probl. Eng.* **2022**, *2022*, e3095122. [CrossRef]

28. Zhang, Q.; Zhang, X.; Im, N. Ship Nonlinear-Feedback Course Keeping Algorithm Based on MMG Model Driven by Bipolar Sigmoid Function for Berthing. *Int. J. Nav. Archit. Ocean Eng.* **2017**, *9*, 525–536. [CrossRef]

29. Min, B.; Zhang, X.; Wang, Q. Energy Saving of Course Keeping for Ships Using CGSA and Nonlinear Decoration. *IEEE Access* **2020**, *8*, 141622–141631. [CrossRef]

30. Borkowski, P. Inference Engine in an Intelligent Ship Course-Keeping System. *Comput. Intell. Neurosci.* **2017**, *2017*, e2561383. [CrossRef]

31. Zirilli, A.; Roberts, G.N.; Tiano, A.; Sutton, R. Adaptive Steering of a Containership Based on Neural Networks. *Int. J. Adapt. Control Signal Process.* **2000**, *14*, 849–873. [CrossRef]

32. Xu, H.-J.; Li, W.; Yu, Y.; Liu, Y. A Novel Adaptive Neural Control Scheme for Uncertain Ship Course-Keeping System. *Sens. Transducers* **2014**, *178*, 282.

33. Zhang, S.; Zhang, Q.; Su, W.; Li, H.; Gai, X. Ship Adaptive RBF Neural Network Course Keeping Control Considering System Uncertainty. In Proceedings of the 2023 IEEE 12th Data Driven Control and Learning Systems Conference (DDCLS), Xiangtan, China, 12–14 May 2023; pp. 1398–1403.

34. Wang, Q.; Sun, C.; Chen, Y. Adaptive Neural Network Control for Course-Keeping of Ships with Input Constraints. *Trans. Inst. Meas. Control* **2019**, *41*, 1010–1018. [CrossRef]

35. Le, T.T. Ship Heading Control System Using Neural Network. *J. Mar. Sci. Technol.* **2021**, *26*, 963–972. [CrossRef]

36. Zhang, Z.; Zhang, X.; Zhang, G. ANFIS-Based Course-Keeping Control for Ships Using Nonlinear Feedback Technique. *J. Mar. Sci. Technol.* **2019**, *24*, 1326–1333. [CrossRef]

37. Wang, C.; Yan, C.; Liu, Z. Leader-Following Consensus for Second-Order Nonlinear Multi-Agent Systems Under Markovian Switching Topologies with Application to Ship Course-Keeping. *Int. J. Control Autom. Syst.* **2021**, *19*, 54–62. [CrossRef]

38. Mohd Tumari, M.Z.; Zainal Abidin, A.F.; Hussin, M.S.F.; Abd Kadir, A.M.; Mohd Aras, M.S.; Ahmad, M.A. PSO Fine-Tuned Model-Free PID Controller with Derivative Filter for Depth Control of Hovering Autonomous Underwater Vehicle. In Proceedings of the 10th National Technical Seminar on Underwater System Technology, Pekan, Malaysia, 26–27 September 2018; Volume 538.

39. Alkhafaji, A.S.; Al-hayder, A.; Hassooni, A. Hybrid IWOPSO Optimization Based Marine Engine Rotational Speed Control Automatic System. *Int. J. Electr. Comput. Eng.* **2020**, *10*, 840–848. [CrossRef]

40. Chen, H.; Xie, J.; Han, J.; Shi, W.; Charpentier, J.-F.; Benbouzid, M. Position Control of Heave Compensation for Offshore Cranes Based on a Particle Swarm Optimized Model Predictive Trajectory Path Controller. *J. Mar. Sci. Eng.* **2022**, *10*, 1427. [CrossRef]

41. Zhu, M.; Sun, W.; Hahn, A.; Wen, Y.; Xiao, C.; Tao, W. Adaptive Modeling of Maritime Autonomous Surface Ships with Uncertainty Using a Weighted LS-SVR Robust to Outliers. *Ocean Eng.* **2020**, *200*, 107053. [CrossRef]

42. Chen, C.; Delefortrie, G.; Lataire, E. Effects of Water Depth and Speed on Ship Motion Control from Medium Deep to Very Shallow Water. *Ocean Eng.* **2021**, *231*, 109102. [CrossRef]
43. Zhu, M.; Tian, K.; Wen, Y.-Q.; Cao, J.-N.; Huang, L. Improved PER-DDPG Based Nonparametric Modeling of Ship Dynamics with Uncertainty. *Ocean Eng.* **2023**, *286*, 115513. [CrossRef]
44. Kijima, K.; Katsuno, T.; Nakiri, Y.; Furukawa, Y. On the Manoeuvring Performance of a Ship with Theparameter of Loading Condition. *J. Soc. Nav. Archit. Jpn.* **1990**, *1990*, 141–148. [CrossRef] [PubMed]
45. Jia, X.; Yang, Y. *Ship Motion Mathematical Model: Modeling Mechanism Modeling and Identification*; Dalian Maritime University Press: Dalian, China, 1999.
46. Broglia, R.; Dubbioso, G.; Durante, D.; Mascio, A.D. Simulation of Turning Circle by CFD: Analysis of Different Propeller Models and Their Effect on Manoeuvring Prediction. *Appl. Ocean Res.* **2013**, *39*, 1–10. [CrossRef]
47. Sun, M.; Zhang, W.; Zhang, Y.; Luan, T.; Yuan, X.; Li, X. An Anti-Rolling Control Method of Rudder Fin System Based on ADRC Decoupling and DDPG Parameter Adjustment. *Ocean Eng.* **2023**, *278*, 114306. [CrossRef]
48. Christofides, P.D.; Scattolini, R.; Muñoz de la Peña, D.; Liu, J. Distributed Model Predictive Control: A Tutorial Review and Future Research Directions. *Comput. Chem. Eng.* **2013**, *51*, 21–41. [CrossRef]
49. Andersen, I.M.V. Wind Loads on Post-Panamax Container Ship. *Ocean Eng.* **2013**, *58*, 115–134. [CrossRef]
50. Zhang, X.; Feng, Y. Control Algorithm of YUPENG Ship Autopilot Based on Tangent Function Nonlinear Feedback. *J. Meas. Sci. Instrum.* **2017**, *8*, 73–78. [CrossRef]
51. Zhang, X. *Simple Robust Control of Ship Motion*; Science Press: Beijing, China, 2012.
52. Kennedy, J.; Eberhart, R. Particle Swarm Optimization. In Proceedings of the ICNN'95—International Conference on Neural Networks, Perth, WA, Australia, 27 November–1 December 1995; Volume 4, pp. 1942–1948.
53. Yasukawa, H.; Yoshimura, Y. Introduction of MMG Standard Method for Ship Maneuvering Predictions. *J. Mar. Sci. Technol.* **2015**, *20*, 37–52. [CrossRef]
54. Zhang, X.; Yang, G.-P.; Zhang, Q. A Kind of Bipolar Sigmoid Function Decorated Nonlinear Ship Course Keeping Algorithm. *J. Dalian Marit. Univ.* **2016**, *42*, 15–19. [CrossRef]

Journal of *Marine Science and Engineering*

MDPI

Article

Experimental Study on Adaptive Backstepping Synchronous following Control and Thrust Allocation for a Dynamic Positioning Vessel

Changde Liu [1,2,*], Yufang Zhang [3], Min Gu [1,2], Longhui Zhang [1,2], Yanbin Teng [1,2] and Fang Tian [1,2]

1 China Ship Scientific Research Center, Wuxi 214082, China
2 Taihu Laboratory of Deepsea Technological Science, Wuxi 214082, China
3 School of Control Technology, Wuxi Institute of Technology, Wuxi 214121, China
* Correspondence: liucd@cssrc.com.cn

Abstract: Cargo transfer vessels (CTVs) are designed to transfer cargo from a floating production storage and offloading (FPSO) unit into conventional tankers. The dynamic positioning system allows the CTV to maintain a safe position relative to the FPSO unit using a flexible cargo transmission pipe, and the CTV tows the tanker during operating conditions. The operation mode can be considered a synchronization tracking control problem. In this paper, a synchronization control strategy is presented based on the virtual leader–follower configuration and an adaptive backstepping control method. The position and heading of the following vessel are proven to be able to globally exponentially converge to the virtual ship by the contraction theorem. Then, the optimization problem of the desired thrust command from the controller is solved through an improved firefly algorithm, which fully considers the physical characteristics of the azimuth thruster and the thrust forbidden zone caused by hydrodynamic interference. To validate the effectiveness of the presented synchronous following strategy and thrust allocation algorithm, a scale model experiment is carried out under a sea state of 4 in a seakeeping basin. The experimental results show that the CTV can effectively maintain a safe distance of 100 m with a maximum deviation of 3.78 m and an average deviation of only 0.99 m in the wave heading 180°, which effectively verifies that the control strategy proposed in this paper can achieve safe and cooperative operation between the CTV and the FPSO unit. To verify the advantages of the SAF algorithm in the thrust allocation, the SQP algorithm and PSO algorithm are used to compare the experimental results. The SAF algorithm outperforms the SQP and PSO algorithms in longitudinal and lateral forces, with the R-squared (R^2) values of 0.9996 (yaw moment), 0.9878 (sway force), and 0.9596 (surge force) for the actual thrusts and control commands in the wave heading 180°. The experimental results can provide technical support to improve the safe operation of CTVs.

Keywords: dynamic positioning; virtual leader–follower; backstepping; synchronization; contraction theory; control allocation; model experiment

Citation: Liu, C.; Zhang, Y.; Gu, M.; Zhang, L.; Teng, Y.; Tian, F. Experimental Study on Adaptive Backstepping Synchronous following Control and Thrust Allocation for a Dynamic Positioning Vessel. *J. Mar. Sci. Eng.* **2024**, *12*, 203. https://doi.org/10.3390/jmse12020203

Academic Editor: Mohamed Benbouzid

Received: 5 December 2023
Revised: 19 January 2024
Accepted: 19 January 2024
Published: 23 January 2024

1. Introduction

With the development of deep-sea oil and gas resource development technology, a brand-new concept of a deep-water dynamic positioning cargo transfer vessel has been proposed. The vessel is composed of a CTV equipped with a dynamic positioning (DP) system and a cargo transfer device onboard the CTV. The CTV performs the same role as a tanker and maintains position within the floating production storage and offloading (FPSO) unit. The offloading hose is passed from the FPSO to the CTV and it is connected. The cargo is then pumped to the tanker via the CTV using booster pumps. A corresponding safe working distance must be maintained between the CTV and the FPSO unit due to the limitations of the transfer device. The design of a strategy to synchronously control the CTV and the FPSO unit under a towing state is important for improving the efficient operation of the CTV. A schematic diagram of the CTV operation is shown in Figure 1.

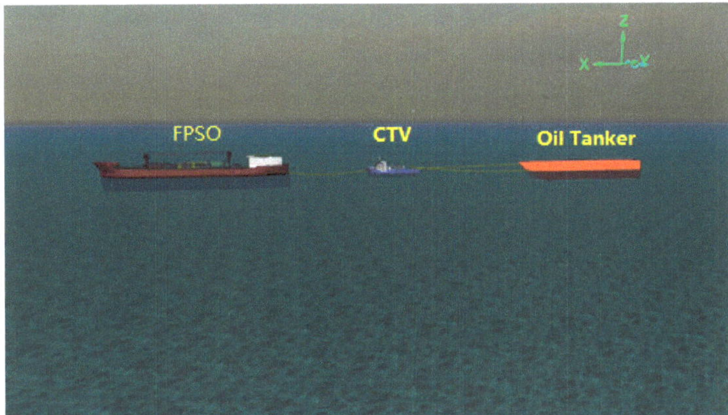

Figure 1. Schematic diagram of CTV operation with an FPSO unit and a tanker.

During cargo transfer between the CTV and the FPSO unit, the whole operation process can be regarded as a formation coordination control problem. At present, formation control strategies are mainly categorized into hybrid behavior-based formation control methods [1], virtual structure-based formation control methods [2] and leader–follower formation control methods. In actual operation processes, leader–follower formation control methods are widely used. Wu et al. [3] combined a leader–follower control strategy with a path planning strategy based on an artificial potential field method to propose a formation motion control method for unmanned surface vehicles (USVs) and designed a set of control laws for underactuated USVs. Wang et al. [4] proposed a fixed-time controller strategy based on the leader–follower mechanism and finite time disturbance observer. Shojaei et al. [5] proposed a leader–follower formation control method for underactuated surface vessels with actuator saturation. Cui et al. [6] proposed a leader–follower formation control for multiple underactuated autonomous underwater vehicles. To improve the tracking performance under uncertain sea conditions, a backstepping control with compensation control was designed [7]. Witkowska et al. [8] designed a course-keeping control system using a genetic algorithm and adaptive backstepping method. In recent years, with the development of intelligent technology, a variety of adaptive swarm formation control methods for unmanned vessels have emerged [9–11].

Most of the above research results are focused on underactuated control systems. Ships equipped with DP systems are generally equipped with tunnel thrusters, azimuth thrusters, and rudder propeller systems. When only used to control the three degrees of freedom in the ship's horizontal plane, the DP system is an overactuated system. Therefore, it can simultaneously control the horizontal, longitudinal, and heading positions in the horizontal plane. Sørensen [12] systematically summarized the DP control methods. In recent years, various improved algorithms for H-infinity control [13], backstepping control [14], sliding mode control [15], and nonlinear model predictive control [16,17] have emerged, and some control strategies have been applied to target point positioning models or full-scale ship experiments [18–20]. To achieve the synchronous control of DP vessels, Erick et al. [21] proposed a synchronization control strategy for underway replenishment, which realized the synchronization control of the supply vessels and the supplied vessels. However, hydrodynamic interactions between the two ships are crucial for the design of their automatic motion control systems; Miller [22] studied the interference of the forces and yaw moments between the two ships through the model tests. To achieve robustness to sea conditions, a leader–follower-based formation control problem was proposed for the collision avoidance study of fully driven USVs, and adaptive control technology was used to estimate the uncertain parameters of environmental disturbances [23].

At present, most of the above existing DP control methods use Lyapunov stability theory to prove the stability of closed-loop systems [15–23]. However, the designed controller method has high requirements for the system equilibrium state information. With the emergence and perfection of contraction theory, the concept of virtual displacement has gradually been used to solve the problem of system stability. Guttorm [24] designed a DP vessel observer based on contraction theory, determined that the system is globally contracting, and verified the robustness of the control system by simulation. Zhang et al. [25] designed a DP control law based on contraction theory to address DP vessels being disturbed by the environment (such as wind, waves, and currents) and verified that the designed adaptive backstepping controller has good robustness to sea conditions through numerical simulation. Alamir et al. [26] proposed a nonlinear model predictive control method based on contraction theory and verified the convergence of the closed-loop system. Traditional DP control methods have shown poor adaptation to the wave direction and sea conditions [20], while the backstepping method has been shown to have strong robustness [25,27].

Due to the overactuated characteristics of the DP propulsion system, for horizontal lateral force, longitudinal force, and yaw moment control commands, there are multiple combinations of forces and thrust directions in the multiple thrusters. Therefore, optimization techniques are usually used to obtain the optimal solution to match the expected control commands generated by the DP controller, thereby ensuring the safe operation of the DP vessels. Many linear and nonlinear optimal thrust allocation methods for DP vessels have been proposed [28]. However, the constraints of the physical characteristics of the thrusters must be considered. Therefore, in engineering applications, control allocation is actually an optimization problem under constraint conditions [29]. The SQP method has been widely used in DP thrust allocation due to its simplicity and efficiency [30–33]. To avoid the reduction in the thrust efficiency due to the thruster–thruster and thruster–ship interference, most researchers reduced the problem of severe thrust loss by setting forbidden zones [30,34,35]. A small number of researchers have introduced an efficiency function to achieve azimuth thruster operation over 0–360 degrees with a modified SQP algorithm [36,37]. However, for azimuth thrusters and rudders, when using the SQP algorithm, Taylor expansion must be performed on the force and torque balance equations to construct the SQP solution form, and second-order and higher-order terms must be truncated. Therefore, the turning angle speed for azimuth thrusters or steering speed for rudders is strictly limited. The recommended value in the literature is generally 1 deg/s for full-scale ships [28], but the actual turning angle speed and rudder turning speed can reach 12 deg/s and 5 deg/s, respectively, making it difficult for azimuth thrusters and rudders to respond quickly to control commands and even posing a risk of reducing the thrust allocation accuracy. Therefore, to improve the accuracy of DP thrust allocation, the use of algorithms (for example, genetic algorithms [38] and particle swarm optimization (PSO) algorithms [39]) has been verified through numerical simulations. The firefly algorithm (FA), which was developed by Xin-She Yang at Cambridge University in 2007 [40], has a strong local search ability and can find the optimal solution in a small area [41]. This algorithm is easy to operate, is simple to implement, has fewer parameters, and there is less impact of parameters on the algorithm. However, due to the high dependence of the search method on excellent individuals, the convergence speed is reduced; therefore, to achieve engineering applications, we make some improvements to enhance the self-adaptive ability.

In view of the new concept of CTV engineering operation requirements, a DP ship synchronous following control strategy and a thrust allocation method are proposed, and relevant model experimental techniques are established. First, we took an FPSO unit as the leader vessel and a CTV as the follower vessel and proposed a synchronous following control strategy for a DP vessel based on virtual leader–follower and adaptive backstepping methods. Next, we achieved feedforward compensation of the external disturbance caused by waves and the towing force, verified the global exponential convergence of the position and heading angle of the DP vessel based on contraction theory, and solved the nonuniversal parameter setting problem of the vessel DP system controller. A self-adaptive firefly (SAF) algorithm was proposed to optimize multiple thrusters considering forbidden zones in thrust allocation. Then, we embedded the control strategy in an independently developed DP control system, established a model experimental setup for a CTV synchronously following an FPSO unit while towing a tanker in irregular waves, and verified the effectiveness of the synchronous following control strategy proposed in this paper.

This article is organized as follows: Section 2 presents the virtual leader–follower strategy, low-frequency DP mathematical model, controller design, and contraction analysis. Section 3 describes the scaled model, experimental environment conditions, and test contents. Section 4 presents the experimental results of the scaled model, and the results are analyzed and discussed. Finally, conclusions are summarized and drawn in Section 5.

2. Mathematical Model and Methods

2.1. Virtual Leader–Follower Strategy

The dynamic positioning of a vessel DP only considers the motion of three degrees of freedom in the horizontal plane: surge, sway, and yaw. To better describe the motion of the DP vessel in the horizontal plane, we established Earth-fixed frame and Ship-fixed frame as the two coordinate frames of the leader vessel, virtual leader vessel, and follower vessel (as shown in Figure 2).

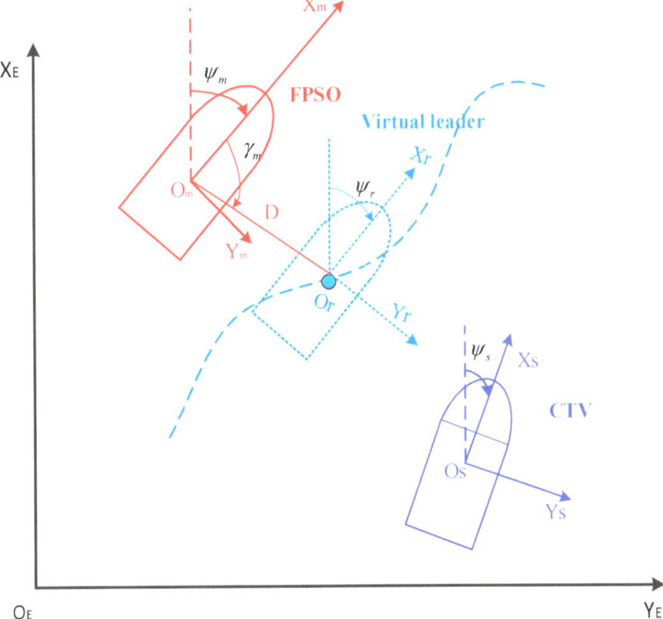

Figure 2. Schematic diagram of the virtual leader–follower strategy.

In Figure 2, $X_E O_E Y_E$ is the Earth-fixed coordinate system, $X_m O_m Y_m$ and $X_s O_s Y_s$ are the Ship-fixed coordinate systems of the leader vessel and the follower vessel, respectively, and $X_r O_r Y_r$ is the Ship-fixed coordinate system of the virtual leader vessel.

The tracking strategy based on the virtual leader–follower method (as shown in Figure 2) assumes a virtual FPSO unit as the leader, and the position of the virtual vessel relative to the leader is used as the feedback information of the closed-loop DP system. By designing the control strategy, the CTV and the FPSO unit maintain a certain target distance D, and the virtual leader–follower method is adopted. The position and heading of the leader vessel and the virtual leader vessel should meet the following requirements:

$$\eta_r = \eta_m + R(\psi_m)L$$

$$R(\psi_m) = \begin{bmatrix} \cos\psi_m & -\sin\psi_m & 0 \\ \sin\psi_m & \cos\psi_m & 0 \\ 0 & 0 & 1 \end{bmatrix} \tag{1}$$

where $\eta_m = [x_m, y_m, \psi_m]^{\mathrm{T}}$ represents the position and heading of the leader vessel, $\eta_r = [x_r, y_r, \psi_r]^{\mathrm{T}}$ represents the position and heading of the virtual vessel, $R(\psi_m)$ is the transformation matrix between the Earth-fixed coordinate system and the Ship-fixed coordinate system, $L = [D\cos\gamma_m, D\sin\gamma_m, 0]^{\mathrm{T}}$, and γ_m is the angle between the longitudinal direction of the leader vessel and the line connecting the centers of gravity of the leader vessel and the virtual vessel.

Then, the speed relationship between the virtual vessel and the leader vessel is as follows [21]:

$$\dot{\eta}_m = R(\psi_m)v_m$$

$$\dot{\eta}_r = \dot{\eta}_m + R(\psi_m)S(r_m)L + R(\psi_m)\dot{L} \tag{2}$$

where $v_m = [u_m, v_m, r_m]^{\mathrm{T}}$ and is composed of the longitudinal and lateral velocities and the heading angular velocity in the hull coordinate system of the leader vessel, and $S(r_m) = \begin{bmatrix} 0 & -r_m & 0 \\ r_m & 0 & 0 \\ 0 & 0 & 0 \end{bmatrix}$.

2.2. Mathematical Model

The CTV operates at low speed when synchronously following the FPSO unit with the DP system, so the nonlinear damping term is ignored, and its mathematical model can be expressed as [12]:

$$\begin{cases} \dot{\eta}_s = R(\psi_s)v_s \\ M\dot{v}_s + D_Lv_s = \tau + T_d \end{cases} \tag{3}$$

where $\eta_s = [x_s, y_s, \psi_s]^{\mathrm{T}}$ is composed of the trajectory and heading angle of the follower vessel in the geodetic coordinate system; $v_s = [u_s, v_s, r_s]^{\mathrm{T}}$ is composed of the surge, sway, and yaw angular velocities of the follower vessel in the hull coordinate system; $\tau = [\tau_x, \tau_y, \tau_z]^{\mathrm{T}}$ is composed of the resultant force and moment under the action of waves and the towing force; T_d is composed of the longitudinal and lateral control forces and the heading control moment generated by multiple thrusters; M is the inertia matrix with added mass $(M = M^T > 0)$; and D_L is the linear damping matrix.

$$M = \begin{bmatrix} m - X_{\dot{u}} & 0 & 0 \\ 0 & m - Y_{\dot{v}} & -Y_{\dot{r}} \\ 0 & -N_{\dot{v}} & I_z - N_{\dot{r}} \end{bmatrix}, \quad D_L = -\begin{bmatrix} X_u & 0 & 0 \\ 0 & Y_v & Y_r \\ 0 & N_v & N_r \end{bmatrix} \tag{4}$$

where m is the mass of the vessel; I_z is the moment of inertia; $X_{\dot{u}}$, $Y_{\dot{v}}$ and $Y_{\dot{r}}$ are added masses; $N_{\dot{r}}$ is the added moment of inertia; X_u, Y_v, Y_r, and N_r are hydrodynamic derivatives.

2.3. Controller Design

Based on contraction theory, a backstepping control strategy for the DP vessel to synchronously follow the virtual leader vessel is designed. The main steps are as follows:

First, define the position and heading error between the follower vessel and the virtual vessel as

$$s_1 = \eta_s - \eta_r \tag{5}$$

From Formula (5), the time derivative of s_1 is obtained as

$$\dot{s}_1 = R(\psi_s)v_s - \dot{\eta}_r = R(\psi_s)v_s - (\dot{\eta}_m + R(\psi_m)S(r_m)L + R(\psi_m)\dot{L}) \tag{6}$$

Let the speed vector v_s of the follower vessel be a virtual control quantity to design a virtual control function:

$$\alpha = R^T(\psi_s)(-K_1 s_1 + \dot{\eta}_r) \tag{7}$$

where $K_1 = \mathrm{diag}(k_{11}, k_{12}, k_{13}) > 0$ and is the design parameter matrix.

Define the speed error between the follower vessel and the virtual vessel as $s_2 = v_s - \alpha$. From Formulas (6) and (7), Formula (8) can be obtained:

$$\dot{s}_1 = R(\psi_s)s_2 - K_1 s_1 \tag{8}$$

Second, according to Formula (3), we can obtain the time derivative of S_2:

$$\dot{s}_2 = -M^{-1}D_L v_s + M^{-1}\tau + M^{-1}T_d - \dot{\alpha} \tag{9}$$

Then, the synchronous following control law for the DP vessel is designed as follows:

$$\tau = M[-K_2 s_2 + M^{-1}D_L v_s - M^{-1}\hat{T}_d + \dot{\alpha} - R^T(\psi_s)s_1] \tag{10}$$

where $K_2 = \mathrm{diag}(k_{21}, k_{22}, k_{23}) > 0$ and $\hat{T}_d$ are the estimated vectors of the disturbance of the wave force and towing force, respectively. Then, Formula (11) can be obtained:

$$\dot{s}_2 = -K_2 s_2 - R^T(\psi_s)s_1 - M^{-1}(\hat{T}_d - T_d) \tag{11}$$

During actual operation, the interference of the DP vessel T_d is bounded. To estimate the bounded interference T_d, the adaptive law is designed:

$$\dot{\hat{T}}_d = Ms_2 \tag{12}$$

Based on contraction theory, the virtual dynamic connection form of Formulas (8), (11) and (12) can be expressed as

$$\frac{d}{dt}\begin{pmatrix} \delta s_1 \\ \delta s_2 \\ \delta \hat{T}_d \end{pmatrix} = \begin{pmatrix} J_{11} & J_{12} \\ J_{21} & 0 \end{pmatrix}\begin{pmatrix} \delta s_1 \\ \delta s_2 \\ \delta \hat{T}_d \end{pmatrix} \tag{13}$$

where the Jacobian matrix $J_{11} = \begin{pmatrix} -K_1 & R(\psi) \\ -R^T(\psi) & -K_2 \end{pmatrix}$, and $J_{21} = -J_{12}^T = [\mathbf{0} \quad -M^{-1}]$.

From the above formula, the matrix J_{11} is uniformly negative definite, and M^{-1} is smooth, so the error variables s_1 and s_2 globally and asymptotically converge to $\mathbf{0}$, and the estimated value of uncertain environmental disturbance is bounded.

Finally, from Formula (A5) in the Appendix A and Formulas (7) and (10), the virtual dynamic formula of the closed-loop control system can be obtained as

$$\frac{d}{dt}\begin{pmatrix} \delta \eta_s \\ \delta v_s \end{pmatrix} = \bar{J}\begin{pmatrix} \delta \eta_s \\ \delta v_s \end{pmatrix} \tag{14}$$

In the formula,

$$\bar{J} = \begin{bmatrix} \bar{J}_{11} & \bar{J}_{12} \\ \bar{J}_{21} & \bar{J}_{22} \end{bmatrix},$$

$$\bar{J}_{11} = \frac{\partial [R(\psi_s)v_s]}{\partial \eta_s^T},$$

$$\bar{J}_{12} = R(\psi_s),$$

$$\bar{J}_{21} = K_2 \frac{\partial \alpha}{\partial \eta_s^T} + \frac{\partial \dot{\alpha}}{\partial \eta_s^T} - \frac{\partial [R^T(\psi_s)(\eta_s - \eta_r)]}{\partial \eta_s^T}, \text{ and } \bar{J}_{22} = -K_2 - \frac{\partial \dot{\alpha}}{\partial \eta_s^T}.$$

If it can be proven that the Jacobian matrix $\bar{J} < 0$, then all solution trajectories of the closed-loop system exponentially converge to the desired trajectory, and the whole system is contracting.

According to the defined state error variable, the coordinate transformation can be obtained as

$$\begin{bmatrix} \delta s_1 \\ \delta s_2 \end{bmatrix} = \Theta \begin{bmatrix} \delta \eta_s \\ \delta v_s \end{bmatrix} \tag{15}$$

In the formula, δs_1 and $\delta \eta_s$ are the virtual displacements, δs_2 and δv_s are the virtual velocities, and the invertible matrix Θ is

$$\Theta = \begin{pmatrix} I_{3\times 3} & 0 \\ -\partial \alpha / \partial \eta_s^T & I_{3\times 3} \end{pmatrix} \tag{16}$$

The transformation matrix $P = \Theta^T \Theta$ is

$$P = \begin{bmatrix} I_{3\times 3} + [\frac{\partial \alpha}{\partial \eta_s}]^T \frac{\partial \alpha}{\partial \eta_s} & -[\frac{\partial \alpha}{\partial \eta_s}]^T \\ -\frac{\partial \alpha}{\partial \eta_s} & I_{3\times 3} \end{bmatrix} \tag{17}$$

Formula (7) indicates that by selecting a reasonable value for the controller parameter K_1, the transformation matrix P is guaranteed to be positive definite.

From Formulas (14) and (17), we obtain

$$\frac{d}{dt}\begin{pmatrix} \delta s_1 \\ \delta s_2 \end{pmatrix} = \tilde{J}\begin{pmatrix} \delta s_1 \\ \delta s_2 \end{pmatrix} \tag{18}$$

where the Jacobian matrix $\tilde{J} = (\dot{\Theta} + \Theta\bar{J})\Theta^{-1} = \begin{bmatrix} \tilde{J}_{11} & \tilde{J}_{12} \\ \tilde{J}_{21} & \tilde{J}_{22} \end{bmatrix}$.

Formula (13) indicates that the error system composed of Formulas (8) and (11) is contracting; that is, if the Jacobian matrix $\tilde{J}$ is uniformly negative definite, then all the solutions of the original closed-loop system exponentially converge to a certain trajectory, and the system is contracting, thus ensuring that the position and heading angle of the follower vessel globally exponentially converge to and remain on those of the virtual vessel η_s.

2.4. Thrust Allocation

The thrust allocation algorithm should be optimized for lower power consumption and wear and tear of the thruster devices. To ensure safe operation, the ability of thrust allocation to always generalize an optimal solution in time is very important, and can be taken as an optimization problem with cost functions and constraints.

The cost functions can be formulated as follows [15].

$$J(\alpha, f, s) = \sum_{i=1}^{m} (f^{\mathrm{T}} P f + (f - f_0)^{\mathrm{T}} M (f - f_0) + (\alpha - \alpha_0)^{\mathrm{T}} Q(\alpha - \alpha_0) + s^{\mathrm{T}} W s) \tag{19}$$

where P is a diagonal matrix, $f \in R^n$ is a vector of thruster forces, $\alpha \in R^n$ is a vector of azimuth angles at sample time k, f_0 and α_0 are the force and angle at time $k - 1$, respectively, and $s \in R^3$ is a vector of slack variables introduced to ensure the existence of a solution. $M \in R^{n \times n}$ and $Q \in R^{n \times n}$ are weighting matrices, and the matrix $W \in R^{3 \times 3}$ is significantly larger to force the optimal solution $s \approx 0$.

The equality constraint is most important to ensure that the controlled forces and moments are produced:

$$s + B(\alpha)f = T \tag{20}$$

where B is the thruster configuration matrix, $T = [F_x, F_y, M_z]^{\mathrm{T}}$, F_x is the surge force, F_y is the sway force, and M_z is the moment in the yaw direction.

Assuming that there are n_f tunnel thrusters and n_p azimuth thrusters and that each thruster k is located at (x_i, y_i), the thruster configuration matrix can be expressed as follows:

$$B(\alpha_{i=1:n_f}) = \begin{pmatrix} 0 \\ 1 \\ -y_i \cdot \cos \alpha_i + x_i \cdot \sin \alpha_i \end{pmatrix}, B(\alpha_{i=n_f+1:n_f+n_p}) = \begin{pmatrix} \cos \alpha_i \\ \sin \alpha_i \\ -y_i \cdot \cos \alpha_i + x_i \cdot \sin \alpha_i \end{pmatrix} \tag{21}$$

The physical constraints on the power limitation, saturation of the RPM input, and turning rate of the azimuth angle can be expressed as follows:

$$\begin{aligned}
f_{\min} - f_0 &\leq \Delta f \leq f_{\max} - f_0 \\
\Delta f_{\min} &\leq \Delta f \leq \Delta f_{\max} \\
\alpha_{\min} - \alpha_0 &\leq \Delta \alpha \leq \alpha_{\max} - \alpha_0 \\
\Delta \alpha_{\min} &\leq \Delta \alpha \leq \Delta \alpha_{\max} \\
s_{x\min} &\leq s_x \leq s_{x\max} \\
s_{y\min} &\leq s_y \leq s_{y\max} \\
s_{y\min} &\leq s_z \leq s_{z\max}
\end{aligned} \tag{22}$$

where $f_{\max} \in R^n$ is a vector of maximum thrust, $f_{\min} \in R^n$ is a vector of minimum thrust, and the maximum $\Delta f_{\max} \in R^n$ and minimum $\Delta f_{\min} \in R^n$ are the rates of change of thrust. The maximum $\Delta \alpha_{\max} \in R^n$ and minimum $\Delta \alpha_{\min} \in R^n$ are the rates of change in the thrust direction; s_i (i = x, y and z) denotes the slack variable.

In this paper, the above formulated optimization problem can be solved by the FA, which was proposed by Yang for solving optimization problems [40].

In the FA, the brightness and attractiveness are two main parameters; brightness guides the direction of fireflies and attractiveness indicates the forward momentum of fireflies with low brightness. Since the brightness of fireflies gradually weakens with distance, absolute and relative brightness are used for characterization.

For any two fireflies i and j in the group, the I_i represents the brightness of *i*th firefly. The relative brightness I_{ij} can be expressed as follows:

$$I_{ij}(r_{ij}) = I_i e^{-\gamma r_{ij}^2}$$

$$r_{ij} - \|\mathbf{x}_i - \mathbf{x}_j\| = \sqrt{\sum_{k=1}^{D} (x_{i,k} - r_{j,k})^2} \tag{23}$$

where γ is light absorption coefficient, r_{ij} is the distance between firefly i and firefly j, D is spatial location dimension, $x_{i,k}$ is the kth component of the spatial location x_i of ith firefly.

The firefly's attractiveness is defined as follows:

$$\beta_{ij}(r_{ij}) = \beta_0 e^{-\gamma r_{ij}^2} \tag{24}$$

where β_0 is attractiveness at $r_{ij} = 0$.

The firefly i will be attracted to another more brighter firefly j, and the spatial location x_i is defined as:

$$x_i^{N_{iteration}+1}(t) = x_i^{N_{iteration}}(t) + \beta_{ij} \times (x_j^{N_{iteration}}(t) - x_i^{N_{iteration}}(t)) + \alpha \times (rand - 1/2) \tag{25}$$

where t is the sampling time, $N_{iteration}$ is the number of generations, and x_i and x_j are the firefly i and j positions in the D-dimensional space, respectively, α is the step size factor, and rand is a uniformly distributed random number.

To reduce the complexity of the FA and avoid the generation of local optimal solutions, the Nth generation of the FA with strong brightness at spatial position x is selected to attract the movement of other fireflies. While ensuring that the firefly search process is affected by the spatial distance, the attraction model is improved to make the search direction more reasonable. The improved attraction model is as follows:

$$\beta_{ij}(r_{ij}) = \theta \cdot \beta_0 e^{-\gamma r_{ij}^2}$$
$$\theta = 1 - \frac{N_{iteration}}{N_{max}} \tag{26}$$

where N_{max} is the maximum number of generations.

In the position update formula of the standard FA, the constant step factor a will not be conducive to the convergence of the optimal solution of the population. Therefore, to improve the convergence of the FA, an adaptive step size factor that varies with the number of iterations is adopted, and its expression is as follows:

$$\alpha_{N_{iteration}} = \alpha_0 \left(1 - e^{-(1 - \frac{N_{iteration}}{N_{max}})}\right) \tag{27}$$

The pseudocode of the thrust allocation-based SAF algorithm is shown in Algorithm 1.

Algorithm 1 SAF algorithm

Inputs:
Maximum number of iterations N_{max};
The population number N_{pop};
The cost functions $f(x) = \min\{J(\alpha, f, s)\}, x = [f_1, f_2, \ldots, f_n, \alpha_1, \alpha_2, \ldots \alpha_{n,}]^T$
Initial boundary limits L_b and U_b; the dimension D, which is twice the number of thrusters.
Initial $x = L_b + (U_b - L_b) \times \text{rand}(D,1)$
Initialize parameters $\alpha_0, \beta_0, \gamma$
Initialize firefly population x_i (i = 1, 2, ..., Npop), number of fireflies.
Calculate the $f(x)$, optimal solution x_{Gbest} is determined for initialize firefly population
While ($N_{iteration} << N_{max}$)
    for i = 1: N_{pop}

        calculate the distance $r_{i,Gbest} = \|x_i - x_{Gbest}\| = \sqrt{\sum_{k=1}^{D}(x_{i,k} - x_{Gbest,k})^2}$

        update the position x_i by Equations (25)–(27)
        boundary treatment $x_i \in [x_{i,L_b}, x_{i,U_b}] \, x_{i,L_b}, x_{i,U_b}$ are the upper and lower bounds
    end

        obtain a new $x_i^{N_{iteration}+1}(t)$
end while
Outputs: the optimal solution $x_{Gbest}(t)$

---

3. Test Overview

3.1. Test Objects

The test object is an 8000-ton CTV. By comprehensively considering the tank testing and wave-making capacity, the installation and arrangement requirements of the model thruster, and the adjustment of the center of gravity and inertia of the test model, the scale ratio of the model is taken as 1:22. The main parameters of the CTV, FPSO unit, and tanker are shown in Table 1. Restricted by the main scale of the basin, the actual scale ratio of the FPSO unit model is 1:68.

Table 1. Model particulars of the CTV, FPSO unit, and tanker.

Parameter	Symbol	Unit	FPSO Model	Tanker	CTV Full-Scale	CTV Model
Length Overall	L_{OA}	m	4.1590	6.8158	89.00	4.0455
Breadth	B	m	0.6820	1.1461	20.00	0.9091
Depth	D	m	0.3320	0.5394	10.50	0.4773
Displacement	$\triangle$	t	0.2270	1.3511	8030	0.7357
Mean draft	T	m	0.1310	0.2994	6.50	0.2955

The CTV is equipped with four azimuth thrusters and one tunnel thruster at the bottom of the bow. The thrusters are simulated according to geometric similarity and power similarity. The main parameters are shown in Table 2. The thruster numbers and layout are shown in Figure 3.

Table 2. Main parameters of the thrusters (scale ratio is 1:22).

Parameter	Unit	T1&T2 Full-Scale	T1&T2 Model	T3&T4 Full-Scale	T3&T4 Model	T5 Full-Scale	T5 Model
Thrust	kN	745.56	0.0700	345.312	0.0324	107.91	0.0101
Diameter	m	3.900	0.1636	2.600	0.1182	1.700	0.0773
Thrust from 0 to T_{max}	kN/s	33.0	0.1454	23.0	0.0101	10.80	0.0048
Azimuth angle from 0 deg to 180 deg	deg/s	12	56.28	12	56.28	/	/

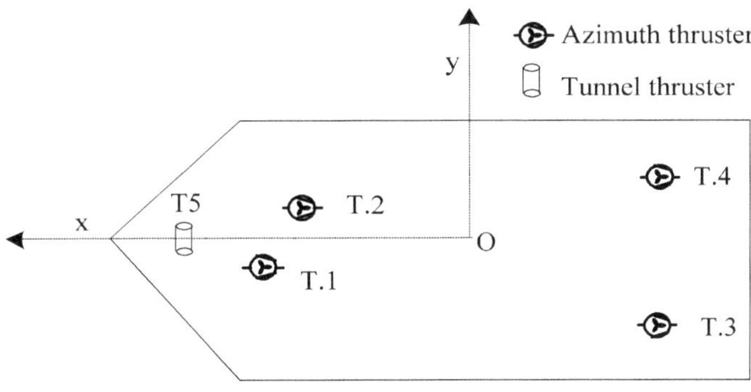

Figure 3. Schematic diagram of the thruster number and layout.

Before the DP model experiment, thruster-hull and thruster-thruster interaction tests were carried out. The thrust forbidden zones are set to $[-60°, 30°]$, $[110°, 150°]$, $[200°, 290°]$ and $[70°, 170°]$, corresponding to thrusters 1, 2, 3 and 4, respectively.

3.2. Simulation of the Flexible Link System with Multiple Floating Bodies

3.2.1. Simulation of the Mooring System of the FPSO Unit

Because the scale ratio of the FPSO unit is different from those of the tanker and the CTV, the stiffness of the mooring system of the FPSO unit cannot be determined according to the scale ratio. Therefore, this test does not simulate the mooring system of the FPSO unit and only simulates the statistical characteristics of the motion of the FPSO unit.

3.2.2. Simulation of the Mooring System of the Tanker

The tanker is connected to the CTV through two mooring cables, which are each composed of a 150 m-long nylon rope and a 9 m-long iron chain. The cables are simulated according to diameter similarity, weight similarity, and stiffness similarity.

3.3. Test Environment Conditions and Contents

The JONSWAP spectrum is adopted in the model test to simulate an irregular wave environment. The JONSWAP spectral density is defined as follows:

$$S_\zeta(\omega) = \alpha H_{1/3}^2 \omega_P^4 \omega^{-5} \exp\left\{-1.25(\omega/\omega_P)^{-4}\right\} \times \gamma^{\exp\left\{-\left(\frac{\omega/\omega_p-1}{\sqrt{2}\sigma}\right)^2\right\}} \tag{28}$$

where $H_{1/3}$ is the significant wave height, ω_p is the spectrum peak frequency, ω is the circular wave frequency, and γ is the peak enhancement factor (for this test, $\gamma = 2.2$). When $\omega < \omega_p$, σ is 0.07, and when $\omega > \omega_p$, σ is 0.09; $\alpha = \frac{0.0624}{0.230+0.0336\gamma-0.185/(1.9+\gamma)}$.

The headings against waves are defined as follows: The wave coming from the bow is defined as 180° (head sea), and that coming from the starboard is defined as 90° (beam sea). The significant wave height and spectral peak period combinations are listed in Table 3.

Table 3. Test contents for synchronous following control of the CTV in the towing state.

Test No.	Wave Heading (°)	$H_{1/3}$		T_p		γ
		Full-Scale (m)	Model (mm)	Full-Scale (s)	Model (s)	
F01	180					
F02	135	2.00	90.91	9.21	1.964	2

Irregular waves are generated by an L-type shaking plate wave generator with approximately 200 oscillating flaps.

Before the experiments, the JONSWAP spectrum of irregular waves was calibrated, and the calibrated results (model scale) are shown in Figure 4. The significant wave height ($H_{1/3}$) and spectral peak period (T_p) are within 5% of the theoretical values under the two headings against waves.

3.4. Test Setup

The scaled experiments are carried out in the seakeeping basin of China Ship Scientific Research Center (CSSRC) with dimensions of 69 m × 46 m × 4 m (length × width × water depth). The initial state of the test is set as the longitudinal tandem mode of the FPSO unit, the CTV, and the tanker. According to the CTV operation requirements, the distance between the stern end of the FPSO unit and the bow end of the CTV should be 4.545 m (corresponding to 100 m at full scale), the distance between the stern end of the CTV and the bow end of the tanker should be 6.818 m (corresponding to at full scale), and the heading angles of the three vessels should all be 0°. A schematic diagram and photographs of the tank layout are shown in Figures 5 and 6, respectively.

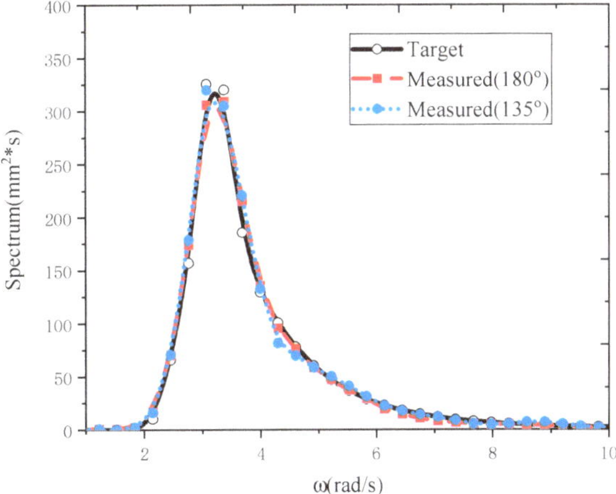

Figure 4. Calibrated wave spectrum.

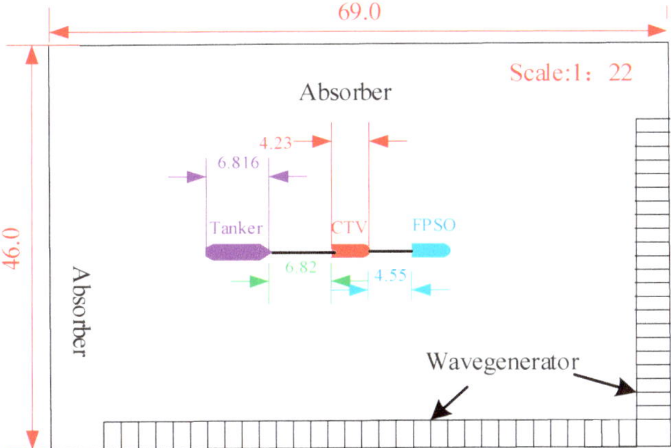

Figure 5. Schematic diagram of the FPSO unit, CTV, and tanker layout.

Figure 6. Photographs of the FPSO unit, CTV, and tanker in the experiment.

4. Experimental Results and Analysis

In the experimental verification, all the results are transformed to full-scale according to Froude number similarity. The added mass and added moment of inertia in the inertia matrix M and the linear damping matrix D_L are obtained by numerical calculation:

$$M = \begin{vmatrix} 8.68907 \times 10^5 & 0 & 0 \\ 0 & 1.44489 \times 10^7 & 0 \\ 0 & 0 & 5.74462 \times 10^9 \end{vmatrix}, \; D_L = \begin{vmatrix} 5.19397 \times 10^3 & 0 & 0 \\ 0 & 3.26129 \times 10^4 & -8.94123 \times 10^3 \\ 0 & 5.98832 \times 10^5 & 1.38939 \times 10^7 \end{vmatrix}$$

4.1. Simulation of the Motion Characteristics of the FPSO Unit

To ensure the successful simulation of the statistical characteristics of the FPSO unit motion, a multifloating body model mooring test was carried out before this test (the scale ratios of the FPSO unit, the CTV, and the tanker were all 1:80). The CTV counteracts the influence of second-order wave forces through horizontal mooring during the model experiment. The statistical results of the sway, surge, and yaw motions of the FPSO unit in the multifloating body model mooring test and the synchronous following test with the DP system are shown in Table 4. As can be seen from Table 4, the motion results of the FPSO unit in this test are slightly larger; Figure 7 shows the time history curves of FPSO for acceleration in the x and y directions and the angular velocity of the heading. Figure 7a,c show the results of the synchronous following control test in the wave headings $0°$ and $135°$, respectively. Figure 7b,d show the multifloating body model mooring test in the wave headings $0°$ and $135°$, respectively. As can be seen from Figure 7, the acceleration and angular velocity of FPSO are obviously larger during the synchronous following test, so the results can more effectively verify the ability of the designed synchronous following control strategy to maintain a safe operating distance between the FPSO unit and the CTV.

Table 4. Statistical motion results of the FPSO unit under the horizontal mooring system.

Test No.	Wave Heading (°)	Item	FPSO (CTV with Mooring)			FPSO (CTV with DP System)		
			Surge (m)	Sway (m)	Yaw (°)	Surge (m)	Sway (m)	Yaw (°)
F01	180	Max	0.815	0.342	0.05	2.442	0.528	1.863
		Min	−1.145	−0.798	−0.206	−1.254	−0.418	−1.147
		Mean	−0.008	0.002	−0.069	0.523	0.065	0.361
		SD	0.543	0.072	0.035	0.710	0.155	0.386
F02	135	Max	2.529	−2.352	0.783	−2.064	−2.184	2.434
		Min	−4.983	−4.314	1.459	−6.533	−5.982	−2.451
		Mean	−1.481	−3.238	0.035	−4.268	−4.025	−0.133
		SD	1.129	0.303	0.268	0.780	0.623	0.700

4.2. Analysis of the Capability of the CTV to Synchronously follow the FPSO Unit

Table 5 shows the statistical values of the sway, surge, and yaw motions of the CTV under different wave heading conditions. All the results in the table are converted into the full-scale vessel state, and the time–history curves of the CTV motion trajectories and heading errors corresponding to different wave heading conditions are shown in Figures 8 and 9.

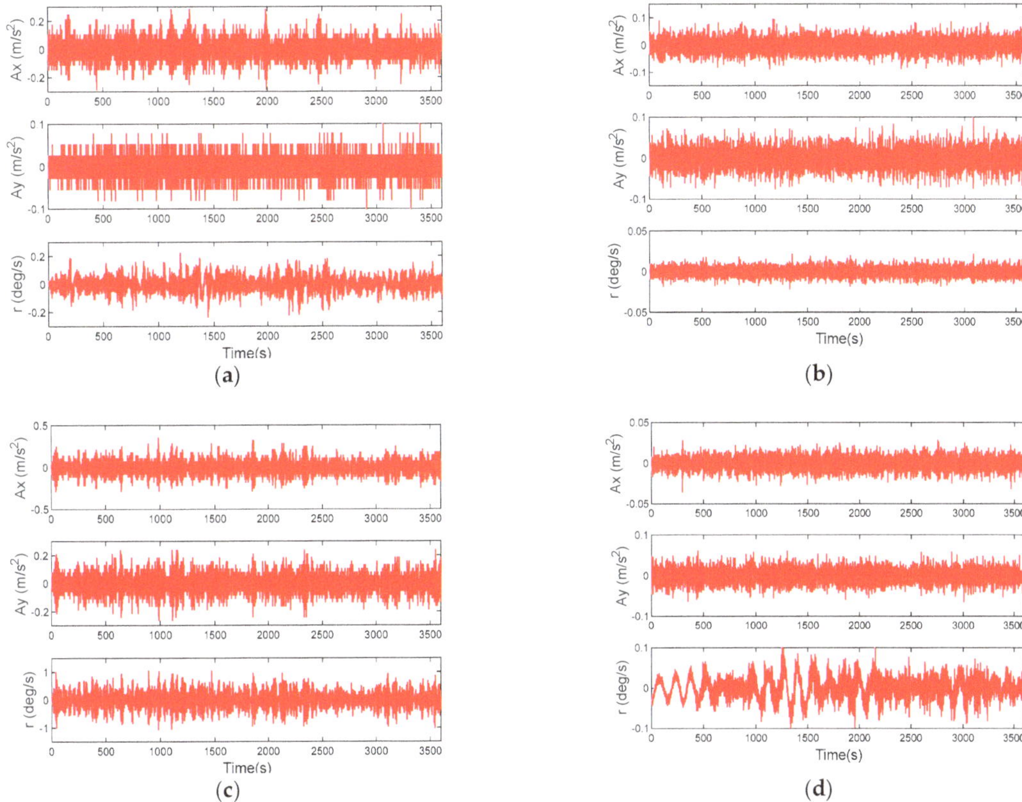

Figure 7. Time–history curves of the FPSO acceleration: (**a**) DP test in wave heading 180°; (**b**) mooring test in wave heading is 180°; (**c**) DP test in wave heading 135°; (**d**) mooring test in wave heading 135°.

Table 5. Statistical results of the motion of the CTV synchronously following the FPSO unit.

Test No.	Wave Heading	Motion	Unit	Mean	Min	Max	Error
F01	180°	Yaw	deg	−0.55	−2.63	2.15	1.65
		Surge	m	−1.52	−3.23	0.29	/
		Sway	m	0.89	−3.28	4.86	/
F02	135°	Yaw	deg	1.28	−5.89	1.04	2.45
		Surge	m	−3.26	−14.70	7.50	/
		Sway	m	−0.72	−13.44	8.68	/

The results in Table 5 show that when the wave heading is 180°, the control effects for surge and sway of the CTV are equivalent, and the average error of the heading control is only −0.55°. At 1400 s and 3400 s in the test (as shown in Figure 9a), the lateral tracking error of the CTV is slightly larger since the CTV is towing the tanker, which leads to a greater deviation in the heading. At this time, the maximum heading error of the CTV following the FPSO unit is 1.65°. When the wave heading is 135°, the capability of the CTV to synchronously follow the position and heading of the FPSO unit is slightly reduced, but the synchronous following error of the heading is still controlled within 2.45°, which effectively verifies that the control strategy in this paper can achieve a synchronous following response of the heading motion and confirms the robustness of the control strategy to different wave heading conditions.

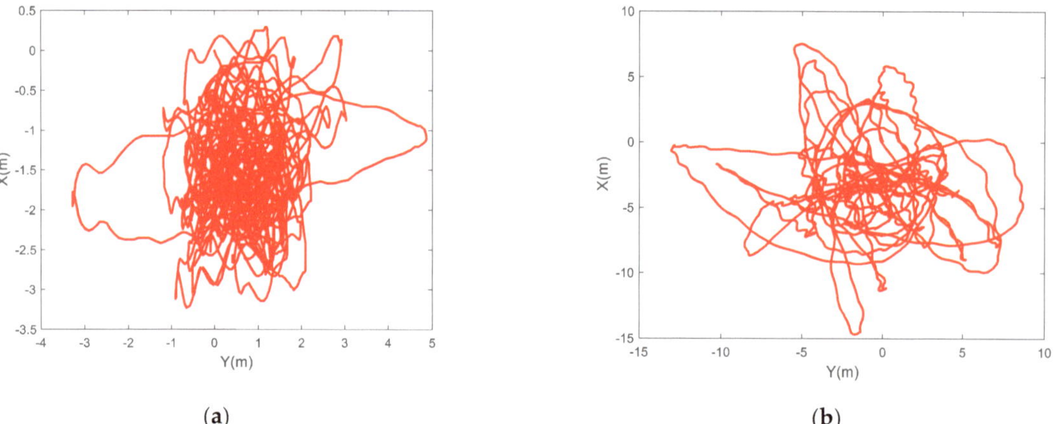

Figure 8. Time–history curves of the CTV trajectory: (**a**) wave heading is 180°; (**b**) wave heading is 135°.

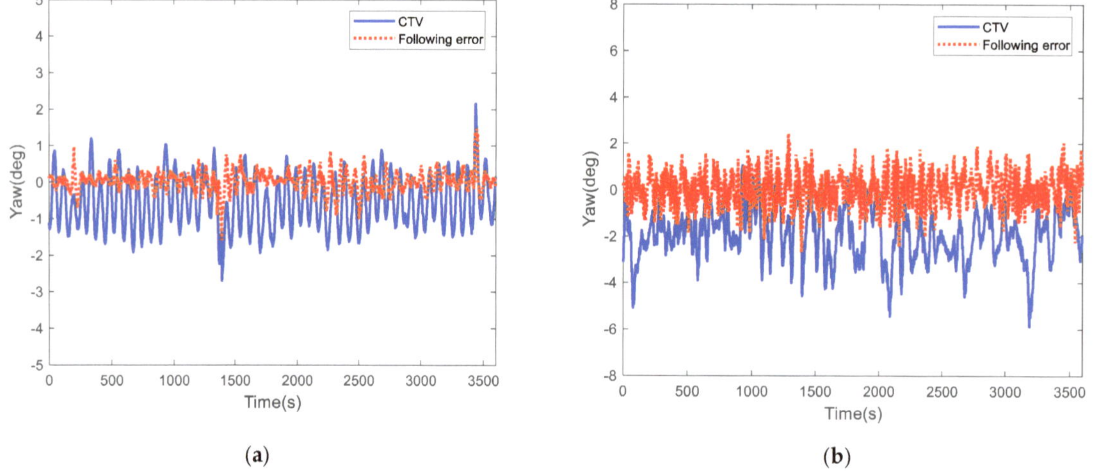

Figure 9. Time–history curves of the heading tracking error of the CTV: (**a**) wave heading is 180°; (**b**) wave heading is 135°.

Figure 10 shows the time–history curves of the distance between the CTV and the FPSO unit under two wave heading conditions, 180° and 135°. As observed in the figure, when the wave heading is 180°, the CTV and the FPSO unit can effectively maintain a safe distance of 100 m with a maximum deviation of 3.78 m and an average deviation of only 0.99 m (95% CI = [0.96, 1.03]); when the wave heading is 135°, due to the reduced heading control accuracy, the distance between the CTV and the FPSO unit is slightly larger, with a maximum deviation of 14.38 m and an average deviation of 2.53 m (95% CI = [2.42, 2.64]), which effectively verifies that the control strategy proposed in this paper can achieve safe and cooperative operation between the CTV and the FPSO unit.

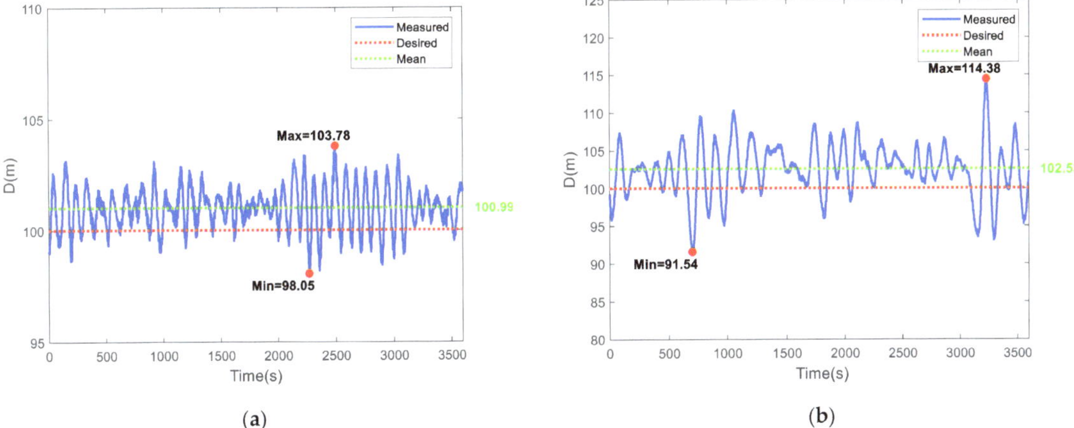

Figure 10. Time–history curves of the distance between the CTV and the FPSO unit: (**a**) wave heading is 180°; (**b**) wave heading is 135°.

To verify the advantages of the SAF algorithm, the SQP algorithm and PSO algorithm are used to compare the experimental results. When using the SQP algorithm, to reduce the truncation error of second-order and higher-order terms produced by Taylor expansions, the rotation speed of the azimuth thruster is selected as 1 deg/s (corresponding to full scale), whereas the rotation speed for the corresponding PSO algorithm is similar to that for the SAF algorithm, which is 12 deg/s (corresponding to full scale).

In order to demonstrate how the firefly algorithm works, consider the maximum number of iterations N_{max} = 150, the population number N_{pop} = 60, and initialization parameters $\alpha_0 = 1$, $\beta_0 = 1$, $\gamma = 0.1$. Figure 11 shows the comparison results of the longitudinal, transverse, and heading thrust allocation commands and control commands for the three degrees of freedom in the horizontal plane under the 180° wave heading. Due to the principle of heading control priority (slack variable matrix W = diag $[10^3, 10^3, 10^4]$), the thrust allocation priority satisfies the heading moment control command. The optimal solution of the yaw moment is almost identical to the control command, and due to the small environmental force under the wave heading of 180°, the lateral and longitudinal control forces are also in good agreement with the control command. Figure 12 shows the comparison results of thrust allocation when the wave heading is 135°. Under this condition, the wave force acting on the CTV in the lateral and yaw directions significantly increases, and its sway, surge, and yaw motions also significantly increase, resulting in a larger error between CTV thrust allocation commands and control commands. However, the optimal solution and control commands for the yaw moment are still better than those in the lateral and longitudinal directions.

To test the effectiveness of the proposed thrust allocation algorithm, R-squared (R^2) is introduced to evaluate the accuracy of the actual thrusts and control commands.

$$R^2 = 1 - \frac{\sum_{i=1}^{N} \left(x\_command(i) - x\_allocation(i) \right)^2}{\sum_{i=1}^{N} \left(x\_command(i) - \hat{x}(i) \right)^2} \tag{29}$$

where $x\_command(i)$ and $x\_allocation(i)$ denote the control command and actual allocation result, respectively, and $\hat{x}(i)$ represents the average of the control commands. When the allocation values perfectly match the control commands, R^2 is 1.

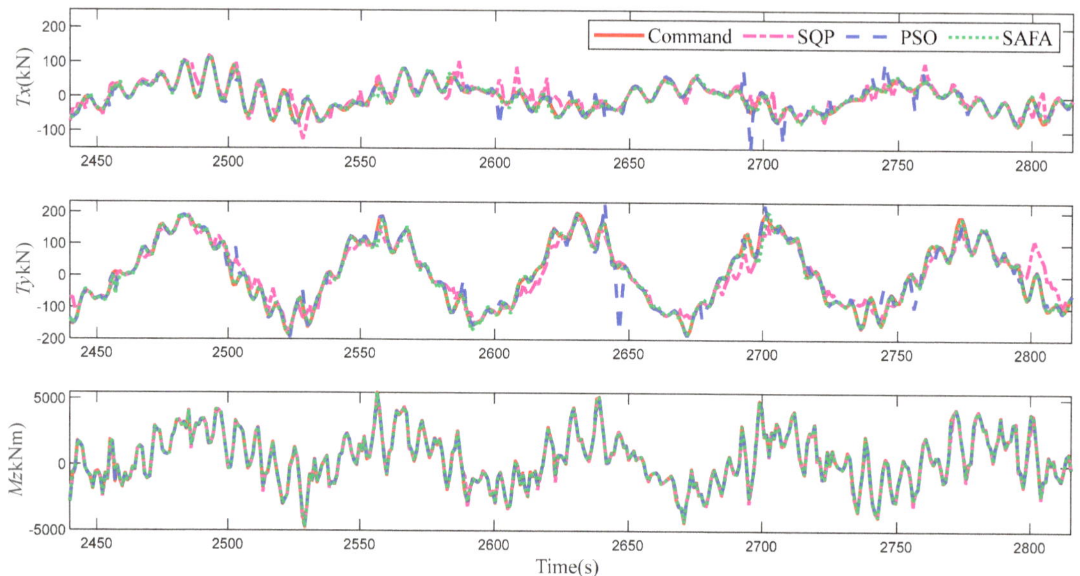

Figure 11. Time–history curves of the control command and thrust allocation under a 180° wave heading.

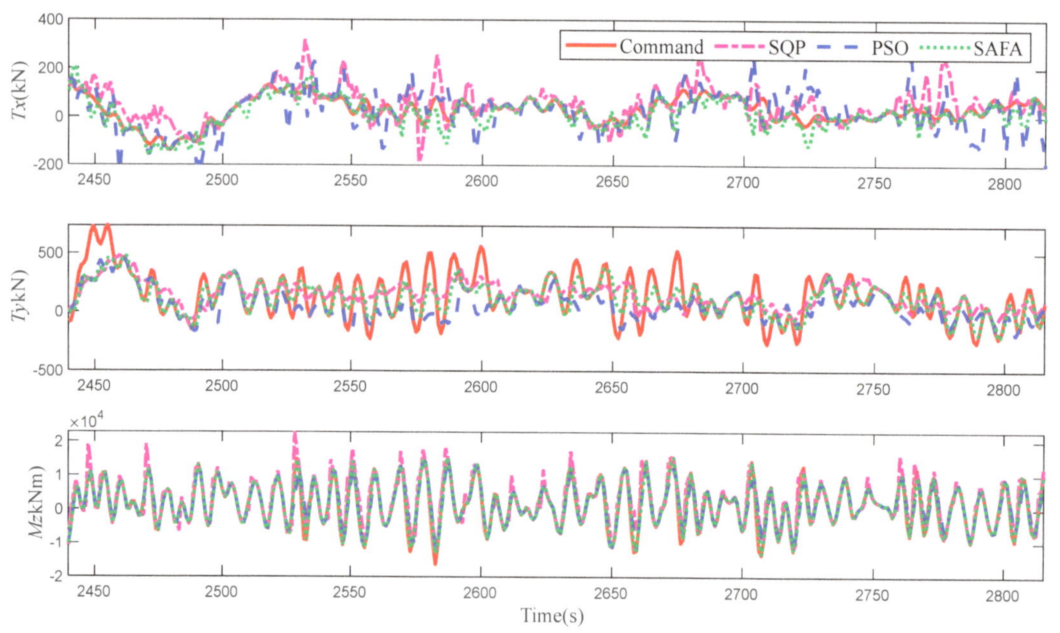

Figure 12. Time–history curves of the control command and thrust allocation under a 135° wave heading.

Figures 13 and 14 show the accuracy and total power consumption comparisons between the SAF algorithm proposed in this paper, the SQP algorithm and the PSO algorithm. According to Figure 13a, when the wave heading is 180°, the actual optimal torque command issued by the propulsion system is almost identical to the control command for the three algorithms, with R^2 values of 0.9727 (SQP), 0.9993 (PSO), and 0.9996 (SAF). The SAF algorithm outperforms the SQP and PSO algorithms in longitudinal and lateral forces. However, compared to the SAF algorithm and PSO algorithm, the SQP algorithm is limited by a rotation angle speed of only 1 deg/s, resulting in a large error in lateral and longitudinal forces. From Figure 13b, the proposed SAF algorithm has higher accuracy under a 180° wave heading, while the power consumption is lower than that for the PSO algorithm. The reason for the lower power consumption of the SQP algorithm is that the smaller force and moment errors result in the first term of Equation (19) being smaller. When the wave heading is 135°, from Figure 8b, Figure 9b, and Figure 12, the wave force and coupled motion significantly increase. Due to the principle of prioritizing heading control, the three allocation algorithms can still meet the control requirements for the yaw torque, and the SQP algorithm even has a slightly higher accuracy in yaw control allocation than the PSO and SAF algorithms, with R^2 values of 0.9876 (SQP), 0.9087 (PSO), and 0.9694 (SAF). However, the accuracy of the SQP algorithm is significantly decreased for lateral and longitudinal forces because the SQP algorithm can only achieve the optimal solution by quickly adjusting the thrust, and the power significantly increases compared to the PSO and SAF algorithms due to the strict limitation of the lower rotation angle speed.

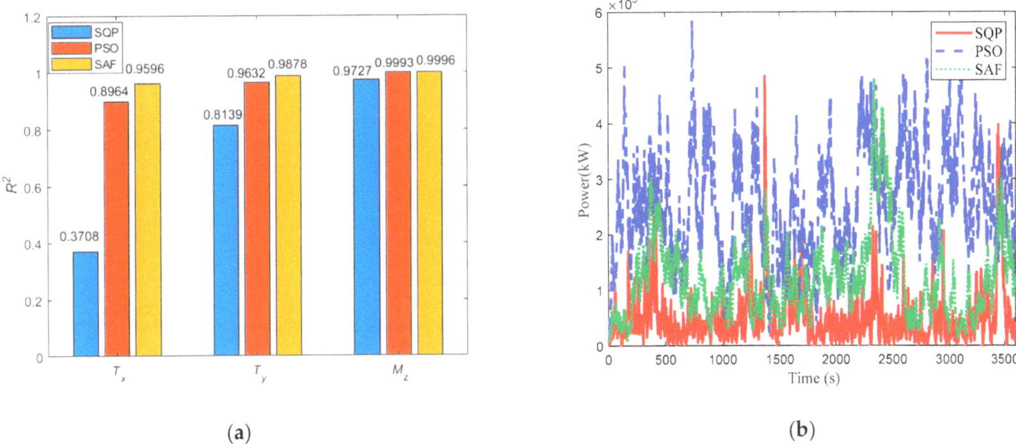

(**a**)   (**b**)

Figure 13. R^2 and total power comparison of different thrust allocation algorithms under a 180° wave heading: (**a**) R^2; (**b**) total power comparison.

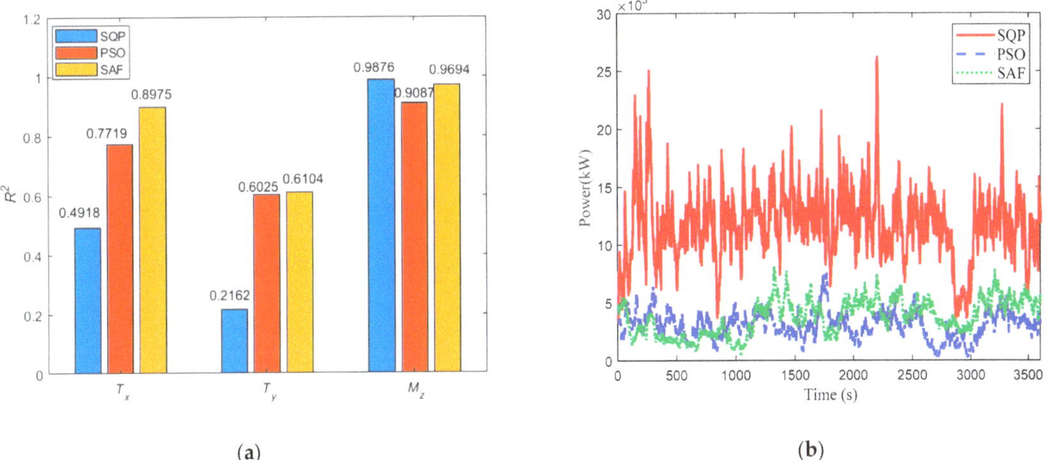

(**a**)   (**b**)

Figure 14. R^2 and total power comparison of different thrust allocation algorithms under a 135° wave heading: (**a**) R^2; (**b**) total power comparison.

From Figures 13b and 14b, we can see that due to limitations of the SQP algorithm for azimuth thrusters in thrust allocation and the different rotation angle speeds, comparing the power consumption of the SQP, PSO, and SAF algorithms is no longer meaningful. Table 6 shows the total power consumption of the CTV. From Table 6, when the wave heading is 180°, compared with the PSO algorithm, the SAF algorithm has lower maximum and average power consumption. When the wave direction is 135°, due to the intensification of coupled motion, to satisfy the yaw moment, the higher thrust allocation accuracy leads to smaller relaxation variable errors. Although the SAF algorithm has higher accuracy, this causes an increase in power compared to the PSO algorithm, with the maximum and average power increased by 8.53% and 7.80%, respectively. Actually, the first and fourth terms in Equation (19) are inherently contradictory, and the allocation accuracy and power consumption trade-off can be solved by adjusting the matrix weight.

Table 6. Statistical results of power consumption for all thrusters of the CTV.

Wave Heading	Control Allocation	Power (kW)			
		Max	Mean	SD	95% Confidence Interval on Mean
180°	PSO	5850	2557	989	[2526, 2588]
	SAF	4802	1193	734	[1170, 1216]
135°	PSO	7541	3526	1155	[3490, 3562]
	SAF	8184	3801	1489	[3755, 3847]

Tables 7 and 8 show the statistical occurrence results for the power utilization of all thrusters under the two wave heading conditions. The results in Tables 7 and 8 show that when the wave heading is 180°, except for thrusters 1 and 2, the power utilization rate of each thruster is basically within 20%; when the wave heading is 135°, due to the increases in the control force, moment, and allocation accuracy, the power utilization (>40%) of all thrusters significantly increases with the SAF algorithm compared to the PSO algorithm.

Table 7. Occurrence of power utilization for each thruster of the CTV under a wave heading of 180°.

Power Utilization	Thrust Allocation	Occurrence (%)				
		No. 1	No. 2	No. 3	No. 4	No. 5
0~20%	PSO	4.72	14.81	90.04	99.95	96.61
	SAF	28.83	52.97	92.18	100.00	89.44
20~40%	PSO	17.41	20.72	8.84	0.05	1.75
	SAF	40.80	33.45	6.67	0.00	5.81
40~60%	PSO	28.55	22.31	1.12	0.00	0.70
	SAF	21.19	9.15	1.15	0.00	2.66
60~80%	PSO	25.31	24.56	0.00	0.00	0.47
	SAF	7.48	2.27	0.00	0.00	1.07
80~100%	PSO	24.01	17.596	0	0	15.98
	SAF	1.69	2.164	0	0	5.18

Table 8. Occurrence of power utilization for each thruster of the CTV under a wave heading of 135°.

Power Utilization	Thrust Allocation	Occurrence (%)				
		No. 1	No. 2	No. 3	No. 4	No. 5
0~20%	PSO	3.96	14.73	59.25	90.75	81.86
	SAF	0.76	31.23	8.79	66.27	30.50
20~40%	PSO	9.38	23.51	28.36	8.06	6.05
	SAF	3.86	17.31	27.48	10.97	16.76
40~60%	PSO	18.20	24.06	9.18	1.12	5.24
	SAF	14.57	11.57	24.50	9.83	13.84
60~80%	PSO	29.35	18.64	2.16	0.08	3.62
	SAF	36.13	19.42	22.81	8.24	19.29
80~100%	PSO	39.1	19.06	1.04	0	18.53
	SAF	44.68	20.46	16.42	4.69	20.26

4.3. Analyzing the Motion of the Tanker When Towed by the CTV

The tanker is connected to the CTV only through the mooring cables during the operation process in which the CTV synchronously follows the FPSO unit. Figures 15 and 16 show the time–history curves of the relative position and motion of the tanker and the CTV when the headings against waves are 180° and 135°, respectively. Figure 15 shows that when the wave heading is 180°, the tanker moves from the initial position along the starboard direction, and the maximum lateral and longitudinal displacements are 30.64 m and 8.16 m, respectively. Figure 17a shows the test photograph of the slow oscillating motion of the tanker about the initial longitudinal position after the lateral position of the tanker is stabilized near 25 m. Figure 16 shows that when the wave heading is 135°, the tanker moves along the larboard direction, and the maximum lateral and longitudinal displacements of the tanker are 76.65 m and 31.88 m, respectively. The tanker finally stabilizes left and posterior to the CTV under the action of waves and the towing force, as shown by the test photograph in Figure 17b.

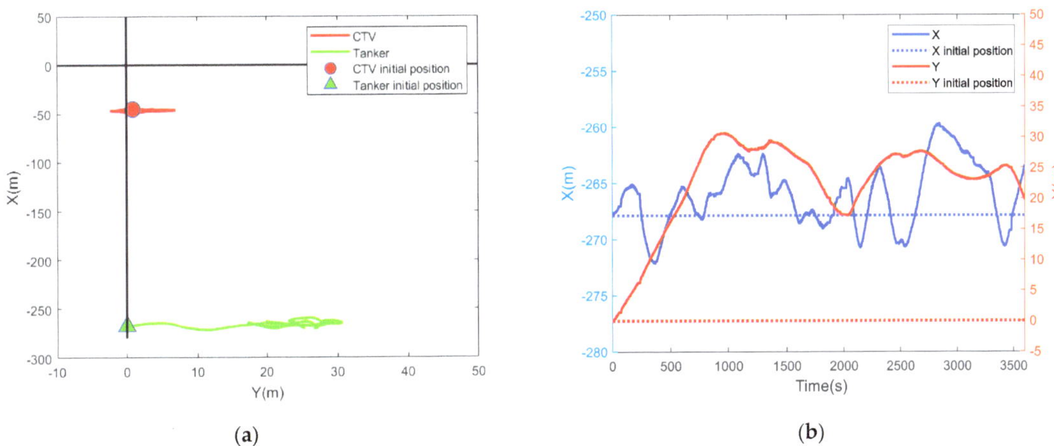

(a)

(b)

Figure 15. Relative position and trajectory of the tanker and the CTV under a wave heading of 180°: (**a**) relative position in the fixed basin coordinate system; (**b**) motion trajectory.

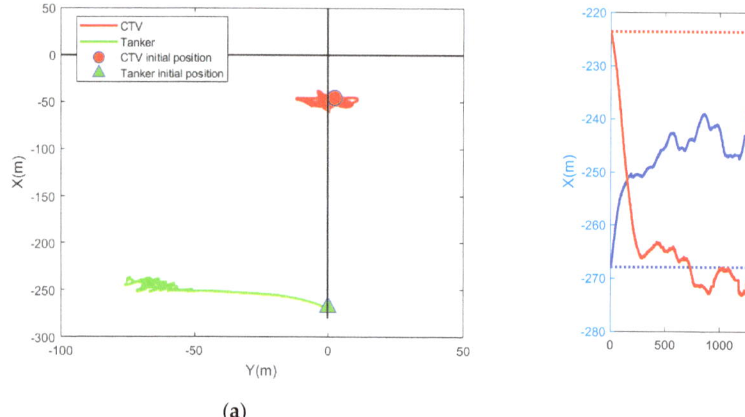

(a) **(b)**

Figure 16. Relative position and trajectory of the tanker and the CTV under a wave heading of 135°: (**a**) relative position in the fixed basin coordinate system; (**b**) motion trajectory.

(a)

(b)

Figure 17. Test photographs of the CTV and the tanker under the towing state: (**a**) wave heading is 180°; (**b**) wave heading is 135°.

5. Conclusions

In this paper, an adaptive backstepping synchronous following control strategy based on a virtual leader–follower was proposed for CTV with DP system. A proof was given of the globally exponentially convergence of the closed-loop DP control system by the contraction theorem. In order to generate forces and moments from the DP controller, the thrust allocation of multiple thrusters was considered with an optimal solution method based on the SAF algorithm.

To validate the effectiveness of the proposed control method, model experiments were conducted in irregular waves with different headings against waves. The results verified the capability of the synchronous following control strategy to maintain a safe distance and heading between the FPSO unit and the CTV and clarified the motion characteristics of the tanker under the towing state. Additionally, the results also verified that the DP adaptive backstepping control strategy had strong robustness to different headings against waves. When the wave heading was 180°, the CTV could better maintain a safe working distance from the FPSO unit compared with a wave heading of 135°.

Then, the optimization problem of the desired control commands from the controller was solved through the SAF algorithm, which fully considered the physical characteristics of the azimuth thruster and the thrust forbidden zone caused by the interference between the thrusters and the ship. The proposed thrust allocation method could fully utilize the rotation rate of the azimuth thruster and effectively solve the high-order truncation error caused in the traditional SQP algorithm through Taylor expansion. When the wave heading was 180°, the actual optimal torque command from the SAF algorithm was almost identical to the control command, with R^2 values of 0.9596 (lateral force), 0.9878 (longitudinal force), and 0.9996 (yaw moment), and the SAF algorithm reduced the maximum and average power consumption compared to the PSO algorithm. When the wave heading was 135°, the wave force and coupled motion significantly increased; however, due to the principle of prioritizing heading control, the SQP, PSO, and SAF allocation algorithms could still meet the control requirements for the yaw moment. Although the power consumption was slightly increased, at this time, the SAF allocation accuracy for the lateral and longitudinal forces significantly increased, with R^2 values of 0.8975 (lateral force), 0.6104 (longitudinal force), and 0.9694 (yaw moment).

Author Contributions: Conceptualization, C.L. and Y.Z.; methodology, C.L. and Y.Z.; software, C.L., L.Z. and Y.T.; validation, C.L., Y.Z. and L.Z.; formal analysis, C.L. and Y.Z.; investigation, C.L. and Y.Z.; resources, C.L., F.T. and Y.Z.; data curation, C.L. and L.Z; writing—original draft preparation, C.L.; writing—review and editing, C.L. and Y.Z.; supervision, M.G. and L.Z.; project administration, C.L.; funding acquisition, C.L. All authors have read and agreed to the published version of the manuscript.

Funding: This research was funded by National Key Research and Development Program of China (grant number 2022YFB3404800) and Natural Science Foundation of Jiangsu Province of China (grant number BK20220222).

Institutional Review Board Statement: Not applicable.

Informed Consent Statement: Not applicable.

Data Availability Statement: Data are contained within the article.

Acknowledgments: We would like to thank to those who helped us in the DP system debug and measurements in scaled model test.

Conflicts of Interest: The authors declare no conflict of interest.

Nomenclature

Abbreviation	Full Name
CSSRC	China Ship Scientific Research Center
DP	Dynamic positioning
CTV	Cargo transfer vessel
FA	firefly algorithm
FPSO	floating production storage and offloading
PSO	particle swarm optimization
R^2	R-squared
SQP	Sequential Quadratic Programming
SAF	self-adaptive firefly
USV	unmanned surface vehicles
Symbol	Description
D_L	linear damping matrix
$H_{1/3}$	significant wave height
I_z	the moment of inertia
M	the mass of the vessel
$N_{\dot{r}}$	added moment of inertia
N_r	hydrodynamic derivative
T_p	spectral peak period
$X_{\dot{u}}$	added masse in x direction
$Y_{\dot{v}}$	added masse in y direction
$Y_{\dot{r}}$	added masses in yaw direction
X_u	hydrodynamic derivative in x direction

Y_v	hydrodynamic derivative in y direction
N_r	hydrodynamic derivative in yaw direction
$N_{\max}$	the maximum number of generations
N_{pop}	the population number
α_0	the step size factor
β_0	attractiveness
γ	light absorption coefficient

Appendix A

Consider a nonlinear system

$$\dot{x} = f(x,t) \tag{A1}$$

where $f(x,t)$ is a continuously differentiable nonlinear function and $x \in \mathrm{R}^n$ is a state vector. Based on the concept of virtual displacement of the system trajectory, if δx denotes the minimum virtual displacement in state x, then the virtual dynamics of system (A1) can be expressed as

$$\delta\dot{x} = (\partial f / \partial x)\delta x \tag{A2}$$

Then,

$$\frac{d}{dt}(\delta x^T \delta x) = 2\delta x^T \frac{\partial f}{\partial x}\delta x \leq 2\lambda_{\max}(x,t)\delta x^T \delta x \tag{A3}$$

where the Jacobian matrix J is $J = \partial f / \partial x$ and $\lambda_{\max}(x,t)$ represents the largest eigenvalue of the matrix $(J^T + J)/2$.

If $\lambda_{\max}(x,t)$ is uniformly negative definite, then by integrating the two ends of Formula (A3), the length of any infinitesimal virtual displacement $\|\delta x\|$ is found to exponentially converge to 0. If $\delta x^T \delta x$ is expressed as the squared distance of the adjacent trajectories of the system, then all solution trajectories of system (A1) exponentially converge to a certain trajectory, independent of the initial conditions of the system.

Definition A1. *For the system $\dot{x} = f(x,t)$, if there is a region in the state space such that $J = \partial f / \partial x$ is uniformly negative definite, then the region is called a contracting region.*

Definition A2. *For the system $\dot{x} = f(x,t)$, if any two trajectories starting from different initial conditions exponentially converge to each other, then the system is said to be contracting.*

Lemma A1 ([42]). *For the system $\dot{x} = f(x,t)$, if there is a positive definite matrix $\Theta^T \Theta$ that makes the Jacobian matrix $(\dot{\Theta} + \Theta(\partial f / \partial x))\Theta^{-1}$ uniformly negative definite, then all solution trajectories of the system converge to a certain trajectory, and the system is contracting.*

The results of contraction theory can be extended to various connection systems, and the contractibility of a whole system can be studied through different connection modes of contracting subsystems. Existing connection modes include feedback connections and hierarchical connections. In this paper, the feedback connection mode is used to design the vessel DP contraction controller, so the feedback connection mode will be briefly introduced.

Consider two systems of different dimensions:

$$\begin{aligned} \dot{x}_1 &= f_1(x_1, x_2, t) \\ \dot{x}_2 &= f_2(x_1, x_2, t) \end{aligned} \tag{A4}$$

If coordinate transformation $\delta\mathbf{z} = \Theta\delta\mathbf{x}$ is used, then the following form is satisfied:

$$\frac{d}{dt}\begin{pmatrix} \delta z_1 \\ \delta z_2 \end{pmatrix} = \begin{pmatrix} F_1 & G \\ -G^T & F_2 \end{pmatrix}\begin{pmatrix} \delta z_1 \\ \delta z_2 \end{pmatrix} \tag{A5}$$

If each subsystem is contracting, then the whole system is contracting.

Lemma A2 ([43]). *If the virtual dynamics of system (A1) satisfy the form*

$$\frac{d}{dt}\begin{pmatrix} \delta x \\ \delta \psi \end{pmatrix} = \begin{pmatrix} \partial f/\partial x & G \\ -G^T & 0 \end{pmatrix}\begin{pmatrix} \delta x \\ \delta \psi \end{pmatrix} \tag{A6}$$

and if $\partial f/\partial x$ is negative definite and G is smooth, then system (A1) is asymptotically stable, and ψ is bounded.

References

1. Tan, G.; Zhuang, J.; Zou, J.; Wan, L. Coordination control for multiple unmanned surface vehicles using hybrid behavior-based method. *Ocean Eng.* **2021**, *232*, 109147. [CrossRef]
2. Yan, X.; Jiang, D.; Miao, R.; Li, Y. Formation Control and Obstacle Avoidance Algorithm of a Multi-USV System Based on Virtual Structure and Artificial Potential Field. *J. Mar. Sci. Eng.* **2021**, *9*, 161. [CrossRef]
3. Wu, T.; Xue, K.; Wang, P. Leader-follower formation control of USVs using APF-based adaptive fuzzy logic nonsingular terminal sliding mode control method. *J. Mar. Sci. Technol.* **2022**, *36*, 2007–2018. [CrossRef]
4. Wang, N.; Li, H. Leader-follower formation control of surface vehicles: A fixed-time control approach. *ISA Trans.* **2022**, *124*, 356–364. [CrossRef]
5. Shojaei, K. Leader–follower formation control of underactuated autonomous marine surface vehicles with limited torque. *Ocean Eng.* **2015**, *105*, 196–205. [CrossRef]
6. Cui, R.; Sam, G.S.; Voon, E.H.B.; Sang, C.Y. Leader–follower formation control of underactuated autonomous underwater vehicles. *Ocean Eng.* **2010**, *37*, 1491–1502. [CrossRef]
7. Bouteraa, Y.; Alattas, K.A.; Mobayen, S.; Golestani, M.; Ibrahim, A.; Tariq, U. Disturbance Observer-Based Tracking Controller for Uncertain Marine Surface Vessel. *Actuators* **2022**, *11*, 128. [CrossRef]
8. Witkowska, A.; Śmierzchalski, R. Designing a ship course controller by applying the adaptive backstepping method. *Int. J. Ap. Mat. Com-Pol.* **2012**, *22*, 985–997. [CrossRef]
9. Sun, Z.; Zhang, G.; Lu, Y.; Zhang, W.D. Leader-follower formation control of underactuated surface vehicles based on sliding mode control and parameter estimation. *ISA Trans.* **2018**, *72*, 15–24. [CrossRef]
10. Sun, Z.; Sun, H.; Li, P.; Zou, J. Formation Control of Multiple Underactuated Surface Vessels with a Disturbance Observer. *J. Mar. Sci. Eng.* **2022**, *10*, 1016. [CrossRef]
11. Er, M.J.; Li, Z. Formation Control of Unmanned Surface Vehicles Using Fixed-Time Non-Singular Terminal Sliding Mode Strategy. *J. Mar. Sci. Eng.* **2022**, *10*, 1308. [CrossRef]
12. Sørensen, A.J. A survey of dynamic positioning control systems. *Ann. Rev. Control* **2011**, *35*, 123–136. [CrossRef]
13. Song, W.; Li, Y.; Tong, S. Fuzzy Finite-Time H∞ Hybrid-Triggered Dynamic Positioning Control of Nonlinear Unmanned Marine Vehicles Under Cyber-Attacks. *IEEE. Trans. Intell. Veh.* **2023**, *99*, 1–11. [CrossRef]
14. Chen, H.; Li, J.; Gao, N.; Han, J.; Aït-Ahmed, N.; Benbouzid, M. Adaptive backstepping fast terminal sliding mode control of dynamic positioning ships with uncertainty and unknown disturbances. *Ocean Eng.* **2023**, *281*, 114925. [CrossRef]
15. Zhang, Y.; Liu, C.; Zhang, N.; Ye, Q.; Su, W. Finite-Time Controller Design for the Dynamic Positioning of Ships Considering Disturbances and Actuator Constraints. *J. Mar. Sci. Eng.* **2022**, *10*, 1034. [CrossRef]
16. Alagili, O.; Khan, M.A.I.; Ahmed, S.; Imtiaz, S.; Zaman, H.; Islam, M. An energy-efficient dynamic positioning controller for high sea conditions. *Appl. Ocean Res.* **2022**, *129*, 103331. [CrossRef]
17. Liu, C.; Sun, T.; Hu, Q. Synchronization Control of Dynamic Positioning Ships Using Model Predictive Control. *J. Mar. Sci. Eng.* **2021**, *9*, 1239. [CrossRef]
18. Shi, Q.; Hu, C.; Li, X.; Guo, X.; Yang, J. Finite-time adaptive anti-disturbance constrained control design for dynamic positioning of marine vessels with simulation and model-scale tests. *Ocean Eng.* **2023**, *277*, 114117. [CrossRef]
19. Fujii, S.; Kato, T.; Kawamura, Y.; Tahara, J.; Baba, S.; Sanada, Y. Invention of automatic movement and dynamic positioning control method of unmanned surface vehicle for core sampling. *Artif. Life Robot.* **2021**, *26*, 503–512. [CrossRef]
20. Ianagui, A.S.S.; De Mello, P.C.; Tannuri, E.A. Robust Output-Feedback Control in a Dynamic Positioning System via High Order Sliding Modes: Theoretical Framework and Experimental Evaluation. *IEEE Access* **2020**, *8*, 91701–91724. [CrossRef]
21. Kyrkjebo, E.; Pettersen, K.Y. A virtual vehicle approach to output synchronization control. In Proceedings of the 45th IEEE Conference on Decision and Control, San Diego, CA, USA, 3–15 December 2006.
22. Miller, A. Interaction Forces Between Two Ships During Underway Replenishment. *J. Navig.* **2016**, *69*, 1197–1214. [CrossRef]
23. He, S.; Wang, M.; Dai, S.-L.; Luo, F. Leader–Follower Formation Control of USVs with Prescribed Performance and Collision Avoidance. *IEEE Trans. Ind. Inform.* **2019**, *15*, 572–581. [CrossRef]
24. Guttorm, T. Nonlinear Control and Observer Design for Dynamic Positioning Using Contraction Theory. Master's Thesis, Norwegian University of Science and Technology, Trondheim, Norway, 2004.
25. Zhang, Y.F.; Liu, C.D. Contraction based adaptive backstepping control of dynamic positioning vessels. *Shipbuild. China* **2020**, *61*, 85–94.

26. Alamir, M. Contraction-based nonlinear model predictive control formulation without stability related terminal constraints. *Automatica* **2017**, *75*, 288–292. [CrossRef]

27. Tomera, M.; Podgórski, K. Control of Dynamic Positioning System with Disturbance Observer for Autonomous Marine Surface Vessels. *Sensors* **2021**, *21*, 6723. [CrossRef] [PubMed]

28. Johansen, T.A.; Fossen, T.I. Control allocation—A survey. *Automatica* **2013**, *45*, 1087–1103. [CrossRef]

29. Lindegaard, K.P.; Fossen, T.I. Fuel-efficient rudder and propeller control allocation for marine craft: Experiments with a model ship. *IEEE. Trans. Contr. Syst. Technol.* **2003**, *11*, 850–862. [CrossRef]

30. Ruth, E. Propulsion control and thrust allocation on marine vessels. Ph.D. Thesis, Norwegian University of Science and Technology, Trondheim, Norway, 2008.

31. Li, X.; Yang, L. Study of constrained nonlinear thrust allocation in ship application based on optimization and SOM. *Ocean Eng.* **2019**, *191*, 106491. [CrossRef]

32. Artyszuk, J.; Zalewski, P. Energy Savings by Optimization of Thrusters Allocation during Complex Ship Manoeuvres. *Energies* **2021**, *14*, 4959. [CrossRef]

33. Zhang, L.H.; Peng, X.Y.; Wei, N.X.; Liu, Z.F.; Liu, C.D.; Wang, F. A thrust allocation method for DP vessels equipped with rudders. *Ocean Eng.* **2023**, *285*, 115342. [CrossRef]

34. Xu, S.; Wang, X.; Wang, L.; Li, B. A Dynamic Forbidden Sector Skipping Strategy in Thrust Allocation for Marine Vessels. *Int. J. Offshore Polar Eng.* **2016**, *26*, 175–182. [CrossRef]

35. Kalikatzarakis, M.; Coraddu, A.; Oneto, L.; Anguita, D. Optimizing Fuel Consumption in Thrust Allocation for Marine Dynamic Positioning Systems. *IEEE. Trans. Autom. Sci. Eng.* **2022**, *19*, 122–142. [CrossRef]

36. Arditti, F.; Souza, F.L.; Martins, T.C.; Tannuri, E.A. Thrust allocation algorithm with efficiency function dependent on the azimuth angle of the actuators. *Ocean Eng.* **2015**, *105*, 206–216. [CrossRef]

37. Tang, Z.; He, H.; Wang, L.; Wang, X. An optimal thrust allocation algorithm with bivariate thrust efficiency function considering hydrodynamic interactions. *J. Mar. Sci. Technol.* **2022**, *27*, 52–66. [CrossRef]

38. Gao, D.; Wang, X.; Wang, T.; Wang, Y.; Xu, X. Optimal Thrust Allocation Strategy of Electric Propulsion Ship Based on Improved Non-Dominated Sorting Genetic Algorithm II. *IEEE Access* **2019**, *7*, 135247–135255. [CrossRef]

39. Hou, M.; Yu, M.; Jiao, L. Research on Ship Thrust Distribution Based on Adaptive Particle Swarm Bee Colony Hybrid Algorithm. *J. Phys. Conf. Ser.* **2023**, *2477*, 012088. [CrossRef]

40. Yang, X.S. *Nature-Inspired Metaheuristic Algorithms*, 2nd ed.; Luniver Press: Frome, UK, 2010; pp. 81–89.

41. Liu, J.; Mao, Y.; Liu, X.; Li, Y. A dynamic adaptive firefly algorithm with globally orientation. *Math. Comput. Simul.* **2020**, *174*, 76–101. [CrossRef]

42. Sharma, B.B.; Kar, I.N. Contraction based adaptive control of a class of nonlinear systems. In Proceedings of the American Control Conference, St Louis, MO, USA, 10–12 June 2009.

43. Mohamed, M.; Su, R. Contraction Based Tracking Control of Autonomous Underwater Vehicle. *IFAC-Pap. OnLine* **2017**, *50*, 2665–2670. [CrossRef]

Journal of
*Marine Science
and Engineering*

Article

Application of Modified BP Neural Network in Identification of Unmanned Surface Vehicle Dynamics

Sheng Zhang *, Guangzhong Liu and Chen Cheng

School of Information Engineering, Shanghai Maritime University, Shanghai 201308, China
* Correspondence: zhangsheng_1990@163.com; Tel.: +86-152-5291-9656

Abstract: Over the past few decades, unmanned surface vehicles (USV) have drawn a lot of attention. But because of the wind, waves, currents, and other sporadic disturbances, it is challenging to understand and collect correct data about USV dynamics. In this paper, the Modified backpropagation neural network (BPNN) is suggested to address this issue. The experiment was conducted in the Qinghuai River, and the receiver collected the data. The modified BPNN outperforms the conventional BPNN in terms of ship trajectory forecasting and the rate of convergence. The updated BPNN can accurately predict the rotational velocity during the propeller's acceleration and stability stages at various rpms.

Keywords: underwater surface vehicle (USV); backpropagation neural network (BPNN); additional momentum method; adaptive learning rate method; vehicle dynamics

1. Introduction

Unmanned surface vehicles (USVs) have recently garnered significant interest from researchers and developers. Compared to traditional manned vehicles, the USV's ability to operate in severely hazardous environments is one of its most remarkable advantages. USVs can perform tasks that conventional ships cannot achieve in terrible ocean environments and other situations. At the same time, these types of ships are of great importance for national defense security and environmental monitoring. However, controlling Unmanned surface vehicles is a notoriously challenging task that remains poorly understood due to various uncontrollable phenomena, including wind, wave, current, and other random disturbances. The unpredictability and dynamic nature of these external forces render the control of USVs particularly intricate. This is further compounded by the complex interactions between these disturbances and the vehicle's dynamic characteristics. To compound the matter even further, USVs are often required to perform precise maneuvers in cluttered environments, which requires a high level of control authority and adaptability. To address these challenges, researchers have been exploring innovative control strategies that can enhance USV performance in hostile environments. These strategies often combine traditional control techniques with advanced algorithms and machine learning methods to develop more robust and adaptive control systems. Additionally, the use of simulation platforms has become an essential tool for testing and validating these control strategies before deploying them in real-world scenarios. In Antonelli's research, it has been shown that using artificial intelligence algorithms for data processing has certain pattern recognition capabilities and can be analyzed without a physical model. However, this method currently lacks a comprehensive scientific explanation and cannot be falsified. However, he believes that data-driven artificial intelligence algorithms are still very worthwhile to study [1].

Numerous studies have been conducted to explore the control methods of Unmanned Surface Vehicles (USVs). The identification of the dynamic coefficient of the unmanned vessel plays a decisive role in the accurate motion control and automatic driving of the

Citation: Zhang, S.; Liu, G.; Cheng, C. Application of Modified BP Neural Network in Identification of Unmanned Surface Vehicle Dynamics. *J. Mar. Sci. Eng.* **2024**, *12*, 297. https://doi.org/10.3390/jmse12020297

Academic Editor: Sergei Chernyi

Received: 18 December 2023
Revised: 3 February 2024
Accepted: 4 February 2024
Published: 7 February 2024

unmanned vessel, especially in the case of external interference. At present, no unmanned vessel can have accurate control ability in complex waters, which is also one of the problems of dynamic identification of hydrodynamic coefficients. The most common technique for handling an indescribable system from input–output data is system identification. In the process of system identification, Nagumo and Noda utilized continuous least squares estimation with error-correcting training [2]. Holzhuter employed recursive least square estimation to identify ship dynamics [3]. Kallstrom and Astrom demonstrated the use of recursive estimation of maximum likelihood in ship steering motion in 1981 [4], yielding well-predicted outcomes. To anticipate motion variables, hydrodynamic force, vehicle speed, and current direction, Yoon and Rhee suggested the extended Kalman filter approach and modified Bryson–Frazier smoother [5]. Shin et al. combined Particle Swarm Optimization (PSO) with an adaptive control technique to anticipate the trajectory of autonomous surface vehicles [6]. Additionally, Selvam described a frequency domain identification system for linear steering equations in ships' maneuvering under calm seas [7].

Daniele's research shows that it is very difficult to establish the dynamics model of both USV and ROV. The system delay problem of the dynamic system and the interference of the external environment lead to the failure of accurate control. Therefore, adaptive control algorithms, especially data-driven algorithms, are needed to solve these problems [8]. The dynamic model of ship dynamics poses a formidable nonlinear challenge due to the influence of wind, ocean currents, and various arbitrary disturbances. The nonlinear issue that emerged last year was successfully addressed through the application of Artificial Neural Networks (ANN). In order to tackle the system identification challenges posed by large oil tankers, Rajesh and Bhattacharyya proposed an artificial neural network approach [9]. The network was trained using the Levenberg–Marquardt algorithm, and multiple hidden neuron densities were evaluated to determine the optimal configuration. Oskin et al. introduced Recurrent Neural Networks (RNN) for identifying both linear and nonlinear behaviors in ship dynamics [10]. Additionally, Pan et al. employed an effective Neural Network (NN) method to track the movements of autonomous surface vehicles with unknown ship dynamics [11]. This series of studies provides crucial insights and innovative approaches for understanding and addressing the complexity of ship dynamics.

This study utilized a Back Propagation Neural Network (BPNN) to address the nonlinear dynamics problem of Autonomous Underwater Vehicles (AUVs). The benefit of the BPNN is that it has a robust non-linear mapping capability and a flexible network topology, making it ideally suited to the problem of vessel dynamic system identification. However, the traditional bp neural network has the ability of self-adaptation and self-learning and has strong nonlinear mapping ability, but it also has some defects such as slow convergence and easy fall into the local optimal solution. The additional inertia method and the adapted learning rate method are also chosen to improve the BPNN method in order to minimize these shortcomings. The training data comes from actual experiments that were conducted in the Qing-Huai river.

In order to study the motion control ability of the BPNN algorithm in USV, the algorithm is designed to take the speed of the left and right thrusters as input and output the surge velocity, swing velocity, rotation velocity, and trajectory through black box calculation, which can realize the motion control solution of USV. The above parameters are selected as input and output because these data can be obtained on the experimental model or can be calculated. This study includes two hidden layers. Through the expert experience method, the weight of the hidden layer is assigned, and the various parameters are correlated.

2. USV Dynamic System

2.1. Deepsea Warriors uBoat (YL1300M)

We propose the following development requirements for the USV to be able to perform algorithm verification in a real environment. The real ship is shown in Figure 1.

Figure 1. The design of the YL1300M.

a.   The body of the YL1300M is shaped to be more streamlined in order to reduce resistance.

b.   The YL1300M must have good stability in order for the USV to function in a freshwater and inshore environment.

c.   The vessel features a sizable storage area for carrying a variety of tools and supplies.

d.   The boat's upper deck features a sizable cover that can be used to disassemble and repair equipment.

e.   There are two electric propulsion modules in the car. Due to the lack of a rudder and gimbaled thruster on the YL1300M, the steering motion can be produced by altering the RPM (revolutions per minute) of the two primary thrusters.

The YuanLi-1300M measures 1.3 m in length, 0.64 m in width, and 0.55 m in height. The vehicle is equipped with various sensors, including two differential GPS units located at the bow and stern, respectively, as well as an electronic compass capable of measuring position, speed, and heading angle.

2.2. USV Dynamic Model

In Marco's paper, the rigid-body dynamics equations for the UUV are described. He described the motion of the vehicle as an equation of six degrees of freedom. When considering the influence of the external environment, the coupling relationship is ignored and only the simplified uncoupling model is considered [12]. In order to characterize the motion of the USV in a simple manner, the three degrees of freedom (surge, sway, and yaw) in horizontal planar motion are taken into account in this work. The USV and its coordinate frame are described in Figure 2 using the nomenclature proposed by Fossen [13].

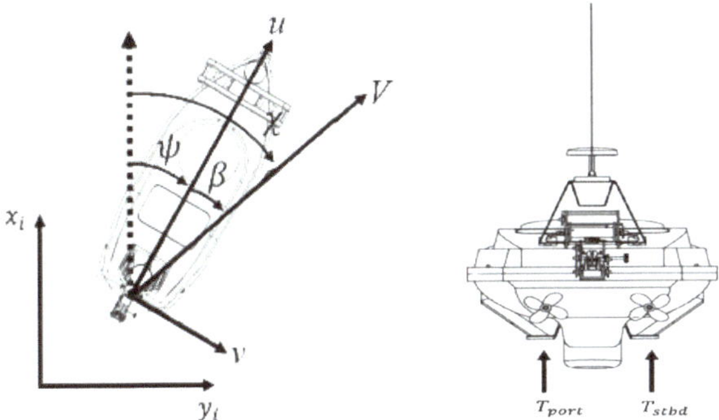

Figure 2. A schematic representation of the USV's differential thrust.

The u, v, and V indicate the upsurge, swing, and overall speed of the USV in a body-fixed frame, respectively while the x_i, y_i denote the north and east direction of the USV in the inertial frame, respectively. In addition, ψ and β denote the heading angle and course angle of the USV, respectively while X represents the side slip angle.

The kinematic model of the USV can be expressed as follows [14]:

$$\dot{\eta} = \mathbf{R}(\eta) \cdot \mathbf{v},\tag{1}$$

$$\mathbf{v} = (u, v, r)^T,\tag{2}$$

$$\eta = (x_i, y_i, \psi),\tag{3}$$

$$\mathbf{R}(\eta) = \begin{bmatrix} \cos(\psi) & -\sin(\psi) & 0 \\ \sin(\psi) & \cos(\psi) & 0 \\ 0 & 0 & 1 \end{bmatrix},\tag{4}$$

where indicates the velocity vector $\mathbf{v}$, the position vector η, and the rotation matrix $\mathbf{R}(\eta)$ that maps the vector from the object's fixed coordinate system to the inertial coordinate system.

The surface vehicle's planar dynamic model can be explained as follows:

$$M\dot{v} + C(v)v + D(v)v = f,\tag{5}$$

where M indicates the mass matrix, which consists of the body's mass and included mass, $C(v)$ indicates the Coriolis and centripetal matrices, $D(v)$ indicates the damping coefficient matrix and the present matrix Equation (6). Given that the YL1300M 's propulsion system is a differential thruster type, Sonnenburg et al. describe it as follows [15]:

$$f = \begin{bmatrix} \tau_X \\ \tau_Y \\ \tau_N \end{bmatrix} = \begin{bmatrix} T_{port} + T_{stbd} \\ 0 \\ (T_{port} - T_{stbd})B/2 \end{bmatrix},\tag{6}$$

where T_{port} and T_{stbd} denote the port side thruster's thrust force and the starboard side thruster, respectively. B the beam of the YL1300M.

Consequently, the following is an illustration of the three degrees of freedom nonlinear dynamic motion equations:

$$(m - X_{\dot{u}})\dot{u} - m(x_G r^2 + vr) + Y_{\dot{v}}vr + \frac{Y_{\dot{r}} + N_{\dot{v}}}{2}r^2 + X_u u + X_{u|u|}|u|u = \tau_X,\tag{7}$$

$$(m - Y_{\dot{v}})\dot{v} + (mx_G - Y_{\dot{r}})\dot{r} + (m - X_{\dot{u}})ur + Y_v v + Y_r r + Y_{|v|v}|v|v + Y_{|r|r}|r|r = \tau_Y,\tag{8}$$

$$(mx_G - N_{\dot{v}})\dot{v} - (I_{zz} - N_{\dot{r}})\dot{r} + mx_G ur - Y_{\dot{v}}uv - \frac{Y_{\dot{r}} + N_{\dot{v}}}{2}ur + X_{\dot{u}}uv + N_v v + N_r r + N_{|v|v}|v|v + N_{|r|r}|r|r = \tau_N,\tag{9}$$

where $X_{(.)}$, $Y_{(.)}$, and $N_{(.)}$ in Equations (7)–(9) represent constant hydrodynamic coefficients, which are partial derivatives of surge, sway force, and yaw moment, respectively.

3. Neural Network

3.1. The Back Propagation Neural Network Principle

In 1986, the back propagation neural network (BPNN) was proposed by Rumelhart and colleagues and McClelland. The input layer, hidden layer, and output layer of the BPNN are depicted in Figure 3. The relationship between the number of the input layer, hidden layer, and output layer may be understood using the empirical formula given by Li et al. [16]:

$$M = \sqrt{N + K} + \varphi,\tag{10}$$

where indicates the hidden layer's unit number, N the input layer's unit number, K the output layer unit number, and φ a constant number that falls within the (1, 10). φ is set to 5.

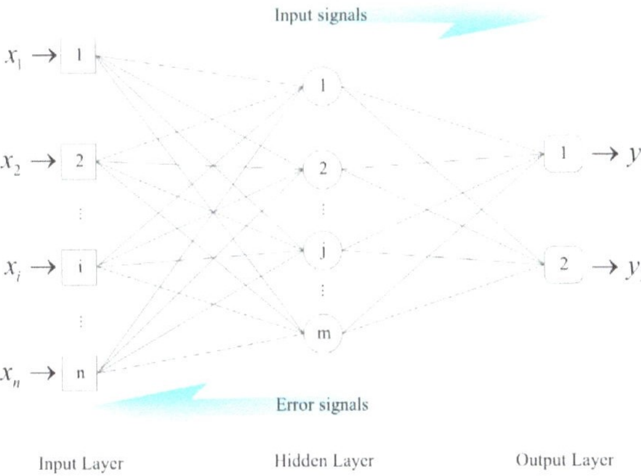

Figure 3. A BP neural network.

Back propagation and forward propagation are two of the methods used in BPNN. The outcome of the neural network can be obtained during this process to be contrasted with the intended data; this discrepancy is regarded as an error and is employed in the backward propagation process to generate the loss function valve. The input data is propagated from the input layer to the output layer. Using optimization techniques, the link weights and threshold can be changed to reduce the loss function.

As can be demonstrated in Figure 4, Assume n input signals are propagated in a neural unit j, and the following is the way the result can be determined:

$$O_j = f\left(\sum_{i=1}^{n} x_i w_{ij} - \theta_j\right), \tag{11}$$

where w_{ij} represents the value of the weight coefficient of friction, θ_j the neural unit's threshold, $f(\cdot)$ an activation function, and the Sigmoid function are used, as shown below:

$$f(X) = \frac{1}{1 + \exp(-X)}, \tag{12}$$

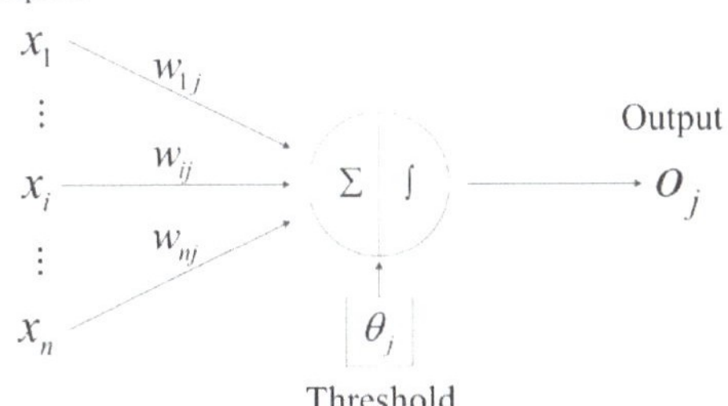

Figure 4. A neural unit.

Substituting Equation (12) into Equation (11) and the output O_j can be expressed as:

$$O_j = \frac{1}{1 + \exp\left(\sum_{i=1}^{n} x_i w_{ij} - \theta_j\right)}. \tag{13}$$

The neural network's i-th output value is compared to the real value, and the difference is known as the unit error of the output layer. Its error function is depicted below:

$$E_i = \frac{1}{2}\sum_{k=1}^{k}(y_{d,k} - y_k)^2,\tag{14}$$

where $y_{d,k}$ and y_k denote expected value and real value of neural unit k, respectively, within the output layer. The entire mistake E for m samples of training is expressed as follows:

$$E = \frac{1}{m}\sum_{i=1}^{i}E_i.\tag{15}$$

The malfunction indicator is obtained layer-by-layer via recursive propagation and the weights will be adjusted to reduce the error. The next will show that how to adjust weights to decrease the error. In the p-th iteration, the mistaken indioncate of the output layer of neuronal cells unit k is as follows:

$$e_k(p) = y_{d,k}(p) - y_k(p).\tag{16}$$

Then the error will be utilized to modulate the weight in the next iteration and can be shown as:

$$w_{jk}(p+1) = w_{jk}(p) + \Delta w_{jk}(p).\tag{17}$$

However, Equation (17) is not able to maintain the accumulation of the learning experience which means the velocity of the convergence is slow. This paper adopts the additional momentum method, which is as follows:

$$w_{jk}(p+1) = w_{jk}(p) + \Delta w_{jk}(p) + \alpha\left[w_{jk}(p) - w_{jk}(p-1)\right],\tag{18}$$

where α denotes the rate of the momentum learning.

The adjusting part of the weight $\Delta w_{jk}(p)$ can be calculated as:

$$\Delta w_{jk}(p) = \eta \times y_i(p) \times \lambda_k(p),\tag{19}$$

where η denotes the learning rate, $y_i(p)$ the output signal of neuron unit j, and $\lambda_k(p)$ the error gradient of neuron unit k.

The standard η is belonging to $(0, 1)$, but the η is difficult to determine, if the η is too large which result in generating the oscillation in the learning process while the η is too small which cause the velocity of convergence become slowly. The adaptive learning rate method is adopted which can be shown as follow:

$$\eta(t) = \eta_{\max} - (\eta_{\max} - \eta_{\min}) \times \frac{t}{t_{\max}},\tag{20}$$

where $\eta_{\max}$ indicates the highest rate of learning, $\eta_{\min}$ the lowest possible rate of learning, $t_{\max}$ the greatest quantity of iteration, and t the current number of iteration.

The error gradient can be calculated as:

$$\lambda_k(p) = \frac{\partial y_k(p)}{\partial X_k(p)} \times \delta_k(p),\tag{21}$$

where $X_k(p)$ denotes the neural unit's weighted input k, $\delta_k(p)$ the error of the neural unit k. If neural unit k is in the output layer,

$$\delta_k(p) = e_k(p).\tag{22}$$

Figure 5 demonstrates the error gradient's back-propagation process for a unit and the error of the neuron can be written as:

$$\delta_K(P) = \sum_{i=1}^{l} (\delta_i(p) \times w_{ki}(p)),\qquad(23)$$

where l denotes the result of the layer's neural unit count.

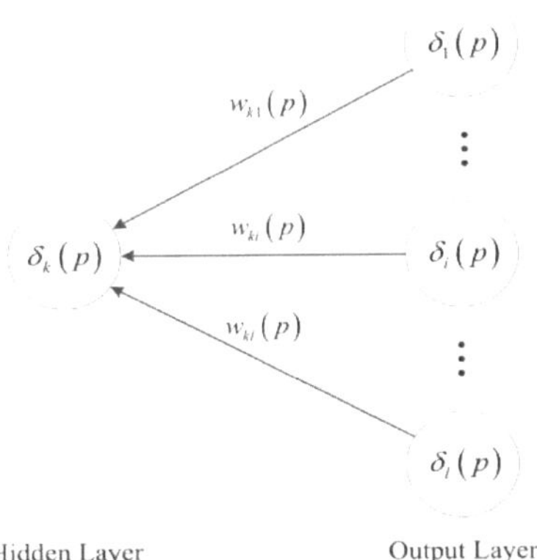

Figure 5. The procedure of mistake the gradient backpropagation for a unit.

In Equation (21):

$$\frac{\partial y_k(p)}{\partial X_k(p)} = \frac{\partial\left\{\frac{1}{1+\exp[-X_k(p)]}\right\}}{\partial X_k(p)} = \frac{\exp[-X_k(p)]}{\{1+\exp[-X_k(p)]\}^2} = y_k(p) \times [1 - y_k(p)].\qquad(24)$$

Substitute Equations (21)–(23) into Equation (24):
When the output layer's neural unit k:

$$\lambda_k(p) = y_k(p) \times [1 - y_k(p)] \times e_k(p).\qquad(25)$$

When the hidden layer's neural unit k:

$$\lambda_k(p) = y_k(p) \times [1 - y_k(p)] \times \sum_{i=1}^{l} \delta_i(p)w_{ki}(p).\qquad(26)$$

3.2. BPNN for Dynamic Model Identification

According to Equations (7)–(9), The thruster force determines the surge velocity, sway velocity, rotational velocity, and trajectory. The RAND function in Matlab is used to assign random weight values and thresholds in the range of -1 to 1. The magnitudes of the thruster forces are contingent upon the rotational velocity of the propeller. Hence, the variables included as inputs in this study are the left RPM and right RPM, whereas the variables considered as outputs are the surge velocity, sway velocity, rotational velocity, and trajectory. The Levenberg–Marquardt (LM) backpropagation algorithm is utilized for the purpose of error minimization. The LM backpropagation algorithm is a very efficient technique utilized in neural networks and has gained significant recognition in academic research (Hagan and Menhaj, 1994; Suri et al., 2002) [17,18]. The threshold values and weight values are updated using the additional momentum method in conjunction with the LM backpropagation algorithm. In the context of computer simulation, a training set consisting of 70% of the available data is employed to train the Backpropagation Neural

Network (BPNN), while a separate testing set comprising 30% of the data is used to evaluate the performance of the trained BPNN.

This article uses the LM algorithm as a calculation method to minimize errors in the improved BPNN algorithm, which is an estimation method for minimizing regression parameters in nonlinear regression. This method is a combination of the steepest descent method and linearization method. When facing the identification of hydrodynamic coefficients for USVs, it has better calculation speed and the ability to solve nonlinear equations.

4. Experimental Setup and Simulation Results

In order to realize the experimental demonstration of the improved BPNN algorithm, we prepared a USV, which is equipped with a propeller that can feedback the speed. At the same time, it is also equipped with a wealth of sensors, including GPS, attitude sensors, electronic compass, and other sensors, which can provide us with track, heading Angle, speed, Roll, Pitch, Yaw, and other data. The USV ship is equipped with a radio communication radio, which can transmit these data to the shore control software at a frequency of 5 Hz in real time and can display and store the data in real time. The topological relationship of the ship control system is shown in Figure 6.

Figure 6. Topology diagram of USV control system.

The investigation was carried out in Nanjing's Qing-Huai River, as shown in Figure 7. It is worth noting that wind and current were consistently present at this particular site. Figure 8 displays the YL1300M expeditions throughout the Qing-Huai river. The turning experiments for the unmanned surface vehicle were conducted by varying the revolution speed of the propeller. The selection of distinct revolution speeds for the left and right propellers is evident in Table 1. Initially, the vehicle came to a halt on the water, utilizing the distinct revolutions per minute (RPM) of both the left and right propellers to push the YL1300M. This enabled the acquisition of rotation performance data under intricate circumstances. The neural network was trained using 70% of the available data, while the remaining 30% was reserved for testing purposes.

Table 1. The different revolution speeds of the propeller were operated in experiments.

Left (RPM)	Right (RPM)	Left (RPM)	Right (RPM)	Left (RPM)	Right (RPM)
-100	0	0	-100	-100	-100
-200	0	0	-200	-200	-200
-300	0	0	-300	-300	-300
-400	0	0	-400	-400	-400
-500	0	0	-500	-500	-500
-600	0	0	-600	-600	-600

Table 1. *Cont.*

Left (RPM)	Right (RPM)	Left (RPM)	Right (RPM)	Left (RPM)	Right (RPM)
−700	0	0	−700	−700	−700
−800	0	0	−800	−800	−800
−900	0	0	−900	−900	−900
−1000	0	0	−1000	−1000	−1000
100	0	0	100	100	100
200	0	0	200	200	200
300	0	0	300	300	300
400	0	0	400	400	400
500	0	0	500	500	500
600	0	0	600	600	600
700	0	0	700	700	700
800	0	0	800	800	800
900	0	0	900	900	900
1000	0	0	1000	1000	1000

Figure 7. The Qing-huai river in google map.

Figure 8. YL1300M expedition on the Qing-Huai River.

Figure 9 provides a visual representation of the YL1300M 's trajectory under varying propeller revolution speeds. The insightful depiction allows us to observe how the left and right propellers contribute to the vehicle's movement at rotational speeds of 100 rpm, 500 rpm, and 1000 rpm. In the left column, a sequence of three images elucidates the trajectory of the left propeller, while the right column concurrently showcases the trajectory of the right propeller at the same rotational speeds. It becomes evident that the YL1300M's path deviates from the anticipated conventional circular trajectory. The images in the left column exhibit the subtle nuances in the trajectory as the propeller revolution speeds increase. At 100 rpm, the trajectory seems relatively stable, maintaining a certain symmetry. However, as the rotational speed escalates to 500 rpm and 1000 rpm, we begin to observe deviations in the circular path, hinting at the influence of external factors or system dynamics. Interestingly, the right column reveals a complementary set of images illustrating the trajectory under negative rotational speeds (-100 rpm, -500 rpm, and -1000 rpm). This adds a layer of complexity to the analysis, as negative rotational speeds might introduce counteracting forces, potentially affecting the overall stability and direction of the YL1300M. The visual representation in Figure 8 not only captures the expected circular trajectory but also unveils subtle deviations and nuances induced by varying propeller revolution speeds. This nuanced analysis provides valuable insights into the dynamic behavior of the YL1300M, shedding light on potential factors influencing its trajectory under different operational conditions.

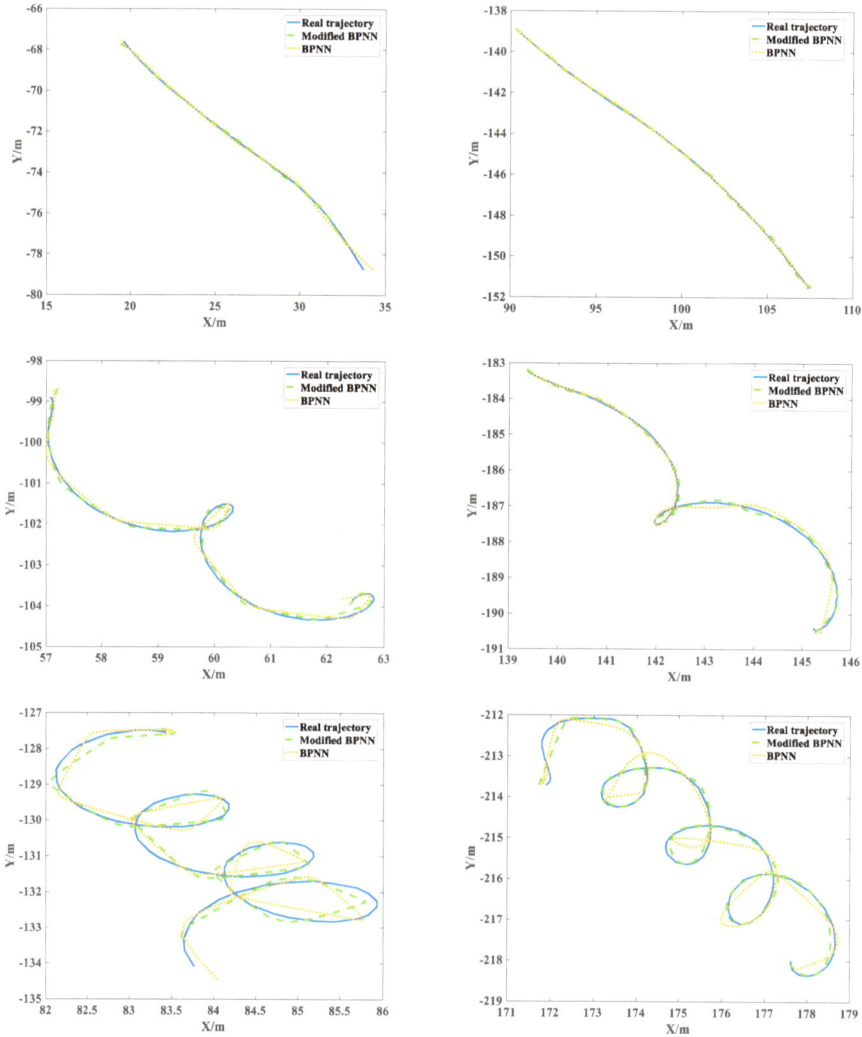

Figure 9. The trajectory of the YL1300M.

Due to the influence of wind and current, the actual trajectory deviates from a circular path and instead takes the form of a linear or spiral trajectory. When the propellant's velocity of rotation is low, the trajectory of the vehicle resembles a straight line. The lateral velocity is significantly smaller in comparison to the velocity of the current, resulting in the current exerting a dominant influence on the vehicle's motion. Consequently, the ship moves in accordance with the current, leading to a linear trajectory. When the rotational velocity of the propeller reaches a high magnitude, the resulting path followed by the ship can be described as a spiral progression. The surge speeds and sway speeds exceed the magnitude of the current velocity, resulting in a ship's motion characterized by a combination of circular trajectory and forward movement. Furthermore, the overall trajectory of the movement aligns with the direction of the current, as depicted in Figure 7. The predictive trajectory determined by the modified BPNN is represented by the blue dotted lines, while the traditional BPNN is represented by the yellow dotted lines. Table 2 demonstrates that the modified Backpropagation Neural Network (BPNN) has superior predictive capabilities for trajectory estimation compared to the conventional BPNN. This approach also exhibits a strong compatibility with the actual trajectory of the vehicle in the presence of wind and current. Furthermore, the modified backpropagation neural network (BPNN) exhibits a higher rate of convergence compared to the original BPNN. Hence, the utilization of the additional momentum method and adaptive learning rate method in enhancing the BPNN yields superior results in predicting the motion of vehicles. There are three types of track types in the experiment: smooth curve without turning, one turning circle, and multiple turning circles. After the experimental demonstration, BPNN can still predict more accurately in the first two working conditions, but when encountering multiple groups of rotation or the greater the curvature, BPNN will completely predict out of control. In contrast, the modified BPNN algorithm has better adaptability, and the effect is far superior to the traditional BPNN.

Table 2. The standard deviation of error at different RPM.

	100 RPM		500 RPM		1000 RPM	
	X (m)	Y (m)	X (m)	Y (m)	X (m)	Y (m)
Traditional BPNN	0.055	0.033	0.139	0.151	0.147	0.325
Modified BPNN	0.048	0.028	0.121	0.147	0.133	0.287
	−100 RPM		−500 RPM		−1000 RPM	
	X (m)	Y (m)	X (m)	Y (m)	X (m)	Y (m)
Traditional BPNN	0.062	0.103	0.118	0.103	0.138	0.289
Modified BPNN	0.044	0.088	0.105	0.096	0.115	0.244

Figure 10 offers a comprehensive visualization of the interplay between empirical observations and predictive data, meticulously acquired through the sophisticated application of a refined Backpropagation Neural Network (BPNN) operating across a spectrum of propeller revolution speeds. The deliberate alignment of the six images in Figure 9 mirrors the arrangement in Figure 8, facilitating a detailed comparative analysis. The discernible rhythmic behavior captured in the data is a consequence of the inherent noise introduced during the meticulous process of data collection. In response to this challenge, the Gaussian filter technique emerges as a pivotal tool, systematically employed to mitigate and filter out noise from the dataset. The outcome is depicted graphically by the red line, unveiling a nuanced narrative delineated by two distinctive phases: acceleration and stability.

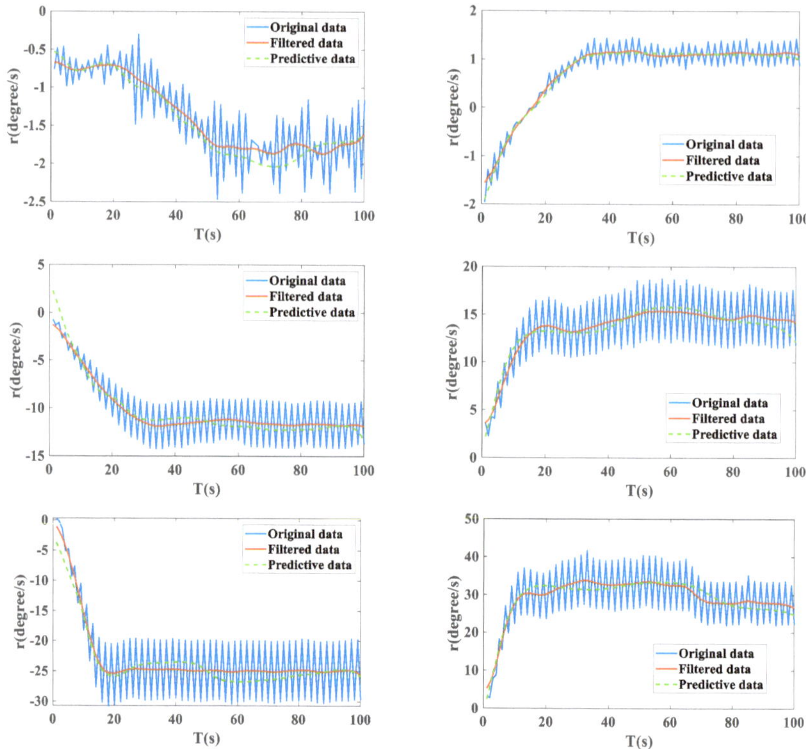

Figure 10. Real and estimated outputs of the different rotation speeds.

Within the intricate interplay of data and model predictions, it becomes evident that the forecasted data elegantly conforms to the intricate dynamics of both acceleration and stabilization stages. However, an analytical eye acknowledges a peak in the error rate, reaching a maximum of 11.33%. This nuanced observation underscores the inherent complexity of the system and hints at potential areas for refinement in the predictive modeling process. Nevertheless, the overall accuracy of the improved Backpropagation Neural Network (BPNN) in delineating the complete rotational speed process across the diverse spectrum of propeller revolution speeds remains notably high, showcasing the model's capacity for capturing and interpreting the intricate dynamics of the propulsion system. The data and curves in Figures 9 and 10 and Table 2 significantly show that the improved BPNN algorithm is relatively reliable in predicting different propeller speeds and track stability.

Transitioning to Figures 11–13, these visual representations intricately dissect the surge velocities and sway velocities observed across varying propeller revolution speeds. The rhythmic patterns discernible in both surge and sway velocities are attributed to the intricate interplay of factors, intertwined with the rotational speed of the propeller. The meticulous projection of these patterns is encapsulated by the green dotted lines, serving as visual overlays that encapsulate the predicted values and patterns extrapolated from the empirical data obtained under distinct propeller revolution speeds. The analysis extends to the amplitude of the predicted values, revealing an admirable alignment with the actual observed values. The meticulous calibration of the predictive model is further emphasized by the recorded maximum inaccuracy rate of 8.9%. This attests to the model's robustness, demonstrating a commendable efficacy in navigating the complexities of forecasting surge and sway velocities even under dynamically intricate conditions. The multifaceted analysis presented in Figures 10–13 encapsulates the synergy between empirical data and predictive modeling, shedding light on the intricate dynamics of the propulsion system under varying operational conditions. The nuances observed in the rhythmic behavior and predictive accuracy offer valuable insights for future refinements, emphasizing the perpetual pursuit of precision and understanding in the realm of marine propulsion systems.

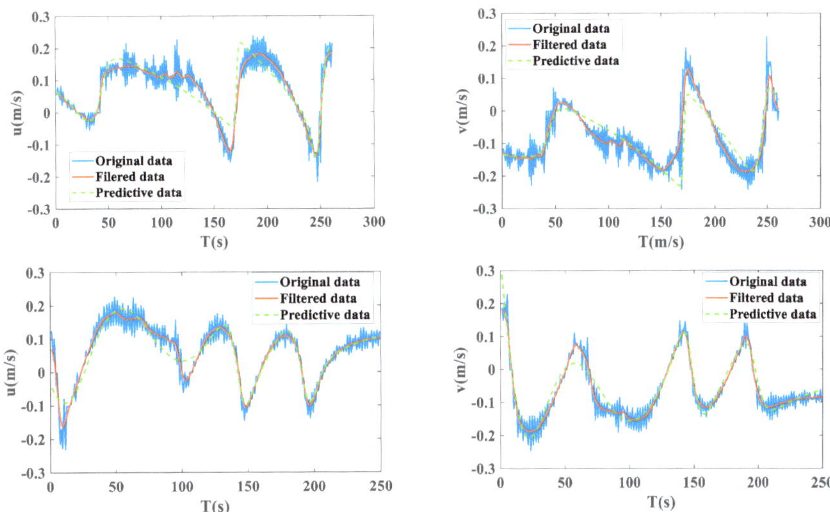

Figure 11. The surge velocities and sway velocities in left propeller and right propeller at 100 rpm and −100 rpm.

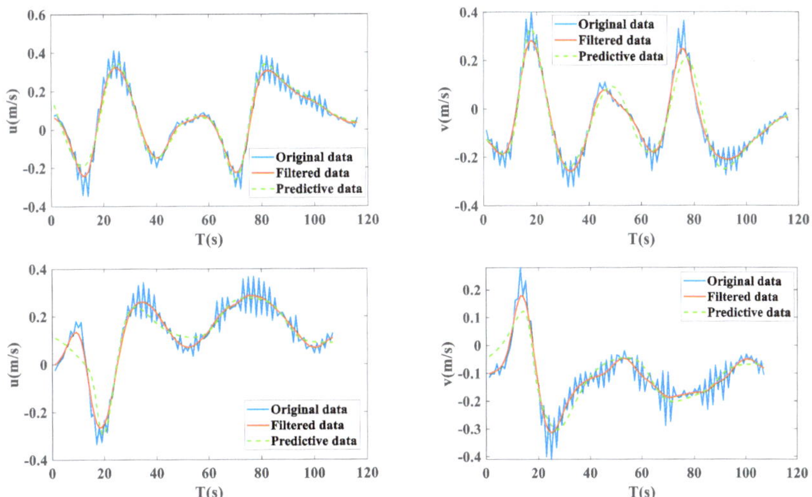

Figure 12. The surge velocities and sway velocities in the left propeller and right propeller are at 500 rpm and −500 rpm.

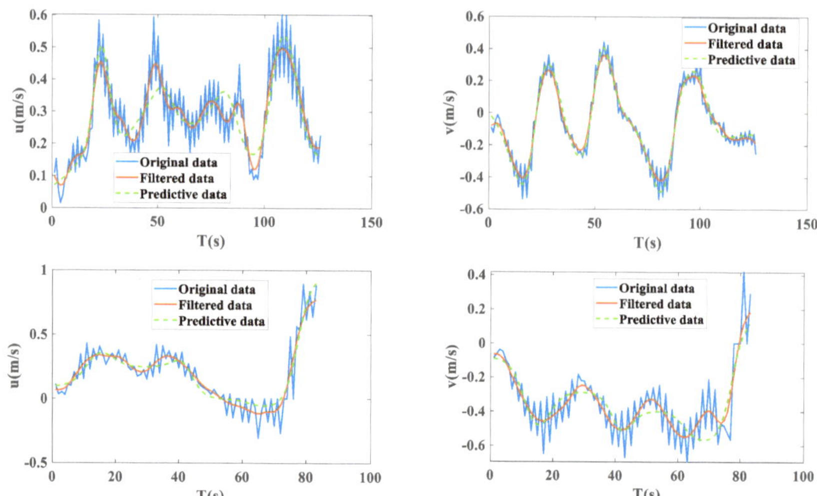

Figure 13. The surge velocities and sway velocities in left propeller and right propeller are at 1000 rpm and −1000 rpm.

5. Conclusions

This study focused on the identification of a three-degree-of-freedom unmanned surface vehicle named YL1300M. Due to the influence of wind, ocean currents, and various unpredictable disruptions, researchers decided to employ a neural network in order to accurately discern and analyze the dynamics of ships. The Backpropagation Neural Network (BPNN) is a valuable approach for addressing complex nonlinear problems, while it does possess certain limitations. The traditional backpropagation neural network (BPNN) was enhanced by using the additional momentum method and the adaptive learning rate approach. Avoiding local minima is a challenging task that can enhance the convergence speed and precision of conventional Backpropagation Neural Networks (BPNNs). The findings indicate:

(1) The modified BPNN exhibits enhanced convergence speed and superior ship trajectory prediction capabilities compared to the conventional BPNN. The calculated trajectories of the propeller at various rotational speeds exhibit a strong correlation with the actual trajectories.

(2) The improved BPNN demonstrates effective estimation capabilities for the rotational velocity throughout both the acceleration and stability stages at various revolutions per minute (rpm) of the propeller. Furthermore, reliable predictions have been made for the acceleration of an increase and sway, in comparison to their actual values.

For the dynamic coefficient of the USV, the trajectory of the receiver and the main engine rotation speed are predicted by the improved BPNN algorithm. Compared with the traditional BNPP algorithm, the proposed algorithm has a larger lift, but still has a higher error when compared with the actual data of the real ship. By analysis, it is thought that there may be unreliable data when acquiring real-time data, or the weight allocation of the algorithm needs to be adjusted. Further studies could be made on this basis in the future. At the same time, the improved idea of this algorithm can be extended to other engineering fields, such as unmanned aerial vehicles, and inverted pendulums, especially in the scenes where the coefficients of nonlinear kinematic equations need to be recognized. The future identification algorithm based on data-driven dataless models will inevitably have more and wider applications.

Author Contributions: Conceptualization, S.Z.; Methodology, S.Z.; Validation, C.C.; Investigation, S.Z.; Data curation, C.C.; Writing—original draft, S.Z.; Writing—review & editing, G.L.; Supervision, G.L.; Project administration, S.Z. All authors have read and agreed to the published version of the manuscript.

Funding: This research received no external funding.

Institutional Review Board Statement: Not applicable.

Informed Consent Statement: Not applicable.

Data Availability Statement: Data are contained within the article.

Conflicts of Interest: The authors declare no conflict of interest.

References

1. Gianluca, A.; Stefano, C.; Paolo, D.L. On data-driven identification: Is automatically discovering equations of motion from data a Chimera? *Nonlinear Dyn.* **2023**, *111*, 6487–6498. [CrossRef]
2. Nagumo, J.I.; Noda, A. A learning method for system identification. *IEEE Trans. Autom. Control* **1967**, *12*, 282–287. [CrossRef]
3. Holzhuter, T. Robust identification scheme in an adaptive track controller for ships. In Proceedings of the 3rd IFAC Symposium on Adaptive System in Control and Signal Processing, Glasgow, UK, 19–21 April 1989; pp. 118–123.
4. Kallstrom, C.G.; Astrom, K.J. Experiences of system identification applied to ship steering. *Automatica* **1981**, *17*, 187–198. [CrossRef]
5. Yoon, H.K.; Rhee, K.P. Identification of hydrodynamic coefficients in ship maneuvering equations of motion by estimation-before-modeling technique. *Ocean Eng.* **2003**, *30*, 2379–2404. [CrossRef]
6. Shin, J.; Kwak, D.J.; Lee, Y.-I. Adaptive path following control for an unmanned surface vessel using an identified dynamic model. *IEEE/ASME Trans. Mechatron.* **2017**, *22*, 1143–1153. [CrossRef]
7. Selvam, R.P.; Bhattacharyya, S.; Haddara, M. A frequency domain system identification method for linear ship maneuvering. *Int. Shipbuild. Prog.* **2005**, *52*, 5–27.
8. Di Vito, D.; Cataldi, E.; Di Lillo, P.; Antonelli, G. Vehicle Adaptive Control for Underwater Intervention Including Thrusters Dynamics. In Proceedings of the 2018 IEEE Conference on Control Technology and Applications (CCTA), Copenhagen, Denmark, 21–24 August 2018; pp. 646–651.
9. Rajesh, G.; Bhattacharyya, S. System identification for nonlinear maneuvering of large tankers using artificial neural network. *Appl. Ocean Res.* **2008**, *30*, 256–263. [CrossRef]
10. Oskin, D.A.; Dyda, A.A.; Markin, V.E. Neural Network Identification of Marine Ship Dynamics. In Proceedings of the 9th IFAC Conference on Control Applications in Marine Systems, Osaka, Japan, 17–20 September 2013.
11. Pan, C.Z.; Lai, X.Z.; Yang, S.X.; Wu, M. An efficient neural network approach to tracking control of an autonomous surface vehicle with unknown dynamics. *Expert Syst. Appl.* **2013**, *40*, 1629–1635. [CrossRef]
12. Bibuli, M.; Zereik, E. Analysis of an unmanned underwater vehicle propulsion model for motion control. *J. Guid. Control Dyn.* **2022**, *45*. [CrossRef]
13. Fossen, T.I. *Marine Control Systems: Guidance, Navigation and Control of Ships, Rigs and Underwater Vehicles*; Marine Cybernetics: Trondheim, Norway, 2002.
14. Woo, T.; Park, J.Y.; Yu, C.; Kim, N. Dynamic model identification of unmanned surface vehicles using deep learning network. *Appl. Ocean Res.* **2018**, *78*, 123–133. [CrossRef]
15. Sonnenburg, C.R.; Woolsey, C.A. Modeling, identification, and control of an unmanned surface vehicle. *J. Field Rob.* **2013**, *30*, 371–398. [CrossRef]
16. Li, J.C.; Zhao, D.L.; Ge, B.F.; Yang, K.W.; Chen, Y.W. A link prediction method for heterogeneous networks based on BP neural network. *Physica A* **2017**, *495*, 1–17. [CrossRef]
17. Hagan, M.T.; Menhaj, M.B. Training feedforward networks with the Marquardt algorithm. *IEEE Trans. Neural Netw.* **1994**, *5*, 989–993. [CrossRef] [PubMed]
18. Suri, R.N.; Deodhare, D.; Nagabhushan, P. Parallel Levenberg_Marquardt-based neural network training on Linux clusters—A case study. In Proceedings of the Third Indian Conference on Computer Vision, Graphics & Image Processing, Ahmadabad, India, 16–18 December 2002.

Journal of
Marine Science and Engineering

Article

A Ship Path Tracking Control Method Using a Fuzzy Control Integrated Line-of-Sight Guidance Law

Bing Han [1,2], Zaiyu Duan [1,*], Zhouhua Peng [3] and Yuhang Chen [1,4]

[1] Shanghai Ship and Shipping Research Institute Co., Ltd., Shanghai 200135, China; han.bing@coscoshipping.com (B.H.); chen.yuhang@coscoshipping.com (Y.C.)
[2] College of Physics and Electronic Information Engineering, Minjiang University, Fuzhou 350108, China
[3] College of Marine Electrical Engineering, Dalian Maritime University, Dalian 116026, China; zhpeng@dlmu.edu.cn
[4] Merchant Marine College, Shanghai Maritime University, Shanghai 201308, China
* Correspondence: duanzaiyu1999@outlook.com

Abstract: A fuzzy control improvement method is proposed with an integral line-of-sight (ILOS) guidance principle to meet the needs of autonomous navigation and high-precision control of ship trajectories. Firstly, a three-degree-of-freedom ship motion model was established with the battery-powered container ship ZYHY LVSHUI 01 built by the COSCO Shipping Group. Secondly, a ship path-following controller based on the ILOS algorithm was designed. To satisfy the time-varying demand of the look-ahead distance parameters during the following process, especially under different navigation conditions, fuzzy logic controllers were designed for different navigation conditions to automatically adjust the look-ahead distance parameters. Thirdly, a controller was applied that uses a five-state extended Kalman filter (EKF) to estimate the heading, speed, and heading rate based on the ship's motion model with the assistance of Global Navigation Satellite System (GNSS) position measurements. This provides the necessary navigational information, reduces the algorithm's dependence on sensors, and improves its generalizability. Finally, path-following experiments were carried out in the MATLAB experimental platform, and the results were compared with different following algorithms. The simulation results showed that the new algorithm has a better following performance, and it can maintain a smooth rudder angle output. The research results provide a reference for the path-following control of ships.

Keywords: path following; ILOS guidance law; fuzzy control; extended Kalman filter

Citation: Han, B.; Duan, Z.; Peng, Z.; Chen, Y. A Ship Path Tracking Control Method Using a Fuzzy Control Integrated Line-of-Sight Guidance Law. *J. Mar. Sci. Eng.* **2024**, *12*, 586. https://doi.org/10.3390/jmse12040586

Academic Editor: Mohamed Benbouzid

Received: 8 March 2024
Revised: 24 March 2024
Accepted: 27 March 2024
Published: 29 March 2024

1. Introduction

With the fast-paced growth of the economy and trade, there has been a surge in demand for freight transportation services. Waterborne transportation plays an indispensable role in efficiently transporting goods due to its cost-effectiveness and large capacity [1]. Container ships and other large vessels, crucial for waterway transportation, are continuously evolving towards digitization, autonomy, and intelligence to meet the ever-increasing demand for trade [2]. Autonomous ship navigation technology represents a fundamental feature of smart ships, and it also embodies the future direction of shipping technology [3]. Autonomous navigation technology needs to control the propulsion power unit according to the current position of the ship so that the ship navigates along the predetermined route, and ship path-following technology is critical to realizing this autonomous navigation of the ship, meaning it has important research significance [4].

In recent studies, there have been several approaches taken to build a simulation model for a real ship. Fossen [5] utilized a first-order model to represent the motion of the vessel. The first-order model [6] simulates the ship's course angle dynamics by mapping the rudder angle to the course angle derived from the data of the ship's maneuverability test. Song [7] employed an integral-type Abkowitz model [8] to describe the ship's motion. The

Abkowitz model approximates ship hydrodynamics by considering the vessel as an entirety and deriving third-order hydrodynamic derivatives from the Taylor expansion of motion equations. Qu [9] used a ship motion model proposed by Fossen [10], which is represented in the state space format and integrates hydrodynamic-component-based modeling with control design models based on vectors and matrices. Sandeepkumar [11] used a ship model of a KVlCC2 tanker. This modeling approach, proposed by the ship maneuvering mathematical model group (MMG) in Japan [12], is characterized by modeling the hull, propeller, and rudder separately and calculating their respective hydrodynamic forces.

In the study of path-following control, several researchers have suggested viable control strategies and addressed the related issues to different extents. Guo Jie [13] developed an Active Disturbance Rejection Controller by using the Fast Non-singular Terminal Sliding Mode. A simulation test was conducted with Dalian Maritime University's "Yulong" ship as the subject, which revealed that the controller could efficiently and accurately follow both straight and curved paths. In [14], a control law for tracking the trajectory of underactuated ships was developed by integrating the output redefinition method, an extended state observer (ESO), and the dynamic inversion control method. The design accounts for uncertainties in dynamics, external disturbances of unknown time-varying nature, and unavailable ship velocities. Ren [15] developed a time-scale decomposition method to solve the RRS control issue in path following. The resulting path-following performance is more stable and smoother. Zhu Kang [16] incorporated a deep reinforcement learning method into the LOS algorithm to suit complex control surroundings. They tested this approach using a 7 m KVLCC2 ship model, achieving a commendable tracking effect even for variable trajectories. Ghommam [17] developed a fuzzy-adaptive observer to estimate the state by solely utilizing the USVs' global position information and local measurement of the orientation angle. Le [18] integrated the Antenna Mutation Beetle Swarm Prediction Learning Algorithm into the line of sight (LOS) algorithm to address the ship parameter uncertainty issue. The algorithm's efficacy was verified through a simulation using a container ship as the test object. Renxiang Bu [19] combined a radial basis neural network with sliding mode control to accurately approximate the total unknown term and achieve precise trajectory tracking control in the presence of wind and wave currents. Huang [20] proposed an observer using internal model control (IMC), to rapidly estimate the sideslip angle in the line-of-sight guidance law, and demonstrated the efficacy of the proposed sideslip angle observer in enhancing the path-following accuracy. Xunwen Liu [21] introduced adaptive neural network and event-triggered control technology to reduce the physical damage of actuators. In recent years, linearized ship models have often been used in studies of ship path following, but actual ships have strong model and disturbance uncertainties [22], meaning that these models do not accurately reflect actual ship navigation. Meanwhile, some control algorithms are designed with idealized control inputs, which assume that theoretical values are equivalent to the real control inputs of the ship. The ship's maneuverability will be influenced by physical constraints, including limitations on the ship's rudder angle and propeller rotation speed during the voyage. Exceeding the working range limit or producing frequent jerks during maneuvering can result in significant physical damage to the ship's control mechanism. However, this approach does not align with actual engineering practice. Most researchers have focused on improving the anti-disturbance capability of an algorithm, but they have neglected the influence of the ship's maneuvering characteristics on the tracking performance under different sailing conditions. For instance, if a ship navigates along a curvilinear or twisting course, an algorithm that functions effectively on a straight trajectory will face issues such as intensified overshooting and biased oscillations, resulting in dreadful tracking performance.

In this paper, an integral line-of-sight navigation method with fuzzy control of the forward-looking distance is proposed to achieve precise path tracking in various sailing conditions. A 700 twenty-foot equivalent unit (TEU) container ship ZYHY LVSHUI 01 that operates on battery power, constructed by the COSCO Shipping Group, is chosen as the control object. Ultimately, simulation and experimental results demonstrate that the motion controller designed for the 700 TEU container ship effectively achieves path-following objectives under various conditions.

The main contributions and the key features of this paper are summarized as follows.

Using line-of-sight (LOS) navigation and fuzzy controllers, a ship motion controller is designed based on the ILOS guidance method with fuzzy control of the variable forward-looking distance. Fuzzy controllers designed for different navigational conditions can improve the performance of the algorithm by correcting the forward-looking distance parameter of the algorithm.

In this paper, a three-degree-of-freedom ship motion model is developed using the sailing data of container ship ZYHY LVSHUI 01. Furthermore, the extended Kalman filter algorithm is developed to accurately estimate speed, heading, and other states utilizing the ship's GNSS position information. This can enhance the general applicability of the control algorithm and decrease its reliance on costly sensors.

The rest of this paper is organized as follows: Section 2 introduces the ship motion model. Section 3 presents the design of the control system, including the introduction of the ILOS navigation method and its improvement. Section 4 illustrates the control algorithm's effectiveness through simulation experiments. Finally, Section 5 presents the conclusion and future work.

2. Preliminaries and Problem Statement

In this paper, a three-degree-of-freedom (DOF) mathematical model for ship maneuvering is presented, which incorporates surge, sway, and heave, based on the parameters of a 700 TEU container ship ZYHY LVSHUI 01. The 700 TEU container ship is equipped with twin engines, twin propellers, and twin rudders. See Table 1 for details of the ship parameters.

Table 1. Ship parameters.

Parameters	Values
Length	119.8 m
Draught	5.5 m
Displacement	12,600,000 kg
Rudder Area	13.02 m^2
Diameter of Propeller	2.8 m
Breadth	23.6 m
Block Coefficient	0.835
Molded Depth	9 m
Aspect Ratio of Rudder	1.355
Propulsion Power	900 kW

The equation for the ship model can be expressed as

$$
\begin{cases}
\dot{x} = u\cos\varphi + v\sin\varphi \\
\dot{y} = u\sin\varphi + v\cos\varphi \\
\dot{\varphi} = r
\end{cases}
$$

$$
\begin{cases}
(m + m_x)\dot{u} - (m + m_y)vr = X_H + X_P + X_R + X_W + X_C \\
(m + m_y)\dot{v} - (m + m_x)ur = Y_H + Y_P + Y_R + Y_W + Y_C \\
(I_{zz} + J_{zz})\dot{r} = N_H + N_P + N_R + N_W + N_C \\
T\dot{\delta} = K\delta_c - \delta
\end{cases}
\tag{1}
$$

where (x, y) are the position coordinates of the ship, φ is the heading angle, m is the ship's mass, m_x, m_y is the added mass component along the respective direction, I_{zz} is the moment of inertia, J_{zz} represents the added moment of inertia, X, Y, and N are the external sway, surge forces, and yaw moments acting on the ship in the body reference frame, and the subscripts H, P, R, W, and C denote the forces and moments of the hull, oars, rudder, wind, and currents applied to the ship, respectively. The kinetic parameters in the equations above were calculated utilizing the empirical formulas supplied in [23]. The forces and moments on the hull are

$$
\begin{cases}
X_H = X(u) + X_{vv}v^2 + X_{vr}vr + X_{rr}r^2 \\
Y_H = Y_v v + Y_r r + Y_{|v|v}|v|v + Y_{|v|r}|v|r + Y_{|r|r}|r|r \\
N_H = N_v v + N_r r + N_{|v|v}|v|v + N_{vvr}v^2 r + N_{vrr}vr^2
\end{cases}
\tag{2}
$$

Table 2 shows the hydrodynamic coefficients calculated with empirical equations.

Table 2. Hydrodynamic coefficients.

Parameters	Values	Parameters	Values
X_{vv}	-0.0519	$Y_{|r|r}$	-0.0126
X_{vr}	-1.3107×10^6	N_v	-0.0737
X_{rr}	-0.065	N_r	-0.0443
Y_v	-0.3509	$N_{|v|v}$	-0.0112
Y_r	-0.0399	N_{vvr}	-0.2879
$Y_{|v|v}$	-0.1937	N_{vrr}	-0.0562
$Y_{|v|r}$	-0.3299		

In this paper, we maintain a constant value for the propeller speed while controlling the ship through the manipulation of the rudder. The rudder characteristics are represented using a first-order system [24]. The recommended rudder angle is indicated by δ_c, while the current rudder angle is δ. K and T represent the control gain and time constant, respectively. The maximum rudder angle is restricted to $\delta \leq \pm 35°$. The forces and moments generated by the rudder are as follows:

$$
\begin{cases}
X_R = (1 - t_R)F_N \sin\delta \\
Y_R = (1 + a_H)F_N \cos\delta \\
N_R = (x_R + a_H x_H)F_N \cos\delta
\end{cases}
\tag{3}
$$

where F_N is the rudder positive pressure and the rudder parameters are as displayed in Table 3.

Table 3. Rudder parameters.

Parameters	Values
t_R	0.1844
a_H	0.8788
x_R	60
x_H	-0.4835

Then, the disturbance force on the hull is divided into two parts, wind and current, and is calculated using empirical equations. The equations below are used to calculate the disturbance forces and moments generated by the wind and the current on the hull.

$$
\begin{bmatrix} X_W \\ Y_W \\ N_W \end{bmatrix} = \frac{1}{2}\rho_a V_w^2 \begin{bmatrix} C_X(\theta_w)A_{Fw} \\ C_Y(\theta_w)A_{Lw} \\ C_N(\theta_w)A_{Fw}L \end{bmatrix}
\tag{4}
$$

$$
\begin{bmatrix} X_C \\ Y_C \\ N_C \end{bmatrix} = \frac{1}{2}\rho L d V_C^2 \begin{bmatrix} C_X(\theta_C) \\ C_Y(\theta_C) \\ C_N(\theta_C)L \end{bmatrix}
\tag{5}
$$

where V_w, V_c is the relative speed of wind and current, θ_w, θ_c is the relative angle of wind and current, ρ_a, ρ is the density of air and water, L is the length of the ship, d is the draft of the ship, A_{Fw} and A_{Lw} are the wind areas of the front and side of the hull, respectively, and

$C_X(\theta_w), C_Y(\theta_w), C_N(\theta_w)$ and $C_X(\theta_C), C_Y(\theta_C), C_N(\theta_C)$ are the wind force and current force coefficient, generally obtained from ship testing results.

The objective of this article is to design an LOS-based path-following control scheme for the target ship that enables it to travel the desired path with high accuracy, regardless of model uncertainty and unknown environmental disturbances.

3. Control System Design

The basic block diagram of the control system is shown in Figure 1. The ship features a GNSS, which obtains the ship's current location in real-time and estimates its condition through an extended Kalman filter. The ship's desired heading is calculated by ILOS with a fuzzy controller. This calculation is based on both pre-set path points and the real-time ship position. Then, the PD controller is utilized to control the rudder rotation, such that the ship can be guided to follow the pre-set path point.

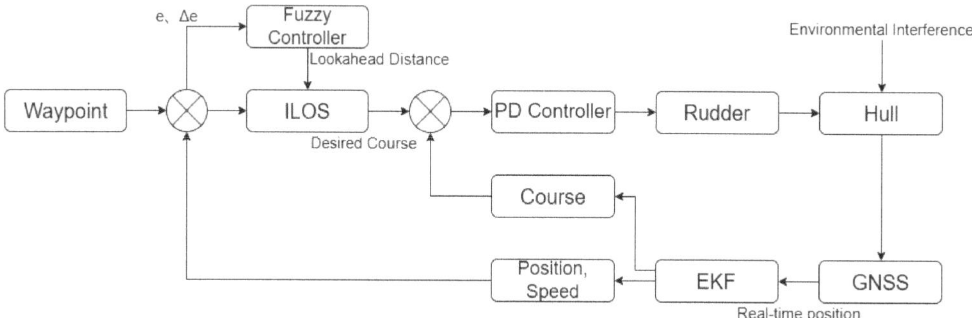

Figure 1. Basic block diagram of control system.

3.1. ILOS Guidance Method

A commonly utilized algorithm for following a path is the line-of-sight (LOS) algorithm. In Figure 2, we indicate some primary variables utilized in the ILOS algorithm.

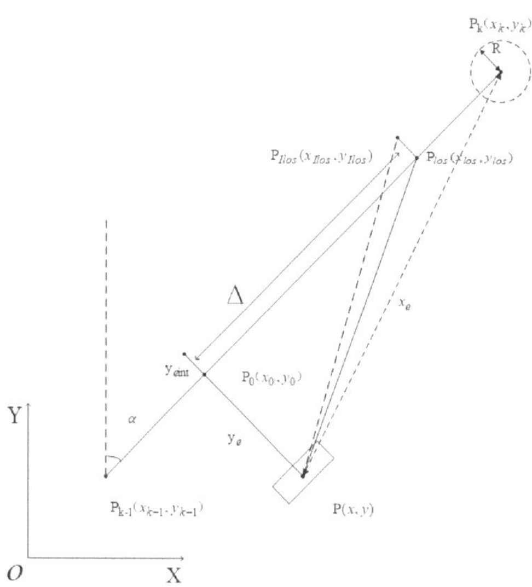

Figure 2. An illustration of the integral line-of-sight guidance.

The straight line between points $P_{k-1}(x_{k-1}, y_{k-1})$ and $P_k(x_k, y_k)$ is the line path to be followed. $P(x, y)$ represents the ship's current location while $P_0(x_0, y_0)$ is the point where the ship's location intersects the path. Then, the direction angle of the path can be calculated using

$$\alpha = \text{atan2}\left(\frac{y_k - y_{k-1}}{x_k - x_{k-1}}\right) \tag{6}$$

Following this, we could compute the along-track and cross-track errors (y_e, x_e) by using

$$\begin{cases} y_e = -(x - x_k)\sin\alpha + (y - y_k)\cos\alpha \\ x_e = (x - x_k)\cos\alpha + (y - y_k)\sin\alpha \end{cases} \tag{7}$$

When the along-track distance x_e is less than R, the LOS algorithm goes to the next waypoint. In the traditional LOS algorithm [10], the desired heading χ is calculated based on

$$\chi = \alpha - \text{atan2}\left(\frac{y_e}{\Delta}\right) \tag{8}$$

where Δ is the looking-ahead distance. However, conventional LOS guidance is not equipped to manage an environmental disturbance, such as wind or current. Accordingly, Borhaug [25] proposed the ILOS algorithm:

$$\chi = \alpha - \text{atan2}\left(\frac{y_e + \kappa y_{eint}}{\Delta}\right)$$
$$\dot{y}_{eint} = \frac{\Delta y_e}{\Delta^2 + (y_e + \kappa y_{int})^2} \tag{9}$$

where $\kappa > 0$ is a designed integral gain. In [26], a different version of the integral LOS algorithm is proposed as follows:

$$\chi = \alpha - \text{atan2}\left(\frac{y_e + \kappa y_{eint}}{\Delta}\right)$$
$$\dot{y}_{eint} = \frac{U y_e}{\sqrt{\Delta^2 + (y_e + \kappa y_{int})^2}} \tag{10}$$

where U is the absolute speed. From Figure 2 with Equations (9) and (10), the integral term indicates that the desired heading angle will be a non-zero constant when $y_e = 0$. This enables the use of a portion of the ship's forward speed to counteract the effects of the flow disturbance. Algorithm 1 provides the pseudo-code for the LOS/ILOS algorithm.

Algorithm 1: LOS/ILOS

Inputs: ship location (x, y); waypoint (wp.x,wp.y)
Output: desired heading angle χ

1. $k \leftarrow 1$ (initialization); set R; set LOS/ILOS parameter Δ, κ
2. Initialization starting point $(xk, yk) \leftarrow (wp.x(k), wp.y(k))$, and end point $(xk\_next, yk\_next) \leftarrow (wp.x(k+1), wp.y(k+1))$
3. Compute the path angle α
4. Compute the along-track and cross-track errors (x_e, y_e)
5. If $x\_e < R\_switch$, then k=k+1, end
6. Compute the desired heading angle χ

If the ship's position is far from the intended path, the accumulation of error can easily lead to integration saturation and result in overshoot. Therefore, this study employs a combination of the ILOS and LOS navigation methods [27], as illustrated in Figure 3. If $y_e < L_{pp}$, the controller employs the ILOS navigation method, and if $y_e \geq L_{pp}$, the controller shall utilize the LOS navigation method.

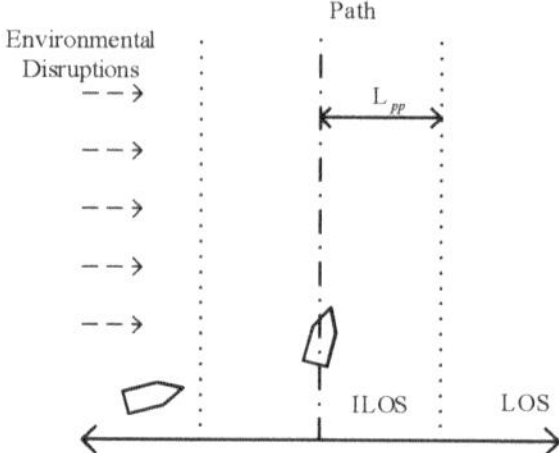

Figure 3. Area schematic.

3.2. Fuzzy-Rule-Based Lookahead Distance Selection Method

The performance of the ILOS algorithm can be enhanced by modifying the parameter for lookahead distance. A shorter lookahead distance typically leads to more aggressive steering and faster attainment of the desired path, but it may also cause unwanted oscillations around it. Conversely, a longer lookahead distance results in smoother steering that prevents such oscillations but has the disadvantage of slower convergence to the path. The fuzzy controller is designed to dynamically adjust the forward-looking distance of the ILOS navigation method by using fuzzy rules based on the deviation e and the difference in the deviation Δe. The process of formulating the mapping from a given input to an output using fuzzy logic is known as the fuzzy inference system (FIS). The fuzzy inference system type utilized in this paper is the "Sugeno Fuzzy Inference System" [28]. The first step is to define the inputs and outputs of the system and determine the degree to which they belong to each corresponding fuzzy set using Gaussian membership functions. In this paper, the inputs to the system include the deviation $e = [-50, 50]$ and the difference in the deviation $de = [-0.3, 0.3]$, while the output is the look-ahead distance $\Delta = [100, 300]$. Then, in the second step, the center-of-area approach, also known as the center-of-gravity method, is the most widely used defuzzification procedure in fuzzy logic control. Essentially, it is

$$u = \frac{\sum_{i=1}^{N} w_i z_i}{\sum_{i=1}^{N} w_i} \tag{11}$$

where N is the number of quantization levels of the output, z_i is the value of the output at quantization level, and w_i represents its membership value in the output fuzzy set. The final step is to define fuzzy rules for different navigational conditions.

Overall, when the ship moves away, we decrease the forward-looking distance to accelerate steering. Conversely, when the ship moves closer, we increase the forward-looking distance to minimize overshooting. In the line condition, the target ship follows a predetermined path on a straight course from a distant position, and the fuzzy controller determines the motion trend of the ship using the deviation e and the difference in the deviation de. If both the deviation e and the difference in the deviation de are positive, it indicates that the ship is moving away from the reference path. In this case, even if e is small, the look-ahead distance Δ needs to be reduced. Conversely, if the deviation e and the difference in the deviation de are in opposite directions, it means that the ship is close to the reference path. Therefore, the value of L needs to be increased appropriately to prevent overshooting. The resulting fuzzy rule table for the line condition is shown in Table 4.

Table 5 shows the design of fuzzy rules for curvilinear conditions, which follows the same logic as that of line conditions. In curvilinear conditions, a ship will make multiple turns. The curvature of the route and the disturbance of the flow will cause a larger deviation. To decrease steering bias, the ship's lookahead distance should be reduced even more, prompting the ship to steer more assertively and ultimately reducing steering bias.

Table 4. Components of fuzzy rules for the line condition.

de	e						
	NB	**NM**	**NS**	**ZO**	**PS**	**PM**	**PB**
NB	PVS	PS	PS	PM	PB	PM	PVS
NS	PVS	PS	PM	PB	PM	PS	PVS
ZO	PVS	PM	PS	PB	PM	PS	PVS
PS	PVS	PS	PM	PB	PM	PS	PVS
PB	PVS	PM	PB	PM	PS	PVS	PVS

Table 5. Components of fuzzy rules for the curvilinear condition.

de	e						
	NB	**NM**	**NS**	**ZO**	**PS**	**PM**	**PB**
NB	PVS	PVS	PVS	PS	PVS	PVS	PVS
NS	PVS	PS	PM	PM	PM	PS	PVS
ZO	PVS	PM	PS	PB	PM	PS	PVS
PS	PVS	PS	PM	PM	PM	PS	PVS
PB	PVS	PVS	PVS	PS	PVS	PVS	PVS

In a turning condition, the ship's heading angle is constantly adjusted, resulting in changing environmental disturbances and deviations that make it difficult for the error to converge to zero. To counter the effects of environmental disturbances, the change in the difference in the deviation *de* determines the magnitude of the disturbances and adjusts the lookahead distance dynamically. The ultimate components of the fuzzy rules are shown in Table 6.

Table 6. Components of fuzzy rules for the turning condition.

de	e						
	NB	**NM**	**NS**	**ZO**	**PS**	**PM**	**PB**
NB	PVS	PVS	PS	PS	PB	PM	PVS
NS	PVS	PS	PM	PM	PM	PS	PVS
ZO	PVS	PS	PS	PB	PS	PS	PVS
PS	PVS	PS	PM	PM	PM	PS	PVS
PB	PVS	PM	PB	PS	PS	PVS	PVS

3.3. Extended Kalman Filter

Researchers sometimes assume that the ship navigation subsystem is available as the perception system, and that the ship control system can access the necessary information directly. To obtain precise navigational information, such as speed and course, ships require costly sensory equipment. So, to improve the generalizability of control algorithms and eliminate the need for sensing equipment, it is necessary to accurately estimate the information required for control. To achieve a precise estimation of variables such as speed and direction and enhance the robustness of the algorithm, in this study, a five-state extended Kalman filter algorithm [29] is employed to estimate the ship's speed, course angle, and other variables utilizing positional data from the GNSS. The dynamics of a ship following a path can be modeled by using a combination of the CV and CA models [30,31] according to

$$\begin{cases} \dot{x}^n = U\cos(\chi) \\ \dot{y}^n = U\sin(\chi) \\ \dot{U} = -\alpha_1 U + \omega_1 \\ \dot{\chi} = \omega_\chi \\ \dot{\omega}_\chi = -\alpha_2 \omega_\chi + \omega_2 \end{cases} \tag{12}$$

where (x^n, y^n) is the north-east position of a ship, ω_1 and ω_2 are Gaussian white-noise processes, and two constants (α_1, α_2) from the Singer model [32] have been incorporated into the model so that U and ω_χ converge to zero during stationkeeping [33]. The discrete form of the equation is

$$\begin{cases} x^n[k+1] = x^n[k] + hU[k]\sin(\chi[k]) \\ y^n[k+1] = y^n[k] + hU[k]\sin(\chi[k]) \\ U[k+1] = (1 - h\alpha_1)U[k] + h\omega_1[k] \\ \chi[k+1] = \chi[k] + h\omega_1[k] \\ \omega_\chi[k+1] = (1 - h\alpha_2)\omega_\chi[k] + h\omega_2[k] \end{cases} \tag{13}$$

where h is the sampling time. Then, the GNSS measurement equations are

$$\begin{cases} y_1 = x^n + \varepsilon_1 \\ y_2 = y^n + \varepsilon_2 \end{cases} \tag{14}$$

where ε_1 and ε_2 are Gaussian white-noise measurement noise. The discrete-time forms are

$$\begin{cases} y_1[k] = x^n[k] + \varepsilon_1[k] \\ y_2[k] = y^n[k] + \varepsilon_2[k] \end{cases} \tag{15}$$

The discrete-time state–space model becomes

$$\begin{aligned} x[k+1] &= A_d x[k] + E_d \omega[k] \\ y[k] &= C_d x[k] + \varepsilon[k] \end{aligned} \tag{16}$$

where $x = [x^n, y^n, U, \chi, \omega_\chi]^T, y = [x^n, y^n]^T, \omega = [\omega_1, \omega_2]^T$, and

$$A_d = \begin{bmatrix} 1 & 0 & \cos(\hat{\chi})h & -h\widehat{U}[k]\sin(\hat{\chi}) & 0 \\ 0 & 1 & \sin(\hat{\chi})h & h\widehat{U}[k]\cos(\hat{\chi}) & 0 \\ 0 & 0 & 1 - h\alpha_1 & 0 & 0 \\ 0 & 0 & 0 & 1 & h \\ 0 & 0 & 0 & 0 & 1 - h\alpha_2 \end{bmatrix}$$

$$E_d = \begin{bmatrix} 0 & 0 \\ 0 & 0 \\ h & 0 \\ 0 & 0 \\ 0 & h \end{bmatrix}, \quad C_d = \begin{bmatrix} 1 & 0 & 0 & 0 & 0 \\ 0 & 1 & 0 & 0 & 0 \end{bmatrix} \tag{17}$$

Extended Kalman filter algorithms based on motion models then become [33]
Initial values:

$$\hat{x}^-[0] = x_0$$
$$\widehat{P}^-[0] = E\left[(x[0] - \hat{x}^-[0])(x[0] - \hat{x}^-[0])^T \right] = P_0 \tag{18}$$

Kalman filter gain:

$$K[k] = \widehat{P}^-[k]C_d^T[k]\left(C_d[k]\widehat{P}^-[k]C_d^T[k] + R_d[k] \right)^{-1} \tag{19}$$

Corrector:

$$\hat{x}[k] = \hat{x}^-[k] + K[k]\left(y[k] - h(\hat{x}^-[k]) \right)$$
$$\widehat{P}[k] = (I - K[k]C_d[k])\widehat{P}^-[k](I - K[k]C_d[k])^T + K[k]R_d[k]K^T[k] \tag{20}$$

Predictor:

$$\hat{x}^-[k+1] = A_d\hat{x}[k] + B_d u[k]$$
$$\widehat{P}^-[k+1] = A_d\widehat{P}[k]A_d^T + E_d Q_d[k]E_d^T \tag{21}$$

where $h\left(\widehat{x}^-[k]\right) = C_d[k]\widehat{x}^-[k]$, and where $Q_d[k]$ and $R_d[k]$ are the process covariance and measurement matrices, respectively.

4. Simulations

In this section, the 700ETU motion model is used as the test object to verify the effectiveness of the modified ILOS algorithm. The motion mathematical model's dynamic parameters are extensively outlined in Section 2. The ship motion model and Kalman filter estimation algorithm were tested using the Zig-Zag and turning tests. Simulation tests were also performed using the traditional LOS algorithm, Borhaug's ILOS algorithm, Lekkas' ILOS algorithm, and the modified ILOS algorithm, respectively, to demonstrate the advantages of the modified ILOS algorithm. The test algorithms utilized the PD controller for heading control, while the other three control algorithms used a fixed lookahead distance parameter, which was set to twice the length of the ship.

4.1. Test Simulation Model

In this section of our work, the Zig-Zag test and turning test were conducted to verify the maneuverability and applicability of the 700 TEU container ship simulation model. In addition, the extended Kalman filter from Section 3 was used for parameter estimation during the test.

First, the $20°/20°$ Zig-Zag test procedure involves the following steps: (1) Initially, the container ship is sailing at a speed of 12 n mile/h (about 6 m/s). After approaching steadily, it rapidly steers $20°$ to starboard and maintains the rudder angle. (2) When the ship's heading is $20°$ off the initial course, the rudder is rapidly turned to the port side at $20°$ and maintained. (3) Finally, this process is repeated until the end of the test. In Figure 4, the test results show that the first overshoot angle is about $4.5°$ which is in accordance with the maneuvering standards. This also confirms that the EKF can estimate the speed and course angle with great precision during the Zig-Zag test.

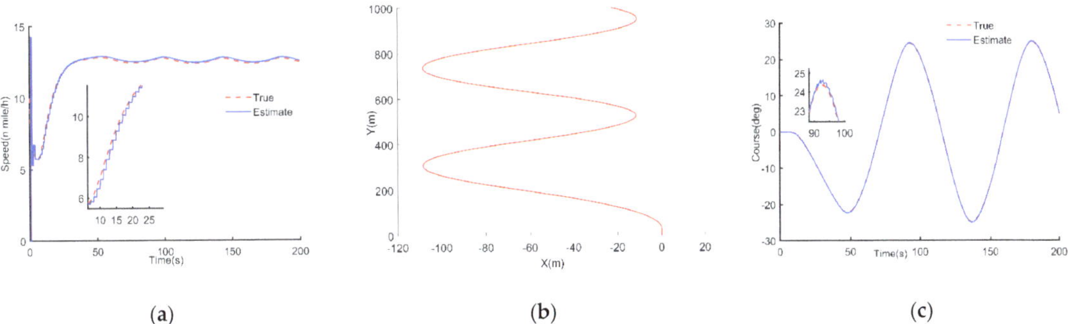

Figure 4. The Zig-Zag simulation test: (**a**) speed estimate; (**b**) ship's trajectory; (**c**) course estimate.

For the turning test, the process begins with the ship maintaining a constant speed of 12 n mile/h (about 6 m/s). Next, the rudder is turned to the maximum right angle of $35°$ and remains in this position until the ship completes a full turning circle beyond $360°$.

Figure 5 shows that in the turning test, the ship's advance distance (Ad) at $90°$ is approximately 323 m, and the tactical diameter distance (DT) at $180°$ is 616 m. It is important to note that the Ad of the turning circle is less than three times the Lpp (length of perpendicular) of the ship, which is about 120 m; and the DT of the turning circle is about six times the Lpp. During the turning test, the EKF obtains accurate estimates of the speed and the course angle.

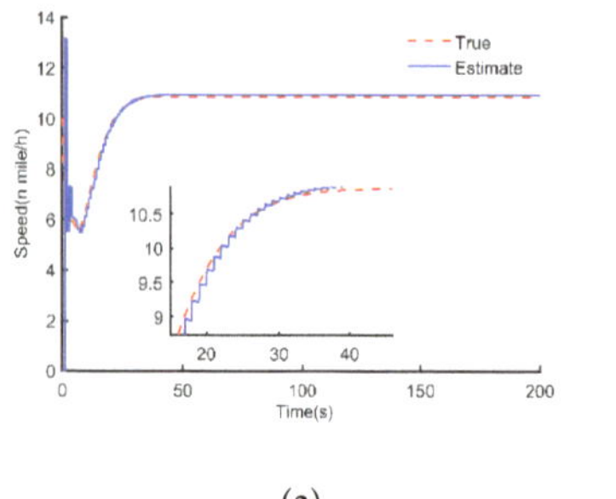
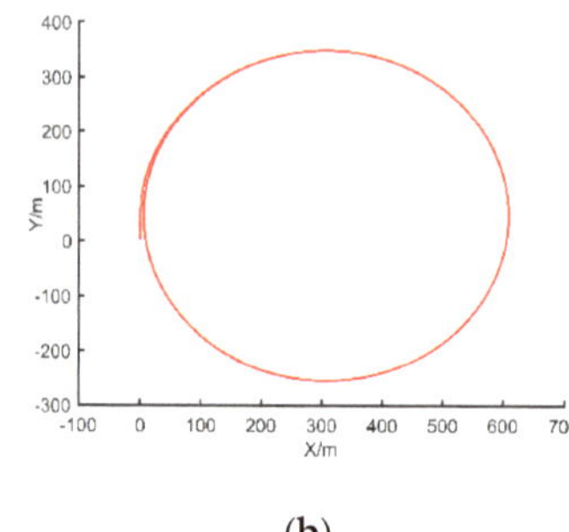
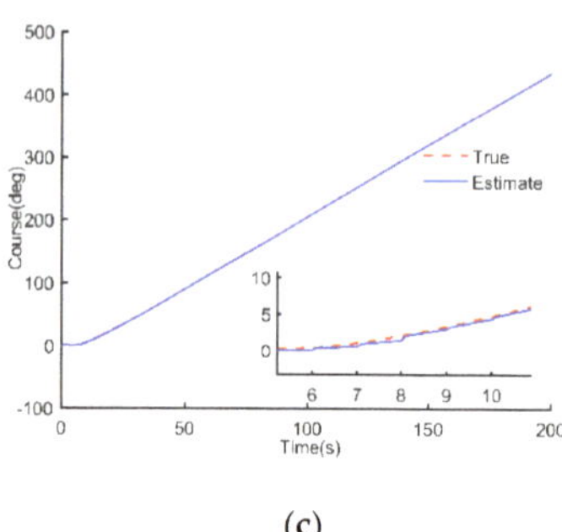

Figure 5. The turning simulation test: (**a**) speed estimate; (**b**) ship's trajectory; (**c**) course estimate.

From the results of the tests above, it can be concluded that the vessel has good maneuverability, while the performance of the EKF is generally acceptable for engineering application.

4.2. Line Path-following Test

In the line path-following test, the desired path begins at the coordinates (1500, 0) and concludes at (1500, 9000). The ship's initial position is (0, 0) and it is heading east. The ship's speed is set at 10 n mile/h, while the wind speed is 1.5 m/s with a wind angle of 45°, and the current speed is 2.2 m/s with a current angle of 45°. The simulation results using the modified ILOS algorithm are illustrated by the red line in Figure 6. Figure 6a,b show that the ship under the modified ILOS algorithm can follow the reference path with satisfactory control performance. The algorithm's final following error is 1 m, with only 5 m of overshoot during the following process. In Figure 6c,d, the controlled ship displays a more reasonable and smooth change in rudder angle and speed. Moreover, the simulation results under the other three algorithms are also shown in Figure 6. Figure 6b illustrates that the traditional LOS algorithm is susceptible to environmental disturbances, resulting in a fixed error of approximately 15 m that cannot be eliminated. Meanwhile, the rest of the ILOS algorithms offset the influence of the environmental interference, and the final convergence error reaches within 1 m. However, the two ILOS algorithms produce overshoots of 117 m and 22 m under the influence of the fixed lookahead parameters and the integral term, respectively. Then, Figure 6c,d show that the control inputs of the other algorithms for the rudder angle have reached the maximum limit of the rudder, resulting in a significant reduction in speed. Consequently, the simulation comparison results indicate that the proposed algorithm can achieve a satisfactory following performance. Moreover, it maintains a reasonable rudder angle and speed. Table 7 compares the performance metrics of different algorithms, including the overshooting, the final following error, and the time required for the error to converge. The modified algorithm has improved convergence speed, reduced tracking error, and significantly decreased overshooting during the convergence process.

Table 7. Comparison of line path-following performance.

	Overshooting/m	Errors/m	Time/s
LOS	none	15	300
Borhaug's ILOS	117	1	190
Lekkas's ILOS	22	1	200
Modified ILOS	5	1	200

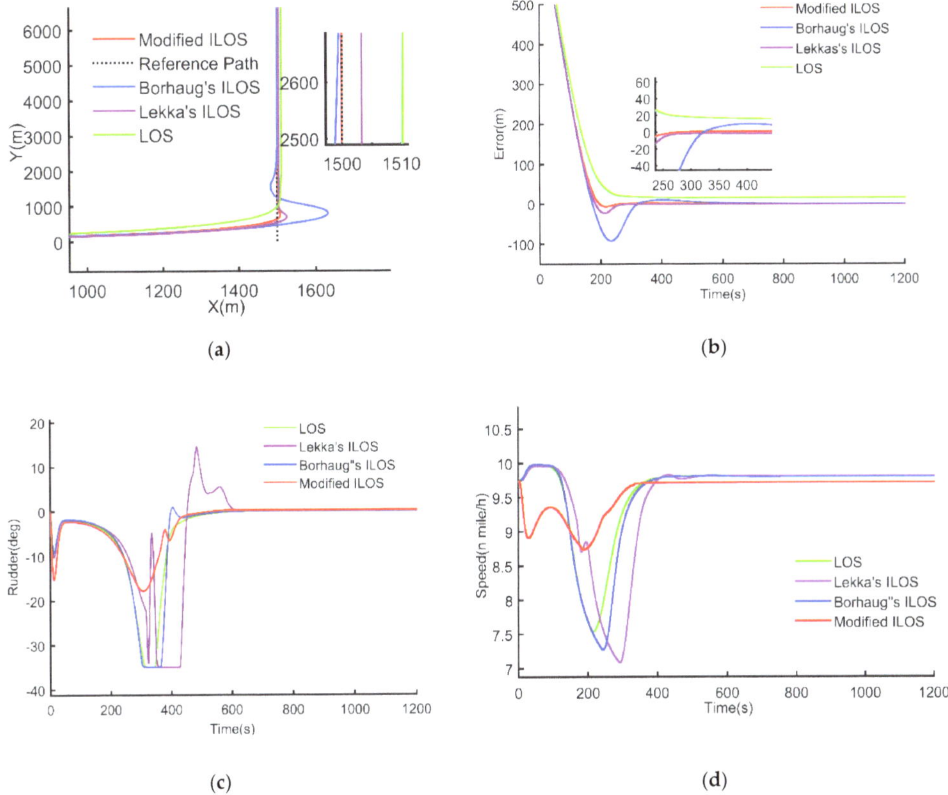

(**a**)   (**b**)

(**c**)   (**d**)

Figure 6. Line path-following simulation results: (**a**) path-following performance; (**b**) cross-tracking errors; (**c**) rudder angles; (**d**) speeds.

4.3. Curvilinear Path-Following Test

In the curvilinear path-following test, the reference path is

$$\begin{cases} X = 500a \\ Y = 500\cos(a) \end{cases} \tag{22}$$

where $a = [0, 4\pi]$, the ship's initial position is $(-100, 500)$, and it is heading east. The ship's speed is set at 10 n mile/h, while the wind speed is 1.5 m/s with a wind angle of $45°$, and the current speed is 2.2 m/s with a current angle of $45°$. The simulation results using the modified ILOS algorithm are illustrated by the red line in Figure 7. Figure 7a shows the performances of the four algorithms, which are all capable of following the reference path. Based on Figure 7b it can be observed that the proposed algorithm has the smoothest convergence process, completing convergence in 220 s with a following error of 1 m. Figure 7c,d exemplify the changes in the rudder angle and ship speed during the following process. The proposed algorithm can maintain stability during the ship's multiple course adjustments by adjusting the lookahead distance through the fuzzy rule.

In Figure 7c, the regulation of the rudder angle displays minimal fluctuations, indicating a consistent and stable control. In Figure 7d, the ship's speed shows a steady cyclic variation. Table 8 presents a comparison of the performance indices of the algorithms. It can be observed that the proposed algorithm has the smoothest convergence process, completing convergence in 220 s with a following error of 1 m. The proposed modified ILOS algorithm can maintain stability during the ship's multiple course adjustments by adjusting the lookahead distance through the fuzzy rule. This contrasts with the other algorithms, which produce large jitter due to heading changes.

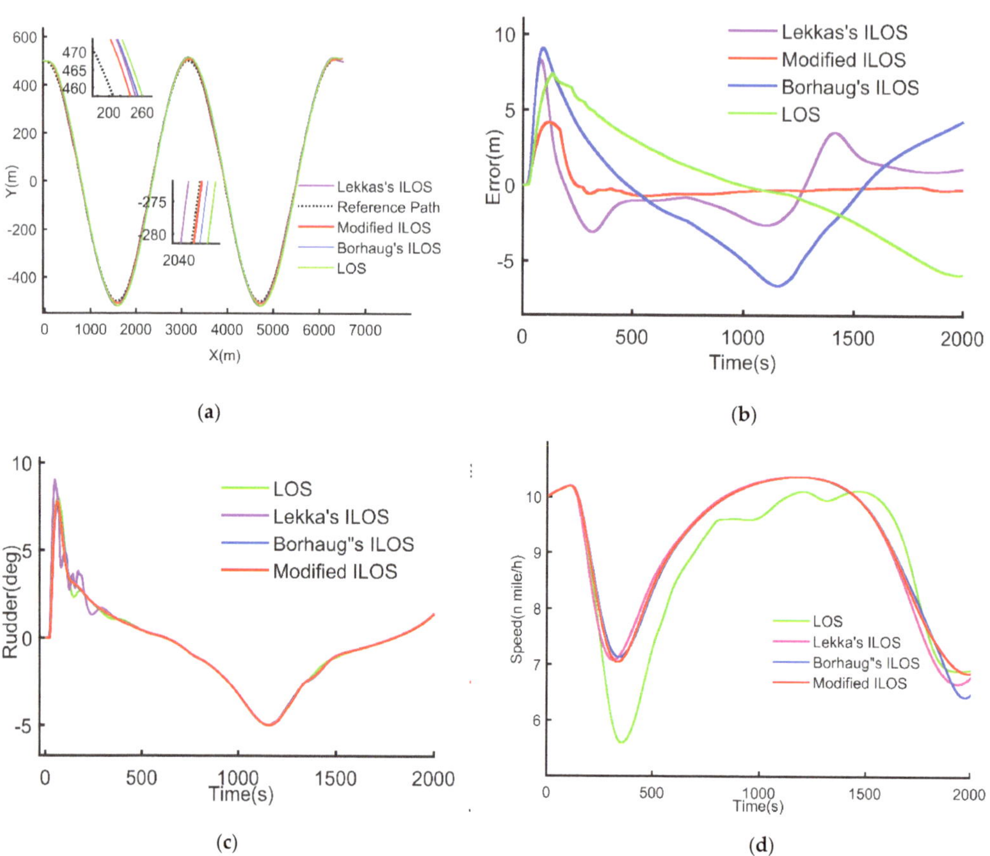

Figure 7. Curvilinear path-following simulation results: (**a**) path-following performance; (**b**) cross-tracking errors; (**c**) rudder angles; (**d**) speeds.

Table 8. Comparison of Curvilinear path-following performance.

	Overshooting/m	Errors/m	Time/s
LOS	7.5	8	1160
Borhaug's ILOS	9.6	8	580
Lekkas's ILOS	8.3	4	190
Modified ILOS	4.2	1	220

4.4. Turning Path-Following Test

In the turning path-following test, the reference path is

$$\begin{cases} X = 1500\sin(a) \\ Y = 1500\cos(a) \end{cases} \tag{23}$$

where $a = [0, 4\pi]$, the ship's initial position is $(-100, 1500)$, and it is heading east. The ship's speed is set at 10 n mile/h, while the wind speed was 1.5 m/s with a wind angle of $45°$, and the current speed is 2.2 m/s with a current angle of $45°$. The simulation results are shown in Figure 8 and the performance quantification indices are summarized in Table 9. Figure 8a shows that the proposed algorithm can drive the ship along the desired path with a high-precision process. Figure 8b shows that the proposed algorithm can converge the error quickly, whereas the other algorithm has an obvious oscillation during the process. Under turning conditions, the ship is subject to continuously changing environmental disturbances and the influence of path curvature. This makes a portion of the following error difficult to converge. The LOS algorithm conventionally has a constant following error of 7 m when following the reference slewing trajectory and is unable to converge. While Borhaug's ILOS algorithm can slowly reduce the following error to 3 m in 2000 s under the effect of the integral term, Lekkas's ILOS algorithm has a much faster error convergence and converges the following error to 1 m in 1500 s. The algorithm mentioned above cannot be adjusted according to the actual navigation situation as it uses a fixed lookahead distance parameter. Therefore, there is still room for improvement in its performance. The fuzzy rule enables the algorithm to use lower lookahead parameters during turning conditions compared to straight line conditions. This prompts the ship to perform more aggressive steering to eliminate errors. The following error is reduced to 1 m at 500 s without generating unstable oscillations. In Figure 8c,d, the controlled ship displays a more reasonable and smooth change in rudder angle and speed. Hence, in this case, the modified ILOS algorithm path-following control method is more effective and robust according to these simulation results.

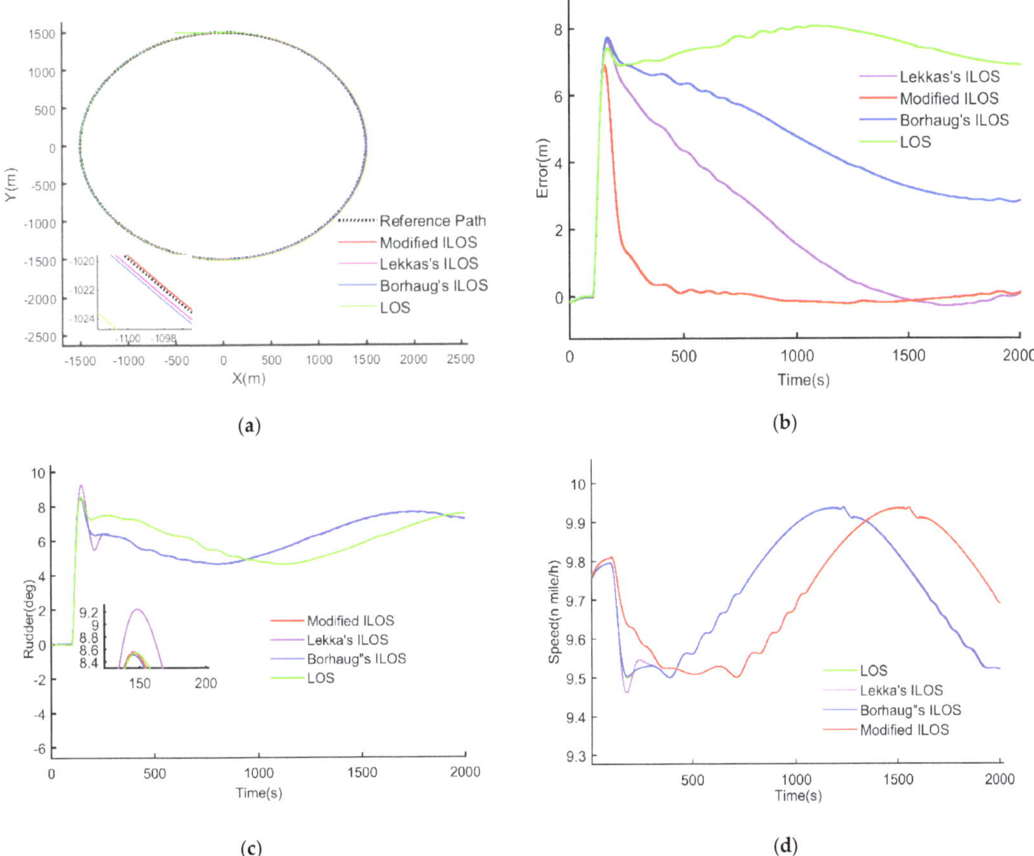

Figure 8. Turning path-following simulation results: (**a**) path-following performance; (**b**) cross-tracking errors; (**c**) rudder angles; (**d**) speeds.

Table 9. Comparison of Turning path-following performance.

	Overshooting/m	Errors/m	Time/s
LOS	7.9	7	2000
Borhaug's ILOS	7.9	3	2000
Lekkas's ILOS	7.8	1	1500
Modified ILOS	7.5	1	500

5. Conclusions

In this paper, a fuzzy-controlled-variable forward-looking-distance ILOS guidance law has been presented to meet the needs of autonomous ship navigation and high-precision ship trajectory control. Fuzzy rules designed for various navigation conditions can improve the accuracy and convergence speed by adjusting the algorithm's lookahead distance parameter. In addition, the algorithm does not rely on accurate ship models or sensing devices, but rather utilizes the EKF algorithm to estimate the ship's state via GNSS position data, making the algorithm both generalizable and universal. Simulation and experimental results have demonstrated that that a ship under the modified ILOS algorithm has satisfactory following results for line, curvilinear, and turning paths, and it performs more reasonable maneuvers compared to when using other algorithms. Meanwhile, it was concluded from Tables 7–9 that the proposed control algorithm can converge the tracking error to within 1 m under the three working conditions with high tracking accuracy, and it also has a faster and better convergence process compared to other comparative algorithms. The results attest to its comprehensive advantages, which are of great significance for achieving autonomous navigation and high-precision control of ship trajectories.

In future research, obstacles and automatic collision avoidance should be considered in path following. Furthermore, since the ship motion model is calculated by using mainly empirical formulas, it can be further optimized to improve the accuracy of the ship motion model.

Author Contributions: B.H.: Conceptualization, Methodology, Supervision, Visualization, Funding Acquisition, Writing—Review and Editing; Z.D.: Conceptualization, Data Curation, Writing—Original Draft, Software, Validation, Writing—Review and Editing; Z.P.: Investigation, Conceptualization, Project Administration, Supervision, Validation, Writing—Review and Editing; Y.C.: Conceptualization, Data Curation, Formal Analysis, Funding Acquisition, Investigation, Project Administration, Writing—Review and Editing. All authors have read and agreed to the published version of the manuscript.

Funding: This research was funded by the National Key R&D Program of China, grant number 2022YFC2807004; the Natural Science Foundation of the Fujian Province of China, grant number 2022J011128; and the Shanghai Science Program for a Shanghai Academic/Technology Research Leader, grant number 23XD1431000.

Institutional Review Board Statement: Not applicable.

Informed Consent Statement: Not applicable.

Data Availability Statement: Data are contained within the article.

Acknowledgments: The authors would like to thank the editor and two reviewers for their useful feedback that improved this paper.

Conflicts of Interest: The authors declare no conflicts of interest.

References

1. Yan, X.P. Implementing self-reliance and scientific and technological innovation to open new horizons for high-quality development of Yangtze River shipping. *China Water Trans.* **2021**, *11*, 12–15.
2. Li, Y.Z.; Zheng, Z.; Wu, T.; Dou, Z.X. Introduction to Development Trend and Challenges of Smart Ships in Era of Industry 4.0. *Ship Eng.* **2023**, *45*, 224–229.
3. Li, Y.J.; Zhang, R.; Wei, M.H.; Zhang, Y. State-of-the-art research and prospects of key technologies for ship autonomous navigation. *Chin. J. Ship Res.* **2021**, *16*, 32–44.

4. Yu, Y.L.; Su, R.B.; Fong, X. Tracking control of backstepping adaptive path of unmanned surface vessels based on surge-varying LOS. *Chin. J. Ship Res.* **2019**, *14*, 163–171.
5. Fossen, T.I. Line-of-sight path-following control utilizing an extended Kalman filter for estimation of speed and course over ground from GNSS positions. *J. Mar. Sci. Technol.* **2022**, *27*, 806–813. [CrossRef]
6. Nomoto, K.; Taguchi, T.; Honda, T.; Hirano, S. On the Steering Qualities of Ships. *J. Zosen Kiokai* **1956**, *99*, 75–82. [CrossRef]
7. Song, L.; Xu, C.; Hao, L.; Yao, J.; Guo, R. Research on PID Parameter Tuning and Optimization Based on SAC-Auto for USV Path Following. *J. Mar. Sci. Eng.* **2022**, *10*, 1847. [CrossRef]
8. Abkowitzm, A. Measurement of hydrodynamic characteristic from ship maneuvering trials by system identification. *Trans. Soc. Nav. Archit. Mar. Eng.* **1980**, *88*, 283–318.
9. Qu, X.; Jiang, Y.; Zhang, R.; Long, F. A Deep Reinforcement Learning-Based Path-Following Control Scheme for an Uncertain Under-Actuated Autonomous Marine Vehicle. *J. Mar. Sci. Eng.* **2023**, *11*, 1762. [CrossRef]
10. Fossen, T.I. *Handbook of Marine Craft Hydrodynamics and Motion Control*; John Wiley & Sons: Hoboken, NJ, USA, 2021.
11. Sandeepkumar, R.; Rajendran, S.; Ranjith, M.; Antonio, P. A unified ship manoeuvring model with a nonlinear model predictive controller for path following in regular waves. *Ocean. Eng.* **2021**, *243*, 110165. [CrossRef]
12. Ogawa, A.; Kasai, H. On the mathematical model of manoeuvring motion of ship. *Int. Ship-Build. Prog.* **1978**, *25*, 306–319. [CrossRef]
13. Guo, J.; Liu, Y.H.; Ma, L.H. Active Disturbance Rejection Control for Ship Trajectory Tracking with Multimodal Fast Nonsingular Terminal Sliding Mode Strategy. *Navig. China* **2020**, *43*, 7–13.
14. Tan, C.; Du, J.L.; Xu, G.X.; Li, D.H. Trajectory Tracking Control Design of Underactuated Ships Based on ESO and Dynamic Inversion. *Ship Eng.* **2020**, *42*, 103–109.
15. Ren, R.-Y.; Zou, Z.-J.; Wang, J.-Q. Time-Scale Decomposition Techniques Used in the Ship Path-Following Problem with Rudder Roll Stabilization Control. *J. Mar. Sci. Eng.* **2021**, *9*, 1024. [CrossRef]
16. Zhu, K.; Huang, Z.; Wang, X.M. Tracking control of intelligent ship based on deep reinforcement learning. *Chin. J. Ship Res.* **2021**, *16*, 105–113.
17. Jawhar, G.; Maarouf, S.; Faisal, M. Prescribed performances based fuzzy-adaptive output feedback containment control for multiple underactuated surface vessels. *Ocean. Eng.* **2022**, *249*, 110898.
18. Le, W.; Li, S.J.; Liu, J.L.; Qing, W. Data-driven path-following control of underactuated ships based on antenna mutation beetle swarm predictive reinforcement learning. *Appl. Ocean. Res.* **2022**, *124*, 103207.
19. Zhang, H.G.; Bu, R.X.; Yu, J.M. RBF neural network sliding mode control for ship path following. *J. Shanghai Marit. Univ.* **2021**, *42*, 7–11.
20. Huang, Y.; Shi, X.; Huang, W.; Chen, S. Internal Model Control-Based Observer for the Sideslip Angle of an Unmanned Surface Vehicle. *J. Mar. Sci. Eng.* **2022**, *10*, 470. [CrossRef]
21. Liu, X.W.; Xu, C.; Chen, Z.F. Event-triggered adaptive PID trajectory tracking control for marine surface vechles. *J. Shanghai Marit. Univ.* **2023**, *44*, 11–17.
22. Li, W.; Xie, H.W.; Han, J.Q.; Meng, F.B. Track control of Under-actuated ship based on Model Free Adaptive PD Control. *Ship Sci. Technol.* **2022**, *44*, 21–25.
23. Jia, X.L.; Yang, Y.S. *Mathematical Model of Ship Motion*; Dalian Maritime University Press: Dalian, China, 1998.
24. Liu, C.; Li, T.S.; Chen, N.X. Dynamic surface control and minimal learning parameter (DSC-MLP) design of a ship's autopilot with rudder dynamics. *J. Harbin Eng. Univ.* **2012**, *33*, 9–14.
25. Borhaug, E.; Pavlov, A.; Pettersen, K.Y. Integral LOS control for path following of underactuated marine surface vessels in the presence of constant ocean currents. In Proceedings of the 2008 47th IEEE Conference on Decision and Control, Cancun, Mexico, 9 December 2008.
26. Lekkas, A.M. *Guidance and Path-Planning Systems for Autonomous Veicles*; Norwegian University of Science and Technology: Trondheim, Norway, 2014.
27. Huang, C.P.; Zou, Z.J.; He, H.W.; Fan, J. Path following and collision avoidance of underactuated ships based on model predictive control and modified particle swarm optimization. *Navig. China* **2023**, *46*, 125–134.
28. Sugeno, M. *Industrial Applications of Fuzzy Control*; Elsevier: Amsterdam, The Netherlands, 1985.
29. Brown, R.A.; Phil, H. *Introduction to Random Signals and Applied Kalman Filtering*; John Wiley & Sons: Hoboken, NJ, USA, 2012.
30. Li, X.R.; Jilkov, V. Survey of Maneuvering Target Tracking. Part I. Dynamic Models. *IEEE Trans. Aerosp. Electron. Syst.* **2003**, *39*, 1333–1364.
31. Siegert, G.; Bany's, P.; Martinez, C.S.; Heymann, F. EKF Based Trajectory Tracking and Integrity Monitoring of AIS Data. In Proceedings of the 2016 IEEE/ION Position, Location and Navigation Symposium (PLANS), Savannah, GA, USA, 11–14 April 2016.
32. Singer, R.A. Estimating Optimal Tracking Filter Performance for Manned Maneuvering Targets. *IEEE Trans. Aerospace Electron. Syst.* **1970**, *AES-6*, 473–483.
33. Fossen, S.; Fossen, T.I. Five-State Extended Kalman Filter for Estimation of Speed over Ground (SOG), Course over Ground (COG) and Course Rate of Unmanned Surface Vehicles (USVs): Experimental Results. *Sensors* **2021**, *21*, 7910. [CrossRef]

Journal of
Marine Science and Engineering

MDPI

Article

Robust Fixed-Time Fault-Tolerant Control for USV with Prescribed Tracking Performance

Zifu Li and Kai Lei *

Navigation College, Jimei University, Xiamen 361021, China; fumeiscjm@jmu.edu.cn
* Correspondence: lkysal@163.com

Abstract: The unmanned surface vessel (USV) is an emerging marine tool with its advantages of automation and intelligence in recent years; the good trajectory tracking performance is an important capability. This paper proposes a novel prescribed performance fixed-time fault-tolerant control scheme for an USV with model parameter uncertainties, unknown external disturbances, and actuator faults, based on an improved fixed-time disturbances observer. Firstly, the proposed observer can not only accurately and quickly estimate and compensate the lumped nonlinearity, including actuator faults, but also reduce the chattering phenomenon by introducing the hyperbolic tangent function. Then, under the framework of prescribed performance control, a prescribed performance fault-tolerant controller is designed based on a nonsingular fixed-time sliding mode surface, which guarantees the transient and steady-state performance of an USV under actuator faults and meets the prescribed tracking performance requirements. In addition, it is proved that the closed-loop control system has fixed-time stability according to Lyapunov's theory. Finally, upon conducting numerical simulations and comparing the proposed control scheme with the SMC and the finite-time NFTSMC scheme, it is evident that the absolute error tracking performance index of the proposed control scheme is significantly lower, thus indicating its superior accuracy.

Keywords: unmanned surface vessel; fixed-time; fault-tolerant control; prescribed tracking performance

Citation: Li, Z.; Lei, K. Robust Fixed-Time Fault-Tolerant Control for USV with Prescribed Tracking Performance. *J. Mar. Sci. Eng.* **2024**, *12*, 799. https://doi.org/10.3390/jmse12050799

Academic Editor: Rafael Morales

Received: 20 March 2024
Revised: 28 April 2024
Accepted: 6 May 2024
Published: 10 May 2024

1. Introduction

In recent years, the unmanned surface vessel (USV) has been extensively utilized in ocean sampling, ocean mapping, and maritime rescue owing to its compact size, exceptional maneuverability, and effective concealment [1]. A requirement of the USV in performing the above tasks is to be able to arrive and remain precisely on the desired trajectory within the specified time. However, there are significant challenges in achieving precise trajectory tracking for the USV. Firstly, the dynamics of the USV exhibit highly nonlinear behavior. Secondly, the exact model of the USV system is unknown, and working conditions are often harshly affected by sea winds, waves, and currents [2]. Accurate trajectory tracking is crucial for the safe and efficient autonomous operation of the USV in different scenarios and thus has important research value and practical significance.

At present, the trajectory tracking control methods of the USV mainly include the following: PID control [3], backstepping control [4,5], fuzzy control [6], adaptive control [7,8], and sliding mode control (SMC) [9–11]. In particular, SMC has been proven to be highly robust to uncertainties and disturbances in nonlinear systems, and it is widely used in ship trajectory tracking control. For example, by introducing a disturbance observer to estimate the disturbance and compensate it in the control law and combining it with SMC, the trajectory tracking of the USV was implemented in [12]. Similarly, Piao et al. [13] proposed an adaptive backstepping SMC strategy that used the adaptive law to estimate the bounds of the external unknown disturbance. But the upper bound of the disturbance is known and constant. Furthermore, Chen et al. [14] designed an adaptive sliding mode controller by combining radial basis function neural networks (RBFNNs), which were used to approximate and compensate modeling uncertainties, and a disturbance observer to estimate and compensate external disturbances. However, these SMC methods all choose a linear sliding surface, which can only guarantee that the tracking error converges to zero asymptotically, and the convergence rate can be adjusted by adjusting the sliding mode surface parameters, while the system tracking error cannot converge to zero in finite time regardless.

In order to speed up the error convergence, finite-time control methods [15–17] have been widely used in the field of USV motion control. Also of relevance, Xu et al. [18] proposed a nonsingular fast terminal SMC scheme based on a finite-time extended state observer (ESO). In [19], the universal approximation property of RBFNNs was used to estimate the uncertainty of the system, and event-triggered control (ETC) and finite-time control were combined to solve the problem of system uncertainty and asymmetric input saturation. Then, Rodriguez et al. [20] designed a finite-time control strategy by combining adaptive boundary estimation and the integral SMC method. Furthermore, Huang et al. [21] used RBFNNs to approximate the uncertain system dynamics and used the minimum learning parameter (MLP) algorithm to reduce the computational complexity. At the same time, by introducing a finite-time SMC algorithm, the finite-time formation control of the USV was realized. Notably, it should be pointed out that the tracking error convergence time in the above control method depends on the control parameters and the initial state of the system, which indicates that when the control parameters are unchanged and the initial error of the system tends to infinity, the error convergence time will also increase unbounded. In order to make the upper bound on the convergence time of the system independent of the initial state, the fixed-time control was first proposed by the researcher Polyakov in [22]. Inspired by this, Yao et al. [23] proposed a fixed-time terminal SMC scheme, and the trajectory tracking error of USV can converge in a fixed time, but this fixed-time terminal sliding mode surface will produce singularity. In view of this, a nonsingular fixed-time terminal SMC strategy was proposed in [24]. Additionally, Chen et al. [25] proposed a nonsingular fixed-time fractional order sliding mode controller and introduced RBFNNs to estimate external disturbances. However, when using neural networks to approximate unknown external disturbances, it is a difficult problem to select the weighting matrix and the number of hidden layer nodes.

In practical engineering applications, it is necessary to consider actuator faults in order to ensure the performance and reliability of USV tracking control. Actuator fault is known as one of the most typical cases of input constraints, which may degrade the control performance, especially for USV motion control systems that require high safety. Therefore, fault-tolerant control (FTC) techniques must be considered. In [26], actuator faults of the USV are addressed by auxiliary systems integrated with adaptive techniques. In addition, Zhang et al. [27] transformed the dynamic model with actuator faults and uncertainties into a nominal model with equivalent disturbances and tracked by a robust compensator. In [28], a fault efficiency estimator is constructed based on a fuzzy-aided nonlinear observer.

The above studies have achieved certain results in improving the steady-state accuracy of USV trajectory tracking, but less consideration has been given to the transient performance and output constraints of trajectory tracking errors, especially for the transient performance (such as overshoot) of trajectory tracking under actuator faults. Heshmati-Alamdari et al. [29] proposed a prescribed performance control (PPC) strategy to achieve trajectory tracking under prescribed transient and steady-state responses without considering the model's uncertainties and external disturbances.

In summary, in order to highlight the difference between this paper and existing related studies, Table 1 indicates that the controller considers multiple factors. Note that if the controller satisfies the factor in Table 1, it is marked by ✓, otherwise, by ✗.

Table 1. Advantages and disadvantages of related literature.

Related Literature	Model Uncertainties	External Disturbances	Actuator Faults	Prescribed Performance	Limited Convergence Time	Convergence Time Is Independent of Initial States
[12]	✗	✓	✗	✗	✗	✓
[13]	✓	✓	✗	✗	✗	✓
[14]	✓	✓	✗	✗	✗	✓
[18]	✓	✓	✗	✗	✓	✗
[19]	✗	✓	✗	✗	✓	✗
[20]	✓	✓	✗	✗	✓	✗
[23]	✓	✓	✗	✗	✓	✓
[24]	✓	✓	✓	✗	✓	✓
[25]	✓	✓	✗	✗	✓	✓
[26]	✗	✓	✓	✗	✓	✗
[27]	✓	✓	✓	✗	✓	✓
[28]	✓	✓	✓	✗	✓	✓
[29]	✗	✗	✗	✓	✗	✓
This paper	✓	✓	✓	✓	✓	✓

In this work, an improved fixed-time disturbances observer-based prescribed performance fixed-time fault-tolerant control (IFxDO-PPFxFC) scheme is proposed to solve the problem that the tracking control results of the USV are susceptible to the initial states of the system, unknown external disturbances, model parameter uncertainties, and actuator faults. Then, the main contributions of this paper are as follows:

(1) An improved fixed-time disturbances observer is proposed. The proposed technique is significant in that it not only provides faster convergence but also effectively reduces the chattering phenomenon by introducing the hyperbolic tangent function. In addition, the bound of the convergence time value can be predicted in advance.

(2) Combining fixed-time SMC, FTC, and PPC theories, a novel IFxDO-PPFxFC scheme is proposed in this paper. Unlike the finite-time stable control scheme [18–20], The proposed control scheme enables the USV to accurately track the desired trajectory in a fixed time, and the convergence time is independent of initial states. Meanwhile, the advantage of this control scheme is its singularity-free. Furthermore, it can guarantee the transient and steady-state performance of output errors of trajectory tracking controller even in the presence of actuator faults; this is of great significance for the safe navigation of USV.

This paper is organized as follows. Section 2 gives the preliminaries and problem formulation. Section 3 presents the controller design and stability analysis. Section 4 verifies it through simulation experiments. Finally, Section 5 gives the conclusions.

2. Preliminaries and Problem Formulation

In this section, in order to facilitate the subsequent controller design and stability proofs, we present the needed notations, definitions, and lemmas for the controller design, introducing the detailed mathematical model of the USV.

2.1. Preliminaries

Notations: $\mathbb{R}^n$ *denotes an* $n \times 1$ *column vector,* $\mathbb{R}^{n \times m}$ *denotes an* $n \times m$ *dimensional matrix,* $\lambda_{\max}(\cdot)$ *and* $\lambda_{\min}(\cdot)$ *denote the minimum and maximum eigenvalue of a matrix,* *diag{·} denotes the diagonal matrix, |·| denotes the modulus of a vector, and ||·|| denotes the 2-norm of a vector or the induced norm of a matrix, respectively. For a vector* $x = [x_1, x_2, \cdots \cdots x_n]^T \in \mathbb{R}^n$ *and a positive scalar a,* $sign(x) = [sign(x_1), sign(x_2) \cdots sign(x_n)]^T$ $[x]^a = [|x_1|^a \, sign(x_1), |x_2|^a \, sign(x_2) \cdots |x_n|^a sign(x_n)]^T$ *and* $\mathbf{x}^a = [x_1{}^a, x_2{}^a \cdots \cdots x_n{}^a)]^T$, *where sign(x) denotes the signum function.*

Definition 1. *Consider the following system:*

$$\dot{y} = f(y), y(0) = y_0, y \in \mathbb{R}^n \tag{1}$$

The origin of the system (1) is said to be fixed-time stable if there exists a bounded constant T_* *which is independent of the initial value such that*

$$\lim_{t \to T_*} \|y\| = 0, \|y\| = 0 \text{ for } t > T_* \tag{2}$$

Lemma 1 ([30]). *For the system (1), assume that there exists a positive definite continuous Lyapunov function that satisfies*

$$\dot{V}(y) \le -\lambda_1 V^\alpha(y) - \lambda_2 V^\beta(y) \tag{3}$$

where $\lambda_1 > 0, \lambda_2 > 0, 0 < \alpha < 1,$ *and* $\beta > 1$. *The system (1) is globally fixed-time stable, ensuring that the convergence time* T_* *satisfies*

$$T_* \le T_*(\alpha, \beta, \lambda_1, \lambda_2) = \frac{1}{\lambda_1(1-\alpha)} + \frac{1}{\lambda_2(\beta-1)} \tag{4}$$

Lemma 2 ([31]). *For the system (1), assume that there exists a positive definite continuous Lyapunov function that satisfies*

$$\dot{V}(y) \le -aV^p(y) - bV^q(y) + \varepsilon \tag{5}$$

where a and b are positive constants, $0 < p < 1, q > 1,$ *and* $0 < \varepsilon < \infty$. *Then, the system (1) is practically fixed-time stable, and the system state can converge to the residual set at fixed time*

$$\left\{ \lim_{y \to T_*} y \middle| V(y) \le \min\left\{ a^{-\frac{1}{p}}\left(\frac{\varepsilon}{1-\theta}\right)^{\frac{1}{p}}, b^{-\frac{1}{p}}\left(\frac{\varepsilon}{1-\theta}\right)^{\frac{1}{q}} \right\} \right\} \tag{6}$$

where θ *is a positive constant and satisfies* $0 < \theta < 1$. *The system convergence time* T_* *satisfies*

$$T_* \le T_*(a, b, p, q) = \frac{1}{a(1-p)} + \frac{1}{b(q-1)} \tag{7}$$

Lemma 3 ([32]). *If* $\varepsilon_1, \varepsilon_2, \ldots, \varepsilon_n \ge 0$, *there exists*

$$\sum_{i=1}^n \varepsilon_i{}^p \ge \left(\sum_{i=1}^n \varepsilon_i\right)^\kappa, 0 < \kappa < 1 \tag{8}$$

$$\sum_{i=1}^n \varepsilon_i{}^p \ge n^{1-p}\left(\sum_{i=1}^n \varepsilon_i\right)^\kappa, \kappa > 1. \tag{9}$$

2.2. USV Mathematical Model

In general, the motion of the USV in the horizontal plane is regarded as the 3-DOF motion, namely surge, sway, and yaw. As shown in Figure 1, the kinematic model parameters of the USV are established based on the earth-fixed OXY and the body-fixed $O_E X_E Y_E$ coordinate frames.

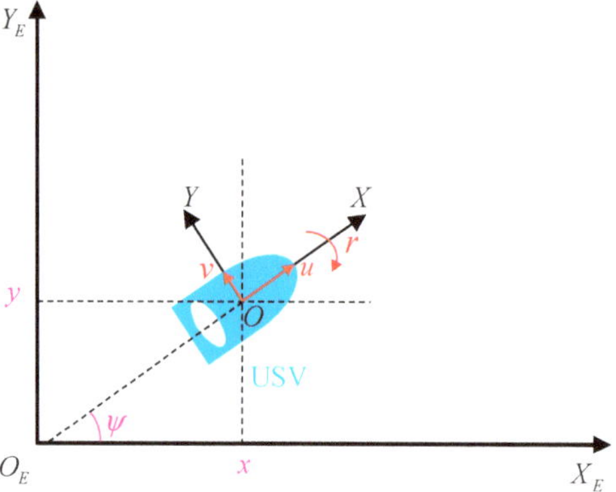

Figure 1. The definition of the earth-fixed OXY and the body-fixed $O_EX_EY_E$ frames.

The kinematic and dynamic models of the USV are defined as follows [33]:

$$\begin{cases} \dot{\eta} = R(\psi)v \\ M\dot{v} = -C(v)v - D(v)v + \tau + b \end{cases} \tag{10}$$

where $\eta = \begin{bmatrix} x & y & \psi \end{bmatrix}^T \in \mathbb{R}^3$ is the position and heading of the USV in the earth-fixed frame. $v = \begin{bmatrix} u & v & r \end{bmatrix}^T \in \mathbb{R}^3$ is the surge, sway, and yaw velocities of the USV in the body-fixed frame. $\tau = \begin{bmatrix} \tau_1 & \tau_2 & \tau_3 \end{bmatrix}^T \in \mathbb{R}^3$ is the control input. $b = \begin{bmatrix} b_u & b_v & b_r \end{bmatrix}^T \in \mathbb{R}^3$ is the external environmental unknown time-varying disturbance. The rotation matrix $R(\psi) \in \mathbb{R}^{3\times3}$ is given as follows:

$$R(\psi) = \begin{bmatrix} \cos\psi & -\sin\psi & 0 \\ \sin\psi & \cos\psi & 0 \\ 0 & 0 & 1 \end{bmatrix} \tag{11}$$

The matrix $R(\psi)$ has the following properties: $\dot{R}(\psi) = R(\psi)S(r)$, $R^T(\psi)S(r)R(\psi) = R(\psi)S(r)R(\psi)^T = S(r)$, $\|R(\psi)\| = 1$. The matrix $S(r)$ is given as follows:

$$S(r) = \begin{bmatrix} 0 & -r & 0 \\ r & 0 & 0 \\ 0 & 0 & 0 \end{bmatrix} \tag{12}$$

$M \in \mathbb{R}^{3\times3}$ is the positive definite matrix, and it satisfies $M = M^T > 0$. Its expression is

$$M = \begin{bmatrix} m_{11} & 0 & 0 \\ 0 & m_{22} & m_{23} \\ 0 & m_{32} & m_{33} \end{bmatrix} \tag{13}$$

where $m_{11} = m - X_{\dot{u}}$, $m_{22} = m - Y_{\dot{v}}$, $m_{23} = mx_g - Y_{\dot{r}}$, $m_{32} = mx_g - N_{\dot{v}}$, and $m_{33} = I_z - N_{\dot{r}}$.

$C(v) \in \mathbb{R}^{3\times3}$ is the Coriolis and centripetal matrix, and it satisfies $C(v) = -C(v)^T > 0$. $D(v) \in \mathbb{R}^{3\times3}$ is the damping matrix. They are expressed as follows:

$$C(v) = \begin{bmatrix} 0 & 0 & c_{13}(v) \\ 0 & 0 & c_{23}(v) \\ -c_{13}(v) & -c_{23}(v) & m_{33} \end{bmatrix}, D(v) = \begin{bmatrix} d_{11}(v) & 0 & 0 \\ 0 & d_{22}(v) & d_{23}(v) \\ 0 & d_{32}(v) & d_{33}(v) \end{bmatrix} \tag{14}$$

where $c_{13}(v) = -m_{11}v - m_{23}r$, $c_{23}(v) = -m_{11}u$, $d_{11}(v) = -X_u - X_{|u|u}|u| - X_{uuu}u^2$, $d_{22}(v) = -Y_v - Y_{|v|v}|v| - Y_{|r|v}|r|$, $d_{23}(v) = -Y_r - Y_{|v|r}|v| - Y_{|r|r}|r|$, $d_{32}(v) = -N_v - N_{|v|v}|v| - N_{|r|v}|r|$, and $d_{33}(v) = -N_r - N_{|v|r}|v| - N_{|r|r}|r|$, where m is the mass of the USV, I_z is the moment of inertia, and x_g is the distance from the center of gravity to the origin on the Y. X_*, Y_*, and N_* are the hydrodynamic parameters acting on the USV.

According to the research of Zhang et al. [27], the control input of the USV with actuator faults can be expressed by the following mathematical method:

$$\tau_i = \tau_{ci} + f_{(t-T_{0i})}((e_i - 1)\tau_{ci} + \overline{\tau}_i) \tag{15}$$

where $\tau_i(i = u, v, r)$ represents the actual control input acting on the USV, τ_{ci} is the desired control forces and moments, $\overline{\tau}_i$ is the additive uncertain fault input, e_i is the thruster efficiency factor with $0 \le e_i \le 1$, and $f_{(t-T_{0i})}$ represents the time-varying fault. Its expression is

$$f_{(t-T_{0i})} = \begin{cases} 0, t < T_{0i} \\ 1 - e^{-a_i(t-T_{0i})}, t \ge T_{0i} \end{cases} \tag{16}$$

where a_i is the unknown fault change rate, and T_{0i} is the fault occurrence time of each degree of freedom.

In summary, the equivalent control forces and moments of the USV propulsion system considering the thruster fault constraint is expressed as follows:

$$\tau = \tau_c + F_{(t-T_0)}\left((E - I_3)\tau_c + \overline{\tau}\right) \tag{17}$$

where $F_{(t-T_0)} = diag\{f_{(t-T_{0u})}, f_{(t-T_{0u})}, f_{(t-T_{0r})}\} \in \mathbb{R}^{3\times3}$ is the time-varying fault matrix of the propulsion system. $E = diag\{e_u, e_v, e_r\} \in \mathbb{R}^{3\times3}$ is the efficiency matrix of the propulsion system, $\tau_c = \begin{bmatrix} \tau_{cu} & \tau_{cv} & \tau_{cr} \end{bmatrix}^T \in \mathbb{R}^3$ is the desired control forces and moments vector, and $\overline{\tau} = \begin{bmatrix} \overline{\tau}_u & \overline{\tau}_v & \overline{\tau}_r \end{bmatrix}^T \in \mathbb{R}^3$ is the additive fault vector.

In addition, consider the model parameter uncertainties:

$$\begin{cases} C(v) = C_0(v) + \Delta C(v) \\ D(v) = D_0(v) + \Delta D(v) \end{cases} \tag{18}$$

where $C_0(v)$ and $D_0(v)$ denote the nominal values of the Coriolis and centripetal matrix and the damping matrix, respectively. $\Delta C(v)$ and $\Delta D(v)$ denote the uncertainty values of the Coriolis and centripetal matrix and the damping matrix, respectively.

Define an auxiliary velocity vector and let

$$\omega = R(\psi)v \tag{19}$$

where $\omega = \begin{bmatrix} \omega_1 & \omega_2 & \omega_3 \end{bmatrix}^T \in \mathbb{R}^3$.

Combining Equations (17) and (18), the mathematical model of the USV can be redefined as follows:

$$\begin{cases} \dot{\eta} = \omega \\ \dot{\omega} = H(\eta, v) + RM^{-1}\tau_c + d \end{cases} \tag{20}$$

where $H(\eta, v) \in \mathbb{R}^3$ is the total nominal component, $d = \begin{bmatrix} d_1 & d_2 & d_3 \end{bmatrix}^T \in \mathbb{R}^3$ is the unknown lumped nonlinearity (model parameter uncertainties, unknown external disturbances, and actuator faults) of the system. They are given as follows:

$$H(\eta, v) = RM^{-1}(MSv - C_0(v)v - D_0(v)v) \tag{21}$$

$$d = RM^{-1}\left(F_{(t-T_0)}\left((E - I_3)\tau_c + \overline{\tau}\right) + b - \Delta C(v)v - \Delta D(v)v\right) \tag{22}$$

Assumption 1. *The unknown lumped nonlinearity **d** (22) is limited by $\|d\| < a$, where a is an unknown constant. The first-order derivative of **d** is bounded by $\|\dot{d}\| < b$, where b is a known positive constant.*

3. Controller Design and Stability Analysis

In this section, Figure 2 presents the trajectory tracking controller framework diagram for the USV. It is mainly made up of the performance function, an improved FxDO, and a PPFxF controller. The performance function provides the tracking error transformation. The improved FxDO can accurately estimate lumped disturbances and faults in a fixed time. The PPFxF controller makes trajectory tracking errors reach zero at a fixed time and ensures that the transient performance and steady-state performance meet the specified requirements.

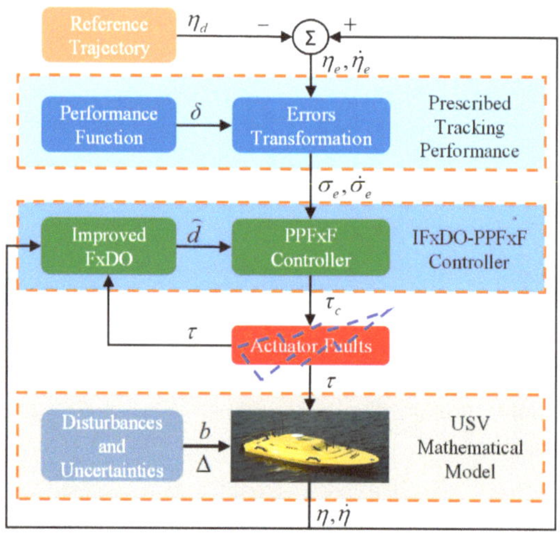

Figure 2. Trajectory tracking controller framework diagram for USV.

3.1. Improved Fixed-Time Disturbances Observer

Notably, when the USV performs its mission in the actual ocean environment, its stability can be seriously affected by unknown disturbances such as wind, waves, and currents. At the same time, the faults of the actuator cannot be ignored. In view of this, by combining the mathematical model of the USV, a fixed-time disturbances observer (FxDO) is applied to estimate and compensate the lumped nonlinearity, including actuator faults, and the observer is designed accordingly as follows:

$$
\begin{cases}
\tilde{\omega} = \omega - \hat{\omega}, \tilde{d} = d - \hat{d} \\
\dot{\hat{\omega}} = H(\eta, v) + RM^{-1}\tau_c + \hat{d} + \varsigma_1|\tilde{\omega}|^{\alpha_1}sign(\tilde{\omega}) + \sigma_1|\tilde{\omega}|^{\beta_1}sign(\tilde{\omega}) \\
\dot{\hat{d}} = \varsigma_2|\tilde{\omega}|^{\alpha_2}sign(\tilde{\omega}) + \sigma_2|\tilde{\omega}|^{\beta_2}sign(\tilde{\omega}) + \gamma sign(\tilde{\omega})
\end{cases}
\tag{23}
$$

Remark 1. *In Equation (23), $\varsigma_1|\tilde{\omega}|^{\alpha_1}sign(\tilde{\omega})$, $\sigma_1|\tilde{\omega}|^{\beta_1}sign(\tilde{\omega})$, $\varsigma_2|\tilde{\omega}|^{\alpha_2}sign(\tilde{\omega})$, and $\sigma_2|\tilde{\omega}|^{\beta_2}sign(\tilde{\omega})$ are continuous regardless of the sign function; however, $\gamma sign(\tilde{\omega})$ is discontinuous, leading to a chattering phenomenon. The fixed-time observer's estimate of the centralized uncertainty will be compensated for in the control input, so the discontinuity term in Equation (23) will cause discontinuities in the control input, which will lead to the chattering phenomenon.*

To avoid the problem of the chattering phenomenon, the signum function is approximated in this paper by the hyperbolic tangent function. An improved FxDO can be expressed as follows:

$$\begin{cases} \widetilde{\omega} = \omega - \widehat{\omega}, \widetilde{d} = d - \widehat{d} \\ \dot{\widehat{\omega}} = H(\eta, v) + RM^{-1}\tau_c + \widehat{d} + \varsigma_1|\widetilde{\omega}|^{\alpha_1}sign(\widetilde{\omega}) + \sigma_1|\widetilde{\omega}|^{\beta_1}sign(\widetilde{\omega}) \\ \dot{\widehat{d}} = \varsigma_2|\widetilde{\omega}|^{\alpha_2}sign(\widetilde{\omega}) + \sigma_2|\widetilde{\omega}|^{\beta_2}sign(\widetilde{\omega}) + \gamma\tanh(\widetilde{\omega}) \end{cases} \tag{24}$$

Theorem 1. *If Assumption 1 is satisfied, the improved FxDO (24) is designed to achieve an accurate estimation of unknown lumped nonlinearity within a fixed time T_d.*

The estimated errors are given as follows:

$$\begin{cases} \dot{\widetilde{\omega}} = -\varsigma_1|\widetilde{\omega}|^{\alpha_1}sign(\widetilde{\omega}) - \sigma_1|\widetilde{\omega}|^{\beta_1}sign(\widetilde{\omega}) + \widetilde{d} \\ \dot{\widetilde{d}} = -\varsigma_2|\widetilde{\omega}|^{\alpha_2}sign(\widetilde{\omega}) - \sigma_2|\widetilde{\omega}|^{\beta_2}sign(\widetilde{\omega}) - \gamma\tanh(\widetilde{\omega}) + \dot{d} \end{cases} \tag{25}$$

where $0 < \alpha_1 < 1, 0 < \alpha_2 < 1, \beta_1 > 1, \beta_2 > 1$, and $\gamma > b$. The options for $\varsigma_1, \varsigma_2, \sigma_1$, and σ_2 are as follows:

$$A = \begin{bmatrix} -\varsigma_1 & 1 \\ -\varsigma_2 & 0 \end{bmatrix}, B = \begin{bmatrix} -\sigma_1 & 1 \\ -\sigma_2 & 0 \end{bmatrix} \tag{26}$$

Matrices A and B satisfy the Hurwitz matrix, and it has been proved in [34] that the estimation error of the system converges to zero at fixed time T_d, and it is given as follows:

$$T_d \leq \frac{\lambda_{\max}^{1-a_1}(P_1)\lambda_{\max}(P_1)}{(1-a_1)\lambda_{\min}(Q_1)} + \frac{\lambda_{\max}(P_2)}{(b_1-1)\lambda_{\min}^{b_1-1}(P_1)\lambda_{\min}(Q_2)} \tag{27}$$

where $1 - c_1 < a_1 < 1, 1 < b_1 < 1 + c_2, 0 < c_1 < 1$, and $c_2 > 0$. P_1, P_2, Q_1, and Q_2 are symmetric positive definite matrixes and satisfy $A_1^T P_1 + P_1 A_1 = -Q_1, A_2^T P_2 + P_2 A_2 = -Q_2$.

3.2. Errors Transformation via Performance Function

The trajectory tracking errors of the USV are defined as follows:

$$\begin{cases} \eta_e = \eta - \eta_d \\ \omega_e = \dot{\eta} - \dot{\eta}_d \end{cases} \tag{28}$$

where $\eta_e = \begin{bmatrix} \eta_{eu} & \eta_{ev} & \eta_{er} \end{bmatrix}^T \in \mathbb{R}^3, \omega_e = \begin{bmatrix} \omega_{eu} & \omega_{ev} & \omega_{er} \end{bmatrix}^T \in \mathbb{R}^3$.

Taking the derivative of Equation (28), the mathematical equation of the tracking errors of the USV can be written as follows:

$$\begin{cases} \dot{\eta}_e = \omega_e \\ \dot{\omega}_e = H(\eta, v) + RM^{-1}\tau_c + d - \ddot{\eta}_d \end{cases} \tag{29}$$

where $\eta_d = \begin{bmatrix} x_d & y_d & \psi_d \end{bmatrix}^T \in \mathbb{R}^3$. η_d is the desired trajectory, $\dot{\eta}_d$ and $\ddot{\eta}_d$ are the first-order and second-order derivatives of the desired trajectory.

Assumption 2. *The desired trajectory of the USV η_d is bounded and derivable, and its derivatives $\dot{\eta}_d$ and $\ddot{\eta}_d$ are smooth and bounded.*

In order to ensure that the transient and steady-state performance of the trajectory error η_{ei} is as shown in Figure 3, a prescribed performance function is introduced to constrain the tracking error. Select the following performance function

$$\delta_i(t) = (\delta_{0i} - \delta_{\infty i})\exp(-\xi t) + \delta_{\infty i}, (i = u, v, r) \tag{30}$$

where ξ is the index of the rate of convergence; researchers can change this index to obtain the prescribed transient and steady-state performance according to the specific needs of the task. δ_{0i} denotes the initial value of the prescribed performance function $\delta_i(t)$, and $\delta_{\infty i}$ is the stabilized value after the convergence of the prescribed performance function $\delta_i(t)$.

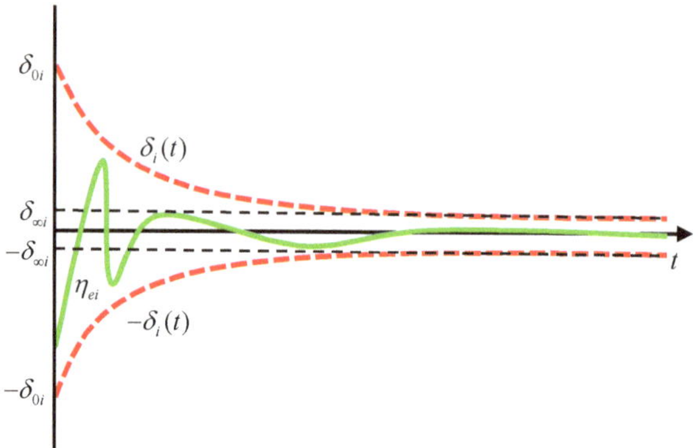

Figure 3. Prescribed performance control diagram.

Then, a boundary function is constructed from the performance function to limit the tracking error of the system as follows:

$$\begin{cases} -f_1\delta_i(t) < \eta_{ei} < \delta_i(t), \eta_{ei}(0) \geq 0 \\ -\delta_i(t) < \eta_{ei} < f_2\delta_i(t), \eta_{ei}(0) \leq 0 \end{cases} (i = u, v, r) \tag{31}$$

From Equation (31), one has

$$\begin{cases} -f_1 < \frac{\eta_{ei}}{\delta_i(t)} < 1, \eta_{ei}(0) \geq 0 \\ -1 < \frac{\eta_{ei}}{\delta_i(t)} < f_2, \eta_{ei}(0) \leq 0 \end{cases} \tag{32}$$

Assumption 3. *([35]). The initial states of the USV satisfy* $-\delta_i(0) < \eta_{ei}(0) < \delta_i(0)$.

The transformation is achieved by rewriting the tracking error to the following equivalent form

$$\eta_{ei} = \delta_i(t)\Gamma(\sigma_{ei}) \tag{33}$$

where $\Gamma(\sigma_{ei})$ is the unconstrained transformation function, σ_{ei} is the transformed error, and $\delta_i(t)$, $\Gamma(\sigma_{ei})$ need to satisfy the following requirements:

- $\delta_i(t)$ is a monotonically decreasing function;
- $\Gamma(\sigma_{ei}) \in (-1, 1)$, $\Gamma(\sigma_{ei})$ is a strictly increasing function;
- $\delta_i(t) > 0, \forall t \geq 0$;
- $\lim_{\sigma_{ei} \to \infty} \Gamma(\sigma_{ei}) = 1, \lim_{\sigma_{ei} \to -\infty} \Gamma(\sigma_{ei}) = -1$.

The transformation function in the form of a hyperbolic tangent is defined as follows:

$$\Gamma(\sigma_{ei}) = \begin{cases} \frac{e^{\sigma_{ei}} - f_1 e^{-\sigma_{ei}}}{e^{\sigma_{ei}} + e^{-\sigma_{ei}}}, \sigma_{ei} \geq 0 \\ \frac{f_2 e^{\sigma_{ei}} - e^{-\sigma_{ei}}}{e^{\sigma_{ei}} + e^{-\sigma_{ei}}}, \sigma_{ei} \leq 0 \end{cases} \tag{34}$$

According to the properties of the hyperbolic tangent function, the transformed error can be expressed as

$$\sigma_{ei} = \Gamma^{-1}\left(\frac{\eta_{ei}}{\delta_i(t)}\right) = \frac{1}{2}\ln\frac{1 + \frac{\eta_{ei}}{\delta_i}}{1 - \frac{\eta_{ei}}{\delta_i}} \tag{35}$$

The first-order derivative and the second-order derivative of σ_{ei} are calculated as

$$\begin{cases} \dot{\sigma}_{ei} = \frac{1}{2}\left(\frac{\dot{\delta}_i + \dot{\eta}_{ei}}{\delta_i + \eta_{ei}} + \frac{\dot{\delta}_i - \dot{\eta}_{ei}}{\delta_i - \eta_{ei}} \right) \\ \ddot{\sigma}_{ei} = \frac{1}{2}\left(\frac{\ddot{\delta}_i(\delta_i + \eta_{ei}) - (\dot{\delta}_i + \dot{\eta}_{ei})^2}{(\delta_i + \eta_{ei})^2} - \frac{\ddot{\delta}_i(\delta_i - \eta_{ei}) - (\dot{\delta}_i - \dot{\eta}_{ei})^2}{(\delta_i - \eta_{ei})^2} \right) + \frac{1}{2}\left(\frac{\delta_i + \eta_{ei}}{(\delta_i + \eta_{ei})^2} - \frac{\delta_i - \eta_{ei}}{(\delta_i - \eta_{ei})^2} \right)\dot{\omega}_{ei} \end{cases}$$

(36)

For the convenience of the subsequent control design, let

$$\ddot{\sigma}_{ei} = \hbar_i + \lambda_i \dot{\omega}_{ei} \tag{37}$$

where

$$\begin{cases} \hbar_i = \left(\left[\ddot{\delta}_i(\delta_i + \eta_{ei}) - (\dot{\delta}_i + \dot{\eta}_{ei})^2\right]/2(\delta_i + \eta_{ei})^2 - \left[\ddot{\delta}_i(\delta_i - \eta_{ei}) - (\dot{\delta} - \dot{\eta}_{ei})^2\right]/2(\delta_i - \eta_{ei})^2 \right) \\ \lambda_i = \left((\delta_i + \eta_{ei})/\left[2(\delta_i + \eta_{ei})^2\right] - (\delta_i - \eta_{ei})/\left[2(\delta_i - \eta_{ei})^2\right] \right) \end{cases} \tag{38}$$

Remark 2. *The error transformation by means of a prescribed performance function and the error η_{ei} also converges if the error σ_{ei} converges. Meanwhile, $-\delta_i(t)$ and $\delta_i(t)$ limit the maximum overshooting of η_{ei}, and the decreasing speed of $\delta_i(t)$ affects the convergence speed of η_{ei}. $\delta_{\infty i}$ and $-\delta_{\infty i}$ limit the steady state error of η_{ei}, so that the prescribed performance control ensures the transient and steady state performance of the error η_{ei} at the same time.*

3.3. Fixed-Time Fault-Tolerant Controller Design

In order to realize the accurate tracking of the desired trajectory in a fixed time, the sliding surface is constructed according to the errors after the transformation of the prescribed performance function. A nonsingular fixed-time sliding surface s is designed as follows [36]:

$$s = \sigma_e + \frac{1}{\lambda_1{}^\alpha}\left[\dot{\sigma}_e + \lambda_2[\sigma_e]^\beta \right]^{\frac{1}{\alpha}} \tag{39}$$

where $s = \begin{bmatrix} s_1 & s_2 & s_3 \end{bmatrix}^T \in \mathbb{R}^3$, $\sigma_e = \begin{bmatrix} \sigma_{eu} & \sigma_{ev} & \sigma_{er} \end{bmatrix}^T \in \mathbb{R}^3$, $1/2 < \alpha < 1, \beta > 1, \lambda_1 > 0$, and $\lambda_2 > 0$.

Theorem 2. *For the nonlinear USV dynamics system (20), in the presence of actuator faults and model parameter uncertainties given by Equations (17) and (18), if the sliding mode surface is designed as (39), then the closed-loop system is stable, and the errors σ_e and $\dot{\sigma}_e$ can converge to zero at a fixed time T_s.*

Proof. When the sliding mold surface $s = 0$ is reached, which yields

$$\dot{\sigma}_e = -\lambda_1[\sigma_e]^\alpha - \lambda_2[\sigma_e]^\beta \tag{40}$$

Select the Lyapunov function as follows:

$$V_1 = \frac{1}{2}\sigma_e{}^T\sigma_e \tag{41}$$

The derivative of Equation (41) is taken as follows:

$$\begin{aligned} \dot{V}_1 &= \frac{1}{2}\sigma_e{}^T\dot{\sigma}_e \\ &= \sigma_e{}^T\left(-\lambda_1[\sigma_e]^\alpha - \lambda_2[\sigma_e]^\beta \right) \\ &= -\lambda_1|\sigma_e|^{\alpha+1} - \lambda_2|\sigma_e|^{\beta+1} \\ &= -\lambda_1 2^{(\alpha+1)/2}V_1{}^{(\alpha+1)/2} - \lambda_2 2^{(\beta+1)/2}V_1{}^{(\beta+1)/2} \\ &\leq -\lambda_1 2^{\alpha/2}V_1{}^{(\alpha+1)/2} - \lambda_2 2^{\beta/2}V_1{}^{(\beta+1)/2} \end{aligned} \tag{42}$$

Due to $1/2 < \alpha < 1, \beta > 1$, according to Lemma 1, $\sigma_e, \dot{\sigma}_e$ will converge to zero at a fixed time T_s, as shown in the following:

$$T_s \leq \frac{2^{1-\alpha/2}}{\lambda_1(1-\alpha)} + \frac{2^{1-\beta/2}}{\lambda_2(\beta-1)} \tag{43}$$

Then, taking the derivative of Equation (39), the following can be obtained:

$$\dot{s} = \ddot{\sigma}_e + \frac{1}{\alpha\lambda_1{}^\alpha} diag\left\{\left\|\dot{\sigma}_e + \lambda_2[\sigma_e]^\beta\right\|^{\frac{1}{\alpha}-1}\right\}\left(\ddot{\sigma}_e + \lambda_2\beta diag\{\|\sigma_e\|^{\beta-1}\}\dot{\sigma}_e\right) \tag{44}$$

Let $F(\sigma_e, \dot{\sigma}_e) = diag\left\{\left\|\dot{\sigma}_e + \lambda_2[\sigma_e]^\beta\right\|\right\}$, substituting Equation (37) into Equation (44), the following can be obtained:

$$\dot{s} = \ddot{\sigma}_e + \frac{1}{\alpha\lambda_1{}^\alpha}F(\sigma_e, \dot{\sigma}_e)^{\frac{1}{\alpha}-1}\left(h + diag\{\lambda_i\}\dot{\omega}_e + \lambda_2\beta diag\{\|\sigma_e\|^{\beta-1}\}\dot{\sigma}_e\right) \tag{45}$$

where $h = \begin{bmatrix} \hbar_u & \hbar_v & \hbar_r \end{bmatrix}^T \in \mathbb{R}^3$, $diag\{\|\sigma_e\|^{\beta-1}\} = diag\{|\sigma_{eu}|^{\beta-1}, |\sigma_{ev}|^{\beta-1}, |\sigma_{er}|^{\beta-1}\}$, and $diag\{\lambda_i\} = diag\{\lambda_u \quad \lambda_v \quad \hbar_r\}$. $\square$

According to Equation (29), Equation (45) can be transformed into

$$\dot{s} = \ddot{\sigma}_e + \frac{1}{\alpha\lambda_1{}^\alpha}F(\sigma_e, \dot{\sigma}_e)^{\frac{1}{\alpha}-1}\left(h + diag\{\lambda_i\}\left(H(\eta, v) + RM^{-1}\tau_c + d - \ddot{\eta}_d\right) + \lambda_2\beta diag\{\|\sigma_e\|^{\beta-1}\}\dot{\sigma}_e\right) \tag{46}$$

Based on Equation (46), the control inputs to the controller can be expressed in terms of the equivalent control input τ_{eq} and the switching control input τ_{sw}, thus $\tau_c = \tau_{eq} + \tau_{sw}$. They are given as follows:

$$\begin{cases} \tau_{eq} = -MR^{-1}\left(H(\eta, v) + \hat{d} - \ddot{\eta}_d\right) - (diag\{\lambda_i\})^{-1}MR^{-1}\left(\lambda_2\beta diag\{\|\sigma_e\|^{\beta-1}\}\dot{\sigma}_e + \alpha\lambda_1{}^\alpha diag\left\{\left\|F(\sigma_e, \dot{\sigma}_e)\right\|^{\frac{1}{\alpha}-1}\right\}\sigma_e + h\right) \\ \tau_{sw} = -(diag\{\lambda_i\})^{-1}MR^{-1}\left(\varphi_1[s]^{\mu_1} + \varphi_2[s]^{\mu_2}\right) \end{cases} \tag{47}$$

where $diag\left\{\left\|F(\sigma_e, \dot{\sigma}_e)\right\|^{\frac{1}{\alpha}-1}\right\} = diag\left\{\left|F(\sigma_{eu}, \dot{\sigma}_{eu})\right|^{\frac{1}{\alpha}-1}, \left|F(\sigma_{ev}, \dot{\sigma}_{ev})\right|^{\frac{1}{\alpha}-1}, \left|F(\sigma_{er}, \dot{\sigma}_{er})\right|^{\frac{1}{\alpha}-1}\right\}$. $0 < \mu_1 < 1, \mu_2 > 1, \varphi_1 > 0$, and $\varphi_2 > 0$.

Remark 3. *In Equation (47), τ_{eq} contains $diag\left\{\left\|F(\sigma_e, \dot{\sigma}_e)\right\|^{\frac{1}{\alpha}-1}\right\}$, but this will not produce singularity; in fact, when $\dot{\sigma}_{ei} = 0, \sigma_{ei} \neq 0$, we can obtain $\left|\dot{\sigma}_{ei}\right|^{1-\frac{1}{\alpha}}\dot{\sigma}_{ei} \geq \dot{\sigma}_{ei}{}^{2-\frac{1}{\alpha}}, 2 - \frac{1}{\alpha} > 0$.*

Theorem 3. *If Assumptions 1–3 are satisfied, the IFxDO-PPFxFC scheme (47) is designed, which can make the USV track the desired trajectory quickly and accurately, ensuring that the position and velocity tracking errors are stable to a small neighborhood around the equilibrium point at a fixed time, and the convergence time is independent of the initial state of the system, and the convergence time is satisfied $T_{\max} \leq T_d + T_s + T_r + T_c$; T_r and T_c are given as follows:*

$$\begin{cases} T_r \leq \frac{2^{1-\mu_1/2}}{\varphi_1 n^{(1-\mu_1)/2}(1-\mu_1)} + \frac{2^{1-\mu_2/2}}{\varphi_2(\mu_2-1)} \\ T_c = (\lambda_1{}^\alpha)^{\frac{\alpha}{\alpha-1}} \end{cases} \tag{48}$$

3.4. Stability Analysis

Proof of Theorem 3. Substituting Equation (47) into Equation (46), the following can be obtained:

$$\dot{s} = \frac{1}{\alpha\lambda_1{}^\alpha}F(\sigma_e,\dot{\sigma}_e)^{\frac{1}{\alpha}-1}\left(-\varphi_1[s]^{\mu_1} - \varphi_2[s]^{\mu_2} + diag\{\lambda_i\}\widetilde{d}\right) \tag{49}$$

Let $\chi = diag\{\chi_1,\chi_2,\chi_3\} = \frac{1}{\alpha\lambda_1{}^\alpha}F(\sigma_e,\dot{\sigma}_e)^{\frac{1}{\alpha}-1} \in \mathbb{R}^{3\times3}$, then one has

$$\dot{s} = \chi\left(-\varphi_1[s]^{\mu_1} - \varphi_2[s]^{\mu_2} + diag\{\lambda_i\}\widetilde{d}\right) \tag{50}$$

Select the Lyapunov function as follows:

$$V_2 = \frac{1}{2}\sum_1^3 s_i{}^2 \tag{51}$$

Take the derivative of Equation (51), then one has

$$\begin{aligned}
\dot{V}_2 &= \sum_1^3 s_i\dot{s}_i \\
&= \chi_i\sum_1^3 s_i\left(-\varphi_1[s_i]^{\mu_1} - \varphi_2[s_i]^{\mu_2} + \lambda_i\widetilde{d}_i\right) \\
&\leq \chi_i\sum_1^3\left(-\varphi_1|s_i|^{\mu_1+1} - \varphi_2|s_i|^{\mu_2+1} + |s_i|\|\lambda_i\|\left\|\widetilde{d}_i\right\|\right) \\
&\leq \chi_i\sum_1^3\left(-\varphi_1|s_i|^{\mu_1+1} - \varphi_2|s_i|^{\mu_2+1} + \Theta_i\right)
\end{aligned} \tag{52}$$

where $\Theta_i = |s_i|\|\lambda_i\|\left\|\widetilde{d}_i\right\|$, $0 < \Theta_i < \infty$.

According to Lemma 3, one has

$$\sum_{i=1}^3|s_i|^{\mu_1+1} \geq n^{(1-\mu_1)/2}\left(\sum_{i=1}^3|s_i|^2\right)^{(\mu_1+1)/2}, \sum_{i=1}^3|s_i|^{\mu_2+1} \geq \left(\sum_{i=1}^3|s_i|^2\right)^{(\mu_2+1)/2} \tag{53}$$

By substituting Equations (51) and (53) into Equation (52), the following can be obtained:

$$\begin{aligned}
\dot{V}_2 &\leq \chi\left(-\varphi_1 n^{(1-\mu_1)/2}\left(\sum_{i=1}^3|s_i|^2\right)^{(\mu_1+1)/2} - \varphi_2\left(\sum_{i=1}^3|s_i|^2\right)^{(\mu_2+1)/2} + \Theta_i\right) \\
&\leq \chi\left(-\varphi_1 n^{(1-\mu_1)/2}2^{(\mu_1+1)/2}V_1{}^{(\mu_1+1)/2} - \varphi_2 2^{(\mu_2+1)/2}V_1{}^{(\mu_2+1)/2} + \Theta_i\right)
\end{aligned} \tag{54}$$

It is worth noting that when $F(\sigma_{ei},\sigma_{ei}) \neq 0$, $\chi_i > 0$. The working state space can be divided into two areas, namely $\Omega1 = \left\{\left(\sigma_{ei},\dot{\sigma}_{ei}\right)\middle|\chi_i > 1\right\}$ and $\Omega2 = \left\{\left(\sigma_{ei},\dot{\sigma}_{ei}\right)\middle|0 < \chi_i < 1\right\}$ for the above two cases, which we will talk about later.

Step 1. Obviously, when the system states meet $\Omega1$, the following can be obtained:

$$\dot{V}_2 \leq -\varphi_1 n^{(1-\mu_1)/2}2^{(\mu_1+1)/2}V_1{}^{(\mu_1+1)/2} - \varphi_2 2^{(\mu_2+1)/2}V_1{}^{(\mu_2+1)/2} + \Theta_i \tag{55}$$

Due to $\mu_1 > 1$ and $0 < \mu_2 < 1$, according to Lemma 2, the system satisfies the fixed-time convergence, and the convergence time T_r is bounded by Equation (48).

Step 2. When the system states meet $\Omega2$, $F(\sigma_{ei},\sigma_{ei}) \neq 0$ can be obtained, and the sliding mode surface s_i will approach $s_i = 0$. We need to prove that $F(\sigma_{ei},\sigma_{ei}) = 0$ is not attractive except for the origin. This is proved in [37], so no matter where the initial state is, As soon as the s sliding mode surface $s_i = 0$ is reached, the system states will reach the origin within the fixed time $T = T_r + T_c$, where the convergence time T_c is given by Equation (48).

The total convergence time of the USV trajectory tracking system is satisfied: $T_{\max} \leq T_d + T_s + T_r + T_c$. □

Remark 4. *In summary, in terms of controller design, this paper reports two novelties compared to recent work in the field. (1) An improved fixed-time disturbances observer is designed. The proposed technique is significant in that it not only provides faster convergence but also effectively reduces the chattering phenomenon. (2) A prescribed performance fixed-time fault-tolerant controller is proposed. In fact, the application of fixed-time control in the field of USV trajectory tracking control is not new, but we consider the case of actuator faults in the model and incorporate the prescribed performance control to ensure the transient and steady-state performance of the output errors while meeting the prescribed tracking performance requirements. The whole proposed control scheme is new to the previous design methods of USV controllers.*

4. Numerical Simulations and Analysis

In this section, the simulation study is performed on CyberShip II [33], a 1:70 scale ship model provided by the Marine Cybernetics Laboratory of the Norwegian University of Science and Technology. The vessel is 1.255 m long, and the detailed parameters are given in Table 2.

Table 2. Hydrodynamic parameters of CyberShip II.

$m = 23.8000$	$Y_v = -0.8612$	$X_{\dot u} = -2.0$
$I_z = 1.7600$	$Y_{|v|v} = -36.2823$	$Y_{\dot v} = -10.0$
$x_g = 0.0460$	$Y_r = 0.1079$	$Y_{\dot r} = -0.0$
$X_u = -0.7225$	$N_{|v|v} = 5.0437$	$N_{\dot v} = -0.0$
$X_{|u|u}|u| = -1.3274$	$N_v = 0.1052$	$N_{\dot r} = -1.0$
$X_{uuu} = -5.8664$		

In the simulation, the model uncertainties are chosen to be $\Delta C(v) = 10\%C_0(v)$ and $\Delta D(v) = 10\%D_0(v)$. $b = \begin{bmatrix} 18\sin(0.9t + \pi/2) & 7\sin(0.8t + \pi/3) & 3\sin(0.5t + \pi/5) \end{bmatrix}^T$ is used to describe the effect of the wind, waves, and current on the USV. The reference trajectories are chosen to be $\eta_d = \begin{bmatrix} 2\sin(0.1t) & 2\cos(0.1t) & 0.1t \end{bmatrix}^T$, and the initial position and initial velocity are chosen as $\eta(0) = \begin{bmatrix} 1 & 1 & \pi/4 \end{bmatrix}^T$ and $v(0) = \begin{bmatrix} 0 & 0 & 0 \end{bmatrix}^T$. The observer gains and parameters are chosen as $\varsigma_1 = \sigma_1 = 16$, $\varsigma_2 = \sigma_2 = 64$, $\alpha_1 = 5/7$, $\beta_1 = 7/5$, $\alpha_2 = 3/7$, and $\beta_2 = 39/35$. The parameters for the controller are chosen as $\delta_{0i} = 2, \delta_{\infty i} = 0.1$, $\xi = 1.3$, $\lambda_1 = 3.5$, $\lambda_2 = 10$, $\alpha = 0.8$, $\beta = 1.4$, $\varphi_1 = \varphi_2 = 10$, $\mu_1 = 4/7$, and $\mu_2 = 7/4$. The actuator faults parameters are chosen as $E = diag\{0.85 \quad 0.85 \quad 0.7\}, \bar{\tau} = \begin{bmatrix} 25 & 25 & 5 \end{bmatrix}^T$, $a = \begin{bmatrix} 15 & 10 & 5 \end{bmatrix}^T$, and $T_0 = \begin{bmatrix} 30 & 30 & 30 \end{bmatrix}^T$.

The FxDO and the improved FxDO are numerically simulated in MATLAB R2018b. Figures 4–6 show that the designed improved FxDO is able to accurately and quickly estimate and compensate for the closed-loop nonlinearity of the system and converge. Even when the actuator starts to fail at $T_{0u} = T_{0v} = T_{0r} = 30$ s, the observer is still able to accurately track the closed-loop disturbances. In addition, the use of the hyperbolic tangent function instead of the sign function reduces the chattering phenomenon from the local zoomed-in plot.

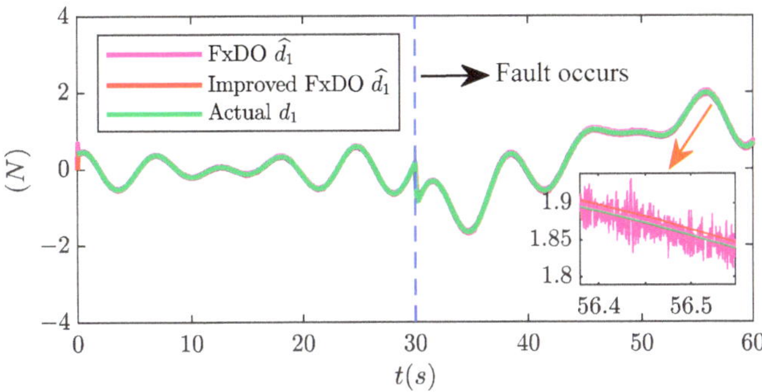

Figure 4. The lumped disturbance d_1 and the observer's estimated output.

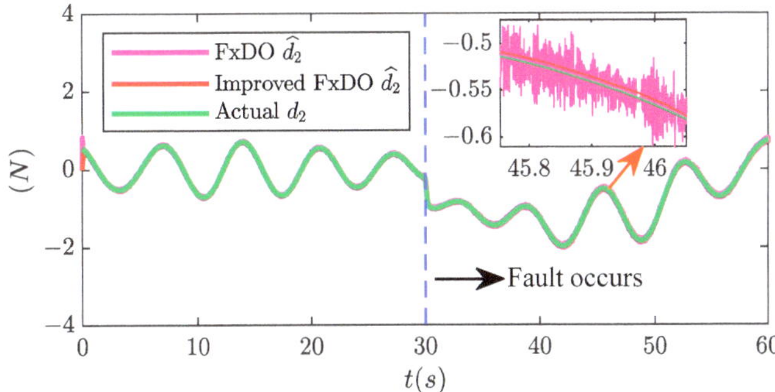

Figure 5. The lumped disturbance d_2 and the observer's estimated output.

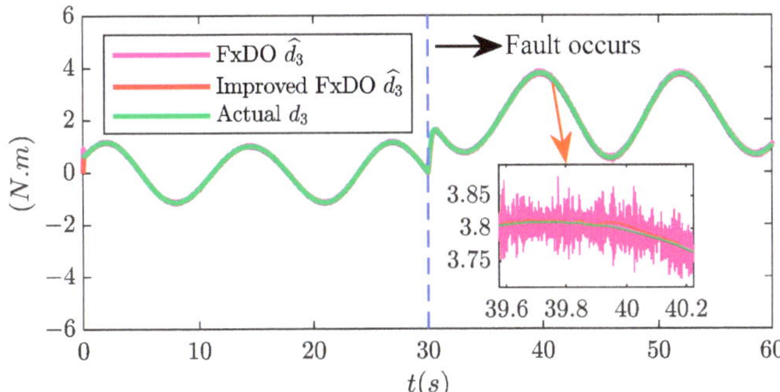

Figure 6. The lumped disturbance d_3 and the observer's estimated output.

In order to better prove the superiority and effectiveness of the proposed IFxDO-PPFxFC scheme, we compare it with Wan et al. [37] proposed SMC scheme and Xu et al. [18] proposed finite-time NFTSMC scheme. The three control schemes are numerically simulated in MATLAB R2018b. First, through the circular trajectory tracking experiment, whether the USV stays on the predetermined circular path with stability ensures the accuracy and stability of navigation. Second, during the autonomous navigation of the USV, it will encounter various external disturbances and changes, such as wind and waves, currents, etc. In addition to this, it may encounter actuator faults, which can test the coping ability and trajectory retention ability of the USV under these conditions. The simulation results are shown in Figures 7–13.

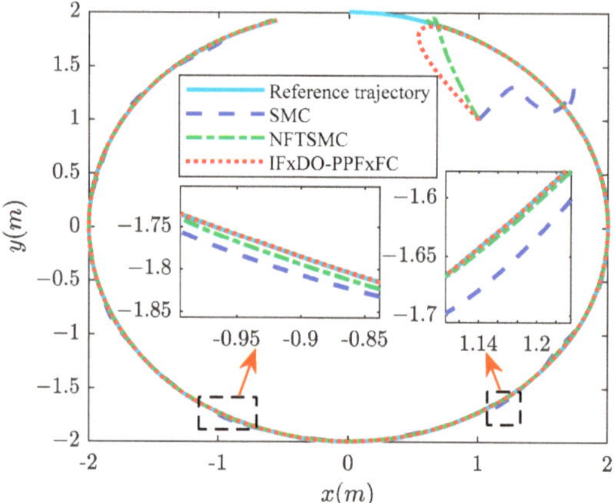

Figure 7. Tracking performance results.

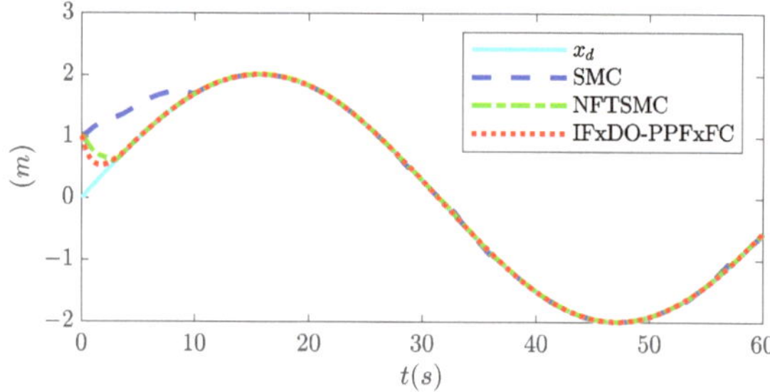

Figure 8. Position x of ship's motion.

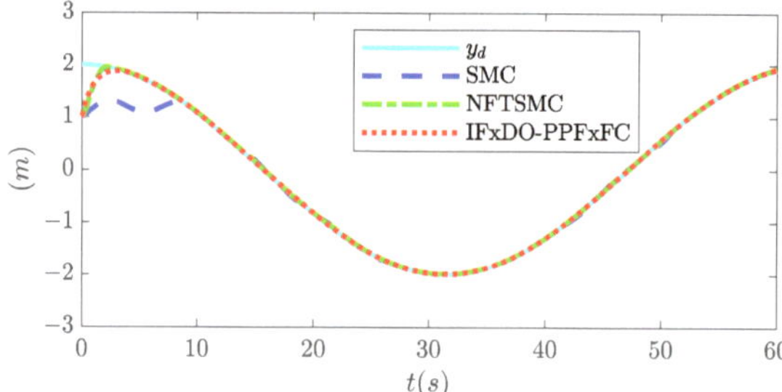

Figure 9. Position y of ship's motion.

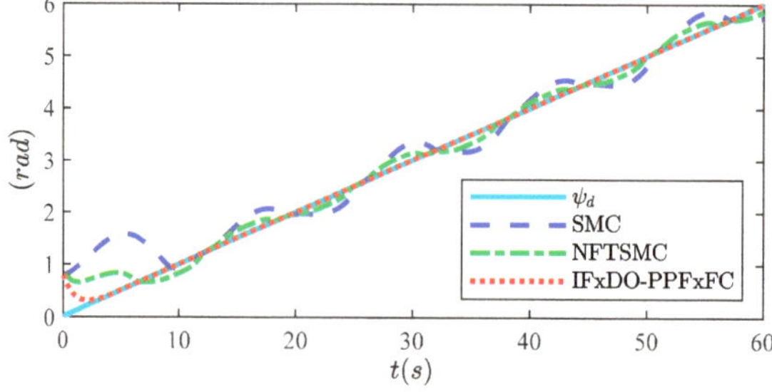

Figure 10. Heading ψ of ship's motion.

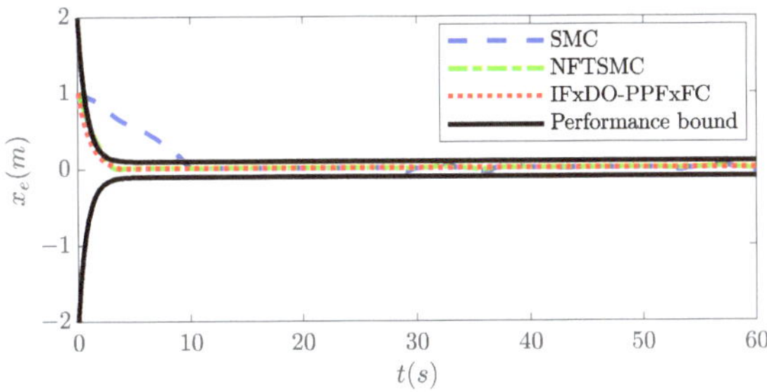

Figure 11. Tracking error of position x.

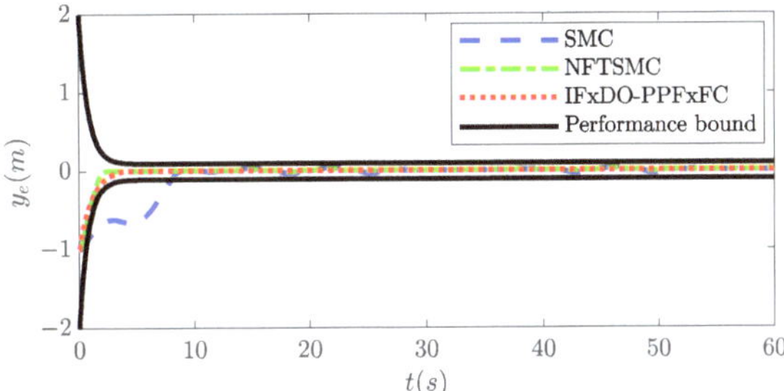

Figure 12. Tracking error of position y.

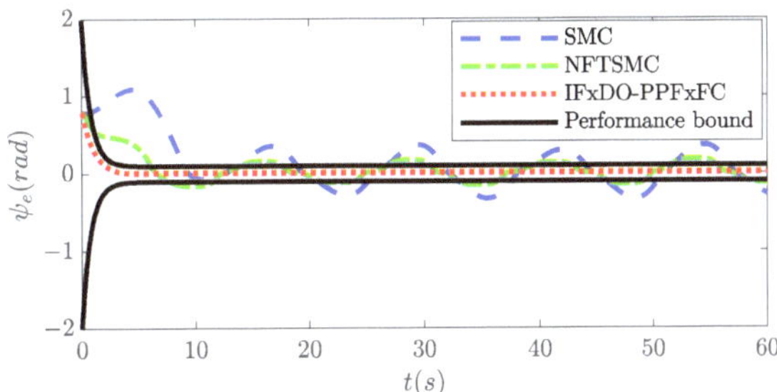

Figure 13. Tracking error of heading ψ.

The tracking performance results of the actual and desired trajectories for the three control schemes are presented in Figure 7. It is evident from the local zoomed-in figure that our proposed control scheme exhibits a faster convergence rate and higher accuracy. The tracking of the actual position and heading of the USV are illustrated in Figures 8–10. All three control strategies successfully achieve a reference position and heading tracking. In comparison to the other two control strategies, the designed control strategy exhibits superior tracking speed and robustness during both position and heading tracking processes, particularly for heading of the USV. The variations in position errors and the heading error with time are illustrated in Figures 11–13. In comparison to the other two control strategies, even when considering lumped nonlinearity, including actuator faults, the tracking error can converge to the specified range more rapidly and meet the designated tracking performance requirements. The convergence curves of the other two control schemes may exhibit slight oscillations and a slower convergence rate.

For further verifying the advantages of the proposed control scheme, IAE indicators are used to quantify the tracking error, which are expressed as follows:

$$\text{IAE} = \int_0^t \left| i_e(\tau) \right| d\tau, \ (i = x, y, \psi) \tag{56}$$

where t denotes the simulation time, and the calculation results are shown in Table 3.

Table 3. Performance index IAE of three control schemes.

Control Scheme	IAE
SMC	IAE (x_e) = 5.7809, IAE (y_e) = 5.4068, IAE (ψ_e) = 17.7117
NFTSMC	IAE (x_e) = 1.4199, IAE (y_e) = 1.0694, IAE (ψ_e) = 7.9170
IFxDO-PPFxFC	IAE (x_e) = 0.9943, IAE (y_e) = 1.0003, IAE (ψ_e) = 0.7680

In order to more intuitively reflect the superiority of the proposed control scheme, we transform the absolute error tracking performance index IAE into the histogram shown in Figure 14.

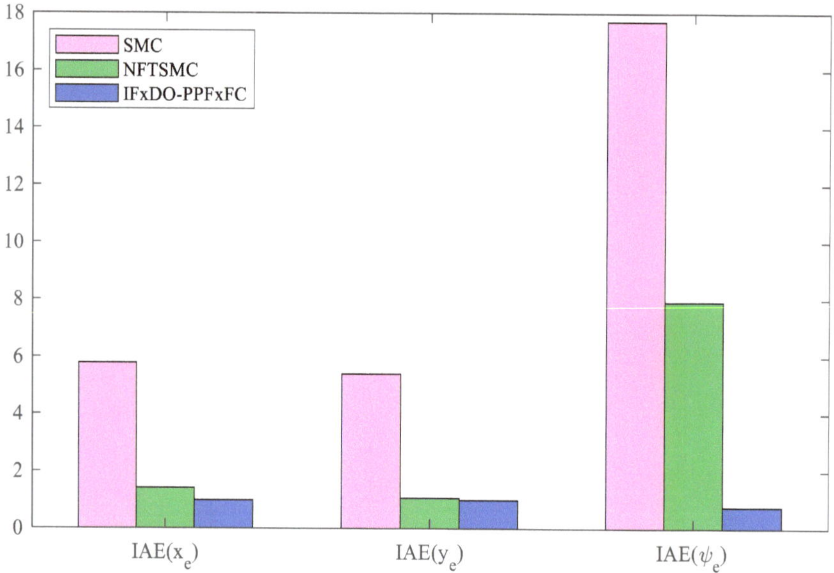

Figure 14. Performance index IAE.

5. Conclusions

The trajectory tracking control of the USV is an important research direction in ship motion control that has important research value and practical significance. In this paper, we studied the problem of the prescribed performance trajectory tracking control of the USV under complex conditions, and a novel IFxDO-PPFxFC scheme was proposed based on the 3-DOF USV mathematical model.

First, we addressed the trajectory tracking problem of the USV under model parameter uncertainties, unknown external time-varying disturbances. We introduced an improved FxDO to realize accurate estimation the lumped nonlinearity of system in a fixed-time, and effectively reduce the chattering phenomenon of the observer's estimation of the lumped nonlinearity compensation in the control input. This technique improved the performance of the controller.

Second, we addressed the trajectory tracking problem of the USV with prescribed tracking performance even in the presence of actuator faults. Based on mathematical model of USV, and by combining fixed-time SMC, FTC, and PPC theories, a prescribed performance fixed-time fault-tolerant controller was designed to ensure accurately and safely tracking of the USV with actuator faults in a fixed time, and the proposed controller was made robust by adding the improved FxDO.

Third, we chose the other two controllers to compare through numerical simulation experiments, from the circular trajectory tracking experiment results, it can be seen that the designed control law had a better control effect than the traditional SMC scheme and finite-time NFTSMC scheme. In addition, for further verifying the advantages of the proposed control scheme, the performance of the controller was described by adding the performance index IAE, which is the lowest compared with the other two methods.

Fourth, the constraints of the control inputs of the system had not been considered yet. in the future, we will consider the case where the input is constrained and extend the method proposed in this paper to an underactuated USV and combine it with other state-of-the art control techniques.

Author Contributions: Conceptualization, K.L. and Z.L.; methodology, K.L. and Z.L.; software, K.L.; validation, K.L.; formal analysis, K.L.; investigation, K.L.; resources, K.L. and Z.L.; data curation, K.L.; writing—original draft preparation, K.L.; writing—review and editing, K.L. and Z.L.; visualization, K.L.; supervision, Z.L.; project administration, Z.L.; funding acquisition, Z.L. All authors have read and agreed to the published version of the manuscript.

Funding: This research was funded by the National Natural Science Foundation of China (No. 51879119), the Key Projects of National Key R&D Program (No. 2021YFB390150), the Natural Science Project of Fujian Province (No. 2022J01323, 2021J01822, 2020J01660, 2023I0019), and the Fuzhou-Xiamen-Quanzhou Independent Innovation Region Cooperated Special Foundation (No: 3502ZCQXT2021007).

Institutional Review Board Statement: Not applicable.

Informed Consent Statement: Not applicable.

Data Availability Statement: The datasets generated during and/or analyzed during the current study are available from the corresponding author on reasonable request.

Conflicts of Interest: The authors declare no conflicts of interest.

References

1. Er, M.J.; Ma, C.; Liu, T.; Gong, H. Intelligent Motion Control of Unmanned Surface Vehicles: A Critical Review. *Ocean Eng.* **2023**, *280*, 114562. [CrossRef]
2. Alvaro-Mendoza, E.; Gonzalez-Garcia, A.; Castañeda, H.; De León-Morales, J. Novel Adaptive Law for Super-Twisting Controller: USV Tracking Control under Disturbances. *ISA Trans.* **2023**, *139*, 561–573. [CrossRef] [PubMed]
3. Song, L.; Xu, C.; Hao, L.; Yao, J.; Guo, R. Research on PID Parameter Tuning and Optimization Based on SAC-Auto for USV Path Following. *J. Mar. Sci. Eng.* **2022**, *10*, 1847. [CrossRef]
4. Dong, Z.; Wan, L.; Li, Y.; Liu, T.; Zhang, G. Trajectory Tracking Control of Underactuated USV Based on Modified Backstepping Approach. *Int. J. Nav. Archit. Ocean Eng.* **2015**, *7*, 817–832. [CrossRef]
5. Tong, H. An Adaptive Error Constraint Line-of-Sight Guidance and Finite-Time Backstepping Control for Unmanned Surface Vehicles. *Ocean Eng.* **2023**, *285*, 115298. [CrossRef]
6. Chen, Y.-Y.; Ellis-Tiew, M.-Z. Autonomous Trajectory Tracking and Collision Avoidance Design for Unmanned Surface Vessels: A Nonlinear Fuzzy Approach. *Mathematics* **2023**, *11*, 3632. [CrossRef]
7. Lu, J.; Yu, S.; Zhu, G.; Zhang, Q.; Chen, C.; Zhang, J. Robust Adaptive Tracking Control of UMSVs under Input Saturation: A Single-Parameter Learning Approach. *Ocean Eng.* **2021**, *234*, 108791. [CrossRef]
8. Qin, H.; Li, C.; Sun, Y. Adaptive Neural Network-based Fault-tolerant Trajectory-tracking Control of Unmanned Surface Vessels with Input Saturation and Error Constraints. *IET Intell. Trans. Sys.* **2020**, *14*, 356–363. [CrossRef]
9. Gonzalez-Garcia, A.; Castaneda, H. Guidance and Control Based on Adaptive Sliding Mode Strategy for a USV Subject to Uncertainties. *IEEE J. Ocean. Eng.* **2021**, *46*, 1144–1154. [CrossRef]
10. Zhao, C.; Yan, H.; Gao, D.; Wang, R.; Li, Q. Adaptive Neural Network Iterative Sliding Mode Course Tracking Control for Unmanned Surface Vessels. *J. Math.* **2022**, *2022*, 1417704. [CrossRef]

11. Wang, R.; Yan, H.; Li, Q.; Deng, Y.; Jin, Y. Parameters Optimization-Based Tracking Control for Unmanned Surface Vehicles. *Math. Probl. Eng.* **2022**, *2022*, 2242338. [CrossRef]
12. Chen, Z.; Zhang, Y.; Zhang, Y.; Nie, Y.; Tang, J.; Zhu, S. Disturbance-Observer-Based Sliding Mode Control Design for Nonlinear Unmanned Surface Vessel with Uncertainties. *IEEE Access* **2019**, *7*, 148522–148530. [CrossRef]
13. Piao, Z.; Guo, C.; Sun, S. Adaptive Backstepping Sliding Mode Dynamic Positioning System for Pod Driven Unmanned Surface Vessel Based on Cerebellar Model Articulation Controller. *IEEE Access* **2020**, *8*, 48314–48324. [CrossRef]
14. Chen, Z.; Zhang, Y.; Nie, Y.; Tang, J.; Zhu, S. Adaptive Sliding Mode Control Design for Nonlinear Unmanned Surface Vessel Using RBFNN and Disturbance-Observer. *IEEE Access* **2020**, *8*, 45457–45467. [CrossRef]
15. He, Z.; Wang, G.; Fan, Y.; Qiao, S. Fast Finite-Time Path-Following Control of Unmanned Surface Vehicles with Sideslip Compensation and Time-Varying Disturbances. *J. Mar. Sci. Eng.* **2022**, *10*, 960. [CrossRef]
16. Yu, Y.; Guo, C.; Li, T. Finite-Time LOS Path Following of Unmanned Surface Vessels with Time-Varying Sideslip Angles and Input Saturation. *IEEE/ASME Trans. Mechatron.* **2022**, *27*, 463–474. [CrossRef]
17. Wang, N.; Gao, Y.; Yang, C.; Zhang, X. Reinforcement Learning-Based Finite-Time Tracking Control of an Unknown Unmanned Surface Vehicle with Input Constraints. *Neurocomputing* **2022**, *484*, 26–37. [CrossRef]
18. Xu, D.; Liu, Z.; Song, J.; Zhou, X. Finite Time Trajectory Tracking with Full-State Feedback of Underactuated Unmanned Surface Vessel Based on Nonsingular Fast Terminal Sliding Mode. *J. Mar. Sci. Eng.* **2022**, *10*, 1845. [CrossRef]
19. Hu, Y.; Zhang, Q.; Liu, Y.; Meng, X. Event Trigger Based Adaptive Neural Trajectory Tracking Finite Time Control for Underactuated Unmanned Marine Surface Vessels with Asymmetric Input Saturation. *Sci. Rep.* **2023**, *13*, 10126. [CrossRef]
20. Rodriguez, J.; Castañeda, H.; Gonzalez-Garcia, A.; Gordillo, J.L. Finite-Time Control for an Unmanned Surface Vehicle Based on Adaptive Sliding Mode Strategy. *Ocean Eng.* **2022**, *254*, 111255. [CrossRef]
21. Huang, B.; Song, S.; Zhu, C.; Li, J.; Zhou, B. Finite-Time Distributed Formation Control for Multiple Unmanned Surface Vehicles with Input Saturation. *Ocean Eng.* **2021**, *233*, 109158. [CrossRef]
22. Polyakov, A. Nonlinear Feedback Design for Fixed-Time Stabilization of Linear Control Systems. *IEEE Trans. Automat. Contr.* **2012**, *57*, 2106–2110. [CrossRef]
23. Yao, Q. Robust Fixed-Time Trajectory Tracking Control of Marine Surface Vessel with Feedforward Disturbance Compensation. *Int. J. Syst. Sci.* **2022**, *53*, 726–742. [CrossRef]
24. Zhang, J.; Yu, S.; Wu, D.; Yan, Y. Nonsingular Fixed-Time Terminal Sliding Mode Trajectory Tracking Control for Marine Surface Vessels with Anti-Disturbances. *Ocean Eng.* **2020**, *217*, 108158. [CrossRef]
25. Chen, D.; Zhang, J.; Li, Z. A Novel Fixed-Time Trajectory Tracking Strategy of Unmanned Surface Vessel Based on the Fractional Sliding Mode Control Method. *Electronics* **2022**, *11*, 726. [CrossRef]
26. Yu, X.-N.; Hao, L.-Y. Integral Sliding Mode Fault Tolerant Control for Unmanned Surface Vessels with Quantization: Less Iterations. *Ocean Eng.* **2022**, *260*, 111820. [CrossRef]
27. Zhang, J.; Yu, S.; Yan, Y.; Wu, D. Fixed-Time Output Feedback Sliding Mode Tracking Control of Marine Surface Vessels under Actuator Faults with Disturbance Cancellation. *Appl. Ocean Res.* **2020**, *104*, 102378. [CrossRef]
28. Wu, W.; Tong, S. Fixed-Time Formation Fault Tolerant Control for Unmanned Surface Vehicle Systems with Intermittent Actuator Faults. *Ocean Eng.* **2023**, *281*, 114813. [CrossRef]
29. Heshmati-Alamdari, S.; Bechlioulis, C.P.; Karras, G.C.; Nikou, A.; Dimarogonas, D.V.; Kyriakopoulos, K.J. A Robust Interaction Control Approach for Underwater Vehicle Manipulator Systems. *Annu. Rev. Control* **2018**, *46*, 315–325. [CrossRef]
30. Wang, Z.; Su, Y.; Zhang, L. Fixed-Time Attitude Tracking Control for Rigid Spacecraft. *IET Control Theory Appl.* **2020**, *14*, 790–799. [CrossRef]
31. Fan, Y.; Qiu, B.; Liu, L.; Yang, Y. Global Fixed-Time Trajectory Tracking Control of Underactuated USV Based on Fixed-Time Extended State Observer. *ISA Trans.* **2023**, *132*, 267–277. [CrossRef] [PubMed]
32. Zuo, Z.; Tie, L. A New Class of Finite-Time Nonlinear Consensus Protocols for Multi-Agent Systems. *Int. J. Control* **2014**, *87*, 363–370. [CrossRef]
33. Skjetne, R.; Fossen, T.I.; Kokotović, P.V. Adaptive Maneuvering, with Experiments, for a Model Ship in a Marine Control Laboratory. *Automatica* **2005**, *41*, 289–298. [CrossRef]
34. Basin, M.; Bharath Panathula, C.; Shtessel, Y. Multivariable Continuous Fixed-Time Second-Order Sliding Mode Control: Design and Convergence Time Estimation. *IET Control Theory Appl.* **2017**, *11*, 1104–1111. [CrossRef]
35. Sui, B.; Zhang, J.; Li, Y.; Liu, Y.; Zhang, Y. Fixed-Time Trajectory Tracking Control of Unmanned Surface Vessels with Prescribed Performance Constraints. *Electronics* **2023**, *12*, 2866. [CrossRef]
36. Van, M.; Ceglarek, D. Robust Fault Tolerant Control of Robot Manipulators with Global Fixed-Time Convergence. *J. Frankl. Inst.* **2021**, *358*, 699–722. [CrossRef]
37. Li, H.; Cai, Y. On SFTSM Control with Fixed-Time Convergence. *IET Control Theory Appl.* **2017**, *11*, 766–773. [CrossRef]

MDPI AG
Grosspeteranlage 5
4052 Basel
Switzerland
Tel.: +41 61 683 77 34

Journal of Marine Science and Engineering Editorial Office
E-mail: jmse@mdpi.com
www.mdpi.com/journal/jmse